# Tools and Trends in Bioanalytical Chemistry

Lauro Tatsuo Kubota ·
José Alberto Fracassi da Silva ·
Marcelo Martins Sena ·
Wendel Andrade Alves
Editors

# Tools and Trends in Bioanalytical Chemistry

Springer

*Editors*
Lauro Tatsuo Kubota
Institute of Chemistry
State University of Campinas
Campinas, Brazil

José Alberto Fracassi da Silva
Institute of Chemistry
State University of Campinas
Campinas, Brazil

Marcelo Martins Sena
Institute of Chemistry
Federal University of Minas Gerais
Belo Horizonte, Brazil

Wendel Andrade Alves
Center for Natural and Human Sciences
Federal University of ABC
Santo Andre - SP, Brazil

ISBN 978-3-030-82383-2       ISBN 978-3-030-82381-8   (eBook)
https://doi.org/10.1007/978-3-030-82381-8

This Springer imprint is published by the registered company Springer Nature Switzerland AG.
The registered company address is: Gewerbestrasse 11, 6330 Cham, Switzerland

# Foreword

Does the very term *Analytical Chemistry* have an absolute scientific meaning or does it only cover a unifying concept encompassing very different fundamental scientific disciplines which nowadays range from chemistry and physics to mathematics, statistics, and science? Or social and environmental sciences through basic chemistry, physics, geology, and biology. Yet even if this variation of the disciplines that make up analytical chemistry and its challenges fluctuates from one country to another and from one era to another according to eminently cultural and historical roots, today there is practically no socioeconomic, technological, scientific, health, or medical field that can be freed from the contributions of chemical analyses.

Does this confer a definition and a particular rank on *Analytical Chemistry* within the scientific corpus? In other words, when the entire socioeconomic, medical, and scientific world uses or produces analytical chemistry data, is it relying on a full-fledged scientific discipline, *Analytical Chemistry*, or is it not? Does it only need and produce chemical analyzes? If so, would chemical analyzes be able to evolve as we observe without the support of a strong scientific field?

Etymologically, chemical analysis involves *"the detailed study of a complex whole by separation into its essential chemical components,"* a meaning which seems to reduce the process of producing chemical analyzes to a simple juxtaposition of technical know-how. In this common acceptance, *Analytical Chemistry* would only be the thesaurus of these skills. There would then be neither science nor discipline, but simply a collection of know-how continuously developing from questions from other scientific or socioeconomic fields and from technological advances made to resolve these questions.

But what do the terms *"any complex,"* *"separation,"* or *"essential chemical components"* on which the preceding definition is based actually mean? Is the notion of essential chemical component the same when it comes to analyzing a chemical element in the sense of Mendeleiev in a mineral, or in a living organism, in which molecules and ions are compartmentalized and present in different chemical forms and at different concentrations depending on the compartments? Does the same notion get involved when it comes to searching for the presence of the toxic enantiomer of an organic molecule whose other enantiomer is a potent drug? Or when a DNA or protein sequence matters?

Do *separation* or *measurement* have the same meanings when it comes to analyzing a single sample, a few milliliters of a biological fluid emanating from a patient or from the winner of a sports competition in order to control, in the first case, the evolution of a medical treatment or, in the second one, the absence of doping? Or when it comes to systematically controlling the dioxin content of another carcinogen or toxin in a product of the agri-food industry? Or to evidence the spread of new variants of a virus as we have just discovered was so important during the COVID-19 pandemics. Do *separation* and *measurement* still have their same meanings when things come to detecting and quantify the presence of fraud on the origins and nature of a product, such as when it comes to verifying the fraudulent origin of a wine or the blend of a natural essence by a synthetic molecule in a great perfume although it is chemically identical to the natural molecule but considerably cheaper and more easily available anywhere than it?

Can apprehending a *complex whole* proceed in the same way when the investigated object is inert like a small fragment of an asteroid or when it is a part of an immensely large and essentially dynamic object like the atmosphere and the oceans? Or a microscopic, very compartmentalized and in rapid and complex kinetic and thermodynamic interactions with its external environment like a living cell? Does a *complex whole* have the same definition when the sample properties result from the interactive superposition of its components as happens when the question relates to a perfume whose odorous properties are only virtual since they only reflect the simultaneous stimulation of a set of neurons by the different components of the *complex odorous whole*?

These different examples clearly show that the usual definition of *chemical analysis*, however crisp and common it may be, is semantically empty. However, these few examples have brought out one of the basic but important qualities of *Analytical Chemistry*. This is deep-rooted on the uncommon ability to develop methods allowing reliable and certified measurements and analyses (metrological science) in order to obtain quantitative chemical information on complex systems intended to solve a problem often posed by another scientific discipline or by the socioeconomic world, nowadays often involving biomedical and health issues.

To answer to the problems which are posed to it today and to those, still unknown, which will be posed to it in the future, perhaps even in the very near future on the occasion of a medical or bio-environmental disaster, *Analytical Chemistry* must continuously invent new concepts, new strategies and develop its own tools. Obtaining the information that is, or will be, necessary to carry out these missions with the right precision and accuracy, within the necessary range of values and for the wide sorts of samples to test and analyze, requires relying on careful fundamental researches prone to introduce new and efficient analytical methods or to adapt known strategies for performing under more and more stringent constraints (reduction in sample volumes, lower detection thresholds, reduced analysis times, increased number of analyzes per day, out-of-laboratories uses, etc.) thanks to new technological openings (biotechnologies, miniaturization, robotization, etc.) as is perfectly illustrated by the recent successful efforts for the continuous advancement of bioanalytical systems.

This is important without any doubt, but it is not enough! *Analytical Chemists* must simultaneously teach their current specialized knowledge and that developed in the best research laboratories in a form that will allow their students and successors in academy and industry, of course, but above all, to be able to adapt their fundamental knowledge and go beyond it in order to resolve the questions that will inevitably be asked of them in the future. This is an important condition for allowing the pursuit of scientific progress and technological advances but also for remedying future problems that will inevitably be emerging in the socio-economic, medical, technological, or environmental worlds.

This new book attempts to respond to these crucial needs at an educational level based on advanced research carried out more specifically in bioanalytical chemistry. This new field of science initially emerged as a subfield of analytical chemistry but quickly developed into a major discipline in itself due to the near-exponential development of its approaches and tools in an interactive dynamic fueled by the increasing demands of our societies for fundamental and practical applications linked to health, medical and environmental biology. This book was planned and organized by Professor Lauro Tatsuo Kubota, Member of the Brazilian Academy of Sciences and General Coordinator of *INCT-Bio*, the *National Institute of Bioanalytical Science and Technology* in Brazil, and by Professors José Alberto Fracassi da Silva, Marcelo Martins Sena, and Wendel Andrade Alves who coordinate each main area of *INCT-Bio*.

*INCT-Bio* is a research network involving several tens of groups hosted in different higher education institutions spread across the whole country and interacting together. This unique institute was created to bring together senior and young researchers in order to merge their specialized training and their expertise in different areas of *Analytical Chemistry* in order to propose and synergistically develop innovative inter/multidisciplinary biosensing tools and methods for clinical diagnosis, biochemical and pharmacological analyzes as well as for solving important emerging problems related to Genomics, Proteomics, Metabolomics, Metallomics, etc., the demands of which are rapidly increasing in our societies.

I was happy to accompany the founders of *INCT-Bio* in the process of giving birth to this remarkable center and to consolidate it over the years because of its dual and intermingled scopes aimed at developing advanced researches and proposing precise and accurate tools likely to be used in advanced centers as well as in the most remote places as well as to teach the new generations of *Analytical Chemists*. I am now happy to be able to welcome the publication of this excellent book. Indeed, it brings *INCT-Bio* educational objectives to a high standard. I am more than sure that it will help many students as well as young researchers around the world to learn and become familiar with *Bioanalytical Chemistry* to the point they can proceed on their own.

Paris, France

Christian Amatore

# Acknowledgments

The reader will see, the book is authored and edited by a select group of scientists that share in common the passion for bioanalytical chemistry. We are members of a virtual research institute, which was formed initially in 2008. In that year, the *Ministry of Science and Technology Innovation and Culture* (MCTIC) through the Brazilian Research Council, the *Conselho Nacional de Desenvolvimento Científico e Tecnológico* (CNPq), released a call for proposals for the creation of institutes of science and technology in all areas of the knowledge.

The National Institutes on Science and Technology Program, the INCTs, had the specific goals of consolidating the program as one of the most important for Brazilian science. The program aimed to aggregate, in an articulated way, the best research groups on the frontier of science and in strategic areas for the sustainable development of the country, driving fundamental scientific research to an internationally competitive level by stimulating the development of cutting-edge scientific and technological research.

The National Institute on Science and Technology in Bioanalytic (INCT-Bio) was initially created in 2008, with the primary objective of consolidating Bioanalytical Chemistry in Brazil. Currently, INCT-Bio consists of a network involving 46 research groups from 18 Brazilian higher education institutions, funded by CNPq and CAPES (*Coordenação de Aperfeiçoamento de Pessoal de Nível Superior*), and by the *State of São Paulo Funding Agency* – FAPESP, approved after a competitive process with more than 200 proposals. We would like to acknowledge the agencies for their financial support.

The INCT-Bio consists of specialists from the main areas of Analytical Chemistry potentially applicable to biological sciences, including researchers with extensive international experience and proven productive capacity, as well as young researchers in groups either not yet consolidated or in institutions that lack tradition in research. Currently, the team of analytical chemists joined forces with researchers from the Biological and Medical Sciences areas, expanding the area of knowledge and scope of our institute.

INCT-Bio is composed of 46 principal investigators and is structured administratively and academically as follows: A general coordinator, Prof. Lauro T. Kubota from the University of Campinas (UNICAMP), a vice-coordinator Prof. Marilia Oliveira Fonseca Goulart from the Federal University of Alagoas (UFAL). For governance, the INCT-Bio relies on the experience of senior research members of an Advisory Board - AB. The AB is composed by Auro Atsushi Tanaka, from Federal University of Maranhão (UFMA), Emanuel Carrilho, from the University of São Paulo (USP), Érico Marlon de Moraes Flores, from Federal University of Santa Maria (UFSM), Lúcio Angnes, from USP, and Marco Aurelio Zezzi Arruda, from UNICAMP. To coordinate the administrative duties, we have three subgroups in bioanalytical chemistry: Separations, Spectroanalysis and Chemometrics, and Sensors & Biosensors. The coordinators are José Alberto Fracassi da Silva, from UNICAMP (Separations), Marcelo Martins de Sena, from Federal University of Minas Gerais (UFMG) replacing Ronei Jesus Poppi, *in memoriam*, from UNICAMP (Spectroanalysis and Chemometrics), and Wendel Andrade Alves, from the Federal University of ABC (UFABC) (Sensors). All members of the INCT-Bio are divided into these three areas, but some members are dedicated to promoting science and chemical education.

We have to stress that this book is an initiative of INCT-Bio and we would like to acknowledge all members of the institute, and especially the authors of the chapters or the contributions, and FAPESP, CNPq, CAPES, and MCTIC for the financial support.

# Contents

# What Is Bioanalytical Chemistry? Scientific Opportunities with Immediate Impact

Vinícius Guimarães Ferreira, Jéssica Freire Feitor,
Mariana Bortholazzi Almeida, Daniel Rodrigues Cardoso, and
Emanuel Carrilho

## 1    Introduction

Chemistry is a fascinating field of study with endless applicability in every aspect of our life. The beauty in this open and diverse universe is the possibility of changing lives for the best. As part of the mission, chemists came to cooperate with the biological and medical sciences and elucidate complex biological systems through chemistry, mostly by using analytical instrumentation. Human curiosity and the necessity of science to explain relevant questions of biological interest continue to push the evolution and improvement of existing analytical tools. In the meantime, analytical chemists still need to be alert about the progress of biological sciences (Horvai et al. 2011). The nineteenth and twentieth centuries were marked by the chemist's efforts on making precise, reliable chemical quantitation a reality. Now, we are living in a biological revolution. These fields, together, will clear obstacles and will bring mutual benefits to researchers working in the field of Bioanalytical Chemistry (Horvai et al. 2011; Wake 2008).

## 2    Bioanalytical Chemistry: Accessing the Chemistry of Life

One of the greatest mysteries of life is to understand its origin. The distinguishment of animate and inanimate matter is surrounded by questions on how and why organic molecules and inorganic compounds might interact through sophisticated molecular recognition, as in a perfect symphony. It is known that chemical reactions are vital to maintaining the structure and the workability of living organisms, guaranteeing their survival and continuous evolution. As an aftereffect, a vast number of different

V. G. Ferreira · J. F. Feitor · M. B. Almeida · D. R. Cardoso · E. Carrilho (✉)
Instituto de Química de São Carlos, Universidade de São Paulo, São Carlos, SP, Brazil
e-mail: emanuel@iqsc.usp.br

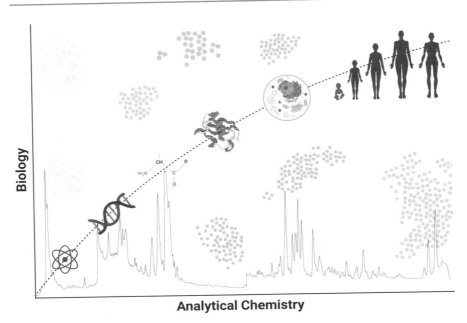

**Fig. 1** Schematic representation of the landscape of bioanalytical chemistry—the integration between analytical chemistry and biology. Created with BioRender (https://biorender.com/)

chemical compounds are repeatedly created as products of those reactions, telling us the history behind each living organism (Pross and Pascal 2013; Datta et al. 2020).

The chemical compounds that originated from a biological organism are called biomolecules. Most biomolecules have organic composition, and some typical examples are nucleic acids, amino acids, peptides, proteins, carbohydrates, and lipids. While the study of life at a molecular level is concerned with biochemistry, bioanalytical chemistry explores the identification, characterization, quantification, and time-dependent monitoring of biomolecules originated from sensitive sample matrix through many specific analytical methods that will be further explored in this book (Fig. 1) (Labuda et al. 2018; Kogikoski et al. 2018; Roat-Malone 2007).

The chemical study beyond biological functions provides the necessary knowledge to perform diagnostics of diseases, to discover novel biomarkers and biomaterials, to improve drug's design by a specific mechanism of action, as well as to develop biosensors for specific target control. All of these purposes carry an enormous responsibility, requiring a high level of accuracy and reliability, those ensured by the analytical procedure and by the analyst (Hersel et al. 2003).

## 2.1 Biomolecules Composition and Properties

Biomolecules are made of some critical elements, such as carbon, nitrogen, oxygen, hydrogen, calcium, phosphorus, and sulfur. Moreover, the presence of sodium, potassium, magnesium, chlorine, and transition metals like iron and copper regulates critical biochemical pathways on biological systems. The chemical interactions provided by those elements dictate the structure of the matter existent in the organism, allowing different physical properties. From an analytical point of view, it is crucial to understand and use chemical and physical properties in favor of methods, to help elucidate the differences between biological molecules and samples.

The discrimination between compounds originated from plants with similar composition yet with different biological properties (Kubinyi 2002; Atanasov et al. 2015), for example, sometimes is only possible due to the separation capability and selectivity (Atanasov et al. 2015; Srikoti et al. 2020; Gault and McClenaghan 2009), the assessment of the exact mass (mass spectrometry) and the determination of their shape and structure by spectroscopic methods (Gault and McClenaghan 2009).

Among biomolecules, nucleic acids and proteins are groups with intricate compositions and functions that play essential roles in life maintenance. In time, as science learns about the specific functions and pathways that these biomolecules conduct, the importance of bioanalysis of these molecules grows continuously. There are two types of nucleic acids, deoxyribonucleic acid (DNA) and ribonucleic acid (RNA). If there is a "code of life," it certainly is the DNA. DNA contains all the genetic information transmitted to the new generation of cells or organisms, while RNA is responsible for the synthesis of proteins. Both are composed of nucleobases (purine and pyrimidine bases), carbohydrates, and phosphoric acid. Covalent binding is responsible for the interactions between the constituents, forming nucleobases and nucleotides chains. The hydrogen bonds are responsible for base pairing, which gives the DNA its 3D structure with a twisted double helix shape (Selzer et al. 2018; Manz et al. 2004).

Proteins, in its turn, consist of amino acids connected by peptide bonds. In living organisms, about 20 different amino acids are arranged in distinct sequences of connections, forming a chain. Then, intermolecular interactions enable the amino acid chain to fold and create a 3D structure and shape, giving the protein a specific activity and function. In general, proteins are responsible for providing the organisms with the structure, tissue support (fibrous characteristic), and specific functions to the immune system and metabolism (antibodies and enzymes) (Manz et al. 2004). The research field responsible for elucidating the proteins and the methods used is called proteomics (Amiri-Dashatan et al. 2018). Other biomolecules also have an essential role in identifying diseases by the –omics sciences (genomics, transcriptomics, proteomics, and metabolomics), each one using a different technology (Bedia 2018).

An important aspect to consider when working with biological samples is the nature of the matrix. Samples originated from animal tissue need different treatment than plant material, for example. Each type of analysis usually has a sample preparation protocol to avoid degradations and analytes loss. At the same time, the targeted compounds need to be extracted from the sample matrix, preventing interference from contaminants or constituents during instrumental analysis. The extraction process is challenging and has to be chosen carefully, taking into account the properties and characteristics of the analyte, cost-benefit, and compatibility with the analytical instrumentation (Bedia 2018; Poole et al. 2016; Clark et al. 2016).

## 2.2    Importance of Bioanalytical Studies

Modern biomedical research is interested in deeply understanding the protein universe since proteins are directly related to the expression of different types of genes and might differ in various conditions and diseases. Nevertheless, to discover a protein and predict its clinical aspects is enough? In science, the answer is usually no. Recently, a new branch has emerged, named Chemical Proteomics. This field of science focuses on integrating synthetic chemistry, cellular biology, and mass spectrometry. Concisely, chemical proteomics uses the affinity chromatography approach to identify protein interactions with other molecules, being the protein uncharacterized or not. This approach has the advantage of using a target fishing method, allowing the identification of not only new proteins but also what kinds of molecules that interact with proteins. Chemical proteomic became an excellent way to test drug response (Amiri-Dashatan et al. 2018; Chen et al. 2020; Eberl et al. 2019). Besides, knowing the proteins involved at receptor-binding domains and understanding its characteristics is crucial when infections need to be overcome, and the race for medicine or a vaccine needs to be fast (Qiao and de la Cruz 2020).

The biomarker discovery, to be clinically useful, needs to present a robust and reproducible method, also helped by a computational process to facilitate the comprehension from scientists to the clinician. That is the goal of translational medicine, to narrow the basic and applied science (Bravo-Merodio et al. 2019; Wang 2012).

## 2.3    Separation Methods for Biomolecules

The study of biomolecules involves a complex manipulation of these analytes. This complexity is due to the researcher's interest in having each biomolecule separated in order to evaluate its physical and chemical properties without interference from other molecules. The complexity in separating biomolecules for its elucidation comes from the fact that many substances have very similar properties and are unquestionably diluted on its substrates. To overcome these conditions, separation and purification methods were developed and enhanced over the past years (Glad and Larsson 1991). The most important of these methods are covered in this book, including electrophoresis, chromatography, and centrifugation (e.g., membrane centrifugation and ultracentrifugation) (Fig. 2).

Electrophoresis is a technique for separating and identifying high molecular weight biomolecules, such as proteins and nucleic acids. This separation occurs through an electric field that moves charged macromolecules depending on their molecular weights, charge magnitude, and tertiary and quaternary structures. The

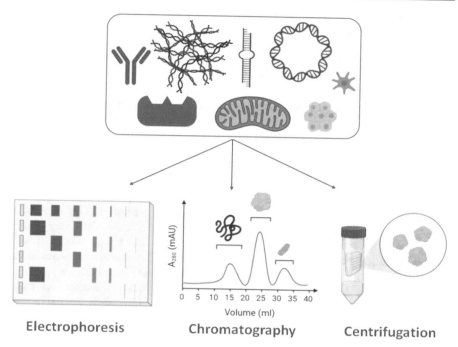

**Fig. 2** The most commonly used separation techniques for biomolecules. Image adapted from BioRender (https://biorender.com/)

basic scheme for electrophoretic separation consists of an anode, a cathode, and a medium filled with a buffer solution and supported by inert material. When voltage is applied, a current is generated by the movement of positively charged molecules to the cathode and negatively charged to the anode (Mikkelsen and Cortón 2016). In electrophoresis, the fundamental parameter, characteristic of each molecule (under given conditions), is the electrophoretic mobility, $\mu$, given by the equation:

$$\mu = v.E = q.f \tag{1}$$

$v$ velocity, $E$ electric field, $q$ net charge, $f$ frictional coefficient.

Over the years, much has been developed in electrophoresis (Campa et al. 2006; Lechner et al. 2020), and mostly, three types are primarily used: the moving-boundary, zone, and steady-state electrophoresis. Choices are based on the best compatibility between the analyte and its substrates and the specificity of the experiment.

Capillary electrophoresis (CE) is the most recent advance in this area, consisting of the same technique mentioned above, with the difference that it is carried out inside of a capillary column or tube. This simple fact makes the resolution of this technique much higher than any other separation technique due to the small volumes supported and the control of the electroosmotic flow rate. Other advantages of this technique, over the conventional electrophoresis, are minor heat dissipation per volume, higher electric fields, faster separations, and smaller diffusional band broadening. This technique also allows the possibility of analyzing small samples such as single cells with high efficiency. Some limitations include low sensitivity due to low selectivity, low limit of detection (LOD) compared to other similar techniques, and high dependency with solution pH and temperature. Also, the requirement of concentrated samples to obtain a clean response is another fragility of this approach (Mikkelsen and Cortón 2016; Fekete and Schmitt-Kopplin 2007).

Chromatography is widely used for separating biomolecules. This technique is based on the flow of a mobile phase (containing the sample) through a stationary phase. There are multiple types of chromatography, but as this chapter concerns biomolecules, this topic focuses on liquid chromatography (LC). In this case, the mobile phase is an aqueous solution, and the stationary phase can be either a particulate solid, a semi-solid porous gel, or a porous monolith. The separation occurs with the mobile phase dragging the analyte through the stationary phase. By the binding strength between the sample contents and the stationary phase, specific retention times are established, and each molecule is released based on this time between the insertion and elution (Mikkelsen and Cortón 2016).

When considering biomolecules separation, it is relevant to mention four main mechanisms: partition, size-exclusion (gel filtration), affinity, and ion-exchange. Partition occurs with the analyte interacting with two liquid phases (mobile and stationary), according to its solubility. The first to elute is the most soluble analytes into the mobile phase. Size exclusion separates molecules by their size and shape, with smaller and compact molecules being retained to the stationary phase and the larger and open-structures ones, eluting first. The affinity mechanism involves particular biomolecules, such as enzymes and antibodies, that have specific interactions with their substrates and antigens, respectively. Ion-exchange separation is specific for charged molecules, in which the stationary phase has an opposite charge compared to the analyte, attracting it (Fig. 3) (Mikkelsen and Cortón 2016).

The most advanced technology involving LC is related to high and ultra-high-performance liquid chromatography (HPLC and UHPLC). The system allows the passage of intense high-pressure flows, increasing the efficiency and the resolution of the analysis. This pressure arises from the tiny size of the stationary phase particles, among other improvements in the whole apparatus, including the coupling with robust detectors such as Mass Spectrometer (MS) (Matuszewski et al. 2003; Yoon et al. 2019).

Among the strengths of this technique, versatility stands out, considering the wide range of molecules that can be separated and analyzed. Also, high resolution, selectivity, specificity, and efficiency are obtained in the most modern chromatographs. Its limitations are the high costs involved and the excessive hazardous waste generated by this technique. Besides, low boiling and high masses ($>10^6$ Da) analytes are better investigated using other techniques (Lottspeich and Engels 2018).

Centrifugation is a consolidated technique that is still broadly used. It consists of the use of centrifugal forces to separate particles from their liquid medium. This

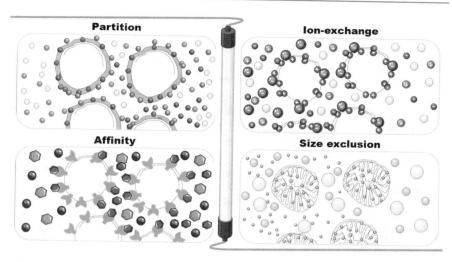

**Fig. 3** Relevant chromatography separation mechanisms for biomolecules: Partition, Ion-exchange, Affinity, and Size exclusion. Images adapted from Biorender (https://biorender.com/) and Servier Medical Art licensed under CC 3.0, https://smart.servier.com

method is based on the principle of gravitational forces that lead particles with higher densities in suspension to deposit downward. This deposition depends on the size, shape, and mass of the particle and viscosity, and density of the supporting fluid. This technique is widely used in biochemical samples such as cells, organelles, DNA, and proteins. Upon using the right exposure time and rotation speed, analytical centrifugation can separate those biomolecules without any harm (Mikkelsen and Cortón 2016; Lottspeich and Engels 2018).

In a centrifuge, the main component is the rotor, which allows high-speed separations. Three types of rotors are used in this technique: swinging-bucket, fixed-angle, and vertical rotors. The difference between them is the direction of the applied forces and, consequently, the angle of deposition. Each rotor is used with different techniques directed to the desired application. Among the techniques, it is possible to mention the differential, zonal, isopycnic, density, and fractional centrifugations. These mechanisms differ according to the exploration of sedimentation, viscosity, and density differences inside of a mixture (Mikkelsen and Cortón 2016; Lottspeich and Engels 2018).

Advances in centrifugation include temperature control, high-speed and ultra-centrifuges (Laue and Stafford 1999). Usually, they are used for specific biological samples that require the control of some factors to avoid damages. Care must be taken with this technique. According to the sample, temperature control is essential to avoid denaturation. Besides, this technique presumably changes the structure and shapes of the particles, including another limitation for some experiments. Undoubtedly, centrifugation is a straightforward and applicable technique, showing its importance for the vast field of biochemical analysis (Mikkelsen and Cortón 2016; Lottspeich and Engels 2018).

## 2.4    Microfluidics: Bringing Solutions to Bioanalytical Chemistry

Microfluidics, also known as micro total analysis systems or lab-on-a-chip, is a growing field offering application in many areas such as sensors, chemical synthesis, biological field, and engineering, among others. Microfluidic devices were initially implemented to study the behavior of fluid through microchannels. The main goal was to miniaturize known processes, enabling the use of lower sample and material volumes, resulting in cheaper and greener processes (Li 2010).

The field of microfluidics was born in 1975, with the fabrication of a silicon-based gas chromatograph (Terry et al. 1979), and it was established with the miniaturization of pumps, valves, and sensors in later works (Hodge et al. 2001; Reyes et al. 2002; Auroux et al. 2002). During the 1990s, not only chromatography and its components were on the track of microfluidics, but also other analytical tools. They evolved toward miniaturization, with the highlight being electrophoresis (Manz et al. 1992; Duarte et al. 2012). Furthermore, and until nowadays, the field keeps having remarkable attention, with the recent development of top technology devices focused on pharmaceutical studies, the so-called organs-on-chip, and diagnosis, with microdevices able to perform the screening of several diseases (Fig. 4) (Reyes et al. 2002; Ingber 2018; Zhang et al. 2018).

Bioanalytical chemistry has always been related to the microfluidics field since there are few conventional methods capable of intensively treating and analyzing biomolecules. Samples such as plasma, saliva, tumor cells, among others, are typically limited to a few microliters, frequently making unfeasible its evaluation

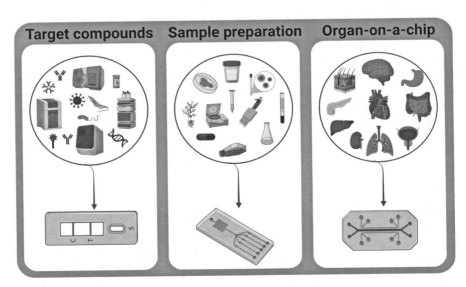

**Fig. 4** Microfluidic devices are applied to target compound analysis, sample preparation, and organs-on-chips. Images adapted from Biorender (https://biorender.com/) and Servier Medical Art licensed under CC 3.0, https://smart.servier.com

through some techniques, e.g., LC-MS and GC-MS. Therefore, microfluidic devices are bringing enormous advances to the sphere of separation (e.g., chromatography and electrophoresis), cell-based assays, nucleic acid, and PCR analysis, immunological and biological reactions, biochemical investigation, protein crystallization, and clinical investigation (Gomez 2010; Imamura et al. 2020; de Oliveira et al. 2020; Bounab et al. 2020; Roman and Kennedy 2007).

The advantages of using microfluidics as a bioanalytical tool include the reduction of materials, reagents, waste, and costs. Improvements in throughput, efficiency, and sensibility are also very convenient in this technique, besides the portability, dynamicity, and disposability it offers. Among its limitations, one can mention the compatibility between materials, samples, and analytical methods, the integration of components in complex systems, and the high investments on microfabrication facilities (Roman and Kennedy 2007).

## 3    Omics Sciences: Guided Bioanalysis

About 20 years ago, the world awoke to a new era. No, we are not talking about the twenty-first century or the internet popularization, but of the human genomics era, guided by the sequencing of the entire human genome achievement of the Human Genome Project (HGP). From now on, scientists believed to have found the perfect tool to answer all the disease-guided questions, an enormous achievement for humanity, the solution for all diseases, and elucidation of thoughtful journeys. Well, reading these beliefs in the 2020s may sound naive, but back then, the scientific belief was guided by the thought that, once DNA holds all the information needed for life, it should also hold all the information related to diseases, behavior, and therapeutic responses (Lander et al. 2001; Emmert-Streib et al. 2017; Gibbs 2020).

Indeed, without DNA, life would not be possible the way we know it. Since Watson-Crick's central dogma statement, DNA became the chest that keeps the necessary information needed to maintain cellular machinery working. Several genes are responsible for different diseases and conditions, for example, the onco-gene p53, known to be an essential gene in several types of cancer (Ozaki and Nakagawara 2011). The p53 gene is responsible for encoding a protein that is directly responsible for checking for DNA damage during the cell cycle. Once the p53 gene is mutated, the p53 protein may lose the ability to check and stop the damaged DNA from being copied, thus, turning possible the proliferation of this flawed cell. This way, after somatic mutations in the DNA, the damaged cell turns into a cancer cell, i.e., a cell with the ability to avoid all the cell cycle checkpoints, multiplying fast and irregularly and becoming a tumor (Pollard et al. 2004; Whibley et al. 2009). Other genes related to different diseases were discovered in the past decades, such as BRCA1 and BRCA2 for breast cancer and the IDDM1 locus for type 1 diabetes (Gomes et al. 2017; Narod and Foulkes 2004).

Life, however, is not simple. Time and immense scientific efforts showed that life is relatively more complex than DNA nucleotides can tell us. Following the idea of

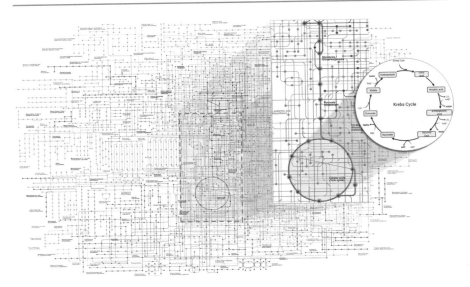

**Fig. 5** Diagram of the metabolic pathways in human physiology. In an analogy of a metro system, each node, or station, represents a metabolite, while each color represents a different train line or metabolic pathway. Lines represent the path traveled by the trains, showing that to a train go from the beginning to the end of the line, it has to pass through all the line stations. Many of the pathways are linked to each other, been the reason that metabolic changes in a specific route can disrupt the entire system. Images adapted from KEGG Pathway Maps (Kanehisa 2019)

the central dogma of biology, proteins also show a crucial role in cellular response, as in viral infections. Here, proteins such as cytokines are responsible for signaling to the immune system cells of the virus presence, and antibodies, also proteins, responsible for marking the virus and bacteria for further elimination by immune system cells or mechanisms (Pollard et al. 2004). Proteins are also responsible for regulating many critical cellular processes. Examples are apoptosis, which is guided by several proteins, including BH3 domain proteins and the cell-division cycle, a process in part controlled by cyclin-dependent kinase (CDK) proteins that regulate each step of the cell cycle (Chipuk et al. 2006; Shamas-Din et al. 2013; Galderisi et al. 2003; O'Connor and Adams 2010).

Although proteins are also not the answer to everything, metabolites enter the field to win, once they are the final products from cellular machinery. Metabolites are responsible for several essential reactions in cellular operation, for example, the ATP and NADH molecules that are responsible for the energy that runs the entire human organism, or even the amino acids, the so-called building blocks of the proteins. Metabolites operate in several different pathways in the cellular workflow, so much that the most recent KEGG (Kyoto Encyclopedia of Genes and Genomes) pathway list, Fig. 5, could be mistaken with a draw of electronic chips given the complexity of the networking and connections, an achievement that would make even the most giant existing subway let behind (Kanehisa 2019). Metabolites are also directly

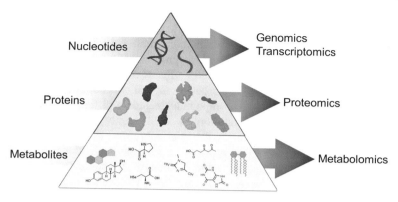

**Fig. 6** Omics sciences applied to molecular biology: from nucleotides to proteins, from proteins to metabolites. Image created with Biorender (https://biorender.com/)

affected by the environment; for example, the diet or the use of pharmaceuticals and drugs directly affects the metabolome—the complete set of metabolites in the organism. It is also highly enjoyable to look in the metabolome for biomarkers of conditions and diseases. As a good example, di-tyrosine and 2-pyrroloylglycine appear to be biomarkers for the evolution of pre-diabetes to diabetes, as cortisol is a biomarker for stress (Zeng et al. 2008; Yan-Do and Macdonald 2017; Hellhammer et al. 2009).

None of the three groups of molecules, however, can, alone, offer a conclusive and final evaluation of all diseases, or be used as a unique tool for explaining human organisms. Nowadays, science realized that no unique perspective of an organism stands alone. The three mentioned groups of molecules work together in a too complex network, where the miss of a single piece of the puzzle may disrupt the entire system or, as humanity calls it: death. This complexity led the scientists to investigate the proteins and metabolites with the same approach as that given to genes. As genes led to genomics, the field of transcriptomics studies RNA, proteomics is the evaluation of the proteome, and metabolomics the investigation of a complete metabolome (de Anda-Jáuregui and Hernández-Lemus 2020; Rahman and Rahman 2018; Joyce and Palsson 2006).

Nowadays, the three sets of biomolecules (nucleotides, proteins, and metabolites) undergo an omics approach for the continued advancement of our knowledge about life. This approach means that using bioanalytical tools, comparisons between different conditions could (1) explain the emerging of a condition or disease, (2) lead to a biomarker, or (3) be employed as a prognostic tool or treatment target (de Anda-Jáuregui and Hernández-Lemus 2020; Karczewski and Snyder 2018). Complexity in the biological matrix can describe the complexity of each of the omics fields as the difficulty of accessing the biomolecules, their range of concentration and localization, and the number of different analytes. Figure 6 represents a visual template of how the omics fields are correlated. Nucleotides (DNA and RNA), besides the complexity of the sequencing, are biomolecules with small molecular diversity and range of concentration, which translate into a smaller analytical challenge. DNA and RNA encode the next group of biomolecules: the proteins, which have a broader range of concentration and more considerable diversity in the same organism than the nucleotides, making proteomics studies more sophisticated from the analytical point of view. Metabolites, the next link in the chain, on the other hand, are present in any organism in a broader range of concentration than the others. They also include a great variety of compounds, including different physicochemical properties, which turns the metabolomics field highly challenging from the bioanalytical analysis perspective. In the next subtopics, the reader will find the main points involving the omics approaches for polynucleotides, proteins, and metabolites, including the most common bioanalytical techniques employed, and the challenges for each one (de Anda-Jáuregui and Hernández-Lemus 2020; Joyce and Palsson 2006; Karczewski and Snyder 2018).

## 3.1   Nucleotides: Genomics and Transcriptomics

Genomics can be considered the precursor of all the omics fields and is defined as the comparison between genomes by sequencing them and checking the similarities and discrepancies. In the past, genomics was one of the most expensive fields of science, with complete genome sequencing costing billions of dollars. Fortunately, sequencing technology made huge progress in the past decades, in such a way that nowadays it is possible to perform DNA analysis at affordable prices, indeed has given start to a new business field for individual DNA analysis (de Anda-Jáuregui and Hernández-Lemus 2020; Raja et al. 2017; Vernon et al. 2019).

As mentioned, "Genomics 101" rests in DNA sequencing for later comparison and analysis of the sequences. During the Human Genome Project – HGP, the sequencing used to be performed by a technology called Sanger sequencing, in reference to the Frederick Sanger sequencing process, Fig. 7, that applied DNA polymerase, deoxynucleotides, and fluorescent marked dideoxynucleotides to synthesize different sizes of DNA chains, therefore separated by capillary gel electrophoresis and read by a fluorescence detector (Carrilho 2000). As electrophoresis separates the polynucleotide chains by size, the different DNA sizes continuously generate the pattern of the DNA sequence. Somehow, we can say that bioanalytical tools made possible the entire field of genomics since, without electrophoresis analysis, no sequencing was possible (Carrilho 2000; Schuster 2008; Dewey et al. 2013; Mamanova et al. 2010; França et al. 2002). More than that, we can say that the analytical chemists saved the Human Genome Project by designing new instrumentation (multiple capillary array electrophoresis), new enzymes (Taq DNA polymerase), new fluorophores (higher quantum yields and convenient spectroscopic properties), and improved separation media (high molecular weight poly(acrylamide)) (Zubritsky 2002).

Nowadays, Sanger sequencing is still widely employed, but the next generation sequencing (NGS) technology took over by storm as it is a faster and cheaper approach than the Sanger sequencing by capillary array electrophoresis. In NGS,

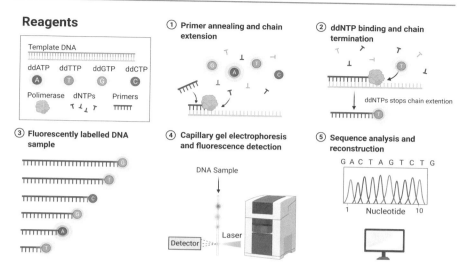

**Fig. 7** Sanger sequencing. The fragmented DNA is combined with a primer, DNA polymerase, deoxynucleotides, and labeled dideoxynucleotides (the Sanger's chain terminator nucleotide) to allow the synthesis of a new DNA strand, originating different sizes of DNA chains. Synthesized DNA chains are later separated by the size of the chain employing capillary electrophoresis using polymer solutions, and read by a four-color fluorescence detector. The electropherogram, combined with the fluorescence measurement, allows the DNA sequence determination. Image created with Biorender (https://biorender.com/)

Fig. 8, the DNA is extracted and purified, fragmented, and reacted with an adapter, a specific sequence of nucleotides. The DNA adapter is later added to a chip, also called a flow cell, containing several molecules of a nucleotide sequence (oligonucleotide) fixed on the surface. This fixed oligonucleotide is complementary to the adapter linked to the DNA, which leads to the pairing of both nucleotides, a process called library hybridization. After this process, DNA polymerase is added to the flow cell, together with deoxynucleotides, leading to a DNA synthesis using the fixed oligonucleotide as a primer. This step is a polymerase chain reaction process (PCR) and is repeated, leading to DNA amplification (Schuster 2008; Dewey et al. 2013; Mamanova et al. 2010; França et al. 2002; Medvedev et al. 2009).

Accordingly, several copies of the DNA fragments are fixed to the surface, and different sequences are located in different clusters on the flow cell. Once the DNA fragment is fixed to the surface, all the previously added deoxynucleotides are washed away, and fluorescence marked terminator deoxynucleotides are added to the flow cell. DNA polymerase then performs the DNA synthesis with the marked deoxynucleotides. After one cycle of synthesis, the equipment read the fluorescence in each of the clusters. Since each marked deoxynucleotide has a different color, the equipment can determine which nucleotide was added in each cluster. After each measurement, chemicals are added to remove the terminator and fluorescence groups from the marked deoxynucleotide, in order to allow DNA polymerase to continue

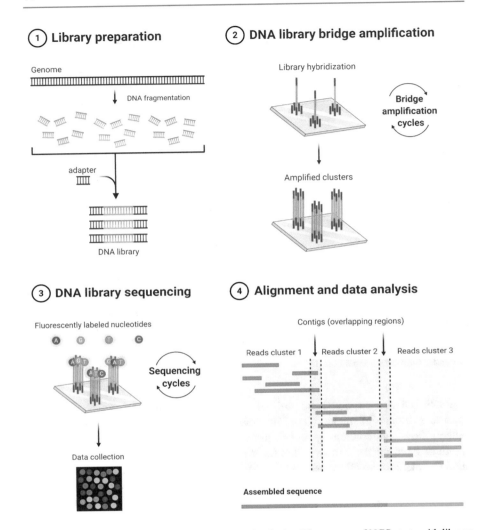

**Fig. 8** One of the next-generation sequencing technologies. The process of NGS starts with library preparation (1) by the DNA fragmentation, followed by the addition of adapters to the DNA sequences. The DNA fragments are inserted into the flow cell, and the DNA is amplified (2). After amplification, fluorescently labeled nucleotides are added in turns, and the sequence of all fragments is acquired (3). As the final step, specific software turns several fragments into a single DNA sequence (4). Image created with Biorender (https://biorender.com/)

the DNA synthesis. The chemicals are then washed away, and the process of adding a newly marked deoxynucleotide is repeated over and over. The equipment measures after each nucleotide is added, turning possible to determine the sequence of all the DNA fragments. Bioinformatics is latter used to gather the fragments into a single DNA sequence (de Anda-Jáuregui and Hernández-Lemus 2020; Vernon et al. 2019;

Carrilho 2000; Schuster 2008; Dewey et al. 2013; Mamanova et al. 2010; Medvedev et al. 2009).

In NGS, electrophoresis separation was abandoned in exchange for a different approach for sequencing, but bioanalytical tools are still turning all the processes possible. The entire NGS process relies on microfluidics devices that afford lower reagents consumption and quicker analysis (França et al. 2002). Therefore, genomics remains holding hands with bioanalytical advances, and further advances in this field will highly benefit genomics technologies (Zubritsky 2002). Despite the sort of technology applied in DNA sequencing, genomics is currently a widely defunded science and permitted several discoveries of how organisms works and the discovery of different genetical markers for disease development, as the use of the BRCA1 gene as a biomarker of breast cancer occurrence probability, and TMSB15A gene as a biomarker for triple-negative breast cancer treatment progress (Davis et al. 2014; Martínez-Jiménez et al. 2020; Carrasco-Ramiro et al. 2017).

Transcriptomics, in parallel, is the evaluation and comparison between the transcriptomes, i.e., the RNA transcripts of the DNA. This field remains highly attractive once RNA transcripts give information about the differentially expressed genes due to a specific condition or treatment. Differently from DNA, which has a vast part of non-coding sequences, RNA transcripts of a given cell are more closely related to the condition and encode information that is going to be used in that specific scenario. Therefore, transcriptomics can be said to be closer to the phenotype (de Anda-Jáuregui and Hernández-Lemus 2020; Sandberg 2014).

Although quite a few techniques permit RNA sequencing or RNA analysis directly, such as the use of microarray technology, transcriptomics is commonly achieved by extraction of RNA, followed by fragmentation of RNA chain and the reverse transcription of the RNA to cDNA (complementary DNA). After the cDNA is obtained, the sequencing is performed by the same techniques explained previously for DNA sequencing. Recently, bioanalytical technologies have been employed in transcriptomics, allowing even the analysis of single-cell transcriptomics. In this approach, organism cells are separated by microfluidic devices and followed by cDNA obtention and sequencing, tuning the science even closer to personalized medicine (de Anda-Jáuregui and Hernández-Lemus 2020; Sandberg 2014; Li et al. 2016).

Nevertheless, the sequencing of nucleotides, both DNA or RNA, still has an enormous impact in different fields as medicine and agriculture. Genomics and transcriptomics were widely developed in crop species in order to understand which loci and genes were responsible for better growth, or better adaptability to a given soil, or even responsible for better response to pathogens. As a result, we have several transgenic species of crop plants, which resulted in better harvests, impacting the food production market (Varshney et al. 2015).

It is interesting to observe that the recent growth of publications, clinical diagnosis, therapies, and agricultural products born from genomics and transcriptomics studies can be closely related to the development of better and advanced bioanalytical tools, that turned possible the advance of these fields and allowed the scientists to transform the world somehow.

## 3.2   Proteins–Proteomics

The term proteomics, published in 1995, is defined as the study of the total protein content able to be encoded by a given genome (Humphery-Smith 2015). The term is still in use, and this field of bioanalytical chemistry seems to be a promising one, with many efforts being put on (Yates et al. 2009; Boggio et al. 2011). The importance of the proteomic study is related to the vital role of proteins in cell functions. Proteins act in a building block scheme for cells, being responsible for numerous functions such as enzymatic reactions, signaling, transcription and translation processes, and structural components, being decisive for cell regulation (Vaz and Tanavde 2019; Aizat and Hassan 2018).

Proteomics can be divided into four different areas: sequence, structural, functional, and interaction and expression proteomics. As the name says, each one of them deals with different properties and aspects of proteins, apart from using different analytical tools for identification and evaluation. Summarizing, sequence proteomics determines amino acid sequences in proteins using chemicals to cleave and tag them. This technique is classically done using Edman sequencing (Edman and Begg 1967). Nowadays, Mass Spectrometry (MS), and Nuclear Magnetic Resonance (NMR), electrophoresis, and HPLC substituted Edman's technique, reducing analysis time. Structural proteomics studies the structure of proteins and the implications in their functions. Many analytical methods are used to determine protein structure, with X-ray diffraction and protein crystallization, and 2D NMR being the most common (Sali et al. 2003; Woolfson 2018; Paganelli et al. 2016). The third area of proteomics studies the functions of proteins and the interactions between proteins and other biomolecules mostly through in vitro analysis and microarrays (LaBaer and Ramachandran 2005; Sutandy et al. 2013). Functional and interaction area is considered the most targeted proteomics of all. Finally, proteomics is capable of elucidating the expression of a protein in a very complex system, with plenty of biomolecules. This elucidation is essential to understand the role of a specific protein in cells, tissues, or entire organisms (e.g., the human body) (Kim et al. 2014), and it is usually analyzed and quantified by MS.

Applications in proteomics are vast and include multiple areas such as diagnosis, drug screening, epidemiology, pathogenicity, oncology, food chemistry, agriculture, and others (Chandrasekhar et al. 2014; Aslam et al. 2017). As every growing field, proteomics brings many challenges. However, as technology develops, new insights emerge, leading to advances in this area.

## 3.3   Metabolites–Metabolomics

Biological sciences are always trying to understand better the organism's conditions, from a healthy functional organism to a disabled one, like an infection, a genetic disease, or cancer. Genomics, transcriptomics, and proteomics have a lot to tell about these conditions. However, there is a set of compounds that are even closer to the phenotype, which means the observable characteristics from the condition, for example, a cough, brain dysfunction, or tumor growth. The group of compounds that are the final product of cellular metabolism and, therefore, the closest step in the relationship of cellular functioning and phenotype is the metabolites (Ahmad 2008).

An array of metabolites is called metabolome and is composed of numerous molecules that have different properties, which can be: inorganic compounds, hydrophilic carbohydrates, organic acids, hydrophobic lipids, complex natural products, volatile molecules, among other (Fiehn 2002). The comparison between the metabolome of a healthy organism and the metabolome of a sick one, known as metabolomics, can lead to the discovery of biomarkers for the detection of several types of diseases in initial stages, allowing the best medical intervention to the patient (Lindahl et al. 2017). In the field of personalized medicine, biomarkers can also be used as prognostic markers, guiding the physicians to the best therapeutic approach (Villas-Bôas and Gombert 2006; Nordström et al. 2006; Wishart 2007).

Metabolomics as a tool is extremely useful for different fields, as in plant science. Here metabolomics can lead to the discovery of different secondary metabolites expelled by the plant in the presence of a pathogen. Alternatively, metabolomics can show the differences in the metabolites of a species after a pesticide application. These discovered compounds can lead to new pharmaceuticals or even the use of less impacting pesticides and are of great interest to natural products science (Funari et al. 2013; Razzaq et al. 2019; Jorge et al. 2016).

Metabolomics studies can be performed in several types of samples, e.g., in vitro (cell culture), in vivo (urine, blood, saliva, and feces), and from different plant parts (Funari et al. 2013; de Zawadzki et al. 2017). Normally, these studies are carried out in three main steps: sample preparation, data acquisition, and data analysis, Fig. 9. Sample preparation, especially the extraction step, is a crucial procedure in non-target metabolomic approaches, that is, when the goal is not a specific set of metabolites, but instead a comprehensive fingerprint of the metabolome (Duportet et al. 2012). Data acquisition can be performed by different analytical techniques, such as liquid chromatography coupled to mass spectrometry (LC-MS), gas chromatography coupled to mass spectrometry (GC-MS), capillary electrophoresis coupled to mass spectrometry (CE-MS), direct infusion to the mass spectrometer (DIMS), ionization and matrix-assisted laser desorption (MALDI) and nuclear magnetic resonance (NMR) (Mesquita et al. 2012; Zawadzki et al. 2018). Each technique has its main strengths and disadvantages, as it will be better discussed in the next chapters of this book. However, it is essential to say that metabolomics studies would not be possible without the recent advances in analytical chemistry (Duportet et al. 2012; Mesquita et al. 2012; Kim and Verpoorte 2010; Ferreira et al. 2016).

Data analysis comprehends statistical and chemometric evaluation of the data to differentiate the groups studied. This evaluation involves, for example, Principal Component Analysis (PCA), Partial Least Square Discriminating Analysis (PLS-DA), Orthogonal Partial Least Square Discriminating Analysis (OPLS-DA), amongst others. Once a metabolite is statistically proved to be able to differentiate the conditions, it is considered to be a potential biomarker. The following step is to identify the compounds using metabolomics databases, such as HMDB and Metlin

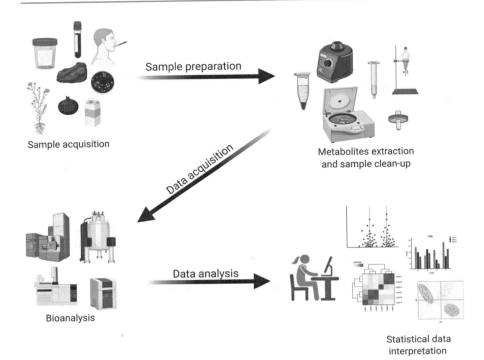

**Fig. 9** Metabolomics workflow. After sample collection, a few steps of sample preparation are processed in order to extract the metabolites and remove interferents. In sequence, the sample is analyzed by analytical tools, such as LC-MS, GC-MS CE-MS, or NMR. At last, the acquired data is analyzed by statistical protocols in order to avoid bias and establish the correct interpretation. Image created with Biorender (https://biorender.com/)

for human metabolome and Nubbe and Dictionary of Natural Products for plant metabolomics (Lindahl et al. 2017; Triba et al. 2015).

Although been one of the newest between omics sciences, metabolomics showed to be extremally useful in clinics already, e.g., in diagnosis and prognosis of different types of cancer, such as breast cancer, where glycerophosphocholine and glucose levels could be used to differentiate tumors from healthy tissue; and prostate cancer, where phosphocholine, lactate, and alanine can be used for diagnosis. Metabolomics discoveries can contribute even harder to physicians by applying the biomarkers in new advanced bioanalytical tools. One of them is the MasSpec Pen, a handheld sample probe that is capable of extract metabolites from tissue directly from the surgery table and sending the metabolites to a mass spectrometer kept inside the surgery room. The MasSpec Pen technology allows the use of biomarkers to differentiate the tumor tissue from healthy tissue, allowing the surgeons to extract less healthy tissue as possible (Spratlin et al. 2009; Zhang et al. 2017).

## 3.4 Bioinformatics and Chemometrics Tools

The vast number of data generated by the advances in analytical instrumentation brought new challenges to perform data analysis and interpretation. Bioinformatics had to be reviewed to differentiate distinct groups of samples, molecules, or methods among themselves. Also, all the design studies had to be properly modeled using statistical tools. Nowadays, chemometrics is useful to help researchers to understand and categorize the obtained data. Moreover, unknown molecules can also be identified through the help of several databases (Bravo-Merodio et al. 2019; Rotroff 2020; Parry-Smith 2019).

Specialized software was developed in the past years to help researchers to perform data analysis, such as Galaxy® for genomics and transcriptomics data, msInspect® for proteomics research, and MetaboAnalyst® for metabolomics data analysis. Nevertheless, software like Statistica® and Minitab® are commonly used for the design of experiments in order to enable the development of better chromatographic separations and comprehensive sample preparation steps. The software, as mentioned earlier, were developed to be user-friendly, not requiring previous programming skills from the user. Although, if the user has programming training, software like R® and MatLab® can be used to refine their data analysis, through the change of the mathematical algorithms, improving the model fitting to their own data (Enot et al. 2011).

Bioinformatics and data analysis protocols vary broadly from lab to lab, and up to today, there is no consensus of the most suitable algorithms for all the research data. Is it suitable to perform a PLS-DA or a neural network with this set of data? Were the results genuinely unbiased? Therein, it is highly important to the analyst take some time to truly understand the statistical models that he is wondering about applying, in order to keep the data interpretation consistent with the acquired data. For this quest, in the next chapters of this book, the reader will find essential information on chemometrics and data analysis.

In summary, all the omics fields have been contributing significantly with the advance of new therapies, diagnostics tools, better crop plantations, food quality control, and even consider contributing to the concept of personalized medicine. It is essential to state that none of the fantastic advances from omics fields would be possible without equal advances in bioanalytical technologies. Improvements in separation techniques, including better chromatographic columns and more durable pumps, evolution in microfluidic devices, and enhancement in detectability and separation on MS and ion-mobility MS and on NMR and chromatographic NMR, were of fundamental importance for the advances in omics sciences; what can exemplify the importance and relevance of the other chapters of this book.

# References

Ahmad S (2008) Plant metabolomics: techniques, applications, trends, and challenges. In: Barh D, Zambare V, Azevedo V (eds) OMICS applications in biomedical, agricultural, and environmental sciences, 1st edn. CRC Press, Boca Raton

Aizat W, Hassan M (2018) Proteomics in systems biology. In: Aizat W, Goh H, Baharum SNS, Azizan KA (eds) Omics applications for systems biology, 1st edn. Springer, Gewerbestrasse

Amiri-Dashatan N, Koushki M, Abbaszadeh H, Rostami-Nejad M (2018) Iran J Pharm Res 17: 1523–1536

Aslam B, Basit M, Nisar MA, Khurshid M, Rasool MH (2017) J Chromatogr Sci 55:182–196

Atanasov AG, Waltenberger B, Pferschy-Wenzig E-M, Linderd T, Wawroscha C, Uhrine P, Temmlf V, Wanga L, Schwaigerb S, Heissa EH, Rollingera JM, Schusterf D, Breusse JM, Bochkovg V, Mihovilovicd MD, Koppa B, Bauerc R, Dirscha VM, Stuppner H (2015) Biotechnol Adv 33:1–18

Auroux PA, Iossifidis D, Reyes DR, Manz A (2002) Anal Chem 74:2637–2652

Bedia C (2018) Experimental approaches in omic sciences. In: Jaumot J, Bedia C, Tauler R (eds) Data analysis for omic sciences: methods and applications, 1st edn. Elsevier, Amsterdam

Boggio KJ, Obasuyi E, Sugino K, Nelson SB, Agar NYR, Agar JN (2011) Expert Rev Proteomics 8:591–604

Bounab Y, Eyer K, Dixneuf S, Rybczynska M, Chauvel C, Mistretta M, Tran T, Aymerich N, Chenon G, Llitjos J-F, Venet F, Monneret G, Gillespie IA, Cortez P, Moucadel V, Pachot A, Troesch A, Leissner P, Textoris J, Bibette J, Guyard C, Baudry J, Griffiths AD, Védrine C (2020) Nat Protoc 15:2920–2955

Bravo-Merodio L, Williams JA, Gkoutos GV, Acharjee A (2019) J Transl Med 17:1–10

Campa C, Coslovi A, Flamigni A, Rossi M (2006) Electrophoresis 27:2027–2050

Carrasco-Ramiro F, Peiró-Pastor R, Aguado B (2017) Gene Ther 24:551–561

Carrilho E (2000) Electrophoresis 21:55–65

Chandrasekhar K, Dileep A, Lebonah DE, Kumari JP (2014) Int Lett Nat Sci 17:77–84

Chen X, Wang Y, Ma N, Tian J, Shao Y, Zhu B, Wong YK, Liang Z, Zou C, Wang J (2020) Signal Transduct Target Ther 5:1–13

Chipuk JE, Bouchier-Hayes L, Green DR (2006) Cell Death Differ 13:1396–1402

Clark KD, Zhang C, Anderson JL (2016) Anal Chem 88:11262–11270

Datta LP, Manchineella S, Govindaraju T (2020) Biomaterials 230:119633

Davis SL, Eckhardt SG, Tentler JJ, Diamond JR (2014) Ther Adv Med Oncol 6:88–100

de Anda-Jáuregui G, Hernández-Lemus E (2020) Front Oncol 10:1–21

de Oliveira GCM, de Souza Carvalho JH, Brazaca LC, Vieira NCS, Janegitz BC (2020) Biosens Bioelectron 152:112016

de Zawadzki A, Arrivetti LOR, Vidal MP, Catai JR, Nassu RT, Tullio RR, Berndt A, Oliveira CR, Ferreira AG, Neves-Junior LF, Colnago LA, Skibsted LH, Cardoso DR (2017) Food Res Int 99: 336–347

Dewey FE, Pan S, Wheeler MT, Quake SR, Ashley EA (2013) Circulation 125:931–944

Duarte GRM, Coltro WKT, Borba JC, Price CW, Landers JP, Carrilho E (2012) Analyst 137:2692– 2698

Duportet X, Aggio RBM, Carneiro S, Villas-Bôas SG (2012) Metabolomics 8:410–421

Eberl HC, Werner T, Reinhard FB, Lehmann S, Thomson D, Chen P, Zhang C, Rau C, Muelbaier M, Drewes G, Drewry D, Bantscheff M (2019) Sci Rep 9:1–14

Edman P, Begg G (1967) Eur J Biochem 1:80–91

Emmert-Streib F, Dehmer M, Yli-Harja O (2017) Front Genet 8:184–186

Enot DP, Haas B, Weinberger KM (2011) Bioinformatics for mass spectrometry-based metabolomics. In: Mayer B (ed) Bioinformatics for omics data, methods and protocols, 1st edn. Humana Press, London

Fekete A, Schmitt-Kopplin P (2007) Capillary electrophoresis. In: Picó Y (ed) Food toxicants analysis, 1st edn. Elsevier, Amsterdam

Ferreira VG, Leme GM, Cavalheiro AJ, Funari CS (2016) Anal Chem 88:8421–8427

Fiehn O (2002) Metabolomics – the link between genotypes and phenotypes. In: Town CD (ed) Functional genomics, 1st edn. Springer, Dordrecht

França LTC, Carrilho E, Kist TBL (2002) Q Rev Biophys 35:169–200

Funari CS, Castro-Gamboa I, Cavalheiro AJ, Bolzani VS (2013) Quim Nova 36:1605–1609

Galderisi U, Jori FP, Giordano A (2003) Oncogene 22:5208–5219

Gault VA, McClenaghan NH (2009) Introduction to biomolecules. In: Gault VA, McClenaghan NH (eds) Understanding bioanalytical chemistry, 1st edn. Wiley-Blackwell, Chichester

Gibbs RA (2020) Nat Rev Genet 21:575–576

Glad M, Larsson PO (1991) Curr Opin Biotechnol 2:413–418

Gomes KFB, Santos AS, Semzezem C, Correia MR, Brito LA, Ruiz MO, Fukui RT, Matioli SR, Passos-Bueno MR, da Silva MER (2017) Sci Rep 7:1–10

Gomez FA (2010) Bioanalysis 2:1661–1662

Hellhammer DH, Wüst S, Kudielka BM (2009) Psychoneuroendocrinology 34:163–171

Hersel U, Dahmen C, Kessler H (2003) Biomaterials 24:4385–4415

Hodge C, Bousse L, Knapp M (2001) Microfluidic analysis, screening, and synthesis. In: Sucholeiki I (ed) High throughput synthesis, 1st edn. Marcel Dekker, New York

Horvai G, Worsfold P, Karlberg B, Andersen JET (2011) Trends Anal Chem 30:422–424

Humphery-Smith I (2015) Proteomics 15:1773–1776

Imamura AH, Segato TP, de Oliveira LJM, Hassan A, Crespilho FN, Carrilho E (2020) Microchim Acta 187:1–8

Ingber DE (2018) Development 145:10–13

Jorge TF, Mata AT, António C (2016) Philos Trans R Soc A Math Phys Eng Sci 374:20150370

Joyce AR, Palsson B (2006) Nat Rev Mol Cell Biol 7:198–210

Kanehisa M (2019) Protein Sci 28:1947–1951

Karczewski KJ, Snyder MP (2018) Nat Rev Genet 19:299–310

Kim HK, Verpoorte R (2010) Phytochem Anal 21:4–13

Kim M-S, Pinto SM, Getnet D, Nirujogi RS, Manda SS, Chaerkady R, Madugundu AK, Kelkar DS, Isserlin R, Jain S, Thomas JK, Muthusamy B, Leal-Rojas P, Kumar P, Sahasrabuddhe NA, Balakrishnan L, Advani J, George B, Renuse S, Selvan LDN, Patil AH, Nanjappa V, Radhakrishnan A, Prasad S, Subbannayya T, Raju R, Kumar M, Sreenivasamurthy SK, Marimuthu A, Sathe GJ, Chavan S, Datta KK, Subbannayya Y, Sahu A, Yelamanchi SD, Jayaram S, Rajagopalan P, Sharma J, Murthy KR, Syed N, Goel R, Khan AA, Ahmad S, Dey G, Mudgal K, Chatterjee A, Huang T-C, Zhong J, Wu X, Shaw PG, Freed D, Zahari MS, Mukherjee KK, Shankar S, Mahadevan A, Lam H, Mitchell CJ, Shankar SK, Satishchandra P, Schroeder JT, Sirdeshmukh R, Maitra A, Leach SD, Drake CG, Halushka MK, Prasad TSK, Hruban RH, Kerr CL, Bader GD, Iacobuzio-Donahue CA, Gowda H, Pandey A (2014) Nature 509:575–581

Kogikoski S, Paschoalino WJ, Kubota LT (2018) Trends Anal Chem 108:88–97

Kubinyi H (2002) J Braz Chem Soc 13:717–726

LaBaer J, Ramachandran N (2005) Curr Opin Chem Biol 9:14–19

Labuda J, Bowater RP, Fojta M, Gauglitz G, Glatz Z, Hapala I, Havliš J, Kilar F, Kilar A, Malinovská L, Sirén HMM, Skládal P, Torta F, Valachovič M, Wimmerová M, Zdráhal Z, Hibbert DB (2018) Pure Appl Chem 90:1121–1198

Lander ES, Linton LM, Birren B, Nusbaum C, Zody MC, Baldwin J, Devon K, Dewar K, Doyle M, Fitzhugh W, Funke R, Gage D, Harris K, Heaford A, Howland J, Kann L, Lehoczky J, Levine R, McEwan P, McKernan K, Meldrim J, Mesirov JP, Miranda C, Morris W, Naylor J, Raymond C, Rosetti M, Santos R, Sheridan A, Sougnez C, Stange-Thomann N, Stojanovic N, Subramanian A, Wyman D, Rogers J, Sulston J, Ainscough R, Beck S, Bentley D, Burton J, Clee C, Carter N, Coulson A, Deadman R, Deloukas P, Dunham A, Dunham I, Durbin R, French L, Grafham D, Gregory S, Hubbard T, Humphray S, Hunt A, Jones M, Lloyd C, McMurray A, Matthews L, Mercer S, Milne S, Mullikin JC, Mungall A, Plumb R, Ross M, Shownkeen R, Sims S, Waterston RH, Wilson RK, Hillier LW, McPherson JD, Marra MA,

Mardis ER, Fulton LA, Chinwalla AT, Pepin KH, Gish WR, Chissoe SL, Wendl MC, Delehaunty KD, Miner TL, Delehaunty A, Kramer JB, Cook LL, Fulton RS, Johnson DL, Minx PJ, Clifton SW, Hawkins T, Branscomb E, Predki P, Richardson P, Wenning S, Slezak T, Doggett N, Cheng JF, Olsen A, Lucas S, Elkin C, Uberbacher E, Frazier M, Gibbs RA, Muzny DM, Scherer SE, Bouck JB, Sodergren EJ, Worley KC, Rives CM, Gorrell JH, Metzker ML, Naylor SL, Kucherlapati RS, Nelson DL, Weinstock GM, Sakaki Y, Fujiyama A, Hattori M, Yada T, Toyoda A, Itoh T, Kawagoe C, Watanabe H, Totoki Y, Taylor T, Weissenbach J, Heilig R, Saurin W, Artiguenave F, Brottier P, Bruls T, Pelletier E, Robert C, Wincker P, Rosenthal A, Platzer M, Nyakatura G, Taudien S, Rump A, Smith DR, Doucette-Stamm L, Rubenfield M, Weinstock K, Hong ML, Dubois J, Yang H, Yu J, Wang J, Huang G, Gu J, Hood L, Rowen L, Madan A, Qin S, Davis RW, Federspiel NA, Abola AP, Proctor MJ, Roe BA, Chen F, Pan H, Ramser J, Lehrach H, Reinhardt R, McCombie WR, De La Bastide M, Dedhia N, Blöcker H, Hornischer K, Nordsiek G, Agarwala R, Aravind L, Bailey JA, Bateman A, Batzoglou S, Birney E, Bork P, Brown DG, Burge CB, Cerutti L, Chen HC, Church D, Clamp M, Copley RR, Doerks T, Eddy SR, Eichler EE, Furey TS, Galagan J, Gilbert JGR, Harmon C, Hayashizaki Y, Haussler D, Hermjakob H, Hokamp K, Jang W, Johnson LS, Jones TA, Kasif S, Kaspryzk A, Kennedy S, Kent WJ, Kitts P, Koonin EV, Korf I, Kulp D, Lancet D, Lowe TM, McLysaght A, Mikkelsen T, Moran JV, Mulder N, Pollara VJ, Ponting CP, Schuler G, Schultz J, Slater G, Smit AFA, Stupka E, Szustakowki J, Thierry-Mieg D, Thierry-Mieg J, Wagner L, Wallis J, Wheeler R, Williams A, Wolf YI, Wolfe KH, Yang SP, Yeh RF, Collins F, Guyer MS, Peterson J, Felsenfeld A, Wetterstrand KA, Myers RM, Schmutz J, Dickson M, Grimwood J, Cox DR, Olson MV, Kaul R, Raymond C, Shimizu N, Kawasaki K, Minoshima S, Evans GA, Athanasiou M, Schultz R, Patrinos A, Morgan MJ (2001) Nature 409:860–921

Laue TM, Stafford WF (1999) Annu Rev Biophys Biomol Struct 28:75–100

Lechner A, Wolff P, Leize-Wagner E, François Y-N (2020) Anal Chem 92:7363–7370

Li PCH (2010) Introduction. In: Li PCH (ed) Fundamentals of microfluidics and lab on a chip for biological analysis and discovery, 1st edn. CRC Press, Boca Raton

Li JR, Sun CH, Li W, Chao RF, Huang CC, Zhou XJ, Liu CC (2016) Nucleic Acids Res 44:D944–D951

Lindahl A, Heuchel R, Forshed J, Lehtiö J, Löhr M, Nordström A (2017) Metabolomics 13:1–10

Lottspeich F, Engels JW (2018) Protein purification. In: Lottspeich F, Engels JW (eds) Bioanalytics—analytical methods in biochemistry and molecular biology, 1st edn. Wiley, Weinheim

Mamanova L, Coffey AJ, Scott CE, Kozarewa I, Turner EH, Kumar A, Howard E, Shendure J, Turner DJ (2010) Nat Methods 7:111–118

Manz A, Harrison DJ, Verpoorte EMJ, Fettinger JC, Paulus A, Lüdi H, Widmer HM (1992) J Chromatogr A 593:253–258

Manz A, Pamme N, Lossifidis D (2004) Biomolecules. In: Manz A, Pamme N, Lossifidis D (eds) Bioanalytical chemistry, 1st edn. Imperial College Press, London

Martínez-Jiménez F, Muiños F, Sentís I, Deu-Pons J, Reyes-Salazar I, Arnedo-Pac C, Mularoni L, Pich O, Bonet J, Kranas H, Gonzalez-Perez A, Lopez-Bigas N (2020) Nat Rev Cancer 20:555–572

Matuszewski BK, Constanzer ML, Chavez-Eng CM (2003) Anal Chem 75:3019–3030

Medvedev P, Stanciu M, Brudno M (2009) Nat Methods 6:S13–S20

Mesquita FS, Alexandri FLD, Scolari SC, Membrive CMB, Papa PC, Cardoso D (2012) Anim Reprod 9:713–722

Mikkelsen SR, Cortón E (2016) Principles of electrophoresis. In: Mikkelsen SR, Cortón E (eds) Bioanalytical chemistry, 2nd edn. Wiley, Hoboken

Narod SA, Foulkes WD (2004) Nat Rev Cancer 4:665–676

Nordström A, O'Maille G, Qin C, Siuzdak G (2006) Anal Chem 78:3289–3295

O'Connor C, Adams JU (2010) Cyclin-dependent kinases regulate progression through the cell cycle. In: O'Connor C, Adams JU (eds) Essentials of cell biology, 1st edn. NPG Education, Cambridge

Ozaki T, Nakagawara A (2011) Cancer 3:994–1013

Paganelli MO, Grossi AB, Dores-Silva PR, Borges JC, Cardoso DR, Skibsted LH (2016) Food Chem 210:491–499

Parry-Smith DJ (2019) Bioinformatics and its applications in genomics. In: Ramesh V (ed) Biomolecular and bioanalytical techniques, 1st edn. Wiley, Chichester

Pollard TD, Earnshaw WC, Lippincott-Schawartz J, Johnson GT (2004) Introduction to cell cycle. In: Pollard TD, Earnshaw WC, Lippincott-Schawartz J, Johnson GT (eds) An introduction to cell biology, 3th edn. Elsevier, Philadelphia

Poole C, Mester Z, Miró M, Pedersen-Bjergaard S, Pawliszyn J (2016) Pure Appl Chem 88:649–687

Pross A, Pascal R (2013) Open Biol 3:120–190

Qiao B, de la Cruz MO (2020) ACS Nano 14:10616–10623

Rahman J, Rahman S (2018) Lancet 391:2560–2574

Raja K, Patrick M, Gao Y, Madu D, Yang Y, Tsoi LC (2017) Int J Genomics 2017:6213474

Razzaq A, Sadia B, Raza A, Hameed MK, Saleem F (2019) Metabolites 9:303–339

Reyes DR, Iossifidis D, Auroux P, Manz A (2002) Anal Chem 74:2623–2636

Roat-Malone RM (2007) Biochemistry fundamentals. In: Roat-Malone RM (ed) Bioinorganic chemistry, 2nd edn. Wiley, Hoboken

Roman GT, Kennedy RT (2007) J Chromatogr A 1168:170–188

Rotroff DM (2020) Chest 158:S113–S123

Sali A, Glaeser R, Earnest T, Baumeister W (2003) Nature 422:216–225

Sandberg R (2014) Nat Methods 11:22–24

Schuster SC (2008) Nat Methods 5:16–18

Selzer PM, Marhöfer RJ, Koch O (2018) The biological foundations of bioinformatics. In: Selzer PM, Marhöfer RJ, Koch O (eds) Applied bioinformatics, 2nd edn. Springer, Berlin

Shamas-Din A, Kale J, Leber B, Andrews DW (2013) Cold Spring Harb Perspect Biol 5:a008714

Spratlin JL, Serkova NJ, Eckhardt SG (2009) Clin Cancer Res 15:431–440

Srikoti M, Bolgar MS, Kazakevich Y (2020) J Chromatogr B Anal Technol Biomed Life Sci 1140:121984

Sutandy FXR, Qian J, Chen CS, Zhu H (2013) Curr Protoc Protein Sci 72:1–16

Terry SC, Herman JH, Angell JB (1979) IEEE Trans Electron Devices 26:1880–1886

Triba MN, Le Moyec L, Amathieu R, Goossens C, Bouchemal N, Nahon P, Rutledge DN, Savarin P (2015) Mol BioSyst 11:13–19

Varshney RK, Kudapa H, Pazhamala L, Chitikineni A, Thudi M, Bohra A, Gaur PM, Janila P, Fikre A, Kimurto P, Ellis N (2015) Crit Rev Plant Sci 34:169–194

Vaz C, Tanavde V (2019) Proteomics. In: Arivaradarajan P, Misra G (eds) Omics approaches, technologies and applications: integrative approaches for understanding OMICS data, 1st edn. Springer, Singapore

Vernon ST, Hansen T, Kott KA, Yang JY, O'Sullivan JF, Figtree GA (2019) Microcirculation 26:e12488

Villas-Bôas SG, Gombert AK (2006) Biotecnologia Ciência e Desenvolvimento 36:59–69

Wake MH (2008) Bioscience 58:349–353

Wang X (2012) Clin Transl Med 1:1–3

Whibley C, Pharoah PDP, Hollstein M (2009) Nat Rev Cancer 9:95–107

Wishart DS (2007) Brief Bioinform 8:279–293

Woolfson MM (2018) Phys Scr 93:032501

Yan-Do R, Macdonald PE (2017) Endocrinology 158:1064–1073

Yates JR, Ruse CI, Nakorchevsky A (2009) Annu Rev Biomed Eng 11:49–79

Yoon D, Choi BR, Kim YC, Oh SM, Kim HG, Kim JU, Baek NI, Kim S, Lee DY (2019) Biomol Ther 9:424–439

Zawadzki A, Alloo C, Grossi AB, do Nascimento ESP, Almeida LC, Junior SB, Skibsted LH, Cardoso DR (2018) Food Res Int 105:210–220

Zeng W, Musson DG, Fisher AL, Chen L, Schwartz MS, Woolf EJ, Wang AQ (2008) J Pharm Biomed Anal 46:534–542

Zhang J, Rector J, Lin JQ, Young JH, Sans M, Katta N, Giese N, Yu W, Nagi C, Suliburk J, Liu J, Bensussan A, Dehoog RJ, Garza KY, Ludolph B, Sorace AG, Syed A, Zahedivash A, Milner TE, Eberlin LS (2017) Sci Transl Med 9:eaan3968

Zhang B, Korolj A, Lai BFL, Radisic M (2018) Nat Rev Mater 3:257–278

Zubritsky E (2002) Anal Chem 74:22A–26A

# Role of Bioanalytical Chemistry in the Twenty-First Century

Rachel A. Saylor and Susan M. Lunte

## 1  Introduction

As we write this introductory chapter, the COVID-19 pandemic is raging around the world. Never has analytical chemistry and specifically, bioanalytical chemistry, been more important. It is involved in the direct detection of the virus as well as the antibodies present against it. Good analytical chemistry is also critical for the validation of vaccine integrity and stability. But what is analytical chemistry? Charles N. Reilley once said that "analytical chemistry is what analytical chemists do." Analytical chemists develop creative new methods to monitor and quantify analytes in all sorts of matrices. Bioanalytical chemists apply analytical chemistry to biological systems to analyze the content of food, drugs, soil, bodily fluids, and tissues, to name a few examples. They even develop methods and instrumentation to look for chemical signatures of life on other planets. Analytical chemists are a versatile and creative group. You can find them in pharmaceutical companies, clinical, environmental, and forensic laboratories, as well in as academia, pushing the frontiers of measurement science. In this chapter, we will highlight exciting, recent advances in analytical chemistry, after providing a brief historical context of the key progress in each technique. Many of the historical developments in analytical chemistry we will discuss have resulted in Nobel Prizes, which are summarized in Table 1.

Much of analytical chemistry in the twentieth century was focused on geochemistry and the determination of metals in ores. Professor Fritz Feigl, who immigrated to Brazil in 1940, was a major driving force in the development of the field of

R. A. Saylor (✉)
Department of Chemistry and Biochemistry, Oberlin College, Oberlin, OH, USA
e-mail: Rachel.Saylor@oberlin.edu

S. M. Lunte
Department of Chemistry, Ralph N. Adams Institute for Bioanalytical Chemistry, University of Kansas, Lawrence, KS, USA

© The Author(s), under exclusive license to Springer Nature Switzerland AG 2022
L. T. Kubota et al. (eds.), *Tools and Trends in Bioanalytical Chemistry*,
https://doi.org/10.1007/978-3-030-82381-8_2

**Table 1** Nobel prizes for analytical methods used in bioanalytical chemistry

| Technique | Prize | Recipient |
| --- | --- | --- |
| Mass spectrometry (isotopes) | Chemistry 1922 | F. Aston |
| Centrifugation | Chemistry 1926 | T. Svedberg |
| Raman | Physics 1930 | C. V. Raman |
| Electrophoresis | Chemistry 1948 | A. Tiselius |
| Liquid chromatography | Chemistry 1952 | A. Martin and R. Synge |
| Electrochemistry | Chemistry 1959 | J. Heyrovsky |
| Laser spectroscopy | Physics 1964 | C. Townes, N. Basov, and A. Prokhorov |
| Immunoassay | Physiology and Medicine 1977 | R. Yalow and S. Berson |
| DNA sequencing | Chemistry 1980 | W. Gilbert and F. Sanger |
| Ion trap | Physics 1989 | H. Dehmelt and W. Paul |
| NMR | Chemistry 1991 | R. Ernst |
| NMR | Chemistry 2002 | Kurt Wüthrich |
| Electrospray ionization | Chemistry 2002 | J. Fenn |
| MALDI | Chemistry 2002 | K. Tanaka |
| Optical tweezers | Physics 2018 | A. Ashkin |

analytical chemistry in Latin America. He developed unique quantitative spot tests for both inorganic and organic compounds that are still widely used today (Schufle and Ionescu 1976). In the early part of the twentieth century, electrochemical and spectroscopic methods were also focused primarily on metal analysis. In fact, many general chemistry labs were focused on identifying the presence of metal ions in an "unknown" sample, a practice that continues even today.

## 2    Electrophoresis and Chromatography

### 2.1    Historical Perspective

Biology and medicine became increasingly important in the twentieth century, as knowledge about medical disorders and diseases grew. Many of the key discoveries in these areas relied on the development of new analytical methods that made the measurements of the basic components of living things possible. To put these developments in perspective, the discovery of biological cells was first reported by Robert Hooke in 1665 using one of the first light microscopes (Rowbury 2012). It was almost 200 years later in 1838 that Gerardus Mulder and Jacob Berzelius discovered and named proteins (Miquel et al. 1838). It took another 100 years for Linus Pauling to elucidate the principles of protein structure (Pauling et al. 1951). In particular, new separation methods for biomolecules developed in the early part of the twentieth century played a key role in these important discoveries in biology and medicine that have had a tremendous impact on the science we do today.

One of these early separation methods was the ultracentrifuge. This technique was developed by Theodor (The) Svedberg and was used to prove conclusively that proteins were unique polymers of amino acids and not colloidal substances (Svedberg 1937). His student, Arne Wilhelm Kaurin Tiselius, subsequently developed electrophoresis, a new analytical method that separated biopolymers based on their charge and size. Tiselius reported the first electrophoretic separation of serum proteins in 1937 (Tiselius 1937). The five bands first reported in his paper are still used in clinical chemistry today, even though each band of proteins can now be separated into many different individual components. Gel and paper electrophoresis followed this initial discovery and today electrophoresis is accomplished in slab, capillary, and microfluidic formats. Recent examples of the use of modern forms of electrophoresis for protein, DNA, and small molecule separations are given throughout this chapter.

Liquid chromatography is another important analytical tool that was first described by Mikhail Tswett at the turn of the century for the separation of plant pigments (Ettre 2003). Over the next 100 years, it evolved into one of the most important and ubiquitous analytical methods used in bioanalysis. This evolution was catalyzed by the work of Archer Martin and Richard Synge who received the Nobel Prize in 1952 for the development of modern liquid chromatography (Martin and Synge 1941). Later, Stanford Moore and William H. Stein combined liquid chromatographic analysis with post-column derivatization to develop a method for determining the amino acid composition of proteins, which became a critical technique for protein identification (Moore and Stein 1954). Liquid chromatography in many different formats is now routinely employed for all types of bioanalytical assays, from food analysis to pharmaceutical development. The development of new types of column supports for protein separations has been a focus more recently and the innovative work of Mary Wirth on the development of highly efficient stationary phases using submicron particles is described below.

## 2.2    Submicron Particles for Protein Separations

Proteins and peptides are large biomolecules, comprised of amino acids, that play a role in many critical functions in the body, including signaling, transporting and storing other molecules, acting as catalysts or enzymes, controlling cell growth and differentiation, among others. Many modern therapeutics are either comprised of proteins or impact protein function in the body. Due to the importance of proteins for biological functions, it is of widespread interest to develop methods capable of interrogating these molecules in complex, often biological, samples.

Often, researchers who are interested in characterizing the protein content of a sample turn to top-down proteomics, where intact proteins are analytically separated using higher-resolution techniques prior to their fragmentation and detection with mass spectrometry (MS). Techniques such as liquid chromatography–mass spectrometry (LC-MS) or tandem MS experiments (MS-MS or $MS^n$) are commonly performed. This approach provides better resolution, dynamic range, and limits of

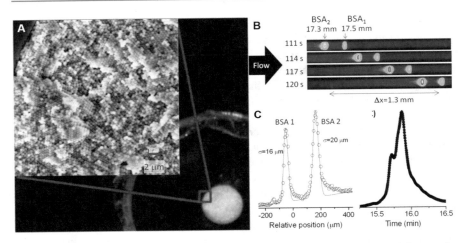

**Fig. 1** Separation of proteins using columns packed with submicrometer particles. (**a**) Image of a cross-sectional area of capillary packed with 470 nm silica particles. (**b**) Separation of two labeled bovine serum albumin (BSA) proteoforms over time using the column in (**a**). (**c**) Comparison of the BSA separation accomplished in (**b**) with one in a commercially available UHPLC. Image adapted with permission from (Wei et al. 2012)

detection, and is a more universal detection strategy than slab gel immunostaining-based methods, which are discussed later in this chapter. However, LC-MS of intact proteins remains challenging due to important, but minute, variations in protein structure caused by post-translational modifications, disulfide bonds, fragmentation, aggregation, oxidation, and other biological and exobiological processes. The various proteoforms of an individual protein can provide information on disease and disorders (Smith and Kelleher 2013), but these and other variations can also cause broad or splitting peaks in chromatographic analysis (Everley and Croley 2008; Wang et al. 2005).

To solve this problem, Mary Wirth and coworkers developed LC columns with uniformly and densely packed submicrometer silica particles to obtain highly efficient protein separations, as seen in Fig. 1 (Wei et al. 2012; Malkin et al. 2010). Submicron particles improve the efficiency of a chromatographic separation in two key ways: (1) the small, uniformly packed silica particles reduce both Eddy diffusion and mass transfer broadening and (2) these particles enable slip flow, or conditions under which the velocity of the mobile phase is non-zero at the wall, enabling a faster overall velocity and a narrower relative velocity distribution. The theoretical basis for these improvements is discussed in more detail in an excellent review (Rogers et al. 2015). Employing these densely packed silica particles with a bonded C18 stationary phase and reverse-phase LC-MS analysis enabled a threefold improvement in chromatographic peak capacities when compared to previous reports (Wu et al. 2014). These separations have been employed to separate proteoforms of *E. coli* (Wu et al. 2014) and bovine serum albumin, as seen in Fig. 1 (Wei et al. 2012), among others.

# 3 Mass Spectrometry

## 3.1 Historical Perspective

Mass spectrometry (MS) was first discovered by Joseph J. Thompson and then further developed by Francis W. Aston in the early part of the last century (Griffiths 2008). Prior to 1980, most mass spectrometry was performed either by direct injection or combined with gas chromatography and was limited to the analysis of volatile compounds. Analysis of biomolecules with GC-MS almost always required derivatization of the analytes to make them volatile. The invention of electrospray ionization by John Fenn made it possible to couple liquid chromatography directly to mass spectrometry (Fenn 1993), revolutionizing bioanalytical chemistry. Fenn was awarded the Nobel Prize in 2002 along with Koichi Tanaka who invented matrix-assisted laser desorption (MALDI) mass spectrometry (and Kurt Wüthrich for his work on nuclear magnetic resonance (NMR) spectroscopy). In the twenty-first century, mass spectrometry has continued to evolve with an emphasis on new types of mass analyzers as well as ionization methods. Of particular interest is the use of mass spectrometry to provide biochemical maps of tissues. A recent example from Amanda Hummon's group investigating drug disposition and metabolism in 3D tumor culture models is discussed below. The use of ambient ionization with mass spectrometry to analyze molecules on surfaces has also been an important area of research, especially in forensic analysis. The development and use of desorption electrospray ionization (DESI) by R. Graham Cooks' group to look for drugs on surfaces is also described in this section.

## 3.2 Biochemical Imaging by MALDI Mass Spectrometry of 3D Cell Cultures

One important aspect of bioanalytical chemistry is in understanding the metabolism and distribution of drugs in the body and its cells. Depending on the application and ease of sampling, an analysis may be performed in humans, animal models, or cell cultures. Cell culture models are often used to evaluate drugs prior to animal or human studies. Most of these experiments rely on 2D, monolayer cell cultures; however, these systems have been shown to provide limited information due to lack of physiological barriers to drug delivery, differences in cell-cell or cell-matrix interactions, and cell proliferation, among other restrictions (Fennema et al. 2013). In contrast, 3D cell cultures more closely mimic in vivo environments. When employing model systems for drug development, it is important to develop methods capable of investigating drug penetration, metabolism, and any resulting metabolomics alterations in the cell(s). Matrix-assisted laser desorption/ionization-mass spectrometry imaging (MALDI-MSI) is a label-free technique that allows for the detection and subsequent visualization of analyte distribution within a tissue section and has been reviewed previously (Liu and Hummon 2015; Spengler 2015; Tobias and Hummon 2020).

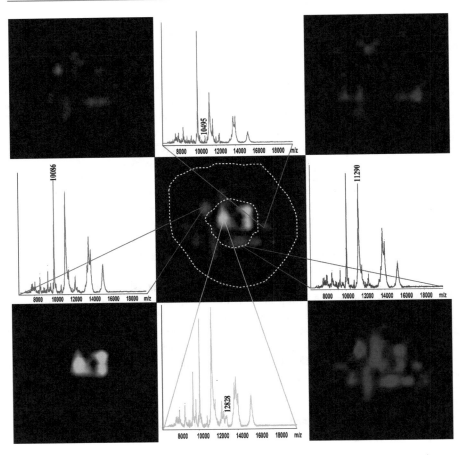

**Fig. 2** MALDI-MS imaging of a spheroid grown with a colon carcinoma cell line. Four species are highlighted with ion intensity maps and mass spectra, showing distinct regions of the cell spheroid are characteristic of different analytes. Image reprinted with permission from (Li and Hummon 2011)

In 2011, the Amanda Hummon lab was the first to apply MALDI-MSI to a 3D cell culture of multicellular tumor spheroids (MCTS), a model of a human solid tumor that possess three distinct concentric zones: necrotic center, quiescent zone, and proliferating outer layer (Li and Hummon 2011). Through MALDI-MSI of a colon carcinoma MCTS, the Hummon lab demonstrated that while some *m/z* values were present throughout the spheroid, there were others that were only present in specific zones, as can be seen in Fig. 2. In later publications, this group expanded their work to include monitoring chemotherapeutic drug delivery and metabolism (Liu et al. 2013). They were able to quantitatively evaluate the proteome of the spheroid in response to chemotherapeutics and identify alterations in several biochemical pathways associated with cancer (Labonia et al. 2018). These methods for analytically investigating 3D cancer cell spheroids provide a viable platform for preclinical drug development and evaluation and highlight the utility of approaching drug development from an analytical perspective.

## 3.3 DESI Mass Spectrometry of Latent Fingerprints

The measurement of chemicals in our environment, for forensic, health, or environmental applications, is an important and ever-growing area of study. However, performing forensic or environmental analysis can be analytically challenging. Depending on the application, sample, and analyte(s) of interest, researchers are often dealing with trace analyte amounts in very complex samples, which can also be incredibly diverse (including soil, water, plant matter, air, documents, hair, tissue, bodily fluids, etc.). In many instances, the analytical method must be non-destructive and/or performed on-site. Most importantly, knowing the validity of the results is imperative, as often health, public safety, and justice are at stake, as in the recent Flint water crisis, or more generally, in crime scene investigations. These analytical challenges can be overcome using a variety of techniques and methods, dependent on the specific application. Mass spectrometry, especially ambient MS, is attractive for environmental and forensic analysis as many different classes of analytes can be detected in complex matrices, and their identities verified using tandem MS. There are previous reviews on this subject (Correa et al. 2016; Cooks et al. 2006). While there are now many influential mass spectrometry techniques, including direct analysis in real time (DART) MS (Cody et al. 2005), easy ambient sonic spray ionization (EASI) MS (Haddad et al. 2006), and paper spray ionization (PSI) MS (Liu et al. 2010), we will turn our attention to the first ambient mass spectrometry method developed in the twenty-first century, desorption electrospray ionization (DESI) mass spectrometry (Takáts et al. 2004).

Until the development of DESI in 2004 by the R. Graham Cooks lab (Takáts et al. 2004), mass spectrometry analysis required samples to be under a vacuum or pretreated in some way (i.e., application of a matrix when employing MALDI mass spectrometry for the analysis of solids). DESI can be considered a combination of electrospray ionization (ESI) with an atmospheric pressure desorption ionization (DI) technique. DESI employs a spray emitter to create charged droplets (similarly to ESI) that bombard the sample; these charged droplets first interact with analyte(s) on the surface of the sample and are subsequently desorbed and propelled into the mass analyzer (similarly to DI) (Takáts et al. 2005). DESI enables samples to be analyzed with mass spectrometry without being placed in a vacuum or undergoing any sample pretreatment; samples can be comprised of solids, liquids, or adsorbed gases and analyzed for many different types of compounds including small molecules, polymers, and proteins, among others. These features allow DESI-MS to be widely applicable in environmental and forensic analysis (Takáts et al. 2005; Morelato et al. 2013; Wójtowicz and Wietecha-Posłuszny 2019), where it has previously been employed in the analysis of inks (Ifa et al. 2007), explosives (Cotte-Rodríguez et al. 2005; Mulligan et al. 2007), and pesticides (Mulligan et al. 2007), among

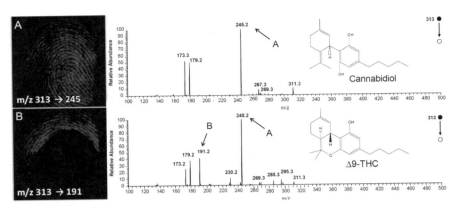

**Fig. 3** DESI-MS imaging used in forensic analysis to distinguish between Δ9-tetrahydrocannabinol (Δ9-THC) and cannabidiol. (**a**) Δ9-THC and/or cannabidiol on paper and corresponding MS/MS transition of *m/z* 313 to *m/z* 245.2. (**b**) Cannabidiol on paper identified by the MS/MS transition of *m/z* 313 to *m/z* 191. Image adapted from (Ifa et al. 2008) with permission

others. This technology was also employed to develop a backpack miniature mass spectrometer for field work (Hendricks et al. 2014).

In one of the first forensic applications of DESI-MS, the Cooks' lab employed this technique to image latent fingerprints (LFPs) (Ifa et al. 2008). The chemical analysis of LFPs potentially allows for both suspect identification, through the particular patterns of the fingerprint, and evidence of the suspect handling explosives, substances of abuse, or other exogenous substances. In this work, DESI-MS imaging was performed by rastering the charged droplet spray across the surface of a fingerprint and collecting a mass spectrum using a linear ion trap mass spectrometer. The researchers were able to construct fingerprints of small amounts of cocaine, Δ9-tetrahydrocannabinol, and the high energy explosive trinitrohexahydro-1,2,4-triazine from individuals whose fingers had been exposed to small amounts of these chemicals (Ifa et al. 2008). Tandem MS was then performed to verify the identity of these chemicals using fragmentation patterns. These chemical images were at a high enough resolution (150μm × 150μm per pixel) to allow specific minutiae of the fingerprint to be detected and compared (Ifa et al. 2008). Select results of the DESI-MS image analysis of a latent fingerprint can be seen in Fig. 3. This report showcases the ability of DESI-MS imaging to identify not only potential suspects but also any (potentially illicit) substances with which they may have come into contact.

# 4    Immunoassays

## 4.1    Historical Perspective

The immunoassay is another important bioanalytical tool that is ubiquitous in biomedical analysis and clinical chemistry; in fact, many of the tests currently used for COVID-19 testing are based on immunoassays. Rosalyn Yalow and Solomon Berson received the Nobel Prize in 1977 for the invention of the radioimmunoassay. This powerful new technique made it possible to detect nanomolar to picomolar quantities of hormones, including proteins and peptides, in complex biological samples. Prior to this advancement, these substances could not be detected by any other method at these biologically relevant concentrations. Important early immunoassays include measurements of insulin and thyroid hormones in the blood (Yalow and Berson 1960). Before the development of immunoassays, these hormones could not be detected in blood, yet they are standard clinical assays today.

Immunoassays are accomplished by targeting specific analytes using an antibody or antigen that also possesses a detectable label; in the case of the radioimmunoassay, the detectable labels are radioactive isotopes. Two important advancements have evolved from the radioimmunoassay: (1) the immunoassay technique was combined with electrophoresis to identify trace proteins in slab gels, leading to huge advances in clinical chemistry and biochemistry and (2) the development of the enzyme-linked immunosorbent assay (ELISA) in 1971 by two Swedish scientists, Eva Engvall and Peter Perlmann. ELISA eliminated the need for the use of radioactivity and made it possible to develop portable colorimetric assays and lateral flow assays, which are ubiquitous today (Engvall and Perlmann 1971). The immunoassay continues to evolve in the twenty-first century and its use in single-cell Western blot analysis is discussed below. Later in this chapter, the use of immunoassays in a microfluidic-based clinical assay is highlighted.

## 4.2    Single Cell Western Blot

One routine immunoassay technique in the biochemist's arsenal for characterizing the protein content of a sample is the Western blot. This technique is characterized by first separating proteins in a sample through electrophoresis (typically an SDS/PAGE gel), transferring the resolved bands onto a blotting membrane, and then employing an antibody stain for the specific protein(s) of interest. Traditional western blots require pooled cell lysate, preventing single-cell heterogeneity from being assessed. Relying simply on an immunoassay analysis of single cells (e.g., without the prior SDS/PAGE separation) can result in errors due to non-specific antibody binding and cross-reactivity. In 2014, the Amy Herr lab developed a microfluidic device capable of single-cell western blotting (scWestern), as seen in Fig. 4 (Hughes et al. 2014). Building on their work in microwestern (μWestern) blotting (Hughes and Herr 2012), the scWestern enables multi-protein detection on single cells, advancing knowledge of cell heterogeneity.

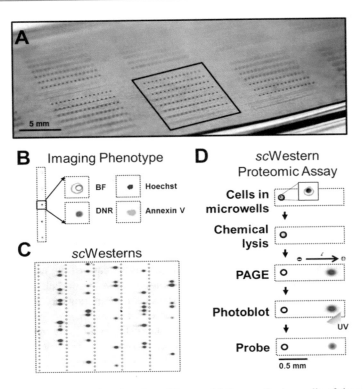

**Fig. 4** Analysis of single cells using the scWestern. (**a**) Image of microwells of the scWestern device, (**b**) phenotype imaging (fluorescence and phase contrast) of single cells in the microwells, (**c**) western blot and (**d**) scWestern blotting workflow. Reprinted with permission from (Kang et al. 2014). Copyright (2014) American Chemical Society. Further permissions related to the material excerpted should be directed to the ACS

To achieve single-cell analysis, each lane in the scWestern array possesses a 20μm microwell, in which cell(s) are seeded via gravity-driven cell settling. After seeding, cells not in wells are washed away, and cells in the microwells are lysed and protein constituents separated via PAGE in a short microchannel adjoining the microwell. The proteins are then immobilized in the gel using a benzophenone methacrylamide co-monomer that has been cross-linked into the PAGE gel. After the application of UV light, the benzophenone methacrylamide hardens, immobilizing the separated bands into the substrate (Hughes et al. 2014). Importantly, this immobilization allows proteins to be probed directly in the device without a blotting transfer step, a step that can add variability and errors in traditional western blotting analysis. Primary and secondary antibodies are then added to probe for protein(s) of interest, and fluorescence measurements taken. Due to the stable covalent protein immobilization, antibodies can be subsequently stripped using a strongly denaturing buffer, and the system re-probed with different antibodies, enabling researchers to study multiple separate proteins in an individual single cell. Each scWestern device is also capable of probing up to 48 protein targets per array (Hughes et al. 2014). Among many other applications, this technology has been employed to investigate cellular signaling and differentiation in neural stem cells (Hughes et al. 2014), investigate chemotherapeutic responses in human glioblastoma cells (Kang et al. 2014), and has since been commercialized by Protein Simple for more widespread use (ProteinSimple 2020).

## 5    DNA Sequencing

DNA is a double-stranded, helical molecule that carries all the genetic information for an organism. Understanding the structure of DNA and the genome has been transformative in fields including biology, medicine, archology, forensics, environmental science, and agriculture. The progress in this area has been exponential since the structure of DNA was elucidated in the late twentieth century (1952) by James Watson, Francis Crick, and Rosalind Franklin (Watson et al. 2011). This was followed by the invention of DNA sequencing, for which Walter Gilbert and Frederick Sanger received the Nobel Prize in 1980. Their sequencing methods were based on the use of radioactive labels and allowed scientists for the first time to explore the genetic code in detail (Karger and Guttman 2009; Shendure et al. 2017).

The field of DNA analysis experienced dramatic growth in the 1980s with the discovery of the polymerase chain reaction (PCR) that made it possible to make many copies of a single strand of DNA prior to sequencing. The huge advances in technology led to the initiation of the Human Genome Project in the 1980s with the ambitious goal to sequence the entire human genome; this feat was achieved early in the twenty-first century (Shendure et al. 2017). The sequencing of the human genome was accomplished, in large part, due to the analytical technique of capillary electrophoresis (Karger and Guttman 2009; Kheterpal and Mathies 1999; Dovichi and Zhang 2000). The sequencing method developed by Sanger was the primary method used by the community in the late 1970s. As mentioned above, this method employs polyacrylamide gel electrophoresis and autoradiography or fluorometry. More importantly, the sequencing process could take up to 24 h for data acquisition and even more time for analysis (Gilbert and Maxam 1973; Sanger et al. 1977; Smith et al. 1986). In addition, a single gel would generate only a few hundred bases of sequence.

The advent of capillary electrophoresis (CE) (Jorgenson and Lukacs 1981; Jorgenson 1986), an analytical technique that separates molecules in a capillary instead of on a gel slab, vastly improved the throughput of DNA sequencing for two primary reasons: (1) the small diameter (typically ~50μm) capillary allows for more efficient heat dissipation than a gel slab, enabling the application of higher field strengths and therefore more efficient separations and, more importantly, (2) the ease of automation of this analytical method. However, despite these improvements, a single capillary gel electrophoresis instrument still did not possess significantly more throughput than a multi-lane gel electrophoresis experiment. Therefore, analytical chemists continued to make advances to improve DNA sequencing: by developing capillary array electrophoresis, where multiple capillaries are employed for a multiplexed separation (Huang et al. 1992); in detection schemes (Swerdlow et al. 1990; Taylor and Yeung 1992); and in temperature control and novel capillary matrices to enable high-resolution separations for long DNA strands (Ruiz-Martinez et al. 1993; Salas-Solano et al. 1998; Goetzinger et al. 1998; Zhou et al. 2000).

These improvements made by analytical chemists in instrumentation have allowed for much faster sequencing of DNA, increasing the amount of data that can be collected by over 10-fold from automated gel electrophoresis instruments (Dovichi and Zhang 2000). These advances enabled the completion of the human genome project in 2003 and provided genomic information on other model organisms, revolutionizing biology. For additional information on DNA sequencing approaches, there are many excellent reviews available (Gilbert and Maxam 1973; Sanger et al. 1977; Smith et al. 1986). In addition, one of the authors of a chapter in this book, Emanuel Carrilho, was intimately involved in this research area (Salas-Solano et al. 1998) and gives more detail on its impact on bioanalytical chemistry in his chapter.

## 6 Microdialysis Sampling

### 6.1 Historical Perspective

Microdialysis is a continuous sampling technique that enables small-molecule monitoring in the extracellular fluid of tissue. Microdialysis sampling stems from an older sampling method, push-pull perfusion, which was first described by J.H. Gaddum in 1960 (Gaddum 1960). Later in the 1960s, Bito and coworkers implanted a dialysis sac in the brain for sample collection (Bito et al. 1966). Developed in its modern form by Urban Ungerstedt in the 1970s for monitoring neurotransmitters in the brain (Ungerstedt and Pycock 1974), microdialysis is now accomplished by implanting a probe with a semipermeable membrane into the tissue of interest. Microdialysis sampling is achieved by pumping perfusate, or a solution with a similar composition (ionic strength and pH) of the extracellular fluid of the tissue of interest, through a probe at a slow flow rate ($0.1–1 \mu L/min$ is typical); molecules diffuse from the extracellular fluid into the probe based on their concentration gradient. Samples are subsequently collected either off-line via a fraction collector or analyzed online by connecting the flow directly to an analysis method.

Two primary advantages of microdialysis sampling for continuous in vivo monitoring of small molecules are: (1) As most membranes have molecular weight cut-offs of ~20 kDa, only salts, small molecules, and some peptides are collected while macromolecules are excluded. (2) Because sampling is based on diffusion, there is no net fluid loss for the tissue of interest, making it possible to sample from a tissue in an awake animal for long periods of time. Microdialysis sampling has impacted the fields of drug discovery and pharmacology, physiology, neuroscience, and biology, among many others, and there are many book chapters and reviews on the subject (Saylor et al. 2017; Perry et al. 2009). Here we will focus on two complimentary areas of analytical advances to the field in the twentieth century: the development of new probe designs to expand microdialysis sampling capabilities outside of the brain and the advancement of separation and detection sciences for analyzing microdialysis samples. We will then highlight a twenty-first century application of microdialysis sampling using microchip electrophoresis with electrochemical detection for on-animal measurements.

One of the pioneers of microdialysis sampling, especially in tissues other than the brain, was Craig Lunte (Lunte et al. 1991). In the 1990s, the C. Lunte group developed and employed several new probe designs that could be used to sample different tissue types for drug metabolism and pharmacokinetic studies. These novel probe designs included a flexible cannula probe for blood sampling (Telting-Diaz et al. 1992) and a shunt probe for sampling bile (Scott and Lunte 1993). Craig Lunte also applied the microdialysis linear probe in novel ways, interrogating soft tissues such as the liver (Davies and Lunte 1996) or skin (Ault et al. 1992, 1994) in drug delivery and drug metabolism studies, respectively. His work in the area of microdialysis substantially advanced the field and showcased the power of microdialysis sampling for investigating many biological processes.

Separation and detection science also advanced substantially in the twentieth century to enable microdialysis samples to be analyzed with higher temporal resolution. When analyzing a microdialysis sample off-line, temporal resolution is dependent on the sampling flow rate (volume of sample), the analyte recovery, and the detection limits of the analytical method being used. With the introduction of capillary electrophoresis by James Jorgenson's group in 1981 (Jorgenson and Lukacs 1981), it became apparent that the low sample volume requirements of capillary electrophoresis were ideally suited for the analysis of microdialysis samples. Dr. Norberto Guzman, the founder of the Latin American Conference on Capillary Electrophoresis (LACE) was one of the first people to use capillary electrophoresis with fluorescence detection to analyze brain microdialysis samples (Hernandez et al. 1993). The Craig and Susan Lunte groups also used capillary electrophoresis with electrochemical detection to monitor electroactive substances in microdialysis samples that were collected off-line (O'Shea et al. 1992).

When microdialysis is directly coupled online to a separation and detection method it yields a "separation-based sensor" that enables the continuous monitoring of multiple analytes simultaneously (Saylor et al. 2017; Saylor and Lunte 2015). An early report of the separation-based sensor approach was from the Craig Lunte group in 1995, in which microdialysis was coupled, online, to microbore liquid chromatography and used to investigate the pharmacokinetics of acetaminophen in awake, freely moving rats (Steele and Lunte 1995). Microdialysis samples have subsequently been coupled, online, to a variety of separation methods including capillary liquid chromatography (Ngo et al. 2017), capillary electrophoresis (Zhou et al. 1995; Hogan et al. 1994; Lada et al. 1997; Bowser and Kennedy 2001), and microchip electrophoresis (Huynh et al. 2004; Li and Martin 2007; Wang et al. 2009); many of these online methods were developed late in the twentieth century and are currently being used in interesting applications today. More thorough reviews of progress in this area have been previously published (Saylor et al. 2017; Saylor and Lunte 2015).

## 6.2   Separation-Based Sensors for On-Animal Measurements

An exciting area of progress in the twenty-first century in the analysis of microdialysis samples is toward developing online separation-based sensors for monitoring biological molecules on-animal. By using microchip electrophoresis and electrochemical detection, the analytical instrument and all associated electronics can be made small, allowing for the entire analysis system to be miniaturized and enabling it to be placed on-animal for freely roaming animal experiments. On-animal, freely-roaming experiments would allow for concentrations of neurotransmitters to be correlated to behavior in awake, freely moving, and behaving animals.

In one example of online analysis of microdialysis samples, a microdialysis–microchip electrophoresis with electrochemical detection (MD-ME-EC) system using telemetric control was developed by the Susan Lunte lab (Scott et al. 2015). This device was comprised of an all-glass microchip with integrated platinum electrodes for detection. In this application, the online separation-based sensor was placed on the back of a sheep to monitor the metabolism of nitroglycerin by skin tissue in near-real time, as seen in Fig. 5 (Scott et al. 2015). While this report showcased the ability to place the entire analysis system on-animal, the long-term goal of this project is to monitor neurochemicals in the brain, including biogenic amines that are present at low concentrations and generate better responses at carbon-based electrodes. Toward this goal, an online MD-ME-EC system, comprised of a poly(dimethyl)siloxane/glass hybrid microchip and integrated detection at a carbon electrode was developed for monitoring analytes in the dopamine metabolic pathway (Saylor and Lunte 2018). This device was demonstrated through monitoring the conversion of L-DOPA into dopamine following L-DOPA perfusion in the brain of a rat (Saylor and Lunte 2018). Later, modifications were made to the construction of the device and the release of dopamine in the rat brain due to potassium stimulation was able to be monitored (Gunawardhana et al. 2020). Future work on these devices aims to lower detection and quantitation limits so that endogenous concentrations of catecholamines can be monitored. Parallel work on the development of new separation-based sensors for near real-time monitoring of energy metabolites and reactive oxygen and nitrogen species is also ongoing and discussed in more detail in the reviews by Saylor (Saylor et al. 2017; Saylor and Lunte 2015) and Schilly (Schilly et al. 2020).

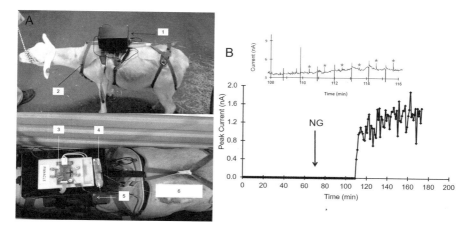

**Fig. 5** On-animal separation-based sensor using microdialysis sampling coupled to microchip electrophoresis with electrochemical detection. (**a**) Analysis device on-animal. (**b**) Production of nitrite in sheep after nitroglycerine (NG) perfusion monitored on-animal with the device. Adapted with permission from (Scott et al. 2015)

# 7 Electrochemistry

## 7.1 Historical Perspective

In 1959 Jaroslav Heyrovsky received the Nobel Prize for the invention of polarography, which employed dropping mercury electrodes primarily for the analysis of inorganic ions. In the 1950s Ralph N. Adams began to explore the use of solid electrodes, and in particular carbon electrodes, to study the electrochemistry of organic compounds (Adams 1958, 1969). The Clark oxygen electrode that is used to measure tissue oxygenation was also invented in the 1950s and commercialized in the 1970s. This sensor employed platinum electrodes and a gas permeable membrane to measure oxygen during cardiovascular surgery (Clark and Lyons 1962; Clark and Clark 1987). The Clark electrode formed the basis for the invention of biosensors that used oxidase enzymes in the presence of a substrate to generate hydrogen peroxide. The introduction of redox mediators (Schuhmann et al. 1991) and the use of dehydrogenase enzymes furthered the development of biosensors based on electrochemical detection (Turner 2013). The major target analyte for the development of biosensors has been the detection of glucose, with the goal of continuous in vivo monitoring for diabetics (Wilson and Hu 2000). Glucose sensors are now commercially available by Abbot and other companies. Below, examples of the uses of fast-scan cyclic voltammetry at carbon electrodes and biosensors to study neurodegenerative diseases are discussed. These exciting research areas would not be possible without the seminal work of scientists like Heyrovsky, Adams, Clark, Turner, Heller, and Wilson, to name a few.

## 7.2 Fast-Scan Cyclic Voltammetry for Brain Physiological Investigations

As mentioned above, one of the early pioneers of the field of bioanalytical chemistry was Ralph N. Adams, who not only studied fundamental electrochemical processes at carbon electrodes but also used electrochemistry as a direct detection strategy for the physiological and pathophysiological measurements of neurotransmitters and neurochemicals in vivo in the 1970s (Kissinger et al. 1973; McCreery et al. 1974a, b). These early efforts were characterized by employing Adams' newly developed solid carbon-paste electrodes to detect various catecholamines, both *in* and ex vivo. Since then, in large part due to Adams' and his students' efforts, electrochemical detection has become faster, more selective toward target analyte(s), and expanded into many different techniques and areas.

Today, one of the most widely employed electrochemical methods for in vivo analysis of chemicals in the brain is fast-scan cyclic voltammetry (FSCV), as developed by R. Mark Wightman who was a post-doctoral scholar in Adam's lab (Millar et al. 1985). Initially employed for monitoring dopamine in vivo, FSCV has since been expanded to include serotonin (Jackson et al. 1995; Hashemi et al. 2009), adenosine (Swamy and Venton 2007), histamine (Samaranayake et al. 2015; Puthongkham et al. 2019), hydrogen peroxide (Sanford et al. 2010), and neuropeptides (Schmidt et al. 2014), to name a few, and applied in cells, tissue cultures, rodents and other animal models, and even humans (Roberts and Sombers 2018; Fox and Wightman 2017; Bucher and Wightman 2015). FSCV is often accomplished in vivo by implanting a carbon fiber working electrode into the brain region of interest and applying an analyte-specific waveform at a high frequency to the electrode; the resulting current from the oxidation and reduction of analyte is monitored. After background-subtraction of non-Faradaic processes, a series of cyclic voltammograms characteristic for a given analyte in a given medium are produced and placed together to form a color plot (Fig. 6c). When a vertical section is taken out of the color plot, the original cyclic voltammogram can be reproduced (current vs. voltage), serving to identify the analyte(s). When a horizontal section is taken, a current vs. time profile is created, which can be transformed into concentration vs. time using calibrations. This concentration vs. time profile provides quantitative information concerning the amount and rate of neurotransmitter released, the process of analyte reuptake back into the cells, and kinetics. Due to the small size (micron) of the electrode and rapid (millisecond) data acquisition, FSCV possesses high spatial and temporal resolution, enabling investigations of physiological and pathophysiological processes.

As an example of employing FSCV to determine physiological function, a recent report from the Wightman lab investigated cross-hemispheric dopamine projections in the rat brain (Fox et al. 2016). Dopamine is an important neurotransmitter

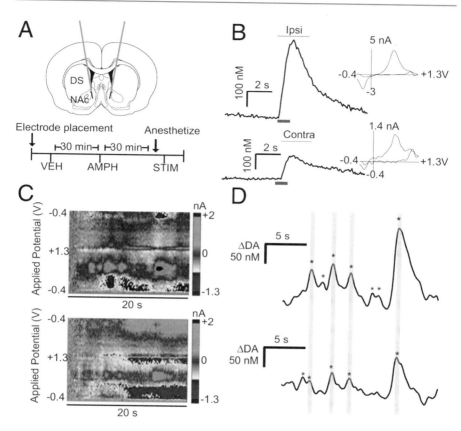

**Fig. 6** Fast-scan cyclic voltammetry detection of dopamine across two brain hemispheres. (**a**) Electrode placement and experimental design. (**b**) Simultaneous stimulated dopamine release in both brain hemispheres. (**c**) Representative FSCV color plots showing synchronous dopamine release transients and (**d**) Dopamine concentration over time in both hemispheres, with asterisks and gray bars indicating synchronous release and asterisks alone indicating asynchronous release. Image adapted with permission from (Fox et al. 2016)

implicated in many processes including locomotion, reward-seeking behavior and addiction, mood disorders, and Parkinson's disease. Previous research into the dopaminergic system provided contradictory results concerning the ability of dopamine neurons to project and function across the two hemispheres in the brain, due in part to the lack of a direct detection of dopamine release in prior research (Nauta et al. 1978; Geisler and Zahm 2005). Wightman's lab demonstrated that dopamine neurons are indeed able to project across to the other hemisphere of the brain. Specifically, these researchers found (Fox et al. 2016): (1) Spontaneous synchronous dopamine release in both hemispheres that increases synchronicity after the administration of a stimulant (amphetamine). (2) Electrical or optical stimulation in one hemisphere of the brain elicits dopamine release in both the ipsilateral (same) and contralateral (opposite) hemispheres, and the magnitude of that release is dependent on the brain area. (3) Dopamine projections are differentially regulated by D2 receptors in different brain regions, as demonstrated using pharmacological manipulations and FSCV measurements of dopamine release. (4) After dopamine neurons have been chemically lesioned in one hemisphere, electrical stimulation of neurons in the contralateral hemisphere can cause the release of dopamine in the lesioned area. Some of these results can be seen in Fig. 6. Taken together, these results indicate that dopaminergic neurons are capable of interhemispheric communication, knowledge that could result in the future development of pharmaceuticals for disorders characterized by dysregulated dopamine signaling, such as Parkinson's (Fox et al. 2016). Importantly, this work showcases the utility of FSCV in physiological investigations of the brain.

## 7.3    Sensors for Monitoring Brain Chemistry in the Clinic

The ability to monitor biomolecules in clinical settings is a vital advancement toward better clinical diagnostics and outcomes. Doctors and patients rely on fast, accurate, and often low-cost testing for various diseases and disorders to make critical judgments about prognosis, improve patient care, and monitor the spread of diseases. There are many analytical challenges associated with this feat, however. Depending on the specific application, developed methods must be robust, able to be used by a non-expert, portable, possess a small footprint, and be low-cost and largely automated. There are many reports of researchers developing instruments and methods capable of point-of-care testing (reviewed here (Rogers and Boutelle 2013; Booth et al. 2018; Chan et al. 2013)). As an example of a modern and important use of electrochemical-based biosensors, we will describe a multisensor device capable of in-clinic monitoring of potassium, glucose, and lactate in the human brain during a spreading depolarization after traumatic brain injury.

Traumatic brain injury is a major health concern that affects thousands of patients, often young adults, per year. Unfortunately, after the initial traumatic injury a cascade of secondary insults, called spreading depolarization, can cause additional damage. Spreading depolarization, or mass depolarization of all brain cells occurring in waves, results in imbalances in ion homeostasis, disruption of blood flow, and large energy demand, all of which can result in additional injury to tissue. Typically, traumatic brain injury patients are monitored through electrocorticography (ECoG) strip electrodes (Strong et al. 2002); however, this method only provides information on the electrical status of the brain and does not give insight into the biochemical processes. To accomplish in-clinic real-time monitoring of spreading depolarization in traumatic brain injury, analytical devices must be minimally invasive, unobtrusive, and compatible with existing care guidelines. Microdialysis sampling, as discussed above, is a minimally invasive technique that allows for long-term monitoring of analytes (Saylor et al. 2017), and has been FDA approved for use in patients.

Coupling microdialysis sampling to an appropriate analysis technique can allow multiple analytes to be monitored, simultaneously. The collaboration of Martyn Boutelle and Anthony Strong has applied continuous online microdialysis sampling coupled to a microfluidic system with glucose, lactate, and potassium electrochemical biosensors to monitor the metabolic effects of spreading depolarization in the brain of patients in the ICU, as seen in Fig. 7 (Rogers et al. 2017). By using a microfluidic platform, these researchers were able to analyze the small volume samples produced by brain microdialysis while keeping the overall footprint of the system small, so as not to interfere with patient care. Additionally, their device incorporated an automatic calibration method for the sensors, ensuring reliable data over multi-day analysis. This point-of-care device was deployed in the ICU to continually monitor lactate, glucose, and potassium for between 24 h and 5 days in 65 patients; the results from one patient can be seen in Fig. 7. Their data shows that during a spreading depolarization in the brain potassium levels increase, glucose levels decrease, and lactate levels increase. The changing levels of glucose and lactate are indicative of the high energy requirements of the tissue in repolarizing; all three biomarkers are thought to be important in determining tissue health (Hutchinson et al. 2015). This report demonstrates the impact of the use of biosensors in a clinical setting for understanding diseases and disorders.

# 8    Micro Total Analysis Systems

## 8.1    Historical Perspective

The use of microfluidics for bioanalysis also started at the end of the twentieth century. H. Michael Widmer introduced the concept of micro total analysis systems in the 1980s (Manz et al. 1990). As part of this effort, the groups of Andreas Manz, D. Jed Harrison (Manz et al. 1992), and J. Michael Ramsey (Jacobson et al. 1994) developed microchip electrophoretic systems. The introduction of soft lithography for microfluidic devices by the George Whitesides group made it possible to produce these systems easily in the laboratory without a cleanroom facility (Duffy et al. 1998). Many Brazilian scientists, including several authors of chapters in this book, have contributed toward the development of novel microfluidic devices for portable analysis systems, especially using polyester toner, plastics, and paper substrates (Coltro et al. 2010). The Susan Lunte group has worked extensively on the development of electrochemical methods of detection for microchip electrophoresis and combining microchip electrophoresis with microdialysis sampling, some of the applications of which were discussed above and in many reviews (Saylor et al. 2017; Saylor and Lunte 2015; Schilly et al. 2020; Gunasekara et al. 2016). Here, we will discuss two applications of micro total analysis systems: point-of-care testing and detecting life on other planets and moons.

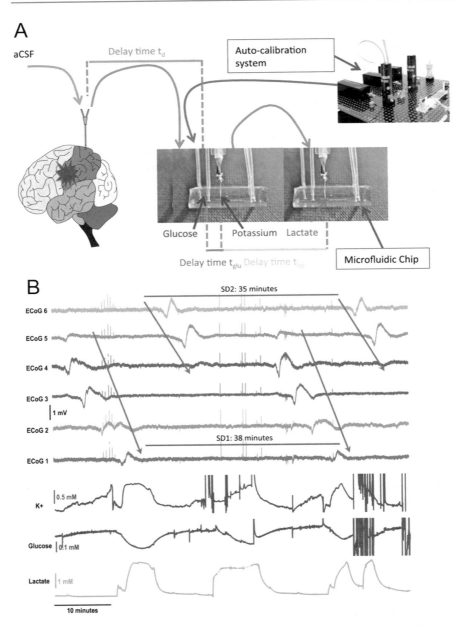

**Fig. 7** Sensors for monitoring the metabolic effects of spreading depolarization in the brain of patients in the ICU. (**a**) Continuous online microdialysis sampling coupled to a microfluidic system with glucose, lactate, and potassium electrochemical sensors. (**b**) ECoG and potassium, glucose, and lactate measurements from a representative patient during two (indicated) spreading depolarization (SD) events. Image adapted with permission from (Rogers et al. 2017)

## 8.2    Detecting Disease with Droplet Microfluidics

We rely on vaccines to stop the outbreak and spread of certain diseases, such as malaria and rubella, and now COVID-19. Unfortunately, in many areas of the world preventable disease outbreaks continue, due in part to low vaccine coverage and reduced access to healthcare. Serological surveys, or the assessment of the presence of immunoglobulin G (IgG) antibodies in a population, can provide a snapshot of the population's immunity. This snapshot informs clinicians about the necessity of implementing a vaccination campaign, especially in situations where individuals have been displaced from their homes. Testing for IgG traditionally involves a blood draw from the patient(s) followed by sending the sample to an established lab, where results are acquired. This process can be cost, time, and logistically prohibitive. Alternatively, surveys in remote settings often rely on vaccine records or an individual's (or their parent's) memory, which can be incomplete or inaccurate. In developing point-of-care analytical devices for this application, in addition to delivering highly accurate data, care must be taken so that the final device is portable, does not rely on external infrastructure, is relatively easy to use by a non-expert, and can process many samples relatively quickly. Digital microfluidic (DMF) systems are capable of addressing many of these analytical challenges. In DMF, droplets of liquid are manipulated using an array of electrodes that have been coated with a hydrophobic insulator (Choi et al. 2012). Through careful application of voltages, droplets are compelled to move, mix, merge, and split; this technology allows for samples to be prepared, processed, and analyzed all on a single platform, without the use of more traditional fluid handling (pumps, valves, mechanical mixers) allowing a more streamlined and less complex final device.

The Wheeler lab has developed a platform for screening individuals for measles and rubella immunoglobulin G (IgG) antibodies using enzyme-linked immunosorbent assays (ELISAs) incorporated into a DMF device, which they called the Measles-Rubella Box (MR box) (Ng et al. 2018). This device was created to be lower cost (<\$2500 USD), portable (25 cm × 20 cm × 28 cm and 4 kg), and not reliant on existing infrastructure (powered by a 12 V laptop power supply), to enable serological surveys in remote locations. DMF was used to carry out all steps in a traditional ELISA assay, from sample introduction, through to chemiluminescent detection. The MR Box was deployed in the Kakuma refugee camp in Kenya for field testing during a vaccination campaign, to detect the presence of measles and rubella IgG in 144 adults and children. When compared to serum samples acquired from study participants that had been sent to a traditional lab, the MR box exhibited an overall agreement of 86% and 84% for measles and rubella, respectively (Ng et al. 2018). This study clearly demonstrates the ability of new and expanding analytical methods to interrogate disease in traditionally difficult to probe locations.

## 8.3 Chemical Laptop to Look for Life on Other Solar Bodies

This chapter has primarily focused on bioanalytical chemistry as it is applied on earth. However, the question of life existing (current or past) on other planets or solar bodies still remains. In some ways, this remains one of the most fundamental questions: What is life, and how can we determine its existence? Shortly after humanity's first explorations into space, scientists began considering how to implement exobiological studies, proposing to investigate the presence of organic molecules, nucleic acids, amino acids and proteins, and/or metabolism (Lederberg 1960). The Viking space missions of the 1970s probed for metabolism using a variety of strategies and, using gas chromatography-mass spectrometry, organic molecules. Taken together, the result was not indicative of life (Dick 2006).

The further development of analytical methods to probe the existence of life in spaceflight missions is understandably challenging and requires several practical instrument design considerations, including:

(1) Employing sampling vs. non-sampling techniques. Non-sampling techniques are typically those that rely on optical methods, including Raman, but are limited both by their low sensitivity to organic compounds and ability to detect only compounds found on the surface. Sampling methods are inherently a bit more complex, as the sample often requires extraction, preconcentration, or other processing procedures prior to analysis; all these steps must be automated and rugged for use remotely. (2) The size, weight, and power requirements, due to the cost and space limitations in sending instrumentation to other planets via spaceflight. (3) Ensuring long-term stability of the instrumentation and chemicals in harsh extraterrestrial environments (Willis et al. 2015).

These design considerations are in addition to what some call the "Ladder of Life Detection," developed by Neveu and coworkers (Neveu et al. 2018), which describes many analytical criteria that must be met to conclusively determine life exists on other planets. These criteria include repeatability, detectability, reliability, and sensitivity of the proposed method, among others (Neveu et al. 2018).

The Peter Willis lab at the Jet Propulsion Laboratory in the USA is tackling these analytical challenges through the use of capillary and microchip electrophoresis with laser-induced fluorescence detection, due to their ability to analyze complex samples for multiple low concentration analytes simultaneously, relatively simple instrumentation requirements, and small overall footprint. One indicator of life on other planets would be the presence of specific types of amino acids. Amino acids can act as biosignatures in three ways (1) their presence, (2) their relative abundance, and (3) their chirality as indicators of abiotic vs. biotic processes may provide evidence for the possibility of life on other planets. Toward this goal, Creamer et al. (Creamer et al. 2017) developed two capillary electrophoresis with laser-induced fluorescence detection methods to separate and detect 17 of the most abundant abiotic and biotic amino acids. These methods were then applied to samples from Mono Lake, a shallow salt brine lake that possesses a high pH and salt concentration, as a model for bodies such as Europa and Enceladus. Detected amino acids and concentrations were then assessed according to the biosignatures listed above. It was determined

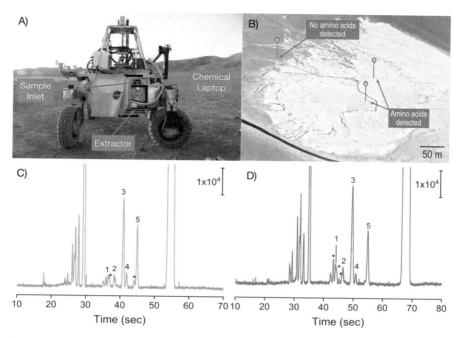

**Fig. 8** Use of the Chemical Laptop for proof-of-concept experiments for exploring life on other planetary bodies. (**a**) Chemical laptop and sample extractor mounted on a rover. (**b**) Sample sites near Yungay Station marked on a satellite image. The analysis of two samples is shown in (**c, d**) for (1) leucine, (2) valine, (3) serine, (4) alanine, and (5) glycine. Image reproduced with permission from (Mora et al. 2020)

that the samples from Mono Lake possessed all three biosignatures, proving the capability of this method for identifying the presence of life (Creamer et al. 2017).

In a separate report from the same group, Mora et al. recently demonstrated the feasibility of employing microchip electrophoresis with laser-induced fluorescence detection, coupled to subcritical water extraction for the analysis of amino acids from soil (Mora et al. 2020). What makes this report particularly noteworthy is that the analysis device, dubbed the Chemical Laptop (CL, seen in Fig. 8), was completely automated: from sample acquisition through sample preparation and labeling, to separation and detection of analytes, all processes were carried out completely remotely. The CL was created by combining and extending previous research, including a subcritical extraction unit, sample handling microchip for sample labeling and dilution, a separations microchip, chemical cartridge for storing liquids needed for analysis, a monolithic pneumatic manifold for fluid control, in addition to optics and electronics (Mora et al. 2020). The completed device was then placed on a rover and deployed in the Atacama Desert in Chile to simulate a Mars mission, where four soil samples taken at various soil depths were sampled and analyzed for amino acid content. Of these four samples, three were found to possess amino acids at concentrations above the blank, as can be seen in Fig. 8. The authors are clear to point out that this report serves only to demonstrate the functionality of the developed device, and was not created to conform to the specific requirements of a mission (Mora et al. 2020). Regardless, this work clearly establishes the feasibility of employing this technology for future missions aimed at seeking life on other planets and moons.

## 9    Conclusions

The discussion and examples above are just a few of the many new methods and techniques that are making an impact in biochemistry and medicine in the twenty-first century. As can be seen by these examples, bioanalytical chemistry continues to evolve as a critical tool for drug discovery and biomedical research. In this chapter, we have focused primarily on the impact of separation and electrochemical methods, as well as mass spectrometry, on bioanalytical research in the twenty-first century. There are many methods that were not discussed such as NMR, surface plasmon resonance, Raman spectroscopy, atomic force microscopy, and confocal microscopy that have also played an important role in bioanalysis. The reader is directed to the chapters in this book on spectroscopic analysis for more information on these methods.

## References

Adams RN (1958) Anal Chem 30:1576
Adams RN (1969) Electrochemistry at solid electrodes. Dekker, New York
Ault JM, Lunte CE, Meltzer NM, Riley CM (1992) Pharm Res 9:1256–1261
Ault JM, Riley CM, Meltzer NM, Lunte CE (1994) Pharm Res 11:1631–1639
Bito L, Levin E, Murray M, Snider N (1966) J Neurochem 13:1057–1067
Booth MA, Gowers SAN, Leong CL, Rogers ML, Samper IC, Wickham AP, Boutelle MG (2018) Anal Chem 90:2–18
Bowser MT, Kennedy RT (2001) Electrophoresis 22:3668–3676
Bucher ES, Wightman RM (2015) Annu Rev Anal Chem 8:239–261
Chan CPY, Mak WC, Cheung KY, Sin KK, Yu CM, Rainer TH, Renneberg R (2013) Annu Rev Anal Chem 6:191–211
Choi K, Ng AHC, Fobel R, Wheeler AR (2012) Annu Rev Anal Chem 5:413–440
Clark LCJ, Clark EW (1987) Int Anesthesiol Clin 25:1–29
Clark LC, Lyons C (1962) Ann N Y Acad Sci 102:29–45
Cody RB, Laramée JA, Durst HD (2005) Anal Chem 77:2297–2302
Coltro WKT, de Jesus DP, da Silva JAF, do Lago CL, Carrilho E (2010) Electrophoresis 31:2487–2498
Cooks RG, Ouyang Z, Takats Z, Wiseman JM (2006) Science 311:1566–1570
Correa DN, Santos JM, Eberlin LS, Eberlin MN, Teunissen SF (2016) Anal Chem 88:2515–2526
Cotte-Rodríguez I, Takáts Z, Talaty N, Chen H, Cooks RG (2005) Anal Chem 77:6755–6764
Creamer JS, Mora MF, Willis PA (2017) Anal Chem 89:1329–1337
Davies MI, Lunte CE (1996) Life Sci 59:1001–1013
Dick SJ (2006) Endeavour 30:71–75
Dovichi NJ, Zhang J (2000) Angew Chem Int Ed 39:4463–4468
Duffy DC, McDonald JC, Schueller OJA, Whitesides GM (1998) Anal Chem 70:4974–4984

Engvall E, Perlmann P (1971) Immunochemistry 8:871–874

Ettre LS (2003) LCGC N Am 21:458–467

Everley RA, Croley TR (2008) J Chromatogr A 1192:239–247

Fenn JB (1993) J Am Soc Mass Spectrom 4:524–535

Fennema E, Rivron N, Rouwkema J, van Blitterswijk C, De Boer J (2013) Trends Biotechnol 31: 108–115

Fox ME, Wightman RM (2017) Pharmacol Rev 69:12–32

Fox ME, Mikhailova MA, Bass CE, Takmakov P, Gainetdinov RR, Budygin EA, Wightman RM (2016) Proc Natl Acad Sci U S A 113:6985–6990

Gaddum JH (1960) Proc Physiol Soc 155:1–2

Geisler S, Zahm DS (2005) J Comp Neurol 490:270–294

Gilbert W, Maxam A (1973) Proc Natl Acad Sci U S A 70:3581–3584

Goetzinger W, Kotler L, Carrilho E, Ruiz-Martinez MC, Salas-Solano O, Karger BL (1998) Electrophoresis 19:242–248

Griffiths J (2008) Anal Chem 80:5678–5683

Gunasekara DB, Wijesinghe MB, Saylor RA, Lunte SM (2016) Principles and strategies for microchip electrophoresis with amperometric detection. In: Arrigan DWM (ed) Electrochemical strategies in detection science. The Royal Society of Chemistry, London, pp 85–124

Gunawardhana SM, Bulgakova GA, Barybin AM, Thomas SR, Lunte SM (2020) Analyst 145: 1768–1776

Haddad R, Sparrapan R, Eberlin MN (2006) Rapid Commun Mass Spectrom 20:2901–2905

Hashemi P, Dankoski EC, Petrovic J, Keithley RB, Wightman RM (2009) Anal Chem 81:9462–9471

Hendricks PI, Dalgleish JK, Shelley JT, Kirleis MA, McNicholas MT, Li L, Chen TC, Chen CH, Duncan JS, Boudreau F, Noll RJ, Denton JP, Roach TA, Ouyang Z, Cooks RG (2014) Anal Chem 86:2900–2908

Hernandez L, Tucci S, Guzman N, Paez X (1993) J Chromatogr A 652:393–398

Hogan BL, Lunte SM, Stobaugh JF, Lunte CE (1994) Anal Chem 66:596–602

Huang XC, Quesada MA, Mathies RA (1992) Anal Chem 64:2149–2154

Hughes AJ, Herr AE (2012) Proc Natl Acad Sci U S A 109:21450–21455

Hughes AJ, Spelke DP, Xu Z, Kang CC, Schaffer DV, Herr AE (2014) Nat Methods 11:749–755

Hutchinson PJ, Jalloh I, Helmy A, Carpenter KLH, Rostami E, Bellander BM, Boutelle MG, Chen JW, Claassen J, Dahyot-Fizelier C, Enblad P, Gallagher CN, Helbok R, Hillered L, Le Roux PD, Magnoni S, Mangat HS, Menon DK, Nordström CH, O'Phelan KH, Oddo M, Barcena JP, Robertson C, Ronne-Engström E, Sahuquillo J, Smith M, Stocchetti N, Belli A, Carpenter TA, Coles JP, Czosnyka M, Dizdar N, Goodman JC, Gupta AK, Nielsen TH, Marklund N, Montcriol A, O'Connell MT, Poca MA, Sarrafzadeh A, Shannon RJ, Skjøth-Rasmussen J, Smielewski P, Stover JF, Timofeev I, Vespa P, Zavala E, Ungerstedt U (2015) Intensive Care Med 41:1517–1528

Huynh BH, Fogarty BA, Martin RS, Lunte SM (2004) Anal Chem 76:6440–6447

Ifa DR, Gumaelius LM, Eberlin LS, Manicke NE, Cooks RG (2007) Analyst 132:461–467

Ifa DR, Manicke NE, Dill AL, Cooks RG (2008) Science 321:805

Jackson BP, Dietz SM, Wightman RM (1995) Anal Chem 67:1115–1120

Jacobson SC, Hergenroder R, Koutny LB, Ramsey JM (1994) Anal Chem 66:1114–1118

Jorgenson JW (1986) Anal Chem 58:743A–758A

Jorgenson JW, Lukacs KDA (1981) Anal Chem 53:1298–1302

Kang CC, Lin JMG, Xu Z, Kumar S, Herr AE (2014) Anal Chem 86:10429–10436

Karger BL, Guttman A (2009) Electrophoresis 30:196–202

Kheterpal I, Mathies RA (1999) Anal Chem 71:31A–37A

Kissinger PT, Hart JB, Adams RN (1973) Brain Res 55:209–213

Labonia GJ, Ludwig KR, Mousseau CB, Hummon AB (2018) Anal Chem 90:1423–1430

Lada MW, Vickroy TW, Kennedy RT (1997) Anal Chem 69:4560–4565

Lederberg J (1960) Science 132:393–400
Li H, Hummon AB (2011) Anal Chem 83:8794–8801
Li MW, Martin RS (2007) Electrophoresis 28:2478–2488
Liu X, Hummon AB (2015) Anal Chem 87:9508–9519
Liu J, Wang H, Manicke NE, Lin JM, Cooks RG, Ouyang Z (2010) Anal Chem 82:2463–2471
Liu X, Weaver EM, Hummon AB (2013) Anal Chem 85:6295–6302
Lunte CE, Scott DO, Kissinger PT (1991) Anal Chem 63:773A–780A
Malkin DS, Wel B, Fogiel AJ, Staats SL, Wirth MJ (2010) Anal Chem 82:2175–2177
Manz A, Graber N, Widmer HM (1990) Sens Actuators B 1:244–248
Manz A, Harrison DJ, Verpoorte EMJ, Fettinger JC, Paulus A, Luedi H, Widmer HM (1992) J
    Chromatogr 593:253–258
Martin AJP, Synge RLM (1941) Biochem J 35:91–121
McCreery RL, Dreiling R, Adams RN (1974a) Brain Res 73:23–33
McCreery RL, Dreiling R, Adams RN (1974b) Brain Res 73:15–21
Millar J, Stamford JA, Kruk ZL, Wightman RM (1985) Eur J Pharmacol 109:341–348
Miquel FAW, Mulder GJ, Wenckebach W (1838) Bull Sci Phys Nat Néerl:104
Moore S, Stein WH (1954) J Biol Chem 211:907–913
Mora MF, Kehl F, Tavares E, Bramall N, Willis PA (2020) Anal Chem 92:12959–12966
Morelato M, Beavis A, Kirkbride P, Roux C (2013) Forensic Sci Int 226:10–21
Mulligan CC, MacMillan DK, Noll RJ, Cooks RG (2007) Rapid Commun Mass Spectrom 21:
    3729–3736
Nauta WJH, Smith GP, Faull RLM, Domesick VB (1978) Neuroscience 3:385–401
Neveu M, Hays LE, Voytek MA, New MH, Schulte MD (2018) Astrobiology 18:1375–1402
Ng AHC, Fobel R, Fobel C, Lamanna J, Rackus DG, Summers A, Dixon C, Dryden MDM, Lam C,
    Ho M, Mufti NS, Lee V, Asri MAM, Sykes EA, Chamberlain MD, Joseph R, Ope M, Scobie
    HM, Knipes A, Rota PA, Marano N, Chege PM, Njuguna M, Nzunza R, Kisangau N, Kiogora J,
    Karuingi M, Burton JW, Borus P, Lam E, Wheeler AR (2018) Sci Transl Med 10:1–13
Ngo KT, Varner EL, Michael AC, Weber SG (2017) ACS Chem Nerosci 8:329–338
O'Shea TJ, Weber PL, Bammel BP, Lunte CE, Lunte SM, Smyth MR (1992) J Chromatogr 608:
    189–195
Pauling L, Corey RB, Branson HR (1951) Proc Natl Acad Sci U S A 37:205–211
Perry M, Li Q, Kennedy RT (2009) Anal Chim Acta 653:1–22
ProteinSimple. Milo single-cell westerns. https://www.proteinsimple.com/. Accessed 20 Oct 2021
Puthongkham P, Lee ST, Venton BJ (2019) Anal Chem 91:8366–8373
Roberts JG, Sombers LA (2018) Anal Chem 90:490–504
Rogers ML, Boutelle MG (2013) Annu Rev Anal Chem 6:427–453
Rogers BA, Wu Z, Wei B, Zhang X, Cao X, Alabi O, Wirth MJ (2015) Anal Chem 87:2520–2526
Rogers ML, Leong CL, Gowers SA, Samper IC, Jewell SL, Khan A, McCarthy L, Pahl C, Tolias
    CM, Walsh DC, Strong AJ, Boutelle MG (2017) J Cereb Blood Flow Metab 37:1883–1895
Rowbury R (2012) Robert Hooke, 1635-1703. Sci Prog 95:238–254
Ruiz-Martinez MC, Belenkii A, Karger BL, Berka J, Foret F, Miller AW (1993) Anal Chem 65:
    2851–2858
Salas-Solano O, Carrilho E, Kotler L, Miller AW, Goetzinger W, Sosic Z, Karger BL (1998) Anal
    Chem 70:3996–4003
Samaranayake S, Abdalla A, Robke R, Wood KM, Zeqja A, Hashemi P (2015) Analyst 140:3759–
    3765
Sanford AL, Morton SW, Whitehouse KL, Oara HM, Lugo-Morales LZ, Roberts JG, Sombers LA
    (2010) Anal Chem 82:5205–5210
Sanger F, Air GM, Barrell BG, Brown NL, Coulson AR, Fiddes JC, Hutchison CA, Slocombe PM,
    Smith M (1977) Nature 265:687–695
Saylor RA, Lunte SM (2015) J Chromatogr A 1382:48–64
Saylor RA, Lunte SM (2018) Electrophoresis 39:462–469

Saylor RA, Thomas SR, Lunte SM (2017) Separation-based methods combined with microdialysis sampling for monitoring neurotransmitters and drug delivery to the brain. In: Wilson GS, Michael AC (eds) Compendium of in vivo monitoring in real-time molecular neuroscience, Microdialysis and sensing of neural tissues, vol 2. World Scientific, Hackensack, pp 1–45

Schilly KM, Gunawardhana SM, Wijesinghe MB, Lunte SM (2020) Anal Bioanal Chem 412:6101–6119

Schmidt AC, Dunaway LE, Roberts JG, McCarty GS, Sombers LA (2014) Anal Chem 86:7806–7812

Schufle JA, Ionescu LG (1976) J Chem Educ 53:174

Schuhmann W, Schmidt HL, Ohara TJ, Heller A (1991) J Am Chem Soc 113:1394–1397

Scott DO, Lunte CE (1993) Pharm Res 10:335–342

Scott DE, Willis SD, Gabbert S, Johnson D, Naylor E, Janle EM, Krichevsky JE, Lunte CE, Lunte SM (2015) Analyst 140:3820–3829

Shendure J, Balasubramanian S, Church GM, Gilbert W, Rogers J, Schloss JA, Waterston RH (2017) Nature 550:345–353

Smith LM, Kelleher NL (2013) Nat Methods 10:186–187

Smith LM, Sanders JZ, Kaiser RJ, Hughes P, Dodd C, Connell CR, Heiner C, Kent SBH, Hood LE (1986) Nature 321:674–679

Spengler B (2015) Anal Chem 87:64–82

Steele KM, Lunte CE (1995) J Pharm Biomed Anal 13:149–154

Strong AJ, Fabricius M, Boutelle MG, Hibbins SJ, Hopwood SE, Jones R, Parkin MC, Lauritzen M (2002) Stroke 33:2738–2743

Svedberg T (1937) Nature 139:1051–1062

Swamy BEK, Venton BJ (2007) Anal Chem 79:744–750

Swerdlow H, Wu S, Harke H, Dovichi NJ (1990) J Chromatogr A 516:61–67

Takáts Z, Wiseman JM, Gologan B, Cooks RG (2004) Science 306:471–473

Takáts Z, Wiseman JM, Cooks RG (2005) J Mass Spectrom 40:1261–1275

Taylor JA, Yeung ES (1992) Electrophoresis. Anal Chem 64:1741–1744

Telting-Diaz M, Scott DO, Lunte CE (1992) Anal Chem 64:806–810

Tiselius A (1937) Trans Faraday Soc 33:524–531

Tobias F, Hummon AB (2020) J Proteome Res 19:3620–3630

Turner APF (2013) Chem Soc Rev 42:3184–3196

Ungerstedt U, Pycock C (1974) Bull Schweiz Akad Med Wiss 30:44–55

Wang Y, Balgley BM, Rudnick PA, Lee CS (2005) J Chromatogr A 1073:35–41

Wang M, Roman GT, Perry ML, Kennedy RT (2009) Anal Chem 81:9072–9078

Watson CJ, Lydic R, Baghdoyan HA (2011) J Neurochem 118:571–580

Wei B, Rogers BJ, Wirth MJ (2012) J Am Chem Soc 134:10780–10782

Willis PA, Creamer JS, Mora MF (2015) Anal Bioanal Chem 407:6939–6963

Wilson GS, Hu Y (2000) Chem Rev 100:2693–2704

Wójtowicz A, Wietecha-Posłuszny R (2019) Appl Phys A Mater Sci Process 125:312

Wu Z, Wei B, Zhang X, Wirth MJ (2014) Anal Chem 86:1592–1598

Yalow RS, Berson SA (1960) J Clin Invest 39:1157–1175

Zhou SY, Zuo H, Stobaugh JF, Lunte CE, Lunte SM (1995) Anal Chem 67:594–599

Zhou H, Miller AW, Sosic Z, Buchholz B, Barron AE, Kotler L, Karger BL (2000) Anal Chem 72:1045–1052

# Sampling and Sample Preparation in Bioanalysis

Ljubica Tasic

## 1 Introduction

The success of any chemical or physical description of the composition of any sample depends greatly on sampling, sample preparation, and the choice of technique that is picked for such tasks. There are many types of samples, which can have different physical states at room (standard) temperature and pressure, such as gas, liquid and solid, or a mix of those like emulsions, suspensions, or aerosols (Pawliszyn 2010). That is why sampling and its representativity are of crucial importance when aimed to the best possible analytical sample description. It is important to say that there is no perfect technique for all types of samples, nor the best all-in-one analysis, and each sample, even if not of complex nature, must be taken as unique in the best effort to achieve as good as possible sample composition description.

All samples are made from elements, molecules, and their mixtures and some are the major, others the minor or trace components (Clark et al. 2016). Thus, it is important to establish the goals previous to any analysis to be performed and develop a suitable strategy to make as few as possible, and the smallest as possible errors, in defining the chemical and physical compositions of the studied samples (Pawliszyn 2010; Clark et al. 2016; Berndt 2020).

L. Tasic (✉)
Chemical Biology Laboratory, Department of Organic Chemistry, Institute of Chemistry, University of Campinas (UNICAMP), Campinas, Sao Paulo, Brazil
e-mail: ljubica@unicamp.br

© The Author(s), under exclusive license to Springer Nature Switzerland AG 2022
L. T. Kubota et al. (eds.), *Tools and Trends in Bioanalytical Chemistry*,
https://doi.org/10.1007/978-3-030-82381-8_3

## 2 Sampling and Biosafety

The number of samples and the samples' physical state requires the use of suitable tools and sampling (Berndt 2020). Sample must be representative (Nahorniak et al. 2015) and sufficient (Suresh et al. 2011) in terms of number and quantity for the analytical method or methods, which are going to be used in sample analyses. Also, sample handling, transportation, storage, and preservation (Nocerino et al. 2005; Tripathi et al. 2020) must be discussed before any analysis. Any sample contamination must be avoided or reduced to the lowest possible chance. It is very important to plan and design a study to avoid data interpretation in a biased manner. One must think about what are the aims of a study and if the designed protocol is the most appropriate to get the right answer—for example, samples are analyzed to determine their constituents and concentrations (Berndt 2020; Nahorniak et al. 2015; Suresh et al. 2011; Nocerino et al. 2005; Tripathi et al. 2020).

First of all, sampling depends on the sample's physical state and there are some standard procedures used for each sample type that must be followed. Herein, bioanalytical samples are discussed, thus many protocols for biosafety (Dickmann et al. 2015; Flemming 2000) and authorized sampling for research use must also be taken into account previous to any sample collection.

Biosafety is defined as—the maintenance of safe conditions in biological research to prevent harm to workers, non-laboratory organisms, or the environment—(World Health Organisation Biosafety). Not all laboratory settings are equipped to work with human samples, pathogenic bacteria, viruses, and so on. There are four levels of laboratory containment protocol (Kimman et al. 2008) expressed as 1–4 in order of danger, with level 4 posing a high risk to individuals and community, working with pathogens that cause serious human or animal diseases and to spread easily, such as deadly viruses (Ebola virus) for humans. A risk group 1 is assigned (Enserink and Du 2004) to no or low individual and community risk, for example, to work and research that employs microorganisms that are unlikely to cause human and animal diseases (World Health Organisation Biosafety) (Bayot and King 2020).

On the other hand, an authorized sampling includes protocols that follow all ethical and legal rights, and when discussing the use of human biosamples (Bayot and King 2020), they must have the consent of donation. It is also important to respect the legal and ethical rights of the people who donate their samples and to guarantee that they will make a personal contribution to the research undertaken with the collected samples (Singh et al. 2020). When discussing the sampling of many types of biological samples, it is also important to obey the protocols on access to genetic resources and the fair and equitable sharing of benefits arising from their use to the convention on biological diversity (CBD) (Mohammedsaleh and Mohammedsaleh 2014). When legal and ethical rights are obeyed, then it is also important to have in mind that after sample collecting, there are two types of samples to be prepared and kept, a laboratory size sample and a sample for the biobank (Fig. 1).

The following sections bring some important protocols for sampling in regard to physical state and type of samples.

**Fig. 1** Illustration of the main steps in a study when discussing sampling in bioanalytics. It is first to define aims and design a study, collect samples and prepare two types of samples from the same cohort—one that is laboratory size and the other to be kept in a biobank

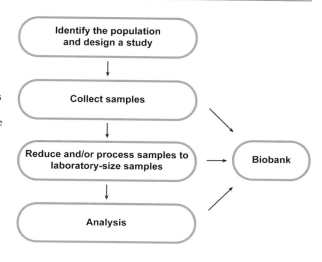

## 3      Sample Size, Composition, and Errors

The sample size is one of the main issues when biological samples are analyzed. For example, when it is a large population researched, getting a sample from everyone is not possible. Therefore, it is important to take a random sample of individuals or species, which represents the population as a whole. An accurate, and statistically significant result, which leads to a successful conclusion of the study, strongly depends on the sample size. If the sample size is too small, it may include a disproportionate number of individuals who are anomalies or outliers, therefore, the obtained data may lead to a biased assessment of the entire population. But if too big, the whole study becomes complex, expensive, long, and even if it brings accurate results, the benefits of such a study are not justified.

Some tips might help to design the study properly concerning how many samples must be used throughout the research, which depends on the size of the population, margin of error, confidence level, and standard deviation to estimate an ideal, or at least a preferable sample size. For example, a preferable sample size for DNA barcoding is at least 20 samples from the same species. Because of the technical and biological variances among samples, there is a need to calculate the sample size and consider: (1) the technique-based basis, (2) laboratory-based level, (3) aims of the study (disease), and (4) the sample type, before deciding how to proceed. Throughout this chapter, and when important, the sample size is discussed, like the number of individuals, or the quantity of the sample in volume or mass units. And when dealing with the mass or volume units, then, it is important to discuss the sample constituents.

Major elements of a biological sample or a mixture are usually considered to be above 1%, while minor elements make between 0.1 and 1% of a sample, and trace elements are below 0.1% and usually measured in ppm. Trace components go from

1 part per billion (ppb) to 100 parts per million (ppm), but there are also ultra-trace components that account for less than 1 ppb. So, there are some concerns about what one intends to measure. When the sample size is taken into account, its components can be considered as a macro if their mass is above 0.1 g, semi-micro if it is made from 0.01 to 0.1 g, micro components are considered those present in the range 0.0001–0.01 g, and ultra-micro if they make up less than 0.0001 g of a sample.

Analytical determination of very small amounts of substances less than 0.01% and going as low as $10^{-8}$% of the sample are called trace analysis and concentration of those substances are expressed in parts per million (ppm) or smaller parts. There are some particular difficulties in trace analysis such as separation and isolation of substances that are masked by many minor or major mixture constituents, then, in the limits of detection of the chosen technique. Trace components are usually obtained by liquid extraction, distillation, sublimation, ion exchange, chromatography, and coprecipitation among others. Also, the best suited analytical technique must be picked wisely to satisfy the limits of detection. Reagents, water, solvents that are used in sample handling and analyses must be carefully purified to avoid contaminants from the laboratory atmosphere and commonly are called chemically pure, spectroscopically pure, or nuclear pure (over 99.9999% pure) reagents. Also, all used glassware and material must be very clean and used carefully. Trace analysis is becoming of great importance in biochemical and ecological studies, as well as in the chemistry and physics of materials. Errors in trace analysis can be positive or negative. Positive errors are introduced accidentally because of the contamination of samples, while negative errors occur due to the loss of the analyte during sample handling.

Errors in sampling can be random if they are associated with unexpected change, for example, due to environmental conditions like high humidity, wind, or extremely cold weather. On the other hand, an instrumental bias or experimenter accounts for systematic errors that can be reduced to minimal if standardized procedures are employed properly. The choice of solvent is very important when the sample is being prepared based on physical and chemical properties and instrumental requirements for posterior sample handling. For instance, the sample must be soluble in the selected solvent and it is advisable to avoid solvents that are not recommended for the instrumental technique chosen for sample characterization. It is also very important to know an accurate sample concentration. Solid samples are weighed on an analytical balance for greater accuracy and if hygroscopic, samples must be dried before weighing in an oven or desiccator until constant weight. Liquid samples are measured by weight or volume and volumetric flasks are preferred to minimize errors. Alternatively, a glass volumetric pipette can be used, which is calibrated to deliver one accurate volume with the last drop remaining in the pipette. Errors can be compensated using an internal standard, which is an inert substance added in a known amount to the sample before any analysis. This way, random errors become the same for the internal standard and the analyte and the response factor is proportional to the ratio of the analyte and standard concentrations in chromatography or spectroscopy measurements. For systematic errors, such as matrix effects in solution, the ratio will be unaffected as long as the matrix effect is equal for the standard and the analyte. Although internal standards provide great benefits, it can be difficult to choose one that is suitable. An internal standard must have a signal that is similar to, but not identical to, the analyte. Also, it cannot affect the measurement of the analyte in any way. Finally, the concentration must be well known. This is achieved by ensuring that the internal standard is not natively present in the sample; thus, the only source of it in solution is one that was added to the sample.

Biological samples are collected and placed into pre-cleaned vials (Pawliszyn 2010; Clark et al. 2016; Berndt 2020). Collection vials, flasks, tubes, and analysis equipment must be cleaned according to the procedures or standard methods based on the type of the sample and analysis to be applied. Sample vials must be clean and sterile for handling biological samples and may consist of a specific material that depends on the sample type and the desired analysis. Vials could be made of Teflon, baked glass, polycarbonate, among other materials, and almost all must be cryogenic resistant. Samples must be frozen as soon as possible after collection, shipped to the laboratory on dry ice, and held at a low temperature of $-20\,^{\circ}$C or $-80\,^{\circ}$C depending on the sample, until processing (Singh et al. 2020; Mohammedsaleh and Mohammedsaleh 2014).

## 4 Gaseous Samples in Bioanalytics

Gaseous samples (Lawal et al. 2017) are rare and commonly sampled in environmental and classical analytical areas of research, but there are also some gaseous biological samples, like air (Mochalski et al. 2014) that is exhaled from the lungs, used in the bioanalytical field of research. Nitrogen, oxygen, carbon dioxide, argon, and water are the major constituents of human exhaled breath, whereas as minor components, some volatile organic compounds (Fig. 2) are found and may be useful in diagnostics, and some of those are present just in trace concentrations. Most of the breath volatile organic compounds of interest have low concentrations expressed in parts per million volume (Marco and Grimalt 2015) (ppmv) or lower. Therefore, their reliable detection and quantification (Phillips 1997) show a great challenge not just because it is necessary to use highly sensitive analytical instruments, but also because of the need to concentrate sample (Pijls et al. 2016) prior to analyses. Some of the most useful methods for concentration purposes of the gaseous samples are thermal desorption devices as solid-phase microextraction (Mochalski et al. 2013) (SPME), then, needle trap devices (Amann et al. 2014) (NTDs).

There is also some caution regarding gaseous sample collection. It is first necessary to define the type of breath to be collected, for example, from which phase of exhalation it is: late expiratory, end-tidal, or mixed phase (Das and Pal 2020). Next, whether single or multiple exhalations (Shehada et al. 2014) is going to be applied in sampling. Also, the breath capture (Hakim et al. 2011) technology choice has to be considered as well as the volume of air that is going to pass through the chosen device (200 mL for example). There are some commercially available breaths collection containers used in clinics that have as low as possible ingress of environmental and egress of breath volatiles (Brusselmans et al. 2018). Finally,

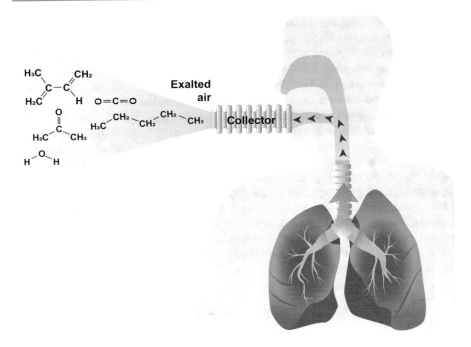

**Fig. 2** Gaseous samples in bioanalytics are mostly used in diagnosis such as exalted air for comparative studies. Exalted air contains valuable but low concentration substances like volatile organic compounds (VOCs) that must be collected and concentrated before analysis. Many different types of devices are used to capture VOCs, with filters that explore solid-phase microextraction (SPME), needle trapping devices (NTDs), or others

during storage, some containers may absorb some gaseous sample constituents from the environment that must be considered when data are analyzed as explained as follows (Brusselmans et al. 2018).

It is important to keep in mind that any device prepared for exalted air sampling must be cleaned, usually with high-purity nitrogen ($N_2$, 99.9999%) in a repeated process (5 times, for example), followed by overnight heating while filled with $N_2$, and evacuation. After being sampled, exhaled air appears as captured VOCs, which must be desorbed, and analyzed following the supplier instructions to avoid mistakes and flaws in results interpretation. But if needle trapping devices (NTDs) are used, because of their small dimensions and low sorbent masses, there is rapid desorption and direct use in the standard inlets of gas chromatographs. Consequently, contrary to traditional sorbent trapping, no additional equipment, such as a thermal desorber is required. It should also be borne in mind that sometimes it is required to subtract unfiltered air and disregard the identified contaminants in our subsequent analysis when air is taken in hospital conditions.

There is a lot of literature on breathomics (Mochalski et al. 2013; Amann et al. 2014; Das and Pal 2020; Shehada et al. 2014; Hakim et al. 2011; Brusselmans et al. 2018) that is still just a research area with the potential to develop into clinical practice application.

# 5    Liquid Samples in Bioanalytics

A liquid sample could be obtained from already liquid, biologically relevant samples such as blood and others (Putcha 2000) or from the dissolution of a solid (McDowall 1989). Bioanalysis in clinical chemistry testing, drug, alcohol, and DNA testing and for diagnostic and therapeutic purposes can be applied on many different liquid samples or it can even use more than one type of sample, for example, sweat, tears, saliva, mucous, urine, blood, plasma, serum, breast milk, and cerebrospinal fluid (CSF) for the analysis of the combined results (Nováková and Vlčková 2009). The quantity in terms of the number of samples and volume depends on the type of sample and analysis that is pretended to be applied and also on the availability of a liquid. If researched in clinical testing, the sample size is around 50 mL for blood tests and complete hemogram (Tiwari and Tiwari 2010), but it can be as small as few microliters (Cassiano et al. 2014) if it is difficult to collect, such as in newborns, dangerous to humans on testing or if the analysis demands do not need great sample volumes. The most common handling of a liquid sample is its freezing immediately after the first pre-treatment to very low temperature (Ali et al. 2008) using liquid nitrogen, and then bio-freezer (Liu and Qiu 2020). Also, the sample must be shipped or transported to the laboratory in dry ice and must not be thawed more than one time before sample preparation for suitable analysis (Serra-Mora et al. 2020). Some samples must be analyzed in short periods after sampling, for example, in 2 weeks in the case of the serum samples (Serra-Mora et al. 2020; Mitra 2004; Amiri 2016).

There is a technology that uses an additive in conjunction with centrifuging, which allows human saliva samples (Nasiria et al. 2020) to be stored at room temperature for up to 6 months. Biological liquid samples that require long-term storage at ambient temperatures require preservation prior to analysis. The preservative (Yi et al. 2008) is composed of sodium benzoate in an amount of at least 0.15% of the sample (weight/volume) and citric acid in an amount of at least 0.025% of the sample (weight/volume). The preservative can be dispensed in different platforms such as coated sample collection tubes (Konoz et al. 2016) like vacutainers and other saliva collection devices, coated preservative discs, compressed tablets, or capsules. Larger volume samples are kept in other devices, such as syringes or dropper bottles for preservation.

Often, the various species of interest must be isolated from the original sample prior to analysis. If the liquid sample is a suspension (blood for example), solid particles are removed using filtration or centrifugation before analysis. On the other hand, if it is necessary to isolate analytes, and extraction, separation, or complexation could be used for sample treatment (Amiri 2016; Nasiria et al. 2020; Yi et al. 2008; Konoz et al. 2016). Usually, sample handlings prior to analysis are named pre-treatments and applied to remove any species that may cause interferences, concentrate analytes, or prepare the sample for the chosen type of analysis.

One of the most applied pre-treatments of the liquid samples is extraction by which the target substance is isolated from a solution (Cassiano et al. 2014). Liquid/ liquid extraction and solid-phase extraction are some of the explored techniques for isolation of the desired analyte from the complex mixtures where affinity differences between two phases are crucial for the process. Immiscible liquids are used in liquid/liquid extractions, such as an aqueous phase of the biological samples and an organic phase (chloroform), which form two layers with a distinct boundary (Serra-Mora et al. 2020; Huang et al. 2019). The affinity of the various components within the sample for each of the two layers is used to separate them. So, the analyte distributes between the two different phases according to its properties that englobe partition coefficient that represents the ratio of the solute's concentration in one solvent to its concentration in the second at given temperature and pressure. A very large or very small partition coefficient is considered desirable to enhance extraction success. Also, multiple extractions may have to be performed to remove most of the solute as fractional partitioning of the solute leads to better results, but never to 100% yield (Lin et al. 2020).

Solid-phase extraction is based on the partitioning of an analyte between the sample's liquid and a solid support because of adsorption. Some trace analysis procedures explore solid-phase microextraction (SPME) in sample preparation for the analysis. This technique is sometimes time consuming.

## 6     Solid and Semisolid Samples

There can be many types of different biological samples that are solids or semisolids, such as hair, tissues, cells, and viruses. Because of that, there are also many different techniques used for their preparation. When samples are in the grind, then obtaining powder from those samples involves grinding a solid sample into a powder using either a mortar and pestle or a ball mill. To ensure that samples are ground and not lost due to sticking to the mechanical apparatus, samples must be dried to expel any water. After reducing particle size and drying, the taken sample has to be homogeneous and representative.

Many analyses require samples that are liquids, so solid samples must be solubilized. Sometimes the sample dissolves easily and is thus ready for analysis or further pretreatment. In other cases, the sample matrix is insoluble in common solvents, and its analytes must be extracted. Analytes may be difficult to remove from the sample insoluble matrix due to adsorption or inclusion phenomena and additional sample preparation steps must be performed to achieve sampling with better accuracy and precision as possible. Besides, smaller particles dissolve faster and are easier to extract because of a greater surface area.

Similar to liquid samples preparation steps before analysis, various extraction methods are used for solid samples, such as solid-liquid extraction, Soxhlet extraction, forced-flow leaching, homogenization, sonication, dissolution, accelerated solvent extraction, supercritical fluid extraction, microwave-assisted extraction, thermal extraction, and others. In solid-liquid extraction, samples are put into closed, or semi-closed containers, and a solvent that can dissolve the targeted analyte, or a desired class of the analytes is added. The solvent is normally kept hot, close to boiling, or refluxed to increase analyte solubilities. After the weight of the extracted solution becomes stable, the solution is dried, filtered, and then analyzed as it is, or further manipulated to obtain an analyte, or analytes in measurable concentrations.

## 7    Sampling and Sample Preparation of RNA and DNA

Nucleic acids, RNA, and DNA (Kresse et al. 2018; Farrosh et al. 2011; Lever et al. 2015; Hedegaard et al. 2014) are fundamental biomolecules for the maintenance of life. Their structure is defined by their primary structure that depicts the sequence of purine and pyrimidine bases, starting from their 5′ to 3′ ends, which is determined using sequencing analysis. The success of sequencing analysis depends on the quality of nucleic acid material. DNA samples are more stable compared to RNA, but sometimes nucleic acids isolated from a material kept at low temperatures ($-80$ °C) for a long might be degraded and difficult to amplify because of degradation or chemical changes that have occurred within the sample (Farrosh et al. 2011). Also, there are many available extraction kits and some efforts have been made to identify optimal RNA and DNA extraction method and kit, nucleic acids quantities, qualities, and the performance of sequencing analysis (Lever et al. 2015). It was observed that for RNA extraction, all protocols gave comparable yields and applicable RNA. But different protocols gave variable quantity and quality of extracted nucleic acids and are to be carefully picked and analyzed prior to any analytical method applied for DNA and RNA characterization (Hedegaard et al. 2014).

Biomedical research samples are in the majority of cases made of tissues that are archived in the form of the formalin-fixed paraffin-embedded (FFPE) blocks. This way, samples from biopsies, surgeries of cancer, and tissues are conserved for long with unchanged morphology (Watanabe et al. 2017; Evers et al. 2011; Howat and Wilson 2014; Williams et al. 1999; Masuda et al. 1999; Head et al. 2014). However, the molecular preservation and quality of the nucleic acids from the FFPE blocks have become critical when discussed the sequencing results. The ability to reliably investigate nucleic acids from FFPE material is thus important, and should also be taken into account in tissue fixation procedures (Gibson et al. 2010). It was evidenced that nucleic acid extraction from FFPE blocks yields a material with an inferior quality compared to one obtained from the fresh frozen tissue (Chory and Pllard 1999). Therefore, determinations of low-frequency genetic variants are very challenging when analyzing only FFPE nucleic acid samples. Nevertheless, the compared data extracted from the paired samples, i.e., fresh frozen and FFPE tissues, showed high concordance in molecular analyses. The nucleic acid can suffer structural changes because of the pH of the fixative, time that spent in the block, the age and storage condition of tissue blocks, and the extraction method used. Because of the chemical reactivity of formaldehyde, which is the active component of formalin, nucleic acids can crosslink with the physically close proteins, or suffer nucleotides' modification (Carini et al. 2016; Eland et al. 2012; Cadillo-Quiroz et al. 2006). Consequently, those factors lead to fragmentation of nucleic acids. In addition, because of the formaldehyde interaction with the key enzymes for sequencing, such as reverse transcriptase, and DNA polymerase among others, can unable molecular analysis. Furthermore, deamination of cytosine to uracil, and other non-reproducible sequence changes that are introduced, can be misinterpreted as mutations (Eland et al. 2012).

DNA and RNA quantifications, generally referred to as nucleic acid quantification (Esteva-Socias et al. 2020), are usually done to determine the average concentration of DNA or RNA in a sample, before proceeding with downstream experiments, and sample purity is fundamental for an accurate sample analysis. There are two optical technologies generally employed to quantify nucleic acids: UV-Vis and fluorescence measurements. An accurate RNA or DNA quantification is a fundamental step in avoiding posterior experimental errors, and the choice of the most suitable methodology for such a task can save time and money (Srivastava et al. 2020).

It is common to analyze nucleic acids using an electrophoretic run in an agarose (highly purified agar) gel (Fig. 3). The speed of nucleic acid migration depends on ion mobility and size, with the smaller molecules moving faster and appearing at the bottom (end) of the gel. This method is able to separate DNA fragments in the range of 0.1–50 kbp, which is far below the size of bacterial genomes, but includes some naturally occurring genetic elements such as bacterial plasmids. A variant of agarose gel electrophoresis, pulsed-field electrophoresis, alternates the orientation of the electrical field in a fashion that allows resolution of much larger DNA fragments (Lee and Tripathi 2020).

The intrinsic absorptivity of aromatic moieties in nucleic acids, DNA and RNA, is explored for their photometric measurement (Eland et al. 2012; Cadillo-Quiroz et al. 2006). When an absorption spectrum is measured, nucleic acids present a characteristic peak at 260 nm (Fig. 3). It is not selective and does not distinguish DNA from RNA (Gong and Li 2014). It presents limited sensitivity with worst detection limits than fluorescence-based methods. But it is very simple with almost no sample preparation, dispenses use of dyes, or standards, and it can provide purity ratios with direct measurements—$A_{260/280}$ and $A_{260/230}$. It can give information about contaminants, and identify non-nucleic acid contamination in samples (proteins, phenol, guanidine salts), and provide correct concentrations (Bueso et al. 2020).

Another method for DNA or RNA characterization explores fluorogenic dyes that bind selectively with the nucleic acids, and permit their fluorometric measurements. The signal is measured by fluorometers. The sample is excited with filtered light at the excitation wavelength, and the emitted light at the emission wavelength is monitored. This way, characterization of double-stranded DNA (dsDNA), single-stranded DNA (ssDNA), and RNA can be specific-performed. The measurements are sensitive, and nucleic acids in low concentrations, such as pg per mL, can be determined. Therefore, it is the recommended method for samples with very diluted nucleic acid. It is accurate despite contamination being present in the sample, including other nucleic acids as contaminants. The major difference between absorbance measurements and fluorescence measurements in determining the concentrations is that fluorophore is used as an indirect way of detecting molecules like DNA in absorbance measurements. A previously measured standard curve is used to determine sample concentrations (Chory and Pllard 1999; Carini et al. 2016;

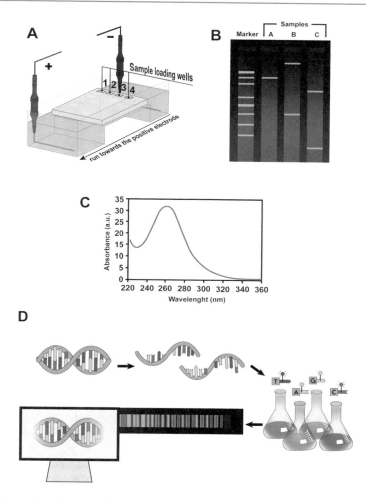

**Fig. 3** Illustration of steps used to evaluate the quality of a nucleic acid sample. Nucleic acids are analyzed using electrophoretic runs (**a**) in agarose (0.8–1%) gels, which are analyzed by comparison with standards (**b**) of known molecular masses, and concentration ng $mL^{-1}$ (sizes are expressed in base pairs, bp, with two standards in greater quantity—1000 bp and 500 bp as shown). (**c**) Sample is checked using UV-Vis, and maximum absorbance at 260 nm. (**d**) If clean (highly pure), and obtained in sufficient quantity (ng $mL^{-1}$), double-stranded DNA (dsDNA) undergoes sequencing

Eland et al. 2012; Cadillo-Quiroz et al. 2006). After combining the standard nucleic acid at least in five different concentrations with the fluorescent dye, a fluorimeter is used to determine the relative fluorescence (RFU).

As already stated, DNA samples can be obtained from different sources such as blood, cells, tissues—frozen, fresh, or stored in paraffin tissue blocks, and so on. DNA extraction and purification involve three steps: (1) breakage of the cells/tissues to expose the DNA, (2) removal of the lipids using a detergent, and

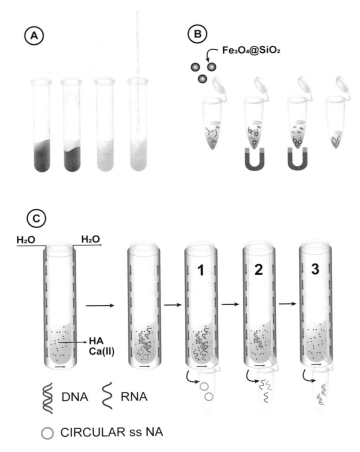

**Fig. 4** Illustration of some of the most important isolation methods used for nucleic acid samples (DNA and RNA). (**a**) Classical DNA extraction method: (1) Cells or tissue samples are lysed with detergent and membranes are disrupted, (2) proteases and RNAase are added and lysis of other that DNA occurs, (3) after centrifugation, DNA is separated in the supernatant, and (4) DNA is precipitated with ethanol and can be spooled as viscous stands. (**b**) (1) Magnetic nanoparticles are added to lysed biosamples and bind with nucleic acids, (2) debris is then washed away using a permanent magnet that keeps magnetic beads loaded with the targeted nucleic acids safe at the bottom of the tube or Eppendorf, (3) nucleic acids are rinsed from the magnetic beads and (4) ready to be used. (**c**) Hydroxyapatite (HA) loaded columns are used for different nucleic acids separations such as DNA (single—ss or double-stranded—ds), RNA (ss, ds), and circular nucleic acids using increased phosphate buffer concentrations (1–3)

(3) precipitation of DNA (Head et al. 2014; Gibson et al. 2010; Chory and Pllard 1999; Carini et al. 2016; Eland et al. 2012; Cadillo-Quiroz et al. 2006; Esteva-Socias et al. 2020; Srivastava et al. 2020; Lee and Tripathi 2020; Gong and Li 2014; Bueso et al. 2020). Also, it would be advisable to remove proteins applying proteases, and RNA using RNases (Fig. 4a). DNA purity increases sample stability if aiming for long-term storage.

If one needs to analyze just the RNA, then there are two protocols to be used, isolating and purifying only the RNA from the source material or isolating the total nucleic acids and add DNase to degrade and take off the DNA from the sample or use liquid chromatography for the separation of different nucleic acids. Sometimes, it is needed to extract DNA or RNA simultaneously from many samples, and then it is needed to simplify, speed up and keep less laborious methods for nucleic acid isolation and purification. And for such extractions, magnetic beads or nanoparticles can be used by exploring solid-phase reversible immobilization (SPRI). Magnetic nanoparticles coated with silica ($Fe_3O_4@SiO_2$) reversibly bind nucleic acid under dehydrating conditions and with the aid of a strong magnet, can be safely immobilized and purified through multiple washes, and manipulation steps (Fig. 4b). Binding DNA to silica on magnetic beads is based on altered affinity of the negatively charged DNA backbone to the silica surface in the presence of chaotropic salts such as guanidinium hydrochloride. By varying chaotropic salt concentration, DNA can be selectively isolated (100 and 3000 bp) and recovered up to 95% of the input DNA. Magnetic bead protocols are inherently scalable since they are independent of centrifugation and the required materials are easy to manufacture on a laboratory scale, and not expensive to purchase. Nevertheless, despite many advantages, magnetic materials are not used often.

If just the RNA is needed, then the starting material is disrupted by adding a chaotropic agent (guanidium thiocyanate, for example), and a reducing agent (beta-mercaptoethanol), vigorously shake or vortex, as to break cystine, and inactivate the contaminant proteins. Phenol and chloroform/ *iso*-amylalcohol are added cold to pass RNA into the aqueous phase. Finally, isopropanol is used to precipitate RNA, which is washed with 75% ethanol to remove impurities. RNA is less stable than DNA, so some extra care is needed for RNA sample handling and posterior use in analysis. It is also possible to prepare just the DNA or the RNA sample from the total nucleic acid fraction, which is divided into: (1) the first fraction that after treated with a DNase 1, recovers the RNA, and (2) the second fraction that after treated with RNase, recovers the DNA. Or total nucleic acids can be purified on gel electrophoresis and purified using open column chromatography (Fig. 4c). Hydroxyapatite (HA) has been employed in the separation of nucleic acids for long. HA exploits two primary binding mechanisms: metal affinity interactions through HA calcium—Ca (II), and cation exchange interactions through HA phosphate. Nucleic acids (ssDNA, RNA, and dsDNA) are generally incubated in the HA column (60 °C) and eluted using a phosphate-containing buffer with concentrations that increase (1–3, Fig. 4c). This way, the circular single-stranded nucleic acids are eluted first (the lowest concentration of phosphate ions), followed by linear ssDNA, and RNA (dsRNA and ssRNA), and the last fraction contains just ssDNA (the highest phosphate concentration). As Ca(II) ions in hydroxyapatite can cause differences in applied runs, it is recommended to optimize the elution of total nucleic acids using a previously known standard.

At last, samples containing pure nucleic acids must be properly stored and handled prior to successful downstream research applications. Nucleic acids can degrade at room temperature, especially RNA samples at neutral or low pH from autolysis, and need to be frozen to maintain their integrity. If the same nucleic acids' sample is going to be used in multiple experiments, it is recommended to make multiple aliquots that will only be thawed once when they are needed. Nucleic acids that are going to be used in a short time frame may be stored at $-20\ °C$ but if not, should be kept in an ultra-low freezer, typically at or below $-80\ °C$. Also, an aliquot of a very important sample can be stored in off-site bio-storage (biobank) as backup material to be kept safe and well monitored.

# 8    Sampling and Sample Preparation of Lipids

Lipids (Granafei et al. 2017) are a very diverse group of molecules that are defined as organic molecules soluble in chloroform. There are many organic compound classes, such as fatty acids and alcohols and their esters, mono-, di-, and triglycerides, sterols, many phospholipids derived from phosphoric acids and esterified with fatty acids, and choline, poly-phosphorylated inositol derivatives, L-serine among others, sphingolipids, and different lipopolysaccharides, which are lipids (Granafei et al. 2017; Abbott et al. 2013; Sajnani and Aribindi 2017; Iverson et al. 2001; Gil et al. 2018). Amphiphilic lipids show dual nature with hydrophilic parts (head group) and lipophilic parts (tails) linked with some polyol (Abbott et al. 2013). Hydrophilic groups could be amino acids, alcohols, amino alcohols, or saccharides. Phospholipids could be mono-, di-, tri- esters, or ethers (Sajnani and Aribindi 2017; Iverson et al. 2001). Even glycerophospholipids with just two fatty acids esterifying the glycerol may show isoforms, with different positions of a given fatty acid on glycerol carbons. Therefore, despite being the smallest biomolecules, lipids are a very complex class of molecules with a considerable number of molecular species that compose lipidome at any given moment. Diverse lipid species and broad application of lipidomics data mean that techniques for isolating the lipid fraction are of considerable interest (Granafei et al. 2017; Abbott et al. 2013; Sajnani and Aribindi 2017; Iverson et al. 2001; Gil et al. 2018; Pizarro et al. 2013; Züllig et al. 2020; Burla et al. 2018; Pellegrino et al. 2014; Bowden et al. 2017).

Lipid isolation processes (Yang et al. 2009; Jiang et al. 2007; Han et al. 2004; Garwolinska et al. 2017; Li et al. 2014) and posterior lipid analyses are of extreme importance for accurate and precise data analyses. For example, if a measured value is very proximate to a standard or known value, the measurement is accurate. On the other side, if two or more measurements are very close to each other, the measurement is precise. Using an example, if a given substance is weighed three times, and gets 2.10008 g each time, then the performed measurement is very precise. If an experiment contains a systematic error, then increasing the sample size increases precision but does not improve accuracy, and the obtained result from inaccurate experiment is just precise. If the systematic error is eliminated, one can improve the accuracy but does not change precision. A measurement system is considered valid if it is accurate and precise (Yang et al. 2009; Han et al. 2004).

The lipophilic fraction of any biological sample comprises all fats, such as triglycerides, fatty acids and alcohols, sterols, waxes, and fat-soluble vitamins

(Garwolinska et al. 2017; Li et al. 2014; Cruz et al. 2016; Correia et al. 2018, 2020; Mor et al. 2020; Satomi et al. 2017). The choice of isolation of this type of lipids must follow the requirements of the desired analyses. Gas chromatography (GC) is the most common technique for identifying fatty acids (FAs). Therefore, lipophilic sample preparation requires isolation of total lipids, separation of the FA compounds from other lipids, followed by transesterification in which the methyl esters of the fatty acids are prepared.

Many different bio-samples can contain lipids, for example, serum, cells, tissues, and viruses (Pellegrino et al. 2014). Also, there are standardized protocols for solid, semi-solid, or liquid samples extraction and handling. Tissues or cells (semi-solid samples), freshly collected or harvested, are the best option for lipid extraction to prevent eventual hydrolysis and oxidation. For example, presence of lysophospholipids, phosphatidic acid, monoacylglycerols, diacylglycerols, or free fatty acids in higher than usual quantities might indicate a possible sample degradation before extraction even started (Bowden et al. 2017; Garwolinska et al. 2017). In some special cases, additional precautions are necessary to minimize rapid degradation of lipid constituents. For example, plant tissues need to be processed immediately after collection by immersion in hot 2-propanol or water (Li et al. 2014). Plant phospholipase D is known to be active even in some solvents. If tissue cannot be immediately handled for extraction in a laboratory (off-site sample collection), then, it should be frozen as quickly as possible in a dry ice or better in a liquid nitrogen and stored in sealed glass containers at $-70\,^{\circ}$C (Cruz et al. 2016; Correia et al. 2020).

Fresh or frozen tissues or cells should be homogenized in the chosen solvent mixture and agitated for a fixed time before elimination of the solid part. Soft tissues may be minced with scissors in the cold and homogenized solvent with an Ultraturax type device, a Waring blender, or a glass-Teflon Potter. Hard tissues (muscle, vessels, bone among others) are best pulverized in a mortar filled with liquid nitrogen. Cells might be treated directly with the solvents (Bowden et al. 2017; Yang et al. 2009; Jiang et al. 2007; Han et al. 2004; Garwolinska et al. 2017; Li et al. 2014; Cruz et al. 2016; Correia et al. 2018, 2020; Mor et al. 2020; Satomi et al. 2017).

The classical Bligh-Dyer method uses chloroform/ methanol/ water solvent mixture but there are some chloroform-free solvent extraction protocols based on 2-methoxy-2-methylpropane also known as methyl *tert*-butyl ether (MTBE) protocol (Bowden et al. 2017; Garwolinska et al. 2017). Replacing chloroform with MTBE is safer, somewhat less carcinogenic and toxic, with lipids dissolved in the upper solvent phase, which allows an easier pipetting, and has the potential for automation (Correia et al. 2020).

It is important to have in mind that lipids must not be stored in dry form for at least two reasons. Oxidative degradation is slower in solutions even when antioxidants are not added. Air and light influences must be diminished. The best storage conditions are bio-freezers ($-70\,^{\circ}$C) for a long storage (up to 1 year) of lipid extracts in chloroform–methanol mixtures that fill well-sealed stopper glass vials (with Teflon liners), the cap is secured with a wide length of self-sticking tape. During long periods of storage, it can be useful to flush vials or tubes with nitrogen before closing to prevent fatty acid oxidation. Also, for the same purpose, a low number of antioxidants, such as butylated hydroxytoluene (BHT) or 2,6-di-tert-butyl-4-methoxyphenol or ethyl gallate (around 50µg BHT/mL), may be added to the solvent, principally because the purified lipid extracts usually contain low quantities of natural antioxidants. The added antioxidant is commonly prepared as a concentrated solution in ethanol (10 mg mL$^{-1}$) (Han et al. 2004; Li et al. 2014; Correia et al. 2018).

Some lipid samples show decrease in esters and plasmalogens, and an increase in free fatty acids, diacylglycerols methyl, or ethyl esters, when stored for a long time (Satomi et al. 2017). These problems can be minimized by using neutral extracts and 2-propanol instead of methanol or ethanol in the solvent mixture. It must be remembered that the best results are obtained with freshly prepared materials.

# 9 Sampling and Sample Preparation of Carbohydrates

Carbohydrates (Sanz and Martínez-Castro 2007) are polar molecules that occur as free, such as mono-, oligo-, and polysaccharides, or as covalently bonded to other biomolecules in the form of bioconjugates, with lipids (glycolipids) or proteins (glycoproteins and proteoglycans) (Sanz and Martínez-Castro 2007; Montañés et al. 2008; Soria et al. 2012a; Melini and Melini 2020; Aoki-Kinoshita et al. 2020). In general, carbohydrates are hygroscopic, but polysaccharides are mostly stable in the solid state, at physiological pH and standard temperature, but labile thermally as they can decompose losing water. Carbohydrates are also grouped into structural and non-structural, and the last ones are abundant in plants (Sanz and Martínez-Castro 2007). If plant material is collected for analysis, then, it is important to avoid sampling degradation effects by diminishing any stress caused (cutting), and inhibit enzymes, using shock-freezing, transportation in coolers (ice or dry ice) to the laboratories, and posterior freeze or oven drying (Ou et al. 2018). Nevertheless, the most important steps to the successful analyses of carbohydrates are not just sampling, and sample handling, but sample preparation, such as the extraction protocol and the analytical methodology used for soluble sugar quantification, and characterization (Melini and Melini 2020; Ou et al. 2018).

There is a fairly large literature on carbohydrates' analysis in food, and many articles and methods have been written or introduced since glycoconjugates have been the objects of study (Soria et al. 2012a; Melini and Melini 2020; A successful analysis of carbohydrates starts with the first step, which is to pick a representative sample, then, proceeds with suitable extraction, and finally, carbohydrates are analyzed.. Aoki-Kinoshita et al. 2020). Also, sample handling and preparation are different if water- or alcohol-soluble carbohydrates, such as mono- or oligosaccharides or polysaccharides, or glycoconjugates are to be analyzed (Chlumská et al. 2014; Landhäusser et al. 2018; Olajos et al. 2008). If the molecules are freely soluble, then the sample preparation is relatively simple, but, still, other biomolecules need to be removed.

Mono- and oligosaccharides are soluble in alcoholic solutions, thus, the semisolid or solid samples, after drying to the constant weight, are treated with 80% ethanol to yield extracts free from proteins and polysaccharides (Melini and Melini 2020; Khatri et al. 2017; Lu et al. 2018; York et al. 2020; Yamada et al. 2020). Such way prepared extracts are still not pure enough for posterior analyses because of other ethanol-soluble molecules—organic and amino acids, vitamins, minerals, pigments from the original sample, and further purification is achieved by clarifying, and use of different chromatography techniques (Chlumská et al. 2014; Landhäusser et al. 2018; Olajos et al. 2008; Khatri et al. 2017; Lu et al. 2018).

For example, thin-layer (TLC), and liquid chromatography (LC) are commonly used to isolate carbohydrates (Melini and Melini 2020; Lu et al. 2018; York et al. 2020; Yamada et al. 2020). Carbohydrates are separated according to their intermolecular interactions with a stationary and a mobile phase, and isolated as bands in TLC and fractions in LC (York et al. 2020; Yamada et al. 2020; Jensen et al. 2012; Trbojević Akmačić et al. 2015; Fomin et al. 2020; Bertagnolli et al. 2014). Carbohydrates can be separated by electrophoresis but only when previously modified into charged species, which after being applied to a gel, show different ion mobility in an electric field (Soria et al. 2012b; Mort and Pierce 2002). Thus, migrate with different speed and are separated based on their size (Dvořáčková et al. 2014), and smaller molecules move faster.

There is no universal procedure for the analysis of most polysaccharides. Therefore, after isolation, and hydrolysis, their constituent monosaccharides are determined (Chlumská et al. 2014; Khatri et al. 2017). Enzymic hydrolysis methods are specific and sensitive but not selective, break up all glycosidic bonds with the specific type—alpha or beta, and used rarely. Dietary fibers, soluble, and total fibers, are composed primarily from non-starch polysaccharides, and analyzed after the removal of the digestible starch with treatment with amylases, proteins are removed with protease, and a nondigestible residue is obtained for further analyses (Melini and Melini 2020).

Sampling and sample pretreatment are very important stages that precede carbohydrate analysis, and depend on the type and composition of the analyzed sample (Soria et al. 2012a; Melini and Melini 2020; Aoki-Kinoshita et al. 2020; Ou et al. 2018; Chlumská et al. 2014; Landhäusser et al. 2018). Sample preparation includes filtration, dialysis, extraction, reverse osmosis, chromatographic fractionation, and biological and chemical treatments. As already explained, the first step is the separation of carbohydrates from the others components, such as lipids and proteins. Ultrafiltration and filtration using membranes are used for sample concentration. Complex biological samples with different classes of carbohydrates in various amounts require the use of preparation techniques before analysis.

Carbohydrates can be extracted using common techniques, such as liquid-liquid and solid-phase extractions (SPE), where the sample affinities between the stationary and mobile phases are explored (Sanz and Martínez-Castro 2007; Melini and Melini 2020). Columns such as C8, C18, porous graphitic carbon, strong cation exchangers are some of the most used as a stationary phase. For instance, mono- and oligosaccharides from many types of honey can be isolated by active carbon and celite columns and ethanol/water as a mobile phase. Nevertheless, the successful results are obtained with ease in LC, the high quantities of solvents, and low analyte concentrations are bottlenecks of these sample preparation methods. Therefore, other methods might be a better choice.

One of such methods relies on extraction with supercritical fluids (SFE) that uses pure fluid or a mixture, such as fluid with the additives, under critical pressure and temperature (Montañés et al. 2008; Melini and Melini 2020). This method explores carbohydrate solubility in a chosen liquid fluid, and diffusivity in the gas under supercritical conditions. Because of the moderate critical temperature and pressure that $CO_2$ has, it is the most commonly used fluid in SFE. Extracts become free of solvent because $CO_2$ turns gaseous under ambient conditions. Finally, the selectivity of the extraction can be controlled with the depressurization steps. However, $CO_2$ has a low polarity, which makes the solubilization of carbohydrates difficult, and different amounts of organic solvents are employed as $CO_2$ modifiers.

Sometimes, it is recommended to apply pressurized liquid extraction (PLE) for carbohydrate sample preparation (Melini and Melini 2020). This method explores just the solvent liquid state at high pressures. Thus, in a short time, it can extract high carbohydrate amounts using lower solvent volumes. PLE is mostly used to extract carbohydrates from food and natural products, and counts on application of various solvents such as water, organic solvents, and water/organic solvents mixtures (Melini and Melini 2020). One interesting example is purification of lactulose from mixtures with lactose using PLE with 70:30 (v/v) ethanol/water mixture (Melini and Melini 2020). Another example is the isolation of di- and trisaccharides from honey by using PLE combined with the use of activated charcoal (Melini and Melini 2020).

Another method for isolating carbohydrates that explores different mobility of the mixture components is field-flow fractionation (FFF) (Melini and Melini 2020). The solution of a mixture of carbohydrates is fractionated when pumped in a laminar flow that suffers a perpendicular effect, which can be gravitational, thermal, magnetic, or electric. Because of the differences in the mobility, components are separated. For example, polysaccharides such as starch, cellulose, pullulan, among others, are commonly isolated from smaller carbohydrates this way (Melini and Melini 2020).

The size exclusion chromatography (SEC) is used to separate oligo- and polysaccharides and elucidate the molecular mass distribution of carbohydrate samples. Another important technique explores ion exchange on anion or cation exchange resins for oligo- and monosaccharides extraction (Sanz and Martínez-Castro 2007; Melini and Melini 2020). On the other hand for large-scale palatinose- and trehalose fractionation, the simulated moving bed chromatography (SMBC) is used (Montañés et al. 2008). Based on adsorbents, ligand-exchange chromatography or size-exclusion chromatography are also employed to raffinate and extract samples for posterior analyses. For example, lactose from a complex sample of human milk oligosaccharides is purified using SMBC (Melini and Melini 2020).

Carbohydrates can also be separated according to their molecular masses exploring dialysis, ultra-, and nanofiltration systems, and different membranes. A sample that contains different carbohydrates can be fractioned using adsorption onto an activated charcoal column, followed by a subsequent elution with ethanol-containing solvents. This way, with the lowest concentration of ethanol (1%) monosaccharides are the first fraction, disaccharides need more concentrated ethanol (5%), and oligosaccharides are obtained just after using the highest concentrated ethanol solution (50%) (Melini and Melini 2020). This is one of the best choices to isolate monosaccharides from a very complex sample. Other sample preparation methods employed in carbohydrate analysis include the use of molecularly imprinted polymers (MIPs). Such MIPs have been demonstrated to be useful in the recognition of different saccharides or the epimeric differentiation of disaccharides.

Different chemical procedures are used to prepare carbohydrates for analysis, starting from their hydrolysis, followed by low molecular mass mono- or oligosaccharides analyses. Poly- and oligosaccharides hydrolysis treatment with strong acid (trifluoroacetic and hydrochloric acid, sometimes, sulfuric acid) whose concentration, as well as temperature, and reaction time must be optimized to complete hydrolysis and avoid sample degradation (dehydration). Sometimes, polysaccharides are just partially hydrolyzed to mono- and oligosaccharides whose analysis might be useful to determine the polysaccharide structure (Olajos et al. 2008; York et al. 2020).

Carbohydrates have very similar polarity, show low volatility, and do not have chromophore (Yamada et al. 2020). Therefore, if wished to be analyzed by GC, carbohydrates should be derivatized to increase the volatility. Monosaccharides are transformed into alditol acetates by reduction and acetylation, while oligosaccharides become volatile if turned into trimethylsilyl or trifluoromethyl ethers. If meant to be detected after a run in HPLC, because carbohydrates can only absorb UV light below 195 nm, amperometric or refractive index detectors must be used. On the other hand, it is advisable to modify their structure with UV absorbing or fluorescent labels with derivatization processes to increase the sensitivity of their detection. The reaction, which is the most usual procedure pre-column derivatization, may take place before or during the analysis (Jensen et al. 2012; Trbojević Akmačić et al. 2015). The most employed approach needs a reductive amination. These procedures are based on the condensation of a carbonyl group with primary amines to produce a Schiff base, which is later on reduced to give an $N$-substituted glycosyl amine. 2-Aminopyridine trisulfonates are among the chemicals more employed. Besides, considering that most carbohydrates are neutral and hydrophilic compounds, derivatization produces not only a sensitivity enhancement but also polarity and electrical charge variations, which can facilitate their separation afterward (Fomin et al. 2020).

At last, the carbohydrate field of research called glycomics is becoming popular and of great interest. Glycomics is a study that maps all carbohydrates of an organism, which may be free or present in bioconjugates, such as glycolipids, glycoproteins, and proteoglycans, and links any deviation from a healthy state to provoked with inherited, and genetic reasons, physiological changes, pathological, and other problems. If analyses of carbohydrate constituents of glycoproteins are in question, then, $50–100 \times 10^{-6}$ g of pure protein in solution is needed. After the release of the N-glycans, labeling and purification steps are held, the sample is dried and resuspended into 20µL and analyzed just 2µL. If a glycoprotein band is analyzed after SDS gel electrophoresis, around 10µg of the sample is needed (Bertagnolli et al. 2014; Soria et al. 2012b; Mort and Pierce 2002; Dvořáčková et al. 2014; Szpakowska et al. 2020; Deng et al. 2004).

## 10    Sampling and Sample Preparation of Proteins

Chemically speaking, proteins (Matejtschuk 2007) are biopolymers made from amino acids linked through peptide bonds in one or more chains of polyamides (polypeptides), which start at the N-terminal and end at the C-terminal. Proteins show a defined number of monomers per chain, high molecular masses expressed in kDa, and characteristic isoelectric points (pI) (Feist and Hummon 2015). Some proteins contain other chemical entities called prosthetic groups that are important for their functions such as heme in hemoglobins, cofactors (e.g., $NAD^+$) in enzymes, among others (Feist and Hummon 2015; Gutstein et al. 2008; Aebersold and Mann 2003). Other proteins appear as bioconjugates like glycoproteins and lipoproteins (Gutstein et al. 2008). They may be functional just if dimeric (insulin receptor, for example) or multimeric (ATPase), and make complex homo- or heteromeric structures. Proteins are important structural components and also have multiple and very important functions as enzymes, as well as take part in cell signalization, transport, DNA replication, and translation, and are molecules with primordial roles that orchestrate life (Feist and Hummon 2015; Gutstein et al. 2008; Aebersold and Mann 2003; Medzihradszky and Chalkley 2015; Scott et al. 2012). They are very diverse and can be rigid and elastic as keratin and silk, respectively, and very sensitive as even a small change in pH, ionic force, and temperature might drastically change their structure. And as all biomolecules, proteins are found in complex mixtures and diverse physical states and can be present in traces or as major components (Feist and Hummon 2015; Gutstein et al. 2008; Aebersold and Mann 2003). Therefore, likewise in other sample handlings, some care must be taken when analyzing proteins, and many of them depend on our target analysis if one wishes to determine and describe all proteins present in a sample (proteomics), or analyze the structure or function of one targeted protein, thus, two different approaches in sample preparation are undertaken (Gutstein et al. 2008; Scott et al. 2012).

If our objective is proteomics, a study of the total protein content of a cell, a tissue, an organ, or a whole organism at a given moment, then proteins must be isolated from other mixture/sample components and analyzed as such. The proteomic sample usually contains hundreds or thousands of proteins in very diverse concentrations. For example, human blood serum is made from 60 to 80 mg/mL of proteins (Hilbrig and Freitag 2003; Doellinger et al. 2020), whose half accounts for human serum albumin (HSA) and almost ¼ to γ-globulins, while others are present in traces, such as cytokines. Thus, even highly sensitive, analytical procedures based on mass spectrometry (MS) or electrophoresis are not capable of detecting all the protein species present in the sample (Hilbrig and Freitag 2003). Without a targeted approach to improving detectability, the low-abundance part of proteomes is in essence inaccessible or difficult to detect (Wiśniewski 2019).

Even if it is our aim to characterize all detectable proteins in a sample, there is a need for sample preparation after collection. The quality of such preparation is critical to successful protein analysis to ensure the best possible analytical results. There are many methods for preparing protein samples for MS analysis. Before deciding on the best method for obtaining the sample, there is some information needed, such as the type, biophysical properties, and quantity of the targeted proteins, their location, and complexity of the samples (Wiśniewski 2019; González-García et al. 2020; Kachuk et al. 2015; Eddhif et al. 2018). Preparation protocols that employ lysis, precipitation, and fractionation of proteins, depletion of high-abundance proteins, then enrichment of selected proteins, and mass tagging tools to accurately identify and quantify the analyzed samples are often used (González-García et al. 2020).

The use of chromatographic systems is a part of sample preparation if aiming to protein fractionation (Visser et al. 2005) in proteomics studies, and sometimes it is applied to isolate and characterize just the targeted protein (Bache et al. 2018). If aiming to study the pure protein and characterize it, then it is necessary to analyze the mixture and determine where the target protein is by applying an electrophoretic run, to evaluate the target protein concentration and its similarity in size with other mixture proteins and choose the best way for posterior purification protocol. The most efficient purification procedure employs affinity liquid chromatography (Fig. 5), usually counting on antibody—antigen or lectin–glycoside interactions, but it is difficult to perform it if the targeted protein is not previously known (Müller et al. 2020). There are many different resins, magnetic and magnetic agarose beads commercially available among which, the ones for the immunoprecipitation (IP) techniques are the most efficient. For example, protein G magnetic beads for immunoglobulins purification from human samples can be used to purify up to 25–30μg human IgG/mg beads (Doellinger et al. 2020; Vuckovic 2013; Zeki et al. 2020; Wawrzyniak et al. 2018). After each purification step, it is necessary to evaluate how efficient was the purification, and electrophoretic runs must be performed. For example, Fig. 5 shows an illustration of the purification of five proteins whose molecular mass can be estimated by comparison of the run that molecular mass marker (M) showed under the same circumstances. Also, before any biophysics method is used for the target protein characterization, its concentration must be determined usually using UV-Vis protein properties (280 nm) or other colorimetric methods based on the application of dyes, like the Bradford method (Coomassie) or fluorescent dyes with many kits sold commercially.

Proteins can be analyzed using electrophoretic runs in 1D and 2D where differences in the molecular masses and isoelectric points are applied to separate proteins. Digest-in gel trypsin lysis for proteomics research is one of the ways to prepare proteins for MS and if the targeted protein exists in a database, only 10–20 ng of the protein is needed. For the identification of proteins in complex mixtures (e.g., separated over a gel lane), the amount of protein should be around or greater than 10μg. For the protein molecular mass determination, at least 10μg at a

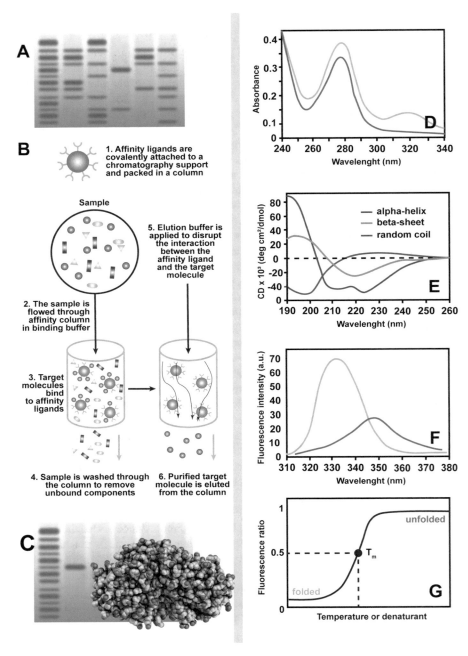

**Fig. 5** Illustration of principle steps needed for the target protein analysis. The target protein is isolated from a complex mixture by precipitation of total proteins. After an electrophoretic run to evaluate concentration and size of the target protein, the best strategy for its purification is proposed,

concentration not lower than 1 mg mL$^{-1}$ is needed (Hughes et al. 2019; Wiśniewski 2019; González-García et al. 2020). There are some important tips to keep in mind, the Coomassie-staining of the gel cannot be done in a microwave or scanned on an overhead projector foil, because microwaving modifies proteins in the gel, and there is no way to extract them, also polymers in the overhead projector foils cause trypsin inhibition (Kachuk et al. 2015).

Proteins in a given sample can be characterized at the proteome or sub-proteome level (Kachuk et al. 2015; Eddhif et al. 2018), for example, one can be interested in determining just the cell phosphoproteome to understand their role and function. The big issue in proteomics studies is linked to separation methods that precede MS analysis. There are some technologies that show great potential for undirected and targeted MS. For example, ion mobility separations, microchip-based, chromatographic nanoscale reversed-phase liquid chromatography, capillary electrophoresis are just some techniques that can be hyphenated with MS.

Another important issue to be considered is protecting proteins from the action of proteases, and other protein-modifying enzymes. It is very important to preserve the real state of the proteins and proteomes from different tissue and cell samples so that analytical results can be trusted (Hughes et al. 2019; Wiśniewski 2019; González-García et al. 2020; Kachuk et al. 2015). If not controlled, degradation and protein modification can potentially take place throughout the complete sample preparation workflow (Eddhif et al. 2018; Bache et al. 2018).

Similar to some biological samples, proteins are sensitive to rough handling, especially if concentrated and purified (Müller et al. 2020). Proteins can degrade during storage due to breakdown, aggregation, and inactivation (Nickerson and Doucette 2020). And while there are many storage options, each has its own tradeoffs. If used briefly or within 2 weeks, concentrated protein samples (1 mg mL$^{-1}$) should be kept in clean, sterilized glassware, or polypropylene tubes at 4 °C or stored in buffers with the appropriate protease inhibitors, antimicrobials, and reducing agents, respectively (Doellinger et al. 2020). In other circumstances, it is recommended to keep protein samples in the freezer, −20 °C with the cryoprotective agents (Vuckovic 2013).

---

**Fig. 5** (continued) being affinity-based bioanalytical method as one of the most efficient if the very pure and low quantity of protein is needed for further analyses. Protein concentration is estimated using UV-Vis measurements, using aromatic acid residues and cystine as protein chromophores or dyes (Coomassie or fluorescent dyes). Biophysical methods that are mostly used to evaluate protein 3D and 2D features are fluorescence and circular dichroism. As proteins exhibit their function only when found in native conformation, their stability (thermal, solvent, like pH and ionic force) might be analyzed by either technique

## 11 Small Molecules Samples

The smallest molecule is the hydrogen molecule ($H_2$), but the term small molecule in bioanalytics refers to molecules found in biological samples that show molecular masses lower than 1.5 kDa (Zeki et al. 2020). There are many powerful analytical techniques and hyphenated ones that are frequently employed in attempts to characterize as many as possible small molecules or metabolites within a given biological sample (metabolomics). Some of those need little or no preparation of samples and others need extensive sample preparation and standardized procedures (Zeki et al. 2020; Wawrzyniak et al. 2018; Roberts et al. 2012; Lin et al. 2007; Wenk 2005; Auslander et al. 2016; Wolthuis et al. 2020; Chong et al. 2018; Coley et al. 2019; Pontes et al. 2016; Pontes et al. 2017). Differently from biomolecules cited in previous chapter sections, small molecule sample preparation is not based on isolation and purification of all of those present in the matrix to be analyzed, but in getting great samples from small molecules based on their polarities. Usually, metabolites are split into polar (Wawrzyniak et al. 2018) and nonpolar (Roberts et al. 2012) isolates and further fractioned (Lin et al. 2007) if needed. In general, those methods are long, require the use of chemicals, such as solvents for extraction, isolation, and purification, other consumables, for example, for water sorption, and prone to errors because of human engagement.

There are two metabolomics approaches, targeted and untargeted (Lin et al. 2007), which are used in biological fluids and tissues analyses. Sample preparation methods for the untargeted metabolomics are different if nuclear magnetic resonance or mass spectrometry as analyzers are applied afterward. It is advisable to keep both sample preparation and handling simple to avoid errors, such as contamination and sample degradation, and quench any sample decomposition to preserve intact metabolome composition as at the time of sampling (Wenk 2005). There are also in vivo methods such as microdialysis (Zeki et al. 2020; Wolthuis et al. 2020), solid-phase microextraction, turbulent flow chromatography, and dried blood spot sampling. Among various sample preparation methods, it is advisable to pick and test the most appropriate one for a given cohort of samples.

For example, in untargeted metabolomics by LC-MS, sample preparation methods include protein precipitation, liquid-liquid extraction, solid-phase extraction (SPE), or ultrafiltration. Protein fraction must be removed using organic solvents or a mixture thereof, followed by centrifugation and supernatant filtration, which are the most frequently used approaches in plasma untargeted metabolomics. The success of extraction in untargeted metabolomics depends on extracting the widest possible range of metabolites from the sample. In targeted metabolomics, the protocol employed for extraction must be efficient for getting just the class of desired metabolites or one metabolite. Thus, tailored or personalized protocols need to be developed.

Independently on the approach applied in metabolomics study, biological replicates and confounders are to be kept in mind when sampling and sample preparation are in question. For example, if human samples are analyzed, investigated individuals must be matched when considering gender, age, health problems, because of the biological variability, and over 50 individuals in each group are recommendable but not limited, and the more the better. If used animal models, then age (weeks), type, diet, exercises must be identical among the studied groups, and at least five plus samples must be investigated. When samples are wild species, then at least five specimens are needed for good statistics that englobe data analysis (Roberts et al. 2012; Lin et al. 2007; Wenk 2005; Auslander et al. 2016; Wolthuis et al. 2020; Chong et al. 2018; Coley et al. 2019; Pontes et al. 2016, 2017).

Alongside the text and in many items, it was pointed that biological samples must be frozen just after being sampled and before transportation to a laboratory, kept at low temperatures and that most of the samples cannot be thawed for more than one time to protect them from degradation or modification of their composition (Lin et al. 2007). Some tissues and cell samples and even biomolecules need to be kept together with the cryoprotectant agents, which are used to prevent ice formation, which causes freezing damage (Lin et al. 2007). Cryoprotectants reduce the ice formation at any temperature by increasing the total concentration of all the solutes present in the system by lowering the melting point of water (Auslander et al. 2016). In order to be used, cryoprotectants must have the capacity to penetrate the cells and show no or very low toxicity. For example, one of the most used cryoprotectants is ethylene glycol, although dimethyl sulfoxide (DMSO), and glycerol are used sometimes (Auslander et al. 2016). Usually, 5–15% of cryoprotectants is enough to allow the isolated cells to survive after freezing and thawing from the liquid nitrogen temperature (Chong et al. 2018; Coley et al. 2019; Pontes et al. 2016; Pontes et al. 2017).

## 12  Final Remarks and Future Perspectives

Sample preparation is one of the most important features in bioanalytical chemistry, and most of the samples are sampled or collected, kept, and transported to the chemical laboratory using standardized procedures and techniques. The first decision that is the key for successful chemical analysis is how to handle and keep the samples to maintain them with the conserved composition until the data are not analyzed and published. If chemical bioanalysis goals are trace molecules and their quantification, then sample handling may need extraction, purification, and concentration techniques and usually, biological samples need the most demanding preparation methods. The laborious methods that need a lot of time to complete are prone to errors and thus laboratory automation can be expected to have a substantial impact on the quality of the sample preparation. Also, improvements in instrumentation and detection limits for many substances may lead to the miniaturization of extraction and purification systems that reduce quantities of samples and chemicals. There are many micro-methods with new materials used as sorbents, micro-devices that may substitute the macro-scale sample preparation. Also, in recent bioanalytics, there are many nanomaterials-based protocols for biomolecules isolation, purification, and concentration. Many employ single-drop micro-extraction, tailored magnetic nanoparticles (magnetic beads) because of their easy separation, molecularly imprinted nanoparticles, polymer-coated nanoclusters, and others. Purification protocols based on nanomaterials explore electrostatic and hydrophobic interactions, but also other specific interactions such as mimetic enzyme–substrate or receptor–ligand interactions, or metal affinity interactions, which may reduce the time needed for sample preparation. Therefore, the development of new nanomaterials to improve the effectiveness of the sample preparation is a new and promising field of research. Some of the proposed protocols still need to be adopted for day-to-day research activities and use, but certainly will enhance sensitivity and selectivity in sample handling. And because they are easy and fast, even if not cheap, they are becoming interesting to apply due to their effectiveness in the extraction, enrichment, and purification of targeted compounds. Trends in this area of science are to use as little as possible of a starting sample, get as much as possible chemical data on it, characterize the sample composition in a short time in an optimized and rather automated way.

# References

Abbott SK, Jenner AM, Mitchell TW, Brown SH, Halliday GM, Garner B (2013) Lipids 48:307–318
Aebersold R, Mann M (2003) Nature 422:198–207
Ali I, Gupta VK, Aboul-Enein HY, Hussain A (2008) J Sep Sci 31:2040–2053
Amann A, Costello Bde L, Miekisch W, Schubert J, Buszewski B, Pleil J, Ratcliffe N, Risby T (2014) J Breath Res 8:34001
Amiri A (2016) TrAC Trends Anal Chem 75:57–74
Aoki-Kinoshita KF, Lisacek F, Mazumder R, York WS, Packer HH (2020) Glycobiology 30:70–71
Auslander N, Yizhak K, Weinstock A, Budhu A, Tang W, Wang XW, Ambs S, Ruppin E (2016) Sci Rep 6:29662
Bache N, Geyer PE, Bekker-Jensen DB, Hoerning O, Falkenby L, Treit PV, Doll S, Paron I, Müller JB, Meier F, Olsen JV, Vorm O, Mann M (2018) Mol Cell Proteomics 17:2284–2296
Bayot ML, King KC (2020) Biohazard levels. In: StatPearls. StatPearls Publishing, Treasure Island
Berndt AE (2020) J Hum Lact 36:224–226
Bertagnolli C, Espindola APDM, Kleinübing J, Tasic L, da Silva MGC (2014) Carbohydr Polym 111:619–623
Bowden JA, Heckert A, Ulmer CZ, Jones CM, Koelmel JP, Abdullah L, Ahonen L, Alnouti Y, Armando AM, Asara JM, Bamba T, Barr JR, Bergquist J, Borchers CH, Brandsma J, Breitkopf SB, Cajka T, Cazenave-Gassiot A, Checa A, Cinel MA, Colas RA, Cremers S, Dennis EA, Evans JE, Fauland A, Fiehn O, Gardner MS, Garrett TJ, Gotlinger KH, Han J, Huang Y, Neo AH, Hyötyläinen T, Izumi Y, Jiang H, Jiang H, Jiang J, Kachman M, Kiyonami R, Klavins K, Klose C, Köfeler HC, Kolmert J, Koal T, Koster G, Kuklenyik Z, Kurland IJ, Leadley M, Lin K, Maddipati KR, McDougall D, Meikle PJ, Mellett NA, Monnin C, Moseley MA, Nandakumar R, Oresic M, Patterson R, Peake D, Pierce JS, Post M, Postle AD, Pugh R, Qiu Y, Quehenberger O, Ramrup P, Rees J, Rembiesa B, Reynaud D, Roth MR, Sales S, Schuhmann K, Schwartzman ML, Serhan CN, Shevchenko A, Somerville SE, St. John-Williams L, Surma MA, Takeda H, Thakare R, Thompson JW, Torta F, Triebl A, Trötzmüller M, Ubhayasekera AJK, Vuckovic D, Weir JM, Welti R, Wenk MR, Wheelock CE, Yao L, Yuan M, Zhao XH, Zhou S (2017) J Lipid Res 58:2275–2288
Brusselmans L, Arnouts L, Millevert C, Vandersnickt J, van Meerbeeck JP, Lamote K (2018) Transl Lung Cancer Res 7:520–536
Bueso YF, Walker SP, Hogan G, Cleasson MJ, Tangney M (2020) Microbiome 8:122

Burla B, Arita M, Arita M, Bendt AK, Cazenave-Gassiot A, Dennis EA, Ekroos K, Han X, Ikeda K, Liebisch G, Lin MK, Loh TP, Meikle PJ, Orešič M, Quehenberger O, Shevchenko A, Torta F, Wakelam MJO, Wheelock CE, Wenk MR (2018) J Lipid Res 59:2001–2017

Cadillo-Quiroz H, Bräuer S, Yashiro E, Sun C, Yavitt J, Zinder S (2006) Environ Microbiol 8: 1428–1440

Carini P, Marsden PJ, Leff JW, Morgan EE, Strickland MS, Fierer N (2016) Nat Microbiol 2:16242

Cassiano NM, Barreito JC, Cass QB (2014) J Braz Chem Soc 25:9–19

Chlumská Z, Janeček Š, Doležal J (2014) Folia Geobot 49:1–15

Chong J, Soufan O, Li C, Caraus I, Li S, Bourque G, Wishart DS, Xia J (2018) Nucleic Acids Res 46(W1):W486–W494

Chory J, Pllard JD (1999) Resolution and recovery of small DNA fragments. In: Ausubel FM et al (eds) Current protocols in molecular biology. Supplementary 45: Unit 2.7

Clark KD, Zhang C, Anderson JL (2016) Anal Chem 88:11262–11270

Coley CW, Wengong J, Rogers L, Jamison TF, Jaakkola TS, Green WH, Barzilay R, Jensen KF (2019) Chem Sci 10:370–377

Correia BSB, Ortin GGD, Mor NC, Santos MS, Torrinhas RS, Val AL, Tasic L (2020) J Braz Chem Soc 12:2531–2543

Correia BSB, Torrinhas RS, Ohashi WY, Tasic L (2018) Analysis tools for lipid assessment in biological assays. In: Baez RV (ed) Advances in lipid metabolism. IntechOpen, London

Cruz M, Wang M, Frisch-Daiello J, Han X (2016) Lipids 51:887–896

Das S, Pal M (2020) J Electrochem Soc 167:037562

Deng J, Davies DR, Wisedchaisri G, Wu M, Hol WG, Mehlin C (2004) Acta Crystallogr Sect D 60: 203–204

Dickmann P, Apfel F, Biedenkopf N, Eickmann M, Becker S (2015) Health Secur 13:88–95

Doellinger J, Schneider A, Hoeller M, Lasch P (2020) Mol Cell Proteomics 19:209–222

Dvořáčková E, Snóblová M, Hrdlička P (2014) J Sep Sci 37:323–337

Eddhif B, Lange J, Guignard N, Batonneau Y, Clarhaut J, Papot S, Geffroy-Rodier C, Poinot P (2018) J Proteomics 173:77–88

Eland LE, Davenport R, Mota CR (2012) Water Res 46:5355–5364

Enserink M, Du L (2004) Science 305:163

Esteva-Socias M, Gómez-Romano F, Carrillo-Ávila JA, Sánchez-Navarro AL, Villena C (2020) Sci Rep 10:3579

Evers DL, He J, Kim YH, Mason JT, O'Leary TJ (2011) J Mol Diagn 13:687–694

Farrosh DW, Andrews-Pfannkoch C, Willlimson SJ (2011) J Vis Exp 55:3146

Feist P, Hummon AB (2015) Int J Mol Sci 16:3537–3563

Flemming DO (2000) Risk assessment of biological hazards. In: Flemming DO, Hunt DL (eds) Biological safety: principles and practices, 3rd edn. ASM Press, Washington

Fomin MA, Seikowski J, Belov VN, Hell SW (2020) Anal Chem 92:5329–5336

Garwolinska D, Hewelt-Belka W, Namiesnik J, Kot-Wasik A (2017) J Proteome Res 16:3200–3208

Gibson JF, Kelso S, Skevington JH (2010) Mol Phylogenet Evol 56:1126–1128

Gil A, Wenxuan Z, Wolters JC, Permentier H, Boer T, Horvatovich P, Heiner-Fokkema MR, Reijngoud D-J, Bischoff R (2018) Anal Bioanal Chem 410:5859–5870

Gong R, Li S (2014) Int J Nanomedicine 9:3781–3789

González-García E, Marina ML, García MC (2020) Sep Purif Rev 49:229–264

Granafei S, Liebisch G, Palmisano F, Carlucci R, Lionetti A, Longobardi F, Bianco G, Cataldi TRI (2017) Food Anal Methods 10:4003–4012

Gutstein HB, Morris JS, Annangudi SP, Sweedler JV (2008) Mass Spectrom Rev 27:316–330

Hakim M, Billan S, Tisch U, Peng G, Dvrokind I, Marom O, Abdah-Bortnyak R, Kuten A, Haick H (2011) Br J Cancer 104:1649–1655

Han X, Yang K, Yang J, Fikes KN, Cheng H, Gross RW (2004) Anal Biochem 330:317–331

Head SR, Komori HK, LaMere SA, Whisenant T, Van Nieuwerburgh F, Salomon DR, Ordoukhanian P (2014) Biotechniques 56:61–64

Hedegaard J, Thorsen K, Lund MK, Hein A-MK, Hamilton-Dutoit SJ, Vang S, Nordentoft I, Birkenkamp-Demtröder K, Kruhøffer M, Hager H, Knudsen B, Lindbjerg Andersen C, Sørensen KD, Pedersen JS, Ørntoft TF, Dyrskjøt L (2014) PLoS One 9:e98187

Hilbrig F, Freitag R (2003) J Chromatogr A 790:79–90

Howat WJ, Wilson BA (2014) Methods 70:12–19

Huang C, Qiao X, Sun W, Chen H, Chen X, Zhang L, Wang T (2019) Anal Chem 91:2418–2424

Hughes CS, Moggridge S, Müller T, Sorensen PH, Morin GB, Krijgsveld J (2019) Nat Protoc 14: 68–85

Iverson SJ, Lang SL, Cooper MH (2001) Lipids 36:1283–1287

Jensen P, Karlsson NG, Kolarich D, Packer NH (2012) Nat Protoc 7:1299–1310

Jiang X, Cheng H, Yang K, Gross RW, Han X (2007) Anal Biochem 371:135–145

Kachuk C, Stephen K, Doucette A (2015) J Chromatogr A 1418:158–166

Khatri K, Klein JA, Haserick JR, Leon DR, Costello CE, McComb ME, Zaia J (2017) Anal Chem 89:6645–6655

Kimman TG, Smit E, Klein MR (2008) Clin Microbiol Rev 21:403–425

Konoz E, Sarrafi AHM, Sahebi H (2016) Can J Chem Eng 94:9–14

Kresse SH, Namløs HM, Lorenz S, Berner J-M, Myklebost O, Bjerkehagen B, Meza-Zapeta LA (2018) PLoS One 13:e0197456

Landhäusser SM, Chow PS, Dickman LT, Furze ME, Kuhlman I, Schmid S, Wiesenbauer J, Wild B, Gleixner G, Hartmann H, Hoch G, McDowell NG, Richardson AD, Richter A, Adams HD (2018) Tree Physiol 38:1764–1778

Lawal O, Ahmed WM, Nijsen TME, Goodacre R, Fowler SJ (2017) Metabolomics 13:110–117

Lee K, Tripathi A (2020) Front Genet 11:374

Lever MA, Torti A, Eickenbusch P, Michaud AB, Šantl-Temkiv T, Jørgensen BB (2015) Front Microbiol 6:476

Li Y, Naghdi FG, Garg S, Adarme-Vega TC, Thurecht KJ, Ghafor WA, Tannock S, Schenk PM (2014) Microb Cell Fact 13:14

Lin Y, Wu H, Tjeerdema RS, Viant MR (2007) Metabolomics 3:55–67

Lin Y-D, Xin GZ, Li W, Liu FJ, Yao ZP, Di XA (2020) Anal Chim Acta 1095:118–128

Liu H, Qiu H (2020) Chem Eng J 393:124691

Lu G, Crihfield CL, Gattu S, Veltri LM, Holland LA (2018) Chem Rev 118:7867–7885

Marco E, Grimalt JO (2015) J Chromatogr A 1410:51–59

Masuda N, Ohnishi T, Kawamoto S, Monden M, Okubo K (1999) Nucleic Acids Res 27:4436–4443

Matejtschuk P (2007) Methods Mol Biol 368:59–72

McDowall RD (1989) J Chromatogr B Biomed Sci Appl 492:3–58

Medzihradszky KF, Chalkley RJ (2015) Mass Spectrom Rev 34:43–63

Melini V, Melini F (2020) In: Galanakis CM (ed) Innovative food analysis. Academic Press, Cambridge

Mitra S (2004) Sample preparation techniques in analytical chemistry. Wiley, Hoboken

Mochalski P, King J, Haas M, Unterkofler K, Amann A, Mayer G (2014) BMC Nephrol 15:43–49

Mochalski P, King J, Klieber M, Unterkofler K, Hinterhuber H, Baumann M, Amann A (2013) Analyst 138:2134–2145

Mohammedsaleh ZM, Mohammedsaleh FA (2014) Global J Health Sci 7:46–51

Montañés F, Corzo N, Olano A, Reglero G, Ibáñez E, Fornarib T (2008) J Supercrit Fluids 45:189–194

Mor NC, Correia BSB, Val AL, Tasic L (2020) J Braz Chem Soc 31:662–672

Mort AJ, Pierce ML (2002) J Chromatogr Libr 66:3–38

Müller T, Kalxdorf M, Longuespée R, Kazdal DN, Stenzinger A, Krijgsveld J (2020) Mol Syst Biol 16:e9111

Nahorniak M, Larsen DP, Volk C, Jordan CE (2015) PLoS One 10:e0131765

Nasiria M, Ahmadzadeha H, Amirib A (2020) TrAC Trends Anal Chem 123:115772

Nickerson JL, Doucette AA (2020) J Proteome Res 19:2035–2042

Nocerino JM, Schumacher BA, Dary CC (2005) Environ Forensic 6:35–44

Nováková L, Vlčková H (2009) Anal Chim Acta 656:8–35

Olajos M, Hajós P, Bonn GK, Guttman A (2008) Anal Chem 80:4241–4246

Ou YM, Kuo S, Lee H, Chang HT, Wang YS (2018) J Vis Exp 137:e57660

Pawliszyn J (2010) Handbook of sample preparation. Wiley, Hoboken

Pellegrino RM, Di Veroli A, Valeri A, Goracci L, Cruciani G (2014) Anal Bioanal Chem 406: 7937–7948

Phillips M (1997) Anal Biochem 247:272–278

Pijls KE, Smolinska A, Jonkers DMAE, Dallinga JW, Masclee AAM, Koek GH, van Schooten F-J (2016) Sci Rep 6:19903

Pizarro C, Arenzana-Rámila I, Pérez-del-Notario N, Pérez-Matute P, González-Sáiz JM (2013) Anal Chem 85:12085–12092

Pontes JGM, Brasil AJM, Cruz GCF, De Souza RN, Tasic L (2017) Anal Methods 9:1078–1096

Pontes JGM, Ohashi WY, Brasil AJM, Filgueiras PR, Espíndola APDM, Silva JS, Poppi RJ, Coletta-Filho HD, Tasic L (2016) Chem Select 6:1176–1178

Putcha L, Nimmagudda RR (2000) Preservation of liquid biological samples. U. S. patent number 6,133,036

Roberts LD, Souza AL, Gerszten RE, Clish CB (2012) Targeted metabolomics. Curr Protoc Mol Biol. Chapter 30, 30.2

Sajnani R., Aribindi. (2017) Lipid Sample Preparation for Biomedical Research. In: Bhattacharya S. (eds) Lipidomics. Methods in Molecular Biology, vol 1609. Humana Press, New York

Sanz ML, Martínez-Castro I (2007) J Chromatogr A 1153:74–89

Satomi Y, Hirayama M, Kobayashi H (2017) J Chromatogr B Analyt Technol Biomed Life Sci 1063:93–100

Scott CT, Caulfield T, Borgelt E, Illes J (2012) Nat Biotechnol 30:141–147

Serra-Mora P, Herráez-Hernández R, Campíns-Falcó P (2020) Sci Total Environ 721:137732

Shehada N, Brönstrup G, Funka K, Christiansen S, Leja M, Haick H (2014) Nano Lett 15:1288–1295

Singh S, Oswal M, Behera BR, Kumar A, Santra S, Acharya R, Singh KP (2020) J Radioanal Nucl Chem 323:1443–1449

Soria AC, Brokt M, Sanz ML, Martínez-Castro I (2012b) Reference module in chemistry. Mol Sci Chem Eng 4:213–243

Soria A, Michal B, Sanz M, Martínez-Castro I (2012a) Comprehensive sampling and sample preparation, vol 4, pp 213–243

Srivastava S, Upadhyau DJ, Srivastava A (2020) Front Mol Biosci 7:582499

Suresh K, Thomas SV, Suresh G (2011) Ann Indian Acad Neurol 14:287–290

Szpakowska N, Kowalczyk A, Jafra S, Kaczyński Z (2020) Carbohydr Res 497:108136

Tiwari G, Tiwari R (2010) Pharm Methods 1:25–38

Trbojević Akmačić I, Ugrina I, Štambuk J, Gudelj I, Vučković F, Lauc G, Pučić-Baković M (2015) Biochemistry (Mosc) 80:934–942

Tripathi R, Khatri N, Mamde A (2020) J Assoc Physicians India 68:14–18

Visser NFC, Lingeman H, Irth H (2005) Anal Bioanal Chem 382:535–558

Vuckovic D (2013) Proteomic and metabolomic approaches to biomarker discovery, pp 51–75

Watanabe V, Hashida S, Yamamoto H, Matsubara T, Ohtsuka T, Suzawa K, Maki Y, Soh J, Asano H, Tsukuda K, Toyooka S, Miyoshi S (2017) Exp Ther Med 14:2683–2688

Wawrzyniak R, Kosnowska A, Macioszek S, Bartoszewski R, Jan Markuszewski M (2018) Sci Rep 8:9541

Wenk MR (2005) Nat Rev Drug Discov 4:594–610

Williams C, Pontén F, Moberg C, Söderkvist P, Uhlén M, Pontén J, Sitbon G, Lundeberg JA (1999) Am J Pathol 155:1467–1471

Wiśniewski JR (2019) Anal Chim Acta 1090:23–30

Wolthuis JC, Magnusdottir S, Pras-Raves M, Moshiri M, Jans JJM, Burgering B, van Mil S, de Ridder J (2020) Metabolomics 16:99

Yamada I, Shiota M, Shinmachi D, Ono T, Tsuchiya S, Hosoda M, Fujita A, Aoki NP, Watanabe Y, Fujita N, Angata K, Kaji H, Narimatsu H, Okuda S, Aoki-Kinoshita KF (2020) Nat Methods 17: 649–650
Yang K, Cheng H, Gross RW, Han X (2009) Anal Chem 81:4356–4368
Yi Z, Mehrotra N, Budha NR, Christensen ML, Meibohm B (2008) Clin Chim Acta 398:105–112
York WS, Ranzinger R, Edwards N, Zhang W, Tiemeyer M (2020) Glycobiology 30:72–73
Zeki ÖC, Eylem CC, Reçber T, Kir S, Nemutlu E (2020) J Pharm Biomed Anal 190:113509
Züllig T, Trötzmüller M, Köfeler HC (2020) Anal Bioanal Chem 412:2191–2209

# UV-Vis Absorption and Fluorescence in Bioanalysis

Erick Leite Bastos

*This chapter is dedicated to the memory of Professor Ronei J. Poppi.*

## 1 Introduction and Scope

The field of bioanalysis started to flourish in the late twentieth century with the development of new technologies for the automatic, ultrasensitive, and robust analysis of biomolecules in complex matrices (Albers 2014; Ramesh 2019). In the broad sense, bioanalysis permeates several other areas of study, including forensic, food, and environmental analyses, clinical chemistry and biodefense (Albers 2014). Several analytical methods have been used independently or combined for bioanalytical purposes. Among them, the monitoring of ultraviolet-visible (UV-Vis) light absorption and fluorescence are fast, scalable, and mostly nondestructive approaches that require relatively simple equipment and minimum technical training (Upstone 2013; Valeur and Berberan-Santos 2012). Hence, these and other photometric methods are ubiquitous in the analysis of species of biological interest, corresponding to more than 90% of all analyses performed in clinical laboratories (Schneider et al. 2014).

UV-Vis light absorption is a useful technique for the detection of *all* chemical species in the sample that are able to absorb photons within a specific wavelength range. Although the analyte of interest in the sample can be quantified by using the

E. L. Bastos (✉)
Departamento de Química Fundamental, Instituto de Química, Universidade de São Paulo, São Paulo, Brazil
e-mail: elbastos@iq.usp.br

Beer-Lambert-Bouguer law, several factors, including optical filter effects, reduce the dynamic range of this technique as compared to spectrofluorimetry (Ramesh 2019). Despite this drawback, several UV-Vis spectrophotometric assays are sensible and specific because they rely on chemical transformations and interactions for producing light-absorbing chromophores. Single- and double-beam arrangements are the most common optical configurations of UV-Vis spectrophotometers and detection is frequently carried out by means of a photomultiplier or a photodiode array. Diffuse reflectance enables the study of solids and semi-solid materials and fiber optic probes made of quartz or Torlon have been developed for clinical and environmental applications. The combination of spectrophotometric detection and separation techniques, such as liquid chromatography and capillary electrophoresis, have been used to develop flow systems, e.g., flow injection analysis (Worsfold 2005).

Analytes promoted to their electronically excited state by the absorption of UV-Vis light enable the monitoring of photophysical, photochemical, and/or photobiological processes (Turro et al. 2010). For practical purposes, in this text UV-Vis light is broadly considered as the electromagnetic radiation with wavelength intervals from 200 to 800 nm range (formally MUV-Vis, 6.5 eV – 1.5 eV; 150–600 kJ mol$^{-1}$) (Akash and Rehman 2020). Although some structural information on the structure of organic chromophores can be inferred from UV-Vis absorption spectra by means of the Woodward-Fieser, Scotts, and Fieser-Kuhn rules (Mistry 2009), the spectrophotometric characterization of species is mostly comparative in nature. Fluorescence is the emission of light accompanying the relaxation of a singlet excited state (mostly $S_1$) to the ground state (Lakowicz 2006). This photophysical process can be affected to different degrees by the environment surrounding the fluorophore and is sensitive enough to study both molecular ensembles and single molecules (Sauer et al. 2011; Rodriguez et al. 2017; Macchia et al. 2020). Fluorescent species, hereafter called fluorophores, fluorogenic compounds, and fluorescence quenchers have been used for the investigation of samples of variable complexity under batch and flow platforms. Fluorescence also makes possible to peering inside cells and other complex structures and matrices, enabling the tracking, visualization, and quantification of species of interest and biochemical processes through a variety of imaging methods (Specht et al. 2017). Knowing the physicochemical, photophysical, and photochemical properties of the fluorophores and/or quenchers as well as the characteristics of the analyte and the basic principles and limitations of the technique chosen is of paramount importance for the development of robust spectrofluorimetric methods and, frequently, for proper interpretation of results (Specht et al. 2017; Rudin and Weissleder 2003; Petryayeva et al. 2013). Overall, both UV-Vis and fluorescence spectroscopies frequently lack specificity and have been less used for the assay of bulk drug materials in pharmacopoeial monographs than chromatographic and titration methods (Siddiqui et al. 2017). However, there are several approaches to enhance the specificity in analyte detection, some of which are discussed in this text.

This chapter is divided into four sections. First, a brief description of selected key concepts in the interaction between light and chemical species in solution is provided for completeness and to stimulate the novice in the field. The following sections present an overview of the classic and contemporary application of UV-Vis absorption and fluorescence spectroscopies for bioanalysis. Major focus is given to the applied aspects of the study of biological and biochemical problems in complex matrices using steady-state and time-resolved spectroscopic platforms. In particular, sensor and biosensor technologies, use of dyes as labels or reporters, and sensitive and selective approaches for bioanalysis are highlighted. A comprehensive review of the use of optical methods for bioanalysis and in-depth description of methods for bioimaging are out of the scope of this chapter and can be found elsewhere (Upstone 2013; Valeur and Berberan-Santos 2012; Turro et al. 2010; Lakowicz 2006; Sauer et al. 2011; Quina 1982; Klán and Wirz 2009; Gauglitz and Moore 2014; Schäferling 2016; Baptista and Bastos 2019).

## 1.1 Fundamental Principles

Both UV-Vis absorption and fluorescence spectroscopies rely on signals generated as a consequence of photoexcitation of chemical species. The absorption of UV-Vis light by species having a chromophore induces electron transitions giving rise to excited state formation. This process is ultrafast ($\sim 10^{-15}$ s, fs timescale), and its probability to occur is proportional to the one-photon cross section ($\sigma_1$) of the analyte at a given wavelength and, therefore, to its molar absorption coefficient ($\varepsilon$). The magnitude of $\varepsilon$ is a measure of the probability of the electronic transition, which depends on the structural and electronic features of both the ground and the excited states (Turro et al. 2012).

The Beer-Lambert-Bouguer law (also known as Beer-Lambert, Lambert-Beer, and Beer's law) states that the absorption of monochromatic light is proportional to the ratio between the intensity of incident and transmitted light and, consequently, to the concentration of the substance, the value of $\varepsilon$, and the absorption path length (Eq. 1).

$$A(\lambda) = \log \frac{I_\lambda^0}{I_\lambda^t} = \varepsilon(\lambda) \cdot c \cdot l \tag{1}$$

where $A$ is the absorbance, $I_\lambda^0$ is the intensity of incident light, $I_\lambda^t$ is the intensity of transmitted light, $\varepsilon(\lambda)$ is the molar absorption coefficient (usually in L mol$^{-1}$ cm$^{-1}$), $c$ is the concentration of the species (in mol L$^{-1}$), and $l$ is the absorption path length (in cm) (Valeur and Berberan-Santos 2012). The UV-Vis absorption spectrophotometric analysis of dyes in solution at room temperature most frequently results in broad spectra formed by the combination of absorption bands instead of discrete lines. The solvent is involved in collisional broadening and can influence the electronic properties of the dye, hence playing a major role in its spectral properties. Water is the main component of all living organisms and the most common solvent in routine bioanalysis. Pure water is virtually transparent from approximately 190 to

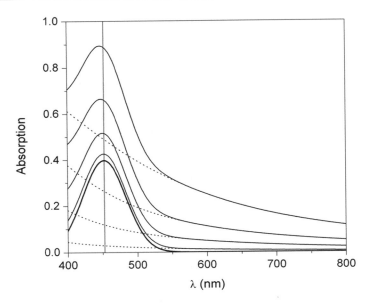

**Fig. 1** Effect of the baseline (dashed line) on the observed absorption and maximum absorption wavelength ($\lambda_{max}$) of a simulated Gaussian curve in a static hypothetical system in which the concentration of the compound of interest is constant and the baseline is deliberately increased. As the baseline increases (by $1/\lambda^4$ order of magnitude), the absorption maximum shifts, and the quantitative measure of absorbance requires correction (Wilson and Walker 2009). Reproduced from Ref. El Seoud et al. (2016) with permission from John Wiley and Sons

850 nm, but most biological samples contain a multitude of water-soluble components. Depending on the amount, the addition of cosolvents, e.g., DMSO, to aqueous solutions may change the spectral cutoff of the system and must be carried out with care (Baptista and Bastos 2019). The material used to construct the recipient used for the measurement (cuvette, microplate, slide, etc.) is also an important source of artifacts in spectrophotometry, and high-quality optical quartz is the most versatile material (Upstone 2013; Baptista and Bastos 2019).

There are many other factors that can influence absorption measurements. As the absorption of the sample increases above three absorbance units, single monochromator instruments may deviate from true linearity due to stray light. Under these circumstances, the sample absorbs 99.9% of the incident light compared to 90% for a sample with absorption of unity. Highly fluorescent samples may also interfere with absorbance measurements, as fluorescence may be added to the transmitted light. For reference, a change of 5% in light intensity corresponds to a variation of 0.02 absorbance units (AU) (Wilson and Walker 2009). Nonlinearity effects originating from light scattering are another important source of experimental artifacts, such as the shift of the absorption maximum wavelength due to the distortion of the baseline. In suspensions, the absorbance can be overestimated if the scattered light lowers the amount of transmitted light reaching the detector. Figure 1 depicts the simulation of the effect of the baseline on the absorption band centered at approximately 450 nm (El Seoud et al. 2016). In some applications, however, changes in the baseline over time are the expected outcome of the process studied, e.g., the analysis of bacterial growth, and the apparent absorbance is reported as optical density (OD).

An important remark when using spectrophotometry in bioanalysis is that light absorption is an additive property, i.e., $A(\lambda) = \Sigma c_i \varepsilon_i(\lambda)$. Thus, the resulting absorption of monochromatic light is the sum of the individual contributions of all chemical species in solution (Goldstein and Day 1954). The background absorption of biological matrices is particularly important because they contain multiple interfering species. For this reason, there are many approaches to minimize background and overlapping effects in UV-Vis spectrophotometry, including the acquisition of absorbance data at several wavelengths, and the use of chemometric tools, such as multivariate calibration or classification, difference spectrophotometry, and derivative absorption spectroscopy.

In samples containing multiple absorbing species, it is desirable that their maximum absorption wavelengths are well separated, but this is rarely the case. It is possible, in principle, to determine the concentration of all $n$ absorbing components in solution by measuring the individual absorbance and molar absorption coefficients at $n$ wavelengths. However, when many overlapping absorptions are present, the accuracy of such determinations is often low. In this case, an alternative approach to increase spectral resolution is the subtraction of one absorption spectrum from another (preferable using a double beam spectrophotometer) to obtain the difference absorption spectrum. Positive and negative variations in absorbance ($\Delta A$) are used to monitor the effect of independent variables, such as time, pH, or temperature variations, on the absorption profile; $\Delta A = 0$ can indicate an isosbestic point and may be used to spot interfering substances (Zuman and Patel 1992). Derivative spectrophotometry also enables the discrimination of species in complex samples (Upstone 2013; El Seoud et al. 2016; Karpinska 2004; Sanchez Rojas and Bosch Ojeda 2009; Vogt 2005). In this technique, the value of $\delta^x \text{Abs}/\delta\lambda^x$ (where $x$ is the derivation order) is plotted against the wavelength, eliminating baseline effects in the spectrum. This approach is useful to separate partially overlapped bands, allowing the determination of the absorption maxima of each component and the degree of band superposition. Second derivative absorption spectroscopy ($x = 2$) is widely used for this goal since the spectrum preserves quantitative information and resembles an inverted absorption spectrum with sharp peaks. Nevertheless, as the derivatization order increases, the signal-to-noise decreases; thus, quality spectral data is required for using this technique. An example of the application of second derivative spectroscopy in food analysis is the kinetic study of the decomposition of betanin in beetroot juice (Goncalves et al. 2012, 2013).

Spectrofluorimetry is theoretically at least two orders of magnitude more sensitive than spectrophotometry (Fersht 1999). The light emission coming from the sample is measured during irradiation because fluorescence ceases shortly after the excitation source is turned off (around $10^{-9}$ s, ns timescale) (Turro et al. 2012; Marcu et al. 2014). Fluorescence uses experimental setups in which the light source and the detector are typically in the right angle or front-face geometries (Bartoloni et al.

2013; Bower 1982; Calderon-Ortiz et al. 2012; Coutinho and Prieto 1993; Barros et al. 2010). The instrumental response is proportional to the intensity of the excitation light, i.e., a drift of 5% in the excitation intensity leads to a 5% change in the detected signal. Therefore, the effect of the background light is often much lower in spectrofluorimetry than in absorbance measurements (Wilson and Walker 2009). Light scattering influences fluorescence, and as excitation light becomes more intense, the Rayleigh scattering (peaks at the excitation wavelength), second-order scattering from the monochromator (observed at twice the excitation wavelength) and Raman inelastic scattering (peaks at fixed energy from the excitation wavelength) by the solvent or matrix become more relevant (Rabek 1982). The fluorescence response of optically dilute samples in homogeneous media is linearly correlated with the concentration of the fluorophore but fluorescence spectroscopy of turbid media, such as human tissue, is complicated due to its highly absorbing and scattering properties. Last, photolysis of the sample must always be considered when performing analyses requiring photoexcitation (Rabek 1982). For pharmaceutical products, photodegradation is a major concern due to the eventual formation of toxic products (Cosa 2004).

The importance of fluorescence measurements in life science applications is due to their ability to provide various simultaneous readouts, such as emission intensity, lifetime of the excited state of the fluorophore, anisotropy, and spectral characteristics. The fluorescence emission spectrum shows the intensity of fluorescence measured over a range of emission wavelengths resulting from excitation at a fixed wavelength, whereas the fluorescence excitation spectrum corresponds to the range of excitation wavelengths leading to emission at a specific wavelength (Lakowicz 2006). Analysis of multiple fluorophores can be performed by using fluorescence excitation–emission matrix (EEM) spectroscopy, in which a contour plot displays the fluorescence intensities as a function of a range of excitation and emission wavelengths (Vishwanath and Ramanujam 2011; Yadav et al. 2019). Several hardware- and software-based approaches are available to remove the signals originating from the excitation source and light scattering from the EEM plot and, some of them, are based on multiway chemometric methods. EEM has been used in combination with parallel factor analysis (EEM-PARAFAC) for characterizing dissolved organic matter and has been applied in a rapidly growing number of studies on drinking water and wastewater treatments (Yang et al. 2015). Also, Multivariate Curve Resolution with Alternating Least Squares (MCR-ALS) have been successfully used to analyze complex EEM data (Dantas et al. 2020). Alternatively, synchronous fluorescence, in which excitation and emission are performed simultaneously at wavelengths that are apart by a fixed value, can be employed to analyze samples containing multiple fluorescent species (Li et al. 2012). This approach leads to spectral simplification, reduction of interference due to light scattering and enhanced selectivity compared to conventional spectrofluorimetry. Synchronous spectrofluorimetry has been used for the nondestructive analysis of food, in particular, to evaluate food adulteration (He et al. 2020a).

For the one-electron excitation of molecules in solution, the emission spectrum is shifted to longer wavelengths than the excitation spectrum due to the loss of energy in intra- and intermolecular processes. The energy difference between the spectral positions of the band maxima (or the band origin) of the absorption and fluorescence arising from the same electronic transition is called Stokes shift (Braslavsky 2007). Small molecular organic dyes that have been widely used in bioimaging, biosensing, medical diagnosis, and environmental detection, such as BODIPY, rhodamine, fluorescein, coumarin, and cyanine derivatives, usually show extensive overlap of the absorption and emission spectra. Consequently, they have very small Stokes shifts (typically <25 nm), which can lead to self-quenching and fluorescence detection errors due to excitation backscattering effects (Fan et al. 2013).

The fluorescence quantum yield ($\phi_{FL}$) is an important figure of merit for a fluorophore. It is a measure of fluorescence efficiency and corresponds to the ratio between the number of photons absorbed and emitted, i.e., $\phi_{FL}$ is less than or equal to unity. The $\phi_{FL}$ allows the comparison of fluorescence data regardless of the instrument used for the measurement, since it can be used to convert relative light units (RLUs) into the absolute number of photons. The $\phi_{FL} \times \varepsilon$ product is the brightness of a fluorophore, which is used to classify fluorescent probes. The $\phi_{FL}$ is related to the lifetime of the singlet excited state ($\tau$, also named fluorescence lifetime), which ranges from approximately 0.1 to 100 ns and depends on internal and environmental factors, being useful for the study of molecular dynamics using pulsed lasers as the excitation source (Berezin and Achilefu 2010). Finally, several fluorophores are *anisotropic,* and their excitation with polarized light is orientation dependent. Although typical fluorescent small molecules, such as fluorescein, do not show polarization in biological assays due to rapid rotation in solution, fluorescence anisotropy allows the study of dynamic processes of biological relevance, such as the monitoring of the association and dissociation of fluorescent ligands to/from G protein-coupled receptors (GPCRs) (Rinken et al. 2018). Note that steady-state fluorescence anisotropy provides information on the mean anisotropy of the system, whereas time-resolved fluorescence anisotropy allows the determination of the decay time, residual anisotropy, and rotational correlation time (Smith and Ghiggino 2015). Both approaches have been used for imaging purposes, though (Vinegoni et al. 2019).

Spectral interference and the occurrence of electron and energy transfer processes must be carefully considered when using fluorescence spectroscopy to analyze unknown samples. The spectrofluorimetric study of turbid samples, such as cell suspensions and tissues, depends on the concentration and distribution of the fluorophore as well as on environmental factors that can alter its fluorescence quantum yield and lifetime. Effective photoexcitation, which can be compromised by nonfluorescent absorbers and scatterers, is required for the analysis of tissues using fluorimetry. In particular, elastic light scattering caused by random spatial variations in the density, refractive index, and dielectric constants of extracellular, cellular, and subcellular components requires special attention (Vishwanath and Ramanujam 2011). The scattering coefficients in living tissues range from 1000 to 10 cm$^{-1}$ from UV to the near-infrared (NIR) spectral regions, decreasing monotonically as the wavelength increases (Vishwanath and Ramanujam 2011). NIR-I-to-NIR-II fluorescence imaging has been used to further improve the image contrast at increased tissue depth. Excitation in the 700–950 nm NIR-I window and emission in the 1000–1700 nm NIR-II window reduces the interference from photon scattering and tissue autofluorescence (Zhao et al. 2018).

Energy transfer between an excited donor species and an acceptor in the ground state can occur in a nonradiative manner at distances considerably longer than the sum of their van der Waals radii. This process is known as Förster resonance energy transfer (FRET), as recommended by the IUPAC (Braslavsky 2007), and has been used to determine the relative position and dynamics of two fluorophores in a variety of systems of biological relevance. The occurrence of efficient FRET requires, among other factors, the superposition of the emission spectra of the donor species and the absorption spectra of the energy acceptor, and the interaction between the transition dipole moments of these entities. Most importantly, no fluorescence is involved in FRET. In bioanalytical applications, FRET is used to turn fluorescence on or off in response to biorecognition events (e.g., ligand–receptor binding, enzyme activity, DNA hybridization) or other stimuli (e.g., pH changes) (Specht et al. 2017). Electronic energy transfer via the Dexter mechanism (Dexter electron transfer, Dexter electron exchange) occurs by electron exchange between the excited donor and the ground state acceptor. The process is nonradiative and occurs at short distances, usually below 1 nm. Singlet excited states can also be deactivated in a nonradiative manner by thermal single-electron transfer following photoexcitation, a phenomenon called photoinduced electron transfer (PeT). Additionally, the occurrence of charge transfer (CT), e.g., intramolecular charge transfer (ICT), metal–ligand charge transfer (MLCT), and twisted intramolecular charge transfer (TICT), are used to rationalize changes in the absorption and fluorescence profiles (Ernst 2015; Misra and Bhattacharyya 2018). Depending on the system, these processes can result in fluorescence quenching or changes in the emission properties. A fluorophore prone to ICT enables the monitoring of ratiometric fluorescence signals. Multifluorophoric systems with energy donor–acceptor architectures are useful for the simultaneous monitoring of multiple emission intensities at different wavelengths. These energy transfer cassettes can provide a built-in correction for environmental effects and supply a facile method for visualizing complex biological processes at the molecular level, such as DNA detection, protein labelling, nucleic acid regulation, and other biomarkers (Fan et al. 2013).

Molecular aggregation is also an important process that is related either to fluorescence enhancement or quenching. Excimers (excited dimers) and exciplexes (excited complexes) have been extensively used in biophysical studies, including the characterization of the properties of surfactants and excited-state dynamics of DNA (Gehlen 2020; Middleton et al. 2009; Takaya et al. 2008). The aggregation-caused emission quenching (ACQ) effect makes highly fluorescent organic dyes virtually non-emissive when in concentrated solutions (Liu et al. 2019). The aggregation-induced emission (AIE) (Wang et al. 2019a; He et al. 2020b), on the other hand, enhances the emission of some dyes that are weakly fluorescent in solution due to the deactivation of the excited state by internal motion (He et al. 2020b; Rodrigues et al. 2020, 2018a; Qi et al. 2020; Li et al. 2020). AIE has become an important tool for biomolecular imaging and has been used in ultrasensitive sensing measurements

(Kaur and Singh 2019; Xu et al. 2020a, b; Zhao et al. 2020a). The occurrence of photochemical reactions, such as photoisomerization and excited-state intramolecular proton transfer (ESIPT), also influences the emission properties, often quenching or shifting the maximum emission wavelength (Rodembusch et al. 2005; Santos et al. 2016).

The application of fluorescent dyes depends on their intrinsic characteristics, and some of them show nonlinear optical properties, being prone to multiphoton excitation (Kang et al. 2020a, b). Two-photon absorption (TPA) can lead to two-photon-excited fluorescence, and the cross section for two-photon absorption, $\sigma_2$ (expressed in units of Goeppert-Mayer, GM), is proportional to the probability of two-photon absorption. Small organic fluorescent dyes with $\sigma_2$ higher than approximately 300 GM in water are often used for two-photon excitation microscopy (Mariz et al. 2017; Rodrigues et al. 2012, 2018b). Species matching the requirements for TPA often can be excited by using red or NIR light and, hence, are frequently used as photodynamic agents and photoactivated drugs, among other applications. NIR light generates less autofluorescence and penetrates deeper into tissues as compared to UV-Vis light. Notably, the Beer-Lambert-Bouger law does not hold for multiphoton absorption; therefore, spot focusing and fluorescence measurements are the preferred techniques for two-photon dyes (Valeur and Berberan-Santos 2012).

## 2 Spectrophotometry

The versatility of spectrophotometry and the development of portable spectrophotometers and automated analytical strategies have contributed to the continuous use of UV-Vis spectrophotometry over the years (Upstone 2013; Worsfold 2005; Zezzi-Arruda and Poppi 2005; Prasada Rao and Biju 2005; Sánchez Rojas and Cano Pavón 2005; Martinez Calatayud 2005; Passos and Saraiva 2019). This technique has a broad scope of application in bioanalysis since most biomolecules absorb UV-Vis light or can be converted into absorbing species. Inorganic species, small molecule organic compounds, proteins, and nucleic acids can be identified and quantified using this technique (Table 1) (Wilson and Walker 2009; Schmid 2001).

### 2.1 Analysis of Blood Components and Related Species

#### 2.1.1 Bilirubin Analysis

Upstone gives an excellent example of why absorption spectroscopy is still relevant in the face of modern methods for bioanalysis in his review on the use of UV-Vis light absorption spectrophotometry in clinical analysis (Upstone 2013). Patients suffering from a suspected subarachnoid hemorrhage, usually resulting from cerebral aneurysm, can be diagnosed using computer-aided tomography (CAT) scanning. Despite being fast, the diagnosis is approximately 97% accurate, and to avoid missing the remaining 3% of cases, a sample of cerebrospinal fluid is analyzed by

**Table 1** Common spectrophotometric assays for biomolecules. Adapted from Ref. Wilson and Walker (2009)

| Analyte | Reagent/method | Wavelength (nm) |
|---|---|---|
| Amino acids | Ninhydrin | 570; Pro at 420 |
| | Cupric salts | 620 |
| Cys and thiolates | Ellman's reagent, (5,5′-dithiobis-(2-nitrobenzoic acid) | 412 |
| Protein | Folin–Ciocalteu | 660 |
| | Biuret | 540 |
| | BCA | 562 |
| | Coomassie Brilliant Blue | 595 |
| | Direct | Tyr, Trp: 278; peptide bond: 190 |
| Coenzymes | Direct | FAD: 438, NADH: 340, $NAD^+$: 260 |
| Carotenoids | Direct | 420, 450, 480 |
| Porphyrins | Direct | ~405 (Soret band) |
| Carbohydrate | Phenol, $H_2SO_4$ | Glucose: 490, xylose: 480 |
| | Anthrone, $H_2SO_4$ | 620 or 625 |
| Reducing sugars | Dinitrosalicylate, alkaline tartrate buffer | 540 |
| Pentoses | Bial (orcinol, ethanol, $FeCl_3$, HCl) | 665 |
| | Cysteine, $H_2SO_4$ | 380–415 |
| Hexoses | Carbazole, ethanol, $H_2SO_4$ | 540 or 440 |
| | Cysteine, $H_2SO_4$ | 380–415 |
| | Arsenomolybdate | 500–570 |
| Glucose | Glucose oxidase, peroxidase, $o$-dianisidine, PBS | 420 |
| Ketohexose | Resorcinol, thiourea, ethanoic acid, HCl | 520 |
| | Carbazole, ethanol, cysteine, $H_2SO_4$ | 560 |
| | Diphenylamine, ethanol, ethanoic acid, HCl | 635 |
| Hexosamine | Ehrlich (dimethylaminobenzaldehyde, ethanol, HCl) | 530 |
| DNA | Diphenylamine | 595 |
| | Direct | 260 |
| RNA | Bial | 665 |
| Sterols and steroids | Liebermann-Burchardt reagent (acetic anhydride, $H_2SO_4$, chloroform) | 415, 625 |
| Cholesterol | Cholesterol oxidase, peroxidase, 4-aminoantipyrine, phenol | 500 |
| ATPase | Coupled enzyme assay, pyruvate kinase, lactate dehydrogenase: ATP→ADP, phosphoenolpyruvate→pyruvate (consumes ADP), pyruvate→lactate (consumes NADH) | NADH: 340 |

UV-Vis spectrophotometry because the products of degradation of red blood cells, in particular bilirubin, are yellow. Although analysis can be performed by visual examination of the cerebrospinal fluid, borderline cases may be misdiagnosed (Upstone 2013).

Bilirubin in cerebrospinal fluid is quantified by the method developed by Chalmers and Kiley (1998) using UV-Vis absorption spectrophotometry. As with most spectrophotometric analyses of biological materials, the main challenge is to reduce the interference of other substances and the effect of light scattering on the baseline. The determination of the net bilirubin absorbance (NBA) at 476 nm and the net oxyhemoglobin absorbance (NOA) at 410–418 nm by this method is performed by measuring the change in absorbance from the absorption maxima to a linear baseline created using two points located before the oxyhemoglobin absorption band and after the absorption band of bilirubin. Diagnosis of xanthochromia, yellowish cerebrospinal fluid, requires NBA of at least 0.007 AU.

Hyperbilirubinemia in neonates is presented as jaundice (icterus neonatorum) in the first days of life. Although it is a benign condition treated with phototherapy, excessive serum bilirubin concentrations can progress to acute bilirubin encephalopathy and kernicterus (chronic bilirubin encephalopathy) with a substantial risk of neonatal mortality and long-term neurodevelopmental impairments (Olusanya et al. 2018). The total bilirubin in the serum can be measured spectrophotometrically at 454 nm and the NBA can be calculated by subtracting the absorbance at 528 nm from that at 454 nm (Kazmierczak et al. 2002). This method is limited to neonates with age below 2–3 weeks due to the prevalence of other forms of bilirubin and chromophores in older children and adults (Ngashangva et al. 2019). Bilirubin concentration in neonates and adults can also be determined from the absorbance in multiple wavelengths (Laterza et al. 2002). The method based on the measurement of the absorbance of a red azodipyrrole produced by the Van den Bergh reaction between bilirubin with diazotized sulfanilic acid has also been extensively used to quantify bilirubin in serum (Kumar and Gill 2008).

Fernandez-Romero and coauthors developed an automatic spectrophotometric method for the determination of total, direct and indirect bilirubins based on flow-injection analysis (Fernández-Romero et al. 1993). A simultaneous injection effective mixing flow analysis (SIEMA) system coupled with spectrophotometric analysis was developed for the fully automated determination of bilirubin and urobilinogen, a colorless by-product of bilirubin reduction, which is produced by bacteria in the intestines (Vichapong et al. 2013). The limit of detection (LOD) for urobilinogen is $1.0$ mg $L^{-1}$ while for bilirubin it corresponds to $0.003$ mg $L^{-1}$. The SIEMA system was also used to determine bilirubin and creatinine in urine samples with LODs of 7 µg $L^{-1}$ and $0.6$ mg $L^{-1}$, respectively (Ponhong et al. 2015).

### 2.1.2 Porphyrins

Porphyria is the name given to nine health disorders due to the buildup of porphyrins in the body, negatively affecting the skin and/or the nervous system (Bechara 1996; Ramanujam and Anderson 2015). Clinical manifestations of porphyria are either neurologic or cutaneous and include severe abdominal pain, vomiting, confusion, personality disorders, paralysis, and sensitivity to light. The subtypes of porphyria are associated with deficiencies related to the enzymes involved in heme biosynthesis and the accurate diagnosis is essential for decision on adequate treatment (Ramanujam and Anderson 2015). UV-Vis spectrophotometry serves as a screening tool because the urine specimens from those suffering from porphyria are generally reddish in color due to the presence of excess porphyrins originated from monopyrrole porphobilinogen (PBG), and may darken further after exposure to light (Upstone 2013; Ramanujam and Anderson 2015; Frank and Parnás 2020). The UV-Vis spectra of porphyrins show characteristic Soret peaks at 400–410 nm and can be used for the quantification and discrimination purposes. For example, the Soret peaks of coproporphyrin and uroporphyrin are centered at 402–403 nm and 406–407 nm, respectively (Upstone 2013). Total porphyrin concentration in urine specimens can be quantified by multiplying the net absorbance at the Soret peak by a factor of 2523 nmol $L^{-1}$/AU (Rimington and Sveinsson 1950).

The biosynthetic porphyrin precursors δ-aminolevulinic acid (ALA) and PBG are water-soluble and can be determined in urine (Woolf et al. 2017). The pyrrolic portion of PBG reacts with 4-dimethylaminobenzaldehyde (Ehrlich's reagent) to produce a red compound that is detected spectrophotometrically (Moore and Labbe 1964). PBG is generally strikingly increased during an attack of acute Porphyria and quantitation, even on a spot sample (rather than a 24 h collection), is highly informative (Frank and Parnás 2020). ALA and PBG are separated from each other by using ion-exchange chromatography and, in the so-called Mauzerall-Granick method, ALA is converted into a pyrrole by reaction with acetylacetone or another β-diketone and derivatized using the Ehrlich's reagent (Mauzerall and Granick 1956). This method has been used to determine the increase in the concentration of ALA in urine upon lead poisoning (Bechara 1996; Woolf et al. 2017; Granick et al. 1973; Costa et al. 1997).

### 2.1.3 Hemoglobin and Carboxyhemoglobin

Hemoglobin (Hb) analysis has been carried out using readings at three distinct wavelengths and the formulas of Harboe, Kahn, Noe, and Fairbanks combined with the Allen method (also referred to as a baseline technique) to correct for background absorption (Paal et al. 2018). The three-point assay of Harboe is based on the specific absorption of porphyrins at the Soret band and has been used for the direct spectrophotometric measurement of hemoglobin (Hb) in nonicteric and nonturbid plasma (Wians et al. 1988). The concentration of hemoglobin can be calculated by measuring the absorbance at 380, 415, and 450 nm, and applying the Allen baseline correction. This method produces unsatisfactory results in the case of hyperbilirubinemia, a frequent condition in patients at risk for intravascular hemolysis (Paal et al. 2018). The Noe or polychromatic formula (absorbance readings at 380, 415, and 470 nm) is a variation of the Harboe assay that has been reported to overestimate the concentration of hemoglobin (Noe et al. 1984). Bilirubin and triglyceride interference in the determination of low levels of hemoglobin (10 mg $dL^{-1}$) are also more evident using the Noe formula. Myoglobin interference was significantly positive only at concentrations not expected to occur under clinical conditions. The determination of free Hb by the method of Kahn requires undiluted specimen and, thus, more sample. However, this method is based on the absorption at 578 nm, which is mostly not influenced by the presence of bilirubin. The Lovibond-Drabkin method has been used for the quantification of Hb in whole blood. It is based on the oxidation and lysis of Hb using Drabkin's reagent, i.e., potassium ferricyanide, potassium cyanide, and potassium bicarbonate (Srivastava et al. 2014). The potassium ferricyanide oxidizes Hb to methemoglobin that is then converted into cyanmethemoglobin in the presence of KCN, which is detected at 540 nm. This spectrophotometric method can measure all hemoglobin forms except sulfhemoglobin, which occurs in blood only in minute concentrations. Although both Drabkin's and Harboe methods are suitable for the analysis of free Hb in red blood cell concentrates (linear response within the Hb concentration range from 0.015 to 220 g $L^{-1}$ Hb), the Harboe method has been found to be more reliable at low concentrations of Hb (Han et al. 2010). Drabkin's reagent has also been used for the study of intracerebral hemorrhage, which is a subtype of stroke with high rates of morbidity and mortality. The method was used to quantify the hemoglobin in both the contralateral and the ipsilateral brain tissue homogenate from mice subjected to complete transcardial perfusion for the removal of intravascular blood (Chang et al. 2011). Second-derivative spectrophotometry has also been used for the quantification of free hemoglobin in both non-icteric and icteric human plasma specimens (Paal et al. 2018). Further information on the methods used for the clinical analysis of Hb can be found elsewhere (Whitehead et al. 2019). Last, carbon monoxide poisoning can be accessed by measuring the concentration of carboxyhemoglobin (COHb) in blood, even postmortem (Egan et al. 1999; Huddle and Stephens 2003).

### 2.1.4 Glucose

Due to the global increase in the incidence of diabetes, glucose monitoring for diabetic care is of major public health and economic importance (Albers 2014). According to the American Diabetes Association, normal fasting plasma glucose (FPG) level tested while fasting range from 3.9 to 5.5 mmol $L^{-1}$ (70–100 mg $dL^{-1}$) and glucose monitoring can provide insight into the impact of diet, physical activity, and medication management on glucose levels for some individuals (American Diabetes 2019). Glucose monitoring may also be useful in assessing hypoglycemia and glucose levels during intercurrent illness. Consequently, glucose monitoring (including glycated hemoglobin, $HbA_{1c}$) is still the largest single-analyte market in diagnostics. Home glucometers are based on electrochemical sensors, but glucose quantification in urine can be accessed by colorimetric methods (Cardosi 2006). The classical colorimetric method for glucose detection in body fluids is based on its reaction with o-toluidine in acidic conditions followed by absorbance measurement at 620–650 nm (Dubowski 1962). However, o-toluidine is a very toxic reagent (occupational exposure limit (OEL) of 9 mg $m^{-3}$; $LD_{50}$ (oral/rat) of 670 mg $kg^{-1}$), and the test is sensitive to the interference of galactose, mannose, lactose, and other carbohydrates. Therefore, this method cannot be used for glucose determination in patients suffering from galactosemia or for those individuals undergoing galactose tolerance tests (Cardosi 2006). Limited specificity to discriminate glucose from other carbohydrates is characteristic of nonenzymatic spectrophotometric methods (Galant et al. 2015). Enzyme assays for the quantification of glucose will be described in the following section.

## 2.2   Protein and Nucleic Acid Quantification

Proteins show a broad absorption band with a maximum absorption wavelength at approximately 280 nm due to aromatic amino acid residues. Therefore, the direct quantification of 20–3000 μg of protein by UV-Vis spectrophotometry is possible, although other biomolecules with intense absorption in the MUV (200–300 nm), such as nucleic acids ($\lambda_{max} = 260$ nm), can make quantification difficult (Noble and Bailey 2009). The molar absorption coefficient of proteins at 280 nm ($\varepsilon_{280}$) can be determined with an accuracy of $\pm 5\%$ using the method proposed by Pace and coauthors (Pace et al. 1995). This method considers the number of residues ($n$) and the $\varepsilon_{280}$ of Trp, Tyr and cysteine involved in disulfide bonds, CysS–S, (5500, 1490, and 125 L mol$^{-1}$ cm$^{-1}$, respectively), according to Eq. (2):

$$\varepsilon_{280}(L\ mol^{-1}\ cm^{-1}) = 5500 \cdot n_{Trp} + 1490 \cdot n_{Tyr} + 125 \cdot n_{CysS-S} \qquad (2)$$

More accurate $\varepsilon_{280}$ values can be calculated by unfolding the proteins with 6 mol L$^{-1}$ guanidinium chloride and using reference $\varepsilon_{280}$ values for Trp, Tyr, and the disulfide chromophore determined in guanidinium chloride solution (Schmid 2001). Pure proteins have a A$_{280}$/A$_{260}$ ratio of approximately 2.0. However, when nucleic acids are present in the sample, the protein concentration (in mg mL$^{-1}$) can be determined using the Warburg-Christian and derived formulas, in which the absorption at 280 nm and 260 nm are weighted and subtracted, i.e., $a$(A$_{280}$) $- b$ (A$_{260}$). The weight variables $a$ and $b$ assume different values depending on the formula; typical values *(a,b)* are: 1.55,0.76; 1.52,0.75; 1.31,0.57 (Manchester 1996). Light scattering from suspended particles in the sample increases the apparent absorbance, and filtration (20 μm) or centrifugation prior to analysis is recommended (Noble and Bailey 2009). Although the intense absorbance of nucleic acids at 260 nm makes difficult the quantification of proteins, it enables the routine quantification of deoxyribonucleic acid (DNA) and ribonucleic acid (RNA) in solution, and the amounts of nucleic acids are often reported in A$_{260}$ units. One A$_{260}$ unit corresponds to 50 mg mL$^{-1}$ double-stranded DNA, 33 mg mL$^{-1}$ of single-stranded DNA, or 40 mg mL$^{-1}$ of single-stranded RNA, meaning an absorbance of one in a cuvette with 1-cm optical path length. In a 1:1 mixture of nucleic acids and proteins, the proteins contribute only about 2% to the total absorbance at 260 nm. Thus, since proteins absorb much more weakly than nucleic acids, contaminating proteins hardly affect the spectrophotometric quantification of nucleic acids (Schmid 2001).

Kits for the analysis of proteins by specific dye-based spectrophotometric methods such as the Bradford, the Lowry, the BCA (bicinchoninic acid), and the Biuret assays, are available from commercial suppliers. These assays generate analytical signals in the visible region of the spectrum, which can be conveniently monitored using simple spectrophotometers. There are several experimental variations of the Biuret (or Piotrowski) assay, but they are all based on the reaction between proteins and $Cu^{2+}$ ions under alkaline conditions to produce a light blue protein–[$Cu^+$] complex having $\lambda_{max}$ at 540 nm. This method can be used to quantify proteins in the 5–160 mg mL$^{-1}$ concentration range by means of a calibration curve obtained with a protein standard, usually bovine serum albumin (BSA). Higher sensitivity is obtained if BCA is added to the protein–[$Cu^+$] complex (which is allegedly in equilibrium with other $Cu^+$ complexes) (Huang et al. 2010) producing a deep purple 2BCA:1$Cu^+$ complex that can be quantified at 562 nm (7700 mol L$^{-1}$ cm$^{-1}$) (Brenner and Harris 1995). The BCA assay allows the quantification of proteins within the 0.0005–2 mg mL$^{-1}$ concentration range (0.2–50 µg range) and is approximately 100 times more sensitive than the Biuret assay. In the Lowry assay, the Folin-Ciocalteu reagent (phosphomolybdate and phosphotungstate) is used instead of BCA, leading to a blue complex that is read at 750 nm. However, the Lowry assay takes longer to complete than the BCA assay and is less sensitive, having a working range of 5–200 µg mL$^{-1}$ (5–100 µg of protein).

The Bradford assay is based on the binding of the protein to the dye Coomassie Brilliant Blue G-250 under acidic conditions, having a working range of 10–2000 µg mL$^{-1}$ (1–50 µg of protein). Upon binding protein, the anionic form of the dye is stabilized, shifting the absorption maximum from 465 to 595 nm. The observed response is mostly due to the interaction of the dye with arginine residues, resulting in the wide protein-to-protein variation characteristic of Bradford assays. Although it is convenient, the Bradford assay is sensitive to interferences from most ionic and nonionic surfactants and glycosylated proteins (Noble and Bailey 2009).

## 2.3 Enzyme Assays

Most enzymes have high substrate specificity and act as efficient biocatalysts under mild experimental conditions, thus promoting their application in bioanalytical methods. The diagnosis of genetic diseases related to enzyme modifications can be performed by measuring the enzyme activity instead of using techniques based on molecular biology, such as DNA amplification. For example, the activity of glucose-6-phosphate dehydrogenase (G6PD, EC 1.1.1.49) is used to diagnose sickle cell disease. The absorption of UV-Vis light is a more common property than fluorescence and, hence, is broadly used for monitoring enzymatic reactions by measuring the absorbance of the product at the end of the reaction, viz., endpoint method, or the reaction rate (Upstone 2013; Sánchez Rojas and Cano Pavón 2005). Nevertheless, the spectrophotometric determination of enzyme activity (Harris and Keshwani 2009) is a niche compared to several automated approaches for enzyme analysis.

Spectrophotometric enzyme assays are often based in the formation of detectable products of reactions mediated by nicotinamide adenine dinucleotide (NAD). NAD is a key enzyme cofactor that can be found in two forms differing in their oxidation state: $NAD^+$ and NADH, which can be discriminated by their absorption maxima at,

**Table 2** Some enzyme-based spectrophotometric assays. Adapted from Ref. Upstone (2013)

| Analyte | Method | Wavelength (nm) |
|---|---|---|
| Acid phosphatase | Kinetic: PNP | 405 |
| Ala aminotransferase | Kinetic: NAD⁺/NADH | 340 |
| Alkaline phosphatase | Kinetic: PNP | 405 |
| Bilirubin | Malloy-Evelyn | 555 |
| Cholesterol | Kinetic: cholesterol oxidase | 500 |
| Creatinine | Jaffe | 510 |
| γ-Glutamyl transferase (GGT) | Kinetic carboxy substrate | 405 |
| Glucose | Hexokinase (NAD⁺/NADH) | 340 |
| Lactate dehydrogenase (LDH) | Kinetic: lactate:pyruvate | 340 |
| Triglycerides | Kinetic: glycerol 3-phosphate oxidase (GPO) | 520 |
| Urea | Kinetic: (NAD⁺/NADH) | 340 |

respectively, 259 nm ($\varepsilon_{259}$ = 16,200 mol L$^{-1}$ cm$^{-1}$) and 340 nm ($\varepsilon_{340}$ ~ 6200 mol L$^{-1}$ cm$^{-1}$) (Ziegler 2013). The monitoring of NAD$^+$/NADH and their phosphate analogues, NADP$^+$/NADPH, in enzymatic transformations have been used in several applications, including photobioredox catalysis (Gonçalves et al. 2019; Goncalves et al. 2019). The activity of NADH-dependent dehydrogenases able to metabolize lactate, malate, and alcohol has been measured directly by the decrease in A$_{340}$ over time (Schmid 2001).

The enzymatic conversion of a colorless reagent into a colored product enables the spectrophotometric monitoring of biotransformations. In some cases, a coupled enzyme process can be used to enable spectrophotometric monitoring. It is essential, however, that in coupled enzyme assays the reaction leading to the product being monitored is faster than the primary reaction (Schmid 2001). Coupled enzyme processes often involve NADH leading to marked changes in absorbance at 340 nm. *p*-Nitrophenol (PNP) derivatives are frequently used as chromogenic substrates since the absorption maximum wavelength of the corresponding phenolate is 405 nm ($\varepsilon_{405}$ 18,000 mol L$^{-1}$ cm$^{-1}$) (Zhang and VanEtten 1991). For example, *p*-nitrophenyl acetate is a substrate for proteases, the phosphate is a substrate for acid and alkaline phosphatases and its glycosylated derivatives can be used to probe amylases or glycosidases. Some common enzyme assays are described in Table 2.

The enzymatic measurement of glucose usually relies on hexokinase (EC 2.7.1.1; ATP: D-hexose 6-phosphotransferase), glucose oxidase (EC 1.1.3.4; β-D-glucose: oxygen 1-oxidoreductase), and glucose dehydrogenase (EC 1.1.1.47; β-D-glucose: NAD(P)$^+$ 1-oxidoreductase), peroxidase (EC 1.11.1.7; phenolic donor: hydrogen peroxide oxidoreductase), and glucose-6-phosphate dehydrogenase (G6PDH, EC 1.1.1.49; D-glucose-6-phosphate:NAD(P)$^+$ 1-oxidoreductase). Glucose can be phosphorylated by ATP/hexokinase and the resulting glucose-6-

phosphate (G6P) is oxidized to 6-phosphogluconate in the presence of G6PDH, which is $NAD^+$-dependent. NADH is produced in equimolar amounts with G6P, enabling the quantification of glucose, providing the reaction is allowed to complete. This endpoint assay results in $A_{340}$ between 0.03 and 1.6 for the analysis of glucose solutions ranging from 0.5 to 50 µg $mL^{-1}$ (2.8–280 µmol $mL^{-1}$) (Cardosi 2006). This assay can be modified to measure the activity of G6PDH by adding iodonitrotetrazolium (INT) and phenazine methosulfate (PMS) to the system. The reaction between NADH and PMS results in $NAD^+$ and PMSH. The reduction of INT by PMSH leads to a formazan dye, which is detected at 520 nm. This method has been used to diagnose G6PDH deficiency, which is an X-linked recessive disorder that predisposes to red blood cell breakdown (Cappellini and Fiorelli 2008).

Glucose has been detected in blood, urine, and cerebrospinal fluid by using a colorimetric assay based on glucose oxidase and peroxidase. The oxidation of β-D-glucose by oxygen in the presence of glucose oxidase produces D-glucono-1,5-lactone, the labile cyclic ester of gluconic acid, and $H_2O_2$. A chromogenic electron donor, such as o-dianisidine or o-toluidine, is oxidized by $H_2O_2$ in the presence of peroxidase producing brown-orange products, which are read at 420–475 nm (Huggett and Nixon 1957; Raba and Mottola 2006). The glucose is stoichiometrically converted into the colored product, but several substances interfere with the method, including catechols, ascorbic acid, cysteine, glutathione, acetylsalicylic acid, L-3,4-dihydroxyphenylalanine (L-DOPA), mercurial diuretics, and tetracycline. The LOD of this method is 25 mg $dL^{-1}$ (Cardosi 2006).

## 2.4    Immunoassays

The strong affinity between an antibody and its target molecule, namely, antigen, is the basis of a wide range of so-called immunoassays, which are used in a plethora of settings ranging from basic research to clinical medicine. Metabolites that can be recognized by antibodies can also be monitored and quantified, enabling several applications such as pathogen identification, tumor marking, and drug discrimination. Immunoassays based on polyclonal and monoclonal antibodies, which recognize the antigen by interaction with multiple or single epitopes, respectively, have been described. When employed in point-of-care testing (PoCT) devices for medical diagnosis, the immunological reaction principle enables the qualitative or quantitative assessment of the concentration of an analyte/measurand. Immunodiagnostics, performed as PoCT, are nowadays used in hospitals, doctor's offices, and by patients themselves (Poschenrieder et al. 2019).

Spectrophotometric immunoassays are less sensitive compared to other detection methods such as spectrofluorimetry. The enzyme-linked immunosorbent assay (ELISA) is a popular method for the analysis of antigens. In the simplest variation, the antigen (or material/specimen containing it) is directly immobilized at a solid surface, usually the bottom of a microplate well, or bind an immobilized antibody. Next, a complimentary biomolecule labeled with a reporter enzyme, such as horseradish peroxidase (HRP) or alkaline phosphatase, interacts with the antigen.

Enzymatic reactions with chromogenic reagents lead to colored products, enabling the qualitative (as in cutoff assays) and quantitative detection of the antigen. Common substrates used for spectrophotometric detections include OPD (*o*-phenylenediamine dihydrochloride), TMB (3,3′,5,5′-tetramethylbenzidine), ABTS (2,2′-azinobis[3-ethylbenzothiazoline-6-sulfonic acid]), whose oxidation by $H_2O_2$ in the presence of HRP outcomes orange ($\lambda^{abs} = 450$ nm), blue ($\lambda^{abs} = 650$ nm) and green ($\lambda^{abs} = 405$ and 734 nm) products, respectively, and PNPP (*p*-nitrophenyl phosphate) that turns yellow by the action of alkaline phosphatase (Wu et al. 2019). Due to its instability and potential carcinogenicity, OPD is less used than TMB. The chromogenic substrate should be easy to prepare, prone to long-term storage, able to react with low concentrations of enzyme and undergo marked color variations. Most importantly, naked eye detection may be misleading, leading to error judgment in visual identification (Wu et al. 2019). There are a large number of ELISA-based assays covering a whole range of applications, including specialized clinical tests, as reported elsewhere (Wu et al. 2019; Aydin 2015; Hosseini et al. 2018). Although mainly used in absorbance, it is possible to adapt ELISA to other techniques such as fluorescence (FELISA) in order to enhance their sensitivity (Upstone 2013).

Although any analyte that is able to specifically bind to an antibody can be used in the development of immunoassays, the simultaneous measurement of multiple analytes, viz., multiplexing for dealing with a range of possible compounds, can be challenging. For example, only specific types of abused drugs can be detected using immunoassays and subtle structural modifications may adversely affect the assay. A high-resolution colorimetric immunoassay platform has been developed based on enzyme-catalyzed multicolor generation and smartphone-assisted signal readout (Xie et al. 2018). The multicolor generation is due to color change from yellow to orange to red of the pH indicator phenol red ($pK_a = 7.9$) induced by the hydrolysis of urea catalyzed by urease, whose activity is indirectly controlled by alkaline phosphatase-mediated ascorbic acid production. The LOD of the model analyte rabbit IgG is 1.73 ng $mL^{-1}$ with a dynamic range from 0 to 18 ng $mL^{-1}$. A colored picture taken using a smartphone can be quantitatively analyzed (Xie et al. 2018). New chromogenic mode and multicolor reagents for ELISA have also been developed by using the localized surface plasmon resonance (LSPR) of metallic nanoparticles such as gold and silver (Guo et al. 2016; Zhang et al. 2015). Enzyme-regulated cascade reactions combined with LSPR results in visual detecting systems with higher sensitivity. Human immunoglobulin G in plasma can be visually detected using an ELISA based on the reduction of iodate to iodine, which etches Au nanorods to Au spherical nanoparticles in the presence of the surfactant CTAB (Zhang et al. 2015). A dual-color response strategy based on $TMB^{2+}$ to etch Au nanoparticles (NPs) relies on color variations from wine red to colorless and then from colorless to yellow in order to quantify prostate-specific antigen (PSA) within a LOD of 9.4 pg $mL^{-1}$ (Guo et al. 2016). The visual detection of the antibiotic cloxacillin in foodstuffs has been achieved by using ELISA based on Au NPs, which are generated by reduction of $HAuCl_4$ by glucose oxidase coupled with the oxidation of glucose (Yu et al. 2017). This Au NPs growth-based ELISA can detect cloxacillin at 11 ng $mL^{-1}$. Molecules with tunable colors, viz., allochroic, have been used to modify carboxyl graphene oxide (cGO), enabling a multicolor detection method called allochroic-cGO linked immunosorbent assay (ALISA) (Li et al. 2016). Combined with chromogenic molecules, such as methyl red, phenolphthalein, and thymolphthalein, this approach can detect three lung cancer biomarkers, carcinoembryonic antigen, neuron-specific enolase, and cytokeratin-19 fragment, in a combined manner and at the pg mL$^{-1}$ level.

Optical biosensors based on label-free (direct) or labeled (indirect) immunosensors have been developed and used for PoCT applications (Poschenrieder et al. 2019). In the label-free approach, the analytical signal is generated directly by the interaction of the analyte with the surface of the transducer. This concept has analytical advantages providing the analyte are in the adequate concentration range. The label-based indirect approach involves the use of a colorimetric, fluorescent, or chemiluminescent tags. Nanoparticles can enhance the performance of optical biosensors mainly by multiplying the number of antibody binding sites due to increased carry capacity (Poschenrieder et al. 2019).

# 3 Spectrofluorimetry

The analytical use of fluorescence is still growing strong, despite the development of myriad other techniques (Vogel 2008). For example, laser-induced fluorescence (LIF) coupled to capillary electrophoresis has been used for DNA sequencing (Tseng et al. 2010; Caruso et al. 2017; Kaneta 2019), and fluorescence in situ hybridization (FISH) is an advantageous method for probing DNA sequences for diagnostic use (Volpi and Bridger 2008; Batani et al. 2019; Moter and Göbel 2000; Wagner et al. 2003), including melanoma testing (Lee and Lian 2019). Fluorescence is also the fundamental process behind modern imaging methods, such as fluorescence correlation spectroscopy (FCS), fluorescence lifetime imaging microscopy (FLIM), and fluorescence intensity distribution analysis (FIDA) (Kask et al. 2002; Luo et al. 2020). Other techniques based on luminescence, in general, are of utmost bioanalytical importance as well. Examples include upconverting phosphor particles (UPTs) (Huang et al. 2019), which have been used for the ultrasensitive detection of DNA and the development of PoCT devices (Vashist 2017), and chemiluminescence- and bioluminescence-based methods (Roda et al. 2000; Roda and Guardigli 2012), which avoid the scattering of excitation light and provide ultrasensitive quantification (LOD as low as zeptomole, $10^{-21}$ mol) of biomolecules, such as alkaline phosphatase and ATP (Baader et al. 2006; Roda et al. 2020).

## 3.1 UV-Vis Fluorescent Dyes

Advances in fluorophore design have changed the way biological systems are studied, enabling the noninvasive study of living organisms and increasing our understanding of the molecular basis of complex biological processes (Cheng et al. 2020). Fluorophores are mainly used as labels, enzyme substrates,

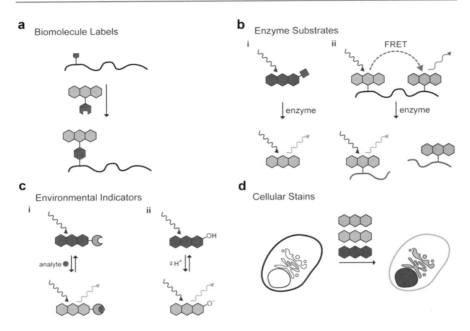

**Fig. 2** Main application of fluorophores for bioanalysis. (**a**) Site-specific labeling of a biomolecule by an orthogonal reaction between two functional groups (red). (**b**) Enzyme substrates (i) enzyme-catalyzed removal of a blocking group (red) elicits a change in fluorescence, (ii) enzyme catalyzes the cleavage of a labeled biomolecule (red) and concomitant decrease in FRET. (**c**) Environmental indicators (i) binding of an analyte (red) elicits a change in fluorescence, (ii) protonation of a fluorophore elicits a change in fluorescence. (**d**) Staining of subcellular domains by distinct fluorophores. Adapted from Ref. Lavis and Raines (2008), Copyright © 2008 American Chemical Society

environmental indicators, and cellular stains (Fig. 2) (Lavis and Raines 2008). Their application is limited by their physicochemical and photophysical properties, such as acidity/basicity, lipophilicity, stability, size, ease of functionalization, spectral features, brightness, fluorescence lifetime, and nonlinear optical properties (Pitre et al. 2016).

Small molecule fluorescent dyes can be tailored to exhibit high brightness and photostability, as well as narrow bandwidth compared to that of fluorescent proteins (Johnson 1998; Tsien 1998). These dyes can be membrane permeable in order to highlight the intracellular environment or membrane impermeable to report on extracellular species, although controlling their localization can be challenging (Zheng and Lavis 2017). Subtle structural modifications, such as deuteration and conformational restriction, can affect the fluorescence quantum yield of a dye and improve its applicability (Grimm et al. 2020, 2015; Wang et al. 2019b; Chi et al. 2020). For example, a better understanding of structure–fluorescence property relationships has led to the development of tracers that are gradually replacing radioactive labels in immunohistochemistry (Jäger et al. 2003).

The judicious choice of fluorophores enables the analysis of biomolecules, organelles, subcellular structures, cells, and tissues in vivo (Petryayeva et al. 2013). Small organic fluorophores have been used in several techniques of clinical and bioanalytical relevance, such as fluorescence-guided surgery in cancer treatment (Lavis and Raines 2008; Zheng and Lavis 2017; Zhang et al. 2017; Hernot et al. 2019; Dijkstra et al. 2019; Pirsaheb et al. 2019; Zheng et al. 2019a) and assays for testing SARS-CoV-2 (Froggatt et al. 2020; Smithgall et al. 2020). Also, sensors based on fluorescent detection have been used to overcome the technical difficulties related to the analysis of hydrogen sulfide ($H_2S$) in its intracellular or subcellular environments. $H_2S$ is volatile and catabolized very quickly. Probes based on nanomaterials, ranging from inorganic nanostructures to organic polymers, have been demonstrated to exhibit highly sensitive and selective detection of this toxic agent both in vitro and in vivo (Luo et al. 2019). Fluorescence bioassays have been optimized to provide enhanced sensitivity, but the use of monoreactive small fluorophores usually limits multiplexing (Macchia et al. 2020; Specht et al. 2017; Jäger et al. 2003; Kairdolf et al. 2017; Goldman et al. 2004). Robust bioanalytical methods based on fluorescence have been updated for enhanced sensitivity and/or automation purposes. For example, the modern microplate fluorometric screening test for phenylketonuria in newborn infants is an updated version of a method developed in 1962 that is based on the detection of the fluorescent product of the reaction between phenylalanine, copper, and ninhydrin (McCaman and Robins 1962; Gerasimova et al. 1989; Fingerhut et al. 1997).

Ideal fluorophores are readily available, photostable, safe, provide reproducible results, are not affected by interfering species, and can be used under mild experimental conditions (Albani 2016). The list of small molecule fluorophores most frequently used in pharmacological, clinical and medical applications include: Alexa Fluor dyes (e.g., Alexa Fluor 488 and 514), BODIPY dyes (Kaur and Singh 2019) (e.g., BODIPY-FL, 8-phenyl BODIPY and BODIPY R6G) (Braun et al. 2016; Wang et al. 2018; Jun et al. 2020; Thapaliya et al. 2017), Cyanine dyes (e.g., Cy 5.5 and Cy 7) (Jun et al. 2020; Kurutos et al. 2020), cypate carbocyanine (Berezin and Achilefu 2010; Achilefu et al. 2005; Ye et al. 2005; Wang et al. 2020a), Fluorescein dyes (e.g., carboxyfluorescein, fluorescein isothiocyanate, FITC, and fluorescein amidite, FAM) (Jun et al. 2020; Shah et al. 2019; Koktysh 2020; Oberbichler et al. 2020), indocyanine green (ICG) (Wu et al. 2020a; Yang et al. 2020; Yeroslavsky et al. 2020; Zhao et al. 2020b), Oregon green (Wang et al. 2019c, 2020b), rhodamine dyes (rhodamine 110, rhodamine 6G, rhodamine X and TAMRA) (Das et al. 2018; Takahashi et al. 2020; Zhang et al. 2020; Lavis 2017), and rhodol dyes (Hong et al. 2020; Poronik et al. 2019). In particular, BODIPY dyes have been used in several applications of bioanalytical interest. Cosa and coworkers designed a BODIPY-$\alpha$-tocopherol adduct for use as an off/on fluorescent antioxidant indicator (Oleynik et al. 2007). BODIPYs are also important labels for fluorescence polarization techniques (Banks et al. 2000). The small Stokes shift of BODIPY dyes leads to self-quenching due to overlabeling. This phenomenon has been explored to analyze protease substrates, because proteolysis of densely labeled proteins increases the fluorescence intensity (Thompson et al. 2000). In addition, the equilibrium between fluorescent and nonfluorescent species can be used for bioanalytical purposes. For example, the nonfluorescent lactone of rhodamine dyes has improved cell permeability and the binding of its fluorescent zwitterion form to biomolecular target enables sensitive detection (Zheng et al. 2019b). Cell permeability of fluorescent dyes is a very important feature for bioanalysis and comprehensive information on the use of fluorescent probes in cell viability assays can be found elsewhere (Tian et al. 2019).

Other fluorophores are emerging for specific bioanalytical applications. In particular, Schiff bases and their metal complexes have become well-known for their biological properties (e.g., antifungal, antibacterial, antimalarial, and antiviral characteristics) and have been used as fluorescent on/off sensors for the determination of diverse analytes, in particular metal cations. As such, they can offer a way to identify toxic ions and/or to provide their speciation in environmental media (Berhanu et al. 2019). Another example is the use of fluorescent and fluorogenic amino acids for tracking protein–protein interactions in situ and for high-resolution imaging of micro- to nanoscopic events in real time (Cheng et al. 2020). Fluorescent amino acids enable the construction of fluorescent macromolecules, such as peptides and proteins, without disrupting their native biomolecular properties (Cheng et al. 2020). Applications include novel methodologies to synthesize building blocks with tunable spectral properties, their integration into peptide and protein scaffolds using site-specific genetic encoding and biorthogonal approaches, and their application to design novel artificial proteins, as well as to investigate biological processes in cells by means of optical imaging (Cheng et al. 2020). Last, advances in the development of multifunctional fluorescent sensors for protein recognition have been achieved by Margulies and coauthors (Margulies and Hamilton 2009; Margulies et al. 2014; Nissinkorn et al. 2015; Unger-Angel et al. 2015; Hatai et al. 2017; Pode et al. 2017; Unger-Angel et al. 2019).

Fluorescent nanomaterials are ubiquitous in bioanalysis (Qing et al. 2020; Keçili et al. 2019). Quantum dots (QDs) are colloidal semiconductor nanocrystals with dimensions between approximately 1 and 10 nm (Petryayeva et al. 2013; Kairdolf et al. 2013; Rosenthal et al. 2011; Molaei 2019). Their high brightness results from high quantum yields ($\phi_{Fl} = 0.1$–$0.9$) and large $\varepsilon$ (between $10^5$ and $10^7$ L mol$^{-1}$ cm$^{-1}$) (Petryayeva et al. 2013; Resch-Genger et al. 2008). In the following sections, the subscript number on the QD description (e.g., QD$_{525}$) denotes the peak photoluminescence wavelength of CdSe/ZnS QDs, except otherwise noted (Petryayeva et al. 2013). QDs are smaller and easier to use for multiplexing compared to silica nanoparticles doped with fluorescent dyes, which are also very bright and photostable (Bae et al. 2012). Nanodiamonds have very high quantum yields and are photostable, but are not very bright and tunable (Mochalin et al. 2011). C-dots are non-blinking small nanoparticles whose emission wavelength depends on the excitation wavelength (Baker and Baker 2010). Carbon nanotubes show NIR emission and excellent resistance to photobleaching but are not very bright (Wu et al. 2010). Lanthanide-based upconversion nanoparticles show multiple narrow emission lines but are larger in size than other nanostructured fluorophores (Wang et al. 2010). New optical materials obtained from fluorescent dyes and macrocycles called small molecule ionic isolation lattices (SMILES) are the brightest fluorescent material to date (Benson et al. 2020).

Fluorescent peptides and proteins expressed as *Aequorea*-derived fluorescent protein (AFP) chimeras can be monitored in cellulo and in vivo (Merola et al. 2010; Kremers et al. 2011). The RNA mimic of the green fluorescent protein (GFP) is called Spinach aptamer and has been used to label RNA in vivo for tracking purposes (Pothoulakis et al. 2014; Trachman and Ferré-D'Amaré 2019). The study of RNA riboswitches stimulated the development of a new class of small molecule and metabolite sensors based on molecule-binding RNA aptamers. Biological nanostructures have been extensively used for developing biosensors (Alves et al. 2011). In particular, photoactive peptide nanotubes are of interest in bioanalysis (Souza et al. 2013, 2017; Pelin et al. 2018). In-depth comparison of the properties of several classes of fluorophores is available in the literature (Petryayeva et al. 2013; Hotzer et al. 2012).

Fluorophores and fluorogenic agents can be easily incorporated into automated analytical methods. Flow cytometry using QD-barcode, for example, offers a rapid, sensitive, and parallel detection of genetic markers for infectious diseases, such as HIV, hepatitis B, hepatitis C, syphilis, and malaria (Giri et al. 2011; Zarei 2018). Further developments in fluorescent nanoparticle biosensors are expected to improve the sensitivity of high-throughput screening (HTS) assays for broad panels of pathogen and disease markers as well as PoCT bioassays (Howes et al. 2014; Zhou et al. 2015; Yang et al. 2019). Fluorescence detection has already been used in HTS systems for drug discovery (Jäger et al. 2003; Blay et al. 2020). Nevertheless, detailed information on biomolecular processes can be obtained by using fluorescence-based assays based on single-molecule detection techniques (Farka et al. 2020; Yokota 1864; Gilboa et al. 2020). These methods are suitable for the direct detection of highly diluted fluorophores by measuring the read-outs inherent to the fluorescence signal and analysis by means of fluctuation methods, such as FCS and FIDA (Luo et al. 2020).

The amounts of fluorophore required for in vivo bioanalysis and bioimaging are usually much lower than the toxic doses reported in the literature. However, toxicity can be an important issue for the use of fluorophores in vivo (Choyke et al. 2009; Luly et al. 2020; Ai et al. 2011). Despite the ready availability of photophysical data of most conventional fluorescent dyes and pigments (Specht et al. 2017; Lavis and Raines 2008; Zheng and Lavis 2017; Grimm et al. 2015), information on the cytotoxicity, tissue toxicity, in vivo toxicity, carcinogenicity, and mutagenicity of non-FDA-approved fluorophores are scarce (Choyke et al. 2009). For example, ICG was approved by the FDA for use as a contrast agent for retinal angiography in the late 1950s. The typical clinical dose of ICG is 3.5 $\mu g\ kg^{-1}$, and the maximum intravenous dose for humans is 5 mg $kg^{-1}$, which is at least 10,000 times lower than the lowest corresponding $LC_{50}$ values (Gandorfer et al. 2003). The phototoxicity of ICG has been considered for use in laser-induced photodynamic therapy against infectious agents (Omar et al. 2008; Jori and Brown 2004) and cancer cells (Baumler et al. 1999). However, ICG has also photodynamic toxicity on the inner retina (Gandorfer et al. 2003; Haritoglou et al. 2005; Narayanan et al. 2005; Kwok et al.

2005; Abels et al. 2000) and, consequently, the use of ICG during macular surgery is not recommended (Hope-Ross et al. 1994). Fluorescent natural products and pseudo-natural derivatives are often regarded as safe, being a potential alternative for the development of safe dyes for bioanalytical applications. For example, the pigments of red beetroots were used to produce a pseudo natural dye for the fluorescence imaging of erythrocytes infected with the malaria parasite (Gonçalves et al. 2013), and a metal-free blue dye (Freitas-Dorr et al. 2020).

## 3.2 NIR Fluorescence Dyes

Fluorophore selection is often guided by the excitation and emission restrictions of the analysis being conducted. Ideally, the absorption and emission maxima of the dye should be separated from those of endogenous substances because autofluorescence decreases the signal-to-noise ratio (Monici 2005; Maltas et al. 2015). Dyes absorbing red and NIR light circumvent some background problems, while being able to penetrate deeper into the tissue compared to UV light (Ding et al. 2018, 2019). Fluorescent bioimaging based on NIR dyes shows the advantages of good contrast and a variety of signal readouts. In particular, near-infrared fluorescent lifetime bioimaging (NIR-FLTB) shows improved signal-to-background ratio due to the suppression of background signals and, therefore, has been used in nanomedicine as a tool for diagnosis, treatment, and imaging (Lian et al. 2019).

There are several subdivision schemes for NIR radiation, e.g., CIE, ISO 20473, astronomical, sensor response, and telecommunication bands. The NIR (IR-A) region is commonly defined as the ~750 to 1400 nm wavelength range and is followed by the short-wavelength infrared (SWIR, IR-B, 1400–3000 nm). However, NIR has been subdivided into NIR-I (~700–1000 nm) and NIR-II (1000–1700 nm) windows. The NIR-II region can be further classified into NIR-IIa' (1000–1300 nm), NIR-IIa (1300–1400 nm), and NIR-IIb (1400–1700 nm) subregions (Zhu et al. 2019). NIR-II fluorescence imaging shows reduced photon scattering, as well as autofluorescence and improved penetration depth compared to traditional and NIR-I (400–900 nm) imaging modalities (Fig. 3) (Ding et al. 2019).

Physiological alterations, such as blood flow and vascular permeability increase, often promote the accumulation of NIR dyes in pathological tissues, where they act as contrast agents in bioimaging applications (Zhao et al. 2020b). NIR fluorescent probes can also be tailored to target specific biological receptors enabling, for example, the noninvasive fluorescence lifetime imaging of tumors (Berezin and Achilefu 2010; Bloch et al. 2005). The same technique has been used for the visualization of drug delivery toward the target cells in cancer immunotherapy (Lian et al. 2019; Tarte and Klein 1999; Almutairi et al. 2008).

Cyanines, squaraines, phthalocyanines, and porphyrin derivatives, rhodamines, and BODIPYs are the most frequently used classes of NIR organic dyes (Lian et al. 2019). Compared to conventional UV-Vis fluorescent dyes, such as FITC, these NIR dyes are more suited to in vivo applications, where noninvasive photoexcitation is desirable, since their absorption and fluorescence have minimal overlap with those

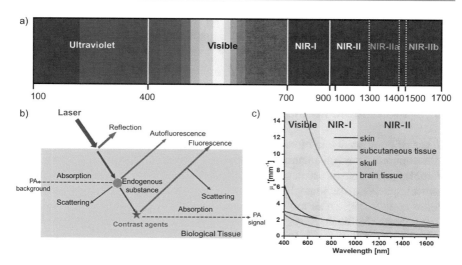

**Fig. 3** Properties of UV-Vis-NIR light. (**a**) The UV-Vis-NIR region of the electromagnetic spectrum. (**b**) Interaction of light and living tissues. PA means photoacoustic. (**c**) Reduced scattering coefficient, $\mu_s$, as a function of the wavelength for various tissue types. Reproduced with permission from Ref. Ding et al. (2019)

of biomolecules in blood and tissues. Cyanines, in particular indocyanine green (ICG) analogues, are NIR optical contrast agents widely used for cancer diagnostics. They can be attached to biomolecules, such as peptides and proteins, to create bright optical bioprobes suitable for fluorescence imaging (Marcu et al. 2014). ICG has the characteristic of being retained in or around the lymph nodes (Zhao et al. 2020b) enabling, for example, the NIR imaging of tumors in canine mammary gland tissue in vivo (Zhang et al. 2017; Reynolds et al. 1999; Newton et al. 2020). Squaraine dyes are a class of organic dyes that are emerging as NIR contrast agents due to their narrow absorption bands and high brightness (Ilina et al. 2020). The squaraine scaffold is susceptible to nucleophilic attack, enabling their conjugation with biomolecules but limiting its broad application. As a consequence, squaraine dyes have been subject to encapsulation and structural modifications to increase their performance in biomedical imaging (Martins et al. 2017). Despite the variety of NIR dyes, many of them suffer from lack of chemical stability, are photosensitive and are not suitable for in vivo applications (Luo et al. 2011). Hence, the development of dyes that overcome these drawbacks yet maintaining high brightness and long fluorescence lifetimes is desirable (Lian et al. 2019; Nolting et al. 2011).

Methods to shift the absorption and fluorescence of dyes to the NIR region have gained considerable interest. The replacement of oxygen atoms by silicon-based groups into xanthene, coumarin, and cyanine scaffolds led to the development of an array of NIR fluorescent dyes (Wang et al. 2019b; Lavis 2017; Pengshung et al. 2020). An interesting example of the potential of NIR dyes are Janelia Fluor

(JF) series of dyes, which were developed by modifying the rhodamine scaffold. JF526-type dyes, for example, are useful ligands for self-labeling tags, stains for endogenous structures, and spontaneously blinking labels for super-resolution immunofluorescence (Zheng et al. 2019b).

## 3.3   FRET-Based Methods

The use of fluorescence in bioanalysis is not restricted to accumulation and binding of fluorophores. Several methods based on FRET enable the study of molecular ensembles and single molecules and have been used for the detection of a number of chemical species of biological interest, including metal ions (Gordon et al. 2016; Bischof et al. 2019), small molecule organic analytes (Thuy et al. 2019; Wu et al. 2020b), drugs (Milligan 2004; Geissler and Hildebrandt 2016; Gong et al. 2020), enzymes (Specht et al. 2017; Pode et al. 2017; Wu et al. 2014; Klockow et al. 2020; Platnich et al. 2020), and toxins and toxicants (Farka et al. 2020; Kattke et al. 2011; Sapsford et al. 2011). FRET enables fluorophores to be turned on/off in response to events of biological interest and, thus, is the principle of hybridization assays (Tavares et al. 2014; Wei et al. 2018), immunoassays (Takkinen and Zvirbliene 2019), study of protein dynamics (Luo et al. 2020; Yokota 1864), and pH sensing (Dennis et al. 2012). Since Stryer and Haugland used FRET to develop a spectroscopic ruler (Stryer and Haugland 1967), FRET-based methods have evolved to a point where it is used in single-molecule methods for structural DNA technology, including the development of sensors and drug delivery vehicles (Platnich et al. 2020; Lacroix et al. 2017).

Energy cassettes based on FRET and NIR fluorescent probes are a valuable tool for bioanalysis, including the sensing of metal cations and small molecules (Fan et al. 2013). In FRET applications, the dyes are usually linked to the species of interest by a nonconjugated spacer and energy transfer is assumed to occur through space. However, depending on how the energy donor and acceptor are connected, through bond energy transfer (TBET) becomes feasible (Fan et al. 2013). A systematic study of these phenomena was undertaken by several research groups, in particular by Verhoeven and coauthors (Oevering et al. 1987, 1988; Reimers and Hush 1989; Hsu 2009). When TBET is faster than non-radiative decay pathways it may not require the donor–acceptor spectral overlap characteristic of conventional FRET pairs (Bajar et al. 2016). Hence, multiplexing with TBET cassettes would not necessarily involve loss of sensitivity at high resolutions (Fan et al. 2013), promoting its applications in chemo- and biosensors (Wang et al. 2019d; Cao et al. 2020; Zhao et al. 2011).

FRET-based fluorescent probes enable ratiometric sensing and dual/multi-analyte responsive systems (Wu et al. 2020b). The pseudo-Stokes shift, viz., the difference between the absorption maximum wavelength of the energy donor and the fluorescence maximum wavelength of the acceptor, of FRET-based energy cassettes is often larger than the Stokes shifts of either the donor or the acceptor dyes. There are several bioanalytical methods relying on FRET. Genetically encoded fluorescent indicators based on FRET have been used for the identification of agonists of human liver X receptors by using ultrahigh-throughput cell-based β-lactamase-dependent assay (Chin et al. 2003), and to monitor various cell functions (Nagai et al. 2004). FRET and two-color global fluorescence correlation spectroscopy (2CG-FCS) have been combined for high-throughput applications in drug screening (Eggeling et al. 2005). Homogeneous time-resolved fluorescence resonance energy transfer (TR-FRET) assays are highly sensitive and robust for the high-throughput quantification of kinase activity (Lundin et al. 2001; Ergin et al. 2016). The performance of ALPHAScreen (Amplified Luminescent Proximity Homogeneous Assay), TR-FRET, and time-resolved fluorescence for the study of cell receptors has been compared (Sepe et al. 2018). Selected examples of tools to monitor cellular dynamics by FRET-based detection are presented in Table 3 (Specht et al. 2017).

The spatial and temporal distributions of $Ca^{2+}$, a second messenger that is ubiquitous in living organisms, can be accessed by FRET-based methods used in combination with other techniques. Synthetic fluorescent indicators have been considered the analytical gold standard for $Ca^{2+}$ detection due to their linear and fast response, and broad dynamic range. However, in several applications, the performance metrics of genetically encoded calcium indicators (GECIs) overcomes those of small molecule fluorescent indicators (Rose et al. 2014). The GCamP platform is a GECI created from the fusion of GFP, calmodulin, and M13, a peptide sequence from myosin light chain kinase. The detection of calcium(II) ion is performed by monitoring changes in fluorescence intensity at a single wavelength or FRET-based ratiometric sensing (Specht et al. 2017; Wang et al. 2019c; O'Banion and Yasuda 2020). However, the comparison of small molecule $Ca^{2+}$ indicators and GCamP6 sensors, an ultrasensitive protein calcium sensor developed for imaging neuronal activity (Chen et al. 2013), revealed that the dyes Cal-520 (EX493/EM515/$\phi_{Fl}[Ca^{2+}$, pH 7.2] = 0.75) and Rhod-4 (EX523/EM551/$\phi_{Fl}$ [$Ca^{2+}$, pH 7.2] = 0.1) are superior for tracking local $Ca^{2+}$ signals (puffs) induced by inositol 1,4,5-trisphosphate (IP3) in cultured human neuroblastoma SH-SY5Y cells (Lock et al. 2015). High-performance GECIs are very sensitive, enabling in vivo fluorescence imaging in a wide range of model organisms (Rose et al. 2014). However, although green GCamPs have been used for the monitoring of neuronal activity through the quantification of $Ca^{2+}$, their use in vivo is limited by the poor tissue penetration of blue excitation light and spectral overlap with optogenetic tools for controlling neuronal signals (Specht et al. 2017; Dana et al. 2016). Consequently, efforts to expand the color palette of such sensors into the red resulted in the development of sensitive indicators based on R-CamP (derived from mRuby) and R-GECO (derived from mApple) (Rose et al. 2014).

## 3.4 Fluoroimmunoassay

Most immunoassays rely on labeling techniques based on fluorescence, such as fluoroimmunoassay and FRET, in addition to the colorimetric enzyme-linked immunosorbent assay (ELISA), electrochemiluminescence, and nonoptical detection

**Table 3** Selected examples of tools to monitor cellular dynamics by FRET-based detection. Reproduced from Ref. Specht et al. (2017)

| Monitored item | Name/description of tool | Type of detection | GE | Key advantages/disadvantages |
|---|---|---|---|---|
| $Ca^{2+}$ | GCamPs/GECIs | Intensity change (single FP sensor) or FRET based | Yes | Plus: Permits long-term measurement of $Ca^{2+}$ transients in transgenic organisms Minus: May perturb endogenous calcium dynamics |
| | Small molecule $Ca^{2+}$ indicators | Intensity change or spectral shift in dye | No | Plus: Fast response time Minus: Intracellular concentration of dye can be very high |
| Kinase activity | KTRs | Change in localization (nucleus versus cytoplasm) of single FP reporter | Yes | Plus: Multiplexing for detection of up to 4 different kinase activities possible |
| Voltage change across a membrane | Genetically encoded voltage sensors (based on conformational or photophysical change of sensing domain) | Intensity change (single FP sensor) or FRET-based | Yes | Plus: Targeting to small pool of neurons in live animals possible Minus: Slow response times compared to small molecule probes |
| | Small molecule-based voltage sensors | Intensity change or FRET-based | No | Minus: Delivery to membrane is difficult; partitioning in other membranes likely |
| Cell cycle stages | Fucci (fluorescent, ubiquitination-based cell cycle indicator) | Cell cycle state-dependent degradation of FP reporters (green/red color change at M–G1 transition, yellow at G1–S transition) | Yes | Minus: Requires delivery of two reporters |
| | CDK2-based localization change | Change in localization (nucleus versus cytoplasm) of single FP reporter | Yes | Plus: Single color allowing it to be multiplexed with other probes |
| NADH/NAD$^+$ | SoNar | Conformational change of single FP fused to NADH/NAD$^+$-sensing domain | Yes | Plus: Insensitive to changes in pH; ratiometric |
| Molecular crowding | GimRET | FRET based | Yes | Plus: Can be targeted to different organelles to monitor crowding |

*GE* genetic encoding

methodologies based on magnetic labels. Due to its high sensitivity, fluorescence is frequently used for signaling with labeled primary or secondary antibodies or labeled tracers, overcoming the typical lack of specific fluorescent dyes and allowing the discrimination of proteins according to their conformation, decomposition pattern, and site modification. Homogeneous immunoassays without bound-from-free separation of immune complexes often utilize fluorescence polarization techniques (Glahn-Martínez et al. 2018), FRET-based methods (Poschenrieder et al. 2019), and evanescent wave fluorescence (Taitt et al. 2016). Furthermore, antibodies labeled with fluorophores enable the visualization of biomolecules and the study of processes of clinical importance in vivo. For example, confocal immunofluorescence microscopy was used to investigate the tight binding of trophozoites of *Giardia lamblia* to host cells, occurring by means of the ventral adhesive disc (Di Genova et al. 2017).

Quantum dots are frequently used in immunoassays due to their unique optical properties and their size-adjustable photophysical properties. QDs are frequently used in FRET systems which can be specifically turned on/off upon interaction with the analyte, enabling sensitive and specific detections (Geissler and Hildebrandt 2016). There are several examples of multiplexing immunoassays based on QD photoluminescence (Ruppert et al. 2015). For example, a sandwich immunoassay based on QD labels ($QD_{510}$, $QD_{555}$, $QD_{590}$, and $QD_{610}$) absorbing at 330 nm enabled the simultaneous detection of cholera, ricin, Shiga-like toxin 1, and staphylococcal enterotoxin B with LODs ranging from 3 to 300 ng mL$^{-1}$ (Goldman et al. 2004). Carcinoembryonic antigen and α-fetoprotein, two tumor markers found in human serum, were detected by using a competitive immunoassay based on $QD_{520}$ and $QD_{620}$ labeling (Tian et al. 2012; Goryacheva et al. 2015). QD–antibody conjugates have been used for labeling fixed cells and tissues. Notably, QD probes were used for immunocytochemistry (ICC) labeling of a variety of subcellular targets, including the HER2 cell surface receptor, cytoskeletal components such as actin and microtubules, and nuclear antigens (Wu et al. 2003).

QD-based spectral imaging has been a valuable tool for studying human cancers. Multispectral imaging and immunohistochemical (IHC) labeling with $QD_{605}$-streptavidin conjugates (via biotinylated secondary antibodies) were used to identify HER2 and the estrogen and progesterone receptors in breast cancer tissue (Chen et al. 2011). The CD15, CD30, CD45, and Pax5 protein biomarkers were simultaneously traced in lymph node biopsy specimens by using $QD_{525}$-, $QD_{565}$-, $QD_{605}$-, and $QD_{655}$-secondary antibody conjugates (Petryayeva et al. 2013; Liu et al. 2010a). This approach enabled the differentiation of Hodgkin's lymphoma from non-Hodgkin's lymphoma and benign lymphoid hyperplasia. A similar approach was used to quantify E-cadherin, high-molecular-weight cytokeratin, p63, and α-methylacyl CoA racemase, which are protein markers associated with prostate cancer (Liu et al. 2010b). An important finding of this study is that progressive changes in benign prostate glands start with a single malignant cell. A fluoroimmunoassay based on QD has also been used for the detection of Zika virus IgG antibodies (Ribeiro et al. 2019).

## 3.5    Sensors and Biosensors

Sensors and biosensors rely on optical, electrochemical, piezoelectric, or calorimetric principles and are inherently related to bioanalysis (Gómez-Hens 2005). Biosensors rely on biochemical recognition systems to produce analytically useful signals (Thevenot et al. 1999). Consequently, the sensing element of a biosensor is of biological origin, for example, an isolated enzyme, an antibody or whole cells, and, ideally, no additional processes, such as separation steps and reagent addition, should be required for the analysis (Albers 2014; Thevenot et al. 1999).

Sensing matrices based on light-absorbing or fluorescent dye compounds have been used for the development of miniature optical sensors, or optrodes (Baldini and North Atlantic Treaty Organization 2006). For the detection of charged species, optical transduction can be achieved by including a lipophilic fluorescent ionophore, viz., fluoroionophore, together with the receptor into the membrane phase. For metal cation detection, for example, the complexation will release a proton and change the fluorescence response of the dye. This approach requires no synthetic modification of the fluorescent dye or the receptor, but the analytical response is pH dependent. Common fluoroionophores with signal-transducing properties include acridine and fluorescein derivatives as well as other small molecule organic fluorescent dyes combined with cavitands, such as calixarenes (Kim and Quang 2007; Kumar et al. 2019). A well-established approach that enables intracellular sensing of various analytes with fluoroionophores is a technique called PEBBLE, namely, Probes Encapsulated By Biologically Localized Embedding (Sumner et al. 2002; Xu et al. 2002; Park et al. 2003; Sumner et al. 2006). Examples of application of this technique include pH and calcium sensing (Clark et al. 1999) and imaging of $Mg^{2+}$ (Park et al. 2003), $Cu^{2+}$ (Sumner et al. 2006; Isarankura-Na-Ayudhya et al. 2010), $Zn^{2+}$ (Sumner et al. 2002), and dissolved oxygen (Cao et al. 2004). In addition, optical probes able to detect free radicals, DNA, and biomarkers at extremely low levels have been developed (Sigaeva et al. 2019). These sensors also have practical significance in clinical assays since they can recognize and discriminate common ions in serum (Albers 2014). Alternatively, the fluorophore can also be an integral part of the receptor in a way that the fluorescence reading is modified by chemical processes (Wu et al. 2017; Wang et al. 2020c).

Optrode technology based on fluorescence detection has been of value for glucose detection. The first optrode for glucose detection has been based on the concanavalin A receptor of glucose and fluorescence detection (Schultz et al. 1982). The update of this technology improved the working range for the detection of glucose to 0.2–3.0 mmol $L^{-1}$ (Ballerstadt and Schultz 2000). In further developments, NIR dyes were introduced in the glucose optrode to reduce background fluorescence from serum samples (Ballerstadt et al. 2004). Although these sensors are capable of long-term and sensitive glucose monitoring, they suffer from interference from fructose (Albers 2014). Also, fluorescent nano-PEBBLE sensors were designed for the imaging of intramolecular glucose (Xu et al. 2002).

Last, molecularly imprinted polymers (MIPs) emerged as promising materials for the recognition and detection of low-molecular-weight analytes, such as steroids, phenethylamines, catecholamines, and a variety of drug compounds (Haupt et al. 2020; Ding et al. 2020; Piletsky et al. 2020; Tarannum et al. 2020). The use of fluorescent functional monomers designed to have specific chemical interactions with the analyte enables the direct monitoring of the binding event via changes in the fluorescence of the MIP (Piletsky et al. 2020; Xiao et al. 2020). By using fluorescent MIPs, methods for the detection of catecholamine, cyclic AMP (cAMP), antibiotics, and bacteria have been developed (Albers 2014; Crapnell et al. 2019). Fluorescent molecularly imprinted polymers were used to replace antibodies and receptors used in ELISA, showing high substrate specificity (Piletsky et al. 2001).

# 4 Concluding Remarks

UV-Vis spectrophotometry is still a small but important tool in bioanalysis, especially in the modern clinical laboratory. The correct interpretation of the acquired data requires some understanding of the analytical system and the limitations of the technique. Data validation and equipment calibration has become increasingly important, mostly due to commercial requirements imposed by the pharmaceutical and food industries. Miniaturization, smaller sampling volumes, high analytical sensitivity, and minimal sample preparation as well as the combination of analytical and separation techniques, are strong trends for the development of new bioanalytical platforms. Device development based on spectrophotometric and spectrofluorimetric measurements aiming clinical applications must be tested in realistic sample matrices. Further, safety is an important, yet often neglected parameter for the effective pharmacological, clinical, and medical application of dyes.

UV-Vis and fluorescence signal detection combined with (bio)sensors, microfluidics, and lab-on-chip devices is a promising alternative on the road to personalized medicine and theranostic applications. Synthetic receptors, including molecularly imprinted polymers and self-assembled molecular layers, are matching the performance of natural receptors. Consequently, immobilization is becoming increasingly integrated with detection schemes. Quantum dot-based assays provide sensitive and reliable methods for bioanalysis, although they have been scarcely used in PoCT devices (Poschenrieder et al. 2019; Lee et al. 2019). Single-molecule sensing platforms are now able to sense at the sub-nanomolar concentration level with minimal background interference. Label-based single-molecule technologies relying on optical transduction have reached ultralow detection limits and are promising alternatives for the analysis of the components of biofluids enabling early diagnostics of myriad pathologies (Macchia et al. 2020).

Monitoring protein dynamics using fluorescence readouts enables the study of complex biological processes from a molecular perspective. Photoresponsive probes enable the monitoring of protein stability and turnover. Data-rich fluorescent techniques with multiplexing capabilities have the potential to revolutionize the field of high-throughput proteomics. Combined nucleic acid imaging, single-molecule detection, and multiplexed labeling are the future of quantitative global analysis of cellular responses. Several classes of fluorophores are providing insight into small molecules, secondary metabolites, metals, and ions. Fluorescence also provides a convenient approach to interrogate cellular dynamics at the RNA level, protein level, and metabolite level, allowing the study of regulatory mechanisms that govern cellular dynamics. Last, spectrophotometry and spectrofluorimetry integrate the group of analytical tools that, combined with automation, analysis, aggregation, visualization, and advanced modeling, such as machine and deep learning, are of fundamental importance for quality by design manufacturing and process analytical technology (PAT) (Wasalathanthri et al. 2020; Fedick et al. 2018).

## References

Abels C, Fickweiler S, Weiderer P, Baumler W, Hofstadter F, Landthaler M, Szeimies RM (2000) Arch Dermatol Res 292:404–411
Achilefu S, Bloch S, Markiewicz MA, Zhong T, Ye Y, Dorshow RB, Chance B, Liang K (2005) Proceedings of the National Academy of Sciences of the U S A 102:7976–7981
Ai J, Biazar E, Jafarpour M, Montazeri M, Majdi A, Aminifard S, Zafari M, Akbari HR, Rad HG (2011) Int J Nanomedicine 6:1117–1127
Akash MSH, Rehman K (2020) In: Akash MSH, Rehman K (eds) Essentials of pharmaceutical analysis. Springer, Singapore, pp 29–56
Albani JR (2016) In: Meyers RA (ed) Encyclopedia of analytical chemistry. Wiley, Chichester, pp 1–25
Albers WM (2014) In: Gauglitz G, Moore DS (eds) Handbook of spectroscopy, vol 3. Wiley-VCH, Weinheim, pp 943–976
Almutairi A, Akers WJ, Berezin MY, Achilefu S, Frechet JM (2008) Mol Pharmacol 5:1103–1110
Alves W, Sousa CP, Amaral H, Liberato MS, Takahashi PM, Silva RF, Kogikoski S Jr, Martins T, Oliveira VX Jr, Alves W (2011) Biosensors based on biological nanostructures. IntechOpen, London
American Diabetes A (2019) Clin Diabetes 37:11–34
Aydin S (2015) Peptides 72:4–15
Baader WJ, Stevani CV, Bastos EL (2006), Chemiluminescence of organic peroxides. In: Rappoport, Z (ed) The chemistry of peroxides. John Wiley and Sons, Chichester, pp. 1211–1278
Bae SW, Tan W, Hong J-I (2012) Chem Commun 48:2270–2282
Bajar BT, Wang ES, Zhang S, Lin MZ, Chu J (2016) Sensors:16
Baker SN, Baker GA (2010) Angew Chem Int Ed 49:6726–6744
Baldini F, North Atlantic Treaty Organization (2006) Optical chemical sensors. Springer, Dordrecht
Ballerstadt R, Schultz JS (2000) Anal Chem 72:4185–4192
Ballerstadt R, Polak A, Beuhler A, Frye J (2004) Biosens Bioelectron 19:905–914
Banks P, Gosselin M, Prystay L (2000) J Biomol Screen 5:329–334
Baptista MS, Bastos EL (2019) In: Pedras B (ed) Fluorescence in industry. Springer, Cham, pp 39–102
Barros TC, Toma SH, Toma HE, Bastos EL, Baptista MS (2010) J Phys Org Chem 23:893–903
Bartoloni FH, Goncalves LCP, Rodrigues ACB, Dorr FA, Pinto E, Bastos EL (2013) Monatsh Chem 144:567–571
Batani G, Bayer K, Boge J, Hentschel U, Thomas T (2019) Sci Rep 9:18618
Baumler W, Abels C, Karrer S, Weiss T, Messmann H, Landthaler M, Szeimies RM (1999) Br J Cancer 80:360–363
Bechara EJ (1996) Braz J Med Biol Res 29:841–851

Benson CR, Kacenauskaite L, VanDenburgh KL, Zhao W, Qiao B, Sadhukhan T, Pink M, Chen J, Borgi S, Chen C-H, Davis BJ, Simon YC, Raghavachari K, Laursen BW, Flood AH (2020) Chemistry 6:1978–1997

Berezin MY, Achilefu S (2010) Chem Rev 110:2641–2684

Berhanu AL, Gaurav, Mohiuddin I, Malik AK, Aulakh JS, Kumar V, Kim K-H (2019) Trends Anal Chem 116:74–91

Bischof H, Burgstaller S, Waldeck-Weiermair M, Rauter T, Schinagl M, Ramadani-Muja J, Graier WF, Malli R (2019) Cell 8:492

Blay V, Tolani B, Ho SP, Arkin MR (2020) Drug Discov Today 25:1807-1821

Bloch S, Lesage F, McIntosh L, Gandjbakhche A, Liang K, Achilefu S (2005) J Biomed Opt 10:054003

Bower N (1982) J Chem Educ 59:975–977

Braslavsky SE (2007) Pure Appl Chem 79:293–465

Braun T, Kleusch C, Naumovska E, Merkel R, Csiszar A (2016) Cytometry A 89:301–308

Brenner AJ, Harris ED (1995) Anal Biochem 226:80–84

Calderon-Ortiz LK, Tauscher E, Bastos EL, Gorls H, Weiss D, Beckert R (2012) Eur J Org Chem:2535–2541

Cao Y, Koo Y-EL, Kopelman R (2004) Analyst 129:745–750

Cao D, Zhu L, Liu Z, Lin W (2020) J Photochem Photobiol C: Photochem Rev 44:100371

Cappellini MD, Fiorelli G (2008) Lancet 371:64–74

Cardosi MF (2006) In: Meyers RA, Evenson MA (eds) Encyclopedia of analytical chemistry. Wiley, Chichester

Caruso G, Fresta CG, Siegel JM, Wijesinghe MB, Lunte SM, Caruso G, Fresta CG, Lunte SM, Siegel JM, Wijesinghe MB, Lunte SM (2017) Anal Bioanal Chem 409:4529–4538

Chalmers AH, Kiley M (1998) Clin Chem 44:1740–1742

Chang CF, Chen SF, Lee TS, Lee HF, Chen SF, Shyue SK (2011) Am J Pathol 178:1749–1761

Chen C, Sun SR, Gong YP, Qi CB, Peng CW, Yang XQ, Liu SP, Peng J, Zhu S, Hu MB, Pang DW, Li Y (2011) Biomaterials 32:7592–7599

Chen T-W, Wardill TJ, Sun Y, Pulver SR, Renninger SL, Baohan A, Schreiter ER, Kerr RA, Orger MB, Jayaraman V, Looger LL, Svoboda K, Kim DS (2013) Nature 499:295–302

Cheng Z, Kuru E, Sachdeva A, Vendrell M (2020) Nat Rev Chem 4:275–290

Chi W, Qi Q, Lee R, Xu Z, Liu X (2020) J Phys Chem C 124:3793–3801

Chin J, Adams AD, Bouffard A, Green A, Lacson RG, Smith T, Fischer PA, Menke JG, Sparrow CP, Mitnaul LJ (2003) Assay Drug Dev Technol 1:777–787

Choyke PL, Alford R, Simpson HM, Duberman J, Craig Hill G, Ogawa M, Regino C, Kobayashi H (2009) Mol Imaging 8:341–354

Clark HA, Kopelman R, Tjalkens R, Philbert MA (1999) Anal Chem 71:4837–4843

Cosa G (2004) Pure Appl Chem 76:263–275

Costa CA, Trivelato GC, Pinto AM, Bechara EJ (1997) Clin Chem 43:1196–1202

Coutinho A, Prieto M (1993) J Chem Educ 70:425–428

Crapnell RD, Hudson A, Foster CW, Eersels K, Grinsven BV, Cleij TJ, Banks CE, Peeters M (2019) Sensors 19:1204

Dana H, Mohar B, Sun Y, Narayan S, Gordus A, Hasseman JP, Tsegaye G, Holt GT, Hu A, Walpita D, Patel R, Macklin JJ, Bargmann CI, Ahrens MB, Schreiter ER, Jayaraman V, Looger LL, Svoboda K, Kim DS (2016) Elife 5:12727

Dantas WFC, Duarte LGTA, Rodembusch FS, Poppi RJ, Atvars TDZ (2020) Methods Appl Fluoresc 8:04006

Das P, Sedighi A, Krull UJ (2018) Anal Chim Acta 1041:1–24

Dennis AM, Rhee WJ, Sotto D, Dublin SN, Bao G (2012) ACS Nano 6:2917–2924

Di Genova BM, da Silva RC, da Cunha JPC, Gargantini PR, Mortara RA, Tonelli RR (2017) J Eukaryot Microbiol 64:491–503

Dijkstra BM, Jeltema HJR, Kruijff S, Groen RJM (2019) Neurosurg Rev 42:799–809

Ding F, Zhan Y, Lu X, Sun Y (2018) Chem Sci 9:4370–4380

Ding F, Fan Y, Sun Y, Zhang F (2019) Adv Healthc Mater 8:e1900260
Ding S, Lyu Z, Niu X, Zhou Y, Liu D, Falahati M, Du D, Lin Y (2020) Biosens Bioelectron 149: 111830
Dubowski KM (1962) Clin Chem 8:215–235
Egan WJ, Brewer WE, Morgan SL (1999) Appl Spectosc 53:218–225
Eggeling C, Kask P, Winkler D, Jager S (2005) Biophys J 89:605–618
El Seoud OA, Baader WJ, Bastos EL (2016) Encyclopedia of physical organic chemistry. Wiley, Chichester, pp 1–68
Ergin E, Dogan A, Parmaksiz M, Elçin AE, Elçin YM (2016) Curr Pharm Biotechnol 17:1222–1230
Ernst HA (2015) Ultrafast photophysics and photochemistry after excited state intramolecular charge transfer in the liquid phase. Logos Verlag, Berlin
Fan J, Hu M, Zhan P, Peng X (2013) Chem Soc Rev 42:29–43
Farka Z, Mickert MJ, Pastucha M, Mikusova Z, Skladal P, Gorris HH (2020) Angew Chem Int Ed 59:10746–10773
Fedick PW, Schrader RL, Ayrton ST, Pulliam CJ, Cooks RG (2018) J Chem Educ 96:124–131
Fernández-Romero JM, Luque de Castro MD, Valcárcel M (1993) Anal Chim Acta 276:271–279
Fersht A (1999) Structure and mechanism in: protein science: a guide to enzyme catalysis and protein folding. W.H. Freeman, New York
Fingerhut R, Stehn M, Kohlschütter A (1997) Clin Chim Acta 264:65–73
Frank EL, Parnás ML (2020) Contemporary practice in clinical chemistry. Elsevier, Amsterdam, pp 885–894
Freitas-Dorr BC, Machado CO, Pinheiro AC, Fernandes AB, Dorr FA, Pinto E, Lopes-Ferreira M, Abdellah M, Sa J, Russo LC, Forti FL, Goncalves LCP, Bastos EL (2020) Sci Adv 6:eaaz0421
Froggatt HM, Heaton BE, Heaton NS (2020) J Virol 94:e01265–e01220
Galant AL, Kaufman RC, Wilson JD (2015) Food Chem 188:149–160
Gandorfer A, Haritoglou C, Gandorfer A, Kampik A (2003) Invest Ophthalmol Vis Sci 44:316–323
Gauglitz G, Moore DS (2014) In: Gauglitz G, Moore DS (eds) Handbook of spectroscopy, 2nd edn. Wiley-VCH, Weinheim
Gehlen MH (2020) J Photochem Photobiol C Photochem Rev 42:100338
Geissler D, Hildebrandt N (2016) Anal Bioanal Chem 408:4475–4483
Gerasimova NS, Steklova IV, Tuuminen T (1989) Clin Chem 35:2112–2115
Gilboa T, Garden PM, Cohen L (2020) Anal Chim Acta 1115:61–85
Giri S, Sykes EA, Jennings TL, Chan WC (2011) ACS Nano 5:1580–1587
Glahn-Martínez B, Benito-Peña E, Salis F, Descalzo AB, Orellana G, Moreno-Bondi MC (2018) Anal Chem 90:5459–5465
Goldman ER, Clapp AR, Anderson GP, Uyeda HT, Mauro JM, Medintz IL, Mattoussi H (2004) Anal Chem 76:684–688
Goldstein JH, Day RA (1954) J Chem Educ 31:417
Gómez-Hens A (2005) In: Worsfold P, Townshend A, Poole C (eds) Encyclopedia of analytical science, 2nd edn. Elsevier, Oxford, pp 170–178
Goncalves LCP, Trassi MAD, Lopes NB, Dorr FA, dos Santos MT, Baader WJ, Oliveira VX, Bastos EL (2012) Food Chem 131:231–238
Goncalves LCP, Di Genova BM, Dorr FA, Pinto E, Bastos EL (2013) J Food Eng 118:49–55
Gonçalves LCP, Tonelli RR, Bagnaresi P, Mortara RA, Ferreira AG, Bastos EL (2013) PLoS One 8:e53874
Gonçalves LCP, Mansouri HR, PourMehdi S, Abdellah M, Fadiga BS, Bastos EL, Sá J, Mihovilovic MD, Rudroff F (2019) Cat Sci Technol 9:2682–2688
Goncalves LCP, Mansouri HR, Bastos EL, Abdellah M, Fadiga BS, Sa J, Rudroff F, Mihovilovic MD (2019) Cat Sci Technol 9:1365–1371
Gong PW, Zhang L, Peng JY, Li SH, Chen JF, Liu XC, Peng HW, Liu Z, You JM (2020) Dyes Pigments 173:107893
Gordon SE, Senning EN, Aman TK, Zagotta WN (2016) J Gen Physiol 147:189–200

Goryacheva IY, Speranskaya ES, Goftman VV, Tang D, De Saeger S (2015) Trends Anal Chem 66: 53–62

Granick JL, Sassa S, Granick S, Levere RD, Kappas A (1973) Biochem Med 8:149–159

Grimm JB, English BP, Chen J, Slaughter JP, Zhang Z, Revyakin A, Patel R, Macklin JJ, Normanno D, Singer RH, Lionnet T, Lavis LD (2015) Nat Methods 12:244–250

Grimm JB, Xie L, Casler JC, Patel R, Tkachuk AN, Choi H, Lippincott-Schwartz J, Brown TA, Glick BS, Liu Z, Lavis LD (2020) bioRxiv 2020.08.17.250027

Guo L, Xu S, Ma X, Qiu B, Lin Z, Chen G (2016) Sci Rep 6:32755

Han V, Serrano K, Devine DV (2010) Vox Sang 98:116–123

Haritoglou C, Yu A, Freyer W, Priglinger SG, Alge C, Eibl K, May CA, Welge-Luessen U, Kampik A (2005) Invest Ophthalmol Vis Sci 46:3315–3322

Harris TK, Keshwani MM (2009) In: Burgess RR, Deutscher MP (eds) Methods in enzymology, vol 463. Academic Press, Cambridge, pp 57–71

Hatai J, Motiei L, Margulies D (2017) J Am Chem Soc 139:2136–2139

Haupt K, Medina Rangel PX, Bui BTS (2020) Chem Rev 120:9554–9582

He Y, Bai XL, Xiao QL, Liu F, Zhou L, Zhang C (2020a) Crit Rev Food Sci Nutr 16:1–21

He X, Xiong L-H, Huang Y, Zhao Z, Wang Z, Lam JWY, Kwok RTK, Tang BZ (2020b) Trends Anal Chem 122:115743

Hernot S, van Manen L, Debie P, Mieog JSD, Vahrmeijer AL (2019) Lancet Oncol 20:e354–e367

Hong JX, Xia QF, Zhou EB, Feng GQ (2020) Talanta 215:120914

Hope-Ross M, Yannuzzi LA, Gragoudas ES, Guyer DR, Slakter JS, Sorenson JA, Krupsky S, Orlock DA, Puliafito CA (1994) Ophthalmology 101:529–533

Hosseini S, Martinez-Chapa SO, Rito-Palomares M, Vázquez-Villegas P (2018) SpringerBriefs in forensic and medical bioinformatics, 1st edn. Springer, Singapore, p 1

Hotzer B, Medintz IL, Hildebrandt N (2012) Small 8:2297–2326

Howes PD, Chandrawati R, Stevens MM (2014) Science 346:1247390

Hsu CP (2009) Acc Chem Res 42:509–518

Huang T, Long M, Huo B (2010) Open Biomed Eng J 4:271–278

Huang H, Huang L, Zhao Y (2019) In: Yang R (ed) Principles and applications of up-converting phosphor technology. Springer, Singapore, pp 81–133

Huddle BP, Stephens JC (2003) J Chem Educ 80:441-443

Huggett ASG, Nixon DA (1957) Lancet 270:368–370

Ilina K, MacCuaig WM, Laramie M, Jeouty JN, McNally LR, Henary M (2020) Bioconjug Chem 31:194–213

Isarankura-Na-Ayudhya C, Tantimongcolwat T, Galla H-J, Prachayasittikul V (2010) Biol Trace Elem Res 134:352–363

Jäger S, Brand L, Eggeling C (2003) Curr Pharm Biotechnol 4:463–476

Johnson I (1998) Histochem J 30:123–140

Jori G, Brown SB (2004) Photochem Photobiol Sci 3:403–405

Jun JV, Chenoweth DM, Petersson EJ (2020) Org Biomol Chem 18:5747–5763

Kairdolf BA, Smith AM, Stokes TH, Wang MD, Young AN, Nie S (2013) Annu Rev Anal Chem 6: 143–162

Kairdolf BA, Qian X, Nie S (2017) Anal Chem 89:1015–1031

Kaneta T (2019) Chem Rec 19:452–461

Kang D, Zhu S, Liu D, Cao S, Sun M (2020a) Chem Rec 20:894–911

Kang G, Nasiri Avanaki K, Mosquera MA, Burdick RK, Villabona-Monsalve JP, Goodson T, Schatz GC (2020b) J Am Chem Soc 142:10446–10458

Karpinska J (2004) Talanta 64:801–822

Kask P, Eggeling C, Palo K, Mets Ü, Cole M, Gall K (2002) In: Kraayenhof R, Visser AJWG, Gerritsen HC (eds) Fluorescence spectroscopy, imaging and probes: new tools in chemical, physical and life sciences. Springer, Heidelberg, pp 153–181

Kattke MD, Gao EJ, Sapsford KE, Stephenson LD, Kumar A (2011) Sensors 11:6396–6410

Kaur P, Singh K (2019) J Mater Chem C 7:11361–11405

Kazmierczak SC, Robertson AF, Catrou PG, Briley KP, Kreamer BL, Gourley GR (2002) Clin Chem 48:1096–1097

Keçili R, Büyüktiryaki S, Hussain CM (2019) Trends Anal Chem 110:259–276

Kim JS, Quang DT (2007) Chem Rev 107:3780–3799

Klán P, Wirz J (2009) Photochemistry of organic compounds: from concepts to practice. Wiley, Chichester

Klockow JL, Hettie KS, La Gory EL, Moon EJ, Giaccia AJ, Graves EE, Chin FT (2020) Sens Actuators B 306:127446

Koktysh DS (2020) Mater Res Bull 123:110686

Kremers GJ, Gilbert SG, Cranfill PJ, Davidson MW, Piston DW (2011) J Cell Sci 124:157–160

Kumar V, Gill KD (2008) Basic concepts in clinical biochemistry: a practical guide. Springer, New York, pp 97–101

Kumar R, Sharma A, Singh H, Suating P, Kim HS, Sunwoo K, Shim I, Gibb BC, Kim JS (2019) Chem Rev 119:9657–9721

Kurutos A, Shindo Y, Hiruta Y, Oka K, Citterio D (2020) Dyes Pigments 181:108611

Kwok AKH, Lai TYY, Yeung CK, Yeung YS, Li WWY, Chiang SW (2005) Br J Ophthalmol 89: 897–900

Lacroix A, Edwardson TGW, Hancock MA, Dore MD, Sleiman HF (2017) J Am Chem Soc 139: 7355–7362

Lakowicz JR (2006) Principles of fluorescence spectroscopy, 3rd edn. Springer, New York

Laterza OF, Smith CH, Wilhite TR, Landt M (2002) Clin Chim Acta 323:115–120

Lavis LD (2017) Annu Rev Biochem 86:825–843

Lavis LD, Raines RT (2008) ACS Chem Biol 3:142–155

Lee JJ, Lian CG (2019) Arch Pathol Lab Med 143:811–820

Lee C, Noh J, O'Neal SE, Gonzalez AE, Garcia HH, P. Cysticercosis Working Group in, Handali S (2019) PLoS Negl Trop Dis 13:e0007746

Li Y-Q, Li X-Y, Shindi AAF, Zou Z-X, Liu Q, Lin L-R, Li N (2012) In: Geddes CD (ed) Reviews in fluorescence 2010. Springer, New York, pp 95–117

Li C, Yang Y, Wu D, Li T, Yin Y, Li G (2016) Chem Sci 7:3011–3016

Li YY, Liu SJ, Ni HW, Zhang H, Zhang HQ, Chuah C, Ma C, Wong KS, Lam JWY, Kwok RTK, Qian J, Lu XF, Tang BZ (2020) Angew Chem Int Ed 59:12822–12826

Lian X, Wei MY, Ma Q (2019) Front Bioeng Biotechnol 7:386

Liu J, Lau SK, Varma VA, Kairdolf BA, Nie S (2010a) Anal Chem 82:6237–6243

Liu J, Lau SK, Varma VA, Moffitt RA, Caldwell M, Liu T, Young AN, Petros JA, Osunkoya AO, Krogstad T, Leyland-Jones B, Wang MD, Nie S (2010b) ACS Nano 4:2755–2765

Liu Z, Jiang Z, Yan M, Wang X (2019) Front Chem 7:712

Lock JT, Parker I, Smith IF (2015) Cell Calcium 58:638–648

Luly KM, Choi J, Rui Y, Green JJ, Jackson EM (2020) Nanomedicine 15:1805–1815

Lundin K, Blomberg K, Nordstrom T, Lindqvist C (2001) Anal Biochem 299:92–97

Luo S, Zhang E, Su Y, Cheng T, Shi C (2011) Biomaterials 32:7127–7138

Luo Y, Zhu C, Du D, Lin Y (2019) Anal Chim Acta 1061:1–12

Luo F, Qin G, Xia T, Fang X (2020) Annu Rev Anal Chem 13:337–361

Macchia E, Manoli K, Di Franco C, Scamarcio G, Torsi L (2020) Anal Bioanal Chem 412:5005–5014

Maltas J, Amer L, Long Z, Palo D, Oliva A, Folz J, Urayama P (2015) Anal Chem 87:5117–5124

Manchester KL (1996) Biotechniques 20:968–970

Marcu L, French PMW, Elson DS (2014) Fluorescence lifetime spectroscopy and imaging: principles and applications in biomedical diagnostics. Routledge, Abingdon

Margulies D, Hamilton AD (2009) Angew Chem Int Ed 48:1771–1774

Margulies D, Rout B, Motiei L (2014) Synlett 25:1050–1054

Mariz IFA, Pinto S, Lavrado J, Paulo A, Martinho JMG, Macoas EMS (2017) Phys Chem Chem Phys 19:10255–10263

Martinez Calatayud J (2005) In: Worsfold P, Townshend A, Poole C (eds) Encyclopedia of analytical science, 2nd edn. Elsevier, Oxford, pp 373–383

Martins TD, Pacheco ML, Boto RE, Almeida P, Farinha JPS, Reis LV (2017) Dyes Pigments 147: 120–129

Mauzerall D, Granick S (1956) J Biol Chem 219:435–446

McCaman MW, Robins E (1962) J Lab Clin Med 59:885–890

Merola F, Levy B, Demachy I, Pasquier H (2010) Springer Ser Fluoresc 8:347–384

Middleton CT, de La Harpe K, Su C, Law YK, Crespo-Hernández CE, Kohler B (2009) Annu Rev Phys Chem 60:217–239

Milligan G (2004) Eur J Pharm Sci 21:397–405

Misra R, Bhattacharyya SP (2018) Intramolecular charge transfer: theory and applications. Wiley

Mistry B (2009) A handbook of spectroscopic data. Chemistry. Oxford Book Company, Jaipur

Mochalin VN, Shenderova O, Ho D, Gogotsi Y (2011) Nat Nanotechnol 7:11–23

Molaei MJ (2019) RSC Adv 9:6460–6481

Monici M (2005) Biotechnology Annual Review, vol 11. Elsevier, Amsterdam, pp 227–256

Moore DJ, Labbe RF (1964) Clin Chem 10:1105–1111

Moter A, Göbel UB (2000) J Microbiol Methods 41:85–112

Nagai T, Yamada S, Tominaga T, Ichikawa M, Miyawaki A (2004) Proceedings of the National Academy of Sciences of the U S A 101:10554–10559

Narayanan R, Kenney MC, Kamjoo S, Trinh TH, Seigel GM, Resende GP, Kuppermann BD (2005) Curr Eye Res 30:471–478

Newton A, Predina J, Mison M, Runge J, Bradley C, Stefanovski D, Singhal S, Holt D (2020) PLoS One 15:e0234791

Ngashangva L, Bachu V, Goswami P (2019) J Pharm Biomed Anal 162:272–285

Nissinkorn Y, Lahav-Mankovski N, Rabinkov A, Albeck S, Motiei L, Margulies D (2015) Chemistry 21:15981–15987

Noble JE, Bailey MJ (2009) Methods Enzymol 463:73–95

Noe DA, Weedn V, Bell WR (1984) Clin Chem 30:627–630

Nolting D, Gore JC, Pham W (2011) Curr Org Synth 8:521–534

O'Banion CP, Yasuda R (2020) Curr Opin Neurobiol 63:31–41

Oberbichler E, Schlogl E, Stangl J, Faschinger F, Gruber HJ, Wiesauer M, Knor G, Hytonen VP (2020) Methods Enzymol 633:1–20

Oevering H, Paddon-Row MN, Heppener M, Oliver AM, Cotsaris E, Verhoeven JW, Hush NS (1987) J Am Chem Soc 109:3258–3269

Oevering H, Verhoeven JW, Paddon-Row MN, Cotsaris E, Hush NS (1988) Chem Phys Lett 143: 488–495

Oleynik P, Ishihara Y, Cosa G (2007) J Am Chem Soc 129:1842–1843

Olusanya BO, Kaplan M, Hansen TWR (2018) Lancet Child Adolesc Health 2:610–620

Omar GS, Wilson M, Nair SP (2008) BMC Microbiol 8:111

Paal M, Lang A, Hennig G, Buchholtz ML, Sroka R, Vogeser M (2018) Clin Biochem 56:62–69

Pace CN, Vajdos F, Fee L, Grimsley G, Gray T (1995) Protein Sci 4:2411–2423

Park EJ, Brasuel M, Behrend C, Philbert MA, Kopelman R (2003) Anal Chem 75:3784–3791

Passos MLC, Saraiva MLMFS (2019) Measurement 135:896–904

Pelin JNBD, Gatto E, Venanzi M, Cavalieri F, Oliveira CLP, Martinho H, Silva ER, Aguilar AM, Souza JS, Alves WA (2018) ChemistrySelect 3:6756–6765

Pengshung M, Neal P, Atallah TL, Kwon J, Caram JR, Lopez SA, Sletten EM (2020) Chem Commun 56:6110–6113

Petryayeva E, Algar WR, Medintz IL (2013) Applied Spectroscopy 67:215–252

Piletsky SA, Piletska EV, Bossi A, Karim K, Lowe P, Turner APF (2001) Biosens Bioelectron 16: 701–707

Piletsky S, Canfarotta F, Poma A, Bossi AM, Piletsky S (2020) Trends Biotechnol 38:368–387

Pirsaheb M, Mohammadi S, Salimi A, Payandeh M (2019) Mikrochim Acta 186:231

Pitre SP, McTiernan CD, Scaiano JC (2016) ACS Omega 1:66–76

Platnich CM, Rizzuto FJ, Cosa G, Sleiman HF (2020) Chem Soc Rev 49:4220–4233
Pode Z, Peri-Naor R, Georgeson JM, Ilani T, Kiss V, Unger T, Markus B, Barr HM, Motiei L, Margulies D (2017) Nat Nanotechnol 12:1161–1168
Ponhong K, Teshima N, Grudpan K, Vichapong J, Motomizu S, Sakai T (2015) Talanta 133:71–76
Poronik YM, Vygranenko KV, Gryko D, Gryko DT (2019) Chem Soc Rev 48:5242–5265
Poschenrieder A, Thaler M, Junker R, Luppa PB (2019) Anal Bioanal Chem 411:7607–7621
Pothoulakis G, Ceroni F, Reeve B, Ellis T (2014) ACS Synth Biol 3:182–187
Prasada Rao T, Biju VM (2005) In: Worsfold P, Townshend A, Poole C (eds) Encyclopedia of Analytical Science, 2nd edn. Elsevier, Oxford, pp 358–366
Qi J, Duan XC, Liu WY, Li Y, Cai YJ, Lam JWY, Kwok RTK, Ding D, Tang BZ (2020) Biomaterials:248
Qing T, Feng B, Zhang P, Zhang K, He X, Wang K (2020) Anal Chim Acta 1105:11–27
Quina FH (1982) In: Adam W, Cilento G (eds) Chemical and biological generation of excited states. Academic Press, San Diego, pp 1–36
Raba J, Mottola HA (2006) Crit Rev Anal Chem 25:1–42
Rabek JF (1982) Experimental methods in photochemistry and photophysics. Wiley, Chichester
Ramanujam V-MS, Anderson KE (2015) Curr Protoc Hum Genet 86:17.20.11–17.20.26
Ramesh V (2019) Biomolecular and bioanalytical techniques: theory, methodology and applications. Wiley, Hoboken
Reimers JR, Hush NS (1989) Chem Phys 134:323–354
Resch-Genger U, Grabolle M, Cavaliere-Jaricot S, Nitschke R, Nann T (2008) Nat Methods 5:763–775
Reynolds J, Troy T, Mayer R, Thompson A, Waters D, Cornell K, Snyder P, Sevick-Muraca E (1999) Photochem Photobiol 70:87–94
Ribeiro JFF, Pereira MIA, Assis LG, Cabral Filho PE, Santos BS, Pereira GAL, Chaves CR, Campos GS, Sardi SI, Pereira G, Fontes A (2019) J Photochem Photobiol B Biol 194:135–139
Rimington C, Sveinsson SL (1950) Scand J Clin Lab Invest 2:209–216
Rinken A, Lavogina D, Kopanchuk S (2018) Trends Pharmacol Sci 39:187–199
Roda A, Guardigli M (2012) Anal Bioanal Chem 402:69–76
Roda A, Pasini P, Guardigli M, Baraldini M, Musiani M, Mirasoli M (2000) Fresen J Anal Chem 366:752–759
Roda A, Arduini F, Mirasoli M, Zangheri M, Fabiani L, Colozza N, Marchegiani E, Simoni P, Moscone D (2020) Biosens Bioelectron 155:112093
Rodembusch FS, Leusin FP, Medina LFC, Brandelli A, Stefani V (2005) Photochem Photobiol Sci 4:254–259
Rodrigues CAB, Mariz IFA, Maçôas EMS, Afonso CAM, Martinho JMG (2012) Dyes Pigments 95:713–722
Rodrigues ACB, Pina J, Dong W, Forster M, Scherf U, Melo JSS (2018a) Macromolecules 51: 8501–8512
Rodrigues ACB, Mariz IFA, Maçoas EMS, Tonelli RR, Martinho JMG, Quina FH, Bastos EL (2018b) Dyes Pigments 150:105–111
Rodrigues ACB, Pina J, Seixas de Melo JS (2020) J Mol Liq 317:113966
Rodriguez EA, Campbell RE, Lin JY, Lin MZ, Miyawaki A, Palmer AE, Shu X, Zhang J, Tsien RY (2017) Trends Biochem Sci 42:111–129
Rose T, Goltstein PM, Portugues R, Griesbeck O (2014) Front Mol Neurosci 7:88
Rosenthal SJ, Chang JC, Kovtun O, McBride JR, Tomlinson ID (2011) Chem Biol 18:10–24
Rudin M, Weissleder R (2003) Nat Rev Drug Discov 2:123–131
Ruppert C, Kohl M, Jacob LJ, Deigner HP (2015) Biosensors J 4:124
Sanchez Rojas F, Bosch Ojeda C (2009) Anal Chim Acta 635:22–44
Sánchez Rojas F, Cano Pavón JM (2005) In: Worsfold P, Townshend A, Poole C (eds) Encyclopedia of analytical science, 2nd edn. Elsevier, Oxford, pp 366–372
Santos FS, Ramasamy E, Ramamurthy V, Rodembusch FS (2016) J Mater Chem C 4:2820–2827

Sapsford KE, Granek J, Deschamps JR, Boeneman K, Blanco-Canosa JB, Dawson PE, Susumu K, Stewart MH, Medintz IL (2011) ACS Nano 5:2687–2699

Sauer M, Hofkens J, Enderlein J (2011) Handbook of fluorescence spectroscopy and imaging: from single molecules to ensembles. Wiley-VCH, Weinheim

Schäferling M (2016) Encyclopedia of analytical chemistry. Wiley, Chichester, pp 1–52

Schmid F-X (2001) eLS

Schneider H, Kurz G, Luppa PB (2014) In: Gauglitz G, Moore DS (eds) Handbook of spectroscopy. Wiley, Weinheim, pp 977–998

Schultz JS, Mansouri S, Goldstein IJ (1982) Diabetes Care 5:245–253

Sepe V, Distrutti E, Fiorucci S, Zampella A (2018) Expert Opin Ther Pat 28:351–364

Shah D, Guo Y, Ocando J, Shao J (2019) J Pharm Anal 9:400–405

Siddiqui MR, AlOthman ZA, Rahman N (2017) Arab J Chem 10:S1409–S1421

Sigaeva A, Ong Y, Damle VG, Morita A, van der Laan KJ, Schirhagl R (2019) Acc Chem Res 52: 1739–1749

Smith TA, Ghiggino KP (2015) Methods and Applications in Fluorescence 3:022001

Smithgall MC, Dowlatshahi M, Spitalnik SL, Hod EA, Rai AJ (2020) Lab Med 51:e59–e65

Souza MI, Jaques YM, de Andrade GP, Ribeiro AO, da Silva ER, Fileti EE, Ávilla ÉdS, Pinheiro MVB, Krambrock K, Alves WA (2013) J Phys Chem B 117:2605–2614

Souza MI, Prieto T, Rodrigues T, Ferreira FF, Nascimento FB, Ribeiro AO, Silva ER, Giuntini F, Alves WA (2017) Sci Rep 7:13166

Specht EA, Braselmann E, Palmer AE (2017) Annu Rev Physiol 79:93–117

Srivastava T, Negandhi H, Neogi SB, Sharma J, Saxena R (2014) J Hematol Transfus 2:1028

Stryer L, Haugland RP (1967) Proc Natl Acad Sci U S A 58:719

Sumner JP, Aylott JW, Monson E, Kopelman R (2002) Analyst 127:11–16

Sumner JP, Westerberg NM, Stoddard AK, Fierke CA, Kopelman R (2006) Sens Actuators B 113: 760–767

Taitt CR, Anderson GP, Ligler FS (2016) Biosens Bioelectron 76:103–112

Takahashi S, Hanaoka K, Okubo Y, Echizen H, Ikeno T, Komatsu T, Ueno T, Hirose K, Iino M, Nagano T, Urano Y (2020) Chem Asian J 15:524–530

Takaya T, Su C, de La Harpe K, Crespo-Hernandez CE, Kohler B (2008) Proceedings of the National Academy of Sciences of the U S A 105:10285–10290

Takkinen K, Zvirbliene A (2019) Curr Opin Biotechnol 55:16–22

Tarannum N, Hendrickson OD, Khatoon S, Zherdev AV, Dzantiev BB (2020) Crit Rev Anal Chem 50:291–310

Tarte K, Klein B (1999) Leukemia 13:653–663

Tavares AJ, Noor MO, Uddayasankar U, Krull UJ, Vannoy CH (2014) Methods Mol Biol 1199: 241–255

Thapaliya ER, Zhang Y, Dhakal P, Brown AS, Wilson JN, Collins KM, Raymo FM (2017) Bioconjug Chem 28:1519–1528

Thevenot DR, Toth K, Durst RA, Wilson GS (1999) Pure Appl Chem 71:2333–2348

Thompson VF, Saldana S, Cong J, Goll DE (2000) Anal Biochem 279:170–178

Thuy TN, Margaret C, Baer RC, Andy F, Chloe G, James G, Allison MD (2019) Proc SPIE:10892

Tian J, Zhou L, Zhao Y, Wang Y, Peng Y, Zhao S (2012) Talanta 92:72–77

Tian M, Ma Y, Lin W (2019) Acc Chem Res 52:2147–2157

Trachman RJ, Ferré-D'Amaré AR (2019) Q Rev Biophys 52:e8

Tseng H-M, Li Y, Barrett DA (2010) Bioanalysis 2:1641–1653

Tsien RY (1998) Annu Rev Biochem 67:509–544

Turro NJ, Ramamurthy V, Scaiano JC (2010) Modern molecular photochemistry of organic molecules. University Science Books, Sausalito

Turro NJ, Ramamurthy V, Scaiano JC (2012) Photochem Photobiol 88:1033–1033

Unger-Angel L, Rout B, Ilani T, Eisenstein M, Motiei L, Margulies D (2015) Chem Sci 6:5419–5425

Unger-Angel L, Motiei L, Margulies D (2019) Front Chem 7:243

Upstone SL (2013) In: Meyers RA (ed) Encyclopedia of analytical chemistry. Wiley, Chichester
Valeur B, Berberan-Santos MN (2012) Molecular fluorescence: principles and applications, 2nd
    edn. Wiley-VCH, Chichester
Vashist SK (2017) Biosensors 7:62
Vichapong J, Burakham R, Teshima N, Srijaranai S, Sakai T (2013) Anal Methods 5:2419–2426
Vinegoni C, Feruglio PF, Gryczynski I, Mazitschek R, Weissleder R (2019) Adv Drug Deliv Rev
    151–152:262–288
Vishwanath K, Ramanujam N (2011) In: Meyers RA (ed) Encyclopedia of analytical chemistry.
    Wiley, Chichester
Vogel HG (2008) Drug discovery and evaluation: pharmacological assays. Springer, Berlin
Vogt F (2005) In: Worsfold P, Townshend A, Poole C (eds) Encyclopedia of analytical science, 2nd
    edn. Elsevier, Oxford, pp 335–343
Volpi EV, Bridger JM (2008) Biotechniques:45. 385–386, 388, 390, 392, 394, 396, 398, 400, 402,
    404, 406, 408–409
Wagner M, Horn M, Daims H (2003) Curr Opin Microbiol 6:302–309
Wang F, Banerjee D, Liu Y, Chen X, Liu X (2010) Analyst 135:1839–1854
Wang M, Vicente MGH, Mason D, Bobadova-Parvanova P (2018) ACS Omega 3:5502–5510
Wang X, Xu M, Huang K, Lou X, Xia F (2019a) Chem Asian J 14:689–699
Wang L, Du W, Hu Z, Uvdal K, Li L, Huang W (2019b) Angew Chem Int Ed 58:14026–14043
Wang W, Kim CK, Ting AY (2019c) Nat Chem Biol 15:101–110
Wang J, Xia S, Bi J, Zhang Y, Fang M, Luck RL, Zeng Y, Chen T-H, Lee H-M, Liu H (2019d) J
    Mater Chem B 7:198–209
Wang S, Yin YPC, Song W, Zhang Q, Yang ZJ, Dong ZL, Xu Y, Cai SJ, Wang K, Yang WL, Wang
    XJ, Pang ZQ, Feng LZ (2020a) J Mater Chem B 8:803–812
Wang LG, Barth CW, Kitts CH, Mebrat MD, Montano AR, House BJ, McCoy ME, Antaris AL,
    Galvis SN, McDowall I, Sorger JM, Gibbs SL (2020b) Sci Transl Med 12:542
Wang P, Xue T, Sheng A, Cheng L, Zhang J (2020c) Crit Rev Anal Chem:1–24
Wasalathanthri DP, Rehmann MS, Song Y, Gu Y, Mi L, Shao C, Chemmalil L, Lee J, Ghose S,
    Borys MC, Ding J, Li ZJ (2020) Biosens Bioelectron
Wei J, Gong X, Wang Q, Pan M, Liu X, Wang F, Liu J, Xia F (2018) Chem Sci 9:52–61
Whitehead RD Jr, Mei Z, Mapango C, Jefferds MED (2019) Ann N Y Acad Sci 1450:147–171
Wians FH, Miller CL, Heald JI, Clark H (1988) Lab Med 19:151–155
Wilson K, Walker JM (2009) Principles and techniques of biochemistry and molecular biology, 7th
    edn. Cambridge University Press, Cambridge
Woolf J, Marsden JT, Degg T, Whatley S, Reed P, Brazil N, Stewart MF, Badminton M (2017) Ann
    Clin Biochem 54:188–198
Worsfold PJ (2005) In: Worsfold P, Townshend A, Poole C (eds) Encyclopedia of analytical
    science, 2nd edn. Elsevier, Oxford, pp 318–321
Wu X, Liu H, Liu J, Haley KN, Treadway JA, Larson JP, Ge N, Peale F, Bruchez MP (2003) Nat
    Biotechnol 21:41–46
Wu H-C, Chang X, Liu L, Zhao F, Zhao Y (2010) J Mater Chem 20:1036–1052
Wu M, Petryayeva E, Algar WR (2014) Anal Chem 86:11181–11188
Wu D, Sedgwick AC, Gunnlaugsson T, Akkaya EU, Yoon J, James TD (2017) Chem Soc Rev 46:
    7105–7123
Wu L, Li G, Xu X, Zhu L, Huang R, Chen X (2019) Trends Anal Chem 113:140–156
Wu HH, He YN, Wu H, Zhou MJ, Xu ZL, Xiong R, Yan F, Liu HM (2020a) Theranostics 10:
    10092–10105
Wu L, Huang C, Emery BP, Sedgwick AC, Bull SD, He X-P, Tian H, Yoon J, Sessler JL, James TD
    (2020b) Chem Soc Rev 49:5110–5139
Xiao D, Su L, Teng Y, Hao J, Bi Y (2020) Mikrochim Acta 187:399
Xie W, Lei L, Tian M, Zhang Z, Liu Y (2018) Analyst 143:2901–2907
Xu H, Aylott JW, Kopelman R (2002) Analyst 127:1471–1477

Xu WH, Zhang ZJ, Kang MM, Guo H, Li YM, Wen HF, Lee MMS, Wang ZY, Kwok RTK, Lam JWY, Li K, Xi L, Chen SJ, Wang D, Tang BZ (2020a) ACS Mater Lett 2:1033–1040

Xu YZ, Li CB, Xu RH, Zhang N, Wang Z, Jing XN, Yang ZW, Dang DF, Zhang PF, Meng LJ (2020b) Chem Sci 11:8157–8166

Yadav M, Aryal R, Short M, Saint C (2019) Water 11:377

Yang L, Hur J, Zhuang W (2015) Environ Sci Pollut Res 22:6500–6510

Yang M, Liu Y, Jiang X (2019) Chem Soc Rev 48:850–884

Yang LJ, Zhang CR, Liu JJ, Huang F, Zhang YM, Liang XJ, Liu JF (2020) Adv Healthc Mater 9: 1901616

Ye Y, Bloch S, Kao J, Achilefu S (2005) Bioconjug Chem 16:51–61

Yeroslavsky G, Umezawa M, Okubo K, Nigoghossian K, Dung DTK, Miyata K, Kamimura M, Soga K (2020) Biomater Sci 8:2245–2254

Yokota H (1864) Biochim Biophys Acta Gen Subj 2020:129362

Yu W, Knauer M, Kunas C, Acaroz U, Dietrich R, Märtlbauer E (2017) Anal Methods 9:188–191

Zarei M (2018) Biosens Bioelectron 106:193–203

Zezzi-Arruda MA, Poppi RJ (2005) In: Worsfold P, Townshend A, Poole C (eds) Encyclopedia of analytical science, 2nd edn. Elsevier, Oxford, pp 351–358

Zhang ZY, VanEtten RL (1991) J Biol Chem 266:1516–1525

Zhang Z, Chen Z, Wang S, Cheng F, Chen L (2015) ACS Appl Mater Interfaces 7:27639–27645

Zhang RR, Schroeder AB, Grudzinski JJ, Rosenthal EL, Warram JM, Pinchuk AN, Eliceiri KW, Kuo JS, Weichert JP (2017) Nat Rev Clin Oncol 14:347–364

Zhang XF, Wang TR, Cao XQ, Shen SL (2020) Spectrochim Acta A 227:117761

Zhao Y, Zhang Y, Lv X, Liu Y, Chen M, Wang P, Liu J, Guo W (2011) J Mater Chem 21:13168-13171

Zhao J, Zhong D, Zhou S (2018) J Mater Chem B 6:349–365

Zhao XJ, Fan ZW, Qiao YQ, Chen Y, Wang S, Yue XM, Shen TL, Liu WT, Yang J, Gao HQ, Zhan XL, Shang LQ, Yin YM, Zhao W, Ding D, Xi RM, Meng M (2020a) ACS Appl Mater Interfaces 12:16114–16124

Zhao SL, Guo X, Taniguchi M, Kondo K, Yamada S, Gu CD, Uramoto H (2020b) Anticancer Res 40:1875–1882

Zheng Q, Lavis LD (2017) Curr Opin Chem Biol 39:32–38

Zheng Y, Yang H, Wang H, Kang K, Zhang W, Ma G, Du S (2019a) Ann Transl Med 7:S6

Zheng Q, Ayala AX, Chung I, Weigel AV, Ranjan A, Falco N, Grimm JB, Tkachuk AN, Wu C, Lippincott-Schwartz J, Singer RH, Lavis LD (2019b) ACS Central Sci 5:1602–1613

Zhou J, Yang Y, Zhang C-Y (2015) Chem Rev 115:11669–11717

Zhu S, Tian R, Antaris AL, Chen X, Dai H (2019) Adv Mater 31:e1900321

Ziegler M (2013) In: Roberts GCK (ed) Encyclopedia of biophysics. Springer, Heidelberg, pp 1710–1712

Zuman P, Patel R (1992) Techniques in organic reaction kinetics. Krieger, Malabar

# High-Resolution Techniques Based on Atomic and Emission Spectrometry Applied to Bioanalytical Purposes

Rodrigo Moretto Galazzi and Marco Aurélio Zezzi Arruda

## 1    Introduction

The observation of the phenomenon of atomic emission is so old than human beings' history. In fact, it probably dates back to the first campfire, which produced a yellow color, probably seen from ancient eyes of hominoids from the vegetal material burnt, due to the emission of the Na, present in the vegetation as well as by carbene ions. Since then, history is prone to visual observations of such events as fireworks, emerging in China over 2000 years ago, or naturally visualized from sunlight, or from auroras Borealis and Australis, among others.

In terms of instrumentation for observing atomic emission phenomena, it is reported the first spectroscope developed in 1859 by Kirchhoff and Bunsen, and in 1860 Kirchhoff postulates the general law on absorption and emission of light and heat by bodies: "For a body of any arbitrary material emitting and absorbing thermal electromagnetic radiation at every wavelength in thermodynamic equilibrium, the ratio of its emissive power to its dimensionless coefficient of absorption is equal to a universal function only of radiative wavelength and temperature." (Goody and Yung 1989). In fact, this law refers to that any material can absorb radiation of the same energy ($\lambda$) in which it emits the radiation.

The first quantitative analysis for Na evaluation was developed by Champion, Pellet, and Grenier in 1873, but it is interesting to emphasize the contribution of Leme in 1918 (Giné and Arruda 2016), a Brazilian who had a seminal contribution in the field of quantitative analysis of minerals using spectrographic analysis,

R. M. Galazzi
Analytical Instrumentation Division, Analytik Jena, an Endress and Hauser Company, São Paulo, SP, Brazil

M. A. Z. Arruda (✉)
Department of Analytical Chemistry, Institute of Chemistry, University of Campinas (Unicamp), Campinas, SP, Brazil
e-mail: zezzi@unicamp.br

© The Author(s), under exclusive license to Springer Nature Switzerland AG 2022          125
L. T. Kubota et al. (eds.), *Tools and Trends in Bioanalytical Chemistry*,
https://doi.org/10.1007/978-3-030-82381-8_5

expanding the methodology for elemental determination in geological samples using a voltaic arc atomic emission spectrograph. Since then, all investigations about the interaction of atoms and light were carried out through atomic emission spectrometry. In fact, at a similar period, a diversity of analysis involving molecular absorption was developed. So, a question was in some minds: Why absorption is applied to molecules and for atoms only atomic emission is available? Then, the birth of atomic spectrometry began with the first proposal of the atomic absorption spectrometer by A. Walsh in 1955 (Walsh 1955).

In 1962, Walsh presents his idea to the Perkin-Elmer company, and in 1963 the first instrument (from 303 series) of flame atomic absorption was then commercially available. Since then, a diversity of developments and progress for both techniques (atomic and emission spectrometry) was carried out allowing the emergence of techniques such as graphite furnace (GF) (Welz and Sperling 1999), hydride generation (HG) (Dedina and Tsalev 1995), cold vapor (CV) (Hatch and Ott 1968), and glow discharge (Grimm 1968), inductively coupled plasma optical spectrometry (ICP OES) (Greenfield et al. 1964), among others.

A diversity of applications were possible with the effective development in the design of burners and atomizers, in sample introduction (with a diversity of nebulizers, and micro nebulizers), in optics (grating monochromators), in torch designs, in electronics (microprocessors to control the instrument and for the collection and processing of data) (Evans et al. 2015), and in hyphenated techniques such as electrothermal vaporization (ETV)-graphite furnace AAS (ETV-GFAAS), ETV-ICP OES, laser ablation (LA)-ICP OES, HG-ICP OES, liquid chromatography (LC)-ICP OES, among others. In fact, all these developments are reflected in the literature so that hundreds of thousands of papers are devoted to AAS or AES, and a diversity of them to bioanalytical applications (Feichtmeier and Leopold 2014; Ozbek and Baysal 2017; Limbeck et al. 2017; Brewer and Marcus 2007; Proch and Niedzielski 2020).

The revolution was so intense that allows the designs of instruments of high-resolution (HR) as HR-continuum source AAS (HR-CS AAS) for atomic absorption proposal (Becker-Ross et al. 1996), and those based on HR-inductively coupled plasma OES (HR-ICP OES) (Krachler and Carbol 2011), for atomic emission ones.

In view of the novelty of the HR techniques, this part of this chapter will cover some characteristics of both techniques, detailing their operational concepts and some applications to (bio)analytical chemistry.

## 2 High-Resolution Continuum Source Atomic Absorption Spectrometry (HR-CS AAS)

Atomic absorption spectrometry (AAS), thanks to its characteristics as fast analysis, robustness, and low cost, has stood as one of the most employed techniques for element determination in several application fields (Welz and Sperling 1999; Ferreira et al. 2018; Welz et al. 2014). During the last few years, additional information was assessed in terms of molecular absorption, multielemental determination, interference management, isotopic evaluation, among others, due to the emergence of HR techniques as HR-CS AAS (Ferreira et al. 2018; Welz et al. 2014). In fact, the development of HR instruments has expanded the applicability of atomic spectroscopy to several research areas, including (bio)analytical chemistry (Oliveira et al. 2010; García-Mesa et al. 2021).

Although the very first experiments in spectroscopy were carried out by using a continuum source, the development of such an area for atomic absorption purposes started to become a reality only from the 1970s. During this time, some researchers pointed out the need for a combination of a continuum source radiation and optics modulation since such continuum source presented a complex emission spectrum. Despite that, some limitations in the development of a suitable continuum source and the lack of HR optics delayed the progress of the CS AAS technique. After approximately 25 years of research, the Becker-Ross group postulated the requirements to make feasible the use of a continuum source for AAS (Becker-Ross et al. 1996; Welz et al. 2005, 2014).

The first commercially instrument for HR-CS AAS was released in 2004, named contrAA$^{®}$ 300 from Analytik Jena GmbH (Jena, Germany), which was available only for flame atomization and chemical vapor generation (CVG), as well as it is based on DEMON research spectrometers (Welz et al. 2005, 2014). Since then, Analytik Jena has devoted efforts to the development of the HR-CS AAS technique by releasing distinct contrAA$^{®}$ Series instruments over the years (Welz et al. 2014). In 2007, a dual atomizer instrument contrAA$^{®}$ 700 was launched to cover the whole application range of the AAS technique and beyond it. A few years later, in 2010, the contrAA$^{®}$ 600 was commercially available for GF atomization. Afterward, the most recent instrument called contrAA$^{®}$ 800, which was released in 2017, presents a possibility of three instruments in one platform, including a fully automated dual atomizer concept (Atomic Absorption, contrAA 800. Analytik Jena 2020).

The main purpose of HR-CS AAS, only provided by contrAA$^{®}$ Series instruments (Analytik Jena GmbH, Jena, Germany), is the combination of conventional AAS advantages as low-cost, simplicity, and robustness, with some benefits from ICP OES technique, as expanded spectral information, flexibility, and multielement determination, including non-metal ones through hyperfine molecular bands. In fact, due to contrAA$^{®}$ features, now, it is feasible for the determination of non-metal elements through the molecular absorption spectrometry (MAS) due to the establishment of HR-CS for absorption spectrometry (Welz et al. 2014; Atomic Absorption, contrAA 800. Analytik Jena 2020). Regarding the instrumentation, although the contrAA$^{®}$ Series in general presents a similar arrangement of the conventional AAS instruments as a source of radiation, atomizer, monochromator optics and detector, some specific modifications in contrAA$^{®}$ components result in several advantages for absorption spectrometry technique (Welz et al. 2005, 2014; Atomic Absorption, contrAA 800. Analytik Jena 2020).

In terms of the radiation source, the conventional AAS instruments comprise a line source (LS) radiation which is specific for each element to be measured. In this sense, as higher is the number of elements greater is the total number of lamps needed. In other words, and considering that ca. 70 elements could be determined by

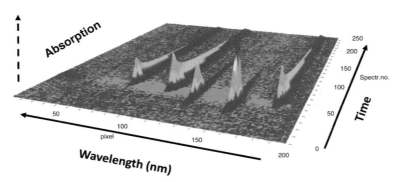

**Fig. 1** 3-D high-definition spectrum obtained from contrAA® Series instrument (Analytik Jena GmbH, Jena, Germany)

AAS, a total of 70 LS-specific lamps are required to cover the complete range of AAS. On the other hand, the contrAA® Series instruments own a single high-intensity Xe short-arc lamp which presents a CS radiation ranging from 185 to 900 nm, thus covering the whole range of interest for AAS. Moreover, the emission intensity of such CS lamps is on average a factor of 100 higher than that usually observed for conventional hollow cathode lamp (HCL) (Welz et al. 2005; Atomic Absorption, contrAA 800. Analytik Jena 2020).

As previously mentioned, for a CS it is mandatory the presence of HR optics to solve the complex emission spectrum from this radiation source and make feasible the element determination by the HR-CS technique. In this sense, contrAA® presents a double monochromator system comprised of a prism pre-monochromator for spectral dispersion and an echelle grating monochromator where the selected spectral range is highly resolved. While in such an instrument the HR is described by the optics, in the conventional AAS the resolution is given by the lamp which is remarkable less than the HR-CS AAS technique. In fact, the combination of a double monochromator in contrAA® instruments results in a resolution of 140,000 which corresponds to a spectral bandpass of 1.6 pm at 200 nm (Welz et al. 2005; Atomic Absorption, contrAA 800. Analytik Jena 2020).

In terms of the detection system, instead of a photomultiplier tube (PMT) detector as in conventional AAS, in HR-CS AAS the entire section of the highly resolved spectrum that contains the analytical line and its environment is reflected onto the detector—a linear charge-coupled device (CCD) array with 200 pixels used for analytical purposes (Welz et al. 2014; Atomic Absorption, contrAA 800. Analytik Jena 2020). In this way, more spectral information is available, and it is possible to obtain a three-dimensional (3-D) spectrum as a function of absorption × wavelength × time (Fig. 1). During the measurement, few pixels are considered to measure the analytical line while the other pixels are used for simultaneous correction purposes, such as variation in lamp emission, continuous background absorption, and to avoid spectral interferences (Welz et al. 2014; Atomic Absorption, contrAA 800. Analytik Jena 2020). Another interesting feature is that related to a dynamic mode where is possible to expand the linear range of concentration through the side pixels evaluation making possible the determination of macro and microelements at the same dilution factor, thus, reducing the time spent with sample preparation (Welz et al. 2014; Oliveira et al. 2010; Atomic Absorption, contrAA 800. Analytik Jena 2020; Heitmann et al. 2007). Finally, despite the intensity of the radiation source has no significant impact on sensitivity in AAS, the HR-CS AAS shows ca. 5 times lower detection limits compared to LS AAS, thanks to the HR optics with a CCD array present in contrAA®, thus, resulting in low noise level and significantly improved signal-to-noise ratios (Welz et al. 2014; Atomic Absorption, contrAA 800. Analytik Jena 2020).

The unique features present in contrAA® provide additional information in terms of multielement determination (fast-sequential or simultaneous) (Oliveira et al. 2010; Vale et al. 2004; Paz-Rodríguez et al. 2015; Pires et al. 2020; Cárdenas Valdivia et al. 2018), interference management (Welz et al. 2014), isotopic evaluation (Wiltsche et al. 2009; Nakadi et al. 2016), determination of non-metals through molecular absorption (Oliveira et al. 2010; Metzger et al. 2019; Welz et al. 2009; Butcher 2013), among others, therefore expanding the range of application of conventional AAS technique. In fact, the capabilities of HR-CS (AAS or MAS) have been pointed out in the literature by several studies related to distinct research fields (Welz et al. 2014; García-Mesa et al. 2021; Pires et al. 2020; Cárdenas Valdivia et al. 2018; Nakadi et al. 2016; Metzger et al. 2019; Shaltout et al. 2020; Abad et al. 2019; Pozzatti et al. 2017).

One of the main advantages of contrAA® is the multielement determination at trace level as reported by Paz-Rodríguez et al. (Paz-Rodríguez et al. 2015) in the application of HR-CS FAAS for trace elements determination in tea and tisanes. For this task, prior to the analysis, the method was validated by obtaining analytical accuracy and precision for fast-sequential multielement determination at $\mu g\ L^{-1}$ level of Ca, Co, Cu, Fe, Mn, Ni, Na, and Zn in the samples. The authors concluded it was possible to distinguish some types of tea/tisanes according to their metallic profile through a fast and low-cost method presenting high analytical performance (Paz-Rodríguez et al. 2015). Oliveira et al. (2010) developed a strategy for the determination of macro and micronutrients in plant leaves by HR-CS AAS/MAS. Taking into account the obtained results, major and minor elements were determined together in one run through the side pixel evaluation. By using this approach, the sensitivities were reduced, thus, increasing the linear working ranges for Ca, Mg, and K. This way, the determination of macro and microelements was performed using the same dilution factor resulting in a fast and simplified strategy for plant leaves routine analysis. Additionally, the wavelength integrated absorbance resulted in increased sensitivities for the determination of B, Mo, P, and S, this last one was determined based on the CS molecular absorption, in plant leaves (Oliveira et al. 2010). In fact, the determination of halogens has been extensively performed by MAS through molecular bands thanks to the distinct capabilities of contrAA Series instruments (Oliveira et al. 2010; Metzger et al. 2019; Welz et al. 2009; Butcher 2013).

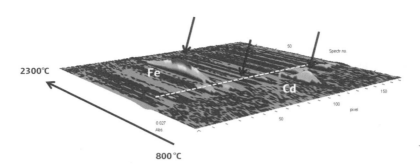

**Fig. 2** Simultaneous determination of Cd and Fe by HR-CS GFAAS in fish sample (Analytik Jena GmbH, Jena, Germany)

In terms of HR-CS GFAAS/MAS, some research papers have demonstrated the advantages of contrAA® for multielement determination in distinct samples, including via direct solid sampling (SS), thus, drastically reducing the total time of analysis (García-Mesa et al. 2021; Cárdenas Valdivia et al. 2018; Wiltsche et al. 2009; Pozzatti et al. 2017), especially when a simultaneous determination is conducted (Fig. 2). The challenges in this combined approach are those related to the development of an optimized temperature program for the elements to be simultaneously determined, and the matrix-matching between the calibration standards and the samples (Nakadi et al. 2016; Adolfo et al. 2019).

In this context, Cárdenas Valdivia et al. (Cárdenas Valdivia et al. 2018) performed a simultaneous determination of V, Ni, Fe, and Ga in fuel fly ash using SS-HR-CS GFAAS. To attain this task, the authors optimized the furnace temperature program and used Ir as a modifier for the simultaneous determination of the analytes by solid sampling. For quantitative purposes, the authors explored a calibration against aqueous standards and such strategy proved to be feasible for the determination of target elements in fuel fly ash since the absorption profile in the aqueous standards was similar to that observed in the solid samples. Analytical recoveries at RSD <10% were obtained by SS-HR-CS GFAAS indicating that such strategy can be suitable for routine analysis since it comprises a fast method where a sample preparation step and reagents are not needed, thus reducing the total cost of the analysis of ashes (Cárdenas Valdivia et al. 2018).

The SS-HR-CS GFAAS was also considered by García-Mesa et al. (García-Mesa et al. 2021) for speciation of $Zn^{2+}$ and ZnO nanoparticles (ZnO-NPs) in cosmetics. The minimization of sample matrix mineralization was achieved through an optimized graphite furnace temperature program, which allowed distinct atomization profiles between the evaluated species. In addition, the determination of $Zn^{2+}$ and ZnO-NPs in eyeshadow samples, through direct solid sampling, was performed over a validated method with an optimized calibration by aqueous standards. Afterward, the proposed strategy, with minimum sample preparation, presented reliable results for speciation of $Zn^{2+}$ and ZnO-NPs in eyeshadow samples since analytical recoveries were obtained when comparing SS-HR-CS GFAAS results, as the sum of $Zn^{2+}$ and ZnO-NPs, with those observed for total Zn digested samples. Moreover, the results show that the total concentration of Zn, as well as its species as $Zn^{2+}$ and ZnO-NPs, is dependent on the eyeshadows composition (García-Mesa et al. 2021).

An interesting investigation for Br isotope determination through monitoring CaBr transitions by isotope dilution (ID)-HR-CS GFMAS was carried out by Nakadi et al. (2016) The methodology was firstly validated, and the ID strategy was required to obtaining accurate results by overcoming Cl interferences in the direct determination of Br in complex matrices by SS. Afterward, this combined approach allowed the simultaneous determination of $Ca^{79}Br$ and $Ca^{81}Br$ since such transitions were spectrally resolved demonstrating the potential of ID-HR-CS GFMAS for direct determination of Br at mg $L^{-1}$ in solid samples (Nakadi et al. 2016).

## 3 High-Resolution Inductively Coupled Plasma Optical Emission Spectrometry (HR-ICP OES)

The use of the plasma as an emission/excitation source has been proposed in the 1960s (Greenfield et al. 1964; Wendt and Fassel 1965) and since then the ICP OES technique has been developed, thus standing out as one of the most employed techniques for elemental determination, including for bioanalytical purposes (Limbeck et al. 2017; Brewer and Marcus 2007; Krachler and Carbol 2011). In the same way of HR-CS for AAS, however, in a smaller level, the HR-ICP OES also presents some advantages in terms of analytical capabilities observed for the ICP OES technique, since the separation of the analytical lines from the interferent ones is critically dependent on the spectral resolution of the spectrometer. The main spectral interferences in the ICP OES technique are those related to a real or partial overlap of the analytical lines with two or more interfering lines, background radiation, light scattering, among others (Ticová et al. 2019; Karadjov et al. 2016). In the case of a non-resolved analytical line overlap, some strategies to solve such limitations can be considered as the selection of alternative spectral lines and mathematical corrections. Although both approaches have been extensively used to overcome some spectral interferences from the ICP OES technique, the use of HR spectrometers reflects in more confident analytical results than low-resolution ones, even those employing correction factors (Ticová et al. 2019; Karadjov et al. 2016).

As demonstrated by Ticová et al. (2019), the PlasmaQuant® (Analytik Jena GmbH, Jena, Germany) instrument which presents a resolution of 2 pm at 200 nm, showed the lowest background equivalent concentration (BEC) values, as well as it provided more reliable results than other evaluated spectrometers (Ticová et al. 2019). Moreover, as depicted by Fig. 3, it is possible to check the reality through an HR instrument by differentiating two analytical lines, e.g., Cd228.802 and As228.812. In this case, such lines can be used for analytical purposes, e.g., single element quantification and/or simultaneous determination of Cd and As since no overlap is seen between these both emission lines. For those spectrometers whose present medium or low-resolution, both lines can be partially or completely overlapped, thus requiring the selection of alternative and fewer intensity lines

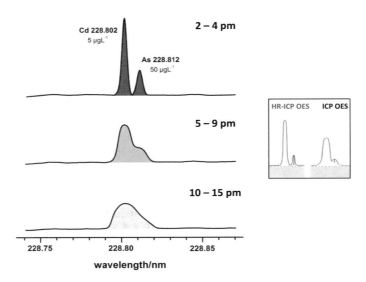

**Fig. 3** Cd228.802 and As228.812 lines at distinct spectral resolutions (Analytik Jena GmbH, Jena, Germany)

which may affect the reliability of the results, as well as the analytical performance of the instrument (Ticová et al. 2019; Velitchkova et al. 2013). Therefore, and since the higher sensitivity analytical lines are available in HR-ICP OES, this technique usually shows improved performance for trace elements determination with lower limits of detection (LOD) than conventional ICP OES (Ticová et al. 2019; Karadjov et al. 2016).

In addition to interference management resulting in greater analytical performance, the HR optics also enables some special applications for the ICP OES technique (Krachler and Carbol 2011; Velitchkova et al. 2013; Krachler et al. 2016; Zolfonoun 2015). In a study developed by Velitchkova et al. (2013), the HR-ICP OES allowed the determination of trace elements in complex environmental samples by assessing the prominent lines, free of interference, for the quantification of several analytes, without preconcentration steps (Velitchkova et al. 2013). In addition, some authors have been considered the HR-ICP OES technique for isotopic analysis, e.g., for the determination of $^6$Li and $^7$Li abundances in synthetic samples (Zolfonoun 2015), and for quality assurance through the identification of $^{234}$U as daughter product of $^{238}$Pu by accurate determination of the $^{234}$U/$^{238}$Pu age dating of Pu materials (Krachler et al. 2016).

Finally, Krachler and Carbol (2011) proposed the HR-ICP OES for isotopic analysis of depleted, natural, and enriched U samples, including scrap metal ones. The method was validated and the HR-ICP OES technique allowed the observation of isotopic splitting of the U emission signal at 411.6 nm for further quantification of $^{235}$U and $^{238}$U intensities. Afterward, the analysis of the $^{235}$U/$^{238}$U ratio by HR-ICP OES provided fast and simple reliable information on the origin of the metal specimens as such on the U enrichment, without chemical separation of the analyte. In addition, the developed strategy allows the estimation of the burn-up of nuclear fuel, since the reliable determination of $^{233}$U in spent $^{232}$Th fuel of thermal breeders was achieved. Finally, the proposed methodology can be considered in nuclear forensics for the identification of U contaminated scrap metal, as well as for the safety assessment of nuclear fuel (Krachler and Carbol 2011).

# 4    Conclusion and Future Trends

In this brief description, we can see the effort of pioneers to construct, brick by brick, the history of emission and absorption spectrometry. Such efforts did allow to reach a stage of HR spectrometers, which could be a tendency in the area, once the molecular absorption/emission can now be explored, under certain circumstances, opening an avenue in terms of applications. Then, the investigation of the role of essential elements in (bio)analytical chemistry can be thinking not only about the total content but taking into account the isotopic distribution. This is of particular importance, once an "isotopic medicine" could be explored. Additionally, in terms of instrumental chemical speciation, the coupling of different separation techniques to the spectrometers is already a reality in terms of applications, but considering HR spectrometers, the resolution is an important figure of merit to be explored in order to increase the accuracy of the methods.

Finally, from the observation of a campfire from the ancient eyes of hominoids to the recent developments nowadays, we can see no limit to the progress. In fact, we will go where the human mind takes us. Then, to infinite and beyond.

# References

Abad C, Florek S, Becker-Ross H, Huang M-D, Lopez-Linares F, Poirier L, Jakubowski N, Recknagel S, Panne U (2019) Spectrochim Acta B 160:105671

Adolfo FR, do Nascimento PC, Leal GC, Bohrer D, Viana C, de Carvalho LM, Colim AN (2019) Talanta 195:745–751

Atomic Absorption, contrAA 800. Analytik Jena. https://www.analytik-jena.com/fileadmin/content/products/contrAA/Br_contrAA800_en.pdf. Accessed 20 July 2020

Becker-Ross H, Florek S, Heitmann U, Weisse R (1996) Fresen J Anal Chem 355:300–303

Brewer TM, Marcus RK (2007) J Anal At Spectrom 22:1067–1075

Butcher DJ (2013) Anal Chim Acta 804:1–15

Cárdenas Valdivia A, Vereda Alonso E, López Guerrero MM, Gonzalez-Rodriguez J, Cano Pavóna JM, García de Torres A (2018) Talanta 179:1–8

Dedina J, Tsalev DL (1995) Hydride generation atomic absorption spectrometry, 1st edn. Wiley, New York. isbn:978-0-471-95364-7

Evans EH, Pisonero J, Smith CMM, Taylor RN (2015) J Anal At Spectrom 30:1017–1037

Feichtmeier NS, Leopold K (2014) Anal Bioanal Chem 406:3887–3894

Ferreira SLC, Bezerra MA, Santos AS, dos Santos WNL, Novaes CG, de Oliveira OMC, Oliveira ML, Garcia RL (2018) Trends Anal Chem 100:1–6

García-Mesa JC, Montoro-Leal P, Rodríguez-Moreno A, López Guerrero MM, Vereda Alonso EI (2021) Talanta 223:121795

Giné MF, Arruda MAZ (2016) Braz J Anal Chem 4:13–15
Goody RM, Yung YL (1989) Atmospheric radiation: theoretical basis, 2nd edn. Oxford University Press, Oxford. isbn:978-0-19-510291-8
Greenfield S, Jones IL, Berry CT (1964) Analyst 89:713–720
Grimm W (1968) Spectrochim Acta B 23:443–454
Hatch WR, Ott WL (1968) Anal Chem 40:2085–2087
Heitmann U, Welz B, Borges DLG, Lepri FG (2007) Spectrochim Acta B 62:1222–1230
Karadjov M, Velitchkova N, Veleva O, Velichkov S, Markov P, Daskalova N (2016) Spectrochim Acta B 119:76–82
Krachler M, Carbol PJ (2011) J Anal At Spectrom 26:293–299
Krachler M, Alvarez-Sarandes R, Rasmussen G (2016) Anal Chem 88:8862–8869
Limbeck A, Bonta M, Nischkauer W (2017) J Anal At Spectrom 32:212–232
Metzger M, Ley P, Sturm M, Meermann B (2019) Anal Bioanal Chem 411:4647–4660
Nakadi FV, da Veiga MAMS, Aramendía M, García-Ruiz E, Resano M (2016) J Anal At Spectrom 31:1381–1390
Oliveira SR, Gomes Neto JA, Nóbrega JA, Jones BT (2010) Spectrochim Acta B 65:316–320
Ozbek N, Baysal A (2017) Spectrochim Acta B 130:17–20
Paz-Rodríguez B, Domínguez-González MR, Aboal-Somoza M, Bermejo-Barrera P (2015) Food Chem 170:492–500
Pires LN, Almeida JS, Dias FS, Teixeira LSG (2020) Food Anal Methods 13:1746–1754
Pozzatti M, Nakadi FV, Vale MGR, Welz B (2017) Microchem J 133:162–167
Proch J, Niedzielski P (2020) Talanta 208:120395
Shaltout AA, Bouslimi J, Besbes H (2020) Food Chem 328:127124
Ticová B, Novotný K, Kanický V (2019) Chem Paper 73:2913–2921
Vale MGR, Damin ICF, Klassen A, Silva MM, Welz B, Silva AF, Lepri FG, Borges DLG, Heitmann U (2004) Microchem J 77:131–140
Velitchkova N, Veleva O, Velichkov S, Daskalova N (2013) J Spectrosc 505871:1–12
Walsh A (1955) Spectrochim Acta 7:108–117
Welz B, Sperling M (1999) Atomic absorption spectrometry, 3rd edn. Wiley-VCH, Weinheim. isbn:978-3-527-28571-6
Welz B, Becker-Ross H, Florek S, Heitmann U (2005) High-resolution continuum source AAS. Wiley-VCH, Weinheim. isbn:978-3-527-3036-4
Welz B, Lepri FG, Araujo RGO, Ferreira SLC, Huang M-D, Okruss M, Becker-Ross H (2009) Anal Chim Acta 647:137–148
Welz B, Vale MGR, Pereira ÉR, Castilho INB, Dessuy MB (2014) J Braz Chem Soc 25:799–821
Wendt RH, Fassel VA (1965) Anal Chem 37:920–922
Wiltsche H, Prattes K, Zischka M, Knapp G (2009) Spectrochim Acta B 64:341–346
Zolfonoun E (2015) J Anal At Spectrom 30:2003–2009

# Vibrational Spectroscopy in Bioanalysis

Mónica Benicia Mamián-López and Vitor H. Paschoal

## 1    Introduction

Implementing an analytical technique as a routine analytical tool for tasks such as assessing if a given foodstuff is suitable for human consumption or as a diagnostic tool for a disease may be challenging. Ideally, such a technique should be fast, with little or no sample processing, and cheap, regarding instrumentation and background training (user-friendliness) costs (Ellis and Goodacre 2006). Some techniques that fulfill these requirements are based on vibrational spectroscopy, such as Fourier transform infrared (FTIR) and (spontaneous) Raman spectroscopies (Ellis and Goodacre 2006). These techniques may be coupled to microscopes giving rise to IR and Raman microscopy, simultaneously correlating spatial and chemical information (Ellis and Goodacre 2006; Krafft 2012).

Both techniques went through a long journey since their discovery, and nowadays, technological advances are being made, allowing their widespread and reliability. For example, infrared (IR) radiation was discovered around 1800 by Herschel and only in 1900, Coblentz registered the first few absorption spectra of different organic compounds (Sala 2008). More recently, quantum cascade lasers were introduced as infrared sources (Beć et al. 2020). Similarly, Raman spectroscopy was discovered in 1928 by Raman and Krishnan (Sala 2008) and independently by Landsberg and Mandelstam (Singh and Riess 2001). Up to the 1950s, the usual Raman spectra were recorded using a mercury lamp arc, prisms, and photographic plates as detectors (Sala 2008). In 1962 Porto and Wood and in 1963, Porto and Kogelnik were among the first ones to use laser sources for Raman spectroscopy

M. B. Mamián-López (✉)
Centro de Ciências Naturais e Humanas, Universidade Federal do ABC, Santo André, SP, Brazil
e-mail: monica.lopez@ufabc.edu.br

V. H. Paschoal
Laboratório de Espectroscopia Molecular, Departamento de Química Fundamental, Instituto de Química, Universidade de São Paulo, São Paulo, Brazil

© The Author(s), under exclusive license to Springer Nature Switzerland AG 2022     135
L. T. Kubota et al. (eds.), *Tools and Trends in Bioanalytical Chemistry*,
https://doi.org/10.1007/978-3-030-82381-8_6

(Sala 2008). And ever since, newer approaches are being developed in the interface between several disciplines (Nafie 2020).

There were also efforts in developing new experiments beyond the linear or spontaneous regimen (Nafie 2020). In general terms, the process of focusing an oscillating electric field of frequency $\omega$ and strength $|E(\omega)|$, such as light, onto a biological sample (or other material), will result in a polarization $P(\omega)$. In general, this polarization can be expressed as power series of $|E(\omega)|$ as shown in Eq. (1) (Boyd 2020):

$$P(\omega) = \varepsilon_0 \left[ \chi^{(1)}(\omega)|E(\omega)| + \chi^{(2)}(\omega)|E(\omega)|^2 + \chi^{(3)}(\omega)|E(\omega)|^3 + \ldots \right] \quad (1)$$

where $\varepsilon_0$ is the vacuum permittivity, and $\chi^{(1)}$, $\chi^{(2)}$, and $\chi^{(3)}$ are the linear (first-order), second-order, and third-order (nonlinear) optical susceptibilities, respectively. The lowest order in this approximation, $\chi^{(1)}$, will be related to the infrared absorption or spontaneous Raman scattering (Boyd 2020).

However, there are advantages in exploring nonlinear effects or even higher-order susceptibilities. In the case of Eq. (1), where we used $|E(\omega)|$, i.e., a scalar, $\chi^{(1)}$, $\chi^{(2)}$, and $\chi^{(3)}$, are also scalar quantities. On the other hand, if the vector nature of the electric field, $E(\omega)$, is considered, $\chi^{(1)}$ becomes a second-order tensor (a matrix), $\chi^{(2)}$ a third-order tensor, $\chi^{(3)}$ a fourth-order tensor, and so on. In each case, the susceptibilities will be constants of proportionality relating the amplitude of $P(\omega)$ and the incident field, $E(\omega)$. One example would be using a high electrical field in the vicinity or between two metallic nanoparticles to increase the intensity of the spontaneous Raman scattering signal, i.e., using surface-enhanced Raman spectroscopy. Another example would be using nonlinear techniques such as vibrational sum-frequency generation (Vib.-SFG, related to $\chi^{(2)}$), which is especially useful for the investigation of proteins at interfaces.[8] Considering even higher terms, coherent anti-Stokes Raman spectroscopy (CARS, related to $\chi^{(3)}$) shows a considerably more intense spectrum than the incoherent (spontaneous) Raman phenomenon (Krafft 2012).

In this chapter, we will provide an overview of different vibrational spectroscopy techniques and their applications. From a qualitative point of view, we will first discuss the fundamentals of infrared absorption and spontaneous Raman scattering, as well as the vibrational frequencies that are expected to be observed in a typical experiment that probes the range of frequencies ($\omega$) from 600 to 4000 cm$^{-1}$. Finally, we will address how nonlinear effects can help in the context of these spectroscopic techniques, first considering surface-enhanced Raman spectroscopy and then very briefly nonlinear polarization effects in Vib.-SFG and CARS.

## 1.1  Basic Aspects of Linear and Nonlinear Techniques and Effects

There are several textbooks on the theory of vibrational spectroscopy considering linear as well as nonlinear techniques. In this chapter, we will only scratch the surface of the theme to provide a minimum background. Furthermore, we will employ mostly a classical approach instead of a quantum mechanical one to attain a broader audience. We kindly invite the interested reader to look up the references mentioned above for a more complete and thorough description (Sala 2008; Boyd 2020; Mukamel 1995; Herzberg 1956, 1963; Long 2002; Le Ru and Etchegoin 2009; Stuart 1997).

In principle, a nonlinear molecule of $N$ atoms will show *3N-6* (or *3N-5* for a linear molecule) fundamental vibrational frequencies associated with the same number of normal coordinates. Not all the normal modes will be present in the spectra due to the selection rules determined mainly by the spectroscopic technique used and by the molecule's symmetry. Each of these normal modes $Q$ represents a set of atomic displacements for which the atoms move at the same frequency with the same phase. Classically, we can assume that the normal coordinate $Q$ takes the form of the solution of the harmonic oscillator, $Q = Q_0 \cos(2\pi\omega t)$, where $\omega$ is the frequency of the mode, usually presented in $cm^{-1}$ ($33.36 \ cm^{-1} \cong 10^{12}$ Hz) and $t$ is time (Sala 2008; Herzberg 1956, 1963; Long 2002).

Suppose we want to measure the frequency of a certain vibration of a mode $v$. Considering a strictly harmonic case, in a quantum mechanical formalism, this would involve a transition from the ground vibrational state, $|v_0 >$ with energy $E_0$ to the first excited state of the same mode $|v_1>$, $E_1$. The energies of these levels are given by the well-known result of the quantum harmonic oscillator $E_n = (n + 1/2)h\omega$, where n is the vibrational level excitation, $h$ is the Planck constant, and $\omega$ would be the mode's frequency. What would be measured experimentally for the fundamental transition would be the difference between these two energy levels, $\Delta E = E_{n\,=\,1} - E_{n\,=\,0} = h\omega$. In a strictly harmonic approximation, the difference between subsequent levels, i.e., 0 and 1, 1 and 2, and so on, is simply $h\omega$. Furthermore, given the quantum mechanical selection rules, the variation of the quantum number $n$ can either be $\pm 1$, i.e., an absorption leading to an excited state or an emission leading to a less excited or the ground state. However, reality imposes itself onto the theory, and anharmonic phenomena may break these rules, resulting, for example, in the separation of levels being smaller than $h\omega$. Other anharmonic effects involve variations of $n$ other than $\pm 1$, giving rise to the harmonic bands where the variations of $n$ become, for example, $\pm 2$ (the first overtone), $\pm 3$ (the second overtone), and the combination bands, when two modes of frequency $\omega_1$ and $\omega_2$ absorb radiation simultaneously and produce a band at frequency $\omega_1 + \omega_2$ (Stuart 1997).

The above summary addresses what is being measured and some of the conditions that allow the processes. We will now dive further into the requirements for the activity of vibrational modes in IR absorption and spontaneous Raman spectroscopies. We will consider as examples of nonlinear techniques vibrational sum-frequency generation (Vib.-SFG, $\chi^{(2)}$) and Coherent Anti-Stokes Raman scattering (CARS, $\chi^{(3)}$). The schematic energy diagrams for each of these techniques are shown in Fig. 1. From a classical perspective, the absorption or emission process of radiation by an oscillating dipole will happen when the radiation frequency matches its oscillation frequency. Within this approximation, $\mu$ can be expanded as a Taylor

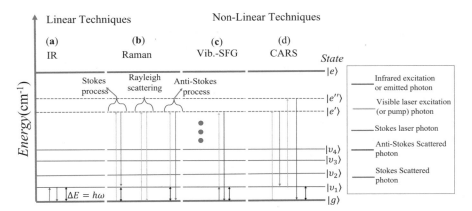

**Fig. 1** Schematic (Jablonski) energy diagram for (**a**) IR absorption, (**b**) spontaneous (or incoherent) Raman, (**c**) vibrational sum-frequency generation (Vib.-SFG) (**d**) coherent anti-Stokes Raman spectroscopy. $|g>$ denotes the ground state, $|v_1>$ the first excited vibrational level and $|v_2>$, $|v_3>$, $|v_4>$ the first four excited vibrational states of this mode. $|e'>$ and $|e''>$ denote some energy level above the ground state and different from the proper excited state $|e>$ of the molecule. The drawing spacing between the lines are not proportional

series as a function of $Q$ around the molecule's equilibrium position according to Eq. (2) (Sala 2008; Herzberg 1956, 1963; Long 2002; Stuart 1997):

$$\mu(Q) = \mu_0 + \left(\frac{d\mu}{dQ}\right)_0 Q + \ldots \tag{2}$$

where $\mu_0$ is the molecule's permanent dipole moment. From the analysis of Eq. (2), the requirements for a peak in an IR measurement to be observed, following the scheme in Fig. 1a, are: (1) the variation of the normal coordinates $Q$ related to the mode $v$ provokes a variation of the molecular dipole (a vector, $\mu$) in short, $d\mu/dQ \neq 0$; (2) the frequency of this variation matches the frequency of the radiation.

The case of Raman spectroscopy, Fig. 1b, differs from IR due to the need for an induced dipole moment ($\mu_{\text{ind}}$), generated by the electrical field of the radiation. The classical induced dipole moment can be written as $\mu_{\text{ind}} = \alpha E$, being $\alpha$ the polarizability tensor (of rank 2, i.e., a matrix) related to the species being studied. We can apply similar reasoning used in Eq. (2) for $\alpha$ as a function of $Q$. If the electrical field $E$ oscillates at frequency $\omega_0$ (with the form of $E = E_0 cos(2\pi\omega_0 t)$) and $Q$ can be written as the solution of the classical harmonic oscillator, it is straightforward to realize that $\mu_{\text{ind}}$ will have the following form (Sala 2008; Herzberg 1956, 1963; Long 2002):

$$\mu_{\text{ind}}(Q) = \alpha_0 E_0 \cos\left(2\pi\omega_0 t\right)$$

$$+ \frac{1}{2}\left(\frac{d\alpha}{dQ}\right)_0 QE_0\{coscos\left(2\pi[\omega_0 - \omega]t\right) + coscos\left(2\pi[\omega_0 + \omega]t\right)\} \tag{3}$$

where $\alpha_0$ is related to the static polarizability of the molecule. It is worth noticing in Eq. (3) what is already anticipated in Fig. 1b: there will be oscillations with the frequency of the incident radiation ($\omega_0$) and other two oscillations, one at higher frequency ($\omega_0 + \omega$, the anti-Stokes side of the spectra) and lower frequency ($\omega_0 - \omega$, the Stokes side of the spectra). It is worth mentioning that there will be a difference in the intensities between the Stokes and anti-Stokes branches, which is related to the fact that excited vibrational states (related to the anti-Stokes branch) are not as much populated in normal conditions, i.e., the system is mainly at its ground state at room conditions (Krafft 2012). The detection of the features related to the last two terms is governed by the variation of the molecule's polarizability with the variation of the coordinate $Q$ (i.e., *$d\alpha/dQ \neq 0$*).

Let us assume that we are probing the carbon-oxygen stretching mode, typical of carbonyl groups of ethers, aldehydes, to name a few, present in lipid bilayers, for example. This group has a typical frequency of *ca.* 1750 cm$^{-1}$ (Beć et al. 2020), i.e., the difference between the ground-state vibrational level and $\Delta E$ is equal to 1750 cm$^{-1}$. In the case of IR, since it is an absorption technique, the amount of energy measured in this case is enough to drive the molecule to $|v_I\rangle$. In the case of spontaneous Raman, since we are dealing with inelastic scattering, assuming an excitation line of 520 nm, this mode would be measured in absolute values of wavenumbers at *ca.* 477 nm in the anti-Stokes and 572 nm Stokes branches of the spectrum. When considering spontaneous Raman, we have dealt so far with the molecule being at the ground-state $|g\rangle$. However, the visible laser excitation could be at sufficient energy to excite the molecule to an excited electronic level, $|e\rangle$, which would lead to the resonant Raman spectroscopy, related to the intensification of specific vibrational modes. The interested reader should refer to the references for further details (Sala 2008; Herzberg 1956, 1963; Long 2002).

The order of magnitude of $\chi^{(1)}$, *i.e.*, for linear techniques, is $10^0$. On the other hand, $\chi^{(2)}$ is of the order of $10^{-12}$ and $\chi^{(3)}$ is of the order of $10^{-23}$ (Boyd 2020). Thus, for nonlinear techniques to be used, the electrical fields in Eq. (1) should be high enough. While spontaneous Raman spectroscopy benefited from lasers' brightness and monochromaticity, for the nonlinear techniques, the use of lasers is fundamental so that high-enough electrical fields are obtained (Ellis and Goodacre 2006; Krafft 2012). In this scenario, where we need strong electrical fields to overcome the increasingly small order of magnitude of the different susceptibilities, one could ask why use nonlinear methods? Aside from the information accessible by the linear techniques, i.e., information related to the inter- and intramolecular interactions and (indirectly) their dynamics, these nonlinear techniques provide the same information but with higher selectivity. There are many $\chi^{(2)}$ and $\chi^{(3)}$ experiments and others that explore even larger orders of susceptibilities. However, we will emphasize two techniques to highlight the increased selectivity that has been widely applied to biomolecules or used in a bioanalytical context (Krafft 2012; Hosseinpour et al. 2020).

The first of these nonlinear techniques that will be discussed is Vibrational sum-frequency generation (Vib. -SFG), which is mainly used to probe interfaces. Although biological processes usually can be thought of as occurring in the bulk of solutions, interfacial processes such as molecular recognition (key to immunological process) happen at a surface, not in the bulk of a solution (Davis et al., 2003). As described in Fig. 1c, if two laser pulses, one in the infrared and one in the visible regions, which are overlapped in space and time on the sample emerge as a single beam, and if the selection rules of these techniques are fulfilled, a signal will be detected with a frequency which is the sum of the frequency of the two pulses. The selection rule for a mode of a sample at the interface combines both the spontaneous Raman and IR absorption selections rules (i.e., $d\mu/dQ{\neq}0$ and $d\alpha/dQ{\neq}0$ simultaneously) (Hosseinpour et al. 2020; Kraack and Hamm 2017), which makes it selective for surfaces since this combination of rules is forbidden for centrosymmetric species in gases, liquids, and crystals, usually broken at surfaces (or interfaces). The IR and visible pulses usually have durations ranging from femto- to picoseconds (Hosseinpour et al. 2020).

The second nonlinear technique discussed is coherent anti-Stokes Raman spectroscopy (CARS), which comes as an alternative to spontaneous Raman spectroscopy, especially Raman microscopy. Spontaneous Raman experiments may yield selective and high-resolution (in the spectroscopic and spatial senses) spectra with low water interference, allowing different biological media measurements (Krafft 2012). However, the main problem with this type of measurement is typically the low probability of the inelastic Raman scattering process to occur, with one photon being scattered inelastically for every $10^3$ or even $10^6$ that interacts with the sample (Krafft 2012). Furthermore, comparing the diagram of Fig. 1b (for the Stokes process) with the diagram for a fluorescence process in Fig. 2 of chapter "UV-VIS Absorption and Fluorescence in Bioanalysis" the vibrational phenomenon may compete with eventual fluorescence emission. Indeed, autofluorescence may easily overcome the Raman signal, which can be $10^6$–$10^8$ weaker than fluorescence. A solution to the fluorescence problem may be in the diagram of Fig. 1b, in the anti-Stokes process, since fluorescence would happen at lower energies than that of the excitation laser. In a CARS experiment, represented in Fig. 1d, two lasers are directed to the same sample simultaneously, the pump laser and the Stokes laser (Krafft 2012). When the difference between the two lasers matches a vibrational mode's energy, then the CARS process occurs, and the vibrational mode signal can be observed. In this type of experiment, the molecular vibrations are coherently excited, in contrast to Raman. Given the nonlinearity and coherence, signal intensity can be much greater in a CARS experiment. There are further considerations over the geometric arrangement of the pumps and Stokes's lasers beams that need to fulfill a phase-matching condition of the incident photons and the CARS scattered photon (that comes from momentum conservation considerations) (Krafft 2012; Boyd 2020). Several different arrangements can be used. Two geometries that are useful for microscopy are the collinear geometry (one of the first demonstrations of CARS microscopy) (Zumbusch et al. 1999) and the epidetected arrangements that allow the detection of samples smaller than the pump and Stokes's wavelength lasers (Volkmer et al. 2001).

As mentioned above, many other nonlinear techniques can be applied to biological systems, such as 2D-IR, stimulated Raman spectroscopy (Liu et al.

2020; Baiz et al. 2013). These techniques also use femto- and picosecond lasers pulses, motivating time-resolved variants of these experiments can be employed (Kraack and Hamm 2017; Buhrke and Hildebrandt 2020). It is also worth noticing that the need for ultrafast lasers can make the instrumentation of nonlinear techniques tricky (Nihonyanagi et al. 2017). From a bird's-eye view, all the mentioned techniques use electromagnetic radiation to probe a vibrational mode. For IR, a classical source would be a heated solid thermal emission spectrum (Stuart 1997), although nowadays, tunable diode and quantum cascade lasers are available (Beć et al. 2020). In the case of Raman, monochromatic laser emissions are used for the measurements (Ellis and Goodacre 2006). More recently, for both techniques, synchrotron light sources are available (Rossi et al. 2020; Martin et al. 2013; Hackett et al. 2013). This intense light source is used to excite the analytes in the sample, and the variation of the amount of light absorbed (in the case of IR absorption) or inelastically scattered (in the case of spontaneous Raman) is the property measured in each experiment. After interacting with the sample, the radiation goes to an interferometer (specifically in the case of FTIR) or a monochromator (spontaneous Raman).

IR spectrometers follow two main designs, starting with a source and ending at a detector: either dispersive, where a monochromator is used to select the frequency to be probed, or interferometric, where an interferometer will work over the whole spectral range in a single shot (Beć et al. 2020). The dispersive design used up to the mid-1940s was based on a prism, and only in the 1950s gratings were introduced. In the interferometric design, on the other hand, the actual frequency information comes from a Fourier transform over a time-domain interferogram (Beć et al. 2020). Although it is usually regarded that the IR spectroscopy is severely impaired by water, this can be overcome to some extent by performing the measurement in the so-called attenuated total reflection (ATR) configuration (Beć et al. 2020). It is worth stressing that the spectra measured in an IR absorption and ATR-IR (or ATR-FTIR) experiments may differ given that in the latter, the band shapes are related to the dispersion of the refractive index of the sample. Similarly to IR, Raman spectrometers can also be dispersive or interferometric. However, Raman spectrometers will rely on lasers sources irrespectively of their setup (Coates 1998; Levin and Lewis 1990).

After this brief description of the techniques, we address what type of modes are detected in the mid-frequency and high-frequency ($600-4000$ $cm^{-1}$) range of usual Raman or IR measurements. While the idea of a normal coordinate that would allow visualizing which atoms are involved in a specific vibration is handy for small molecules, in the case of a large protein, for example, it may not be straightforward to obtain such a normal coordinate or even to provide a clear visualization of the motions taking place. In these situations, based on some previous chemical knowledge of the species she or he is studying, it is possible to interpret the spectra based on the characteristic vibrational frequencies of given chemical groups, often provided in tables such as Table 1 (Krafft 2012; Beć et al. 2020; Stuart 1997; Türker-Kaya and Huck 2017; Movasaghi et al. 2007).

**Table 1** Examples of characteristic frequencies observed in IR and Raman spectra and the related chemical groups

| Frequency range (cm$^{-1}$) | Examples of biochemical species |
|---|---|
| *IR* | |
| 3200–3500 | O-H or N-H streches in carbohydrates, proteins, alcohols, etc. |
| 2800–3020 | C-H stretches in carbohydrates |
| 1630–1660 | C=C stretch in phenolic compounds |
| 1600–1620 | Amide I (C=O stretch) in proteins |
| 1540–1560 | Amide II (C=N and N-H stretches) in proteins |
| 1455–1460 | Amide III (aromatic C-C vibrations) in proteins |
| 1370–1380 | C-H sym. bending of CH$_2$ and CH$_3$ of lipids and proteins |
| 720–1120 | Rocking CH$_2$ of proteins |
| 860–900 | CO ring stretch, CCC in plane ring bending and ring deformations in polysaccharides |
| *Raman* | |
| 3200–3500 | O-H and N-H vibrations |
| 2850–2965 | CH$_2$ and CH$_3$ sym. and asym. stretches in lipids |
| 2550–2580 | S-H symmetric stretch in methionine |
| 1600–1620 | Amide I (C=O stretch) in proteins, or ring deformations in aromatic compounds |
| 1540–1560 | Amide II (C=N and N-H stretches) in proteins |
| 1455–1460 | Amide III (aromatic C-C vibrations) in proteins |
| 790–1200 | Phosphodiester bands in DNA and RNA |
| 760–780 | Ring deformations in nucleobases |
| 620–650 | C-C twisting modes in proteins |

*sym.* symmetrical, *asym.* asymmetrical

We focus on this frequency range for two main reasons. The first one is that modes in the near-infrared region of the spectra ($>4000$ cm$^{-1}$), probed by near-infrared spectroscopy (NIRS), usually are related to combination bands and harmonic bands that are usually broad and less specifically relatable to a chemical group, which is usually not an issue for quantification of substances and classification of samples, but lacks specificity (Ellis and Goodacre 2006). The low-frequency range, on the other hand, especially for crystalline samples, shows features that are related to collective modes (phonons), which are specific to crystal phases and sensitive to impurities and defects. While these are very important information, the need for specialized instrumentation to reach the low-frequency range might not allow the application of this technique as a cheap, routine experiment in spite of recent results showing the analytical capabilities of reaching the THz range (Baxter and Guglietta 2011).

The frequencies compiled in Table 1 should not be considered as absolute values but as typical frequencies where these modes are observed. It is also worth mentioning that changes in the sample, e.g., from a crystalline solid to a solution or protein denaturation, may result in frequency shifts and other spectroscopic modifications.

### 1.1.1 Nonlinear Effects: Surface-Enhanced Raman Spectroscopy

In the previous section, we discussed the basics of linear and the advantages of nonlinear techniques related to high-intensity electrical fields considering the higher-order terms of Eq. (1) (Boyd 2020). An alternative can be found exploring plasmonic effects. These effects arise from strong light-matter couplings, which in its turn leads to enhanced linear and nonlinear optical effects, especially in the case of nanostructured noble metal particles. Our focus in this section is to outline the basics of plasmonic effects and their interplay on enhancing the spontaneous Raman signal resulting in the so-called Surface-Enhanced Raman Spectroscopy, or SERS for short. There are other signal intensification approaches in different experiments that rely on plasmonic effects, for example, SEIRAS (Surface-Enhanced Infrared Absorption Spectroscopy) (Le Ru and Etchegoin 2009; Neubrech et al. 2017), SEFS (Surface-Enhanced Fluorescence Spectroscopy) (Le Ru and Etchegoin 2009; Li et al. 2017), TERS (Tip-Enhanced Raman Spectroscopy) (Le Ru and Etchegoin 2009; Zhang et al. 2016), Enhanced CARS (Ichimura et al. 2004), to name a few. However, we will focus on SERS, given its importance in bioanalytical applications.

The relationship between plasmonics and SERS is mutually beneficial. On the one hand, the continuous development and characterization of different substrates lead to better SERS applications. On the other, the work done by spectroscopists attempting to probe the plasmonic effects through the intensification of modes with different molecules and conditions may point to relevant characteristics of the substrates as well as challenge mechanisms and hypotheses behind the theories of the intensification of the Raman scattering and the interactions between molecules and nanostructures (Le Ru and Etchegoin 2009; Langer et al. 2020).

Although some recent works show the possibility of using semiconductor nanoparticles to enhance the Raman signal (Ji Zhao and Ozaki 2016), noble metal nanoparticles, especially coinage metals such as gold, silver, and copper, are usually employed for SERS substrates (Le Ru and Etchegoin 2009). The common trait between these metals is the presence of almost free electrons in the conduction bands. The connection between these electrons and the plasmonic effects is not necessarily straightforward, but it can be qualitatively understood within the context of the Drude model. Shortly, the consequences of this model within the context of SERS using visible excitation sources are that these metals show values that are small, positive, values for the imaginary part of the wavelength (or conversely, the frequency) dependent dielectric coefficient, $\varepsilon(\lambda)$, ($Im[\varepsilon(\lambda)] = \varepsilon''(\lambda)$), and negative values for the real part of ($Re[\varepsilon(\lambda)] = \varepsilon'(\lambda)$) of $\varepsilon(\lambda)$, the latter being crucial for plasmon effects (Le Ru and Etchegoin 2009). We illustrate the usual dependence of $\varepsilon''(\lambda)$ and $\varepsilon'(\lambda)$ for gold in the panels (a) and (b) of Fig. 2 (Babar and Weaver 2015). Given all the exposed above, one question remains, what are the plasmons in plasmonics?

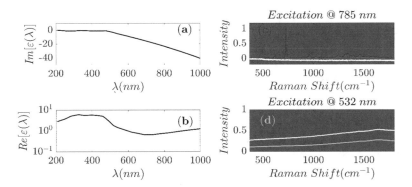

**Fig. 2** The imaginary ($Im[\varepsilon] = \varepsilon$") and real ($Re[\varepsilon] = \varepsilon$') parts of the dielectric function of gold are shown in panels (**a**) and (**b**). Panels (**c**) and (**d**) illustrate the SERS (red curves) intensification outside and close to the plasmon resonance region for adenine in comparison to their spontaneous Raman. $\varepsilon$" and $\varepsilon$' were taken from reference (Babar and Weaver 2015)

Similarly, plasmons are a type of collective excitation in many-body systems, as in the case of phonons mentioned in the previous section. In these types of excitations, the system displays a collective, wavelike motion that is not centered around a single atom or molecule but is spread throughout the sample. Briefly (and not rigorously), phonons are collective vibrations observed in a crystal. Plasmons naturally arise from our depiction of the metals mentioned above, where they present almost free electrons subject to a periodic potential due to the nuclei of the metals. In this context, "almost free" implies gas-like behavior, and from this, a charged gas-like system resembles a plasma (Le Ru and Etchegoin 2009). When an oscillating electrical field of high-enough energy, such as a laser, hits a thin metallic foil, it may induce oscillations of frequency $\omega_P$ on the electronic density (the number of electrons per site) in the sample (Boyd 2020). This is known as a plasma wave, and like photons have energy $\hbar\omega$, the excitations associated with the collective plasma waves, plasmons, have energy $\hbar\omega_P$ (Boyd 2020).

In the scenario above, the coupling between the plasmon fluctuations and the radiation's electric field in a thin film gives rise to the so-called surface plasmon-polariton, polariton being the collective excitation associated with the coupling of the oscillating electric field and the plasma oscillations. Instead of a thin foil, suppose a sphere of dimensions much smaller than those of the laser's excitation wavelength (Le Ru and Etchegoin 2009). In this case, we can represent the system as a sphere inside a dielectric continuum embedded in a uniform electric field. In this framework, using the classical electrostatic toolset and assuming our Drude model for the description of the complex frequency dielectric function, one can show that the absorption and scattering coefficients of these systems, we will show resonances, i.e., at specific frequencies the optical response properties of the particle will be large. The above scenario describes the case where we will observe a localized surface plasmon resonance (LSPR) (Le Ru and Etchegoin 2009). Furthermore, considering visible laser excitations, we can promptly understand the need for nanostructures. In this resonance condition, the electric field will be enhanced, and therefore, the Raman signal will be intensified. This intensification will be largest at the vicinity of the surface of the particle (Le Ru and Etchegoin 2009).

A phenomenon that can enhance even further both the electric field strength and the Raman signal is when we consider the region between two particles, given an appropriate distance between the two surfaces, then the so-called hotspot can be formed. The enhancement of the emitted field can be up to $10^3$ in the hotspot compared to a single particle's surface (Le Ru and Etchegoin 2009). However, the hotspots' contribution to the overall signal is small given the low probability of one hotspot being probed during a measurement (Zong et al. 2018). Furthermore, spite the enhancement factor within a hotspot is such that the single-molecule regimen can be reached, peak intensity fluctuations, "blinking," frequency shifts, among other effects, can turn SERS analysis challenging (dos Santos et al. 2019). These difficulties have led to the development of different measurement protocols and highlight the importance of chemometric tools for the correct assessment of experimental data, either as an effort better to understand the SERS phenomenology (dos Santos 2020), or as an applied strategy for diagnosis (Almehmadi et al. 2019). The mechanism exposed so far mostly deals with the so-called electromagnetic enhancement. Another possibility that is highly dependent on the substrate's nature and of the molecule being probed is the chemical mechanism that relies on charge transfer between molecule and substrate. The importance of these two mechanisms has been recently discussed in the literature (Langer et al. 2020).

We illustrate the SERS effect on a solution of adenine of concentration 1 $10^{-5}$ mol $L^{-1}$ using excitations of 785 and 532 nm in panels (c) and (d) of Fig. 2. In panels (c) and (d), the spectra in white correspond to the pure adenine solution and the spectra in red (panel (c)) and green (panel (d)) are due to mixtures between adenine and gold nanoparticles. As it can be noticed, the panel (c) shows a more intense spectrum for the SERS case, under the same conditions, and no bands can be seen for the pure solutions. The relationship between the excitation wavelength and the enhancement condition is highlighted in panel (d). No intensification is observed at 532 nm, even though the absorption spectrum shows a maximum at this wavelength. This is related to the so-called dark plasmon modes discovered and demonstrated by Schatz and Van Duyne (Langer et al. 2020).

Just as a side note, Van Duyne was one of the first researchers along with Jeanmarie to report the SERS phenomena in 1977 (Jeanmaire and Van Duyne 1977). The first report of this phenomenon was published by Fleischmann, Hendra, and McQuillan in 1974 (Fleischmann et al. 1974).

The main take-away message, especially in the context of SERS, is that these plasmonic effects are affected by the chemical nature of the metal, its shape (i.e., if we are dealing with nanospheres, nanorods, nanoprisms, nanopyramids, among other shapes), the size of these particles and the necessity of nanostructured substrates, what is the wavelength of the laser excitation, and where this particle is embedded (that is, if it is a solid substrate or a colloidal suspension in some solvent and what that solvent is), that is, the nature of the substrate. We will explore this last aspect further.

Whether for fundamental studies focusing on plasmonics or applied studies trying to quantify and detect analytes, the efforts in obtaining substrates have the same goals: reproducibility, uniformity of geometries and size for the nanoparticles (Langer et al. 2020). The reproducibility of the synthesis procedure ensures reproducibility of measurements and enables these substrates' applications. As suggested from the exposed characteristics of LSPR excitations and the SERS phenomena, uniform geometries and sizes ensure comprehensive optical properties, and within the context of SERS, a comprehensible intensification behavior (Langer et al. 2020). Bearing these characteristics in mind, colloidal nanoparticles in suspension, especially aggregated colloidal nanoparticle suspensions, are among those substrates that deliver the best cost-benefit ratio (Langer et al. 2020). Although colloidal nanoparticle suspensions were not the first substrate used to study SERS (Van Duyne and Jeanmarie (Jeanmaire and Van Duyne 1977) and Fleischmann et al. (Fleischmann et al. 1974) carried out studies on roughened silver electrodes), this type of substrate was used in the first efforts to reach the single-molecule regimen (Le Ru and Etchegoin 2009). There are several different protocols for the synthesis of metallic nanoparticles, to name some of the most common, the Lee-and-Meisel (1982), and the Turkevich-Frens methods (Turkevich et al. 1951), both to obtain *quasi*-spherical (gold or silver) nanoparticles by using these metals salts and then reducing them to their metallic form with sodium citrate, the main difference between these methods being the size of the nanoparticles, the stability and hotspot distribution of the aggregate forms of the colloids (Mamián-López and Temperini 2019). There are other types of reducing agents that can be used to speed up or slow down the reduction process, surfactants, and two-step approaches where a nanoparticle seed is used to induce particle nucleation and growth aiming to control the dispersion of nanoparticle sizes (Le Ru and Etchegoin 2009). However, the classical synthesis methods usually show good performance and reliability. Surfactants and seeded procedures may also be used to synthesize anisotropic particles such as nanorods (Lohse and Murphy 2013), nanostars, or even more exotic shapes (Langer et al. 2020).

This type of substrate can be applied as a colloidal suspension. However, this suspension can be dried out to generate 2D-like substrates, *i.e.*, with a 2D arrangement of the hotspots, contrary to the 3D arrangement of metallic nanoparticle aggregates (Le Ru and Etchegoin 2009). Alternatively, solids substrates can be grown using techniques such as lithography or other microfabrication techniques, and in general, reproducibility and size dispersion are far-well better controlled within this context. However, the fabrication of such substrates is considerably more complicated and expensive than the bottom-up approach (Langer et al. 2020). Other efforts to control the arrangement of hotspots is through supporting the nanoparticles in different materials, *e.g.*, paper (Hong et al. 2017), biopolymers (Langer et al. 2020), or aluminum (Silveira et al. 2019).

The specificity and stability can be improved through different strategies revolving around surface modification. The nanoparticle's surface can be modified with molecules that work as labeling agents that are linked to the metal surface of these materials. The resulting material is called a SERS nanotag, which can be made, for example, from a gold nanoparticle and a modified antibody that is able to attach to its surface (Langer et al. 2020; Zhang et al. 2018). This type of device is suited bioimaging in vivo or ex vivo and as biosensors ranging from metals to pathogens (Zhang et al. 2018). Another improvement that comes from surface modifications is when the metallic nanoparticles' surface is covered by a thin, transparent layer (shell) of an oxide, for example, $SiO_2$. This type of modification prevents aggregation and direct exposition of the nanoparticle surface to the molecule or environment being probed, the resulting application of these modified materials is referred to as SHINERS, shell-isolated nanoparticle-enhanced Raman spectroscopy (Krajczewski and Kudelski 2019).

The synthesis of substrates lives in the interface between plasmonic fundaments (which has already spread out through different areas of knowledge) and spectroscopy. This is an essential aspect of the cycle mentioned above that moves the understanding of SERS phenomenology and drives plasmonic substrate and device design efforts. Since this is a broad topic, the interested reader is encouraged to seek the relevant literature on the topic (Le Ru and Etchegoin 2009) and a recent review of the current status of the field (Langer et al. 2020).

## 2 Bioanalytical Applications

Various instrumental techniques that have accompanied biomedical and bioanalytical research, such as chromatography, nuclear magnetic resonance, mass spectrometry, among others, have their limitations in two main situations of remarkably high interest and significance in this field. Firstly, acquiring images when high spatial resolution is not a typical characteristic of those techniques, and secondly, in many cases, they fail to allow in vivo studies. Hence, we could say that in these aspects, vibrational spectroscopy has found one of its main strengths as a bioanalytical tool. We would add one additional and remarkable advantage: minimal sample preparation is needed for most of the analysis.

Nevertheless, besides reaching this level of detail of the systems under study, vibrational spectroscopy techniques have also shown to be very versatile, suitable to be exploited for solving a wide variety of problems, from very fundamental to in vivo studies. Examples of applications range through quality control of foods and pharmaceuticals, biomarkers sensing, dynamic processes, imaging, and diagnosis.

Performing a unique or absolute classification of all bioanalytical applications is not a simple task, mainly due to the complexity and variability of most biological systems and because designing a bioanalytical protocol requires to consider, all the dynamics involved, i.e., biochemical processes occurring in the "background" of each sample under study.

This section aims to outline relevant applications of the techniques explained in the previous section, emphasizing their strength, also depicted in Fig. 3. We would like to emphasize that although a univariate approach can be used for bioanalytical data processing, much of the improvement has been made in this field thanks to

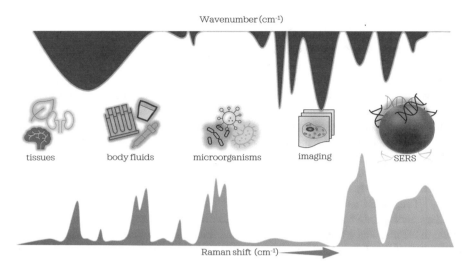

**Fig. 3** Main research subfields applications of vibrational spectroscopies discussed in this section

multivariate methods. We encourage the reading of chapter "Chemometrics in Bioanalytical Chemistry" to learn more about this topic.

## 2.1 Infrared

IR spectroscopy is a very mature technique for analytical purposes in general. Near-Infrared spectroscopy (NIRS), for example, had its very first analytical applications around 1940–1950 (Ellis and Bath 1938; Evans et al. 1951), and nowadays has become an attractive technique that offers non-invasive, non-destructive, precise, and rapid possibilities of analysis.

### 2.1.1 Structural and Fundamental Studies

As chemical specificity is one of IR spectroscopy's main strengths, searching for structural information through this technique is one of the main objectives. Currently, there is plenty of information about typical IR bands in the literature, including those related to biochemical species, as we presented in Table 1. It is worth mentioning that those frequencies in biomolecules may have slightly different wavenumber values because the associated groups can be affected by their chemical surroundings, and significant overlap between modes can also be observed—furthermore, the influence of the complexity of the biological matrix can cause frequency shifts. Thus, better identification of biomolecules is made from a univariate approach when well-resolved, non-overlapped bands are selected. An example is the case of isolated DNA nucleobases, with typical peaks at $1050$–$1070$ $cm^{-1}$ (C-O-C stretch), $1100$–$1040$ $cm^{-1}$ (sym. stretch of phosphate groups of phosphodiester linkages), 1220–1240 cm$^{-1}$ (asym. stretch PO$_2^-$), 1666 and 1717 cm$^{-1}$ (C=O stretch of pyrimidine and purine bases, respectively). The chemical structure of lipids, another quite common kind of biomolecules, are identifiable through the bands centered at 1600–1800 cm$^{-1}$ (C=O stretch), 2852 (sym. CH$_2$ stretch in acyl chain), 2874 cm$^{-1}$ (CH$_3$ sym. stretch in acyl chain), 2922 cm$^{-1}$ (CH$_2$ asym. stretch in acyl chain) and 2956 cm$^{-1}$ (CH$_3$ asym. stretch in acyl chain).

One of the most challenging problems in fundamental research associated with bioanalysis relies on the understanding of the hydrogen bonding (H-bonding) nature, a problem that has captivated interest in research for more than 100 years now (Goymer 2012), and the understanding of this type of interaction is still being modified to the point that recently two more elements, Se and S, start to be considered taking part of this interaction. This has been addressed through IR absorption, taking advantage of its typical effect on spectra, i.e., significant frequency shifts and intensity variations (Fornaro et al. 2015).

Mundlapati et al. (2017) used IR-UV double resonance spectroscopy along with quantum chemistry calculations to study the strength of N − H···Se H-bonds in proteins and other biomolecular systems, in a search for evidence about their strength, when compared to amide−N − H···O H-bonds in nucleobases, for example. It was found that amide−N − H···Se H-bonds are as strong as classical amide−N − H···O and amide−N − H···O=C H-bond. This kind of structural findings is useful for example, for protein design, either from the sequence and structure of a known protein or through de novo design (completely from scratch) (Yang et al. 2021).

The vibrational properties of proteins have also been studied with vibrational spectroscopy, exchanging hydrogen by deuterium, Brielle and Arkin (2018), measured the impact of the hydrogen exchange through frequency shifts in the vibrational mode of amide I of viral (influenza A and SARS coronavirus) membrane peptides in its native environment. Other interesting examples related to structural studies in biomolecules, using IR and Raman spectroscopies, can be found in the review by Wiercigroch et al. (2017), devoted to carbohydrates and in the work of S. Olsztyńska-Janus et al. (2018)

## 2.1.2  Biological Fluids and Biomarkers

The analysis of biological fluids deserves special considerations, given the intrinsic complexity and variability of this kind of analytical matrix and due to its uncontestable impact on our life (Lovergne et al. 2016). Fluids commonly used in diagnosis and bioanalysis can be classified from the point of view of their accessibility since sampling and sample preparation are critical stages when analytical methodologies are designed (see chapter "Sampling and Sample Preparation in Bioanalysis"). Samples such as urine, blood, saliva, and tears are acquired in a minimal or not invasive way, drastically simplifying the overall procedure. Less commonly analyzed fluids whose sampling is invasive include synovial, amniotic, cerebrospinal fluid, and bile. Furthermore, these are analyzed only when a very specific biomarker or signature is sought.

One of the high-interest species analyzed via IR is glucose, mainly due to the well-known fact that changes in its levels in the blood can significantly affect our quality of life. Researchers are continuously pursuing a methodology that allows continuous and non-invasive monitoring (Yadav et al. 2015) to replace invasive or minimally invasive methods, for example, fingerstick and subcutaneous biosensors, and Raman and IR spectroscopy have played a significant role here. NIRS has shown somewhat promising results with some advantages such as high sensitivity and low cost. However, various variables, including temperature, humidity, skin hydration, and carbon dioxide presence, can interfere with the measurement, bringing up common problems found in such a complex analytical matrix analyzed via IR absorption.

Lately, molecular analysis of blood samples has started to show promising results for non-invasive tests of neurodegenerative disorders. However, contrary to vibrational spectroscopies, these tests are time-consuming and costly therefore not suitable for clinical or routine testing. In the work of Paraskevaidi et al. (2018a) NIR was applied to the analysis of blood plasma for diagnosis of Alzheimer's disease (AD). The method was minimally invasive, with a low complexity sample preparation procedure. They reached figures as high as 92%, 87.5 and 96.1% of accuracy, sensitivity, and specificity, respectively. Moreover, it was found through multivariate analysis that most of the spectral features that allowed discrimination of AD patients were related to changes in proteins that could be related to the most studied biomarkers for this disease (building-up of Aβ plaques and neurofibrillary tangles, primarily consisting of tau protein).

In a recent work by Yu et al. (2017) FTIR was used to analyze urine, another widely studied biomarker reservoir where vibrational spectroscopy has also been entered as an excellent option for analysis (Paraskevaidi et al. 2018b; Takamura et al. 2019; De Bruyne et al. 2018). In this case, three potential renal biomarkers were found in dried urine samples, specifically for inflammatory and crescentic glomerulonephritis (GN). The bands located at $1668 \text{ cm}^{-1}$, $1545 \text{ cm}^{-1}$ were assigned to amide I (probably from urea and creatinine), amide II (from peptide bonds of urinary protein), respectively, and a peak centered at $1033 \text{ cm}^{-1}$, with a not straightforward assignment due to the high overlapping.

Following the search for biomarkers, studies of salivary fluid have shown that this easily accessible fluid, rich in carbohydrates, proteins, lipids, vitamins, not to mention its DNA content, is a very promising tool for biomedical diagnosis (Paluszkiewicz et al. 2020; Sun et al. 2015). Hormones, immunoglobulins, mucins, enzymes are among the most common and useful biomarkers in saliva. Ferreira et al. (2020), using ATR-FTIR on saliva samples, showed it was possible to differentiate breast cancer patients from those with benign disease and healthy or control patients. They found two potential salivary biomarkers bands, the first one located at $1041 \text{ cm}^{-1}$ assigned to the $PO_2^-$ symmetric stretching mode (from glycogen and nucleic acids) and the second one corresponding to the $COO^-$ symmetric stretching mode (lipids and proteins). It is worth mentioning that centrifugation (for sample recovering) and lyophilization were the only sample processing required.

To illustrate how IR is an optimal tool to study fluids that are not easily accessible, Graça et al. (2013) made the metabolic profiling of amniotic fluid using

ATR-FTIR, analyzing the mid-IR frequency range, allowing the identification of modifications associated with fetal malformations (FM) and preterm delivery (pre-PTD). The authors proposed assignments of the MIR spectrum of amniotic fluid associated with each disorder studied. Furthermore, the loadings from PLS-DA analysis showed the main bands at 1170, 1020, and 990 cm$^{-1}$ (related to C–C and C–O stretches in glucose and NH aromatic vibrations) associated with FM and the band located at 1039 cm$^{-1}$ (due to C–C and C–O from α-oxoisovalerate, citrate, glucose, alanine, tyrosine, among others), related to pre-PTD. Details about studies involving these not so common (yet) fluids can be found, for example, in work by Yonar et al. (2018)

### 2.1.3   Single Living Cells and Tissues

Contrary to biological fluids, spatial information becomes a priority in most cases when dealing with living cells and tissues. For such samples, the possibility to construct chemical images of undisturbed living systems or even in their native habitat may provide helpful insights (Fang et al. 2017). To obtain these chemical images, the appropriate experimental methodologies and equipment must be used, *i.e.*, proper sample conditioning and the availability of a microscope coupled to the IR spectrometer. This does not mean that this is the only way of extracting relevant or useful information from this kind of sample through a microscopic resolution technique, as reported previously (Boydston-White et al. 2005). Depending on the goals, samples can be analyzed without consideration of the spatial information, for example, as shown by Fale et al. (2015), where ATR-FTIR was used to study the response of living cells to doxorubicin treatment. Here, three different cell lines were treated with this anticancer drug, and spectral analysis could track chemical changes during the treatment. Three different cell lines were treated with the anticancer drug doxorubicin, and spectral analysis allowed for chemical changes during the treatment. HeLa cells, for example, were monitored for 24 h, and significant changes were observed in the bands located at 1645, 1550, and 1085 cm$^{-1}$, associated with amide I and II, and phosphate mainly from the DNA backbone, respectively. Overall, the procedure showed in this work demonstrated that FTIR is a potential cost-effective tool for in vitro drug analyses. Similar works have been recently reported by Hu et al. (2020), Altharawi et al. (2019) and Kelp et al. (2019).

   The application of IR spectral imaging brings out a substantial amount of spatial information associated with a given molecule's chemical distribution. For example, Pereira et al. (2020) followed biochemical changes in the thyroglobulin protein in goiter and normal tissues. Their methodology demonstrated it is possible to identify changes in the secondary structure, thyroglobulin iodination, and glycosylation. Spectral regions located at *ca.* 1600–1700 cm$^{-1}$ associated with amide I (α-helix and β-sheet conformations), and 1030–1074 cm$^{-1}$ associated with carbohydrate features, showed to be more important to differentiate the tissues. Bands related to amide II had subtle changes. Similar works on IR imaging of human tissues (cancer of different organs, normal tissues) have notoriously strengthened the application of vibrational spectroscopy in the biomedical community, as thoroughly described in the review by Diem et al. (2016).

Reaching the single-cell scale in imaging through vibrational spectroscopy includes a series of challenges associated with the behavior of light-matter interactions at microscopic scales. Conversely, information obtained from single cells is unique and allows access to details such as the cell behavior itself and related to its surroundings by identifying biochemical modifications or processes that cannot be "seen" by analyzing a tissue or average spectra of cells. Nevertheless, reaching this scale with the ease that Raman imaging techniques can offer remains challenging, mainly because the spatial resolution is not enough to image intracellular details (*e.g.*, organelles, vacuoles). FTIR imaging approaches in single-cell systems (Perez-Guaita et al. 2016; Morrish et al. 2019) have been reported, for example, in the work of Perez-Guaita et al. (2016), who developed a methodology to discriminate erythrocytes infected with malaria parasites from healthy ones. Images in this work reached 0.66μm of pixel resolution.

In general, for these applications, the main problem relies on the low signal-to-noise ratio (S/N) because of the confinement of IR beam using small apertures and focusing a small spot. Here, the highly collimated synchrotron IR light can overcome this limitation, reaching up to 1000 times higher brightness. Vongsvivut et al. (2019) showed advances in synchrotron macro-ATR-FTIR using red blood cells, individual neurons, and a Eucalyptus leaf as models. However, nowadays, Raman spectroscopy is still a more suitable option for single-cell scale, as discussed in the following topic.

## 2.2    Raman

From a historical perspective, Raman spectroscopy was, for many years, an unviable technique for analytical applications in general, mainly because of the intrinsically low scattering signal, as discussed before. It was only after the introduction of lasers, charged-coupled device detectors, among other technologies, around 1960 that an outbreak of papers applying this technique was registered. Later in the mid-1980s, NIR laser lines in FT-Raman equipment were introduced, allowing reducing fluorescence, mainly in biological matrices or samples. This discovery widened its possibilities to the point that fields such as biomedicine or imaging have notoriously benefited. However, studies involving quantifications only started to grow later, around 2010, and nowadays, Raman is a suitable alternative for analytical and bioanalytical applications (Kuhar et al. 2018).

Although molecules like carbohydrates, lipids, proteins, nucleic acids, among others, have characteristic Raman signatures, there is also a probability of peak overlap, making it difficult in many cases to perform a complete, precise band assignment, a problem partially minimized by multivariate analysis.

### 2.2.1    Fundamental and Structural Studies

The structural aspects of biomolecules studied using Raman have been widely reported in the literature. Throughout this time, some characteristic bands have been thoroughly described (De Gelder et al. 2007). The study of proteins is highly benefited from the amount of information that Raman offers, details of the three-dimensional architecture, decisive in its biological function can be analyzed mainly through amide I and amide III modes (secondary structure), also the ratio between 1360 and 1340 $cm^{-1}$ band, a hydrophobicity marker for tryptophan (tertiary structure). The C=O stretch in lipids found around 1750 $cm^{-1}$, as well as the peaks at 1453 $cm^{-1}$ (C–H bending), 1443 $cm^{-1}$ (CH$_2$ and CH$_3$ deformation of in lipids and triglycerides), 1430 $cm^{-1}$ (CH$_2$ scissoring) allows estimating properties such as the degree of unsaturation and have been used as markers in various studies. This high relative specificity found in lipids and proteins is considerably lower in carbohydrates (saccharides). The Raman bands of these compounds can be assigned along the entire fingerprint region, even in the high-frequency region (2800–4000 $cm^{-1}$). Similar behavior comes from Raman spectra of polysaccharides whose bands are broader than those from smaller saccharides such as trisaccharides, with typical sharp, narrow bands.

Nucleic acids have also been widely studied through Raman and SERS spectroscopies, as we will discuss later. They present characteristic features related to the ring breathing modes in the region between 600 and 800 $cm^{-1}$ and different backbone conformations (Okotrub et al. 2015), which can be studied using the frequencies of the amide I to III modes.

Aside from the biomolecules mentioned above, an honorable mention must be made. Water is an active solvent in biology and essential for living processes and organisms and hence can offer extensive insights when studied through vibrational spectroscopy. Although the water molecule is a weak Raman scatterer, with characteristic bands in the region of 3400–3600 $cm^{-1}$ (sym. and asym. stretches) and 1600 $cm^{-1}$ (bending mode), when interacting with biomolecules through hydration shells, its properties drastically differ from bulk water (Raschke 2006), and Raman spectroscopy allows detecting these modifications. The concept of bound and unbound (free), essential for studying water behavior in biological environments (Leikin et al. 1997), is not new and refers us to the very beginning of water's structural studies using Raman carried out by Walrafen (1964).

Various papers have used water Raman signals in the biological environment, such as tissues (Ghita et al. 2020), single-cells (Takeuchi et al. 2017), proteins (Verville et al. 2021), even hydration shells of DNA (Lee et al. 2013; Marini et al. 2017). For instance, in the papers by Fields et al. (2017) Unal et al. (2019), and Téllez-Soto et al. (2021), Raman spectroscopy was used to study bound water in cartilage as a whole and also in collagen fibers. The relation between water content in the tissue, collagen conformation can be probed in this type of tissue through these high-frequency modes of water but also through the proline-hydroxyproline and amide I bands (Nguyen et al. 2013).

In the paper by Téllez-Soto et al. (2021), a in vivo study involving confocal Raman measurements of the human dermis of 73 female participants allowed detecting differences in bound water through analysis of regions between 3350 and 3550 $cm^{-1}$, related to free and bound water vibrational modes, respectively. It is worth saying that Raman spectroscopy has a great potential for in vivo measurements, as discussed in other works (Unal and Akkus 2018).

## 2.2.2 Detection and Identification of Microorganisms

The application of Raman spectroscopy in the detection and identification of microorganisms has notoriously grown in the past 20 years to the point that today we have "spectral fingerprints" to identify them, as exemplified in Fig. 4 (Rebrošová et al. 2017).

Although a Raman spectrum of microorganisms has contributions of proteins, lipids, carbohydrates, and nucleic acids, it also has phenotypic signatures determined by the expression of parts of the genome that define the cell composition granting specificity to the Raman signal.

Microorganisms are easily found in food, environmental samples, inside cells, and different biological fluids, and thus the analytical matrices can be quite diverse. However, here it is possible to see the potential of Raman spectroscopy, going from the food industry to clinical applications and diagnostics, describing each of these sample environments. Furthermore, the speed of Raman analyses and availability of portable equipment, compared to traditional microorganism identification methods (time-consuming microbiological cultures), are additional advantages that contribute to the potential of the technique. The application of Raman in microbiology has already reached detection and identification of pathogens in biological fluids such as urine (Kloß et al. 2013), ascitic fluid (Kloß et al. 2015), sputum (Rusciano et al. 2013), also in tissues such as skin and nails (Smijs et al. 2014; Kourkoumelis et al. 2018). At a higher microscopic resolution, it was possible to identify microorganisms intracellularly via Raman imaging (Harrison and Berry 2017; Große et al. 2015). Recently, this subfield has become closer to genomics since it has shown to be a tool to assist the connection between bacterial phenotypes and their physiology (Hong et al. 2021).

## 2.2.3 Raman Imaging

Raman imaging (RI) has a central role in solving bioanalytical problems in research, as evidenced by the scientific literature. Medical and biological applications take great advantage of the benefits that the technique offers. Compared with IR imaging, Raman can achieve lower spatial resolution, even reaching the single-cell scale (human (Wang et al. 2020), vegetable (Kneipp et al. 1991; Tamamizu and Kumazaki 2019), and animal (Høgset et al. 2020)) with relative ease. Furthermore, water presence is not a problem here since the probability of overlapping with water's vibrational modes is low given its small Raman cross-section. Raman equipment availability in configurations coupled to confocal microscopes allows measurements in fixed volumes or slices, making microscopic chemical mappings in 3D possible. Despite several characteristics that classify RI in an essential place among techniques able to generate chemical maps, a challenging aspect is the time required for analysis, mainly for in vivo measurements and because of practical reasons (for example, long time for routine procedures or in cases when there is a high number of samples to be analyzed), thus forcing the sacrifice of spatial resolution.

During the first years of Raman imaging, univariate approaches dominated the spectral analysis and are still used when the interest lies in a specific signal (for example, cytochrome c, DNA, or even proteins/lipids specific modes in cells)

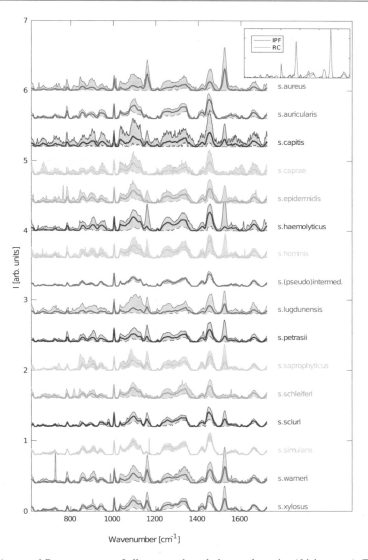

**Fig. 4** Averaged Raman spectra of all measured staphylococcal species (thick curves). The gray area depicts the variations of measured spectral intensities corresponding to a given wavenumber. Border curves of this interval correspond to 0.1st (dashed) and 99.9th percentiles, respectively. Fluorescence spectral background was removed by the IPF method. The top-right inset compares both background removal methods (IPF and RC) on a single randomly selected spectrum of *S. aureus*. Reproduced from reference (Rebrošová et al. 2017)

(Hosokawa et al. 2014; Palonpon et al. 2013) or when a Raman marker is introduced. However, after multivariate analysis met the needs of RI, most data analysis follows this approach, gaining an entirely different perspective and obtaining highly informative chemical maps. Today, there are plenty of algorithms capable of processing and extracting information from RI datasets. For more details on the theory of processing, we recommend the review by Mitsutake et al. (2019), Bedia et al. (2020), and Byrne et al. (2016).

RI of tissues has shown to be promising in biomedical and clinical applications (Ralbovsky and Lednev 2019), Raman information can reveal biochemical changes associated with diverse processes involved in the disease. An illustration of this can be appreciated in the work of Kopec et al. (2019), where the vibrational signatures of glycosylated and non-glycosylated proteins could be distinguished in normal and cancer tissues, as depicted in Fig. 5. Also, the work assessed the spatial distribution of lipids, nucleic acids, glycans, and proteins and modifications of glycan structures associated with oncogenic transformations.

## 2.3    SERS

A journey through the history of SERS development, from the enhancement effect discovery, back in 1974 to the current days, clearly shows how its scope has notoriously changed and is widespread. Among the various research fields that have been permeated by SERS spectroscopy, bioanalysis has taken great advantage of the unique properties of the technique, such as its ultrahigh sensitivity, fingerprint information, tunability of plasmonic properties through morphology changes in nanostructures. Of course, its successful introduction to bioanalysis as a very suitable tool has been accompanied by technological developments in optics, electronic devices (for biosensing), and digital processing and data mining algorithms.

As the metal surface is the critical component in the enhancement process, several morphologies and synthesis procedures are available after ca. 40 years of SERS. Despite gold, silver, and copper nanoparticles being used to enhance Raman scattering regularly, in bioanalytical applications, gold and silver have been preferred. Gold-based nanostructures are more suitable for measurements at excitation wavelengths higher than 600 nm and a better option for in vivo studies, given their lower toxicity. On the other hand, silver offers higher enhancement factors allowing reaching a lower limit of detection (LOD) values and is a better option for in vitro analysis. Besides, measurements on this surface are performed at excitation sources longer than 400 nm. Although several morphologies are available and their syntheses are well established, the simpler spherical morphology is one of the most popular, especially attractive for intracellular interrogation. Other morphologies, such as nanorods, allow tuning their LSPR, making them very versatile, also widely used in bioanalytical applications (Bi et al. 2020). Nanostars, nanowires, and even combinations of both metals (core-satellite or core-shell configurations) in a single nanoparticle have also been applied.

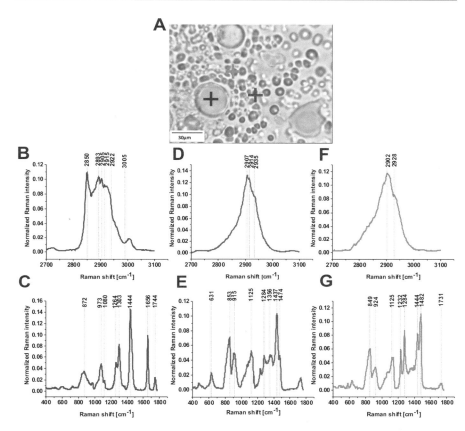

**Fig. 5** Typical vibrational Raman spectra in normal breast tissue from the tumor margin (for the same patient P157 as in Fig. 3), lipid-rich region (blue cross) and glycan-rich regions (red and green cross) in the white light microscopy image (**a**)) in the high frequency (**b, d, f**) and in the fingerprint (**c, e, g**) spectral regions of human breast normal tissue, integration time for a single spectrum 2 s, 10 accumulations, laser power 10 mW, patient P157. Reproduced from reference (Kopec et al. 2019)

Solid substrates are used in SERS applications, but as explained before, the higher cost of preparing them precludes their application for routine biosensing. Instead, several studies pursue low-cost sensors based on less expensive materials. More complex assemblies of metal nanostructured surfaces avoid the direct or label-free route. In all the cases mentioned above, focusing on a more specific sensing method, i.e., labeled SERS detection (SERS nanotags) were choosing the best Raman reporter for a given system is a key step.

SERS' bioanalytical applications will be discussed in this section, starting from fundamental studies of biomolecules up to overly complex systems such as living cell imaging or in vivo diagnosis.

### 2.3.1 The Simplest Case: A Biomolecule Anchored to a Nanoparticle

The simplest scenario for illustrating a bioanalytical methodology using SERS includes a biological-related molecule anchored to a nanoparticle surface. These more fundamental structural studies focus on understanding the biomolecules' interaction with a noble metal nanostructure, allowing the design of more effective, robust, and reproducible methodologies aiming at more complex systems (cells, fluids, tissues) where these molecules can be found (Villa et al. 2020; Tzeng and Lin 2020).

In this kind of application, one of the most representative that, at the same time, opened doors to innumerous studies, derivatives from the fact that the fundamental pieces of DNA and RNA, the nucleobases, have a particular affinity (lower for thymine and uracil) towards nanostructured noble metal surfaces (Kneipp et al. 1991; Bell and Sirimuthu 2006). The behavior of nucleobases and DNA on those surfaces and factors governing it has been extensively studied (Jang 2002; Muntean et al. 2019; Pergolese et al. 2005; Mamián-López et al. 2016), even on not-so-conventional enhancing surfaces such as aluminum (Tian et al. 2017). Still, some aspects of their behavior remain intriguing and controversial (Harroun 2018).

In the work of Li et al. (2019a), modified silver nanoparticles were used to study and distinguish DNA strands (with the same base composition) forming random coils or hairpin conformations. The features associated with the Watson-Crick hydrogen bond in base pairs allowed assessing the guanine-cytosine content in the DNA hairpins. A scheme of the procedure is reproduced in Fig. 6.

### 2.3.2 SERS and Body Fluids

A high number of analytes can be detected in biological fluids through SERS spectroscopy, from biomarkers to metabolites, for example, from drug administration or due to other types of exposure. Also, the motivations are diverse, such as diagnosis, forensics, and pharmacological treatments. The first condition here is that the species of interest can be efficiently attached to a metal nanoparticle, perhaps the most straightforward part of the entire process until detection. A second, more difficult step is to optimize the interaction between nanoparticle and analytical matrix conditions without losing SERS enhancement. When the enhancer is in colloidal form, factors such as analyte to colloid ratio and sample concentration are some of the most critical factors because a poor adjustment of these conditions can easily cause strong nanoparticle aggregation, drastic lowering, or even complete losing enhancement effect.

When using solid nanostructures, a common procedure is dropping a fixed volume of samples. The main problem with this strategy is the volume that should be deposited to effectively "attached" molecules to the enhancing surface, avoiding registering noise signals or Raman bands from the dried material's external layers. Usually, sample conditioning does not require complex experimental procedures. Filtration, centrifugation, and dilution are among the most common. It is important to remark that optimization of analyte-nanoparticle-analytical matrix assembly is more complex and time-consuming when functionalized or decorated nanoparticles are used.

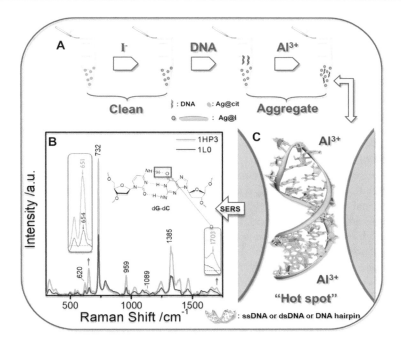

**Fig. 6** (**a**) Schematic diagram of the experimental procedure for DNA SERS detection. (**b**) SERS spectra of 1HP3 (DNA hairpin) (green) and1L0 (ssDNA) (blue) normalized by the intensity of the phosphate band at 1089 cm$^{-1}$. Insert: Structural diagram of the GC base pair. (**c**) Hotspots guided by aluminum ions on Ag@I. Ag@I: iodide-modified Ag nanoparticles. Ag@cit: silver citrate nanoparticles. Reprinted with permission from reference (Li et al. 2019a) Copyright (2019) American Chemical Society

The work of Panikar et al. (2019) shows a label-free methodology for detecting two drug compounds: paclitaxel and cyclophosphamide, with blood serum as an analytical matrix, reaching LOD values as low as $1.5 \times 10^{-8}$ and $5 \times 10^{-9}$ mol L$^{-1}$, respectively. In this case, L-cysteine functionalized star-like nanoparticles were used on a solid substrate where samples were directly deposited. In another work by Moisoiu et al. (2019), a relatively simple SERS methodology was carried out as a screening strategy for breast cancer, with urine as an analytical matrix. As a SERS substrate, silver nanoparticles were synthesized and then activated with Ca$^{2+}$ to promote anionic purine metabolites chemisorption. Regarding the sample conditioning, centrifugation was applied as the unique procedure before the spectral acquisition.

### 2.3.3 SERS Imaging

The introduction of nanoparticles into a given medium or material for illuminating it and acquiring SERS imaging is as challenging as fascinating. Properties such as nanoparticle size, stability in the "new" environment, chemical nature of its surface, or the necessity of modifying it, among other factors (Wang et al. 2018), must be carefully considered. These considerations are especially significant in tissues, single-cell imaging, and intracellular drug delivery, where SERS imaging has recently gained prominence. While RI is an exceptionally suited technique for cells and tissues, as discussed above, specific features or species in low concentrations (as in the case of intracellular drugs or highly specific biomarkers) are almost impossible to be probed without causing damage or death in in vivo analyses.

Research work devoted to studying the interactions of metallic nanoparticles in living cells for sensing has been carried out for a while, showing the significance of size, time of the interaction, type of metal, and morphology. Nowadays, we are able to tune nanoparticles size, morphology, composition, among other properties (Kneipp et al. 2002). In general, it is possible to say that gold is preferred over silver and spherical morphologies over nanorods, nanoprisms, and nanostars. Most used excitation radiations are around 630 and 785 nm, where fluorescence does not interfere with the spectral response.

The nanoparticles' chemical nature is critical in biological and biomedical applications since their stability could be compromised when entering a cell or tissue. Nanoparticles coming from typical syntheses (Lee-and-Meisel (1982) or Turkevich-Frens methods (Turkevich et al. 1951)) are stabilized in colloidal form by small molecules like citrate. However, in the intracellular environment, ions, or bigger biomolecules make them very susceptible to aggregation. Useful procedures are available for modifying their surface by coating (with polymers or lipids and more recently, multidentate ligands), bioconjugation (covalent bonding through amino groups, bioconjugation to antibodies exploiting thiol group affinity), or non-specific binding, i.e., binding of biomolecules to the metal surface.

Choi et al. (2020) used Au-Ag in hollow nanospheres, then encapsulated with a polyethylene glycol layer for identifying multiple biomarkers in breast cancer cells. Their SERS nanotags showed to be highly stable under different pH, temperature, and salt conditions, and three cancer biomarkers could be identified in SERS chemical maps.

Metal nanoparticles are also playing the leading role in drug-delivery studies, in the identification of pathogens, and analysis of biofilms. With quite simple experimental procedures for most cases, including the direct interaction of nanoparticles with microorganisms without prior induced aggregation (Akanny et al. 2020). The main disadvantage here may be low reproducibility, which can be partially managed by taking a high number of spectra to assess the mean spectral feature. Indirect SERS detection of bacteria (or other pathogens) is also an option by using a selective molecule or Raman reporter attached to the SERS substrate (Stöckel et al. 2016).

### 2.3.4 Nonlinear Spectroscopic Techniques

Among nonlinear techniques, stimulated Raman spectroscopy (SRS) and CARS are particularly interesting in solving bioanalytical problems. These techniques can drastically decrease the imaging acquisition time. They are mostly free of interference by autofluorescence and offer linear concentration dependence, an attractive attribute from the analytical perspective. This characteristic is a particular advantage of SRS compared to CARS and Raman. With SRS, it is possible to perform in vivo quantitative measurements, a task not trivial with normal RI or CARS.

Species such as proteins, cholesterol, DNA, and lipids have already been imaged through SRS. Li et al. (2019b), using SRS imaging, showed a linear relationship of signals and fatty acid concentration to determine its variation in an in vivo system (*C. elegans*). The non-invasive methodology also allowed assessing lipids desaturation in vivo. A more comprehensive landscape of SRS advances and possibilities can be found in the article by Hu et al. (2019).

The significant signal improvement offered by CARS (compared to Raman) is shown, for example, in the study by Nuriya et al. (2019), who imaged water at high resolution in live mammalian cells, characterizing it intra- and extra-cellularly. They analyzed C-H stretching ($2930–2940$ cm$^{-1}$) and -O-H ($3420–3430$ cm$^{-1}$) regions, and results showed important differences in the strength of hydrogen bonds. Furthermore, a saccharide (mannitol) was introduced to mimic a pharmacologic intervention, causing an increase in osmolality. Changes in -O-H signal could be observed, giving information about this compound's effect on cell hydration.

Other techniques in this category, such as hyper Raman, time-resolved 2D IR, vibrational sum-frequency generation, time-resolved Raman, have also been applied in the study of biomolecules. However, these techniques have not become so widespread in bioanalysis yet. Efforts using nonlinear techniques were recently reviewed in the literature (Liu et al. 2020; Kawata et al. 2017).

## 3  Concluding Remarks

Vibrational spectroscopies and their variations are analytical techniques that for sure are always up-to-date. Their versatility and unique advantages, especially their non-invasiveness and ability to manage water presence in almost all applications, make them highly appropriate for biological systems. It is not unrealistic to say that any biosystem could be analyzed through infrared or Raman, and scientific literature reveals this.

The high output of IR (absorption and its variations), Raman, SERS applications also reveals a trend towards medicine and clinical diagnostics. Advances in portable spectrometers (mainly Raman) could eventually reach point-of-care testing. SERS spectroscopy, as the most popular strategy for intensifying Raman scattering, has gained countless applications. Although many research groups still use the most classical substrate synthesis methodologies, its development shows that it has transcended successfully to analytical applications with quantitative purposes. Furthermore, it is fair to say that concerns about reproducibility are being managed with a combination of experimental and computational strategies. Structural and fundamental studies of biomolecules are also highly benefited from SERS due to its well-known ultra-sensitivity.

Vibrational spectroscopic imaging has become almost a research field on its own, and its possibilities have notoriously spread out over time. Its ability to map in vivo processes and increase spatial resolution is undoubtedly its main attractive point. Its broad use in tissues, single cells, and drug-delivery studies carry the technique towards routine use in biomedical research and clinical medicine. While nonlinear techniques have been shown to be able to overcome various disadvantages of spontaneous Raman, they have not yet been popularized since, so far, the cost and complexity inherent to these techniques is their main drawback.

Finally, many aspects remain unexplored or poorly explored, for example, the detection of volatile biomarkers, compounds that are highly informative of biological processes and systems. Probably, optimization of sensing devices and portable equipment could boost research in this field.

**Acknowledgments** The authors are indebted to FAPESP (Grants 2016/21070-5 and 2019/00207-0).

# References

Akanny E, Bonhommé A, Commun C, Doleans-Jordheim A, Farre C, Bessueille F, Bourgeois S, Bordes C (2020) J Raman Spectrosc 51:619–629
Almehmadi LM, Curley SM, Tokranova NA, Tenenbaum SA, Lednev IK (2019) Sci Rep 9:12356
Altharawi A, Rahman KM, Chan KLA (2019) Analyst 144:2725–2735
Babar S, Weaver JH (2015) Appl Optics 54:477–481
Baiz CR, Reppert M, Tokmakoff A (2013) An introduction to protein 2D IR spectroscopy. In: Fayer MD (ed) Ultrafast infrared vibrational spectroscopy. CRC Press, Boca Raton, p 475
Baxter JB, Guglietta GW (2011) Anal Chem 83:4342–4368
Beć KB, Grabska J, Huck CW (2020) Anal Chim Acta 1133:150–177
Bedia C, Sierra À, Tauler R (2020) Anal Bioanal Chem 412:5179–5190
Bell SEJ, Sirimuthu NMS (2006) J Am Chem Soc 128:15580–15581
Bi L, Wang X, Cao X, Liu L, Bai C, Zheng Q, Choo J, Chen L (2020) Talanta 220:121397
Boyd RW (2020) Nonlinear optics. Academic Press, London
Boydston-White S, Chernenko T, Regina A, Miljkovic M, Matthäus C, Diem M (2005) Vib Spectrosc 38:169–177
Brielle ES, Arkin IT (2018) J Phys Chem Lett 9:4059–4065
Buhrke D, Hildebrandt P (2020) Chem Rev 120:3577–3630
Byrne HJ, Knief P, Keating ME, Bonnier F (2016) Chem Soc Rev 45:1865–1878
Choi N, Dang H, Das A, Sim MS, Chung IY, Choo J (2020) Biosens Bioelectron 164:112326
Coates J (1998) Appl Spectrosc Rev 33:267–425
Davis SJ, Ikemizu S, Evans EJ, Fugger L, Bakker TR, van der Merwe PA (2003) Nat Immunol 4: 217–224
De Bruyne S, Speeckaert MM, Delanghe JR (2018) Crit Rev Clin Lab Sci 55:1–20
De Gelder J, De Gussem K, Vandenabeele P, Moens L (2007) J Raman Spectrosc 38:1133–1147
Diem M, Ergin A, Remiszewski S, Mu X, Akalin A, Raz D (2016) Faraday Discuss 187:9–42
dos Santos DP (2020) J Phys Chem C 124:6811–6821
dos Santos DP, Temperini MLA, Brolo AG (2019) Acc Chem Res 52:456–464
Ellis JW, Bath J (1938) J Chem Phys 6:108–108
Ellis DI, Goodacre R (2006) Analyst 131:875-858
Evans A, Hibbard RR, Powell AS (1951) Anal Chem 23:1604–1610
Fale PL, Altharawi A, Chan KLA (2015) Biochim Biophys Acta Mol Cell Res 1853:2640–2648
Fang Y, Chen W, Shi W, Li H, Xian M, Ma H (2017) Chem Commun 53:8759–8762
Ferreira ICC, Aguiar EMG, Silva ATF, Santos LLD, Cardoso-Sousa L, Araújo TG, Santos DW, Goulart LR, Sabino-Silva R, Maia YCP (2020) J Oncol 2020:4343590
Fields M, Spencer N, Dudhia J, McMillan PF (2017) Biopolymers 107:e23017

Fleischmann M, Hendra PJ, McQuillan AJ (1974) Chem Phys Lett 26:163–166
Fornaro T, Burini D, Biczysko M, Barone V (2015) J Phys Chem A 119:4224–4236
Ghita A, Hubbard T, Matousek P, Stone N (2020) Anal Chem 92:9449–9453
Goymer P (2012) Nat Chem 4:863–864
Graça G, Moreira AS, Correia AJV, Goodfellow BJ, Barros AS, Duarte IF, Carreira IM, Galanho E, Pita C, Almeida M, Gil AM (2013) Anal Chim Acta 764:24–31
Große C, Bergner N, Dellith J, Heller R, Bauer M, Mellmann A, Popp J, Neugebauer U (2015) Anal Chem 87:2137–2142
Hackett MJ, Borondics F, Brown D, Hirschmugl C, Smith SE, Paterson PG, Nichol H, Pickering IJ, George GN (2013) ACS Chem Nerosci 4:1071–1080
Harrison JP, Berry D (2017) Front Microbiol 8:675
Harroun SG (2018) ChemPhysChem 19:1003–1015
Herzberg G (1956) Molecular spectra and molecular structure II. Infrared and Raman spectra of polyatomic molecules. D. Van Nostrand Company Inc., Princeton
Herzberg G (1963) Molecular spectra and molecular structure I. Spectra of diatomic molecules. D. Van Nostrand Company Inc., Princeton
Høgset H, Horgan CC, Armstrong JPK, Berholt MS, Torraca V, Chen Q, Keane TJ, Bugeon L, Dallman MJ, Mostowy S, Stevens MM (2020) Nat Commun 11:6172
Hong KY, de Albuquerque CDL, Poppi RJ, Brolo AG (2017) Anal Chim Acta 982:148–155
Hong J-K, Kim S, Lyou ES, Lee TK (2021) J Microbiol 59:249–258
Hosokawa M, Ando M, Mukai S, Osada K, Yoshino T, Yamaguchi H, Tanaka T (2014) Anal Chem 86:8224–8230
Hosseinpour S, Roeters SJ, Bonn M, Peukert W, Woutersen S, Weidner T (2020) Chem Rev 120: 3420–3465
Hu F, Shi L, Min W (2019) Nat Methods 16:830–842
Hu R, Zong C, Lin L-X, Wei T, Cheng Q, Jiang Y-X, Lin C-J, Ren B, Tian Z-Q (2020) Vib Spectrosc 109:103068
Ichimura T, Hayazawa N, Hashimoto M, Inouye Y, Kawata S (2004) Phys Rev Lett 92:220801
Jang N-H (2002) Bull Korean Chem Soc 23:1790–1800
Jeanmaire DL, Van Duyne RP (1977) J Electroanal Chem Interfacial Electrochem 84:1–20
Ji W, Zhao B, Ozaki Y (2016) J Raman Spectrosc 47:51–58
Kawata S, Ichimura T, Taguchi A, Kumamoto Y (2017) Chem Rev 117:4983–5001
Kelp G, Li J, Lu J, DiNapoli N, Delgado R, Liu C, Fan D, Dutta-Gupta S, Shvets G (2019) Lab Chip 20:2136–2153
Kloß S, Kampe B, Sachse S, Rösch P, Straube E, Pfister W, Kiehntopf M, Popp J (2013) Anal Chem 85:9610–9616
Kloß S, Rösch P, Pfister W, Kiehntopf M, Popp J (2015) Anal Chem 87:937–943
Kneipp K, Pohle W, Fabian H (1991) J Mol Struct 244:183–192
Kneipp K, Haka AS, Kneipp H, Badizadegan K, Yoshizawa N, Boone C, Shafer-Peltier KE, Motz JT, Dasari RR, Feld MS (2002) Appl Spectrosc 56:150–154
Kopec M, Imiela A, Abramczyk H (2019) Sci Rep 9:166
Kourkoumelis N, Gaitanis G, Velegraki A, Bassukas ID (2018) Med Mycol 56:551–558
Kraack JP, Hamm P (2017) Chem Rev 117:10623–10664
Krafft C, Dietzek B, Popp J, Schmitt M (2012) J Biomed Opt 17:040801
Krajczewski J, Kudelski A (2019) Front Chem 7:410
Kuhar N, Sil S, Verma T, Umapathy S (2018) RSC Adv 8:25888–25908
Langer J, Jimenez de Aberasturi D, Aizpurua J, Alvarez-Puebla RA, Auguié B, Baumberg JJ, Bazan GC, Bell SEJ, Boisen A, Brolo AG, Choo J, Cialla-May D, Deckert V, Fabris L, Faulds K, García de Abajo FJ, Goodacre R, Graham D, Haes AJ, Haynes CL, Huck C, Itoh T, Käll M, Kneipp J, Kotov NA, Kuang H, Le Ru EC, Lee HK, Li J-F, Ling XY, Maier SA, Mayerhöfer T, Moskovits M, Murakoshi K, Nam J-M, Nie S, Ozaki Y, Pastoriza-Santos I, Perez-Juste J, Popp J, Pucci A, Reich S, Ren B, Schatz GC, Shegai T, Schlücker S, Tay L-L, Thomas KG, Tian

Z-Q, Van Duyne RP, Vo-Dinh T, Wang Y, Willets KA, Xu C, Xu H, Xu Y, Yamamoto S, Zhao B, Liz-Marzán LM (2020) ACS Nano 14:28–117
Le Ru E, Etchegoin P (2009) Principles of surface enhanced Raman spectroscopy and related plasmonic effects. Elsevier, Amsterdam
Lee PC, Meisel D (1982) J Phys Chem 86:3391–3395
Lee SA, Tao N-J, Rupprecht A (2013) J Biomol Struct Dyn 31:1337–1342
Leikin S, Parsegian VA, Yang W-H, Walrafen GE (1997) Proc Nat Acad Sci 94:11312–11317
Levin IW, Lewis EN (1990) Anal Chem 62:1101A–1111A
Li J-F, Li C-Y, Aroca RF (2017) Chem Soc Rev 46:3962–3979
Li Y, Gao T, Xu G, Xiang X, Han X, Zhao B, Guo X (2019a) J Phys Chem Lett 10:3013–3018
Li X, Li Y, Jiang M, Wu W, He S, Chen C, Quin Z, Tang BZ, Mak HY, Qu JY (2019b) Anal Chem 91:2279–2287
Liu X, Liu X, Rong P, Liu D (2020) Trends Anal Chem 123:115765
Lohse SE, Murphy CJ (2013) Chem Mater 25:1250–1261
Long DA (2002) The Raman effect. Wiley, New York
Lovergne L, Bouzy P, Untereiner V, Garnotel R, Baker MJ, Thiéfin G, Sockalingum GD (2016) Faraday Discuss 187:521–537
Mamián-López M, Temperini M (2019) Quim Nova 42:1084–1090
Mamián-López MB, Corio P, Temperini MLA (2016) Analyst 141:3428–3436
Marini M, Allioni M, Torre B, Moretti M, Limongi T, Tirinato L, Giugni A, Das G, di Fabrizio E (2017) Microelectron Eng 175:38–42
Martin MC, Dabat-Blondeau C, Unger M, Sedlmair J, Parkinson DY, Bechtel HA, Illman B, Castro JM, Keiluweit M, Buschke D, Ogle B, Nasse MJ, Hirschmugl CJ (2013) Nat Methods 10:861–864
Mitsutake H, Poppi R, Breitkreitz M (2019) J Braz Chem Soc 30:2243–2258
Moisoiu V, Socaciu A, Stefancu A, Iancu SD, Boros I, Alecsa CD, Rachieriu C, Chiorean AR, Eniu D, Leopold N, Socaciu C, Eniu DT (2019) Appl Sci 9:806
Morrish RB, Hermes M, Metz J, Stone N, Pagliara S, Chahwan R, Palombo F (2019) Front Cell Dev Biol 7:141
Movasaghi Z, Rehman S, Rehman IU (2007) Appl Spectrosc Rev 42:493–541
Mukamel S (1995) Principles of nonlinear optical spectroscopy. Oxford University Press, London
Mundlapati VR, Sahoo DK, Ghosh S, Purame UK, Pandey S, Acharya R, Pal N, Tiwari P, Biswal HS (2017) J Phys Chem Lett 8:794–800
Muntean CM, Biter T-L, Bratu I, Toşa N (2019) J Mol Model 25:162
Nafie LA (2020) J Raman Spectrosc 51:2354–2376
Neubrech F, Huck C, Weber K, Pucci A, Giessen H (2017) Chem Rev 117:5110–5145
Nguyen TT, Happillon T, Feru J, Brassart-Passco S, Angiboust J-F, Manfait M, Piot O (2013) J Raman Spectrosc 44:1230–1237
Nihonyanagi S, Yamaguchi S, Tahara T (2017) Chem Rev 117:10665–10693
Nuriya M, Yoneyama H, Takahashi K, Leproux P, Couderc V, Yasui M, Kano H (2019) J Phys Chem 123:3928–3934
Okotrub KA, Surovtsev NV, Semeshin VF, Omelyanchuk LV (2015) Cytometry A 87:68–73
Olsztyńska-Janus S, Pietruszka A, Kiełbowicz Z, Czarnecki MA (2018) Spectrochim Acta A Mol Biomol Spectrosc 188:37–49
Palonpon AF, Sodeoka M, Fujita K (2013) Curr Opin Chem Biol 17:708–715
Paluszkiewicz C, Pięta E, Woźniak M, Piergies N, Koniewska A, Ścierski W, Misiołek M, Kwiatek WM (2020) J Mol Liq 307:112961
Panikar SS, Ramírez-García G, Sidhik S, López-Luke T, Rodríguez-Gonzales C, Ciapara IH, Castillo PS, Camacho-Villegas T, de la Rosa E (2019) Anal Chem 91:2100–2111
Paraskevaidi M, Morais CLM, Freitas DLD, Lima KMG, Mann DMA, Allisop D, Martin-Hirsch PL, Martins FL (2018a) Analyst 143:5959–5964
Paraskevaidi M, Morais CLM, Lima KMG, Ashton KM, Stringfellow HF, Martins-Hirsch PL, Martin FL (2018b) Analyst 143:3156–3163

Pereira TM, Diem M, Bachman L, Bird B, Miljković M, Zezell DM (2020) Analyst 145:7907–7915

Perez-Guaita D, Andrew D, Heraud P, Beeson J, Anderson D, Richards J, Wood BR (2016) Faraday Discuss 187:341–352

Pergolese B, Bonifacio A, Bigotto A (2005) Phys Chem Chem Phys 7:3610–3613

Ralbovsky NM, Lednev IK (2019) Spectrochim Acta Part A Mol Biomol Spectrosc 219:463–487

Raschke TM (2006) Curr Opin Struct Biol 16:152–159

Rebrošová K, Šiler M, Samek O, Růžička F, Bernatová S, Holá V, Ježek J, Zemánek P, Sokolová J, Petráš P (2017) Sci Rep 7:14846

Rossi B, Bottari C, Catalini S, D'Amico F, Gessini A, Massiovecchio C (2020) Synchrotron-based ultraviolet resonance Raman scattering for material science. In: Molecular and Laser spectroscopy. Elsevier, Amsterdam, pp 447–482

Rusciano G, Capriglioni P, Pesce G, Abete P, Carnovale V, Sasso A (2013) Laser Phys Lett 10: 075603

Sala O (2008) Fundamentos da Espectroscopia Raman e no Infravermelho. Editora Unesp, São Paulo

Silveira RL, Mamián-López MB, Rubim JC, Temperini MLA, Corio P, Santos JJ (2019) Anal Bioanal Chem 411:3047–3058

Singh R, Riess F (2001) Notes Rec R Soc Lond 55:267–283

Smijs TG, Jachtenberg JW, Pavel S, Bakker-Schut TC, Willemse-Erix D, de Haas ERM, Sterenborg H (2014) J Eur Acad Dermatol Venereol 28:1492–1499

Stöckel S, Kirchhoff J, Neugebauer U, Rösch P, Popp J (2016) J Raman Spectrosc 47:89–109

Stuart B (1997) Biological applications of infrared spectroscopy. Wiley, New York

Sun Y, Du W, Zhou C, Zhou Y, Cao Z, Tian Y, Wang Y (2015) IEEE Trans Nanobioscience 14: 167–174

Takamura A, Halamkova L, Ozawa T, Lednev IK (2019) Anal Chem 91:6288–6295

Takeuchi M, Kajimoto S, Nakabayashi T (2017) J Phys Chem Lett 8:5241–5245

Tamamizu K, Kumazaki S (2019) Biochim Biophys Acta Bioenerg 1860:78–88

Téllez-Soto CA, Silva MGP, dos Santos L, Mendes TO, Singh P, Fortes SA, Favero P, Martin AA (2021) Vib Spectrosc 112:103196

Tian S, Neumann O, McClain MJ, Yang X, Zhou L, Zhang C, Nordlander P, Halas NJ (2017) Nano Lett 17:5071–5077

Türker-Kaya S, Huck C (2017) Molecules 22:168

Turkevich J, Stevenson PC, Hillier J (1951) Discuss Faraday Soc 11:55

Tzeng Y, Lin B-Y (2020) Biosensors 10:122

Unal M, Akkus O (2018) J Biomed Opt 23:015008

Unal M, Akkus O, Sun J, Cai L, Erol UL, Sabri L, Neu CP (2019) Osteoarthr Cartil 27:304–313

Verville GA, Byrd MH, Kamischke A, Smith SA, Magers DH, Hammer NI (2021) J Raman Spectrosc 52:788–795

Villa JEL, Afonso MAS, dos Santos DP, Mercadal PA, Coronado EA, Poppi RJ (2020) Spectrochim Acta A Mol Biomol Spectrosc 224:117380

Volkmer A, Cheng J-X, Xie XS (2001) Phys Rev Lett 87:023901

Vongsvivut J, Pérez-Guaita D, Wood BR, Heraud P, Khambatta K, Hartnell D, Hackett MJ, Tobin MJ (2019) Analyst 144:3226–3238

Walrafen GE (1964) J Chem Phys 40:3249–3256

Wang F, Wen S, He H, Wang B, Zhou Z, Shimoni O, Jin D (2018) Light Sci Appl 7:18007

Wang D, He P, Wang Z, Li G, Majed N, Gu AZ (2020) Curr Opin Biotechnol 64:218–229

Wiercigroch E, Szafraniec E, Czamara K, Pacia MZ, Majzner K, Kochan K, Kaczor A, Baranska M, Malek K (2017) Spectrochim Acta A Mol Biomol Spectrosc 185:317–335

Yadav J, Rani A, Singh V, Murari BM (2015) Biomed Signal Process Control 18:214–227

Yang C, Sesterhenn F, Bonet J, van Aalen EA, Scheller L, Abriata LA, Cramer JT, Wen X, Rosset S, Georgeon S, Jardetzky T, Krey T, Fussenegger M, Merkx M, Correia BE (2021) Nat Chem Biol 17:492–500

Yonar D, Ocek L, Tiftikcioglu BI, Zorlu Y, Severcan F (2018) Sci Rep 8:1025

Yu M-C, Rich P, Foreman L, Smith J, Yu M-S, Tanna A, Dibbur V, Unwin R, Tam FWK (2017) Sci Rep 7:4601

Zhang Z, Sheng S, Wang R, Sun M (2016) Anal Chem 88:9328–9346

Zhang W, Jiang L, Piper JA, Wang Y (2018) J Anal Test 2:26–44

Zong C, Xu M, Xu L-J, Wei T, Ma X, Zheng X-S, Hu R, Ren B (2018) Chem Rev 118:4946–4980

Zumbusch A, Holtom GR, Xie XS (1999) Phys Rev Lett 82:4142–4145

# Biological Systems Investigated by Small-Angle X-Ray Scattering

Barbara B. Gerbelli and Cristiano L. P. Oliveira

## 1    Introduction

The organization of biomolecules depends on the molecular composition as well as the environment in which they are immersed. In order to investigate supramolecular assemblies, it is necessary to use/combine several experimental techniques, such as spectroscopy, scattering, mechanics, calorimetry, microscopies, among others. In this section, practical tools based on scattering methods will be briefly described and discussed, with examples of their use in biophysics and correlated fields.

Scattering techniques are widely used for the structural characterization of systems at the nanoscale. One of the main advantages of these methods is the fact they are relatively simple (in most cases, a transmission experiment) and can be applied to almost any type of sample, including liquids, gels, gases, and solids. Several scattering techniques depend on the specific characteristics of the beam, energy, geometry, and setup of the desired experiment. It is outside the scope of this chapter to describe all details of the experimental procedure; more information can be found in the literature (Jones 1999; Xu 2015; Chu and Liu 2000; Rambo and Tainer 2013; Rahimabadi et al. 2020; Jacrot 1976; Petoukhov and Svergun 2006).

A typical setup for scattering is shown in Fig. 1. The source can be an X-ray tube, synchrotron radiation, laser, or a waveguide from a neutron reactor/spallation source. Beam optics and collimation are necessary to define the photons/neutrons' energy and beam size/shape. The most common scattering experiment is performed on a transmission mode, where the beam passes through the sample. The photons will interact with the electron clouds for X-rays, whereas neutrons may interact with

B. B. Gerbelli
Centro de Ciências Naturais e Humanas, Universidade Federal do ABC, Santo André, SP, Brazil

C. L. P. Oliveira (✉)
Instituto de Física, Universidade de São Paulo, São Paulo, SP, Brazil
e-mail: crislpo@if.usp.br

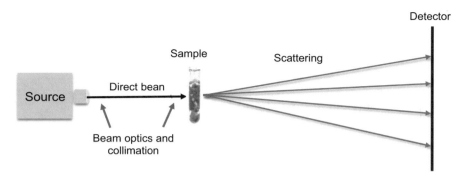

**Fig. 1** Typical setup for scattering techniques

nuclei scattering lengths. Lasers will also interact with the materials' electron structure, and the overall process is successfully described by the polarizability characteristics of the system, which can be expressed in terms of the refraction index (Clausius-Mossotti equation (Talebian and Talebian 2013)).

In all cases, the scattered intensity is collected as a function of the scattering angle $2\theta$ or, for greater utility, as a function of the modulus of reciprocal space momentum transfer vector, which is defined as $q = (4\pi\, n\, \sin\theta)/\lambda$, where $\lambda$ is the wavelength, $2\theta$ is the scattering angle and $n$ is the refraction index. For X-rays and neutrons, $n$ is close to unity, but for laser light, the use of the correct value is critical. The wavelength $\lambda$ defines the type of interaction between the radiation and the sample.

For static and dynamic light scattering (DLS and SLS, respectively), the wavelengths used are in the range of visible light (from 400 to 700 nm), where one has several high-power lasers commercially available. Lasers have a high beam coherence and polarizability, which can be explored in the experiments. DLS takes advantage of this coherence since it allows the construction of autocorrelation curves for the scattering intensity, which is directly linked to diffusion properties of the particles (Chu and Liu 2000). For macromolecules in solution, these diffusion coefficients can be associated with the apparent hydrodynamic radius using the Stokes-Einstein equation (Miller 1924). Advanced modeling of DLS autocorrelation function using methods like Contin (Scotti et al. 2015) or NNLS (Roig and Alessandrini 2007) can even provide the size distribution of the particles, which makes this technique very versatile, fast, and reliable. On the other hand, SLS experiments measure the dependency of the scattering intensity as a function of the scattering angle. Such an approach provides information on intensity-average molecular weight (MW) and particle characteristic dimensions (for instance, the radius of gyration $R_g$), as well as details on the intraparticle interactions. DLS and SLS may give information from few nanometers up to microns.

Typical small-angle X-ray and neutron scattering experiments (SAXS and SANS, respectively) use wavelengths from 0.1 to 1 nm. For X-rays, this corresponds to photon energies of ~7–12 keV (Oliveira 2011), and for neutrons, these are the so-called cold/thermal neutrons (Heller 2010). In both cases, the scattering can be considered elastic, i.e., fulfilling the first Bohr approximation. SAS techniques with these characteristics allow the observation of particles and correlations up to tens of nanometers. In all cases, the scattering is generated by the so-called scattering length contrast between the particles and the medium. For X-rays, this effect depends on the differences in electron densities among the particles and medium. The contrast depends on differences in the nuclear scattering lengths density between the particles and the medium for neutrons. For light, it is related to the difference in refraction index among the particles and the medium. Besides small details, SAXS, SANS, and SLS can be described with the same theoretical framework (Svergun and Koch 2003; Zemb and Lindner 2002).

In the last decades, the use of SAXS and SANS has proven to be a fantastic tool for the investigation of biomolecules in solution (Rambo and Tainer 2013; Jacrot 1976; Petoukhov and Svergun 2006; Heller 2010; Glatter 1977; Gerbelli et al. 2018), from proteins and protein complexes to RNA, DNA, and other types of macromolecular assemblies, aiming the determination of both structural and dynamical characteristics of these systems (Nasta et al. 2019; Oliveira et al. 2009; Rojas et al. 2019; Silva et al. 2018). There are many examples in the literature, and the reader is invited to consult good reviews in this field (Chu and Liu 2000; Rambo and Tainer 2013; Lombardo et al. 2020; Orthaber et al. 2000; Mahieu and Gabel 2018). A few interesting examples will be presented in this chapter focusing on the application of SAXS technique.

## 2    Examples of SAXS Applications

Hemoglobin is a well-known protein responsible for transporting oxygen in the blood and a potentially harmful component to tissues due to its highly reactive heme groups. During intravascular hemolysis, common when the individual has malaria or other hemoglobinopathies, massive hemoglobin amounts are released into the blood. As a response, the body increases haptoglobin protein expression, which captures part of these hemoglobins and promotes a protective action for this acute phase. In a fascinating article by Andersen et al. (2012), it was obtained, for the first time, the crystallographic structure of the hemoglobin–haptoglobin complex at the atomic level (resolution of 2.9 Å). SAXS investigations were critical in that work since they guided the expression and purification of the proteins in several trials and buffers to have a monodisperse system, which, ultimately, enabled the crystallization of the complex. An example of the obtained SAXS results is shown in Fig. 2, revealing that the protein complex is composed of a haptoglobin homodimer in which two hemoglobin dimers are covalently linked (Fig. 2b). These analyses led to the comprehension of fundamental processes related to the interaction between hemoglobin and haptoglobin. The scavenger receptor action that the protein CD163 has with the binding to the hemoglobin-haptoglobin complex allows the clearance of this complex from the body (Andersen et al. 2012).

A wide range of proteins can be used as biosensors of enzymatic activities in cells. An example of this application is the CYNEX4 probe (Fig. 3a, left), composed

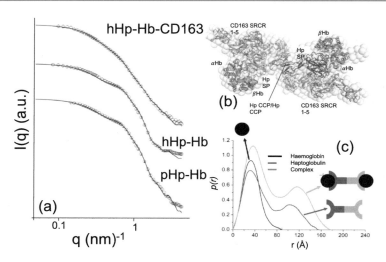

**Fig. 2** (**a**) SAXS data of porcine Hp–Hb (pHp–Hb), human Hp–Hb (hHp–Hb), and CD163 bound to hHp–Hb. The lines correspond to the theoretical intensities calculated from the structure of pHp–Hb. (**b**) used to fit the experimental data. (**c**) Pair distance distribution function, p(r), of the proteins and the complex system. Adapted with permission from (Andersen et al. 2012) Copyright 2020, Nature Publishing Group

of an annexin-A4 nucleus (A4) with an N-terminal domain, which is sensitive to phosphorylation in the presence of calcium, and two fluorescent proteins (donor ECFP and acceptor EYFP). The quantification of calcium ions in cells can be used as a potential pharmacological target in cystic fibrosis disease (Piljic and Schultz 2006). When the annexin-A4 nucleus interacts with calcium ions, the fluorescent proteins become closer to each other (Fig. 3a, right). This conformational change of the probe promotes an increase in intracellular fluorescence. Mertens et al. (2012) proposed a protocol to understand how mutations in the core sector of the probe could affect its 3D conformation and efficiency in detecting the calcium ions. The authors calculated the theoretical fluorescence intensity as a function of the distance between donor and acceptor proteins and demonstrated that by increasing this distance the detection of calcium would decrease. From Foerster resonance energy transfer (FRET) assays, it was observed the decrease of fluorescence intensity for the mutant probe T266D compared to the native CYNEX4 (Fig. 3c). Given these facts, Mertense et al. used SAXS to obtain structural information on these probes with and without mutation (Fig. 3b). From the analyses of the curves at low-q (Guinier region), it was possible to obtain the radius of gyration ($R_g$) as well as the molecular weight (MW) of these complexes (Fig. 3d). T266D exhibited larger $R_g$ and MW compared to CYNEX4. The envelope of these protein complexes, obtained by ab initio modeling (Fig. 3e), also presents differences in the 3D conformation. In conclusion, the main consequence of the mutation was to increase the length of sector A4, promoting the increase of the distance between donor and acceptor proteins, which then led to

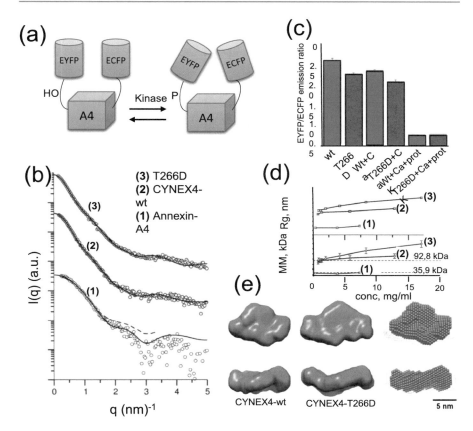

**Fig. 3** (**a**) Schematic representation of the CYNEX4 biosensor (left), composed of the domain annexin-A4 and the fluorescent donor (ECFP) and acceptor (EYFP) proteins. In the presence of calcium, there is a conformational change of the probe (right). (**b**) SAXS data of annexin-A4, native probe (CYNEX4) and mutant probe (T266D). (**c**) FRET assays in the presence and absence of calcium ions for native probe (wt) and mutant probe (T266D). (**d**) Structural parameters radius of gyration (Rg) and molecular weight (MW) obtained from the Guinier region analyses as a function of concentration for (1) annexin-A4, (2) native probe (CYNEX4) and (3) mutant probe (T266D). (**e**) Ab initio modeling results for CYNEX4 (left) and T266D (middle). A shape comparison is also presented (right). Adapted with permission from (Mertens et al. 2012) Copyright 2020, Elsevier

the decrease in fluorescence observed in the FRET assays. The structural results from SAXS were fundamental for the understanding of the biosensor performance in the biological assays.

Another system extensively studied via scattering techniques is the one that mimics the cell membrane. The high interest in exploring these systems is related to their importance on several biological functions, acting as a substrate for various biomolecules, including glycoproteins, glycolipids, proteins, and short amino acid sequences (Singer and Nicolson 1972; Capito et al. 2008; Goni 1838). Scattering methods are suitable for these investigations due to the possibility of study directly in solution, with minimum interaction and disturbance to the delicate thermodynamic

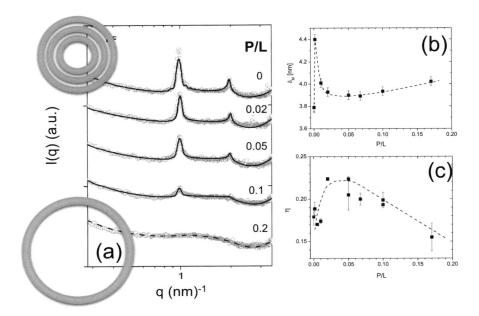

**Fig. 4** (a) SAXS data obtained varying the molar ratio peptide/lipid (P/L). The solid lines represent the fitting of experimental data considering a periodic stacking of bilayers, whereas the dashed lines represent a fitting considering the scattering of spherical shells. Structural parameters of lecithin vesicles as a function of FF content: (**b**) membrane thickness, $\delta_M$, and (**c**) Caillé parameter, $\eta$

equilibrium of these systems (Oliveira 2011; Luzzati et al. 1957). Several works in the literature provided the understanding of the interaction mechanisms and organization of biomolecules and model membranes (Gerbelli et al. 2018; Pabst et al. 2007). Gerbelli et al. (2018) studied the interaction between the short peptide L, L-diphenylalanine (FF) and membranes of vesicles and lamellar phases composed by soy lecithin. The self-association property of FF has been recognized as a critical element for the formation of amyloid fibers involved in neurodegenerative pathologies such as Alzheimer's and Parkinson's disease. By using SAXS experiments (Fig. 4a), the supramolecular organization of lipids membrane was obtained, where a multilamellar-unilamellar transition induced by the increase of the FF concentration was observed. An advanced SAXS data modeling (Oliveira et al. 2012) was used for the determination of relevant structural parameters, such as membrane effective thickness $\delta_M$, electron density contrast profile ($\Delta\rho$) within the membrane, and the Caillé parameter (Caille 1972) ($\eta$) that characterizes the elasticity of the lamellar structure. The authors observed changes in the elastic properties of the membranes of both vesicles and lamellar phases (Fig. 4b, c) after the incorporation of the FF peptide. These experimental results contributed to a better understanding of the mechanisms underlying peptide aggregation and stability of lipid membranes, providing insights into biological implications of such processes in the development of neurodegenerative diseases associated to amylogenic fibers.

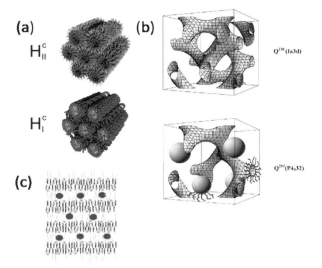

**Fig. 5** (a) Schematic representation of the inverse lipid hexagonal phase $H_{II}^C$ containing DNA fragments interspersed within the tubes formed by the lipids, and the $H_I^C$ phase, with the fragments in the interstitium of the lipid phase. (b) Cubic phases composed by monoolein with the incorporation of protein cytochrome c. (c) Representation of a "sandwich" structure, in which the lamellar phase consists of lipids and the DNA molecules are inserted in the aqueous channels between the bilayers. Adapted with permission from (Ewert et al. 2006) the American Chemical Societ, (Lendermann and Winter 2003) Copyright 2020 Royal Society of Chemistry

Lipids and fatty acids can be used as carrier systems of DNAs, proteins, and small drugs. For DNA, since it is a highly anionic molecule, an often adopted strategy consists in the use of a host lamellar system formed by cationic lipids (Radler et al. 1997; Koltover et al. 1998; Ewert et al. 2006). Radler et al. (1997) formulated these complexes using a mixture of DOTAP-DOPC lipids and applied X-ray diffraction (XRD) and fluorescence microscopy to characterize the system. By indexing the peaks, the authors observed the formation of a lamellar phase with DNA fragments (48,502 and 4361 pb) interspersed, organized in a two-dimensional smectic phase. Hexagonal arrangements were also observed in different lipid mixtures (DOPE/DOTAP), where the DNA fragments are interleaved between columnar lipid aggregates (Fig. 5a) (Koltover et al. 1998; Ewert et al. 2006). More recently, new systems using zwitterionic host lipids have been studied since cationic lipids have high cytotoxicity. Silva et al. proposed a lamellar host formed by a mixture between soy lecithin and the commercial co-surfactant Simulsol (Silva et al. 2011). This co-surfactant is composed of short chain fatty acids that have the function of increasing the temperature range for which the fluid lamellar phase is stable, therefore providing more flexibility to the lamellar structure. In addition, the fatty acids incorporation resulted in the increase of the separation distance between adjacent lecithin bilayers, allowing the incorporation of DNA (Gerbelli et al. 2013). SAXS and wide-angle X-ray scattering (WAXS) were used to characterize the systems that exhibited great polymorphisms as a function of DNA concentration (Fig. 5c). Other supramolecular structures can also be observed using fatty acids. Lendermann et al. studied cubic phases composed by monoolein (Fig. 5b) with the incorporation of protein cytochrome c (cyt c) (Lendermann and Winter 2003). The authors verified that the presence of the protein promotes a host phase transition never seen before, going from a cubic to a putative phase, i.e., composed of cubic structures of type Ia3d where the aqueous channel is constituted by reverse spherical micelles (Fig. 5b) (Lendermann and Winter 2003).

DNA/RNA biomolecules are another system used in many applications as drug delivery systems and biosensors due to their specificity and programmed assembly (Oliveira et al. 2010). Chuaychob et al. (2020) proposed a rapid and straightforward colorimetric sensor using G4-DNA and AuNPs. The G4-DNA has four guanines, and its structure is composed of a four-stranded helical structure with stacks of planar guanine tetrads stabilized by a monovalent cation inserted between the quartets. In this work, the authors chose to detect a biomolecule used as a powerful antineoplastic drug (cisplatin). The cisplatin in high concentration (800 mM) induces cell necrosis, giving rise to severe side effects, although at low concentration (8 mM) can cause cell apoptosis (Lieberthal et al. 1996). Due to these dangerous effects, there is a concern about this molecule's excretion in the patients' urine and local sewage networks contamination. In this way, rapid detection of this molecule is of great importance. X-ray scattering investigations were fundamental for the proposition of the self-assembly process of the colorimetric biosensor and the interaction with cisplatin (Chuaychob et al. 2020). Fig. 6a is shown the scattering intensity curves of G4-DNA/AuNPs as a function of cisplatin concentration. The authors observed that when increasing the percentage of salt in the sample (Fig. 6b), the low-q region presents a diffuse peak that becomes more prominent with the increase in the biomolecule concentration. This information is indicative of ordering in the sample with an increase in the amount of cisplatin. Also, the experimental curves were fitted with a spherical shape factor from the scattering of gold nanoparticles. With this information, it was possible to propose an interaction mechanism between the biosensor and cisplatin (Fig. 6c) where it would occur the deformation of G4 by the formation of a cisplatin-monoadduct inducing the non-crosslinking aggregation of the colloidal particles (Chuaychob et al. 2020).

Polymer is another class of biomaterials widely used. Folkertsma et al. combined in situ SAXS and impedance spectroscopy experiments to study the changes in porous membrane composed by poly(ferrocenylsilane) and poly(ionic liquid)s during the redox process (Folkertsma et al. 2017). They observed, at the low-q region of the SAXS curves (Fig. 7a), an increase of the slope as the potential is elevated. This is an indication of the formation of large structures, being corroborated by scanning electron microscopy images (Fig. 7d) (Folkertsma et al. 2017).

The SAXS curves were fitted with a empiral model taking into account an effective polymer correlation length ($\zeta$). In Fig. 7b is shown the behavior of $\zeta$ as a function of the number of cycles at different applied potentials. The authors observed that $\zeta$ is independent of the number of cycles, suggesting a morphologically reversible system (Fig. 7c). In addition, $\zeta$ is smaller (the polymer is more compressed) for

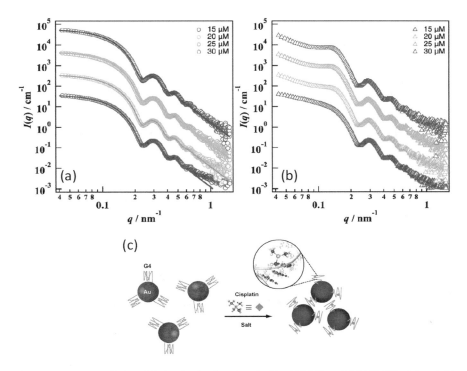

**Fig. 6** SAXS data of 35-nt G4–AuNPs in phosphate buffer (pH 5.0) at 0.1 M NaNO$_3$ (**a**) and (**b**) 1.0 M NaNO$_3$. The solid lines in (**a**) and (**b**) are the fitting curves of form and structure factors. (**c**) Schematic illustration of cisplatin detection by G4–AuNPs. The deformation of G4 by the formation of a cisplatin-monoadduct induces the non-crosslinking aggregation of the colloidal particles. Adapted with permission from (Chuaychob et al. 2020) Copyright 2020, Royal Society of Chemistry

the oxidized state compared to the reduced one (where the polymer is more stretched) (Folkertsma et al. 2017).

Sarode et al. carried out SAXS experiments on a block copolymer varying the system temperature as well as the presence of different ions on the water uptake in order to investigate temperature-dependent structural changes and electrochemical energy conversion on anion-exchange membranes (AEMs) (Sarode et al. 2017). In Fig. 8a is shown the SAXS curve for the block polymers composed of polyethylene and poly(vinyl benzyl trimethylammonium bromide (PE-b-PVBTMA) in the presence of bromite ions. From the analysis of the scattering curves, the authors propose a representative model of the polymeric system (Fig. 8b) based on the decaying law $I(q) \propto q^{-4}$, characteristic of spherical particles, and on the presence of a peak (star) and a shoulder (two stars), both indicating interesting internal organization of the polymer hydrophilic regions with different amounts of water promoted by the presence of ions (Sarode et al. 2017).

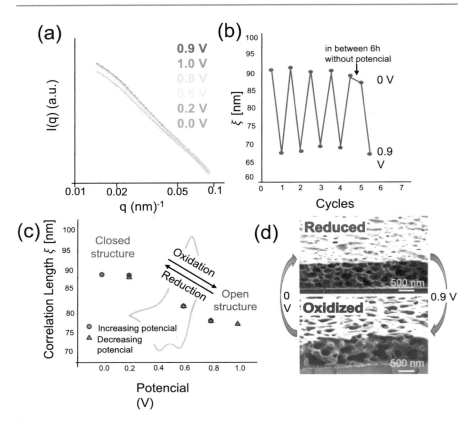

**Fig. 7** (**a**) SAXS data of membranes composed by poly(ferrocenylsilane) and poly(ionic liquid)s. (**b**) Effect of switching the potential between 0 and 0.9 V on the correlation length ($\xi$) as a function of the number of cycles. (**c**) Behavior of $\xi$ as a function of redox potential. Adapted with permission from (Folkertsma et al. 2017) Copyright 2020, American Chemical Society

There are many other examples in the literature regarding the use of scattering techniques, which could be used in large-scale facilities like synchrotrons and neutron centers or laboratory-based systems. As shown in this chapter, the possibility of study several systems, particularly biomolecular assemblies directly in solution with an easy variation of environmental conditions, is one of the main reasons why these methods are widely spread in different areas and become essential tools for structural and dynamic investigations.

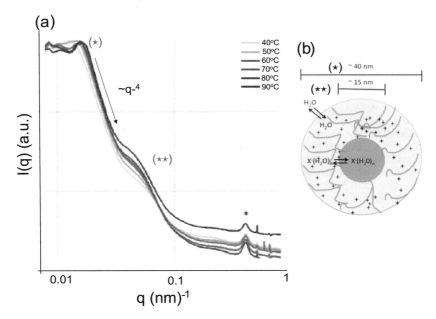

**Fig. 8** (**a**) SAXS data of PE-b-PVBTMA in the presence of bromite ions at different temperatures. (**b**) Model representation of the hydrophilic polymer regions whose diameters were obtained by SAXS experiments. Adapted with permission from (Sarode et al. 2017) Copyright 2020, American Chemical Society

# References

Andersen CBF, Torvund-Jensen M, Nielsen MJ, Oliveira CLP, Hersleth HP, Andersen NH, Pedersen JS, Andersen GR, Moestrup SK (2012) Nature 489:456–459

Caille A (1972) Comptes Rendus Hebdomadaires Des Seances De L Academie Des Sciences Serie B 274:891

Capito RM, Azevedo HS, Velichko YS, Mata A, Stupp SI (2008) Science 319:1812–1818

Chu B, Liu TB (2000) J Nanopart Res 2:29–41

Chuaychob S, Thammakhet-Buranachai C, Kanatharana P, Thavarungkul P, Buranachai C, Fujita M, Maeda M (2020) Anal Methods 12:230–238

Ewert K, Evans H, Zidovska A, Bouxsein N, Ahmad A, Safinya CR (2006) J Am Chem Soc 128: 3998–4006

Folkertsma L, Zhang KH, Czakkel O, Boer HL, Hempenius MA, van den Berg A, Odijk M, Vancso GJ (2017) Macromolecules 50:296–302

Gerbelli BB, Rubim RL, Silva ER, Nallet F, Navailles L, Oliveira CLP, Oliveira EA (2013) Langmuir 29:13717–13722

Gerbelli BB, Silva ER, Soares BM, Alves WA, Oliveira EA (2018) Langmuir 34:2171–2179

Glatter O (1977) J Appl Cryst 10:415–421

Goni F (1838) BBA-Biomembranes 2014:1467–1476

Heller WT (2010) Acta Crystallogr Sect D Biol Crystallogr 66:1213–1217

Jacrot B (1976) Rep Prog Phys 39:911–953

Jones AR (1999) Prog Energy Combust Sci 25:1–53

Koltover I, Salditt T, Radler J, Safinya CR (1998) Science 281:78–81
Lendermann J, Winter R (2003) Phys Chem Chem Phys 5:1440–1450
Lieberthal W, Triaca V, Levine J (1996) Am J Physiol Ren Fluid Electrolyte Physiol 270:F700–
    F708
Lombardo D, Calandra P, Kiselev MA (2020) Molecules 25:5624–5661
Luzzati V, Mustacchi H, Skoulios A (1957) Nature 180:600–601
Mahieu E, Gabel F (2018) Acta Crystallogr Sect D Struct Biol 74:715–726
Mertens HD, Piljic A, Schultz C, Syergun DI (2012) Biophys J 102:2866–2875
Miller CC (1924) Proceedings of the Royal Society of London Series A, containing papers of a
    mathematical and physical character, vol 106, pp 724–749
Nasta V, Vela S, Gourdoupis S, Ciofi-Baffoni S, Svergun DI, Banci L (2019) Sci Rep 9:18986
Oliveira CLP (2011) Current trends in X-ray crystallography. InTech, London, pp 367–392
Oliveira CLP, Behrens MA, Pedersen JS, Erlacher K, Otzen D (2009) J Mol Biol 387:147–161
Oliveira CLP, Juul S, Jorgensen HL, Knudsen B, Tordrup D, Oteri F, Falconi M, Koch J,
    Desideri A, Pedersen JS, Andersen FF, Knudsen BR (2010) ACS Nano 4:1367–1376
Oliveira CLP, Gerbelli BB, Silva ER, Nallet F, Navailles L, Oliveira EA, Pedersen JS (2012) J Appl
    Cryst 45:1278–1286
Orthaber D, Bergmann A, Glatter O (2000) J Appl Cryst 33:218–225
Pabst G, Danner S, Podgornik R, Katsaras J (2007) Langmuir 23:11705–11711
Petoukhov MV, Svergun DI (2006) Eur Biophys J Biophys Lett 35:567–576
Piljic A, Schultz C (2006) Mol Biol Cell 17:3318–3328
Radler JO, Koltover I, Salditt T, Safinya CR (1997) Science 275:810–814
Rahimabadi PS, Khodaei M, Koswattage KR (2020) X-Ray Spectrom 49:348–373
Rambo RP, Tainer JA (2013) Annu Rev Biophys 42:415–441
Roig AR, Alessandrini JL (2007) Part Part Syst Charact 23:431–437
Rojas J, Gerbelli BB, Ribeiro A, Nantes-Cardoso I, Giuntini F, Alves WA (2019) Biopolymers 110:
    e23245
Sarode HN, Yang Y, Motz AR, Li YR, Knauss DM, Seifert S, Herring AM (2017) J Phys Chem C
    121:2035–2045
Scotti A, Liu W, Hyatt JS, Herman ES, Choi HS, Kim JW, Lyon LA, Gasser U, Fernandez-Nieves
    A (2015) J Chem Phys 142:234905–234918
Silva ER, Oliveira EA, Fevrier A, Nallet F, Navailles L (2011) Eur Phys J E 34:83
Silva ER, Listik E, Han SW, Alves WA, Soares BM, Reza M, Ruokolainen J, Hamley IW (2018)
    Biophys Chem 233:1–12
Singer S, Nicolson G (1972) Science 175:720
Svergun DI, Koch MHJ (2003) Rep Prog Phys 66:1735–1782
Talebian E, Talebian M (2013) Optik 124:2324–2326
Xu RL (2015) Particuology 18:11–21
Zemb T, Lindner P (2002) Neutrons, X-rays and light: scattering methods applied to soft condensed
    matter. Elsevier, Amsterdam

# High-Resolution Optical Fluorescence Microscopy for Cell Biology Studies

Fernando Abdulkader, Richard P. S. de Campos, José A. F. da Silva, and Fernanda Ortis

## 1    Introduction

The cell is the smallest functional and structural unit of all living organisms. Interestingly, this name was coined by the English scientist Robert Hooke in 1665, upon observing cork sections with the first described composed microscope (Hooke 1665/2014). The observed polyhedric structures, nowadays known as plant cells, reminded him of either honeycomb cells or monastic cells in cloisters.

Thus, all living organisms are composed by cells, as they appear in endless forms from unicellular to more complex multicellular organisms, such as humans (Nelson and Cox 2017). These cells, even in multicellular organisms, have specific individual

F. Abdulkader
Instituto Nacional de Ciência e Tecnologia em Bioanalítica, INCTBio, Campinas, SP, Brazil

Department of Physiology and Biophysics, Institute of Biomedical Sciences (ICB), University of São Paulo (USP), São Paulo, SP, Brazil

R. P. S. de Campos
Department of Analytical Chemistry, Institute of Chemistry, University of Campinas (UNICAMP), Campinas, SP, Brazil

Current address: Nanotechnology Research Centre, National Research Council of Canada, Edmonton, AB, Canada

J. A. F. da Silva
Instituto Nacional de Ciência e Tecnologia em Bioanalítica, INCTBio, Campinas, SP, Brazil

Department of Analytical Chemistry, Institute of Chemistry, University of Campinas (UNICAMP), Campinas, SP, Brazil

F. Ortis (✉)
Instituto Nacional de Ciência e Tecnologia em Bioanalítica, INCTBio, Campinas, SP, Brazil

Department of Cell and Developmental Biology, Institute of Biomedical Sciences (ICB), University of São Paulo (USP), São Paulo, SP, Brazil
e-mail: fortis@usp.br

L. T. Kubota et al. (eds.), *Tools and Trends in Bioanalytical Chemistry*,
https://doi.org/10.1007/978-3-030-82381-8_8

characteristics important for their specific functions and proper maintenance of the organism's characteristics. In fact, cellular malfunctioning is at the basis of all diseases (Price and Culbertson 2007).

All cells are bounded by the plasma membrane that, as a selectively permeable barrier, separates the extracellular from the intracellular environment (Karp 2010; Alberts et al. 2014). The most important components for cellular structure and function are found in the intracellular compartment. These cellular components may be divided into two groups, and the first one is composed by supramolecular complexes with specific functions, known as organelles. Mitochondria, endoplasmic reticulum, the Golgi, ribosomes, vacuoles, and chloroplasts are examples of cellular organelles (Alberts et al. 2014). The remaining cellular components—proteins, nucleic acids, amino acids, metabolites, coenzymes, and ions—are found in suspension in the cytosol or in the nucleus, which is only found in eukaryotic cells (Alberts et al. 2014).

Most descriptions of cellular structure, particularly those of organelles, came from microscopy and classic functional studies, which consist in isolating the organelles or cytosolic components and evaluating their function by various techniques, including the use of optical and electron microscopy (Nelson and Cox 2017). Indeed, the increase in the understanding of cell biology can be historically correlated to the improvement in microscopy technologies and equipment. We can define microscopy as a group of techniques that allow the observation of structures that are not visible by the naked eye. Human curiosity to observe the microenvironment is ancient, and there are descriptions of amplifying lens use by the philosophers Seneca (4 BC–65 AD) and Pliny (23–79 AD). Many improvements have been achieved since the description of the first observations by Hooke in 1665 using a composed microscope. With the field evolution, we now have phase, fluorescence, and electronic microscopes with multiple functions that allow for the observation of specific structures with high precision, magnification, and resolution (Fig. 1).

Although of extreme importance, functional in vitro studies of organelles and biomolecules isolated from the cells may not totally reflect their real function as when these molecules are working in the cell in vivo (Nelson and Cox 2017). Indeed, the ability of cells to respond to the environment is related to complex signaling systems, in which many biomolecules may be involved and interacting among themselves. Thus, the use of systems with intact cells, in vivo, such as modern optical microscopy techniques, may allow a better understanding on the behavior of groups of biomolecules or even of groups of organelles in their own environment (Tinoco and Gonzalez 2011).

Herein we discuss some highly efficient contemporary fluorescence-based microscopy techniques for eukaryotic cell studies. We focus on *Confocal Microscopy* that allows the capture of images with high resolution and 3D reconstruction of intact cells or tissues samples as well as real-time sequential image capture from living cells. In addition, the use of fluorescence, based on its high intrinsic sensibility, has become a great ally for the technological development of other microscopy techniques, such as *Förster resonance energy transfer (FRET)* and *Total Internal Reflection Fluorescence (TIRF)* that are also discussed here, which are aimed at monitoring intracellular molecules, reaching enough sensitivity and specificity to

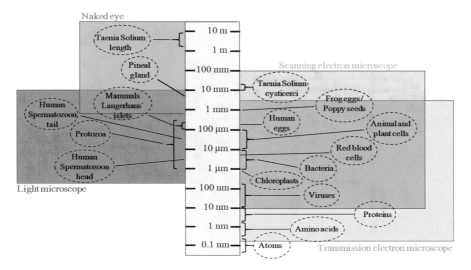

**Fig. 1** Diagram showing the range of object sizes that can be visualized by different methods. The figure shows selected organisms, organelles, and molecules to exemplify the resolution power of different microscopes. The range for the naked eye resolution is shown in green, for light microscopes in red, for scanning electron microscopy in green and for transmission electron microscopy in yellow. Some techniques have overlapping ranges, what is shown in the figure by color overlapping (courtesy of Dr. Alexandre Z. Carvalho, UFABC-Brazil)

study the behavior of individual molecules in vitro (Toomre and Bewersdorf 2010; Neto et al. 2012).

## 2 Introduction to Biological Applications of Fluorescence

The absorption of electromagnetic radiation brings molecules from a fundamental electronic state ($E_0$) to an excited higher energy electronic state ($E_1$). The subsequent loss of this energy may occur by radiating or non-radiating pathways. In the radiating pathway, part of the energy is lost as light, with photon emission, through which the molecule then returns to its fundamental energetic state. If the excited chemical species spontaneously emits radiation while retaining its spin multiplicity, this phenomenon is called *fluorescence*, and the molecular entity that emits fluorescence is thus called a *fluorophore*.

Fluorescence emission of light usually occurs within nanoseconds after the absorption of a shorter wavelength (higher energy) light (Clegg 2012; Skoog et al. 2013). Its efficiency is defined by the ratio between the number of emitted photons and the total number of excitation photons. In other words, it is the difference between the exciting and emitted wavelength, known as Stokes' shift, that makes fluorescence a powerful tool, which allows the specific visualization of the object that is fluorescent. Light emission that occurs after electron spin change (e.g., from

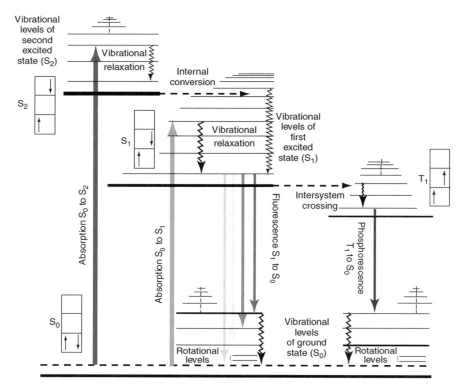

**Fig. 2** Simplified Perrin-Jablonski diagram illustrating the electronic states of a molecule and the transition between them. $S$ represents the *singlet* state and $T$ the *triplet* state. The $0$ represents the fundamental electronic state, and $1$ and $2$ represent the first and second excited electronic states, respectively. Reproduced with permission from Ref. (Clegg 2012)

the singlet to triplet state) has a prolonged duration in a phenomenon denominated phosphorescence. The processes that occur during molecular excitation and radiation emission are represented in the simplified Perrin-Jablonski diagram (Fig. 2).

As mentioned earlier, fluorescence has a high intrinsic sensibility compared to other techniques that rely on light absorption measurements. This occurs because the excitation and emission wavelengths are different, and thus it is feasible to completely filter out the excitation light without blocking the emitted fluorescence. Therefore, only the fluorescent object is imaged (Tinoco and Gonzalez 2011). This way, fluorescence has been used in many areas of study in which low detection limits are necessary, such as individual cell analysis (Wu et al. 2004; Huang et al. 2007), and biomolecule (Liu et al. 2013; Simplicio et al. 2010; Su et al. 2011), food (Tran et al. 2013; Zhou et al. 2013), or drug analyses (Kolberg et al. 2009; Chang and Yu 2011), among others.

Molecular fluorescence spectrometry is an analytical technique based on the detection of the emitted light from a molecule during the fluorescence event

(Skoog et al. 2013). Likewise, fluorescence microscopes are light microscopes that use the fluorescence or phosphorescence phenomenon to generate an image or a measurable signal. The resulting image can be fairly simple, as the ones acquired by epifluorescence microscopes, or more complex, like the ones generated by confocal microscopes, using optical sections to obtain a better resolution of the fluorescent image, as further discussed below.

It is important to point out that not all molecules have the capacity to generate fluorescence, since their structure may preferably allow for energy loss in a non-radiative pathway after light excitation. Fluorescent molecules possess fluorophore groups in their structure, which are responsible for the light absorption and emission properties. More effective fluorescence effects are obtained using compounds that have resonant ring structures (such as aromatic molecules) (Skoog et al. 2013). These fluorescing structures can be made available within biological systems either by chemical derivatization of molecular probes with fluorophore moieties or, as it is increasingly more frequent, via the genetic expression of fluorescent protein sensors. Thus, many of the detection methods are performed indirectly, when a non-fluorescent molecule is analyzed after being detected upon specifically interacting with a fluorophore dye. These dyes can be used alone due to their affinity for specific molecules. For example, ethidium bromide intercalates in DNA and RNA molecules, generating a measurable signal. However, most dyes are conjugated with other molecules, such as antibodies, allowing for the detection of the target structure or molecule.

The most common synthetic fluorescent dyes used as conjugates are rhodamine-B, cyanine (Cy3 and Cy5), fluorescein isothiocyanate (FITC), tetramethyl rhodamine-isothiocyanate (TRITC) and variations of Alexa Fluor® (Liu et al. 2013; Lamichhane et al. 2013) (Table 1). As mentioned above, another important contemporary tool in cell studies are the fluorescent proteins, most of them derived from the green fluorescent protein (GFP), which was the first fluorescent protein to be described and isolated. Using biotechnology techniques, we now have many varieties of fluorescent proteins that have fluorescent color emissions that range from green to blue, red, and yellow (Table 2) (Snapp 2009).

As observed in Tables 1 and 2, each fluorophore and fluorescent protein has a specific excitation and emission wavelength. Thus, it is crucial to previously know this information to design the experiments and even to be sure that the equipment is prepared to detect a specific dye. Of note, many microscopy experiments may require evaluation of different molecules in the same sample, thus double or triple labeling may be used. In this case, it is important to use dyes with different excitation and emission wavelengths. Other important characteristics that need to be taken into account when choosing a dye are its photostability and phototoxicity (Coelho et al. 2013; Shaner et al. 2005). The first refers to its resistance to photobleaching (i.e., loss of the fluorescence property) after cycles of excitation and emission. Phototoxicity is a phenomenon that leads to cell damage by either exposure to low- or high-wavelength lights used to excite fluorophores, which can also release reactive oxygen species that are also potentially dangerous for cellular structures, and/or heat the sample as well (Dailey et al. 2006; Laissue et al. 2017).

**Table 1** Common fluorophores used in fluorescence microscopy

| Fluorescent label | Absorption wavelengths (nm) | Emission wavelengths (nm) | visible color |
|---|---|---|---|
| Hydroxycoumarin | 325 | 386 | Blue |
| DyLight™ 405 | 400 | 420 | Blue |
| Alexa Fluor 350 | 346 | 442 | Blue |
| Aminocoumarin | 350 | 445 | Blue |
| DyLight™ 488 | 493 | 518 | green |
| Fluorecein (FITC) | 495 | 519 | green |
| ATTO 488 | 501 | 523 | green |
| Alexa fluor 430 | 433 | 541 | green |
| ATTO 532 | 532 | 553 | Yellow |
| Alexa fluor 532 | 532 | 553 | yellow |
| Cy3™ | 554 | 568 | Yellow |
| DyLight™ 550 | 562 | 576 | Yellow |
| Alexa fluor 555 | 555 | 565 | Orange |
| Rhodamine (TRITC) | 550 | 570 | Orange |
| Alexa fluor 546 | 556 | 573 | Orange |
| ATTO 550 | 554 | 576 | Orange |
| Alexa fluor 568 | 578 | 603 | red |
| Texas Red | 596 | 620 | red |
| ATTO 594 | 601 | 627 | red |
| DyLight™ 594 | 593 | 618 | red |
| DyLight™ 649 | 646 | 647 | far-red |
| Allophycocyanin (APC) | 650 | 660 | far-red |
| Cy5™ | 650 | 667 | far-red |
| ATTO 655 | 663 | 684 | far-red |
| Cy5.5™ | 678 | 703 | near infra-red |
| Alexa fluor 750 | 749 | 775 | near infra-red |
| DyLightTM 755 | 754 | 776 | near infra-red |
| DyLightTM 800 | 777 | 794 | infra-red |

**Table 2**  Common fluorescent protein used in cellular biology studies

| Fluorescent protein | Absorption wavelengths (nm) | Emission wavelengths (nm) | visible color |
|:---:|:---:|:---:|:---:|
| ECFP | 439 | 476 | cyan |
| Cerulean | 433 | 475 | cyan |
| mTurquoise | 434 | 474 | cyan |
| CyPet | 435 | 477 | cyan |
| AmCyan | 458 | 489 | cyan |
| mTFP1 (teal) | 462 | 492 | cyan |
| EBFP | 383 | 445 | Blue |
| EBFP2 | 383 | 448 | Blue |
| Azurite | 384 | 450 | Blue |
| mTagBFP | 399 | 456 | Blue |
| EGFP | 484 | 507 | green |
| Emerald | 487 | 509 | green |
| Azami Green | 492 | 505 | green |
| T-Sapphire | 399 | 511 | green |
| Tag GFP | 482 | 505 | green |
| EYFP | 514 | 527 | Yellow |
| Topaz | 514 | 527 | Yellow |
| mCitrine | 516 | 529 | yellow |
| ZsYellow1 | 529 | 539 | Yellow |
| mBanana | 540 | 553 | Yellow |
| Kusabira Orange | 548 | 559 | Orange |
| mOrange | 548 | 562 | Orange |
| dTomato | 554 | 581 | Orange |
| DsRed | 563 | 582 | Orange |
| mTangerine | 568 | 585 | Orange |
| mRuby | 558 | 605 | red |
| mCherry | 587 | 610 | red |
| mApple | 568 | 592 | red |
| mStrawberry | 574 | 596 | red |
| mRaspberry | 598 | 625 | red |

# 3    High-Resolution Fluorescence Microscopy

The high detection power of fluorescence combined with the possibility of selectively labeling molecules or specific regions of macromolecules with fluorescent tags makes fluorescence a prime technique for the optical detection of molecules in biological samples. In fact, multiple variations of fluorescence spectroscopy have been developed to monitor cellular processes in vivo (Tinoco and Gonzalez 2011; Toomre and Bewersdorf 2010; Coelho et al. 2013): wide-field fluorescence microscopy, total internal reflection fluorescence (TIRF) microscopy, highly inclined and laminated optical sheet (HILO) microscopy, selective plane illumination microscopy (SPIM), fluorescent speckle microscopy (FSM), photoactivation and photoconversion (PA and PC) fluorescence, photoactivated light microscopy (PALM), stochastic optical reconstruction microscopy (STORM), stimulated emission depletion (STED) and Förster (or fluorescence) resonance energy transfer (FRET). The special resolution of some of these techniques is so high, such as PALM, STORM, and STED, that they are aptly considered super-resolution imaging techniques (Alberts and Davidson 2013). Though they are not the scope of this chapter, these techniques are worth mentioning since they are increasingly becoming more affordable, despite still requiring, for instance, high laser power, special photoswitchable fluorophores, outstanding photodetection devices, and/or massive digital processing.

Wide-field fluorescence microscopy is the simplest application of fluorescence microscopy to biological samples. It involves the illumination of a sample with an excitation wavelength and collection of the emitted light from in-focus and out-of-focus planes (thus the name wide-field). As the elicited emission happens in all directions regardless of the path of the excitation light, it can be detected either opposite from the source of excitation, in what is called transillumination, or by collecting the emitted fluorescence through the same objective lens through which the excitation light was delivered to the sample. This second means of fluorescence detection, in the same path as that of the excitation light, is called epifluorescence (Fig. 3a). When this illumination comes from a beam directed upwards from the objective, as in inverted microscopes, the lens can usually be brought much closer to the sample, thus the efficiency of illumination is maximized.

In wide-field microscopy, however, the image can be blurred by the collection of fluorescence from both in-focus and out-of-focus planes, and further improvements in the design of microscopes and illumination schemes are desired. In fact, spatial resolution, i.e., the ability of a microscope to distinguish two close objects in a focal plane, is limited by the diffraction of light coming from the sample in the aperture of the objective lens. This can be increased by the use of oil immersion lenses, as the oil occupies the space between the sample and the objective that would be filled by air, whose refraction index is very different from both the sample and the objective ones. This also allows for an increased angle of light coming from the sample that can be captured by the lens, and thus also more photons can be collected. This efficiency of the lens in collecting light from a single focal point at different angles is quantified in what is called the *numerical aperture* of the lens (Alberts and Davidson 2013).

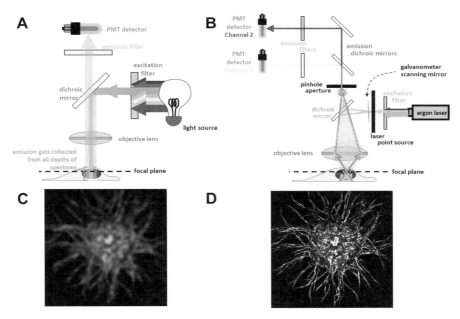

**Fig. 3** Confocal microscopes and confocal images compared to regular fluorescence microscopes. (a) Basic components of a conventional fluorescence microscope. Note that the emitted fluorescence is collected from both in-focus and out-of-focus depths of the specimen. (b) Illustration of the components of a laser confocal scanning microscope. The light orange spot below the focal plane also emits fluorescence, but that does not converge to the pinhole aperture, and so is not detected by the photomultiplier sensors located behind the pinhole. (c) Example of a blurred image that would be obtained if collected in a conventional fluorescence microscope. (d) The actual image, which was obtained in a laser confocal scanning microscope (3D culture of HUVEC cells expressing GFP grown in fibrin gel, 10x magnification—courtesy of Dr. Vanessa M. Freitas. University of São Paulo)

Despite that, conventional wide-field microscopy cannot achieve single-molecule imaging even with the use of high numerical aperture lenses. However, techniques such as TIRF and HILO have dramatically increased the spatial resolution of the images to limits much lower than those found in conventional microscopes. In some cases, resolution can be brought to 25–80 nm (Toomre and Bewersdorf 2010).

The various applications of the high-resolution fluorescence microscopies demonstrate their potential for the detection of multiple cellular components both in vitro and in vivo (Coelho et al. 2013). Among them, laser scanning confocal microscopy has been widely used to obtain images from very thin sections and combine them in highly accurate 3D reconstructions of the sample. Additionally, the analysis of single molecules by FRET (smFRET) has been praised for its capacity to detect, with a high temporal resolution, stochastic behaviors that are inherent to the chemical reactions and macromolecular conformational changes that are characteristic of the intracellular dynamic processes (Lamichhane et al. 2013; Wang et al. 2013). Likewise, the development of TIRF microscopy has allowed the acquisition of fast and reliable data with high resolution and reduced background fluorescence, which has boosted the applications of smFRET in recent years (Ha 2001; Lin and Hoppe 2013). The following sections will discuss these three modes of high-resolution fluorescence microscopy in more detail.

# 4 Confocal Microscopy

Confocal Microscopy, usually in the form of confocal laser scanning microscopy (CLSM), also known as laser confocal scanning microscopy (LCSM), is a powerful technique that revolutionized cell biology studies by providing optimal resolution of tridimensional images. Confocality can be defined as the capacity to select light coming from a specific depth in a sample by filtering out light emitted by or scattered from other depths. Confocal microscopy allows for the spatial localization of structures and molecules within the cell, using living and fixed cells as well as tissue sections. These 3-D images are obtained by computational reconstruction and combination of sectioned 2-D image data from the studied sample; thus, the term section microscopy is also used to describe this technique.

Another particularly important feature of confocal microscopes that make them much more potent than conventional wide-field optical and fluorescence microscopes is their increased optical resolution due to the elimination of background pick-up caused by out-of-focus light and scatter (Conchello and Lichtman 2005; Paddock 2000). This permits the captured sectioned images to be sharp, with great spatial precision. In conventional fluorescence microscopy, for instance, the illumination of the sample elicits fluorescence in the whole depth of the sample, rendering the majority of the cell volume out of focus, independently of the focus being set vertically on the sample (a cell, for example) (Fig. 3). In addition, this out-of-focus light coming from the image may suffer diffraction, reflection, and refraction by the sample, before being captured by the objective lens of the microscopy. Thus, the captured light emission will appear to be coming from the last point that was scattered and not from the real point from where it was emitted. This is highly affected by sample depth. Therefore, the use of confocality to capture images of thin slices from thick objects, this way removing the influence of out-of-focus light in each slice, allows the acquisition of extremely well-defined images (Paddock 2000). However, there is still a limit for the depth penetration of these microscopes, and for studies of thicker samples (e. g. large spheroids, organoids, and small animals), other microscopy approaches should be used, such as two-photon microscopy or LSFM.

The use of fluorescence is an important component for image detection in confocal microscopy, thus the principles of excitation and emission, signal loss, diffraction, autofluorescence, and fluorophores must be considered in any application of this technique.

The first known confocal microscope was built by Marvin Minsky in 1955, during his post-doctoral training at Harvard University, to study images of neural nets from brain living tissues (Minsky 1988). This technology, patented in 1957, is still used in modern confocal microscopes, with few improvements. Modern confocal microscopes have the same basic components: (1) a *light source* to be projected onto the sample (the *laser* beam is the ideal light source since it contains all its energy in a collimated coherent plane wave). (2) A *scanner*, also known as Z *control*, allows the laser beam focus to be moved line-by-line on the sample, and in this way, the image can be acquired in sections in the X-Y (lateral coordination) and Z (depth coordination) axes. (3) A detector, typically a *photomultiplier tube* (PMT), will detect emission and reflected photons from the sample. This type of detector has a low noise and fast response, being able to detect even single photons. (4) The *pinhole*, which is fundamental for the discrimination of the X and Y axis depth and position, allowing a focused spotlight to reach the detector. The pinhole position, always in front of the detector, is crucial to remove the background coming from the out-of-focus light. (5) The *beam splitter*, composed by dichroic mirrors and emission filters. (6) *Objectives*, responsible for the optical image formation, which determine the image quality properties, as well as the resolution in the X, Y, and Z axes (Fig. 3b–d).

The confocal specific principle is that the image is scanned by individual and sequential illumination of specific regions of the sample, while at the same time avoiding the detection of light coming from other regions (out-of-focus light). The resolution quality of the image (due to blocking of scattered light capture) is inversely proportional to the size of the scanned region at each moment. Thus, the laser beam, which provides a light focus sharply at a single point, decreasing the beam thickness, crosses the sample at each specific section before being deflected by the scanner to the adjacent section. In confocal fluorescence microscopy, the laser promotes fluorophore excitation, and the wavelength of the laser is specific for the fluorophore to be imaged (Pawley 1991; Webb 1999). Thus, depending on the required spectral line coverage needed for each study, there is a variety of lasers for confocal microscopy; the most common are the argon-ion lasers, which can emit from UV (230 nm) to green (514 nm). There are also the helium-neon (He-Ne) and Krypton-ion (that can also be combined with argon, Ar-Kr) and helium-cadmium (He-Cd) lasers, which provide lines from the blue to the red, as well as zinc-selenium (Zn-Se) diode lasers (Gratton and vandeVen 2006).

When the emitted light returns from the sample, it passes through the pinhole in a plane that allows the illumination spot and the pinhole aperture to be focused simultaneously at the same point, from which comes the name confocal scanning. This position allows the pinhole to detect only the light coming from the focal plane, rejecting the scattered light coming from around the illuminated point and part of the background light collected by the objectives. The diameter of the pinhole is important for its efficiency: while the background is minimal in exceedingly small pinholes, this compromises signal capture, especially when the light level is limited. Thus, the pinhole size should be between 60 and 80% of the diameter of the diffraction-limited spot of the image one intends to acquire (Conchello et al. 1994; Sandison et al. 1995).

The specific light is then detected with the help of the photomultipliers, and the information is processed using computers that reconstruct the several scanned images from the different points of the sample. To produce these images, the biological samples are immobilized and the light beam moves from each point to the next for the scanning to be performed. One common method used in cell biology studies uses a laser scanner with the help of two or three oscillating mirrors, which deflect the angle of the light beam going into the sample and of the emitted light coming from the sample (respectively described as scanning and "descanning") across a fixed pinhole and detector (Conchello and Lichtman 2005). This way, each mirror scans the illumination along two axes (X and Y), creating a 2-D image. By moving the focus vertically ($z$-axis) it is possible to create stack images used for computational 3-D reconstructed imaging (Conchello and Lichtman 2005; Paddock 2000). This approach builds the image sequentially one pixel at a time, thus the capture of scanned images is not as fast as for conventional microscopy. This slow speed can be a problem since the time and the intensity in which the sample is illuminated can increase photobleaching and phototoxicity (better described below). The latter is especially important when live cells and organisms are studied. Although the photomultiplier tube has high sensitivity, being able to capture even single photons, they are not very efficient, detecting only 10% or less fluorescence signals coming from the pinhole, thus increasing scanning speed can generate loss of image quality (Paddock 2000).

Taking this into account, it is important to increase the velocity of the scanning, but in a way that preserves both the quality and accuracy of the image. For studies that involve only 2D images, the use of fast scan mirrors is acceptable. However, for 3D images a multiplex approach is commonly used. Thus, the image is formed by the illumination of several pixels and collection of light simultaneously. Commonly this is achieved using a disk with several pinhole apertures that move to different regions of the sample (spinning disk) (Paddock 2000; Xiao et al. 1988; Maddox et al. 2003). One disadvantage of the spinning-disc confocal scanning microscope is the decrease in sensitivity. To address this issue, another design was created by Yokogawa Electric Corporation in 1992, using a dual spinning-disk system (also known as micro-lens enhanced confocal scanning microscope). In this system, every pinhole is associated with a micro-lens, by the placement of a second spinning-disk containing micro-lenses in front of the pinhole spinning-disk (Maddox et al. 2003; Toomre and Pawley 2006). This fit significantly increases the amount of light directed to each pinhole, helping the capture of a broadband of light. In both cases, the light that passes through the pinhole is imaged in a detector that is typically a charge-coupled device (CCD) camera (Conchello and Lichtman 2005; Paddock 2000).

Despite being a powerful technique that revolutionized studies involving measurements in cells and tissues, as well image acquisition with a particular accuracy for spatial localization studies, there are some potential problems that compromise the quality of the information obtained by confocal microscopy. Most of the limits of the technique can be avoided by better understanding the intended specific applications and thus choosing a more suitable equipment, or configuration of the parameters and biological sample preparation. If the results can be analyzed by a 2-D image, the use of confocal microscopy with multiple mirrors (confocal laser scanning microscopes) is acceptable. However, if a 3-D image is necessary, the use of a spinning-disk CSM would be preferred, as it decreases the photobleaching effect.

Photobleaching is the irreversible destruction of the fluorochromes after repeated cycles of excitation and emission, leading to loss of signal (Lichtman and Conchello 2005). Since each fluorochrome can stand a determined quantity of excitation and emission cycles, knowing this is important to achieve the maximum efficiency of the specific fluorochrome in each analysis. In addition to using more efficient confocal scanning microscopes that reduce time and exposure intensity to light as well as increase the speed of signal recording (as discussed above), it is also important to choose more stable fluorochromes and use antifading conjugates that protect the sample from photobleaching (Dailey et al. 2006). Interestingly, the photobleaching principle helped the development of important methodologies used in biology studies, such as fluorescence loss in photobleaching (FLIP) and fluorescence recovery after photobleaching (FRAP), as ways to study mobility and molecular diffusion in cell systems (Dailey et al. 2006).

The duration of light exposure is also especially important when live samples are being analyzed, since, in addition to photobleaching, phototoxicity is also prohibitive, as discussed above. Of note, these characteristics are of great concern when using time-lapse imaging. This is a powerful technique in which a living cell (or organism) image can be recorded in sequence over long time periods, allowing the study of cellular dynamics within a living cell (Dailey et al. 2006; Pawley 1991).

Another important feature to be considered in biological studies is that some structures can produce their own fluorescence, which is known as autofluorescence. This can lead to confusion between the emission coming from the excited fluorophore and that from autofluorescence. This phenomenon cannot be overlooked since it can negatively affect the interpretation of the results. Thus, it is always important to use unlabeled samples (without fluorophore addition) as negative controls. In addition, a special attention is needed when preparing the biological sample for microscopy analyses, since some fixator solutions can induce autofluorescence.

Despite these few shortcomings that can be easily overcome by prior and careful knowledge of the equipment, the sample, the fluorochrome and the structures aimed at, confocal microscopy is a very powerful technique to obtain reliable and high-resolution images of cells and tissues. The widespread availability of confocal microscopes in research institutions and their increasing affordability testify to that.

## 5    TIRF Microscopy as a Technique for Selective Fluorophore Excitation in Cell Studies

TIRF microscopy is a technique based on the principles of light refraction and reflection. When an incident beam of light travels from a material with higher refractive index ($n_1$) to a material with lower refractive index ($n_2$), a change in the light angle relative to the normal is observed. Refraction then occurs and can be calculated by applying Snell's law: $n_1 \sin (\theta_1) = n_2 \sin (\theta_2)$, where $\theta_1$ and

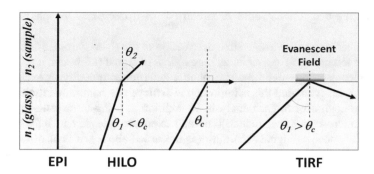

**Fig. 4** Schematic representation of the light refraction in the techniques epifluorescence (EPI), highly inclined thin illumination (HILO), and total internal reflection fluorescence (TIRF). In this scheme, $n_1 > n_2$, $\theta_1$ is the incident beam angle and $\theta_c$ is the critical angle

$\theta_2$ represent the angles of incidence and refraction, respectively. Therefore, if the angle of incidence is substantially increased, the angle of refraction will decrease proportionally, reaching a point where the refracted light will travel in a tangent between both surfaces with different refractive indexes. The angle in which this condition occurs is known as critical angle and can be calculated by $\theta_c = \sin^{-1}\left(\frac{n_1}{n_2}\right)$. At any angle above $\theta_c$, the inciding light will be totally reflected back to the material with index $n_1$. These behaviors are depicted in Fig. 4.

In TIRF microscopy, the condition of total reflection is leveraged. The sample is illuminated by a light beam with an angle of incidence $>\theta_c$ through a glass slide with $n$ higher than the fluorophore solution. When light is totally reflected, a weak local electromagnetic field known as the evanescence field is generated. The evanescent field propagates to the material of lower $n$ and can excite fluorophores of interest present in this medium. The intensity of the evanescent field decays exponentially with penetration distance (the distance point in relation to the interface between the glass slide and liquid medium), allowing for a selective excitation in this small depth range, which is typically between 70 and 300 nm (Tinoco and Gonzalez 2011; Saffarian and Kirchhausen 2008). Due to this selective illumination, lower background fluorescence from out-of-focus molecules is generated. Thus, high sensitivity and signal/noise ratio can be obtained, allowing for the detection of extremely low fluorescent signals, such as for the detection of single molecules in vitro and in vivo (Lamichhane et al. 2013; Wang et al. 2011). The instrumentation for TIRF microscopy is fairly simple and usually requires illumination through an oil immersion objective or through a fused silica prism positioned under the sample compartment. Detection is often performed with a CCD camera, which becomes the limiting factor for the temporal resolution of this technique.

In an application example, Sgro et al. (2013) coupled TIRF and microfluidics to detect specific proteins present in neurotransmitter vesicles by isolating and labeling single vesicles with fluorescently labeled antibodies. The detection was possible because vesicles were immobilized directly onto the surface of the polydimethylsiloxane (PDMS) device. In fact, to ensure that events close to the cell membrane or organelles can be visualized with TIRF it is necessary to immobilize them on the microscope slide (or microfluidic device) surface to be imaged (Roy et al. 2008). The main disadvantage of TIRF, when applied to biological systems, is due to the depth of the evanescent field—the target molecule must be located in close proximity to the surface and cannot participate in dynamic processes that would cause it to diffuse out of the detection region. The free vesicle movement in endocytosis is an example of an in vivo process that would hardly be compatible with TIRF. On the other hand, processes that occur with limited diffusion can possibly be monitored. For instance, the spatial source of vesicles for regulated exocytosis, whether from vesicles previously bound to the membrane or recruited from the cytosol, can be studied using TIRF microscopy (Ohara-Imaizumi et al. 2009).

There are, however, methods that can be used to increase imaging depth. Illumination modes with alternative angles of incidence can be used depending on the application. Examples include epifluorescence and highly inclined thin illumination (HILO) microscopy. As discussed before, in epifluorescence, the sample is illuminated at an angle of incidence of 90°, generating a theoretically infinite depth (Fig. 4). However, in applications that require detection of low abundance molecules (or that have low fluorescence intensity in general), this technique can only be effectively applied if the background fluorescence and the total amount of fluorescent molecules is low. Unlike TIRF, epifluorescence excites the whole depth of the sample and molecules of the focal plane can contribute to the background fluorescence, effectively decreasing the signal/noise ratio and possibly increasing photobleaching (Coelho et al. 2013).

In HILO, the angle of incidence is set to a value slightly lower than the critical angle (Fig. 4). The light travels to the solution with lower $n$ and is refracted with a high inclination angle, increasing the illumination depth up to 20 μm (Coelho et al. 2013). This illumination depth increase allows for the visualization of fluorescent molecules localized in regions further away from the glass/sample interface, but also decreases the signal/noise ratio when compared to TIRF. Konopka and Bednarek (2008) applied HILO for the visualization of the cortical cytoskeleton, membrane proteins, and organelles (mitochondria and Golgi complex close to the plasma membrane) of epidermal plant cells. In that work, the authors evaluated and exemplified HILO as a technique suitable for fluorescent imaging inside plant cells, which can have a wall thickness in the order of 0.5 μm. Although less sensitive, that work illustrates an example in which HILO is more suitable than TIRF.

Furthermore, the use of different illumination methods for the detection of biomolecules in cells has been discussed in the literature. For instance, Wang et al. (2010) evaluated the effect of extracellular stimuli on protein translocation between the cell membrane and cytosol using TIRF, epifluorescence and a microfluidic device (flow cytometry analogous) coupled with TIRF. In this work, the Syk kinase protein was fluorescently labeled, and its translocation was successfully visualized using TIRF when the molecule reached the evanescent field close to the cell membrane. Thus, the translocation was monitored by the increase in fluorescence signal observed 10 and 20 min after the stimulus. The epifluorescence technique generated an image with a lower fluorescence intensity that was only evident 60 min after the stimulus. The microfluidic flow cytometry device coupled with TIRF was used to monitor single cells passing through the microfluidic channel. Each cell was individually compressed under the detection regions to ensure proper evanescent field illumination, and each fluorescence signal was measured per cell after different time points following the extracellular stimulus. The data indicated that protein translocation occurred between 10 and 20 min after stimulus. In this case, the use of microfluidic flow cytometry device coupled to TIRF for this application has the advantage to be less susceptible to intensity losses, due to photobleaching, under a constant illumination source. However, the disadvantage of using this methodology is that it is impossible to perform real-time fluorescence monitoring due to the transient nature of the experiment. It also highlights that the flow cytometry experiment was favored by the use of a photomultiplier tube rather than a CCD camera due to its higher data acquisition. The authors, however, do not comment on the common drawback of having cells adhering to the walls of PDMS microfluidic devices (Lee et al. 2004). More importantly, the article represents a good example of study on how different microscopy techniques may affect the interpretation of the results obtained.

As an additional alternative to minimize the specific limitations of TIRF (as well as of HILO and epifluorescence), the detection of fluorescently labeled biomolecules can be performed in multi-angular illumination systems, i.e., a system in which the light beam angle of incidence is varied over time. In their work, Saffarian and Kirchhausen (2008) highlighted this approach by proposing a method for the monitoring of two sets of fluorophores using sequential TIRF and epifluorescence acquisition. In that work, the authors were able to monitor the axial separation between clathrin protein and its adaptor complex AP-2 during the formation of vesicles on the membrane of live BSC-1 cells.

In conclusion, TIRF, HILO, and epifluorescence are powerful techniques capable of generating fluorescent images with high resolution and provide means for several applications related to cell studies. Together with other high-resolution fluorescence techniques, they can help the scientific community to understand specific interactions in the intracellular medium. TIRF, especially when associated with Förster resonance energy transfer (FRET) microscopy, has provided excellent tools to detect dynamic cellular processes.

## 6  Monitoring Intracellular Biochemical Interactions and Macromolecules Conformation Using Förster Resonance Energy Transfer Microscopy

As discussed in detail in Chap. 4 "UV-Vis Absorption and Fluorescence in Bioanalysis," Förster Resonance Energy Transfer (FRET) is useful to evaluate the relative positions of two fluorophore groups. When a molecule containing a fluorophore is excited (donor), the released energy can be non-radiatively transferred to a second molecule (acceptor), depending on the distance between them (Tinoco and Gonzalez 2011). For FRET to occur, it is necessary that the emission band of the donor overlaps the absorption band of the acceptor (Roy et al. 2008). In Chemistry

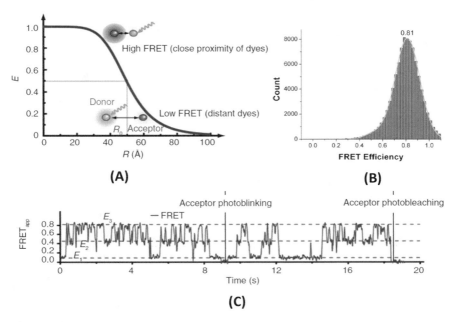

**Fig. 5** (**a**) The efficiency in FRET as function of the distance between the donor and acceptor fluorophores; (**b**) Example of histogram of event counting in FRET; (**c**) Typical signal for $E_{FRET}$ monitored as function of time, containing information on null efficiency ($E_1$), low FRET ($E_2$), and High FRET ($E_3$). Reproduced from Refs. (Roy et al. 2008; Lamboy et al. 2013) with permission

and Biochemistry, this phenomenon can be explored to monitor the stochastic nature of a given chemical reaction, in which both reagents are labeled with fluorophores, and the interaction generates a measurable signal through FRET. Also, macromolecules such as DNA, RNA, and proteins can be labeled in specific positions to detect conformational changes of individual molecules (Sakon and Weninger 2010).

The efficiency in FRET ($E_{FRET}$) is defined as the fraction of the transferred energy that is absorbed by the acceptor and is a function of the sixth power of the distance between donor and acceptor ($R$):

$$E_{FRET} = \frac{1}{1 + R^6/R_0^6} = \frac{I_A/Q_A}{I_A/Q_A + I_D/Q_D}$$

Where $R_0$ is the Förster distance (distance between donor and acceptor at 0.5 $E_{FRET}$), $I_A$, $I_D$, $Q_A$, and $Q_D$ are the intensities of fluorescence and quantum yields of the acceptor and donor, respectively (Tinoco and Gonzalez 2011).

$E_{FRET}$ ranges from 0 to 1, corresponding to bigger and smaller distances, respectively. Usually, $E_{FRET}$ is plotted as a time-dependent signal or by means of a histogram (Fig. 5). The resolution is about a few nanometers, although the absolute distances are difficult to measure.

Either confocal or epifluorescence microscopes can be used to obtain FRET images. These techniques present high compatibility with a wide range of fluorophores, have high image acquisition rate and do not require post-processing to obtain real images (Toomre and Bewersdorf 2010). When compared to TIRF, the main disadvantages are the lower spatial resolution and stronger photobleaching. This way, the choice of the labeling fluorescent probe is of utmost importance in FRET. Most authors use fluorescent proteins derived from GFP (green fluorescent protein) or synthetic dyes (Tables 1 and 2), although other methods are also employed (Roy et al. 2008; Wegner et al. 2013).

Due to its characteristic mechanism of fluorescence generation, the applicability of FRET on biomolecules interaction studies is broad, providing information not available from other techniques. Lamichhane et al. (2013), using TIRF microscopy, employed single-molecule (sm) FRET to understand the mechanisms involved in the DNA polymerase $3'$-$5'$ exonuclease activity, in vitro. Many DNA polymerases possess two domains spatially separated, the polymerase domain (*pol*), responsible for the addition of nucleotides (complementary to the DNA template) to a primer, and the exonuclease domains (*exo*), involved in the proofreading of the newly synthesized chain. Using Alexa Fluor dyes-labeled oligonucleotides and genetically modified DNA polymerase, it was possible to investigate the mechanisms and dynamics for DNA substrate switching between the *pol* and the *exo* sites, since the binding of the primer to the exonuclease domain was projected to reduce $E_{FRET}$ (Fig. 6). In fact, the authors demonstrated that the binding of the primer to the DNA polymerase can occur initially in both domains and that the change in domain proceeds through dissociation and diffusion of either the primer or the DNA polymerase. The retention time of the FRET state evidenced the applicability of this technique to study the biding equilibrium between primer and DNA polymerase domains (Fig. 6).

Regarding conformational analyses, Wang et al. (2013) also applied the concept of smFRET to study the bending of DNA by the HIV-1 nucleocapsid protein (NC). DNA strands labeled with Cy3 and Cy5 (Table 1) probes were immobilized on glass coverslip surfaces so that the bending of the DNA would cause the energy transfer effect (Fig. 7). The results indicate that the DNA strand stays in an unfolded form in the absence of NC given the small values of $E_{FRET}$. In the presence of NC, $E_{FRET}$ presents a heterogeneous distribution, suggesting a dynamic change in the DNA chain bending, which may be related to the incorporation of the viral DNA to the genetic material of the host that is observed in vivo.

In a broader approach, Su et al. (2013) showed the possibility of monitoring two biomolecular interactions using biosensors based on FRET and employing two pairs of fluorophores, measured simultaneously. This way, the authors monitored increases in $[Ca^{2+}]_{ic}$ and Src kinase activity produced in the presence of endothelium growth factor (EGF). Measurements were done in the blue-green region for Src kinase and yellow-orange for $Ca^{2+}$. As a result, it was possible to determine the kinetics of these two biomolecular events in the same sub-cellular site (Fig. 8). Upon EGF stimulus, $Ca^{2+}$ was released in the cytosol and returned to basal levels after 8 min, while Src levels remained high during the whole time of acquisition of the

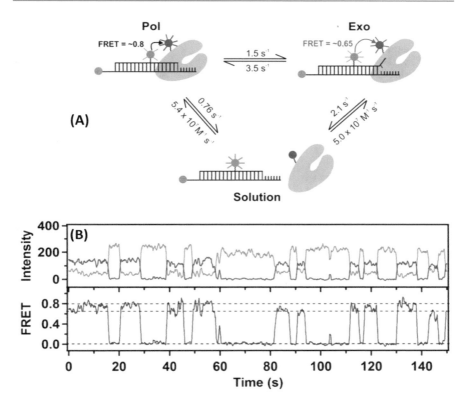

**Fig. 6** Dynamic domain change in DNA polymerase. (**a**) Proposed mechanism of domain change and values of retention times and kinetic constants; (**b**) $E_{FRET}$ signal showing the change in the domain at 0.80 and 0.65 and monitoring of the fluorescence intensities for the donor (green) and acceptor (red). Reproduced with permission from Ref. (Lamichhane et al. 2013). Copyright 2013 American Chemical Society

images. Other biosensors have been proposed with a similar approach, such as for cAMP and cGMP monitoring (Sprenger and Nikolaev 2013).

The comparison of different illumination techniques is important to decide the best methodology to apply. Lin and Hoppe (2013) compared the images and FRET signals by using both TIRF and epifluorescence modes. Gag genes from HIV virus units were labeled with the fluorescent proteins CFP *mCerulean* and YFP *mCitrine* (derived from GFP, Table 2) to exhibit FRET when these viruses interact with the plasma membrane of COS 7 cells. Due to the high resolution of the experiments, it was possible to detect events of high $E_{FRET}$ in TIRF that were not observed in epifluorescence, which were related to the presence of individual viruses at the cell membrane.

Taken together, these examples show how FRET can be used to obtain effectively and uniquely spatial, conformational, or kinetic information in biological systems that cannot be assessed with other light-based techniques. Moreover, the use of high-

**Fig. 7** Schematic representation of FRET effect produced through the DNA bending by HIV-1 nucleocapsid protein. Reproduced from Ref. (Wang et al. 2013) with permission. Copyright 2013 American Chemical Society

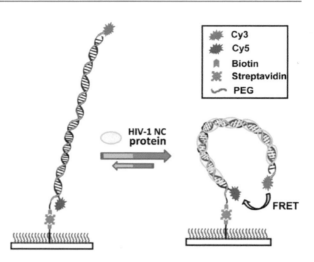

resolution microscopy techniques, such as TIRF, in combination with single molecular FRET would allow interesting research possibilities.

# 7 Concluding Remarks

The evolution of high-resolution fluorescence techniques is dependent on the current technological development of its different instruments, such as lasers, detectors, and data computational processing, as well as on the development of more stable and specific fluorescent probes. The detection methods for cellular components discussed in this chapter have an important common characteristic, which is that all can also be used for in vivo studies.

Laser scanning confocal fluorescence microscopy has been widely used to study cell structure and function with high resolution, and despite its limitations regarding photobleaching and phototoxicity, it still has a place in various biological studies. Also, TIRF microscopy is a valuable tool to visualize fluorescent molecules in the vicinities of the cellular plasma membrane, as well as to detect immobilized components in the glass/sample interface of the microscope. Additionally, TIRF is particularly suitable for applications based on FRET, which allows for the monitoring of inter and intramolecular interactions in a time-resolved manner.

Although FRET resolution is much greater using a TIRF microscope, this technique can also be applied using confocal or conventional epifluorescence microscopy, with satisfactory results for biomolecule analyses in vivo. Thus, TIRF allows FRET applications for analyses of biomolecules close to the plasma membrane, while the other techniques can be used to evaluate other cellular regions. In addition, combining these high-resolution fluorescence techniques with microfluidics devices improves the evaluation of single cells, eliminating heterogeneity with high sensitivity.

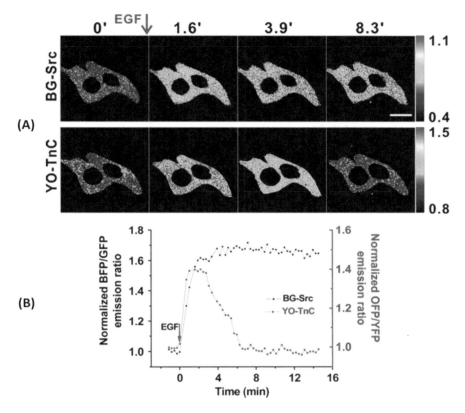

**Fig. 8** Dual FRET biosensor. (**a**) Overlapped fluorescence intensities for the simultaneous monitoring of Src kinase (blue-green Src) and $Ca^{2+}$ (yellow-orange TnC) in HeLa cells; (**b**) Normalized emission ratios for both pairs of fluorophores. EGF: endothelium growth factor. Reproduced from Ref. (Su et al. 2013) with permission

Therefore, the use of these different high-resolution fluorescence methods, each one with specific characteristics that extend the applicability and versatility of fluorescence use to biomolecule and organelle detection, combined with other analytical techniques, can increase the quality of the results obtained, improving the understanding of the multiple biological processes in the cellular context.

# References

Alberts DB, Davidson MW (2013) Fundamentals of light microscopy and electronic imaging, 2nd edn. Wiley-Blackwell

Alberts B, Johnson AD, Lewis J, Morgan D, Raff M, Roberts K, Walter P (2014) Molecular biology of the cell, 6th edn. W. W. Norton and Company, New York

Chang X, Yu Z (2011) J Braz Chem Soc 22:1246–1252

Clegg RM (2012) Fluorescence and FRET: theoretical concepts 101. In: Egelman EH (ed) Comprehensive biophysics, vol 1. Elsevier, pp 592–617

Coelho M, Maghelli N, Tolic-Norrelykke IM (2013) Integr Biol 5:748–758

Conchello JA, Lichtman JW (2005) Nat Methods 2:920–931

Conchello JA, Kim JJ, Hansen EW (1994) Appl Opt 33:3740–3750

Dailey ME, Manders E, Soll DR, Terasaki M (2006) Confocal microscopy of living cells. In: Handbook of biological confocal microscopy. Springer, Berlin

Gratton E, vandeVen MJ (2006) Laser sources for confocal microscopy. In: Handbook of biological confocal microscopy. Springer, Berlin

Ha T (2001) Methods 25:78–86

Hooke R (1665/2014) Micrographia: or some physiological descriptions of minute bodies made by magnifying glasses, with observations and inquiries thereupon. Courier Dover Publications, Mineola, p 113. https://books.google.com/books?id=0DYXk_9XX38C&q=Micrographia+honeycomb&pg=PA113

Huang B, Wu HK, Bhaya D, Grossman A, Granier S, Kobilka BK, Zare RN (2007) Science 315:81–84

Karp G (2010) Cell and molecular biology: concepts and experiments, 6th edn. Wiley, Hoboken

Kolberg DIS, Presta MA, Wickert C, Adaime MB, Zanella R (2009) J Braz Chem Soc 20:1220–1226

Konopka CA, Bednarek SY (2008) Plant J 53:186–196

Laissue PP, Alghamdi RA, Tomancak P, Reynaud EG, Shroff H (2017) Nat Methods 14:657–661

Lamboy JA, Kim H, Dembinski H, Ha T, Komives EA (2013) J Mol Biol 425:2578–2590

Lamichhane R, Berezhna SY, Gill JP, Van der Schans E, Millar DP (2013) J Am Chem Soc 135:4735–4742

Lee JN, Jiang X, Ryan D, Whitesides GM (2004) Langmuir 20:11684–11691

Lichtman JW, Conchello JA (2005) Nat Methods 2:910–919

Lin J, Hoppe AD (2013) Microsc Microanal 19:350–359

Liu HP, Wang Y, Li H, Wang ZL, Xu DK (2013) Dyes Pigments 98:119–124

Maddox PS, Moree B, Canman JC, Salmon ED (2003) Methods Enzymol 360:597–617

Minsky M (1988) Scanning 10:128–138

Nelson DL, Cox MM (2017) Lehninger principles of biochemistry, 7th edn. W. H. Freeman, New York

Neto BAD, Corrêa JR, Carvalho PHPR, Santos DCBD, Guido BC, Gatto CC, de Oliveira HCB, Fasciotti M, Eberlin MN, da Silva Jr EN (2012) J Braz Chem Soc 23:770–781

Ohara-Imaizumi M, Aoyagi K, Nakamichi Y, Nishiwaki C, Sakurai T, Nagamatsu S (2009) Biochem Biophys Res Commun 385:291–295

Paddock SW (2000) Mol Biotechnol 16:127–149

Pawley JB (1991) Scanning 13:184–198

Price AK, Culbertson CT (2007) Anal Chem 79:2614–2621

Roy R, Hohng S, Ha T (2008) Nat Methods 5:507–516

Saffarian S, Kirchhausen T (2008) Biophys J 94:2333–2342

Sakon JJ, Weninger KR (2010) Nat Methods 7:203–205

Sandison DR, Piston DW, Williams RM, Webb WW (1995) Appl Opt 34:3576–3588

Sgro AE, Bajjalieh SM, Chiu DT (2013) ACS Chem Neurosci 4:277–284

Shaner NC, Steinbach PA, Tsien RY (2005) Nat Methods 2:905–909

Simplicio FI, Seabra AB, de Souza GFP, de Oliveira MG (2010) J Braz Chem Soc 21:1885–1895

Skoog DA, West DM, Holler FJ, Crouch SR (2013) Fundamentals of analytical chemistry, 9th edn. Cengage Learning, Boston

Snapp EL (2009) Trends Cell Biol 19:649–655

Sprenger JU, Nikolaev VO (2013) Int J Mol Sci 14:8025–8046

Su JJ, Wang LY, Zhang XH, Fu YL, Huang Y, Wei YS (2011) J Braz Chem Soc 22:73–79

Su T, Pan ST, Luo QM, Zhang ZH (2013) Biosens Bioelectron 46:97–101

Tinoco I, Gonzalez RL (2011) Genes Dev 25:1205–1231

Toomre D, Bewersdorf J (2010) Annu Rev Cell Dev Biol 26:285–314

Toomre D, Pawley JB (2006) Disk-scanning confocal microscopy. In: Handbook of biological confocal microscopy. Springer, Berlin

Tran DT, Knez K, Janssen KP, Pollet J, Spasic D, Lammertyn J (2013) Biosens Bioelectron 43: 245–251

Wang J, Fei B, Geahlen RL, Lu C (2010) Lab Chip 10:2673–2679

Wang B, Ho J, Fei JY, Gonzalez RL, Lin QA (2011) Lab Chip 11:274–281

Wang H, Musier-Forsyth K, Falk C, Barbara PF (2013) J Phys Chem B 117:4183–4196

Webb RH (1999) Confocal Microsc 307:3–20

Wegner KD, Lanh PT, Jennings T, Oh E, Jain V, Fairclough SM, Smith JM, Giovanelli E, Lequeux N, Pons T, Hildebrandt N (2013) ACS Appl Mater Interfaces 5:2881–2892

Wu HK, Wheeler A, Zare RN (2004) Proc Natl Acad Sci U S A 101:12809–12813

Xiao GQ, Corle TR, Kino GS (1988) Appl Phys Lett 53:716–718

Zhou BL, Xiao JW, Liu SF, Yang J, Wang Y, Nie FP, Zhou Q, Li YG, Zhao GH (2013) Food Control 32:198–204

# Nuclear Magnetic Resonance Spectroscopy in Analyses of Biological Samples

Danijela Stanisic, Lucas G. Martins, and Ljubica Tasic

## 1    Introduction

Since very remote times, which correspond to the first scientific reports, studies on the content and functions of different biological samples have been carried out by researchers. Currently, science has an infinite number of techniques and analytical methods that make it possible to carry out these studies at the levels of cells, organelles, macrostructures, molecules, and atoms. The choice of the appropriate methodology depends on the results one wants to obtain, the origin of the sample, and the physicochemical properties of the molecules present in the sample. For example, when considered plant samples, we should know that majority of the natural products used in the development of drugs are their secondary metabolites. However, it is necessary to take into account that plants have many other components, such as chlorophyll, cellulose, hemicellulose, lignin, pectin, all the primary metabolites essential for their development, among others. The extraction and isolation of the compounds of interest need to be compatible with the chosen analytical technique (Sardans et al. 2011). But biological samples can have many different origins, such as metabolites and proteins from single-cell organisms such as bacteria and some fungi, biological samples from humans or other animals (such as many tissues, blood, urine, cells, cell cultures, saliva, cerebrospinal fluid, etc.) and many others (Gosetti et al. 2013; Emwas et al. 2013). Each type of sample can be analyzed by different methodologies; however, every methodology, in turn,

D. Stanisic · L. Tasic (✉)
Chemical Biology Laboratory, Department of Organic Chemistry, Institute of Chemistry, University of Campinas (UNICAMP), Campinas, Sao Paulo, Brazil
e-mail: ljubica@unicamp.br

L. G. Martins
Chemical Biology Laboratory, Department of Organic Chemistry, Institute of Chemistry, University of Campinas (UNICAMP), Campinas, Sao Paulo, Brazil

Facultad de Ingenieria Industrial, Universidad de Lima, Distrito de Santiago de Surco, Lima, Peru

© The Author(s), under exclusive license to Springer Nature Switzerland AG 2022                203
L. T. Kubota et al. (eds.), *Tools and Trends in Bioanalytical Chemistry*,
https://doi.org/10.1007/978-3-030-82381-8_9

demands a specific treatment, or pretreatment, which in result gives the perfectly prepared sample for analysis.

From a chemical point of view, studies of biological samples are based on the structure and dynamics of their different components. To specify which analytical technique should be used in an analysis, one must consider the sample's origin and prior knowledge about the physicochemical properties of the molecules of interest. For example, the study of the biochemical pathways of fatty acid molecules and their connections in living organisms is called lipidomics (Zhao et al. 2014). The choice of the appropriate methodology must be based on knowledge on physical-chemical properties, such as lipid solubility and the viability of using the picked solvent for the chosen analytical technique, size of the molecules, organic functions that can be found, etc. On the other hand, the interest of study can be based on proteins and enzymes of a certain organism, which we can call proteomics, and depend on those biomolecules physicochemical properties and it is necessary to use other analytical methodologies when compared to lipidomics (Zhao et al. 2014; Gevaert and Vandekerckhove 2000; Shin et al. 2008). Last but not least, most studies of biological samples aim to qualitatively and quantitatively explore the primary and secondary metabolites found in specific samples of a given organism. This can be an exploratory study or directed specifically to elucidate some biochemical process of interest. This type of scientific study is called metabolomics (Smolinska et al. 2012; Mahrous and Farag 2015; Pontes et al. 2017a).

An analytical tool that can be applied to all these types of studies is Nuclear Magnetic Resonance (NMR). NMR is one of the most widely used spectroscopic techniques for structural studies of organic compounds. It presents several advantages in the studies of the structure and dynamics of molecules with different molecular masses, providing structural details at the atomic level. This analytical methodology can be applied to liquid, semisolid and solid samples, depending on the origin of the sample and the pretreatment carried out with it. For metabolomics, metabonomics, and lipidomics studies, different NMR experiments are combined with statistical tools, such as chemometrics (Madsen et al. 2010), to explore and maximize the results obtained. Based on what has been presented, this chapter has the objective of helping the reader to understand the procedures necessary for carrying out analyses of biological samples applying NMR. Here we will discuss the sample pretreatment (plant extracts, microorganism extract, blood serum, blood plasma, lipid isolation, urine), a sample preparation (liquid preparation, semisolid preparation), some introduction regarding NMR techniques (liquid NMR, HR-MAS), and the presentation of some NMR experiments (different $^1$H NMR, CPMG, diffusion filter, and the most common 2D NMR techniques) and their uses. Post-analyses processing (spectra processing, statistical analysis) and metabolites' identification will also be covered.

## 2    Liquid-State NMR and Semisolid-State NMR (HR-MAS)

In a very simplistic way, NMR signals result from the interaction of electromagnetic waves (on the radiofrequency scale) with the magnetic moment of the nuclear spins, in the presence of an external magnetic field ($B_0$). There are several nuclei that are observable by NMR, such as $^1H$, $^{13}C$, $^{15}N$, $^{31}P$, among others. However, the most observed nucleus is that of hydrogen ($^1H$) because it has a high natural abundance (99.98%) and a large gyromagnetic ratio ($\gamma$), which results in greater sensitivity in detection. For organic compounds, the non-quadrupole nuclei $^{13}C$ and $^{15}N$ are also widely used, especially in structural elucidation (Claridge 2009).

When a magnetic atomic nucleus is exposed to an external magnetic field ($B_0$), the ground state of energy is split into different energy levels, which are proportional to the strength of the magnetic field and the gyromagnetic ratio of the nucleus in question. This splitting, resulting from the interaction of the nucleus magnetic moment with the external magnetic field, is called the Zeeman interaction, it has a magnitude on the order of hundreds of MHz, and it is useful for identifying different types of nuclei placed in magnetic fields (Levitt 2007a). However, this is not the only interaction that occurs. There are some perturbations to the Zeeman interaction resulting from other magnetic and electronic interactions with the nucleus, which are responsible for providing information regarding the structure and dynamics of the molecules whose nucleus belongs to. These interactions are: quadrupolar coupling, which occurs in the order of MHz; chemical shift, which occurs in the order of kHz; dipolar coupling, also observed in the order of kHz and scalar coupling or $J$ coupling, which occurs in the order of Hz. The quadrupole coupling comes from the interaction of the quadrupole moment with electric field gradients (EFG) in nuclei with spin $>1/2$. However, for the nuclei most observed in biological sample studies, such as $^1H$, $^{13}C$, $^{15}N$, $^{31}P$, which contain spin $= 1/2$, there is no quadrupolar coupling (Smith 1971).

The chemical shift interaction is the most investigated phenomenon in NMR since, as a result of this interaction, the nuclei of the same chemical element that are chemically non-equivalent have slightly different frequencies, providing the specific spectral window of each nucleus. Given non-equivalence results from local electronic environment surrounding the nucleus, which are the result of bonding to different atoms, bond lengths and angles, or any shielding or de-shielding effect of the observed nucleus. Through this interaction, it is possible to differentiate $^1H$ from a methylene group, $^1H$ from an olefin, or $^1H$ from an aromatic compound in the $^1H$ NMR spectrum, for example. Chemical shift interaction has two components, one isotropic and the other anisotropic. The last one has the direct effect of broadening the line shapes in the spectrum. In liquid state NMR the contributions to the chemical shift anisotropy are averaged due to the molecular reorientation, resulting in narrow spectral line shapes. However, in solid or semisolid-state NMR, in which molecular reorientation does not occur or occurs partially, respectively, the spectra show broader lines, thus losing resolution (Levitt 2007b; Massiot et al. 2002).

Dipolar coupling results from interactions between the local magnetic fields generated by the nuclei. This interaction can occur through nuclei of the same chemical elements (homonuclear dipolar coupling) or between nuclei of different elements (heteronuclear dipolar coupling). As a consequence of being an interaction that occurs through space, the dipolar coupling provides information about the distance between the coupled nuclei. The closer spatially the nuclei are coupled, the stronger is the observed interaction. Another factor that contributes to the intensity of the coupling is the natural abundance of the magnetic nuclei, since the more abundant the nucleus is, the greater the probability that two of these nuclei are close enough for the coupling to be observed. Dipolar coupling creates a disturbance of the Zeeman interaction, generating variations in the spin nuclear levels, which result in loss of resolution in the NMR spectrum. In liquids, this coupling is significantly minimized by the movement of the molecules and the distance between them. However, in solids and semisolids, in which the molecules are arranged more closely together, dipolar coupling has a significant effect on the spectral linewidth (Levitt 2007c; Boesch and Kreis 2001; Cohen-Addad 1974).

Finally, we introduce the perturbation of the Zeeman interaction of less magnitude, the scalar coupling or $J$ coupling. This interaction is defined as the intramolecular coupling of two magnetic nuclei, which can be homo- or heteronuclear, but which is necessarily propagated through the bonding electrons. Unlike other interactions, $J$ coupling does not cause an enlargement of the NMR spectrum lines, but rather in splitting the frequencies of each coupled magnetic nucleus into $N + 1$ peaks (Levitt 2007d). This division of the peaks carries structural information within the molecule. This phenomenon is constantly explored in 2D NMR experiments for structural determination (Keeler 2010; Eberstadt et al. 1995).

Although the NMR experiments carried out using probes designed for liquids result in spectra with good resolution for fluid samples, mostly, the same does not occur with solids and semisolid samples at the same probe. Biological samples of tissues, cells, and others that are classified as solid or semisolid exhibit an extremely broad NMR signal due to extensive homo- and heteronuclear dipole coupling, chemical shift anisotropy (CSA), and quadrupole interactions. The absence of molecular reorientation in non-fluid samples results in the emergence of an orientation component that scales as $(3\cos2\theta - 1)$ in the Hamiltonians that describe dipolar interactions, CSA interactions, and magnetic susceptibility (Hennel and Klinowski 2004). In other words, the restriction in the orientation of molecules in solids and semisolids results in spatial dependence on the orientation of their nuclear spins, which is mathematically depicted by the term $(3\cos2\theta - 1)$ in the equations that describe the disturbances of the Zeeman interaction. Therefore, to improve the resolution of the spectra, it was necessary to propose a methodology that could mimic, even partially, the effect of molecular reorientation for nuclear spins. At the end of the 1950s, two independent studies proposed the acquisition of NMR experiments with high sample rotation to improve the resolution of the spectrum (Andrew et al. 1958; Lowe 1959). These studies showed that the best results are obtained when the angle ($\theta$) between the sample rotation axis and the external magnetic field ($B_0$) is 54.74°. When $\theta = 54.74°$, the term $3\cos2\theta - 1 = 0$, minimizing the effects of dipolar, chemical shift anisotropy, and quadrupolar interactions in the NMR spectrum. The $\theta$ angle became known as the *magic angle*

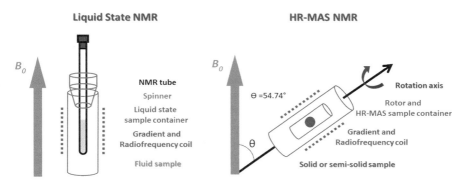

**Fig. 1** Representation of NMR probe heads oriented with respect to the external magnetic field. On the left is the representation of the liquid sample NMR probe, parallel to $B_0$, in which are indicated its main components (such as sample container and gradients and radiofrequency coil) and the spinner coupled with the NMR tube. On the right is the representation of the HR-MAS NMR probe with an orientation of the sample container at an angle ($\theta$; magic angle) of 54.74° between the axis of rotation and $B_0$. Inside the sample container is the rotor with the sample

and the technique was denominated *Magic Angle Spinning* (MAS). When spectra are obtained with spinning at $\theta = 54.74°$ the nuclear dipole–dipole interaction between magnetic moments of nuclei averages to zero. The chemical shift anisotropy, a nuclear-electron interaction, averages to a non-zero value while the quadrupolar interaction is only partially averaged by MAS (Hennel and Klinowski 2004).

In the early 1990s, an NMR technique emerged called high-resolution magic angle spinning (HR-MAS) which extended the applications of high-resolution NMR to samples whose molecular reorientation is limited, as is the case of semisolids and liquid crystals. Essentially, HR-MAS combines the "high resolution" for liquid-state NMR with the "narrowing of the spectral lines" used in solid-state NMR. Basically, HR-MAS attenuates the gradient of magnetic susceptibility and residual dipolar coupling inherent in samples whose molecular and spin reorientation is restricted, which results in semisolid NMR spectra with a spectral resolution comparable to liquid state NMR spectroscopy (Wong and Lucas-Torres 2018). Recent technological advances have enabled the design of HR-MAS probes, which differ in operability when compared to probes designed for liquid-state NMR. In addition to the HR-MAS probe head differing from the probe for the liquid state, as shown in Fig. 1, the technique for semisolids also depends on a pneumatic unit to promote the rotation of the sample properly (Simon et al. 2015).

Technological advances in the development of a specific NMR probe for liquids and HR-MAS have allowed studies of metabolites and biomolecules present in the most diverse biological samples, such as plant metabolites and agro-foods (Mazzei and Piccolo 2017), in the diagnosis of innumerable human diseases (Emwas et al. 2013), among others. Although most studies are carried out by $^1$H NMR, other nuclei such as $^{13}$C, $^{15}$N, $^{19}$F, and $^{31}$P can also be used in metabolomics studies, both in liquid state and HR-MAS techniques. More recent studies have also shown the use of 2D NMR techniques for metabolomics studies from biological samples (Emwas et al. 2019; Kruk et al. 2017). NMR techniques can also be used for studies of metabolomics of diseases in plants (Pontes et al. 2016), to find biomarkers of pathogenic microorganisms in the blood of affected animals (Pontes et al. 2017b), as well as they can help to understand the mechanism of bactericidal or bacteriostatic action in pathogenic bacteria when evaluating, in vitro, the variation of the metabolites in face of different anti-mycotic treatments (Stanisic et al. 2018).

From a clinical point of view, the metabolomics performed with fluids and tissues provide a lot of important information regarding diseases, treatments, or changes in the organism's homeostasis. Among the numerous examples to be cited are studies of metabolites found in human blood and which are related to psychiatric diseases, such as schizophrenia (Tasic et al. 2017a) and bipolar disorder (Tasic et al. 2019). It is also possible to evaluate the effects of crack cocaine addiction (Costa et al. 2019) and differentiate users of that drug and patients with schizophrenia (Tasic et al. 2017b), since the behavior of both is similar but the metabolic response in the body is different. It is possible to evaluate orlistat-induced modulation of plasma metabolites in overweight patients who are undergoing treatments with this drug (Lopes et al. 2013). In the studies of cancer biomarkers, a topic of extreme interest in modern medicine, there are several studies applying NMR and we can mention two reviews that deal with breast cancer tissue metabolomics (Gogiashvili et al. 2019) and solid embryonic tumors in children (Escobar et al. 2019). There are even studies on the effects of antipsychotics applied to worms specifically designed to act as an animal putative model for studying the pathways correlated to schizophrenia (Monte et al. 2019).

## 3    What NMR Experiment Should I Perform with My Biological Sample?

The choice of the NMR experiment to be carried out with a given biological sample depends on some factors, which are: type of sample (solid, liquid, semisolid), molecular components of the sample (presence of water, presence of proteins, presence of any self-assembling structures), amount of overlap of the $^1$H NMR signals on the spectrum and the objective of the study or the hypothesis to be proven. One of the main drawbacks of unpurified biological samples is the presence of water, which in most cases is present in large quantities and the residual water signal in the NMR spectrum covers all other signals.

Since the creation of gradients field and selective radiofrequency pulses, many methods for water signal suppression have been created, from methods based on signal saturation, such as PRESATURATION, 1D-NOESY-PRESATURATION to methods based on the destruction of magnetization, such as the classic ones WET and WATERGATE, among many others (Price 1999). All methods and pulse sequences developed for water suppression in NMR spectra are constantly being improved, so that several of these methods can be optimized and applied for quantitative studies (Giraudeau et al. 2015; Barding and Salditos 2012). As a standard experiment, the WATERGATE method makes it possible to obtain good results. The original method and some updates showed $J$ modulation distortions, because the evolution under the chemical shift is refocused by a spin echo, but the effects of scalar coupling are not. Recently, a small modification in the pulse sequence was proposed to solve the distortions caused by $J$ modulation (Adams et al. 2013).

Another very common problem in spectra of unpurified biologic samples is the presence of molecules with very similar structures, which have a high range of many overlapping signals, making it very difficult to identify the metabolites and perform statistical analysis. There are several experiments capable of eliminating homonuclear couplings, also called pure shift. One of the most promising methods is PSYCHE (Foroozandeh et al. 2018), which can be applied in 1D (Santacruz et al. 2020; Bo et al. 2019) and 2D (Foroozandeh et al. 2016) experiments. It is also possible to overcome the problems of signal overlap in the NMR spectrum using the 2D $J$-resolved technique, which is used to separate the effects of chemical shift and $J$-coupling into two independent dimensions (Ludwig and Viant 2010).

The improvement in the evaluation of NMR signals from samples that show a lot of overlapping can also be achieved by applying 2D techniques, resulting from homo- and heteronuclear correlations, such as COSY and HSQC, respectively. Although obtaining any of these types of spectra requires more acquisition time, there are fast 2D techniques that require relatively short acquisition times and offer good results (Le Guennec et al. 2014).

The choice of the appropriate NMR experiment also depends on the type of molecules to be observed. Metabolites, in general small molecules, have short correlation times and long transverse relaxation times ($T_2$) and diffuse very well in solutions. On the other hand, high molecular weight molecules, such as proteins, have long correlation times and short transversal relaxation times ($T_2$) and have little diffusion in the solution. As a consequence, it is possible to edit the NMR pulse sequences as a function of transversal relaxation ($T_2$) or diffusion in order to specifically observe molecules of certain molecular weight ranges. The $^1$H-NMR-Diffusion-edited experiment eliminates, through pulsed field gradient, the NMR signals of the molecules that diffuse more in the solution, maintaining the signals of those molecules that diffuse less. This is convenient to eliminate the signals from the metabolites in order to visualize the medium and large molecules. This methodology is very suitable for lipidomics studies, for applications of lipid studies in tissues by HR-MAS (Rooney et al. 2003), as well as for any application to biological samples (Smith et al. 2007). In contrast, the $^1$H-NMR-$T_2$ filter-edited, equipped with a pulse sequence block called CPMG (Carr–Purcell–Meiboom–Gill), is an experiment that through a spin trap allows the magnetization of large molecules to lose coherence while magnetization of small molecules remains in coherence. As a result, signals from large molecules are eliminated and signals from small molecules are maintained. This methodology is essential when studying fluids such as blood, which contain a large amount of proteins whose NMR signals overlap the signals of metabolites (Barbosa et al. 2018).

Although the $^1$H nucleus is the most observed in studies of biological samples by NMR, this nucleus is not the only one that can be studied in this type of sample. There are several experiments based on the observation of $^{13}$C, $^{15}$N, and $^{31}$P nuclei (Emwas et al. 2019).

Finally, it is possible to identify and elucidate the metabolite structure in non-purified biological samples, whenever this metabolite has its signals partially assigned in the $^1$H NMR spectrum, and the correlations existing in the 2D spectra can be verified. Two-dimensional NMR spectroscopy can be used for many applications, including molecular identification and structural elucidation. 2D NMR allows to overcome the problem of overlapping resonances, spreading the peaks in a second dimension. Additionally, the homonuclear correlations observed in different 2D NMR experiments, such as Total Correlation Spectroscopy (TOCSY) (Sandusky and Raftery 2005) and Correlation Spectroscopy (COSY) (Feraud et al. 2015) can provide information about the spin systems and the spatial proximity of different parts of the metabolite structure, which helps a lot in the identification and elucidation. The 2D NMR experiments based on heteronuclear correlations, such as Heteronuclear $^1$H-$^{13}$C Single Quantum Coherence ($^1$H-$^{13}$C-HSQC) and Heteronuclear Multiple Bond Correlation Experiments (HMBC) have been used in metabolomics studies and are essential for the elucidation of metabolites structures (Beckonert et al. 2007; Nicholson et al. 1995; Yuk et al. 2010). Together, all NMR data forms a big puzzle and needs to be analyzed and organized to provide the desired results (Parkinson et al. 2019; Kruk et al. 2017).

## 4 NMR Data Preprocessing: Matrices Preparation

The accurate metabolomics spectral analyses can be divided into five main segments: (1) spectral data collection, (2) data preprocessing—preparation of the matrices, (3) statistical analyses of the spectral data, (4) validation, and (5) interpretation of the results (Fig. 2). We have already talked about spectral data collection, so the second segment count on Data preprocessing or Preparation of the matrices, which include the transformation of the obtained $^1$H NMR spectra (1D: suppressed

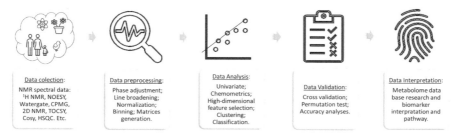

**Data colection:**
NMR spectral data:
$^1$H NMR, NOESY,
Watergate, CPMG,
2D NMR, TOCSY,
Cosy, HSQC. Etc.

**Data preprocessing:**
Phase adjustment;
Line broadening;
Normalization;
Binning; Matrices
generation.

**Data Analysis:**
Univariate;
Chemometrics;
High-dimensional
feature selection;
Clustering;
Classification.

**Data Validation:**
Cross validation;
Permutation test;
Accuracy analyses.

**Data Interpretation:**
Metabolome data
base research and
biomarker
interpratation and
pathway.

**Fig. 2** The accurate metabolomics spectral analyses are divided into five main segments: spectral data collection, data preprocessing—preparations of the matrices, statistical analyses of the spectral data, data validation, and interpretation of the results

$^1$H, NOESY, WATERGATE, CPMG, and other sequences) into convenient data for further analyses (Barbosa et al. 2018). Spectral alignment, normalization, binning, and fitting are crucial steps for further analytical processing. All these steps can be performed using the TopSpin and/or MestreNova, or other programs for spectral analyses and afterward data can be saved and exported as matrices (Emwas et al. 2018). These steps are important for obtaining high-quality NMR spectral data for metabolomics and metabonomics identification.

All obtained $^1$H NMR spectra should have spectral and baseline phases adjusted. Before applying any processing technique, vertical peak intensities should be increased as much as possible. Phasing is also one of the important steps, since any small phase error or misalignment can affect further spectral alignment, spectral binning and for sure measured peak chemical sift and area. Metabolomics spectra always should be phase corrected manually, despite that the automatic phase correction option is enough in processing NMR spectra. A phase distortion in the solvent region spectra (typically from HDO ~4.7–4.8 ppm) can affect peaks near this region. Therefore, the manual phase correction in metabolomics gives better results (Emwas et al. 2018). The solvent peak region is always excluded from NMR spectra, but this will be explained in the following section.

Correct spectral baselines are significant for proper alignment and peak integration. Baseline correction is an inevitable part of the processing technique for removing incomplete digital sampling, removing spectral artifacts that can be consequences from electronic distortion, and removing inadequate digital filtering (Smolinska et al. 2012). NMRPipe, TopSpin, Chenomx NMR Suite and MestreLab Inc's MNova provide software packages that can perform high-quality baseline correction.

There are several manners to align a spectrum, one of them is using singlet (0.00 ppm) of TMSP (99.9 atom %D, contains 0.05 wt.% 3-(trimethylsilyl) propionic-2,2,3,3-*d4* acid) or its salt in deuterium oxide ($D_2O$) or TSM (99.8 atom D, contains 0,0.3–1% v/v tetramethylsilane) in deuterated chloroform ($CDCl_3$). 4,4-dimethyl-4-silapentane-1-sulfonic acid (DSS) is also widely used as a chemical shift reference standard (Pontes et al. 2017a; Emwas et al. 2018; Smolinska et al. 2012; Dona et al. 2016) When there is no addition of organosilicons, the spectra can be aligned with lactate **CH**$_3$- doublet signal (s, 1.33 ppm, $J = 7$ Hz) our other metabolite known signal (Zheng et al. 2020; Chong et al. 2019). It is important to mention that the TSP is pH-sensitive, because of the possibility to interact with proteins and lipoproteins, and for correct chemical shift alignment is sometimes preferable to use sodium acetate or sodium formate (Emwas et al. 2018). The signal of HDO (~4.8 ppm for urine, plasma, and serum samples dissolved in $D_2O$) is not recommendable as a standard peak because of the rapid exchange with $D_2O$. Proper chemical shift assignment is important for further correct identification of the peaks.

Binning (also called bucketing) refers to the dividing of aliened NMR spectra into small regions that are wide enough to include one or more NMR peaks and is a principal step in preprocessing, especially because of the sizes of the acquired spectra. The NMR spectra are composed of thousands time-domain data points that after Fourier transformation are presented in the frequency domain (Claridge 2009). Binning literally means the division of the spectra in histograms and reduction of the points that identified the peaks. One peak can be presented with two to three data points, depending on its width, instead of ten or more (Liland 2011). The most common method of equidistant binning is dividing the spectra into spectral width of 0.04 ppm, sometimes 0.05 ppm (van den Berg et al. 2006).

The normalization is the preprocessing spectra step where the collected data are transformed into comparable values. There are several proposed methods to normalize data before performing any chemometrics analyses, such as Centering, Autoscaling, Pareto scaling, Range, Level and Logarithmic scaling, Logarithmic transformation, and Power transformation (Emwas et al. 2018). For example, in NMR spectra the intensity of peaks is different for different types of metabolites, for example, lactate when compared to tyrosine, therefore it is necessary to perform normalization of spectra before transposing the data into the matrix for chemometrics analyses. Normalization using centering the peak intensities will convert all values into fluctuation around the new "zero," while mean centering will adjust values around the mean data area. These methods are used in combination with all other normalization techniques since sometimes are not enough for high-quality PCA and PLS-DA performance (Emwas et al. 2018; van den Berg et al. 2006). Autoscaling and Pareto scaling are methods that each value of dependent variable is subtracted by their mean and divided by standard deviation or by square root of standard deviation. In range scaling method, each dependent variable is adjusted to the mean and then divided with the minimum and maximum of that particular value. Level scaling makes rough variance equal since each value is adjusted to the mean and then divided by the same mean. Logarithmic and power transformation (pseudo-scaling) are used when there are sub-populations that have different values of variabilities from others (Emwas et al. 2018).

## 5 Univariate and Multivariate Statistic Analyses

Data analysis may include two or more processing techniques, and they can be divided into: Univariate statistic modeling (t-test, Volcano plots, Fold change analysis); Chemometrics multivariate analysis (Principal Component Analysis—PCA, Partial Least Squares-Discriminant Analysis—PLS-DA); High-dimensional feature selection (Significant analysis of microarrays—SAM, Empirical Bayesian analysis of microarrays—EBAM); Cluster analysis (K-means, Self-Organizing Maps—SOM, Dendrogram, Heatmap), and Classification or as is called nonlinear method (Support Vector Machine—SVM and Random Forests) (Madsen et al. 2010; Xia and Wishart 2016).

Univariate analysis is used for potential biomarker identification, where the probability value ($p$-value) is the standard measure of significance. Multiple testing is susceptible to false positives, unless a corrected significance limit is used (Madsen et al. 2010). The statistical analysis is useful to locate a significant difference among metabolites and different conditions. Basically, a statistical test is used when it is necessary to compare two hypotheses: the null and the alternative hypothesis. The null hypothesis presumes that there is no difference and the alternative hypothesis is reverse. Performing the statistical test, the obtained $p$-value is evidence against the null hypothesis, evidently if the $p$-value is $<0.05$ than it is considered that there is a significant difference, and if it is $<0.01$ it is considered that there is a very significant difference. Student $t$-test is an example of a parametric test where the normal distribution of the data is taken into analysis (Winkler 2020). The univariate method is useful to discover new metabolites and compare specific candidates when control and case are explored in different conditions. A fold-changes is represented as the ratio between two values. In spectroscopy dataset when it comes to peak, a fold-change is used to compare mean intensities between two classes or groups of metabolites, by dividing them. If the value of the fold-change is 2 (twofold increase, $\log_2$ fold-change $= 1$) it means that the intensities of the peaks from the first class are twice as big as intensities from the second class, and if the fold-change is 0.5 (twofold decrease, $\log_2$ fold-change $= 0.5$) than the peaks from second class are twice as big as mean intensities of the first-class peaks (Winkler 2020). The univariate methods are not robust; they cannot identify outliers and may produce misleading results in the presence of irregular concentrations of metabolites. Moreover, these abnormal levels of metabolites may lead to the wrong assumption of normality in metabolomic datasets. The Volcano plot is performed to identify different metabolites using a combination of Fold-change and $t$-test. The data are plotted as $\log_2$(Fold-change) on the $x$-axis against the $-\log_{10}(p$-value) on the $y$-axis. The classical Volcano plot is heavily influenced by outliers and Kumar and colleagues (2018) performed a new robust volcano plot for NMR analyses. The robust Volcano test use Kernel weighted average and variance, but also plot $\log_{10}(p$-value) on the $y$ axis and $\log_2$(fold-change) on the $x$-axis (Kumar et al. 2018).

Chemometrics has been defined as the application of statistical and mathematical techniques to evaluate chemical measurements and extract not so obvious correlations between dependent and independent variables (Kowalski 1980). The power of each of diagnostic statistics is investigated in terms of its ability to provide a statistically significant measure of the discrimination between two classes of subjects (e.g., the cases and the controls) when known multivariate effects of different magnitudes are present in the dataset (Szymańska et al. 2012).

PCA is a chemometric technique, in which weighted average or optimal representative variable of whole data is formally called linear combination. Therefore, original data or variables are statistically treated in a linear way when we consider PCA (Bro and Smilde 2014). It is used to reduce the multidimensionality of dataset (latent variables), maintaining as much variance as possible (Scholz and Selbig 2007). PCA transforms dimensional sample vector $x = (x1, x2, \ldots, xd)$T into usually lower-dimensional vector $y = (y1, y2, \ldots, yk)$T, where d represents the number of latent variables and $k$ is a number of selected components, while $y$ is a new vector in the same space as the $x$ variables (Bro and Smilde 2014; Scholz and Selbig 2007).

PLS-DA is a chemometric technique used to handle multiple dependent variables and optimize separation between different groups of samples (Gromski et al. 2015). In PLS-DA models, a relationship between the metabolomics data and the categorical variable y is developed in such a way that categorical variable values can be predicted for samples of unknown origin given the metabolomics data (Szymańska et al. 2012). It is a technique that extends covariance between the independent variables ($^1$H NMR spectral data in case of metabolomics by NMR) and the corresponding dependent variable (groups or classes) by finding a linear sub-space on a planar variable (Gromski et al. 2015). Herein, PLS is a regression method with a special binary "dummy" y-variable and in metabolomics study is used for classification purposes and biomarker selection. The selection of the optimal model complexity (optimal number of latent variables—LV) and the assessment of the overall quality of the model are critical for creating PLS-DA model.

PLS-DA is very widespread, and usually, authors would prefer a PLS scores plot over a PCA scores plot, but many of them are often unaware that this way easily can overfit data. Separating groups with a PLS-DA can be obtained even with a randomly grouped dataset. A PLS-DA method particular issue is when it is used for two or more groups/classes with a small number of samples per class. There are many papers and presentations presenting this method for multi-class problems. Validation of PLS-DA analyses is rarely reported in sufficient detail to determine exactly the data steps in model evaluation and without that detail quite different and some erroneous results can be obtained (Xia et al. 2009).

Q2 is an estimate of the predictive ability of the model and is calculated via cross-validation (CV). In each CV, the predicted data are compared with the original data, and the sum of squared errors is calculated. The prediction error is then summed over all samples (Predicted Residual Sum of Squares or PRESS). For convenience, the PRESS is divided by the initial sum of squares and subtracted from 1 to resemble the scale of the R2. Good predictions will have low PRESS or high Q2. It is possible to have negative Q2, which means that your model is not at all predictive or is overfitted (Szymańska et al. 2012).

Significance Analysis of Microarrays and/or Metabolites (SAM) and Empirical Bayesian Analysis of Microarrays (EBAM) are methods for the identification of significant features in high-dimensional data. SAM has purpose false discovery rate problems (FDR) when running multiple tests on high-dimensional data. First, it assigns significance to the result on each variable basis of its change in relation to the standard deviations of repeated measurements. It then selects variables with scores greater than the adjustable threshold and compares its relative difference in the distribution of estimated random permutations by class labels. A certain percentage of the variables in the permutation set will be found to be significant by chance, for each threshold. The number is used to calculate the FDR. This way one is able to perform permutation analyses, that MetaboAnalyst *t*-tests is not able. The EBAM can be defined as a variation of the SAM method. It uses a modified *t*-statistic to calculate the score. The SAM and EBAM plots are designed to help users in choosing the best parameters and viewing the results (Xia et al. 2009).

Clustering method is widely used in different fields, not only in analytic assays. It is a technique or method that provides finding subgroups, or clusters in data set (James et al. 2013). It could be useful to determine the outliers in the data set. In metabolomics, proteomics and genomics analyses are used three main clustering approaches: *K*-means, Dendrogram, and Heatmap. *K*-mean is an unsupervised technique that groups a dataset in a subgroup by the defined distance similarities. *K*-mean method applies several distance measures such as correlation coefficient, Euclidean distance, or Manhattan distance (James et al. 2013; Wang et al. 2009). Hierarchical clustering, as opposed to *K*-mean where the number of expected clusters must be stated, does not require a defined choice of subgroups. The sample of the hierarchical clustering method is Dendrogram and is preferably used in genomics and proteomics (James et al. 2013). SVM uses rigorous statistical learning theory and it treats variables as input and outputs as classification or a regression vector. Using a Kernel method, it maps input vectors into dimensional characteristics of space using the core function. The training procedure leads to the discovery of higher planes where two class vectors are optimally separated. In metabolomics, when a new characteristic vector (row or sample) is input, his classes affiliation is predicted from which side of the plane is mapped (Wang et al. 2009).

# 6    Metabolite Identification

The final steps in metabolomics data analyses are validation of the chemometrics analyses, interpretation of the obtained results, and proposing the list of identified metabolites. Variations depend on the type of sample, for example, if the sample is from human source (urine, serum, plasma, cell, and tissue), plants (extracts from leaf, root, rhizome, fruit), and/or from microorganism (whole cells our bacteria lysate, yeast, worm, etc.), interpretation of the obtained results must follow biochemical and chemical data available for the given species. There are several databases related to the specific sample type and analytical technique that enable the metabolite data to be searched and linked to specific metabolic pathways. The database are usually open access platforms available for any user, and the name and links of several databases useful for NMR data interpretation are listed below:

- Human metabolome data base (HMDB): https://hmdb.ca/
- BioMagResBank (BMRB): http://www.bmrb.wisc.edu/
- Madison Metabolomics Consortium Database (MMCD): http://mmcd.nmrfam.wisc.edu/
- Birmingham Metabolite Library Nuclear Magnetic Resonance database (BML-NMR): http://www.bml-nmr.org/
- Livestock Metabolome Database (LMDB): http://lmdb.ca/
- Serum Metabolome database (SMDB): https://serummetabolome.ca/
- BiGG Models (BiGG): http://bigg.ucsd.edu/
- NMRShiftDB: https://nmrshiftdb.nmr.uni-koeln.de/

In metabolomics analysis, besides 1D $^1$H NMR spectra, homonuclear and heteronuclear 2D NMR, *J*-resolved spectra are often used for the identification of molecules. 2D NMR homonuclear spectra are used in most cases, such as correlation (COSY) and total correlation (TOCSY) spectroscopies that give spin–spin coupling connectivities and provide information on which hydrogens in a molecule are close neighbors in chemical bond term, up to four bonds in TOCSY. 2D $J$-resolved NMR spectroscopy is performed for confirmation and determination of signals chemical assignment because $J$-couplings are insensitive to physiological conditions (Smolinska et al. 2012).

There are also several softwares for peak detection: Bruker AMIX (1D AND 2D), Chenomx rNMR Suite (1D), Bayesil (1D), Metabominer MVAPack, COLMAR (2D TOCSY AND HSQC), Data Exchange Formats, which assist in final identification of NMR chemical shifts. Recently, online platforms for the deposition of metabolites are constructed, such as MetaboLights and Metabolomics Workbench (Ellinger et al. 2013; Brereton 2015).

Figure 3 and Table 1 illustrate just the main biomarkers that can be found in serum sample using $^1$H NMR spectra assignment.

Lactate doublet (-$CH_3$) at 1.33 ppm ($J = 7.0$ Hz) is easily recognizable and always present in $^1$H NMR spectra of blood serum or plasma, cells, and tissues. This doublet is frequently used as an internal standard if there is no standard addition (TSP, TMSP, DSS). Also, two doublets at 4.63 ppm ($J = 7.98$ Hz) and 5.22 ppm ($J = 3.80$ Hz) are assigned to glucose and always present in above-mentioned samples. There are many differences among metabolites in biological samples, but before mentioned biomarkers (Table 1) and their chemical assignments are always present in human and animal sera and plasma spectra. The urine, sera, plasma, and tissue spectra depend on various factors, such as age, diet, use of licit or illicit drugs, lifestyle, presence of chronic disease, gut microflora, and physical activity (Takis 2019). Nevertheless, the list of metabolites that can be observed and qualified in blood serum samples by NMR goes up to 75–80 metabolites. Also, databases count on 800 to 1000 metabolites and need to be greatly improved as to compete with other databases and methodologies used for metabolomics research (Wishart 2019).

NMR-based metabolomics approach demonstrated to be a reliable and quick tool for possible biomarkers discovery, which found application in cancer, cardiovascular, rheumatic, pulmonary, psychiatric, different animal model research, as well in plant, and food-industry research. If taken into account that the sample preparation is fast and without the application of additional extractive methods, except for lipidomics assays, then liquid and semisolid-state NMR could be considered as the most reliable methods for studying and determining biomarkers without any possible interference on their structure by sample preparation and handling.

# 7 Final Remarks

NMR spectroscopy is one of the most used bio-screening tools in bioanalytical samples analyses with $^1$H as the most useful nucleus for metabolomics, metabonomics, and/or lipidomics analyses. NMR is a highly reproducible, reliable, and quantitative technique that does not destroy the sample, which might be liquid, such as blood serum, cerebrospinal fluid, urine, their extracts or cells'extracts, or semisolid, like cells, and tissues, and is very useful for quantitative analyses for

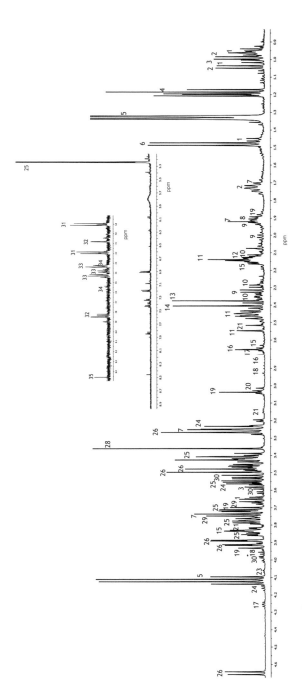

**Fig. 3** Example of $^1H$ NMR spectral data from the blood serum from Wistar rat ($D_2O$, 500 MHz) with top 35 metabolites assignment. Lower panel shows the 0.8–2.8 ppm region followed by 2.7–4.7 ppm, from left to right. Upper panel shows the 5.0–7.1 ppm region and enlarged (eightfolds) 6.7–8.6 ppm region

**Table 1** Chemical shifts, spectral peaks multiplicities, and coupling constants of some of the most important biomarkers (1–35) as depicted in Fig. 3 found in blood serum (source: HMDB)

| Metabolite | N° | Chemical shift (ppm), spectral peaks multiplicities and coupling constant |
|---|---|---|
| Isoleucine | 1 | 0.926 t ($J = 8$ Hz); 0.997 d ($J = 9$ Hz); 1.248 m; 1.457 m; 1.968 m; 3.661 d ($J = 4$ Hz) |
| Leucine | 2 | 0.948 t ($J = 8$ Hz); 1.71 m; 3.72 m |
| Valine | 3 | 0.978 d ($J = 8$ Hz); 1.029 d ($J = 7$ Hz); 2.261 m ($J = 14, 7, 4.41$ Hz); 3.601 d ($J = 4$ Hz) |
| 3-Hydroxy butyrate | 4 | 1.204 d ($J = 6.26$ Hz); 2.314 m; 2.414 m; 4.160 m |
| Lactate | 5 | 1.33 d ($J = 7$ Hz); 4.12 q ($J = 6.93$ Hz) |
| Alanine | 6 | 1.47 d ($J = 7.28$ Hz); 3.77 q ($J = 7.28$ Hz) |
| Arginine | 7 | 1.68 m ($J = 7$ Hz); 1.90 m ($J = 7.28$ Hz); 3.23 t ($J = 6.93$ Hz); 3.76 t ($J = 6.11$ Hz) |
| Acetate | 8 | 1.91 s |
| Proline | 9 | 1.99 m ($J = 3$ Hz); 2.06 m ($J = 4$ Hz); 2.34 m ($J = 4$ Hz); 3.33 dt ($J = 14.02, 7.11$ Hz); 3.41 dt ($J = 11.65, 7.02$ Hz); 4.12 dd ($J = 8.63, 6.42$ Hz) |
| Glutamate | 10 | 2.04 m; 2.119 m; 2.341 m; 3.748 dd ($J = 7.186, 4.724$ Hz) |
| Glutathione | 11 | 2.15 m; 2.54 m; 2.97 dd ($J = 14.23, 9.47$ Hz); 3.78 m; 4.20 q ($J = 7.14$ Hz) |
| Glutamine | 12 | 2.125 m; 2.446 m; 3.766 t ($J = 6.18$ Hz) |
| Pyruvate | 13 | 2.46 s |
| Succinate | 14 | 2.39 s |
| Methionine | 15 | 2.157 m; 2.631 t ($J = 7.587$ Hz); 3.851 dd ($J = 7.1, 5.382$ Hz) |
| Aspartate | 16 | 2.66 dd ($J = 17.45, 8.85$ Hz); 2.80 dd ($J = 17.45, 3.72$ Hz); 3.89 dd ($J = 8.82, 3.75$ Hz) |
| Malate | 17 | 2.35 dd ($J = 15.37, 10.24$ Hz); 2.66 dd ($J = 15.37, 2.99$ Hz); 4.29 dd ($J = 10.23, 2.98$ Hz) |
| Asparagine | 18 | 2.84 m; 2.94 m; 4.00 dd ($J = 7.69, 4.26$ Hz) |
| Lysine | 19 | 1.46 m; 1.71 m; 1.89 m; 3.02 t; 3.74 t ($J = 6.09$ Hz) |
| Creatine | 20 | 3.02 s; 3.92 s |
| Ethanolamine | 21 | 3.13 t; 3.83 t |
| Citrate | 22 | 2.525 d; 2.655 d ($J = 15.16$ Hz) |
| Choline | 23 | 3.189 s; 3.507 dd ($J = 5.816, 4.162$ Hz); 4.056 ddd |
| O-Phosphocholine | 24 | 3.21 s; 3.58 m; 4.16 m |
| α-Glucose | 25 | 3.23 dd ($J = 9.41, 7.98$ Hz); 3.40 m; 3.46 m; 3.52 dd ($J = 9.82, 3.77$); 3.73 m; 3.82 m; 3.88 dd ($J = 12.30, 2.23$ Hz); 4.63 d ($J = 7.98$); 5.22 d ($J = 3.80$) |
| β-Glucose | 26 | 3.25 m; 3.49 m; 3.50 m; 3.88 m; 3.91 m; 4.66 d |
| Betaine | 27 | 3.25 s; 3.89 s |
| Methanol | 28 | 3.33 s |
| Mannitol | 29 | 3.649 dd ($J = 11.76, 6.26$ Hz); 3.729 m; 3.771 d; 3.840 dd ($J = 11.87, 2.86$ Hz) |
| Myo-inositol | 30 | 3.268 t ($J = 11.76$ Hz); 3.524 dd ($J = 9.978, 2.867$); 3.613 t ($J = 9.702$ Hz); 4.053 t ($J = 2.839$ Hz) |

(continued)

**Table 1** (continued)

| Metabolite | N° | Chemical shift (ppm), spectral peaks multiplicities and coupling constant |
|---|---|---|
| Tyrosine | 31 | 3.03 dd ($J = 14.55$, 8.01 Hz); 3.34 dd ($J = 14.53$, 4.68 Hz); 4.04 dd ($J = 8.03$, 4.68 Hz); 6.94 m; 7.20 dd ($J = 7.95$, 1.51 Hz); 7.24 td ($J = 7.76$, 1.71 Hz) |
| Histidine | 32 | 3.16 dd ($J = 15.55$, 7.75 Hz); 3.23 dd ($J = 16.10$, 4.93 Hz); 3.98 dd ($J = 7.73$, 4.98 Hz); 7.09 d ($J = 0.58$ Hz); 7.90 d ($J = 1.13$ Hz) |
| Phenylalanine | 33 | 3.19 m; 3.98 dd ($J = 7.88$, 5.31 Hz); 7.32 d ($J = 6.96$ Hz); 7.36 m; 7.42 m |
| Tryptophan | 34 | 3.292 dd ($J = 15.353$, 8.076 Hz); 3.472 dd ($J = 15.38$, 4.796 Hz); 4.046 dd ($J = 8.104$, 4.851 Hz); 7.194 m; 7.274; 7.310 s; 7.531 d ($J = 8.159$ Hz); 7.723 d ($J = 7.993$ Hz) |
| Formate | 35 | 8.447 s |

molecules whose concentrations are $>1\mu$mol $L^{-1}$. Different samples can be analyzed, and their preparation is minimal for most NMR techniques (liquid NMR, or semisolid HR-MAS NMR), also some most common NMR experiments (CPMG, diffusion filter, 2D NMR techniques) could greatly improve the post-analyses processing and metabolites' identification. For these reasons, NMR is explored in bioanalytics for decades (almost 1/2 century) and it has provided solutions (biomarkers discovery) for diagnostics, prognostics, and other biomedical applications; but, also for biotechnology, drug design, forensic studies, food and fuel research, and adulterations, etc. Although NMR can be used for many purposes, stand out its unique potential as a diagnostic tool for body fluids with a great medical application that is starting to be explored worldwide with in situ clinical applications. In a meantime, there are some issues to be solved, like improvements in software and hardware, construction of better libraries with richer databases, for example, today, NMR libraries cover around 800–1000 compounds, improving its sensitivity, lowering instrumental and maintaining costs, discovery and employment of new probes and sequences that can reduce the time of acquisitions.

# References

Adams RW, Holroyd CM, Aguilar JA, Nilsson M, Morris GA (2013) Chem Commun 49:358–360

Andrew ER, Bradeury A, Eades RG (1958) Nature 182:1659

Barbosa BS, Martins LG, Costa TBBC, Cruz G, Tasic L (2018) In: Guest P (ed) Methods in molecular biology, vol 1735. Humana Press, New York, pp 365–379

Barding R, Salditos CK, Larive CK (2012) Anal Bioanal Chem 404:1165–1179

Beckonert O, Keun HC, Ebbels TMD, Bundy J, Holmes E, Lindon JC, Nicholson JK (2007) Nat Protoc 2:2692–2703

Bo Y, Feng J, Xu J, Huang Y, Cai H, Cui X, Dong J, Ding S, Chen Z (2019) Food Res Int 125: 108574

Boesch C, Kreis R (2001) NMR Biomed 14:140–148

Brereton RG (2015) Chemom Intell Lab Syst 149:90–96

Bro R, Smilde AK (2014) Anal Methods 6:2812–2831

Chong J, Wishart DS, Xia J (2019) Curr Protoc Bioinformatics 68:e86

Claridge TDW (2009) In: Backvall JE, Baldwin JE, Williams RM (eds) High resolution NMR techniques in organic chemistry, 2nd edn. Elsevier, Oxford

Cohen-Addad JP (1974) J Chem Phys 60:2440–2453

Costa TBBC, Lacerda ALT, Dal Mas C, Brietzke E, Pontes JGM, Marins LAN, Martins LG, Nunes MV, Pedrini M, Carvalho MSC, Mitrovitch MP, Hayashi MAF, Saldanha NL, Poppi RJ, Tasic L (2019) J Proteome Res 18:341–348

Dona AC, Kyriakides M, Scott F, Shephard EA, Varshavi D, Veselkov K, Everett JR (2016) Comput Struct Biotechnol J 14:135–153

Eberstadt M, Gemmecker G, Mierke DF, Kessler H (1995) Angew Chem Int Ed 34:1671–1695

Ellinger JJ, Chylla RA, Ulrich EL, Markley JL (2013) Curr Metabolomics 1:28–40

Emwas AM, Salek RM, Griffin JL et al (2013) Metabolomics 9:1048–1072

Emwas A-H, Saccenti E, Gao X, McKay RT, dos Santos VAP, Roy R, Wishart DS (2018) Metabolomics 14:31

Emwas A-H, Roy R, McKay RT, Tenori L, Saccenti E, Gowda GAN, Raftery D, Alahmari F, Jaremko L, Jaremko M, Wishart DS (2019) Metabolites 9:123–162

Escobar MQ, Maschietto M, Krepischi ACV, Avramovic N, Tasic L (2019) Biomol Ther 9:843–859

Feraud B, Govaerts B, Verleysen M, de Tullio P (2015) Metabolomics 11:1756–1768

Foroozandeh M, Castañar L, Martins LG, Sinnaeve D, Dal Poggetto G, Tormena CF, Adams RW, Morris GA, Nilsson M (2016) Angew Chem Int Ed 55:15579–15582

Foroozandeh M, Morris GA, Nilsson M (2018) Chem Eur J 24:13988–14000

Gevaert K, Vandekerckhove J (2000) Electrophoresis 21:1145–1154

Giraudeau P, Silvestre V, Akoka S (2015) Metabolomics 11:1041–1055

Gogiashvili M, Nowacki J, Hergenröder R, Hengstler JG, Lambert J, Edlund K (2019) Metabolites 9:19–46

Gosetti F, Mazzucco E, Gennaro MC, Marengo E (2013) J Chromatogr B 927:22–36

Gromski PS, Muhamadali H, Ellis DI, Xu Y, Correa E, Turner ML, Goodacre R (2015) Anal Chim Acta 879:10–23

Hennel JW, Klinowski J (2004) Top Curr Chem 246:1–14

James G, Witten D, Hastie T, Tibshirani R (2013) An introduction to statistical learning with application in R, vol 10. Springer Texts in Statistics, Berlin, p 385

Keeler J (2010) In: Keeler J (ed) Understanding NMR spectroscopy, 2nd edn. John Wiley and Sons, West Sussex, pp 159–226

Kowalski BR (1980) Anal Chem 52:112–122

Kruk J, Doskocz M, Jodłowska E, Zacharzewska A, Łakomiec J, Czaja K, Kujawski J (2017) Appl Magn Reson 48:1–21

Kumar N, Hoque MA, Sugimoto M (2018) BCM Bioinformatics 19:128

Le Guennec A, Giraudeau P, Caldarelli S (2014) Anal Chem 86:5946–5954

Levitt MH (2007a) In: Levitt MH (ed) Spin dynamics: basics of nuclear magnetic resonance, 2nd edn. John Wiley and Sons, West Sussex, pp 5–15

Levitt MH (2007b) In: Levitt MH (ed) Spin dynamics: basics of nuclear magnetic resonance, 2nd edn. John Wiley and Sons, West Sussex, pp 171–206

Levitt MH (2007c) In: Levitt MH (ed) Spin dynamics: basics of nuclear magnetic resonance, 2nd edn. John Wiley and Sons, West Sussex, pp 211–217

Levitt MH (2007d) In: Levitt MH (ed) Spin dynamics: basics of nuclear magnetic resonance, 2nd edn. John Wiley and Sons, West Sussex, pp 217–223

Liland KH (2011) Trends Anal Chem 30:827–841

Lopes TIB, Martins LG, Nagassaki S, Geloneze B, Marsaioli AJ (2013) J Chem Chem Eng 7:547–555

Lowe IJ (1959) Phys Rev Lett 2:285–287

Ludwig C, Viant MR (2010) Phytochem Anal 21:22–32

Madsen R, Lundstedt T, Trygg J (2010) Anal Chim Acta 659:23–33

Mahrous EA, Farag MA (2015) J Adv Res 6:3–15

Massiot D, Fayon F, Capron M, King I, Le Calvé S, Alonso B, Durand J-O, Bujoli B, Gan Z, Hoatson G (2002) Magn Reson Chem 40:70–76

Mazzei P, Piccolo A (2017) Chem Biol Technol Agric 4:11–24

Monte GG, Nani JV, Campos MR, Dal Mas C, Marins LA, Martins LG, Tasic L, Mori MA, Hayashi MAF (2019) Prog Neuropsychopharmacol Biol Psychiatr 92:19–30

Nicholson JK, Foxall PJD, Spraul M, Farrant RD, Lindon JC (1995) Anal Chem 67:793–811

Parkinson JA, Lindon J, Nicholson J, Holms E (2019) NMR spectroscopy methods in metabolic phenotyping. In: The handbook of metabolic phenotyping. Elsevier, Cambridge, pp 53–96

Pontes JGM, Ohashi WY, Brasil AJM, Filgueiras PR, Espındola APDM, Silva JS, Poppi RJ, Coletta-Filho HD, Tasic L (2016) ChemistrySelect 6:1176–1178

Pontes JGM, Brasil AJM, Cruz GCF, de Souza RN, Tasic L (2017a) Anal Methods 9:1078–1096

Pontes JGM, Santana FB, Portela RW, Azevedo V, Poppi RJ, Tasic L (2017b) Metabolomics 7:1–7

Price WS (1999) Annu Rep NMR Spectrosc 38:289–354

Rooney OM, Troke J, Nicholson JK, Griffin JL (2003) Magn Reson Med 50:925–930

Sandusky P, Raftery D (2005) Anal Chem 77:2455–2463

Santacruz L, Hurtado DX, Doohan R, Thomas OP, Puyana M, Tello E (2020) Sci Rep 10:5417–5429

Sardans J, Peñuelas J, Rivas-Ubac A (2011) Chemoecology 21:191–225

Scholz M, Selbig J (2007) Metabolomics: methods and protocols, vol 87. Human Press

Shin J, Lee W, Lee W (2008) Expert Rev Proteomics 5:589–601

Simon G, Kervarec N, Cérantola S (2015) In: Stengel DB, Connan S (eds) Natural products in marine algae. Methods and protocols. Springer, New York, pp 191–205

Smith JAS (1971) J Chem Educ 48:39–48

Smith LM, Maher AD, Cloarec O, Rantalainen M, Tang H, Elliott P, Stamler J, Lindon JC, Holmes E, Nicholson JK (2007) Anal Chem 79:5682–5689

Smolinska A, Blanchet L, Buydens LMC, Wijmeng SS (2012) Anal Chim Acta 750:82–97

Stanisic D, Fregonesi NL, Barros CHN, Pontes JGM, Fulaz S, Menezes UJ, Nicoleti JL, Castro TLP, Seyffert N, Azevedo V, Duran N, Portela RW, Tasic L (2018) RCS Adv 8:40778–40786

Szymańska E, Saccenti E, Smilde AK, Westerhuis JA (2012) Metabolomics 8:S3–S16

Takis PG, Ghini V, Tenori L, Turano P, Luchinat C (2019) Trends Anal Chem 120:115300

Tasic L, Pontes JGM, Carvalho MS, Cruz G, Dal Mas C, Sethi S, Pedrini M, Rizzo LB, Zeni-Graiff M, Asevedo E, Lacerda ALT, Bressan RA, Poppi RJ, Brietzke E, Hayashi MAF (2017a) Schizophr Res 185:182–189

Tasic L, Pontes JGM, Souza RN, Brasil AJM, Cruz GCF, Asevedo E, Dal Mas C, Poppi RJ, Brietzke E, Hayashi MAF, Lacerda ALT (2017b) ChemistrySelect 2:2927–2930

Tasic L, Larcerda ALT, Pontes JGM, da Costa TBBC, Nani JV, Martins LG, Santos LA, Nunes MFQ, Adelino MPM, Pedrini M, Cordeiro Q, Santana FB, Poppi RJ, Brietzke E, Hayashi MAF (2019) J Psychiatr Res 119:67–75

van den Berg RA, Hoefsloot HCJ, Westerhuis JA, Smilde AK, van der Werf MJ (2006) BMC Genomics 7:142

Wang T, Shao K, Chu Q, Ren Y, Mu Y, Qu L, He J, Jin C, Xia B (2009) BMC Bioinformatics 10:83

Winkler R (2020) Processing metabolomics and proteomics data with open software: a practical guide. Royal Society of Chemistry, Croydon

Wishart DS (2019) J Magn Reson 306:155–161

Wong A, Lucas-Torres C (2018) In: Keun HC (ed) NMR-based metabolomics. The Royal Society of Chemistry, Cambridge, pp 133–150

Xia J, Wishart DS (2016) Curr Protoc Bioinformatics 55:14.10.1–14.10.91

Xia J, Psychogios N, Young N, Wishart DS (2009) Nucleic Acids Res 37:W652–W660

Yuk J, McKelvie JR, Simpson MJ, Spraul M, Simpson AJ (2010) Environ Chem 7:524–536

Zhao YY, Cheng XL, Lin RC (2014) Int Rev Cell Mol Biol 313:1–26

Zheng H, Baijun D, Ning J, Shao X, Zhao L, Jiang Q, Ji H, Cai A, Xue W, Gao H (2020) Clin Chim Acta 501:241–251

# SPR Sensors: From Configurations to Bioanalytical Applications

Dênio E. P. Souto, Jaqueline Volpe, and Denys R. de Oliveira

## 1 Fundamental Aspects

Collective electronic excitations on metallic surfaces are well known to play an important role in many areas, comprising chemistry, physics, material sciences, and even biology (Souto et al. 2019; Damos et al. 2004; Adzhri et al. 2016). According to Pitarke et al. (2007), the studies performed by Tonks and Langmuir (1929), Ruthemann (1948), Lang (1948), Pines and Bohm (1952), Pines (1956), and Ritchie (1957) were essential for Powell and Swan (1959) to demonstrate the existence of collective excitation on metallic surfaces. This phenomenon was then named surface plasmon (SP) by Stern and Ferrell (1960). Specifically, SPs are charge oscillations (electrons in general) at the metal-dielectric interface that generate evanescent electromagnetic fields which exponentially decay in a perpendicular direction to the metallic surface. Different approaches have been used to describe the relation of dispersion for the SP (equation below) (Stegeman et al. 1983a, b; Burke et al. 1986), which consists of the dependence of the angular frequency ($\omega$) to the propagation constant for the SP ($\beta_{SP}$) on a metal-dielectric interface (Eq. 1):

$$\beta_{PS} = \frac{\omega}{c} \sqrt{\frac{\epsilon_r \epsilon_a}{\epsilon_r + \epsilon_a}} = \frac{2\pi}{\lambda} \sqrt{\frac{\epsilon_r \epsilon_a}{\epsilon_r + \epsilon_a}}. \tag{1}$$

where $c$ is the speed of light in the vacuum; $\lambda$ is the wavelength in the vacuum; $\epsilon_r$ and $\epsilon_a$ are the real part of the dielectric constant of the metal and environment, respectively. Later these studies became essential to assure the exploration of the SPR phenomenon through optical sensors for the evaluation of interfacial and surface processes. Fixing $\epsilon_r$ and $\lambda$ (see Eq. 1), it is possible to obtain a direct relationship between $\beta_{SP}$ and $\epsilon_a$, which can variate due to changes in the refractive index from the

D. E. P. Souto (✉) · J. Volpe · D. R. de Oliveira
Department of Chemistry, Federal University of Paraná (UFPR), Curitiba, Paraná, Brazil
e-mail: denio.souto@ufpr.br

© The Author(s), under exclusive license to Springer Nature Switzerland AG 2022
L. T. Kubota et al. (eds.), *Tools and Trends in Bioanalytical Chemistry*,
https://doi.org/10.1007/978-3-030-82381-8_10

media (relation established from the Maxwell boundary conditions). It allows, for example, that interactions between a bioreceptor and an analyte lead to changes in the refractive index, which can generate a signal proportional to the concentration of the analyte. All of these aspects are the basis for the use of SPR-based sensors in bioanalytical applications.

By definition, a chemical sensor with an optical transducer is a detection device that optically converts a chemical information into a signal based on characteristics of the light, which is originated from the interaction between the analyte and a receptor (Homola 2008). This enables several methods to be used in optical sensor platforms, including the variation of refractive index, absorbance, and other measurements based on spectroscopy (Fan et al. 2008). When the optical sensor is based on the SP excitation, they are referred to as SPR sensors (Tudos and Schasfoort 2008). In these systems, variations of the refractive index near an interface (such as metal-dielectric) are followed by variations in the SP propagation constant. These variations result in a change in the conditions of coupling between the incident wave and the plasmonic wave. Therefore, depending on the type of alteration observed in the incident wave, SPR sensors are classified as a sensor of angular modulation (Matsubara et al. 1988), wavelength (Zhang and Uttamchandani 1988), intensity (Nylander et al. 1982), or phase (Brockman et al. 2000).

Typically, SPR optical sensors can be divided into two different categories, propagating SPR and localized SPR (LSPR), being both related to SP oscillations. LSPR is based on metallic nanoparticles, which nonpropagating surface plasmons are excited and its resonance can be adapted by size, shape, and composition of nanoparticles. In this category, after an applied electric field, the conduction of electrons runs through collective harmonic oscillations, generating a dipolar response. In turn, propagating SP are excited on continuous metal films, through different electromagnetic wave couplers (prism, diffraction gratings, and wave guides, for example) and the SPR effect occurs along the metal–dielectric interface. In this chapter, different propagating SPR configurations applied to bioanalytical applications have been discussed (Souto et al. 2019; Damos et al. 2004; Adzhri et al. 2016).

## 2 SPR-Based Sensors Construction: Timelapse on Configurations

Since the first report of a chemical sensor based on SPR, made by Nylander et al. (1982), this field of research has gained significant technological and applicational progress greatly due to the development of optical devices, such as couplers. Even though some new configurations have emerged recently (Tokel et al. 2014), the physical phenomenon of attenuated total internal reflection (ATIR) employing prisms as couplers or optical waveguides, and the principle of light diffraction using a diffraction grating, are the main configurations used so far in commercial SPR-based sensors, see schematic representation (Fig. 1) (Abdulhalim et al. 2008; Fathi et al. 2019).

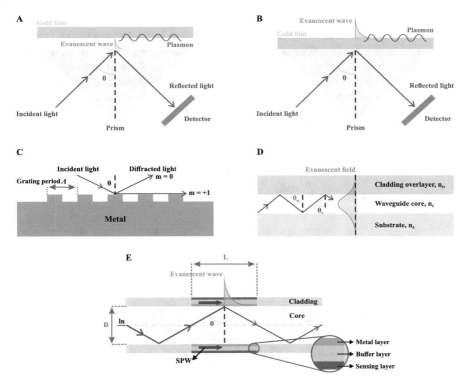

**Fig. 1** Principle of prism coupler-based SPR sensor operating by Otto (**a**) and Kretschmann (**b**) configurations. Grating coupler-based SPR sensor (**c**), fiber-optic coupler-based SPR sensor (**d**), and basic waveguide coupler-based configuration (**e**) illustrating the light confined in the waveguide core by the phenomenon of ATIR; the sensing length of fiber is represented as L

## 2.1 Prism Coupling

Otto in 1968 (Otto 1968; Teng and Stern 1967) first introduced the attenuated total internal reflection (ATIR) coupling method for SP excitation, using a prism in a glass-air-film arrangement (Fig. 1a) (Danlard and Akowuah 2020). However, the need to maintain an air gap of sub-wavelength range between the prism base and the metal surface made this configuration practically ineffective (Tabassum and Kant 2020). In the same year, Kretschmann and Raether updated the configuration proposed by Otto and demonstrated the excitation of surface plasmons using an optical configuration composed of a glass-film-air arrangement (Fig. 1b) (Kretschmann and Raether 1968), which became the most popular configuration on current research and commercial SPR instrumentation (Homola 2008; Gupta and Kant 2018; Wang et al. 2019). Although these configurations are subtly different, the idea behind their operation is similar (Danlard and Akowuah 2020; Tabassum and Kant 2020). To satisfy the boundary conditions necessary for SP excitation, the incident light must be p-polarized, since in this condition the surface plasmon wave

(SPW) present only one electric field component, which is normal to the surface (Kim and Koh 2006). When a $p$-polarized light reaches a prism with a high refractive index, this light is totally reflected at a specific angle with the prism's base. Although the light is totally reflected, a component of its radiation, wave or evanescent field, penetrates the prism–metal interface (Carvalho et al. 2003). In a certain angle of incidence, when the propagation constant of the surface plasmon wave equals the propagation constant of the evanescent wave ($\kappa_{SP} = \kappa_{EV}$) part of the radiation couples with the free electrons oscillating on the metallic film, resulting in the surface plasmon resonance. The resonance condition can be mathematically expressed through the Eq. (2) (Carvalho et al. 2003; Wijaya et al. 2011):

$$\kappa_{SP} = \kappa_0 \left( \frac{\varepsilon_S \varepsilon_m}{\varepsilon_s + \varepsilon_m} \right)^{1/2} = \kappa_{EV} = \kappa_0 \sqrt{\varepsilon_P} \, \sin \, \theta_{\text{res}}. \tag{2}$$

where $\kappa_0$ describes the propagation constant of the incident light of wavelength $\lambda$, and $\varepsilon_m$ and $\varepsilon_S$ are the wavelength-dependent dielectric constants of metal and dielectric, respectively. $\varepsilon_P$ represents the dielectric constant of the prism and $\theta_{\text{res}}$ is the resonant angle (Gupta and Kant 2018).

Despite being the most explored configuration to date in bioanalytical applications (discussed later), an unfavorable characteristic is in relation to the bulky size of the prism along with several operational mechanical parts of the sensor, which limit its versatility and ability to integrate into miniaturized arrays and lab-on-a-chip platforms (Tabassum and Kant 2020; Rossi et al. 2018).

## 2.2    Grating Coupling

The moment of the incident optical wave can be intensified and applied to the excitation of the SP using a grating or metallic diffraction grid with dimensions smaller than the wavelength of the incident light (Gupta and Kant 2018). Conceptually, if a metal-dielectric interface is periodically irregular, the incident wave can suffer diffracttion, forming a series of beams that diverge at different angles starting from the interface (Fig. 1c) (Čtyroký et al. 1999). In this case, the resonance conditions are provided by diffraction of the incident light (Rossi et al. 2018). The scatters of the incident light wave by the diffraction grating modified the original component x of the incident photons wave vector ($K_x$) by an integer multiple of the grating wave number ($G$), depending on the diffraction order ($m$) (Wijaya et al. 2011; Zhao et al. 2019) to match the propagation constant of the surface plasmon wave (Gupta and Kant 2018). The resonance condition for excitation of surface plasmons grating coupling is expressed through Eq. (3) (Gupta and Kant 2018; Wijaya et al. 2011; Zhao et al. 2019):

$$K_d = K_x + mG = \frac{2\pi}{\lambda} n_d \sin \sin \, \theta + m \frac{2\pi}{\Lambda} \tag{3}$$

where $K_d$ is the propagation constant of the diffracted light, $\lambda$ is the wavelength of the incident light, $n_d$ represents the dielectric refractive index, and $\Lambda$ is the grating period (Gupta and Kant 2018; Wijaya et al. 2011; Zhao et al. 2019). In structural terms, SPR sensors based on diffraction grating are diverse, operating in transmission and reflection mode. In transmission mode, the incident light is directed through one extremity and the spectrum of transmitted light is recorded on the other extremity (Byun et al. 2007). In the reflection mode, the photodetector is on the same side of the light source with respect to the metal film, which allows an extremely compact biosensing apparatus (Byun et al. 2007). The grating-coupling-based devices present reduced cost when compared to the prism-based mechanism and offer much more features, like miniaturization and integration, thus arousing great interest in applications for lab-on-a-chip (Rossi et al. 2018; Borile et al. 2019). However, its current application in sensors is not as widespread as the use of prism, partly due to its lower compared sensitivity (Rossi et al. 2018; Čtyroký et al. 1999; Dai et al. 2018). A sensitivity improvement can be achieved by working in "conical mounting," a configuration where the scattering plane is rotated of an azimuthal angle with respect to the grating wave vector (Rossi et al. 2018; Borile et al. 2019). The breaking of symmetry with the grating rotation allows the excitation of SP with combined polarized waves of p and s, overcoming the restrictions of a single polarization mode (Kim 2005). Another interesting aspect is that the diffraction grating can be combined in SPR sensors with modulation of intensity, wavelength, or angular (Homola et al. 1999).

## 2.3 Waveguide Coupling

Several SPR detection devices using slab waveguides, rectangular waveguides, channel waveguides, and circular waveguides have been developed (Čtyroký et al. 1999; Zhang and Hoshino 2014). In general, in waveguide coupling the excitation of the SP is carried out in an analogous way to the Kretschmann and Otto coupling through the phenomenon of ATIR, the difference, therefore, is the replacement of the prism by a waveguide (Gupta and Kant 2018; Homola et al. 1999). In these systems, the light propagates in a waveguide in a confined way and the evanescent wave penetrates the metal covering a portion of the waveguide to interact with the dielectric sensing medium, culminating in SPR phenomenon at the resonance condition (Gupta and Kant 2018; Homola et al. 1999). Among the existing waveguides, we can highlight those based on optical fiber (Fig. 1d). Therefore, a dielectric waveguide is composed basically of a core, which is located between the substrate and the covering medium (cladding and/or sensed analyte layer) (Fig. 1e) (Li and Ju 2013).

The first SPR sensors based on optical fibers were reported in the early 1990s (Homola 1995; van Gent et al. 1990). Ever since, numerous plasmonic fiber-optic structures based on experimental and simulation studies have been reported in the literature (Hu and Ho 2017), including tapered fibers (Fig. 2a) (Navarrete et al. 2014; Verma et al. 2007), U-bent (Fig. 2b) (Verma and Gupta 2008), D-shaped fibers

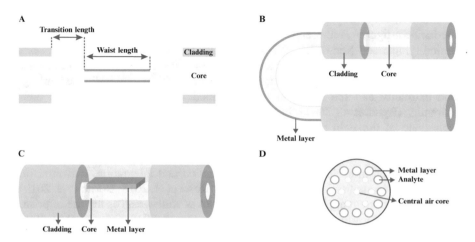

**Fig. 2** Structural representation of several optical fiber detection probes: (**a**) tapered, (**b**) U-bent, (**c**) D-shaped, and (**d**) photonic crystal fiber based

(Fig. 2c) (Wang et al. 2006), among others (Gupta and Kant 2018; Rifat et al. 2017; Klantsataya et al. 2017). To manufacture an SPR detection probe based on optical fiber, usually, a small part of the cladding around the core of the optical fiber is removed and coated with a thin metal layer, with a typical thickness of the order of 50 nm (Gupta and Kant 2018; Sharma et al. 2018). This metal layer can also be covered with a buffer layer, which is finally covered with the sensing layer (analyte sample) (Sharma et al. 2018). SPR sensors based on optical fibers have a high potential for constructing point-of-care (POC) type devices (Tokel et al. 2014; Caucheteur et al. 2015). The simple, slim compact, and flexible design enables their application in small spaces, as well as real-time detection in situ and in vivo measurements (Gupta and Kant 2018; Sharma et al. 2018; Zhao et al. 2019).

## 2.4    SPR Sensors Based on Recent Couplers

Photonic Crystal Fiber Surface Plasmon Resonance (PCF SPR) is a highly sensitive detection technology that has attracted a lot of attention in recent years (Rifat et al. 2017). Structurally, the PCF can be produced by fused quartz (Danlard and Akowuah 2020), polymer (Dash and Jha 2014), or plastic (Cennamo et al. 2013) as background material, with a periodic or aperiodic arrangement of small air holes that extend longitudinally through the fiber and manage the light propagation (Fig. 2d) (Danlard and Akowuah 2020; Rifat et al. 2017; Caucheteur et al. 2015). Through the modification of air hole geometry and by altering the number of rings, the light propagation can be controlled, thus demonstrating exclusive resources to overcome the prism and optical fiber-based SPR sensors problems (Rifat et al. 2017).

PCF can be divided into three groups: solid-core PCF, where the light is guided by the ATIR phenomenon; hollow-core PCF, in which the light is guided by a photonic bandgap; and hybrid PCFs, which carry out the propagation of light using both mechanisms (ATIR and bandgap) simultaneously (Danlard and Akowuah 2020; Caucheteur et al. 2015). PCF-based SPR sensors are promising devices that provide fast, ultra-sensitive, and miniaturized platforms of the point-of-care type (Danlard and Akowuah 2020), enjoying advantages, such as small size, ease in light launching, single-mode propagation, and ability in controlling evanescent field penetration (Rifat et al. 2017). However, the reported PCF-based SPR sensor structures are complex from the manufacturing point of view (Rifat et al. 2017; Zhao et al. 2019). As a result, PCF SPR sensors are applied only in laboratory environments, and the commercial application of these devices still remains a great challenge (Rifat et al. 2017; Zhao et al. 2019). Another development trend for PCF SPR sensors is the portable and rapid laboratory tests for diagnosis at the point of care (Rifat et al. 2017). The implementation of the fiber-optic SPR (FOSPR) detection mechanism in smartphones has increased the relevance of the highly efficient portable optical detection (Gupta and Kant 2018; Chen et al. 2019). For example, Liu et al. (2015) reported the smartphone-combined SPR biosensor construction using this system, in which part of the cladding around the core of optical fiber was removed and coated with 50 nm gold film. When the sample was injected into the flow cell, the light interacting with the sensing region was absorbed because of the SPR phenomenon and the phone camera measures a corresponding intensity change of the light coming out of the lead-out fiber. The smartphone SPR system is a field to be explored with magnificent potential for application.

# 3 Bioanalytical Applications

The numerous possible configurations for the development of SPR sensors with the goal of studying interactions involving biomolecules made it a promising tool for bioanalytical applications, also supported by its high sensitivity detection. In the last decades, many pharmaceutical and biomedical approaches resorted to SPR-based biosensors as an evaluation tool and checked its applicability in drug discovery processes, quality control of pharmaceutical compounds, studies of different receptor-ligand interactions, kinetic and thermodynamic analysis of biomolecular interactions, food quality control, biogases detection and the development of new methods of diagnostic (Pattnaik 2005). Besides, the development of miniaturized SPR equipment offers a promising possibility in point-of-care applications (discussed later) (Cassedy et al. 2020). Fig. 3 summarizes the main applications of SPR biosensors in the bioanalytical field.

## 3.1 SPR-Based Biosensors: Definitions and Classifications

Biosensor, according to IUPAC (International Union of Pure and Applied Chemistry), is a class of sensors where the unity of recognition (receptor) is a biological element, like organisms, tissues, cells, enzymes, antibodies, etc. It is important that

**Fig. 3** Overview of current bioanalytical applications for SPR-based sensors

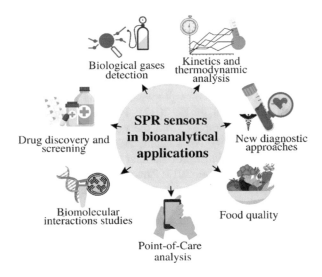

the analytical interaction is strong and selective to generate a detectable electric signal (Hulanicki et al. 1991; Faria et al. 2015). In SPR-based biosensors, the biomolecular interaction studied causes an alteration of the refractive index on the SPR-biosensor surface (Wijaya et al. 2011). Optical biosensors based in SPR offer, besides a label-free and real-time response, a high sensitivity, low electromagnetic interference, low volume of sample required, fast response, decreased number of experimental steps, and the possibility of surface sensor regeneration (Souto et al. 2019; Olaru et al. 2015). This tool has been one of the most useful techniques to characterize biomolecular interaction through kinetic (association and dissociation rates) and thermodynamic (equilibrium dissociation constant) parameters (Kamat et al. 2020). Depending on the nature of the biological element (unity of recognition) immobilized on the transducer's surface, biosensors can be classified into different types, as exemplified with current applications in Table 1. Even though SPR offers many advantages over other transducers biosensors, there are some challenges and limitations related to the technique. One of them is the detection of small molecules, in which it is difficult to differentiate the signal from the noise, since the change in the angle of reflected light is also proportional to the analytes' molecular mass. Another important point to be considered is the immobilization of the unity of recognition onto the transducer's surface, because biomolecules are sensitive to their environment and sometimes it is difficult to immobilize them without affecting their functionality (Pattnaik 2005). Since there are many interactions that can be explored by SPR biosensors, involving different molecular properties, there is not a general immobilization method (Homola 2006). In other words, one of the critical parts of constructing an SPR-based biosensor is the way of immobilizing the biomolecule at the surface's sensor, which can directly affect the performance of the device (Wijaya et al. 2011).

**Table 1** Interactions explored on SPR-based biosensors and recent bioanalytical applications

| Type | Kind of interaction | SPR application |
|---|---|---|
| Immunosensor | Antigen–antibody | Detection of a specific tuberculosis well-established biomarker (HspX—heat shock protein X) for possible application in a POC analysis for the disease's diagnosis. The analyte interacts with a specific monoclonal antibody (bioreceptor) and offers a LOD of 0.63 ng mL$^{-1}$ and LOQ of 2.12 ng mL$^{-1}$ (Pelaez et al. 2020) |
| Genosensor | DNA/RNA–complementary strand | New bioanalytical platform for MicroRNA-21 (cancer biomarker) quantification, with LOD of 0.3 fM and LOQ of 1.0 fM (Mujica et al. 2020) |
| Enzymatic | Enzyme–substrate | Bromocriptine (important Parkinson's disease drug) therapeutic monitoring on plasma, binding the drug to a laccase active site. The kinetic affinity was calculated ($K_D = 5.4$ nM), and the LOD of the analytical method was 0.001 ng mL$^{-1}$ (Jabbari et al. 2017) |
| Peptide-based | Peptides–different targets (membrane, antibody, antigen, protein) | Characterization of the potential of peptide ligands for detection of human IgG, and kinetics studies of adsorption and desorption for evaluation of peptides as alternative biosensors (Shen et al. 2014) |
| Aptasensor | Aptamer–different targets (ions, cells, protein, antibody, etc.) | Several aptamers were evaluated by their selectivity through the dissociation constant, to valuation of PCR detection of *B. cereus* spores in milk (Fischer et al. 2015) |
| Cells-based | Cell–different targets (virus, protein, etc.) | *Escherichia coli* was used for the detection of phages for application in quality control on modern fermentation industries, achieving sensibility of 1 10$^7$ PFU mL$^{-1}$ (Changqing et al. 2012) |

## 3.2 Biosensors Development and Current Approaches to Sensitivity Enhancement

It is important to be aware of some practical requirements when producing SPR biosensors to avoid inadequate responses. The quality of reagents, control of binding that are nonspecific, cleanliness of the instrument, base line's stability, and proper washing steps are factors that must be considered (Pattnaik 2005). The metal of choice as SPR substrate can also be crucial for the biosensor operation since it must exhibit, among other characteristics, a free movement of electrons behavior and good stability (Rich and Myszka 2000). Gold is more often used due mainly to its biocompatibility, stability, and low toxicity (Sohrabi and Hamidi 2016). Metal oxides can also be explored since they are low-cost, biocompatible, resistant, and safe materials to be used for in vivo biosensing, but they do not have a significant

SPR effect, often being necessary to associate them with other metals. Different metals, as silver, copper, and aluminum, for example, are commonly used as plasmonic material (Mei et al. 2020).

Regarding the methods for immobilization of the bioreceptor, since metal substrates such as gold and silver spontaneously adsorb proteins and some other molecules at their surface, early bioanalytical applications were simply based on physical adsorption. However, this method results in a loss of the bioactivity of the receptor, due to a reorganization of the molecule to a more favorable thermodynamic state (Homola 2008). Likewise, some adsorbent materials can be attached to the gold's surface, such as glass, silica gel, and collagen, for adsorption of biocomponents. Still, the interaction forces are not strong enough and the stability of the biosensor may be compromised (Monošík et al. 2012). To better anchor the receptor on the SPR sensor, surface modifications started to take place, offering better immobilization conditions by covalent attachment, as well as stability and the possibility of regeneration of the sensor. Commercial functionalized SPR chips produced by Biacore (GE Healthcare Life Sciences, now Cytiva), for example, uses different surface modifications for different applications. One of the most popular is based on thin layers of hydrogel-like polymer, carboxylated dextran (CM-dextran), formed mainly by unbranched glucose units, which allow the receptor molecules to move freely at its structure, promoting less binding interference than other methods. CM-dextran films supply a good number of attachment points, but hydrogel can hinder molecular diffusion through its structure, which may alter kinetics measurements (Wijaya et al. 2011; Homola 2006). Biacore also offers other commercial functionalized surfaces, for distinct applications. For example, for large analytes a short dextran surface (F1) can be found, or for the capture of liposomes a lipophilic dextran (L1) is offered, and also a streptavidin surface (SA) functionalization surface is available for biotin conjugation (Rich and Myszka 2000).

The use of previously functionalized chips can be practical, but they do not allow adaptations based on user necessity and, for most cases, their surface regeneration can be complicated, which tends to make the analysis more expensive. The introduction of other surface modifications, convenient for diverse surface functionalities, became immensely popular over the last decades. In particular, the formation of self-assembled monolayers (SAMs), became one of the most explored methods for covalent bioreceptor attachment, offering good reproducibility and durability over other methods (Wijaya et al. 2011). SAMs are usually formed by functionalized long-chain aliphatic molecules that can assemble on the metal surface forming a highly organized nanometer-thick layer of molecules with a preferred orientation, which makes it especially useful to biosensor applications. Its formation is also easy to obtain by simply dipping the substrate in a dilute solution of the organic molecule for a certain time, followed by washing steps to take off the excess of the reagent. For the formation of SAMs on gold surfaces, like the SPR sensor chip, usually, thiols are used due to the high affinity between sulfur and gold. Furthermore, SAMs have as terminal groups -OH, -COOH, and $-NH_2$, which are explored to future covalent attachment of the bioreceptor depending on the desired surface property. The size of the chain relates to the structuring of the monolayer, a consequence of hydrophobic interactions, like van der Waals, between the hydrocarbon long-chain part of the molecules, which induces a certain level of organization (Samanta and Sarkar 2011; Homola 2008).

The possibility of exploring the method of immobilization for sensitivity enhancement can offer optimized properties for specific applications. 3D structures, for example, have offered extra accessible areas for biological interactions, improving the sensitivity of SPR devices by immobilizing more bioreceptors than classical methodologies (Souto et al. 2019). For example, we have compared the use of dendrimers and dendrons, supramolecular three-dimensional structures, combined or not with SAMs of alkanethiols. It was possible to observe an improvement in the sensibility, directly related to the higher capacity of biomolecules immobilization on their 3D structure, if compared to the SPR sensor functionalized with just a SAM (Souto et al. 2015; Luz et al. 2016).

The use of nanomaterials has also been extremely explored for the enhancement of sensibility on SPR-based biosensors in bioanalytical applications. Metallic nanoparticles (MNPs) present localized surface plasmon resonance (LSPR), a phenomenon related to the electromagnetic waves around the MNPs which, in association with the high surface area of these materials, have been used for sensitivity improvement of SPR-based biosensors. Different parameters can be explored and manipulated for their application, as the metal of choice, size, and shapes of the NPs (Kaushal et al. 2020). These different parameters were explored by Liu and coworkers on ultrasmall (from 1 to 2 nm) gold and silver nanoparticles, and the biosensing sensibility of the enzyme papain by BSA were evaluated. The results showed that depending on the metal and its size, greater sensitivity can be achieved, compared to a control sensor (without nanoparticles) (Liu et al. 2020). Nanocomposites of MNPs with different materials, such as magnetic nanoparticles, can also be explored and achieve greater sensibilities by synergetic effects (Ferhan et al. 2018).

Graphene is another handy material, due to the high transmittance of the graphene-sheets, and propagation of the surface plasmonic polaritons, making it possible to enhance the sensibility of the SPR-based sensor. A genosensor for detection of a cancer biomarker in urine samples was explored by Mujica et al., by employing graphene oxide and polydiallydimethylammonium chloride on a gold surface modified with 3-mercaptopropane sulfonate (Mujica et al. 2020). It was possible to obtain femtomolar quantification of the prostate-specific antigen (PSA). The authors indicated that compared to the established method of functionalization with 11-MUA, the cysteamine/graphene oxide functionalized sensor presented higher sensibility (Miyazaki et al. 2019).

Some other methodologies may be explored to achieve the most functional sensor surface, such as other polymeric films (Kyprianou et al. 2009), or overcoating with a wide variety of materials like silica, oxides, metals, and carbon-based material (Wijaya et al. 2011). In the last decade, it was possible to observe a growth of published papers in this field, especially with the soaring development of materials science and nanotechnology, which aids in the improvement of sensitivity in these sensors (Qu et al. 2020). Furthermore, the investment toward the miniaturization of the SPR-based devices offers perspectives to make use of these platforms in POC applications in the medicinal field (Cassedy et al. 2020).

## 3.3 SPR Toward Point-of-Care Applications

Traditional methods of diagnosis are usually complex, relatively expensive, and require clinical analysis and hospital structure. To overcome these problems, miniaturized biosensors are being explored to portabilize biosensing, aiming for the use in point-of-care (POC) applications (Zhao et al. 2020). POC offers a simpler alternative for complicated tests, designing devices to fit the user, thus not requiring specialized training for usage, and enabling their application in different environments like ambulances, doctor's offices, emergency rooms, and even in the user's house (Mauriz et al. 2019). For specific applications as diseases diagnosis, there are some challenges to overcome (e.g., sensitivity, simplicity, and portability) to enable the use of SPR biosensors as diagnosis devices. Considering portability, SPR devices have been developed in the last decade for on-site detection, and some companies provide commercial portable-SPR instruments, such as Texas Instrument, K-MAC, and Seattle Sensor Systems. However, their resolution is impaired if compared to a high-performance laboratory instrument, and further improvement is still necessary (Breault-turcot and Masson 2012).

Peláéz and collaborators (Pelaez et al. 2020) worked on a POC SPR-based biosensor for Tuberculosis diagnosis (detection of the heat shock protein X). The SPR device was homemade designed and incorporates a gold sensor chip between microfluidics subsystems ($20 \times 20$ cm$^2$) and a trapezoidal prism. For plasmons excitement, a collimated halogen light is used at a fixed incidence angle of $70^0$ and the reflected light (carried by optic fiber) is detected by a CCD spectrophotometer. The limit of quantification (LOQ) of the method was 2.12 ng mL$^{-1}$, reporting a low-cost and portable biosensor (Pelaez et al. 2020).

For an even more portable and user-friendly device, smartphones can offer a new perspective in the development of portable biosensors (Zhang and Liu 2016). Combining the strong SPR physical phenomenon, with a smartphone camera, it is possible to develop smartphone-based SPR devices, applicable on point-of-care platforms (Lertvachirapaiboon et al. 2018). Smartphones have good computing capability, great image resolution, and an operational system possible to program. These characteristics can be associated with biosensors, offering screen display with the possibility of data analysis, and also detectors in one small size equipment (Zhang and Liu 2016). In Fig. 4, it is possible to observe two different approaches to the development of a smartphone-based SPR biosensor, one by Preechaburana et al. (2012) (Fig. 4a), and the second one by Sanjay et al. (2020) (Fig. 4b).

Exemplified by Fig. 4a, the portable biosensor was based on the light from the screen (red rectangle on image sufficient to promote SPR) toward the chip, and the SPR image was directed to the camera. As for the device responsible for the light enhancement, acting as the prism of more classical SPR equipment, it was made by

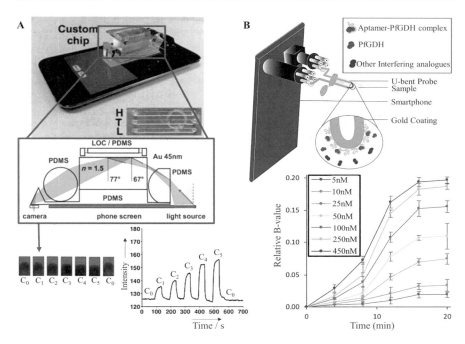

**Fig. 4** SPR-based POC biosensors by (**a**) Preechaburana et al. (2012), and its respective response detected by a frontal camera, light source promoted by the screen, and light coupling by the PMDS based configuration specified; (**b**) Sanjay et al. (2020) based on U-bent probe optic fiber configuration, light source promoted by the LED flash, detection by the camera plus its response based on relative B-value

polydimethylsiloxane (PDMS), which has the same refracting index of glass ($n = 1.5$). This configuration is disposable, leaving the phone intact after the measurement. The polymer interface is terminated by the glass with a thin layer of gold, where the biomolecular interaction occurs. The resulting images show dark patterns related to the analyte's concentration, which require appropriate software for measurements acquisition and interpretation. The system was tested and compared with a commercial test chip for detection of $\beta_2$ microglobulin (biomarker for cancer, inflammatory diseases, etc.), reaching a LOD of $0.1\mu g\ mL^{-1}$. The second approach, illustrated in Fig. 4b, lies in the use of an optic fiber platform, and through this work was possible to detect *Plasmodium falciparum* glutamate dehydrogenase (applied for Malaria disease diagnosis). The fiber probe was u-bent and sputtered with gold, for later functionalization with thiol aptamer and blocked with 6-mercapto 1-hexanol. One of the sides of the fiber was connected to the flash (light source) and the other to the camera (detector). For the SPR phenomenon to happen by a required wavelength, an overlapping blue, green, and orange filter was added after the flash. Also, macro lenses were added to the camera, and the pictures were captured on manual mode with focus to macro. The concentrations were related to the difference in the blue value of the LED flash fluctuations, since the intensity of light in the fiber change according to the refractive index of that region. LOD of the method in diluted

**Table 2** Summary of smartphone-based SPR devices and their analytical parameters

| Coupler | LOD | Linear response range | Bioanalytical application |
|---------|-----|----------------------|---------------------------|
| Prism derived (Preechaburana et al. 2012) | $0.1\mu g\ mL^{-1}$ | $(0.132–1.32)\mu g\ mL^{-1}$ | Detection of $\beta_2$ microglobulin (Cancer biomarker) |
| Optic Fiber (Sanjay et al. 2020) | $352\ pmol\ L^{-1}$ | $(5–450)\ nmol\ L^{-1}$ | Detection of *Plasmodium falciparum* glutamatedehydrogenase (malaria disease diagnosis) |
| Diffraction Grating (Zhang et al. 2018) | $32.5\ ng\ mL^{-1}$ | – | Detection of lipopolysaccharides from *Klebsiella pneumonia* |
| Diffraction Grating (Pan et al. 2018) | $2.43\mu g\ mL^{-1}$ | – | BSA/anti-BSA |

serum was 352 pM. In Table 2 are disposed the analytical parameters of the examples of smartphone-based SPR biosensors disposed in Fig. 4, and some other examples found in literature. Through these studies it is possible to evidence the potential of smartphone-based SPR devices, offering new insights into the field of biosensors.

# 4 Conclusions and Future Trends

In the past few decades, SPR sensors have achieved significant technological progress and gained new applications through new configurations. SPR devices based on optical fibers and photonic crystal fibers have shown high potential for the construction of miniaturized SPR devices, providing fast, ultra-sensitive, and portable platforms. Despite the progress made with the new configurations, the use of prisms as couplers is still the main arrangement used so far in commercial SPR-based sensors.

The numerous configurations for developing SPR biosensors for the study of interactions involving biomolecules made it a powerful tool for bioanalytical applications, which is widely explored in drug discovery processes, quality control of pharmaceutical compounds, studies of different receptor–ligand interactions, food quality control, biogases detection, and the development of new methods of diagnosis. Considering portability, the resolution of SPR devices developed for on-site detection is up to this time lower compared to a high-performance laboratory instrument, so further improvement is still necessary.

As future trends, we strongly believe that research activities related to SPR biosensors will be focused on the development of robust portable systems with high resolution and feasibility to be applied in the point-of-care analysis. For an even more portable and easy-to-use device, we believe that smartphones should offer new advances in the area. It is expected the implementation of the detection mechanism using fiber-optic SPR in smartphones will allow real-time detection for in situ and in vivo analysis in clinical diagnostics.

# References

Abdulhalim I, Zourob M, Lakhtakia A (2008) Electromagnetics 28:214–242

Adzhri R, Arshad KM, Gopinath SCB, Ruslinda AR, Fathil MFM, Ayub RM, Nor MNM, Voon CH (2016) Anal Chim Acta 917:1–18

Borile G, Rossi S, Filippi A, Gazzola E, Capaldo P, Tregnago C, Pigazzi M, Romanato F (2019) Biophys Chem 254:106262

Breault-turcot J, Masson J-f (2012) Anal Bioanal Chem 403:1477–1484

Brockman JM, Nelson BP, Corn RM (2000) Annu Rev Phys Chem 51:41–63

Burke JJ, Stegeman GI, Tamir T (1986) Phys Rev B 33:5186–5201

Byun KM, Kim SJ, Kim D (2007) Appl Opt 46:5703–5708

Carvalho RMD, Rath S, Kubota LT (2003) Quim Nova 26:97–104

Cassedy A, Mullins E, O'Kennedy R (2020) Biotechnol Adv 39:107358

Caucheteur C, Guo T, Albert J (2015) Anal Bioanal Chem 407:3883–3897

Cennamo N, D'Agostino G, Donà A, Dacarro G, Pallavicini P, Pesavento M, Zeni L (2013) Sensors (Switzerland) 13:14676–14686

Changqing X, Fenglei J, Bo Z, Ran LI, Yi L (2012) Sci China Chem 55:1931–1939

Chen Y, Liu J, Yang Z, Wilkinson JS, Zhou X (2019) Biosens Bioelectron 144:111693

Čtyroký J, Homola J, Lambeck PV, Musa S, Hoekstra HJWM, Harris RD, Wilkinson JS, Usievich B, Lyndin NM (1999) Sens Actuators B 54:66–73

Dai Y, Xu H, Wang H, Lu Y, Wang P (2018) Opt Commun 416:66–70

Damos FS, Mendes RK, Kubota LT (2004) Quim Nova 27:970–979

Danlard I, Akowuah EK (2020) Opt Fiber Technol 54:102083

Dash JN, Jha R (2014) IEEE Photon Technol Lett 26:595–598

Fan X, White IM, Shopova SI, Zhu H, Suter JD, Sun Y (2008) Anal Chim Acta 620:8–26

Faria AR, de Castro Veloso L, Coura-Vital W, Reis AB, Damasceno LM, Gazzinelli RT, Andrade HM (2015) PLoS Negl Trop Dis 9:e3429

Fathi F, Rashidi M-R, Omidi Y (2019) Talanta 192:118–127

Ferhan AR, Jackman JA, Park JH, Cho NJ (2018) Adv Drug Deliv Rev 125:48–77

Fischer C, Hu T, Jarck J-h, Frohnmeyer E, Kallinich C, Haase I, Hahn U, Fischer M (2015) J Agric Food Chem 63:8050–8057

Gupta BD, Kant R (2018) Opt Laser Technol 101:144–161

Homola J (1995) Sens Actuators B Chem 29:401–405

Homola J (2006) Surface plasmon resonance based sensors. Springer, Berlin

Homola J (2008) Chem Rev 108:462–493

Homola J, Yee SS, Gauglitz G (1999) Sens Actuators B Chem 54:3–15

Hu DJJ, Ho HP (2017) Adv Opt Photon 9:257

Hulanicki A, Glab S, Ingman F (1991) Pure Appl Chem 63:1247–1250

Jabbari S, Dabirmanesh B, Shahriar S, Amanlou M (2017) Sens Actuators B 240:519–527

Kamat V, Rafique A, Huang T, Olsen O, Olson W (2020) Anal Biochem 593:113580

Kaushal S, Nanda SS, Samal S, Yi DK (2020) Chembiochem 21:576–600

Kim D (2005) Appl Opt 44:3218–3223

Kim S-H, Koh K (2006) In: Kim S-H (ed) Functional dyes, pp 185–213

Klantsataya E, Jia P, Ebendorff-Heidepriem H, Monro TM, François A (2017) Sensors 17:12

Kretschmann E, Raether H (1968) Z Naturforsch A Phys Sci 23:2135–2136

Kyprianou D, Guerreiro AR, Chianella I, Piletska EV, Fowler SA, Karim K, Whitcombe MJ, Turner APF, Piletsky SA (2009) Biosens Bioelectron 24:1365–1371

Lang W (1948) Optik 3:233

Lertvachirapaiboon C, Baba A, Shinbo K, Kato K (2018) Anal Methods 10:4732–4740
Li B, Ju H (2013) Biochip J 7:295–318
Liu Y, Liu Q, Chen S, Cheng F, Wang H, Peng W (2015) Sci Rep 5:1–9
Liu X, Zhang Y, Wang S, Liu C, Wang T, Qiu Z, Wang X, Waterhouse GIN, Xu C, Yin H (2020)
    Microchem J 155:104737
Luz JGG, Souto DEP, Machado-Assis GF, de Lana M, Luz RCS, Martins OA, Damos FS, Martins
    HR (2016) Clin Chim Acta 454:39–45
Matsubara K, Kawata S, Minami S (1988) Appl Spectrosc 42:1375–1379
Mauriz E, Dey P, Lechuga LM (2019) Analyst 144:7105–7129
Mei GS, Menon PS, Hegde G (2020) Mater Res Express 7:012003
Miyazaki CM, Camilo DE, Shimizu FM, Ferreira M (2019) Appl Surf Sci 490:502–509
Monošík R, Streďanský M, Šturdík E (2012) Acta Chim Slovaca 5:109–120
Mujica ML, Zhang Y, Bedioui F, Gutierrez F, Rivas G (2020) Anal Bioanal Chem 412:3539–3546
Navarrete MC, Díaz-Herrera N, González-Cano A, Esteban Ó (2014) Sens Actuators B Chem 190:
    881–885
Nylander C, Liedberg B, Lind T (1982) Sens Actuators B Chem 3:79–88
Olaru A, Bala C, Jaffrezic-Renault N, Aboul-Enein HY (2015) Crit Rev Anal Chem 45:97–105
Otto A (1968) Z Phys 216:398–410
Pan MY, Lee KL, Lo SC, Wei PK (2018) Anal Chim Acta 1032:99–106
Pattnaik P (2005) Appl Biochem Biotechnol 126:079–092
Pelaez EC, Estevez MC, Mongui A, Menendez MC, Toro C, Herrera-Sandoval OL, Robledo J,
    Garcia MJ, Portillo PD, Lechuga LM (2020) ACS Infect Dis 6:1110–1120
Pines D (1956) Rev Mod Phys 28:184–198
Pines D, Bohm D (1952) Phys Rev 85:338–353
Pitarke JM, Silkin VM, Chulkov EV, Echenique PM (2007) Rep Prog Phys 70:1–87
Powell CJ, Swan JB (1959) Phys Rev 116:81–83
Preechaburana P, Gonzalez MC, Suska A, Filippini D (2012) Angew Chem Int Ed 51:11585–11588
Qu J-H, Dillen A, Saeys W, Lammertyn J (2020) Anal Chim Acta 1104:10–27
Rich RL, Myszka DG (2000) Curr Opin Biotechnol 11:54–61
Rifat AA, Ahmed R, Yetisen AK, Butt H, Sabouri A, Mahdiraji GA, Yun SH, Adikan FRM (2017)
    Sens Actuators B Chem 243:311–325
Ritchie RH (1957) Phys Rev 106:874–881
Rossi S, Gazzola E, Capaldo P, Borile G, Romanato F (2018) Sensors 18:1621
Ruthemann G (1948) Ann Phys 2:135–146
Samanta D, Sarkar A (2011) Chem Soc Rev 40:2567–2592
Sanjay M, Singh NK, Ngashangva L, Goswami P (2020) Anal Methods 12:1333–1341
Sharma AK, Pandey AK, Kaur B (2018) Opt Fiber Technol 43:20–34
Shen F, Gurgel PV, Rojas OJ, Carbonell RG (2014) Biosens Bioelectron 58:380–387
Sohrabi F, Hamidi SM (2016) Eur Phys J Plus 131:221
Souto DEP, Damos FS, Andrade HM, Kubota LT, Barragan JTC, Fonseca AM, Luz RDCS (2015)
    Biosens Bioelectron 70:275–281
Souto DEP, Volpe J, Goncalves CC, Ramos CHI, Kubota LT (2019) Talanta 205:120122
Stegeman GI, Burke JJ, Hall DG (1983a) Opt Lett 8:383–385
Stegeman GI, Wallis RF, Maradudin AA (1983b) Opt Lett 8:386–388
Stern EA, Ferrell RA (1960) Phys Rev 120:130–136
Tabassum R, Kant R (2020) Sens Actuators B Chem 310:127813
Teng YY, Stern EA (1967) Phys Rev Lett 19:511
Tokel O, Inci F, Demirci U (2014) Chem Rev 114:5728–5752
Tonks L, Langmuir I (1929) Phys Rev 33:195–210
Tudos AJ, Schasfoort RBM (2008) Handbook of surface plasmon resonance. The Royal Society of
    Chemistry
van Gent J, Lambeck PV, Kreuwel HJM, Gerritsma GJ, Sudhölter EJR, Reinhoudt DN, Popma TJA
    (1990) Appl Opt 29:2843

Verma RK, Gupta BD (2008) J Phys D Appl Phys 41:095106

Verma RK, Sharma AK, Gupta BD (2007) IEEE Photon Technol Lett 19:1786–1788

Wang SF, Chiu MH, Chang RS (2006) Sens Actuators B Chem 114:120–126

Wang D, Loo J, Chen J, Yam Y, Chen S-C, He H, Kong S, Ho H (2019) Sensors 19:1266

Wijaya E, Lenaerts C, Maricot S, Hastanin J, Habraken S, Vilcot JP, Boukherroub R, Szunerits S (2011) Curr Opinion Solid State Mater Sci 15:208–224

Zhang JXJ, Hoshino K (2014) Molecular sensors and nanodevices. Academic Press, pp 233–320

Zhang D, Liu Q (2016) Biosens Bioelectron 75:273–284

Zhang LM, Uttamchandani D (1988) Electron Lett 24:1469–1470

Zhang J, Khan I, Zhang Q, Liu X, Dostalek J, Liedberg B, Wang Y (2018) Biosens Bioelectron 99: 312–317

Zhao Y, Tong RJ, Xia F, Peng Y (2019) Biosens Bioelectron 142:111505

Zhao W, Tian S, Huang L, Liu K, Dong L, Guo J (2020) Analyst 145:2873–2891

# Introduction to Electroanalysis

Lúcio Angnes

## 1  Historical Remarks

The period of the discovery of electricity is not clear, but ancient Egyptian texts, dating from 2750 BC describe the effects of shocks from electric fish from the Nile River. At 600 BC Thales of Miletus described static electricity and until ~1600 AC, it remained as an intellectual curiosity, when William Gilbert presented a critical study, demonstrating the relation between electricity and magnetism. In the following century, the contributions of Otto von Guericke, Robert Boyle and especially from Benjamin Franklin were noteworthy. In 1791, the work of Luigi Galvani was fundamental to understanding the mechanism of transmission of impulses in living bodies. Many authors consider this experiment the first electrochemical experimentation. 1800 was the year that the physicist Alessandro Volta presented for the first time his batteries. In the same year, the English chemists William Nicholson and Johann Wilhelm Ritter demonstrated the electrolysis of water, generating hydrogen and oxygen. In sequence, Ritter also demonstrated the electroplating of metals. In 1829, Becquerel developed a constant current cell, a predecessor of the Daniell cell, which was fundamental for Michael Faraday to start in 1832 the studies that resulted in the establishment of its two laws. Faraday experiments gave origin to electrogravimetry and also to coulometry, which are considered the first electroanalytical methods. Equally older are conductivity measurements. The Nernst equation (1889) was the base for the potentiometric measurements and in 1906 the glass electrode was introduced by Cramer. Many advances gave origin to different applications and in parallel the theory advanced to explain the new phenomena. Advancing to 1922, the starting of polarography is noteworthy, the first voltammetric technique developed, which used a dropping mercury electrode and

L. Angnes (✉)
Department of Fundamental Chemistry, Institute of Chemistry, University of São Paulo (USP), São Paulo, SP, Brazil
e-mail: luangnes@iq.usp.br

© The Author(s), under exclusive license to Springer Nature Switzerland AG 2022
L. T. Kubota et al. (eds.), *Tools and Trends in Bioanalytical Chemistry*,
https://doi.org/10.1007/978-3-030-82381-8_11

was able to quantify concentrations as low as $10^{-5}$ mol $L^{-1}$ of analytes, a fantastic achievement for that time. Voltammetric techniques advanced drastically, and almost one century later, progress has been remarkable. Important remarks from the development of polarography were described some years before (Barek and Zima 2002).

## 2    Electrogravimetry Coulometry and Conductometry

Electrogravimetry and coulometry are techniques directly related with the two Faraday laws. In the middle of the nineteenth century, these techniques were fundamental for basic applications. It is important to realize that the most accurate instrument at this time was the analytical balance and electrogravimetry was based on weight electrodes before and after the deposition of such metal. In this way, the determination of the atomic weight of many elements was achieved. Measurements of current were imprecise until the development of the Weston meter (1886), which results were really precise (Matsumoto 1920). With this new meter, scientists were able "to count electrons," a condition of great importance for applications where deposition on the electrodes does not occur.

These techniques gained importance for many applications, such as purifications, separations, pre-concentrations and even for electrosynthesis. Conductometry was also started in the early eighteenth century, first using continuous current. Friedrich Kohlrausch introduced the alternated current in 1860 and applied the technique for measurements of ions in water, acids, and other solutions. But the major applications were not the direct measurements, but the titrations involving the conductance measured during the procedure (Lubert and Kalcher 2010). The equivalence point of many acid-base, precipitation, and complexation reactions can be followed by conductometric measurements, with good precision. Unfortunately, many authors of didactic books eliminated the conductometry chapter from their texts.

## 3    Potentiometry

Potentiometry involves the measurement of the difference of potential between an indicator electrode and a reference electrode, using a high impedance voltmeter, a condition where almost any current flows. The potential of the indicator electrode is dependent on the activity of the analyte, while the potential of the reference electrode remains constant. This technique allows the use of a great number of different electrodes, deserving emphasis on the use of the glass electrode, considered the most used sensor worldwide (at least until the development of the zeta sensor, in use in all cars). The metallic electrodes also provide important applications and are classified as electrodes of the first, second, and third orders. Additionally, crystalline membrane electrodes, noncrystalline membrane electrodes, selective electrodes, enzymatic electrodes, etc. were developed for numerous applications. The simplicity of the instrumentation and the easy operation of the electrode are attractive for the use of this technique. Even being a simple technique, potentiometry allows detection limits of 100 nanomoles per liter of the total of the ions present in a solution, although down to ten picomolar differences in concentration can be measured (Brett and Brett 1998).

# 4   Polarography

Polarography is considered an initial technique, responsible for providing the basis for voltammetric techniques. It was developed for almost a century by Professor Jaroslav Heyrovsky, who carried out the pioneering experiment which used a two-compartment cell, being the working electrode a dropping mercury electrode and a large area reference electrode which actuates simultaneously as the counter electrode too. The mercury electrode was constituted by a mercury reservoir connected with a flexible tube to a glass capillary (hole with $\emptyset = 10$–$100$ μm). The mercury drops last for 3–6 s, when attains the weight to break up, then the electrode is renewed. A typical reference electrode was constituted by a mercury pool, covered by a calomel layer and a saturated KCl solution. A porous plate separated the two compartments. Along the time, polarography advances significantly and, from the technique that allowed quantifications in the $10^{-5}$ mol L$^{-1}$ at the beginning, the limits of determination decreased to $10^{-7}$–$10^{-8}$ mol L$^{-1}$ with the advent of differential pulse polarography (Zhang 1998), $10^{-9}$ mol L$^{-1}$ using square wave voltammetry (Zhang et al. 2007) and $10^{-12}$ mol L$^{-1}$ utilizing hanging mercury electrodes and stripping voltammetry associated with adsorptive processes (Wang et al. 1999). To reach these limits, the evolution of instrumentation plays a predominant role. Unfortunately, there is a worldwide movement against the use of mercury electrodes (classified as "almost pathologic" by Bard and Zotski (Bard and Zotski 2000) with which this author fully agrees), so that the best electrode ever developed is falling out of use.

# 5   Voltammetry

Voltammetry techniques are based on the application (of a program) of potential to an electrode and measure the resulting current that flows through the working electrode. Polarography is a particular case of voltammetry and it was fundamental for establishing the principles of the technique. In polarography, mercury electrodes present elevated overvoltage to hydrogen reduction, a characteristic that favors the reduction of many metals, as well as organic molecules. In addition, its intrinsic qualities of mercury favor his obtention with elevated purity (Wilkinson 1972) and the surface of the electrodes (=drops) are perfect at the atomic level. Unfortunately, mercury does not work well in the anodic region (at potentials above ~+0.2 (vs. Ag/AgCl), mercury is oxidized). This fact leads the scientist to search for other electrodes capable of working in a larger potential window in the anodic region. The paper "Voltammetry at Solid Electrodes" published by R. E. Parker and R. N. Adams in 1956 (Parker and Adams 1956), probably was related to the use of solid electrodes. At that time, the authors used a rotating platinum electrode and defined the experiment as "the solid electrode polarography."

The development of a potentiostat with three electrodes occurred in 1942, but its use was spread only in the 1960s and 1970s with the advances in instrumentation and the introduction of computers to control systems and perform the data acquisition. In parallel, theory also advanced significantly in this period. The use of carbon paste electrodes (Adams 1958) was introduced in 1958. The main advantage of these solid electrodes is the possibility of work in the anodic region, in potentials that cannot be attained by the mercury electrode. Solid electrodes in that period were few, only noble metals and pyrolytic graphite were explored. In the following years, the number of electrodes grew significantly, with detach to glassy carbon and diamond electrodes, and experienced a notable expansion with the many ways of modification of the surface of the solid (or paste) electrodes.

## 6     Linear Sweep Voltammetry (LSV)

In LSV, the potential is altered linearly in the function of time and the current that flows in the working electrode is monitored. The rate of voltage change over time during this phase is known as the experiment's scan rate (V/s). This procedure is analogous to the one adopted in polarography. The scan can be done in the anodic or cathodic direction and generally, this technique is utilized in preliminary studies.

This technique is generally used in exploratory studies, once its sensibility is not elevated due to the fact that during the application of the ramp of potential, the charging of the electrical double layer is responsible for a non-negligible current. As it is not easy to discriminate the capacitive from the faradaic current, for low concentrations this technique is not appropriate.

## 7     Cyclic Voltammetry (CV)

In CV, similar to what is done in LSV, the potential is ramped linearly versus time, but when it attains the extreme value, is scanned in the opposite direction, usually returning to the initial potential. In CV, the current is monitored in both steps. A plot of current versus applied potential provides important information about the system under investigation. Analogous to LSV, CV also has low sensitivity, affected by the charging process of the electric double layer produced by the variation of the applied potential. Even so, CV is a fantastic tool for many studies and is widely used for several applications. Using CV is possible to determine the formal reduction potential of such an analyte, if a process is reversible or irreversible, get information about the reaction products and verify if some intermediate species are formed. CV can also be used to have information about the kinetics of such a reaction, determine the diffusion coefficient of an analyte, the number of electrons involved in a redox process, and obtain information about adsorption processes. For not so low concentrations (typically $10^{-5}$ to $10^{-3}$ mol $L^{-1}$), the current is proportional to concentration and CV can be utilized for quantitative analysis.

# 8    Amperometric Methods

Amperometry involves the application of a fixed potential between two electrodes. This technique was originally explored to detect the equivalence point in titrations executed under stirring (Kolthoff and Langer 1940; Laitinen and Kolthoff 1941). In the 1960s, amperometry was explored in association with chromatography, detection being done at the end of the column. The association of amperometry with flow injection analysis (FIA) or batch injection analysis (BIA) is a very favorable condition because combines the advantage of elevated transport of analyte to the electrode and on this condition, the contribution of the charging of the electrical double layer is virtually zero (to get this condition, the analyte and the carrying solution should have the same electrolyte). Alternatively, pulsed amperometry can be an efficient alternative for the regeneration of the electrode surface in complex samples (Silva et al. 2013) or even to favor the quantification of mixtures with close oxidation potentials (Freitas et al. 2016), as demonstrated in the quantification of three electroactive compounds. Chronoamperometric methods involve measurements of current in the function of time, under potentiostatic control. For this, the potential of the working electrode is stepped from an initial potential to the final potential and the current is measured as a function of time. In a very short time, the contribution of the capacitive current is elevated, many times much larger than the faradaic current. Fortunately, capacitive current decays very rapidly and in relatively short periods. After the decline of the capacitive current, the remaining current is (almost all) the product of the faradaic process, governed by the diffusion of the analyte to the working electrode. Chronoamperometry was developed in the early 1960s (Beilby and Budd 1962) and at the initial experiments, the application of the potential step was applied manually and the results recorded in mechanical recorders. With the advent of more modern instrumentation and their connection to computers, this and the following pulsed techniques became faster and more accurate. The possibility of executing different techniques in multi-electrodes brings many advantages for getting more and most precise results (Fava et al. 2019).

# 9    Pulse Techniques

The application of potential steps is the base of pulse techniques. The creation of the steps was initially developed for the dropping mercury electrode. The main idea was to synchronize the pulses with the growth of the drop of mercury, thus minimizing the capacitive current at the end of the drop's life. This concept led to the development of tast polarography. In this technique, the current is measured for a short time near the end of the life of the mercury drop and in sequence, a knocker detaches the mercury drop (from the capillary) and a new drop starts to grow. Near the end of the life of this new drop, the current is more one time sampled and in sequence the knocker detaches this drop and so on. In this way, a better distinction between the faradaic and capacitive current is achieved. Normal pulse voltammetry (NPP), differential pulse voltammetry (DPP), and square wave voltammetry (SWV) were developed in the 1970s. In all these techniques, pulses are applied and after a fixed time the current is sampled (Bond 1980).

In NPP, a base potential is maintained along with all the experiments and increasing steps are applied. The current is sampled after a fixed time after the application of each step. Compared to tast polarography, NPP has the advantage that the faradaic process occurs only during the application of the pulses. During the period in which there is no application of the pulses, there is no consumption of analyte and thus the material previously consumed (in the vicinity of the electrode) can be replaced. In this way, NPP achieves larger signals compared to tast polarography.

DDP is similar to NPP, with two basic differences, responsible for its higher sensibility: a) The base potential is not maintained at a fixed value, but varies between the pulses; b) Current is measured just before the application of the pulse and also just before the end of the pulse. The differences between both signals generate a sinusoidal curve. The potential where the maximum current is attained is situated in the $E_{1/2}$ region for reversible species. For more irreversible species, the curve is enlarged and the peak does not coincide with the $E_{1/2}$ region. In the NPP, the differences between the currents before and after each pulse virtually cancel the effect of the capacitive current present after the application of the peak of potential in each step.

SWV was proposed in 1952 by Barker and Jenkins (Backer and Jenkins 1952), but at that time the instrumental limitations were a great barrier. The evolution of instrumentation and their association with microcomputers makes this technique popular. In a simplified way, SWV is a form of linear potential sweep voltammetry that uses the combination of square wave and staircase potential applied simultaneously to the electrode. During the same experiment, signals can be collected in the forward scan and in the back scan. The peaks corresponding to the oxidation and reduction of electroactive species can be obtained in the same experiment and by subtraction, the difference produces an increase in signal. This fact is very important in analytical applications for the analysis of very low concentrations, since in addition to increasing the signal, it also removes the contribution of dissolved oxygen. SWV is considered one of the most advanced and versatile of the pulse voltammetric techniques (Mirceski et al. 2018). SWV combines the advantages of pulse techniques with regard to high sensitivity and cyclic voltammetry, which allows obtaining information about reaction mechanisms. An additional advantage is the speed of analysis, an aspect of great importance when a fast analysis is required, such as in the detection of eluents in HPLC.

There are other pulse techniques, developed for specific applications, such as reverse pulse voltammetry (RPV), differential normal pulse voltammetry (DNPV), and double differential pulse voltammetry (DDPV). The applications of pulse techniques are not restricted to the ones here discussed. This task is now much easier with access to microprocessors and even commercial equipment, which allow the development of new waveforms for particular applications.

# 10   Impedance Measurements and Pre-Concentration Techniques

Impedance measurements started in the nineteenth century and, for more than half a century, were used mainly for electrical measurements. The first article that explored the electrochemical impedance at a mercury electrode/solution interface was published by Graham in 1947 (Graham 1947). Electrochemical Impedance Spectroscopy (EIS) is based on introducing a perturbation in the system being studied by using a sine wave current or small amplitude potential. The instrument senses the resulting changes in the form of an impedance diagram that provides useful data. Due to the multiplicity of variables that can be present in complex systems, EIS is considered by some researchers one of the most complex techniques in electrochemical research. A search in the literature shows a narrow number of papers until 1990, when an almost exponential growth of publications started. The main reasons for this expressive growth were the popularization of the technique and the advances of instrumentation. In addition, the manufacturers of commercial potentiostats began to include the technique in their equipment and the corresponding softwares were improved and turned friendly to use. EIS found a great number of applications in the most different interfaces, from electrochemistry applications to photovoltaic systems, from batteries to applications in life sciences. In the electrochemical area, impedance is important for many practical applications, such as corrosion and metallic surface protection (weight loss, inhibitors persistence, presence of adsorbed intermediates), so as to quantify diffusion coefficients, electron transfer rates, adsorption mechanisms, charge transfer resistances or determination of particles and/or pore sizes (Randviir and Banks 2013).

Pre-concentration is a very practical way of increasing the detection sensitivity of such an analyte. Solvent extraction possibly is the oldest form of pre-concentration of certain species and already in the early twentieth century solvent extraction was explored for analytical purposes. A huge number of methodologies have been developed for the pre-concentration of species of interest for analytical purposes. The content of analytes existing not only in aqueous media but also in air and solids became of interest to analytical chemists. In the electroanalytical field, many processes of pre-concentration of trace metals were also developed, in some cases allowing attain concentrations of $10^{-14}$ mol $L^{-1}$ (Freire and Kubota 2004). In the field of biosensors, the use of immunosensors allowed the attainment of even lower concentrations and limits of detection in the $10^{-19}$ (Wong et al. 2016), $10^{-20}$ (Rahi et al. 2016) or even $10^{-21}$ mol $L^{-1}$ (Posha et al. 2018).

## 11    Measurements Under Flowing Regime

The classical voltammetric techniques, performed under a quiescent regime, suffer from the fact that the measured signal is dependent on the diffusion of the electroactive species to the electrode. Typical diffusion coefficients (D) of common species are situated between $10^{-5}$ and $10^{-6}$ cm$^2$ s$^{-1}$, in relatively diluted solutions. The value of D decreases with the increase of concentration and viscosity of the medium and in the case of ionic liquids it can decrease to $10^{-10}$ cm$^2$ s$^{-1}$. To overcome this situation, the use of rotating electrodes was the first way encountered to significantly increase the transport of material to the electrode surface, increasing the sensitivity achieved in a similar extension. In reality, the first register of rotating electrodes encountered in the literature dates from 1904 (Amberg 1904). The first amperometric register was done by Kolthoff and Jordan, who used a rotating gold electrode to quantify traces of dissolved oxygen (Kolthoff and Jordan 1952). The single rotating electrode was followed by the rotating ring-disc electrodes (Bruckenstein and Feldman 1965), which were an important tool for understanding a series of aspects involved in many voltammetric reactions (reaction mechanism, formation of intermediate species, parallel reactions, etc.). In the 1960s and 1970s, obtaining electrodes with "perfect" geometry and motors with variable and precise speeds was a great challenge.

There are many other ways to move the solution toward the electrode surface. Wall jet electrodes were proposed in 1973 (Yamada and Matsuda 1973). Two years later, flow injection analysis (FIA) was presented by Ruzicka and Hansen (Ruzicka and Hansen 1975). The first studies involved exclusively spectrophotometry and, in some cases, selective electrodes. The first register that describes the association of flow injection analysis and voltammetric sensors was done by Pungor et al. (Pungor et al. 1978) the first full paper was presented by Strohl and Curran (1979). Flow injection analysis is a strong area and a search shows almost 32.000 articles. Initially, spectrophotometric methods predominated, but in the following years, the potential of the voltammetric techniques became evident. High sensitivity, good reproducibility, and the possibility of miniaturization, associated with portability are some of the advantages of voltammetric methods. Under a flowing regime, almost all the procedures done in the laboratory can be performed. The association with reactors containing reagents, catalysts (between them, enzymes) amplifies the scope of possible reactions.

In the 1990s, two other "sisters" of FIA were proposed: Sequential Flow Injection (SIA) and Batch Injection Analysis (BIA). SIA is based on a multi-way central valve which is computer-controlled, enabling the selection of aliquots of sample and reagent(s) solutions to be sequentially aspirated to and coil (Ruzicka and Marshall 1990). The mixture of the reagents is attained by reverse and forward movements of this mixture, which in sequence, is directed to the detection region. The advantages of this technique are the precise temporization of the reactions, and the use of small volumes of reagents and analytes, the consumption of reagents, and the waste generated is smaller in comparison with FIA. Disadvantages are the necessity of a relatively expensive valve and the frequency of analysis is lower than the frequencies normally utilized in FIA. BIA is much simpler than FIA and SIA, and basically requires an automatic pipette, a measurement cell, where the transducer is positioned inside. The main advantages are the absence of external injectors, tubes, and pumps (Wang and Taha 1991). The motor of the pipette is responsible for the transport of the analyte from the pipette tip to the surface of the transductor. The possibilities of exploring reactions that occur along the flowing process, such as those performed in FIA and SIA are more restricted in BIA. However, the technique is quite effective for the direct detection of electroactive analytes, providing analysis frequencies of up to $360 \ h^{-1}$, with great sensitivity and excellent reproducibility.

## 12 Voltammetry + Biosensors: A Powerful Association

In 1956, Clark (Clark 1956) described the development of the $O_2$ electrode, able to monitor the dissolved oxygen in solutions. Six years later, Clark and Lyons presented the first sensor for glucose (Clark and Lyons 1962), based on the entrapment of glucose oxidase on its oxygen electrode, using a dialysis membrane. In 1963, Katz and Rechnitz described the direct potentiometric determination of urea after urease hydrolysis (Katz and Rechnitz 1963). In 1969, Guilbault e Montalvo introduced the potentiometric urea electrode (Guilbault and Montalvo Jr 1969). Four years later, Racine, Mindt, and Schlaepf presented the lactate biosensor (Racine et al. 1973). In the same year, Yellow Spring launched their first commercial glucose sensor, which was re-launch with success in 1975. In the following years, scientists demonstrated that biosensors can be constructed not only with enzymes and the utilization of microbes, fungus, organelles, cells, vegetal and animal tissues, proteins, antigens, antibodies, aptamers, DNA, RNA, etc. was demonstrated. For the immobilization of the biological material on the electrodes, there are many possibilities, but adsorption, entrapment, covalent bonding, and cross-linking are responsible for the majority of the processes. According to Mohamad and co-workers (Mohamad et al. 2015), enzyme immobilization methods can be divided into two general classes: Chemical and physical methods. Physical methods generally are weaker, involving weaker interactions such as hydrogen bonds, hydrophobic interactions, van der Waals forces, affinity binding, ionic binding of the enzyme to the electrode. Chemical methods involve stronger interactions, by the formation of covalent bonds, achieved by amide, carbamate, ether of thio-ether bonds of the enzyme to the electrode. The selection of the appropriate immobilization method plays a very important role and the choice of the way to fix the enzymes on the electrodes is complex due to the nature of the tertiary protein structure and also the characteristics of the electrodes.

The main detection methods utilized in these studies are electrochemical, optical, piezoelectric, and calorimetric. The use of electrochemical detection is growing due to the facility of miniaturization, simplified instrumentation, and elevated sensibility. The association of electron transfer mediators was a very important way for lowering the reaction potential and consequently avoid interferences in electrochemical measurements involving complex samples such as blood. Glucose in blood was

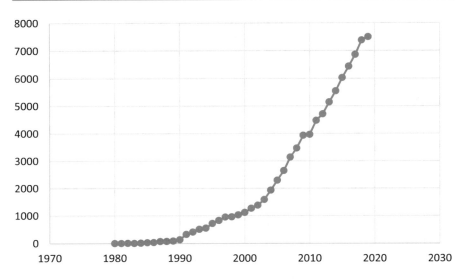

**Fig. 1** Evolution of the number of papers published in the area in the 1980 to 2018 period, using the word "biosensor*". Data obtained from the Web of Sciences database

(and still) the principal market driver in the biosensors area, whereas the applications of biosensors are now significantly broad. Applications in food and beverages quality control (wines production, milk industry), water quality management (metals, pharmaceuticals, hormones), environmental monitoring (pesticides, marine toxin monitoring, pathogenic bacteria) defense applications (neurotoxic gases, explosives), drug discovery, soil quality monitoring, clinical applications (early diagnosis of infections, lung cancer biomarkers), and industrial process control (synthesis, fermentation) are some of the areas where biosensors are gaining increasing importance. The first biosensors have brought the laboratory to the bedside of the patient and in sequence for homecare. The next step consisted of the amplification of the number of species and now scientists are looking for continuous monitoring, preferentially applying non-invasive methods and as wearable sensors (Gonzales et al. 2019). The association of the biosensing units with carbon allotropic forms (fullerenes, nanotubes, nano-diamonds, nano-horns, graphene and its derivatives, carbon dots, etc.) organic materials (nanofilms, nanogels, dendrimers, polymer nanocomposites) inorganic materials (quantum dots, magnetic nanoparticles, gold and silver nanoparticles, nano-shells, nanowires, nanocages) are some of the materials that can improve significantly the performance of biosensors.

As will be seen in the next chapters, the world of biosensors is very exciting. A search using the web of science database showed a total of 88,520 results for a search using "biosensor*" as keyword (Fig. 1).

The evaluation year after year of the published papers does not let any doubt about the impressive growth of this field. From 2003 to 2018 occurred an almost linear growth and this tendency should be seen in the following years.

# References

Adams RN (1958) Anal Chem 30:1576–1576
Amberg R (1904) Phys Chem 10:385–386
Backer GC, Jenkins IL (1952) Analyst 77:685–696
Bard AJ, Zotski CG (2000) Anal Chem 72:346A–352A
Barek J, Zima J (2002) Electroanalysis 15:467–472
Beilby AL, Budd AL (1962) Anal Chem 34:493–495
Bond AM (1980) Modern polarography methods in analytical chemistry. Marcel Dekker, New York
Brett CMA, Brett AMO (1998) Electroanalysis. Oxford, p 3
Bruckenstein S, Feldman GA (1965) J Electroanal Chem 9:395–399
Clark LC (1956) Trans Am Soc Artif Internal Organs 2:41–48
Clark LC, Lyons C (1962) Ann N Y Acad Sci 102:29–45
Fava EL, Silva TA, Prado TM, Moraes FC, Faria RC, Fatibello-Filho O (2019) Talanta 203:280–286
Freire RS, Kubota LT (2004) Electrochim Acta 22–23:3795–3800
Freitas JM, Oliveira TC, Gimenes DT, Munoz RAA, Richter EM (2016) Talanta 146:670–675
Gonzales WV, Mobashsher AT, Abbosh A (2019) Sensors 19:1–45
Graham DC (1947) Chem Rev 41:441–501
Guilbault GG, Montalvo JG Jr (1969) J Am Chem Soc 91:2164–2165
Katz S, Rechnitz GA (1963) Fresenius J Anal Chem 196:248–251
Kolthoff IM, Jordan J (1952) Anal Chem 24:1071–1072
Kolthoff IM, Langer A (1940) J Am Chem Soc 62:211–218
Laitinen HA, Kolthoff IM (1941) J Phys Chem 7:1079–1093
Lubert K, Kalcher K (2010) Electroanalysis 22:1937–1946
Matsumoto E (1920) Recueil des Travaux Chimiques des Pays-Bas 39:280–302
Mirceski V, Skrzypek S, Stojanov L (2018) ChemTexts 4:text number 17
Mohamad NR, Marzuki NHC, Buang NA, Huyop F, Wahab RA (2015) Biotechnol Biotechnol Equip 29:205–220
Parker RE, Adams RN (1956) Anal Chem 28:828–832
Posha B, Nambiar SR, Sandhyarani N (2018) Biosens Bioelectron 101:199–205
Pungor E, Feher Z, Nagy G, Toth K (1978) Abstr Pap Am Chem Soc 176:72–72
Racine P, Mindt W, Schlaepf P (1973) J Electrochem Soc 120:C115–C117
Rahi A, Sattarahmady N, Heli H (2016) Anal Biochem 510:11–17
Randviir EP, Banks CE (2013) Anal Methods 5:1098–1115
Ruzicka J, Hansen EH (1975) Anal Chim Acta 78:145–157
Ruzicka J, Marshall GD (1990) Anal Chim Acta 237:329–343
Silva IS, Capovilla B, Freitas KHG, Angnes L (2013) Anal Methods 5:3546–3551
Strohl AN, Curran DJ (1979) Anal Chem 51:1045–1049
Wang J, Taha Z (1991) Anal Chem 63:1053–1056
Wang J, Czae M, Lu J, Vuki M (1999) Microchem J 62:121–127
Wilkinson MC (1972) Chem Rev 72:575–625
Wong A, Silva TA, Vicentini FC, Fatibello-Filho O (2016) Talanta 161:333–341
Yamada J, Matsuda H (1973) J Electroanal Chem 44:189–198
Zhang GR (1998) Analyst 123:749–751
Zhang H, Xu L, Zheng J (2007) Talanta 71:19–24

# Amperometric Detection for Bioanalysis

Eduardo M. Richter and Rodrigo A. A. Munoz

## 1    Introduction

The electrochemical technique known as chronoamperometry or conventional amperometry (constant potential) is often used in initial analytical tests due to characteristics such as ease of understanding (simple theory), ease of using (choice of potential and current acquisition time), and the possibility of working with very simple and low-cost equipment. These features are interesting to be used with portable analytical methods (Serra et al. 2007). Figure 1a illustrates a conventional amperometric test that involves the application of two sequential potential pulses. Usually, the first potential is selected in a potential region where no faradaic current occurs and, at the second one, in a potential region where a redox event (oxidation or reduction) occurs at the electrode surface. The current-time response obtained for a stationary working electrode and steady-state solution is shown in Fig. 1b. The result shown in Fig. 1c can be obtained in different ways (hydrodynamic amperometry), such as using either stationary working electrode in stirred solutions, or rotating disk electrode, or yet a working electrode coupled to a flow-through electrochemical cell.

In the result shown in Fig. 1b (steady-state solution and planar electrode), the limiting current comes only from the mass transport by diffusion and the current decays with the inverse of the square root of time as predicted by the Cottrell equation (Cottrell 1902; Wang 2000).

$i(t) = i_d(t) = \frac{nFAD^{1/2}C}{(\pi t)^{1/2}}$ Cottrell equation where $i$ is the current (A), $n$ is the number of electrons transferred in the half-reaction, $F$ is Faraday's constant, $A$ is the electrode area (cm$^2$), $D$ is the diffusion coefficient (cm$^2$/s), $C$ is the concentration in bulk solution (mol/cm$^3$), and $t$ is the time (s).

E. M. Richter · R. A. A. Munoz (✉)
Institute of Chemistry, Federal University of Uberlandia (UFU), Uberlandia, MG, Brazil
e-mail: munoz@ufu.br

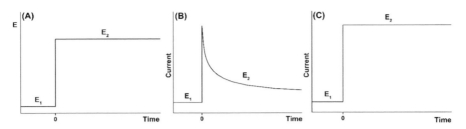

**Fig. 1** Conventional amperometric experiment: (**a**) potential-time waveform; (**b**) current-time response for stationary working electrode and steady-state solution; (**c**) current-time response for hydrodynamic amperometry

As can be observed in the Cottrell equation, the faradaic current (proportional to the concentration) is time dependent. This can be inconvenient if manual commercial instruments are used because the current used in analytical methods (calibration and sample analysis) always needs to be measured at the same time. Non-Cottrell behavior is observed in some conditions such as after long times of amperometric experiments (natural convection effects), when microelectrodes with high perimeter-to-area ratios are used as working electrodes and yet under hydrodynamic conditions. Under constant hydrodynamic conditions, the thickness of the Nernst diffusion layer is constant, and the detected current is invariant with time. This phenomenon can be observed in the data shown in Fig. 1c (hydrodynamic amperometry). In addition, if hydrodynamic amperometry is used, better sensitivity is frequently achieved due to the high rate of mass transport to the electrode surface. Therefore, from the point of view of analytical chemistry, hydrodynamic amperometry is preferably used. In the literature, equations can be found that described the performance (limiting current response) of amperometric flow-through detectors for various electrode geometries, such as tubular, planar (parallel and perpendicular flow), thin-layer, and wall-jet configurations (Wang 2000; Hanekamp and Jong 1982). In addition to the use for analytical purposes, chronoamperometry can also be used for measuring the diffusion coefficient of electroactive species (Wang et al. 2011; Nowinski and Anjo 1989) and mechanisms of electrode processes (Herman and Blount 1969). If the current from a potential step is integrated (charge measurement), the procedure is known as chronocoulometry. Among various applications, chronocoulometry is useful for the study of absorbed species on electrode surfaces (Lauer and Osteryoung 1966; Anson 1966) and measuring the surface area of working electrodes (Fragkou et al. 2012).

## 2    Hydrodynamic Amperometry

The positioning of the three electrodes (working, reference, and auxiliary) in a simple electrochemical cell with solution (supporting electrolyte) inside the cell under constant stirring is the simplest amperometric arrangement frequently used by analytical electrochemists (Njagi et al. 2007). A constant potential is applied to

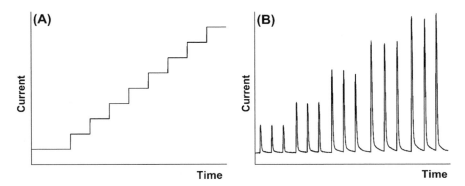

**Fig. 2** Typical current responses under hydrodynamic experiments using constant potential amperometric detection. (**a**) Simple electrochemical cell containing stirred solution (supporting electrolyte) and successive additions of the target analyte at a constant concentration (final concentration of analyte in the cell increases after each addition); (**b**) Flow system (FIA or BIA) with triplicate injections of solutions with increasing concentrations of the target analyte

the working electrode and the resulting current is measured as a function of time (Fig. 2a). This arrangement is useful to test new electrodes, but it has some disadvantages for quality control methods, such as low signal-to-noise ratio (stirring with magnetic bars generates noise background current), low analytical frequency (new sample can only be analyzed after cleaning of the electrochemical cell), and a greater tendency to electrode contamination (full-time contact of the electrode with the analyte).

Figure 2b shows a typical current-time response (transient peaks) obtained with flow-injection analysis (FIA) or batch-injection analysis (BIA) systems with amperometric detection (Felix and Angnes 2010; Rocha et al. 2018; Stefano et al. 2016). For the use of FIA or BIA systems, prior investments with accessories, such as tubes and pumps (FIA), and also injection device and electrochemical flow cell (both FIA and BIA) must be provided, which is a disadvantage compared to the current-time experiment shown in Fig. 2a (stirred solution). However, the use of FIA or BIA systems also have advantages such as high analytical frequency (~60 injections per hour), less contamination of the working electrode surface (shorter contact time with the analyte), and use of reduced volume of reagents and samples by analysis (<150 μL). The signal shape detected in these techniques (peak shape) is also similar to that obtained in several other well-known analytical or characterization techniques.

In addition to flow systems such as FIA and BIA, conventional amperometry is also used as an electrochemical detection mode in separation techniques, including liquid chromatography and capillary electrophoresis (Fedorowski and LaCourse 2015; Trojanowicz 2009, 2011; Islam et al. 2019). The amperometric technique is especially important in separation techniques for a variety of electroactive compounds not detectable with the most commonly used detector (UV-Vis spectroscopy). Despite the selectivity (use of proper potential and/or electrode material), high sensitivity, and instrumental simplicity, the conventional amperometry with solid electrodes has a disadvantage which is the surface contamination with solution impurities and/or by-products from electrochemical reactions (Islam et al. 2019). In order to avoid or minimize this phenomenon, constant potential amperometric detection can be replaced by pulsed amperometric detection (PAD) (Siangproh et al. 2009). This technique involves the application of a set of potential pulses (waveform) to the working electrode as a function of time. The analyte is detected in one of the potential pulses of the waveform and the other potential pulses are used as cleaning and/or activation steps of the electrode surface. Various schematic diagrams of PAD waveforms have already been used and shown in the literature (Islam et al. 2019). The number of potential pulses and applications times can vary depending on the material used as working electrode and the target analyte and/or sample composition. A considerable improvement in the response stability of solid electrodes is usually achieved if conventional amperometry is replaced by the PAD technique. However, if the PAD technique is used, the detected non-faradaic (capacitive) current may not be negligible, especially if potential pulses with very different values are used. Usually, the detected non-faradaic current (background current) is constant in the absence or presence of the analyte and can be easily subtracted.

Finally, a slightly different version of PAD is known as multiple pulse amperometry (MPA) or multiplex-pulsed amperometric detection (MPAD) (dos Santos et al. 2008; Medeiros et al. 2010). Similar to PAD, two or more potential pulses are applied to the working electrode as a function of time; however, the current is monitored at all applied potentials. This strategy makes possible simultaneous determinations of two (dos Santos et al. 2008; Medeiros et al. 2010) or three (Freitas et al. 2016) compounds using a single working electrode without the use of previous separation step (selectivity improvement). If conventional amperometric detection (constant potential) is used, simultaneous determinations are only possible if more than one working electrode is used (Matos et al. 2000). This technique (MPA) also allowed for the use of the internal standard method in FIA (Gimenes et al. 2010) and BIA (Gimenes et al. 2012) systems with amperometric detection.

## 3    Applications of Amperometric (Bio)sensors for Bioanalysis

The most popular amperometric sensor applied for bioanalysis is the glucose biosensor. The main reason is the high demand for glucose measurements in blood for the diagnosis of diabetes mellitus worldwide, especially considering millions of diabetics that need to measure glucose levels daily. The first glucose sensor was developed in 1962 using a Clark oxygen electrode on which a glucose oxidase was immobilized with the aid of a porous dialysis membrane (Clark and Lyons 1962). In this pioneering approach, oxygen consumption by the enzymatic reaction in the presence of glucose was monitored by amperometric detection on a platinum electrode involving the electrochemical reduction of oxygen. Later in 1973, Guilbault and Lubrano proposed the electrochemical biosensing of glucose based on the amperometric detection (oxidation) of hydrogen peroxide, which is generated by the enzymatic reaction (Guilbault and Lubrano 1973). These pioneer works opened up the possibility of immobilizing other enzymes to convert electroinactive compounds into electroactive products, such as hydrogen peroxide. Other examples are listed in Table 1, including analytical characteristics of the proposed amperometric enzymatic biosensors.

Table 1 shows some examples of amperometric biosensors using 12 different enzymes immobilized over different electrodes and mostly applied for bioanalysis. Major effort by researchers has been focused on glucose biosensors with more than 1000 papers published with this aim to date (data from Web of Science®). First studies in the literature involved the use of glassy-carbon, platinum, or gold electrodes as a substrate to immobilize different enzymes, while in the subsequent years the development of electrochemical biosensors using disposable screen-printed electrodes has appeared in combination with metal nanoparticles and carbon nanostructures due to electrocatalytic properties for the oxidation or reduction of enzymatic products. Table 1 lists these examples using recent references. The stability of the proposed biosensors and the evaluation of analytical characteristics were predominantly investigated by performing experiments using stirred solutions. On the other hand, some of these papers demonstrated the feasibility of biosensors under flow conditions whose advantages were previously highlighted in the text (Vidal et al. 2008; Guerrieri et al. 2019). In fact, the immobilization of an enzyme as a biorecognition element on the electrodic transductor allows for the development of highly selective sensors. Nevertheless, the main *"Achilles' heel"* of such biosensors is their shelf-life and pH stability inherent from enzymes.

Chemically modified electrodes have been developed to obtain similar selectivity of biosensors but with improved stability characteristics. Probably, the most successful example is the use of iron hexacyanoferrate as a chemical modifier, commonly known as Prussian blue, which enables the electrochemical reduction of hydrogen peroxide at low potential values (close to zero versus Ag/AgCl). This compound can be incorporated by electrodeposition on different substrates or incorporated within carbon paste or screen-printed electrodes.

Such a condition enables the amperometric detection of hydrogen peroxide free from the interference of species commonly found in biological samples. Due to such unique properties, Prussian blue is called artificial peroxidase (Karyakin and Karyakina 1999). Once hydrogen peroxide is a common product of several enzymatic reactions used for biosensors, the immobilization of different enzymes on Prussian blue-modified electrodes enables the production of highly sensitive and selective amperometric biosensors. Another example is the use of metallic oxides to the direct low-potential detection of species, such as the determination of L-tyrosine using copper oxide nanoparticles as a chemical modifier free from the interference of species found in biological samples (Razmi et al. 2011).

Another potential application of amperometric biosensors involves the development of wearable sensors. Considering the previous example of Prussian blue-modified electrodes and their combination with enzymes, Kim et al. (2015) developed a mouthguard biosensor using flexible PET substrates at which Prussian blue inks were screen-printed, followed by uricase immobilization. The enzyme electrode

**Table 1** Examples of amperometric biosensors and respective analytical characteristics

| Analyte | Biorecognition element (enzyme) | Base electrode | Linear range (mol L$^{-1}$) | LOD (mol L$^{-1}$) | Sample | Ref. |
|---|---|---|---|---|---|---|
| D-alanine | D-amino acid oxidase | Au nanofilm/MWCNTs/GCE | $2.5 \times 10^{-7}$ to $4.5 \times 10^{-6}$ | $20 \times 10^{-9}$ | Human serum | (Shoja et al. 2017) |
| Organophosphorus and triazine | Tyrosinase | Nafion/GCE | $8.0 \times 10^{-7}$ to $1.0 \times 10^{-5}$ | $7.0 \times 10^{-9}$ | Water | (Vidal et al. 2008) |
| Cholesterol | Cholesterol oxidase | Au NPs/MWCNTs/GCE | $2.0 \times 10^{-3}$ to $8.0 \times 10^{-3}$ | $1.0 \times 10^{-4}$ | Human serum | (Alagappan et al. 2020) |
| Choline | Choline oxidase | Poly-pyrrole/poly-aminophenol/GCE | $1.0 \times 10^{-6}$ to $1.0 \times 10^{-4}$ | $1.0 \times 10^{-6}$ | Human serum and dialysate | (Guerrieri et al. 2019) |
| Dopamine and spironolactone | Laccase | MWCNTs/GCE | $3.0 \times 10^{-6}$ to $3.9 \times 10^{-5}$<br>$3.0 \times 10^{-6}$ to $3.9 \times 10^{-5}$ | $1.3 \times 10^{-7}$<br>$9.4 \times 10^{-7}$ | Urine Cerebrospinal fluid | (Coelho et al. 2019) |
| Glucose | Glucose oxidase | Polyaniline/PtNPs/Pt | $1.0 \times 10^{-4}$ to $1.9 \times 10^{-3}$ | $1.0 \times 10^{-7}$ | Human blood | (Zheng et al. 2020) |
| Glutamate | Glutamate dehydrogenase | Co$_3$O$_4$ NPs/SPE | $1.0 \times 10^{-5}$ to $6.0 \times 10^{-4}$ | $1.0 \times 10^{-5}$ | Blood | (Hu et al. 2020) |
| Hydrogen peroxide | Horseradish peroxidase | AuNPs/3D-printed poly-latic acid/graphene | $2.5 \times 10^{-5}$ to $1.0 \times 10^{-4}$ | $9.1 \times 10^{-6}$ | Human serum | (Marzo et al. 2020) |
| L-lactate and pyruvate | Lactate dehydrogenase and pyruvate oxidase | Two working Pt electrodes | $5.0 \times 10^{-6}$ to $1.0 \times 10^{-3}$<br>$1.0 \times 10^{-5}$ to $1.0 \times 10^{-3}$ | $5.0 \times 10^{-6}$<br>$5.0 \times 10^{-6}$ | Human serum | (Kucherenko et al. 2019) |
| Urea | Urease | MWCNTs/poly-o-toluidine/SPE | $1.0 \times 10^{-10}$ to $1.1 \times 10^{-8}$ | $3.0 \times 10^{-11}$ | Blood | (Hassan et al. 2018) |
| Uric acid | Uricase | Graphene oxide/Prussian blue/chitosan/GCE | $1.0 \times 10^{-4}$ to $2.0 \times 10^{-3}$ | $2.3 \times 10^{-5}$ | Human urine | (Zheng et al. 2019) |

*GCE* glassy-carbon electrode, *LOD* limit of detection, *MWCNTs* multi-walled carbon nanotubes, *NPs* nanoparticles, *SPE* screen-printed electrodes

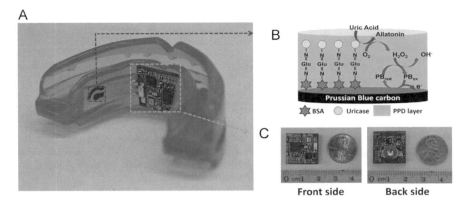

**Fig. 3** (**a**) Image of the mouthguard biosensor integrated with a wireless amperometric circuit board. (**b**) Reagent layer on the chemically modified printed Prussian-Blue carbon working electrode containing uricase for uric acid detection. (**c**) Images of the wireless amperometric circuit board (front and back side) (with permission) (Kim et al. 2015)

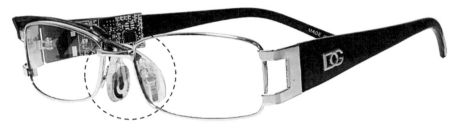

**Fig. 4** Real image of the eyeglasses biosensor system integrated with a wireless circuit board along the arms (with permission) (Sempionatto et al. 2017)

was attached to the mouthguard and the real-time amperometric measurement of uric acid in saliva was performed with the aid of a miniaturized printed circuit board attached to the mouthguard for wireless data collection. Figure 3 shows the image of the mouthguard biosensor (Fig. 3a), the chemical composition of the working electrode modified with Prussian Blue carbon and uricase (Fig. 3b), and an image of the wireless amperometric circuit board (Fig. 3c).

Using similar miniaturized electronics placed on the arms of an eyeglasses frame, a wearable amperometric biosensor was developed to monitor lactate in sweat (Sempionatto et al. 2017). Screen-printing of the electrodes on a flexible PET was also performed using a Prussian blue ink at which a lactate oxidase was immobilized. The flexible amperometric biosensor was attached to the nose-bridge pads of the eyeglasses, integrated with the wireless electronics at the eyeglass's arms, for real-time monitoring of lactate in sweat. Figure 4 shows the fully integrated eyeglasses wireless multiplexed chemical sensing platform for simultaneous real-time detection of lactate in sweat electrolytes. This approach represents an attractive wearable

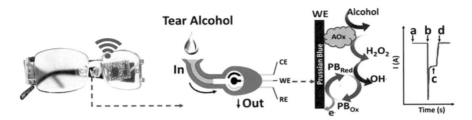

**Fig. 5** Real image and schematics of the microfluidic device and wireless electronics integrated into the eyeglasses approach along with a representation of enzymatic alcohol detection and signal transduction: (a) baseline; (b) current change due to the captured tear; (c) alcohol signal and (d) drying of the device (with permission) (Sempionatto et al. 2019)

platform where amperometric detection was selected due to its simplicity in fabrication and data collection.

A third example of a wearable amperometric biosensor was demonstrated for the real-time analysis of tears (Sempionatto et al. 2019). Glucose and alcohol were amperometrically monitored using biosensors fabricated over a polycarbonate substrate containing a fluidic collector of tears. This device was placed on the arm of the eyeglasses frame, which was assembled with wireless electronics. This wearable biosensor was demonstrated for the real-time detection of glucose and ethanol using glucose oxidase and alcohol oxidase enzymes, respectively, which were immobilized over the Prussian-blue screen-printed electrode over the polycarbonate platform. Figure 5 shows the wearable eyeglasses-based platform with an alcohol biosensor flow detector mounted onto the nose-bridge pad for alcohol detection in tears.

Most examples of the application of amperometric detection for bioanalysis involve the application of a single potential for monitoring the electroactive species generated by enzymatic reaction (biosensor) or by a chemically modified electrode. LaCourse demonstrated that PAD detection can be applied for the determination of sugars (glucose, fructose, and sucrose) in tobacco from cigarettes and cigars, peptones (enzymatically or chemically digested organisms) derived from different samples (soy, pea, yeast, and meat) and carbohydrate in bacterial polysaccharides (LaCourse 2004). PAD detection using gold or platinum electrodes combined with a separation technique, such as HPLC, can measure in single run glucose, fructose, and sucrose in tobacco extracts. Other examples shown in the same work are the chemical profile of peptones that are enzymatically digested organisms and the determination of carbohydrates in bacterial polysaccharides.

As previously mentioned, MPA detection enables the simultaneous determination of analytes using a single working electrode. MPA combined with BIA was proposed for the simultaneous determination of uric acid and nitrite in biological fluids using a screen-printed electrode modified with carbon nanotubes (Caetano et al. 2018). This work demonstrates the potential of BIA-MPA for fast analysis of multiple analytes in biological samples using complete portable instrumentation, a low-cost electrochemical cell, and disposable electrodes.

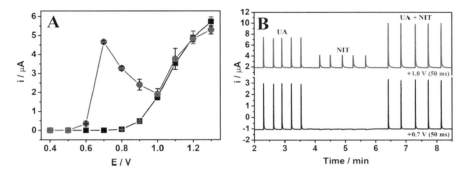

**Fig. 6** (**a**) Hydrodynamic voltammograms obtained by plotting peak current values as a function of the corresponding applied potential pulses (70 ms each pulse). The solutions contained $10 \ \mu mol \ L^{-1}$ uric acid (●) and nitrite (■). (**b**) Amperometric responses (n = 5) of solutions containing only uric acid ($20 \ \mu mol \ L^{-1}$), only nitrite ($20 \ \mu mol \ L^{-1}$) and uric acid + nitrite ($20 \ \mu mol \ L^{-1}$ each). BIA parameters: dispensing rate of $280 \ \mu L \ s^{-1}$ and injection volume of $100 \ \mu L$. Supporting electrolyte: $0.1 \ mol \ L^{-1}$ Britton–Robinson buffer solution (pH 2) (with permission) (Cardoso et al. 2020b)

The potential of a BIA system connected to screen-printed carbon electrodes was also evaluated for amperometric detection UV-induced DNA degradation in the solution phase (Svitková et al. 2019). In this study, sensitive amperometric detection was used to evaluate the structural changes in double-stranded DNA (dsDNA) after UV-C irradiation. A time-dependent response (decrease of amperometric signal up to 58% after 10 min of the irradiation) was found for the detection of damage to low molecular weight salmon sperm dsDNA. According to the authors, the advantages of this low-dimensional and low-cost system are also promising for examining other types of DNA damages.

Additive manufacturing or three-dimensional (3D) printing has generated a great revolution in several areas, including medical, civil engineering, electronics, and aerospace. In chemistry, 3D printing technology has also brought several advances, especially in (Bio)Analytical Chemistry (Palenzuela and Pumera 2018). Graphene-polylactic acid (G-PLA) composite platforms can be quickly fabricated under a low cost (less than $1.00 each sensor) by fused deposition modeling 3D-printing and potentially applied for bioanalysis (Cardoso et al. 2020a). The simultaneous determination of uric acid and nitrite in biological fluids (saliva and urine were diluted in supporting electrolyte) on 3D-printed G-PLA electrodes using BIA with MPA detection was demonstrated (Cardoso et al. 2020b).

The simultaneous determination free from the interference of the other analyte is feasible because both compounds are oxidized on the 3D-printed G-PLA electrode at different oxidation potentials, separated by at least 200 mV. Figure 6a shows two hydrodynamic voltammograms obtained for uric acid (red circles) and nitrite (black squares) using BIA with amperometric detection for constantly applied potentials (at each potential indicated in the x-axis, the current for a triplicate injection was registered) using the 3D-printed G-PLA electrode. The plot of current as a function of applied potential presented a similar behavior verified for cyclic voltammetric experiments in which uric acid oxidized at around +0.7 V while nitrite oxidation started at +0.9 V (both potentials vs Ag/AgCl/KCl). Hence, MPA detection was feasible by the application of the potentials of +0.7 V (for uric acid detection) and +1.0 V (nitrite detection after subtraction of the contribution of uric acid oxidation at this potential), as demonstrated in Fig. 6b. This experiment was performed by applying +0.7 and +1.0 V sequentially (50 ms at each potential) and Fig. 6b shows that at +0.7 V only uric acid presented a current free from possible interference from nitrite. When a mixture of both compounds was injected, current variation was observed at both potentials. The current obtained at +0.7 V can be used to quantify uric acid while the current at +1.0 V required the subtraction of uric acid contribution. This is the first work that demonstrated the MPA detection applied to a 3D-printed electrode for bioanalytical applications. Moreover, the authors verified that the G-PLA electrodes required a surface treatment, which consisted of a mechanical polishing followed by chemical treatment in dimethylformamide, to improve the electrochemical activity of such electrodes.

Another relevant application of these platforms was the fabrication of amperometric biosensors. Glucose oxidase was successfully linked to the G-PLA surface by glutaraldehyde to produce an amperometric biosensor for glucose mediated by ferrocenomethanol, which was applied for blood plasma analysis after simple dilution in an electrolyte (Cardoso et al. 2020b). Similarly, the development of a glucose biosensor on a 3D-printed electrode in which a conductive PLA filament containing carbon black (instead of graphene) was achieved, and the amperometric detection of glucose in biological fluids was presented (Katseli et al. 2019). The direct electron transfer between the 3D-printed G-PLA surface and the enzyme was reported and the absence of a crosslinker to immobilize the enzyme on the polymeric surface was a great advancement in comparison with the previous reports (Marzo et al. 2020). Therefore, these reports show great promises of 3D printing for the development of novel amperometric sensors and biosensors for the detection of several other species of biological interests.

# References

Alagappan M, Immanuel S, Sivasubramanian R, Kandaswamy A (2020) Arab J Chem 13:2001–2010

Anson FC (1966) Anal Chem 38:54–57

Caetano LP, Lima AP, Tormin TF, Richter EM, Espindola FS, Botelho FV, Munoz RAA (2018) Electroanalysis 30:1870–1879

Cardoso RM, Kalinke C, Rocha RG, dos Santos PL, Rocha DP, Oliveira PR, Janegitz BC, Bonacin JA, Richter EM, Munoz RAA (2020a) Anal Chim Acta 1118:73–91

Cardoso RM, Silva PRL, Lima AP, Rocha DP, Oliveira TC, do Prado TM, Fava EL, Fatibello-Filho O, Richter EM, Muñoz RAA (2020b) Sens Actuators B Chem 307:127621

Clark L Jr, Lyons C (1962) Ann N Y Acad Sci 102:29–45

Coelho JH, Eisele APP, Valezi CF, Mattos GJ, Schirmann JG, Dekker RFH, Barbosa-Dekker AM, Sartori ER (2019) Talanta 204:475–483

Cottrell FG (1902) Z Phys Chem 42:385–431

dos Santos WTP, Almeida EGN, Ferreira HEA, Gimenes DT, Richter EM (2008) Electroanalysis 20:1878–1883

Fedorowski J, LaCourse WR (2015) Anal Chim Acta 861:1–11

Felix FS, Angnes L (2010) J Pharm Sci 99:4784–4804

Fragkou V, Ge Y, Steiner G, Freeman D, Bartetzko N, Turner APF (2012) Int J Electrochem Sci 7: 6214–6220

Freitas JM, Oliveira TC, Gimenes DT, Munoz RAA, Richter EM (2016) Talanta 146:670–675

Gimenes DT, dos Santos WTP, Munoz RAA, Richter EM (2010) Electrochem Commun 12:216–218

Gimenes DT, Pereira PF, Cunha RR, da Silva RAB, Munoz RAA, Richter EM (2012) Electroanalysis 9:1805–1810

Guerrieri A, Ciriello R, Crispo F, Bianco G (2019) Bioelectrochemistry 129:135–143

Guilbault G, Lubrano G (1973) Anal Chim Acta 64:439–455

Hanekamp HB, Jong HG (1982) Anal Chim Acta 135:351–354

Hassan RYA, Kamel AK, Hassan HNA, Abd El-Ghaffar MA (2018) J Solid State Electrochem 22: 1817–1823

Herman HB, Blount HN (1969) J Phys Chem 73:1406–1413

Hu F, Pang J, Chu Z, Jin W (2020) Sens Actuators B Chem 306:127587

Islam MA, Mahbub P, Nesterenko PN, Paull B, Macka M (2019) Anal Chim Acta 1052:10–26

Karyakin AA, Karyakina EE (1999) Sensors Actuat B Chem 57:268–273

Katseli V, Economou A, Kokkinos C (2019) Electrochem Commun 103:100–103

Kim J, de Araujo WR, Warchall J, Valdés-Ramirez G, Paixao TRLC, Mercier PP, Wang J (2015) Biosens Bioelectron 74:1061–1068

Kucherenko IS, Soldatkin OO, Topolnikova YV, Dzyadevych SV, Soldatkin AP (2019) Electroanalysis 31:1608–1614

LaCourse WR (2004) Proc SPIE Int Soc Opt Eng 5261:103–112

Lauer G, Osteryoung RA (1966) Anal Chem 38:1106–1112

Marzo AML, Mayorga-Martinez CC, Pumera M (2020) Biosens Bioelectron 151:111980

Matos RC, Augelli MA, Lago CL, Angnes L (2000) Anal Chim Acta 404:151–157

Medeiros RA, Lourenção BC, Rocha-Filho RC, Fatibello-Filho O (2010) Anal Chem 82:8658–8663

Njagi J, Warner J, Andreescu S (2007) J Chem Educ 84:1180–1182

Nowinski SA, Anjo DM (1989) J Chem Eng Data 34:265–268

Palenzuela CLM, Pumera M (2018) Trends Anal Chem 103:110–118

Razmi H, Nasiri H, Mohammad-Rezaei R (2011) Microchim Acta 173:59–64

Rocha DP, Cardoso RM, Tormin TF, de Araujo WR, Munoz RAA, Richter EM, Angnes L (2018) Electroanalysis 30:1386–1399

Sempionatto JR, Nakagawa T, Pavinatto A, Mensah ST, Imani S, Mercier P, Wang J (2017) Lab Chip 17:1834–1842

Sempionatto JR, Brazaca LC, Garcia-Carmona L, Bolat G, Campbell AS, Martin A, Tang G, Shah R, Mishra RK, Kim J, Zucolotto V, Wang J (2019) Biosens Bioelectron 137:161–170

Serra PA, Rocchitta G, Bazzu G, Manca A, Puggioni GM, Lowry JP, O'Neill RD (2007) Sens Actuators B Chem 122:118–126

Shoja Y, Rafati AA, Ghodsi J (2017) Enzyme Microb Technol 100:20–27

Siangproh W, Leesutthipornchai W, Dungchai W, Chailapakul O (2009) J Flow Inject Anal 26:5–25

Stefano JS, Cordeiro DS, Marra MC, Richter EM, Munoz RAA (2016) Electroanalysis 28:350–357

Svitková V, Labuda J, Vyskočil V (2019) Electroanalysis 31:2001–2006

Trojanowicz M (2009) Anal Chim Acta 653:36–58

Trojanowicz M (2011) Anal Chim Acta 688:8–35

Vidal JC, Bonel L, Castillo JR (2008) Electroanalysis 20:865–873

Wang J (2000) Analytical electrochemistry, 2nd edn. Wiley-VCH, New York
Wang Y, Limon-Petersen JG, Compton RG (2011) J Electroanal Chem 652:13–17
Zheng L, Ma H, Ma Y, Meng Q, Yang J, Wang B, Yang Y, Gong W, Gao G (2019) Int J Electrochem Sci 14:9573–9583
Zheng H, Liu M, Yan Z, Chen J (2020) Microchem J 152:104266

# Potentiometric Biosensors

Orlando Fatibello-Filho

## 1 Introduction

Potentiometric methods are those based on voltage ($E_{cel}$) measurements of a galvanic cell. The potential difference between two electrodes such as $E_{bios}$ and $E_{ref}$ (Eq. 1):

$$E_{cel} = E_{bios} \quad - E_{ref} + E_j \tag{1}$$

immersed in an analyte solution occurs under a nearly zero-current condition using high input impedance equipment (potentiometer or pH meter). The biosensor ($E_{bios}$: biosensor (indicator electrode)), combining a biorecognition element (e.g., enzymes, antibodies, antigens, organisms, animal and plant tissues, cells, organelles, and so on) with a transducer, which converts biochemical events into electrical signals (Guilbault et al. 1991; Fatibello-Filho and Capelato 1992). The obtained cell potential ($E_{cel}$) being logarithmically correlated with the analyte activity ($a_i$) (Nernst equation), since the reference electrode ($E_{ref}$) and liquid junction ($E_j$) potentials are constants. Figure 1 shows a general scheme of a typical galvanic cell used in potentiometric analysis using a biosensor.

The Nernst equation (Eq. 2) for a biosensor can be represented by:

$$E_{bios} = K \pm \frac{2.303RT}{nF} \log a_i \tag{2}$$

where $E_{bios}$ is the biosensor potential (V), $K$ is a constant potential term, $R$ is the universal gas constant (8.314 J K$^{-1}$ mol$^{-1}$), $F$ is the Faraday constant (96,485 C mol$^{-1}$), $n$ is the number of electrons or the charge of the analyte, and $a_i$ is the analyte activity.

O. Fatibello-Filho (✉)
Department of Chemistry, Federal University of São Carlos, São Carlos, SP, Brazil
e-mail: bello@ufscar.br

**Fig. 1** General scheme of a galvanic cell in potentiometric determinations, where $E_{Ref}$ is the reference electrode, $E_j$ is the liquid junction potential and Bio. elem. is the biorecognition element immobilized on a selected potentiometric electrode surface (biosensor)

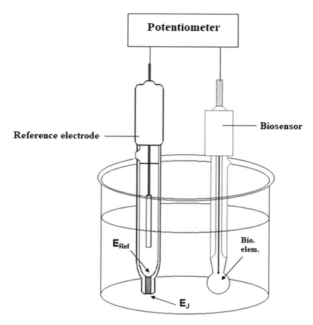

The galvanic cell notation or line notation is shown in Eq. (3):

$$E_j$$
$$Ag \mid AgCl \ (3.0 \ mol \ L^{-1}KCl) \parallel (analyte) \mid E_{bios} \tag{3}$$

where $E_{bios}$ is the biosensor, Ag|AgCl (3.0 mol L$^{-1}$ KCl) is the reference electrode ($E_{ref}$) and $E_j$ is the liquid junction potential.

## 2    First Potentiometric Biosensor

Guilbault and Montalvo (1969) constructed the first potentiometric urea biosensor by immobilizing the urease enzyme on a Nylon membrane-supported polyacrylamide matrix directly on an $NH_4^+$ selective glass electrode.

When the electrode is immersed in a solution containing urea, the urea diffuses into the gel layer containing the *urease* enzyme (*urea aminohydrolase*, EC 3.5.1.5), catalyzing the hydrolysis reaction of this substrate, as shown in Eq. (4):

$$NH_2CONH_{2(aq)} + 2H_2O_{(l)} + H^+_{(aq)} \xrightarrow{urease} 2NH_4^+{}_{(aq)} + HCO_3^-{}_{(aq)} \tag{4}$$

After 30–60 s, the system reaches equilibrium and the electrode potential is measured. The glass electrode responds to the $NH_4^+$ cation produced in the enzymatic reaction (Eq. 4), which under conditions of fixed enzyme concentration, is a linear function of log [urea] in the sample solution. The analytical curve was linear in the urea concentration range from $4.84 \times 10^{-5}$ to $1.16 \times 10^{-2}$ mol L$^{-1}$ with a limit of detection (LOD) of $1.67 \times 10^{-5}$ mol L$^{-1}$.

There are at least five other possibilities to determine urea using potentiometric biosensors. Urea electrodes based on pH measurement were developed by immobilization of urease directly to a glass electrode and onto metal-metal oxides electrodes (Guilbault et al. 1991; Fatibello-Filho and Capelato 1992). Ammonium cations produced in the enzymatic reaction can be detected by means of an $NH_4^+$ selective liquid-membrane electrode (Benco et al. 2003; Guilbault 1982) or an $NH_3$ gas electrode can be used at pH higher than 7.7–8.0 (Guilbault et al. 1991; Fatibello-Filho and Capelato 1992). Alternatively, the $CO_3^{2-}$ ion can be monitored either directly with a carbonate-sensitive liquid-membrane electrode (Guilbault et al. 1991; Fatibello-Filho and Capelato 1992) or after conversion into $CO_2$ with a carbon dioxide gas electrode (Guilbault et al. 1991; Fatibello-Filho and Capelato 1992).

## 3    Selected Potentiometric Biosensors

Since the first potentiometric biosensor (Guilbault and Montalvo 1969), a few hundred new potentiometric biosensors have been designed and employed in the analysis of various analytes in biological processes such as biochemistry, clinical, forensic, pharmaceutical, physiologic, and so on. There are not much commercially available biosensors, which use ion-selective electrodes (ISE), glass, metal-metal oxide, and gas electrodes. Nonetheless, there is a large number of such devices, which can be easily self-made. For further details, the reader is referred to the excellent review articles and/or books (Guilbault et al. 1991; Fatibello-Filho and Capelato 1992; Guilbault 1982; Nery and Kubota 2013; Ravalli and Marrazza 2018; Koncki 2007; Scheller and Schubert 1992). Table 1 shows a selected list of these potentiometric biosensors.

As can be seen from Table 1, many base electrodes can be used for the fabrication of a potentiometric biosensor. The pH-glass electrode is easily available and many enzymatic reactions produce pH changes (hydrogen ion concentration changes) at the enzymatic layer. This explains the vast amount of enzyme-based pH-sensing devices (not shown in Table 1). In this kind of biosensor, it is particularly important to control the buffer capacity of the sample solution to prevent the fluctuation of pH with time. For those cases where the pH changes are verified, differential measurements using a second identical electrode but without the immobilized enzyme are recommended. In addition to pH glass electrodes, a plethora of metal-metal oxide electrodes and polymeric membranes have been used as base electrodes (Guilbault et al. 1991; Fatibello-Filho and Capelato 1992; Zhang et al. 2015; Psychoyios et al. 2013; Ismail and Adeloju 2014).

Polycrystalline membrane and cation-selective glass ion-selective electrodes, such as iodide, chloride, fluoride, ammonium, and others, have been successfully used in the construction of several potentiometric biosensors (Guilbault et al. 1991; Fatibello-Filho and Capelato 1992).

**Table 1** Configurations and some characteristics of potentiometric biosensors

| Analyte | Biorecognition element | Base electrode | Linear range $(\text{mol L}^{-1})$ | LOD | Ref. |
|---|---|---|---|---|---|
| Acetylcholine (Ach) | Acetylcholinesterase | Ach ISE | $1.0 \times 10^{-9}$ to $1.0 \times 10^{-3}$ | $8.0 \times 10^{-10}$ | (Mousavi et al. 2018) |
| Adenosine | Adenosine deaminase | $NH_3$ gas | $7.0 \times 10^{-7}$ to $1.0 \times 10^{-2}$ | $2.0 \times 10^{-7}$ | (Deng and Enke 1980) |
| L-amino acids (general)[a] | L-amino acid oxidase | $NH_4^+$ cation | $1.0 \times 10^{-4}$ to $1.0 \times 10^{-2}$ | $8.0 \times 10^{-5}$ | (Guilbault 1982) |
| L-asparagine | L-asparaginase | $NH_4^+$ cation | $5.0 \times 10^{-5}$ to $1.0 \times 10^{-2}$ | $2.0 \times 10^{-5}$ | (Fatibello-Filho and Capelato 1992; Guilbault 1982) |
| L-asparagine | L-aspartase | $NH_3$ gas | $1.6 \times 10^{-5}$ to $1.5 \times 10^{-3}$ | $1.3 \times 10^{-5}$ | (Fatibello-Filho et al. 1989a) |
| L-aspartame | Carboxypeptidase A + L-aspartase | $NH_3$ gas | $5.2 \times 10^{-4}$ to $8.1 \times 10^{-3}$ | $3.4 \times 10^{-4}$ | (Fatibello-Filho et al. 1988) |
| L-aspartate | L-aspartase | $NH_3$ gas | $7.0 \times 10^{-4}$ to $2.0 \times 10^{-2}$ | $3.1 \times 10^{-4}$ | (Fatibello-Filho et al. 1989b) |
| Cardiac troponin I (cTnI) | Rabbit polyclonal antibodies to cTnI | Gold nanoparticle-functionalized screen-printed graphite electrode (SPGE) | 10 pg/mL to 10 ng/mL | 7.3 pg/mL | (Ni et al. 2020) |
| Cardiac biomarker (troponin I-T-C (Tn I-T-C) complex | Sandwich ELISA | CGE/(PANI/DNNSA) | (>6 order of magnitude) | < 5 pg/mL | (Zhang et al. 2015) |
| Cardiac biomarkers (myoglobin (myo) and cardiac troponin I (cTnI)) | Anti-Myo and anti-cTnI | CdS NW/Au/ITO and SnNb2O6 NS/Au/ITO | 5.0 pg/mL to 50 ng/mL for both biomarker | 2.0 pg/mL (myo) and 2.5 pg/mL (cTnI) | (Cao et al. 2020) |
| Cholesterol | Cholesterol oxidase | Al/ZnO-nanowalls | $1.0 \times 10^{-6}$ to $1.0 \times 10^{-3}$ | $4.0 \times 10^{-7}$ | (Psychoyios et al. 2013) |

| | | | | | |
|---|---|---|---|---|---|
| Creatinine | Creatinase | $NH_4^+$ cation | $8.8 \times 10^{-5}$ to $1.0 \times 10^{-2}$ | $8.8 \times 10^{-5}$ | (Kobos and Ramsey 1980) |
| Enterovirus 71 (EV71) | Polyclonal mouse anti-EV71 antibody | Silver ($Ag^+$) ion-selective electrode (ISE). | 0.3–300 ng/mL | 0.058 ng/mL | (Sun 2018) |
| Flavin adenine dinucleotide (FAD) | Alkaline phosphatase + adenosine deaminase | $NH_3$ gas | $8.0 \times 10^{-5}$ to $4.0 \times 10^{-3}$ | $8.0 \times 10^{-5}$ | (Rashid et al. 1984) |
| Glucose | Glucose oxidase | $W/WO_3$ (pH) | $2.4 \times 10^{-4}$ to $1.7 \times 10^{-3}$ | $9.0 \times 10^{-5}$ | (Magna et al. 1993) |
| Oxalate | Oxalate decarboxylase | $CO_2$ gas | $2.0 \times 10^{-4}$ to $1.0 \times 10^{-2}$ | $4.0 \times 10^{-5}$ | (Thompson and Rechnitz 1974) |
| Pathogen bovine herpes virus-1 (BHV-1) | Recombinant BHV-1 gE, | FET | $10^{-4}$ to $10^{-1}$ anti-gE in serum dilutions | Anti-gE in serum dilution $<2.10^{-3}$ | (Tarasov et al. 2016) |
| Penicillin | Penicillinase | Pt/PVA | $7.5 \times 10^{-6}$ to $2.8 \times 10^{-4}$ | $1.7 \times 10^{-6}$ | (Ismail and Adeloju 2014) |
| Penicillin | Penicillinase | pH glass | $1.0 \times 10^{-4}$ to $1.0 \times 10^{-2}$ | $7.0 \times 10^{-5}$ | (Guilbault 1982) |
| Recombinant human myelin basic protein (rhMBP) | Anti-hMBP antibody | Plasticized PVC–COOH membrane electrode | $0.10$–$20.00$ µg/mL | 50.00 ng/mL | (Al-Ghobashy et al. 2019) |
| Salicylate | Salicylate hydroxylase | $CO_2$ gas | $3.3 \times 10^{-5}$ to $2.0 \times 10^{-3}$ | $2.0 \times 10^{-5}$ | (Fonong and Rechnitz 1984) |
| Salmonella typhimurium | Salmonella monoclonal antibody | PEDOT:PSS filter paper modified with PVC-COOH based membrane | $12$ to $12 \times 10^3$ cells/mL | 5 cells/mL | (Silva et al. 2019) |
| Urea | Urease | pH glass | $5.0 \times 10^{-5}$ to $5.0 \times 10^{-3}$ | $8.0 \times 10^{-6}$ | (Guilbault 1982) |
| Triglycerides | Lipase | ISFET[b] | $5.0 \times 10^{-3}$ to $30 \times 10^{-3}$%[c] | – | (Vijayalakshmi et al. 2008) |

(continued)

**Table 1** (continued)

| Analyte | Biorecognition element | Base electrode | Linear range (mol $L^{-1}$) | LOD | Ref. |
|---|---|---|---|---|---|
| Urea | Urease | $NH_4^+$ | $2.0 \times 10^{-6}$ to $8.0 \times 10^{-5}$ | $1.0 \times 10^{-6}$ | (Guilbault 1982) |
| Urea | Urease | $NH_3$ gas | $5.0 \times 10^{-5}$ to $5.0 \times 10^{-2}$ | $5.0 \times 10^{-5}$ | (Guilbault 1982) |
| Urea | Urease | $CO_2$ gas | $1.0 \times 10^{-4}$ to $1.0 \times 10^{-2}$ | $8.0 \times 10^{-5}$ | (Guilbault 1982) |
| Uric acid | Urate oxidase (uricase) | $CO_2$ gas electrode | $1.0 \times 10^{-4}$ to $2.5 \times 10^{-3}$ | $5.0 \times 10^{-5}$ | (Kawashima and Rechnitz 1976) |

[a]Electrode responds to L-cysteine, L-leucine, L-tyrosine, L-tryptophan, L-phenylalanine and L-methionine
[b]Ion-selective field effect transistor (ISFET)
[c]Concentration in % v/v; PVA: poly(vinyl alcohol)

Nowadays the development of enzyme field-effect transistor (EnFET)) is becoming common. ISFET is a classical metal/oxide/semiconductor (MOS) field-effect transistor (FET) with a gate formed by a separated reference electrode and attached to the gate area via a solution. These semiconductor FETs have an ion-sensitive surface. The surface electrical potential changes due to the interaction between ions and the semiconductor and can be subsequently measured (Vijayalakshmi et al. 2008; Yoon 2013).

In the first biosensor generation, gas permeable electrodes ($NH_3$, $CO_2$) were frequently used. Currently, they are still employed due to the high selectivity of the microporous polymeric membrane and the facility of enzyme immobilization. Moreover, there is a tendency in using solid-state composites, polymeric films modified with enzymes together with carbon nanoparticles, metal and/or metal oxide nanoparticles, and so on. An interesting example of that is an ultrasensitive immunosensor based on potentiometric ELISA for the detection of a cardiac biomarker, troponin I-T-C (Tn I-T-C) complex. In this work, a glassy carbon (GC) working electrode was first coated with emulsion-polymerized polyaniline/dinonylnaphthalenesulfonic acid (PANI/DNNSA) and the coated surface was utilized as a transducer layer on which sandwich ELISA incubation steps were performed. The analytical signal was achieved using open circuit potentiometry (OCP). The proposed immunosensor exhibited an excellent detection limit ($<5$ pg $mL^{-1}$) with a wide dynamic linear range ($>6$ order of magnitude) (Zhang et al. 2015).

A critical review discussing point-of-care (POC) (bio)electrochemical sensors including potentiometric detection was recently presented by Noviana et al. (2020)

# 4    Conclusions

In conclusion, potentiometric biosensors have a well-established position for the determination of several analytes in biological processes, being the biosensor for urea the most used. The advantages of potentiometric biosensors are their simplicity and clear principles of operation, sensitivity, easy miniaturization, and low cost. The applications of carbon and metal nanomaterials in potentiometric biosensors are a promising research area because several nanostructured materials could be developed. Smartphone-based potentiometric biosensors for point-of-care (POC) applications and mobile health were recently proposed (Huang et al. 2018). Wearable potentiometric tattoo biosensor for on-body detection of G-type nerve agent simulants is already a reality (Mishra et al. 2018) and many other wearables potentiometric biosensors are found, and others should be developed in the next years.

# References

Al-Ghobashy MA, Nadim AH, El-Sayed GM, Nebsen M (2019) ACS Sens 4:413–420

Benco JS, Nienaber HA, McGimpsey WG (2003) Anal Chem 75:152–156

Cao J-T, Ma Y, Lv J-L, Ren S-W, Liu Y-M (2020) Chem Commun 56:1513–1516

Deng I, Enke C (1980) Anal Chem 12:1937–1940

Fatibello-Filho O, Capelato MD (1992) Quim Nova 15:28–39

Fatibello-Filho O, Suleiman AA, Guilbault GG, Lubrano GJ (1988) Anal Chem 60:2397–2399

Fatibello-Filho O, Suleiman AA, Guilbault GG (1989a) J Macromol Sci A Chem 26:1261–1269

Fatibello-Filho O, Suleiman AA, Guilbault GG (1989b) Biosensors 4:313–321

Fonong T, Rechnitz GA (1984) Anal Chim Acta 158:357–362

Guilbault GG (1982) Ion-Select Electrode Rev 4:187–231

Guilbault GG, Montalvo JG (1969) J Am Chem Soc 91:2164–2165

Guilbault GG, Suleiman AA, Fatibello-Filho O, Nabirahni MA (1991) Immobilized bioelectrochemical sensors. In: Wise DL (ed) Bioinstrumentation and biosensors. Marcel Dekker, Inc, New York, pp 659–692

Huang X, Xu D, Chen J, Liu J, Li Y, Song J, Ma X, Guo J (2018) Analyst 143:5339–5351

Ismail F, Adeloju SB (2014) Electroanalysis 26:2701–2709

Kawashima T, Rechnitz GA (1976) Anal Chim Acta 83:9–17

Kobos RK, Ramsey TA (1980) Anal Chim Acta 121:110–118

Koncki R (2007) Anal Chim Acta 599:7–15

Magna A, Capelato MD, Fatibello-Filho O (1993) J Braz Chem Soc 4:72–75

Mishra RK, Barfidokht A, Karajic A, Sempionatto JR, Wang J, Wang J (2018) Sens Actuat B 273:966–972

Mousavi MPS, Abd El-Rahman MK, Mahmoud AM, Abdelsalam RM, Bühlmann P (2018) ACS Sensors 12:2581–2589

Nery EW, Kubota LT (2013) Anal Bioanal Chem 405:7573–7595

Ni E, Fang Y, Ma F, Ge G, Wu J, Wang Y, Lin Y, Xie H (2020) Anal Methods 12:2914–2921

Noviana E, McCord CP, Clark KM, Jang I, Henry CS (2020) Lab Chip 20:9–34

Psychoyios VN, Nikoleli G-P, Tzamtzis N, Nikolelis DP, Psaroudakis N, Danielsson B, Israr MQ, Willandere M (2013) Electroanalysis 25:367–372

Rashid A, Yatim M, Riechel TL (1984) Anal Lett 17:835–855

Ravalli A, Marrazza G (2018) Electrochemical-based biosensor technologies in disease detection and diagnostics. In: Altintas Z (ed) Biosensors and nanotechnology-applications in health care diagnostics. Wiley, Hoboken, pp 95–124

Scheller F, Schubert F (1992) Biosensors. Elsevier, Amsterdam

Silva NFD, Almeida CMR, Magalhães JMCS, Gonçalves MP, Freire C, Delerue-Matos C (2019) Biosens Bioelectron 141:111317.

Sun A-L (2018) Analyst 143:487–492

Tarasov A, Gray DW, Tsai M-Y, Shields N, Montrose A, Creedon N, Lovera P, O'Riordan A, Mooney MH, Vogel EM (2016) Biosens Bioelectron 79:669–678

Thompson H, Rechnitz GA (1974) Anal Chem 46:246–249

Vijayalakshmi A, Tarunashree Y, Baruwati B, Manorama SV, Narayana BL, Johnson REC, Rao NM (2008) Biosens Bioelectron 23:1708–1714

Yoon J-Y (2013) Introduction to biosensors—from electric circuits to immunosensors, 2nd edn. Springer, Cham

Zhang Q, Prabhu A, San A, Al-Sharab JF, Levon K (2015) Biosens Bioelectron 72:100–106

# Electrochemical Methods Applied for Bioanalysis: Differential Pulse Voltammetry and Square Wave Voltammetry

Sergio A. Spinola Machado and Fernando Henrique Cincotto

## 1      Introduction

Electrochemical methods, as chronoamperometry and voltammetry, have been conveniently applied for the analysis of biological molecules in several media. In such applications, they are accurate, sensitive, selective, and fast. However, the most important characteristic is the possibility to obtain analytical devices miniaturized to micro or even nano dimensions, thus making possible the use in biological media as within cells or inside living beings or microorganisms. In such applications, they provide a unique tool to investigate the role of biological molecules in the homeostasis of such organisms.

The electrochemical techniques evaluate the current signal obtained after the perturbation of an electrode surface by a defined program of potential. However, these current signals contain two components, i.e., a faradaic current added to a capacitive one. Only the faradaic portion is proportional to the analyte concentration and can be used to quantify. When working with traces amount of analyte, faradaic components are, frequently, much smaller than the capacitive one; thus, the interference could be severe. So, pulse techniques have been developed to minimize the capacitive contribution to the total current. This is achieved considering the different rates of decay of each contribution with time. The capacitive current decays exponentially while the faradaic one (in a diffusion-controlled process) decays with $\text{time}^{-1/2}$ so the rate of decay of capacitive currents is faster than the faradaic one. In this way, the capacitive contribution to current becomes negligible after approximately $5R_uC_{dl}$ after the potential pulse, where $R_u$ is the ohmic resistance and $C_{dl}$ is

S. A. Spinola Machado (✉)
São Carlos Institute of Chemistry, University of São Paulo, São Carlos, SP, Brazil
e-mail: sasmach@iqsc.usp.br

F. H. Cincotto
Federal University of Rio de Janeiro, Institute of Chemistry, Rio de Janeiro, RJ, Brazil

© The Author(s), under exclusive license to Springer Nature Switzerland AG 2022
L. T. Kubota et al. (eds.), *Tools and Trends in Bioanalytical Chemistry*,
https://doi.org/10.1007/978-3-030-82381-8_14

**Fig. 1** (**a**) Potential
perturbation of the electrode
in a differential pulse
voltammetry experiment. (**b**)
Scheme of a differential pulse
voltammogram

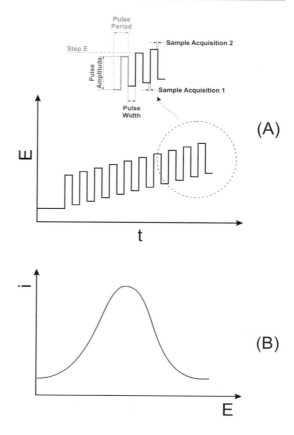

the double layer capacity. Among several pulse techniques, differential pulse
voltammetry and square wave voltammetry are, by far, the most common, and the
difference between both is just the form of the programming of potential.

In differential pulse voltammetry (DPV), the electrode is perturbed by several
potential pulses with the same amplitude superposed to a crescent staircase potential
variation. The perturbation of potential is illustrated in Fig. 1a, where the experimental parameters are defined.

The current sample acquired at the foot of the pulse (sample acquisition 1) is
subtracted from that acquired at the end of the pulse width (sample acquisition 2),
and the result is plotted against the electrode potential (defined by the staircase
perturbation), resulting in the gaussian voltammogram (Fig. 1b). The subtraction
eliminates the capacitive contribution that is approximately constant in both acquisition points. By solving the appropriate differential equations, it is possible to
demonstrate that the peak current ($i_p$) is proportional to the analyte concentration
(Wang 2001):

**Fig. 2** Square wave voltammetry responses: (1) pulses potential perturbation, (2) staircase perturbation, (3) square wave perturbation, (4) current responses, (5) sampled current signals, (6) differential currents, and (7) total currents vs. potential. Based on (Souza et al. 2003)

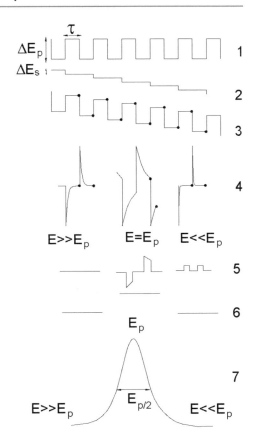

$$i_p = \frac{nFAD^{1/2}C}{\sqrt{\pi t_m}} \left(\frac{1-\sigma}{1+\sigma}\right) \tag{1}$$

where $t_m$ is the time interval between sample acquisition and sample acquisition and $\sigma$ is exp. ($nF/RT \times \Delta E/2$) with $\Delta E$ being the pulse amplitude. Equation (1) shows that a calibration curve is possible to be made by graphing $i_p$ vs. $C$ for several known values of the analyte concentration, considering that, now, $i_p$ is free from the influence of capacitive contributions.

Alternatively, in square wave voltammetry (SWV) the electrode is perturbed by a more complex waveform. Here, the symmetrical square wave is superimposed to the staircase potential program, as exemplified in Fig. 2. Symmetrical pulses in a square shape, with height, equal to $\Delta E_p$ (2–1), are superimposed in the staircase program, where the potential step is defined as $\Delta E_s$ (2–2), in a way that the first pulse is set in the direction of the potential scan while the second one in the opposite direction (2–3) (Souza et al. 2003). The current response originated after each pulse shows the regular exponential decay, as illustrated in (2–4).

The response after the pulse in the direction of the scan is called forward current ($i_{\text{forward}}$), while the other, in the opposite direction, backward current ($i_{\text{backward}}$). Both current signals are sampled at the end of the corresponding pulse duration (2–5 and 2–6). They are subtracted from each other (maintaining the negative signal of one of them) and the result is plotted against the staircase potential as a Gaussian curve (2–7).

In the SWV case, solving the differential equations can yield more complex solutions. For example, the peak current for reversible electrode reactions involving adsorption of reagent or reagent and product is described by the following equations, respectively (Souza et al. 2003):

$$I_p = K\, n^x \Delta E_s f \Delta E_p^y t_0^{1/2} C^* \qquad \text{for } 1 < x < 2; y < 1 \tag{2}$$

and

$$I_p = K\, n^x f^{1/2} \Delta E_p^y t_0^{1/2} C^* \qquad \text{for } 1 < x < 2; y < 1 \tag{3}$$

However, despite the complexity of such derivations, the linear dependence of $I_p$ and the analyte concentration in the solution ($C^*$) suggests that calibration curves can also be obtained similarly as with differential pulse voltammetry and used to quantify analytes in different media.

SWV is characterized by an excellent sensitivity, mainly since the total current is larger than the forward or backward components (since it is a sum of both components), being higher than for other pulse techniques. Moreover, with its lack of sensibility to capacitive currents, very low detection limits can be achieved, even lower than $1 \times 10^{-8}$ mol $L^{-1}$ (Wang 2001).

Because of their unique analytical properties, DPV and SWV have been extensively used in bioanalysis. Some of these several applications are described below.

## 2     Applications in Bioanalysis with Electrochemical Sensors

Wang (2001) developed a disposable ITO-based analytical device consisted of an ITO modified with Au nanoparticles working electrode, which was applied in the non-enzymatic detection of extracellular $H_2O_2$ release from promyelocytic leukemia NB4 cells and in the evaluation of sodium selenite induced apoptosis. In this study, the authors employed DPV, reaching a limit of detection of 0.08 mM for hydrogen peroxide in a physiological environment. Here, DPV was selected as the electrochemical technique due to its improved sensitivity. The results are presented in Fig. 3 (Wang et al. 2016).

Ruiyi et al. (2016) proposed a sensing platform composed of nitrogen-doped multiple graphene aerogel/gold nanostar for ultrasensitive detection of circulating free DNA in human serum by DPV. The authors postulated that a large number of DNA molecules are released into peripheral blood in a tumor necrosis process, resulting in an increased cfDNA concentration in blood. The tumor cfDNA

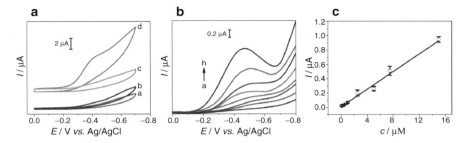

**Fig. 3** (**A**) Steady-state cyclic voltammograms of ITO electrode (curves a and b) and AuNPs/ITO (curves c and d) in 0.1 M PBS containing 0 (curves a and c) and 10 mM $H_2O_2$ (curves b and d), with a scan rate of 100 mV s$^{-1}$. (**B**) Differential pulse voltammetry curves for $H_2O_2$ in pH 7.4 PBS, at different concentrations (a–h): 0, 0.1, 0.5, 1.0, 2.5, 5.0, 7.5 and 15 μM. (**C**) Calibration curves for $H_2O_2$. From Ref. (Wang et al. 2016) with permission

containing mutations can be identified in the plasma of cancer patients. They used Au nanoparticles due to their electrocatalytic activity and surface immobilization in biological assays. Here, the DPV curves showed a characteristic current peak at approximately 0.28 V which were linear with dsDNA concentrations in the range of $1.0 \times 10^{-21}$ to $1.0 \times 10^{-16}$ g mL$^{-1}$ leading to an absurd limit of detection of $1.0 \times 10^{-22}$ g mL$^{-1}$ (Ruiyi et al. 2016).

Ashrafi et al. (2018) reported the electrodeposition of dopamine and the biocompatible polymer chitosan on the surface of a PDA modified glassy carbon electrode. On this modified surface, gold nanoparticles were deposited, generating a POLY (DA-CS)-AuNP organic–inorganic hybrid surface with excellent sensitivity for the detection of some benzodiazepines by DPV and SWV techniques. According to Ref. (Ashrafi et al. 2018), the DPV and SWV are effective and rapid electrochemical techniques with allow to obtain low limits of detection. They are employed to the determination of selected drugs in the standard human plasma sample, generating well-defined oxidation peaks whose peak current presented a good linear relationship with the selected drugs concentration.

Chen et al. (2019) stated that porphyrinic metal-organic frameworks (porph-MOFs) are being investigated as electrodes due to their redox activity mainly in the porphyrin subunit. So, they developed a core-shell hybrid material, PMeTh, containing the poly(3-methyophene) conducting polymer coated on an iron-based porph-MOFs PCN-222(Fe) which showed an excellent electrochemical response toward the levodopa detection. The authors claimed that the sensor exhibited a low limit of detection of 2 nmol L$^{-1}$, as well as excellent stability after 120 cycles in 10 μmol L$^{-1}$ levodopa solution. The sensor was tested in human urine samples by standard addition with DPV as the experimental technique.

Zhang et al. (2020) synthesized a PdNPs@ZnO-Co$_3$O$_4$-MWCNT nanocomposite to be used as a novel electrode material to construct a non-enzymatic electrochemical sensor for high sensitivity quantification of tanshinol, one kind of Chinese herbal medicine, widely used to prevent and treat hyperlipidemia, coronary disease, and cerebrovascular disease. The sensor, using the DPV technique, shows two well

**Fig. 4** Simultaneous determination of epinephrine (EA) and dopamine (DA) using the AgNP/SiO$_2$/GO/GCE electrode in PBS buffer, pH 7.0. The calibration curve of current (i) vs. EA and DA concentrations (inset figure). From Ref. (Cincotto et al. 2014), with permission

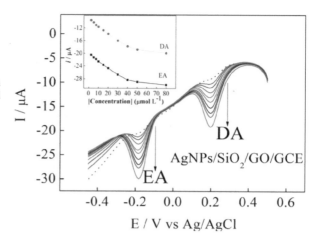

linear relationships between the peak current and the tanshinol concentration in the range of 0.002–0.69 mM and 0.69–3.75 mM with a limit of detection of 0.019 μM.

Cincotto et al. (2014) synthesized a SiO$_2$/graphene oxide decorated with Ag nanoparticles as a new hybrid material. A glassy carbon electrode modified with AgNP/SiO$_2$/GO was used in the development of a sensitive electrochemical sensor for the simultaneous determination of epinephrine (EA) and dopamine (DA) by electrocatalytic reduction, using SWV. Well-defined reduction peaks were observed in the PBS buffer at pH 7 (Fig. 4).

No significant interference was observed for mainly biological interferents, such as uric acid and ascorbic acid, in the detection of EA and DA. The study demonstrated that the electrode modified with the synthesized hybrid material was extremely sensitive for the simultaneous determination of dopamine and epinephrine, with the detection limits obtained of 0.26 and 0.27 μmol L$^{-1}$, respectively.

Cesarino et al. (2015) modified a glassy carbon electrode using reduced graphene oxide (rGO) with antimony nanoparticles (SbNPs) for the determination of estriol in natural water. The synergistic combination of rGO sheets and Sb nanoparticles resulted in a limit of detection of 0.5 nmol L$^{-1}$ (0.14 μg L$^{-1}$) for estriol hormone. A natural water sample was analyzed by the proposed device and high-performance liquid chromatography (HPLC), and the authors obtained a percentage of 98% in agreement between both techniques. The proposed device was successfully applied in the determination of hormone in a natural water sample and was a promising platform for a simple, rapid, direct, and ultrasensitive analysis of estriol.

## 3    Applications in Bioanalysis with Electrochemical Biosensors

Kong et al. (2011) developed a new immunosensor composed of nanocomposites of gold nanoparticles (GNPs) with DNA and methylene blue (MB) labeled with the secondary CEA antibody (Ab$_2$) for highly sensitive bioanalysis of carcinoembryonic

**Fig. 5** SWV recordings for the immunosensor in 0.01 mol $L^{-1}$ PBS solution with CEA concentrations from 0 to 2.0 pg $mL^{-1}$ under optimized experimental conditions. The inset is the respective analytical curve. Adapted from Ref. (Kong et al. 2011), with permission

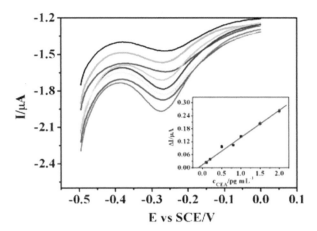

antigen (CEA) using SWV to oxidize the labeled MB to detect the CEA. The receptor-analyte complex was formed by the biorecognition of MB labeled $Ab_2$ with the primary CEA antibody ($Ab_1$) immobilized on the surface of the biosensor. The immunosensor showed a linear response in the range of 0.10–2.0 pg $mL^{-1}$ CEA concentrations with a limit of detection of 0.05 pg $mL^{-1}$. The SWV conditions used were 5–50 Hz for the square wave frequency, 0.01–0.05 V for the amplitude, 0.001–0.010 V for the scan increment and $-0.4$ to 1.0 V for initial scan potential. The results are presented in Fig. 5, together with the respective calibration curve.

Ding et al. (2018) proposed an electrochemical biosensor to quantify the p53 gene by the "in situ" deposition of polyaniline (PANI) catalyzed by G-quadruplex/hemin horseradish peroxidase-mimicking DNAzyme. For that, the target p53 DNA hybridizes with the capture DNA deposited on a gold electrode. Then the released sequences in capture DNA trigger the hybridization chain reaction (HCR) between two hairpin DNA probes, which catalyzes the oxidation of aniline to PANI with $H_2O_2$ leading to a measurable "turn-on" electrochemical signal. This signal was detected with differential pulse voltammetry with an exceptionally low detection limit (0.5 fM). This advanced electrochemical biosensor showed selective recognition capacity to discriminate base-mismatched sequences.

Chen et al. (2018) developed an autocatalytic strand displacement amplification (ASDA) strategy which was coupled with a hybridization chain reaction (HCR) event for label-free detection of nucleic acid. With the current cascade ASDA and HCR strategy, the quantification of target DNA was performed with a limit of detection as low as 0.16 fM and good selectivity. The developed biosensor used DPV to detect the electrochemical signal exhibiting the advantage of good flexibility and simplicity in probe design and biosensor fabrication and label-free electrochemical detection being adequate to be applied in bioanalysis and clinical biomedicine. The electrochemical detection of target DNA is presented in Fig. 6 (Kong et al. 2011):

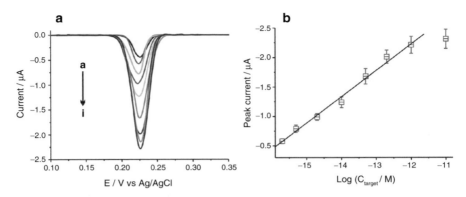

**Fig. 6** (**A**) DPV signals for different concentrations of target DNA by using the cascade ASDA and HCR strategy, where DNA concentrations are: (a) 0 M and (i) 10 pM. (**B**) The calibration curve for results of (A). Adapted from (Kong et al. 2011), with permission

Shoja et al. (2019) adopted a new approach for detection and assessment of Gemcitabine (GEM) as an anti-cancer drug using its interaction with EGFR exon 21-point mutant gene. An electrochemical biosensor was developed by a new molecularly bioimpressed siloxane polymer (MBIS) procedure, where the EGFR exon 21 acts as an identification probe. The biosensor also contained multi-walled carbon nanotubes and Ag nanoparticles to amplify the signal. The electrochemical technique used was DPV, which was able to detect the oxidation signal of adenine and guanine in linear range in the concentration window of GEM from 1.5 to 93 μM, with a limit of detection of 12.5 nM.

Cincotto et al. (2016) developed an electrochemical immunosensor for ethinylestradiol hormone (EE2) using diazonium salt grafting onto silver nanoparticles-silica–graphene oxide hybrids. With such a device, an advanced immunoassay method was proposed aiming at the quantification of the hormone, employing a peroxidase-labeled ethinylestradiol (HRP-EE2), which electrochemical response was acquired amperometrically at $-200$ mV, using hydroquinone (HQ) as redox mediator. The analytical curve for EE2 presented a linear behavior with hormone concentration in the range between 0.1 and 50 ng/mL, allowing to obtain a detection limit as low as 65 pg/mL, Fig. 7.

The authors had also provided an interference study, using some hormones with similar structures as EE2, which revealed the lack of sensitivity of this method toward the action of interfering species. Moreover, the long-term stability of the biosensor modified with anti-EE2 was demonstrated by maintaining it at least for 15 days stored at 4 °C under humid conditions. The developed electroanalytical method was then applied to the quantification of the hormone in spiked urine with excellent results.

Povedano et al. (2017) produced a new nanocomposite material constituted of reduced graphene oxide and rhodium nanoparticles, which were synthesized by a one-pot reaction method. For that, they used the simultaneous reduction of $RhCl_3$

**Fig. 7** Calibration plot for EE2 at the anti-EE2-Phe-AgNPs/SiO$_2$/GO/GCE immunosensor. Adapted from Ref. (Cincotto et al. 2016), with permission

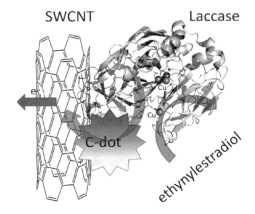

**Fig. 8** Pictorial view of the SWCNT/C-dot/Lac catalysis. Adapted from Ref. (Canevari et al. 2016), with permission

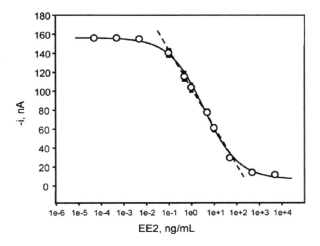

and graphene oxide by NaBH$_4$, allowing the in-situ deposition of the nanoparticles on top of the 2D carbon nanomaterial planar sheets. This nanocomposite was used as a nanostructured substrate for cross-linking the enzyme laccase with glutaraldehyde, generating a voltamperometric biosensor for 17β-estradiol hormone quantification in the 0.9–11.0 pM range. This biosensor presented excellent analytical activity, showing a high sensibility of 25.7 A $\mu$M$^{-1}$ cm$^{-1}$ and an exceptionally low detection limit of 0.54 pM with high selectivity. The device was applied to a rapid and low-cost determination of the hormone in spiked synthetic and real samples of human urine.

Canevari et al. (2016) developed an efficient electrochemical biosensor for ethynylestradiol based on the laccase enzyme supported on single-walled carbon nanotubes decorated with nanocrystalline carbon quantum dots (SWCNT/C-dots), as schematically shown in Fig. 8.

This modifying film was applied to a glassy carbon electrode surface and employed, using the differential pulse voltammetry technique, to quantify the enhanced electroanalytic response for 17a-ethynylestradiol. This enhanced response was associated, by the authors, with an increase in the active area of the biosensor surface, which they believed to be promoted by the C-dots. However, the authors stated that the best biosensor performance was achieved after the immobilization of laccase enzyme on CGE/SWCNT/C-dots, point that the quantum dots may improve the electron transport between the substrate, SWCNTs, and the copper ions inside the enzyme active sites. Such an advanced device was a very efficient bioelectrochemical sensor for 17 $\alpha$-ethynylestradiol endocrine disruptors, with a detection limit of 4.0 nmol $L^{-1}$ for real samples.

# References

Ashrafi H, Hasanzadeh M, Ansarin K, Ozkan SA, Jouyban A (2018) Int J Biol Macromol 120: 2466–2481

Canevari TC, Cincotto FH, Nakamura M, Machado SAS, Toma HE (2016) Anal Methods 8:7254–7259

Cesarino I, Cincotto FH, Machado SAS (2015) Sens Actuators B Chem 210:453–459

Chen Z, Liu Y, Xin C, Zhao J, Liu S (2018) Biosens Bioelectron 113:1–8

Chen Y, Sun X, Biswas S, Xie Y, Wang Y, Hu X (2019) Biosens Bioelectron 141:111470

Cincotto FH, Canevari TC, Campos AM, Landers R, Machado SAS (2014) Analyst 139:4634–4640

Cincotto FH, Martínez-García G, Yáñez-Sedeño P, Canevari TC, Machado SAS, Pingarrón JM (2016) Talanta 147:328–334

Ding L, Zhang L, Yang H, Liu H, Ge S, Yu J (2018) Sens Actuators B Chem 268:210–216

Kong F-Y, Zhu X, Xu M-T, Xu J-J, Chen H-Y (2011) Electrochim Acta 56:9386–9390

Povedano E, Cincotto FH, Parrado C, Díez P, Sánchez A, Canevari TC, Machado SAS, Pingarrón JM, Villalonga R (2017) Biosens Bioelectron 89:343–351

Ruiyi L, Ling L, Hongxia B, Zaijun L (2016) Biosens Bioelectron 79:457–466

Shoja Y, Kermanpur A, Karimzadeh F, Ghodsi J, Rafati AA, Adhami S (2019) Biosens Bioelectron 145:111611

Souza D, Machado SAS, Avaca LA (2003) Química Nova 26:81–89

Wang J (2001) Analytical chemistry, 2nd edn. Wiley-VCH, New York, p 69

Wang Q, Li W, Qian D, Li Y, Bao N, Gu H, Yu C (2016) Electrochim Acta 204:128–135

Zhang C, Ren J, Xing Y, Cui M, Li N, Liu P, Wen X, Li M (2020) Mater Sci Eng C 108:110214

# Impedimetric Immunosensors for Clinical Practices: Focus on Point-of-Care Diagnostics

Blanca A. G. Rodríguez, Paula A. B. Ferreira, and Rosa Fireman Dutra

## 1    Introduction

In recent years, attending the demand for rapid, practical, and low-cost tests for diagnostic of numerous diseases has been a major challenge for public health, since the conventional laboratory tests are operationally complex and require specialized laboratory and skilled professionals, appropriate physical and machine infrastructures, etc. On the other hand, the time taken to deliver the results of these tests is often long as compared to the emergency treatment, which usually results in treatment failure and unnecessary hospitalizations. The continuous and urgent monitoring of diseases that result in sepsis, heart or kidney failure, as well as other therapeutic procedures, could be enhanced, resulting in a strong, positive impact on health. Electrochemical Impedance Spectroscopy (EIS) stands out as a powerful technique for application in electroanalytical devices for immunosensors, especially due to the possibility of label-free detections. The possibility to produce miniaturized systems, in situ measurements, low relative cost, and impedimetric sensors has attracted attention, such as the possibility of point-of-care testing (POCT). POCTs allows increases the quickness of diagnostic response, favoring fast reference for medical assistance, decentralization of medical services, and short-time response at the beginning of treatment, among other benefits. Many efforts have been driven toward the development of quantitative POCTs that offer the exact concentration of relevant information or target analytes.

A biosensor is classically defined as a device that incorporates a biological recognition element (enzymes, antibodies, etc.) in close contact with a transducer element. This integration ensures a convenient conversion of biological events into a quantifiable electrical signal. When focused on immunoassay, they are called the immunosensors being based on the affinity with antigen–antibody interaction, i.e.,

B. A. G. Rodríguez · P. A. B. Ferreira · R. F. Dutra (✉)
Biomedical Engineering Laboratory, Federal University of Pernambuco, Recife, PE, Brazil
e-mail: rosa.dutra@ufpe.br

© The Author(s), under exclusive license to Springer Nature Switzerland AG 2022
L. T. Kubota et al. (eds.), *Tools and Trends in Bioanalytical Chemistry*,
https://doi.org/10.1007/978-3-030-82381-8_15

devices that involve the coupling of immunochemical reactions to appropriate transducers (Duffy and Moore 2017). Immunosensors have been applied in different areas, assisting in environmental and pollutants control, hormone detection in the food industry and veterinary, medicine, among others; but it is in healthcare that these biosensors have gained highest impact, due to its great effectiveness in quick and convenient diseases diagnosis and monitoring of chronic diseases, or in cases in which the treatment needs an effective intervention, when the route used is not suitable, as is the case with degenerative diseases such as cancer, AIDS, among others. The selectivity of immunosensors is mainly due to the antigen–antibody interactions. When antibodies are immobilized, their paratopes interact with the epitopes from specific antigens, ensuring the selectivity. Hydrophobic or electrostatic interactions, van der Waals force, and hydrogen bonding are established between them. The strength of a single antigen–antibody bond is called the antibody affinity (Kumagai and Tsumoto 2002). The antigen–antibody reaction is reversible, and owing to the relative weakness of the forces holding the antibody and antigen together, the complex formed would dissociate according to the environment (e.g., pH and ion strength). This same transduction principle can be used to detect interaction between DNA–RNA, DNA–DNA, aptamer–proteins, that is, between affinity bonds.

Different transduction systems are used for the electrical conversion measurements of the antigen–antibody interactions. It is well known that the bonds between these antigenic molecules occur without their modification or transformation, either generation of ionic species, unlike the enzyme–substrate reaction. Using oxidoreductase enzymes, such as glucose oxidase and peroxidase, the interaction results in the generation of ionic species that can be employed to promote the output current. Thus, given these limitations, transducers that allow direct detection are preferably chosen in comparison to those that require antigen–antibody tracers for the response, then a second antibody or antigen is conjugated to enzymes, fluorophores, chemiluminescent compounds, nanoparticles, etc. These tracers can indirectly monitor the analytical response, being termed labeled immunosensor. Many electrochemical transducers were initially developed using antibodies conjugated to the peroxidase enzyme for the generation of current response (Gomes-Filho et al. 2013). However, because they are more time consuming and involve more than one incubation step, which makes point-of-care monitoring difficult, these sandwich assays were being replaced by direct and more attractive assays and several label-free immunosensors were developed. The first successful immunosensors that were developed using label-free detection were those focused on optical transducing by Surface plasmon resonance—SPR (Jena et al. 2019) and the Localized Surface Plasmon Resonance (LSPR) (Adegoke et al. 2017; Camara et al. 2013). SPR immunosensors have reached the clinical routine in the market for laboratory analysis, but SPR technologies are not yet capable of portable detection, even though SPR immunosensors with piezoelectric transducers, based on quartz crystal microbalance, have been widely studied (Ertekin et al. 2016), they have not been successful for applications due to the constant need for calibrations and the high cost of quartz electrodes. For all this, electrochemical transduction is a promising, revolutionary technique for the production of electroanalytical immunosensors. Using electrochemical transductions, electrical detection of antigen–antibody interactions has been widely carried out by square wave voltammetry (SWV) (Silva et al. 2016), differential pulse voltammetry (DPV), and the electrical impedance spectroscopy (EIS) technique (Yao and Fu 2014).

EIS makes it possible to characterize antigen–antibody using label-free detections, since neither tracer nor enzymes are required to generate the analytical responses. SWV and DPV techniques are normally used for studies of faradaic currents, but it is also possible to also use them for detecting non-faradaic currents. Impedimetric immunosensors can use faradaic and non-faradaic currents (Faria et al. 2019). The knowledge of specific parameters of faradaic and non-faradaic currents is important for the in-depth understanding of the mechanisms and fundaments involving the sensing phenomena, in order to achieve better results and optimal conditions (Biesheuvel and Dykstra 2018). Considering that the demand for point-of-care immunosensors has recently been increasing, and taking into account that EIS is a feasible technique to develop label-free and portable immunosensors, it is reasonable that this technique will be widely studied and applied from the medical clinic in different electrode systems, including screen printed electrodes with three-electrode system, interdigitated electrodes, among others.

## 2 Basic Concept for Impedimetric Immunosensor

Immunosensors based on the EIS measurements enable the quantification of antigenic species from antibodies or antigens immobilized at the sensor's surface. They have the advantage of being able to offer responses in real time since the impedance changes in the sensor interfacial layer can be monitored. Impedance variations come from changes in the physical interfacial properties since that most of the biological molecules adsorbed by the electrochemical surface have an insulating nature, which promotes an increase in electrical impedance. Some biological molecules, however, may decrease electrical impedance, depending on their conformation and complexity, and also may generate faradaic currents on the electrode surface when modified, with compounds of a redox nature, for example. In the EIS measure, two parameters can be evaluated, the charge transfer resistance at the electrode/electrolyte interface and the interfacial capacitance. Thus, EIS can determine both the resistive and the capacitive (dielectric) properties of materials on the electrode surface.

A constant direct potential (DC voltage) is superimposed by an AC voltage that excites the interface potential also feeds the perturbation of the steady state on the surface, driving the charge transfer to the electrode-electrolyte and vice versa. Thus, when the system is perturbed by an AC voltage, depending on the electrical impedance, it is possible to observe changes in the phase constant and the amplitude module of the output AC voltage (Zhang et al. 2018a; Prodromidis 2010). This process is constantly repeated over a range of frequencies, thus yielding the impedance spectra at each frequency. Both the phase and the peak-to-peak voltage vary in function of the applied voltage sinusoidal, according to the material's impedance;

**Fig. 1** Schematic representation of AC voltage perturbation on the impedimetric immunosensor. Displacement in phase angle and current amplitude AC by perturbing on the electrode surface. AC current changes by antigen–antibody interactions

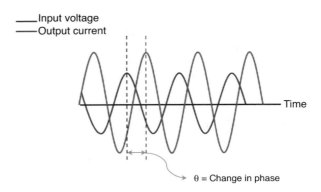

then the ratio $V(t)/I(t)$ is defined as the Impedance ($Z$) and can be mathematically expressed as:

$$Z(\omega) = \frac{Vosin(\omega)}{Iosin(\omega + \theta)}$$

where, $Vo$ is the maximum voltage and $Io$ is the maximum current signals, "$\omega$" is the excitation frequency in rad. $s^{-1}$ ($\omega = 2\pi f$, $f$ is frequency in Hz), "$t$" is time, "$\theta$" is the phase shift between the voltage–time and current–time functions and "$Z$" is the complex conductance that is inversely proportional to the obtained impedance.

Electrochemical Impedance is a complex value affected by modifications on the electrode interface, like the antigen–antibody coupling, which can be described either by changes on the $Z$ magnitude ($|Z|$) and the phase angle ($\theta$) or alternatively as a combination of real ($Zre$) and imaginary ($Zim$) parts of the impedance. When the applied AC voltage and AC current are in phase ($\theta = 0$), the system behaves like a pure ohmic resistor, and in this case, there is no modifying on the surface. Simple perturbation on the electrode surface may cause changes in phase angle and $|Z|$ (Grieshaber et al. 2008; Kokkinos et al. 2016). Fig. 1 represents changes on both the phase angle and the $Z$ magnitude when the antibody or recognizing element interacts with its respective antigens or targets. It is important to point out that impedance measurements are obtained on each swept frequency, and in the respective frequency it is possible to observe changes on the phase angle and magnitude of the impedance through modifications on the sensor layer by antigen–antibody interactions, this change being proportional to the concentration or number of species linked to the sensor surface. In general, the disturbances of the systems must be performed by AC signals of low peak-to-peak amplitude (generally <50 mV) so that the observed disturbance is nondestructive in nature and the system can still be operated microscopically in a linear range. On the other hand, the frequency spectrum modulates the phenomena of capacitance and these changes in phase angle are, therefore, better observed at low frequencies, instead high frequencies that favor the observation of phenomena related to electrical resistance. The impedance data in a complex plane is displayed according to frequency, in which the time constant-dependent due to the relaxation phenomena allows the separation of each contribution: (1) high-frequency region is due to electrolyte resistance, (2) intermediate frequency due to the interface electrode/electrolyte, and (3) low-frequency region due to material–electrode interface. The most popular formats for evaluating EIS data are the Nyquist plots (Zim vs. Zre) and Bode [log|Z|, $\theta$ vs. log($f$)] (Chu et al. 2017). Other designations to Zim and Zre can be $Z''$ and $Z'$ (Grieshaber et al. 2008; Lisdat and Schäfer 2008).

## 3   Faradaic and Non-faradaic EIS

In an electrochemical system, the disturbance by an AC voltage leads to electron transfers through the electrolyte/electrode, as well result in a transfer of charges of species that come from oxide/redox reactions occurred in the electrolyte to electrode or on the electrode. In the second case, for a charge transfer it is necessary to introduce electroactive species, which is done commonly through redox probes, being widely used the [Fe(CN)$_6^{3-}$/$^{4-}$] ferri/ferrocyanide as redox probe (Kokkinos et al. 2016; Lindholm-Sethson et al. 2010). However, the redox phenomenon is not only caused by the redox probes, but also due to charge transfers that can also occur along some conducting polymers by doping phenomena (West et al. 1993) or other tracers for conducting charges (Cecchetto et al. 2015). Thus, it is possible to perform measurements using EIS faradaic and non-faradaic methods, when redox probes or others are used together with the support electrolyte, respectively.

In the faradaic EIS is possible a configuration using two- or three-electrodes immersing in an electrochemical cell. However, it is recommended the electrochemical configuration with three-electrodes due to the fact that in a two-electrode system there are potential changes at the electrode–electrolyte interface, caused by the leakage current leading to instability in the electrochemical system. The third electrode as the counter electrode is essential to avoid current variations and to stabilize the system (Ianeselli et al. 2014). In a faradaic EIS using two electrodes (working and counter electrodes), the bias potential at the working electrode is fixed according to the open circuit potential and electrical contact is done between the working electrode and the solution. Thus, it is more difficult to obtain a stable potential, since the ionic strength of the solution, material of electrode, and the working electrode–solution interface can be variable, affecting the calibrations during the measurements. Thus, AC potential must be fixed between the working electrode and the reference electrode, and it must be kept at a constant distance from the working electrode so that the calibrations are stable. The working electrode should also be kept as close as possible to the auxiliary electrode. Then, working, reference and auxiliary electrodes should be immersed in an electrochemical cell in the presence of a redox probe, as for instance, the hexacyanoferrate (II)/(III), and AC potential should be applied to the system, acting in a DC potential that promotes the development of electrochemical processes by disturbance AC (Fig. 2a). The current that is generated by the diffusion of charge comes from the redox probe, obeying Ohm's Law. The diffusion of species to the electrode surface depends on changes on dielectric constant that varies according to antibodies or antigens on the sensor

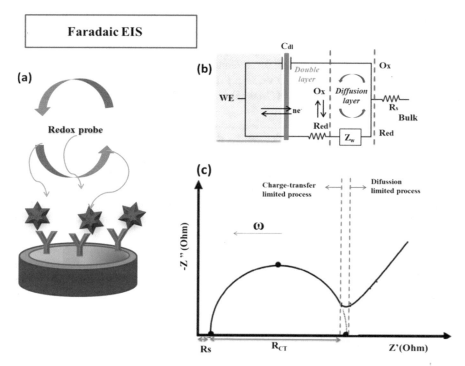

**Fig. 2** Schematic representation of faradaic EIS with a redox probe (**a**) Equivalent electric circuit (**b**) Nyquist plot to estimate the Rct (**c**)

surface, thus it will be possible to report the analytical response by the amplitude of the current proportional to the number of antibodies or antigens bound to the electrode surface. Electrical parameters of charger transfer resistance (Rct) and double-layer capacitance (Cdl) and resistance of solution (Rs) and Warburg impedance (W) participate in the reaction (Mundinamani and Rabinal 2014; Randles 1947).

Equivalent circuits are used in order to approximate the experimental impedance data with these ideal, and the elements are arranged in series and/or in parallel; the Randles circuit is the most used equivalent circuit (Heinze et al. 1980; Silva et al. 2010). In the presence of a redox probe in the measuring solution, the term charge-transfer resistance (Rct) is commonly used instead of resistance of polarization, while an extra component, the Warburg impedance (Zw) is also included (Fig. 2b). A Nyquist plot is exhibited for a faradaic EIS, in which, at low frequency, the diffusion of charge transfer is limited and it is possible to estimate the Rct (Fig. 2c).

In a non-faradaic EIS, there is no presence of a redox probe, and the contribution of the faradaic current has to be minimal, so the presence of redox species at the electrode/electrolyte interface has to be avoided; pinholes, and the use of conductive polymers combined with nanomaterials that can derive catalytic species, should both also be avoided. This may explain why these systems work better at lower

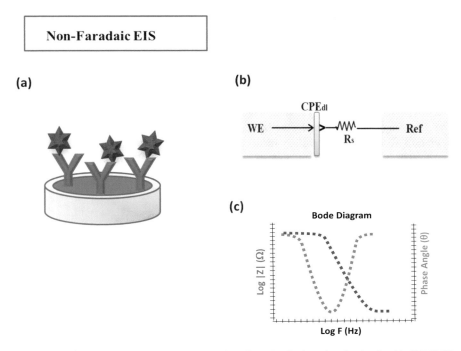

**Fig. 3** Schematic representation of Non-faradaic EIS (**a**) Equivalent electric circuit with CPEdl (**b**) Bode diagram to estimate the phase angle and |Z| (**c**)

frequencies and with AC voltage pulses of low peak-to-peak amplitudes (<10 mV). In non-faradaic EIS, two-electrodes are sufficient, and the system behaves like a biochemical capacitor and is described by the equation:

$$C = \frac{\varepsilon \cdot \varepsilon o \cdot A}{d}$$

where "$\varepsilon$" is the dielectric constant of the medium between the plates, "$\varepsilon_o$" is the permittivity of free space, "$A$" is the surface area of the plates, and "$d$" is the thickness of the insulating layer.

When an AC potential is applied between the two electrodes immersed in an electrochemical cell, the two capacitor plates are ideally formed, one represented by the electrode surface (work electrode) and the other by the electrolyte (bulk). Thus, the double layer (dl) will be formed on the electrode/electrolyte interface, and the so-called dielectric capacitance will be used for measurements of antigen–antibody interactions on the working electrode interface (Fig. 3a). Thus, when an AC potential is applied between the two electrodes immersed in an electrochemical cell, the AC current response is lagged and the magnitude of the phase shift is measured. The variation of the phase angle will be proportional to the concentration of target (antigen or antibody) that interacts with its respective immobilized receptor, and the capacitance is due mainly to the changes in the double layer since the proteins have an insulating nature (Silva et al. 2010) and contribute increasing the ε. The equivalent circuit is simplified, in comparison to the faradaic EIS, consisting of a serial combination of the solution resistance and electrode/solution and double-layer capacitances of antibody, antigen and chemical surface modification (Kokkinos et al. 2016). However, it is important to note that the impedance of the solid electrode/electrolyte interface usually differs from purely capacitive behavior. Surface effects like the roughness of metal electrodes, surface inhomogeneity, etc. can provoke deviation from ideal capacitive behavior. For experimental spectra fitting of the interfacial capacitance the Constant Phase Element (CPEdl) is better adjusted (Brosel-Oliu et al. 2019) Fig. 3b. The Bode diagram representing the phase angle and |Z| is shown in Fig. 3c.

## 4    Electrode Designs

The most used reference electrode is the silver–silver chloride electrode due to its stable hysteresis properties in the adopted frequency range (Ianeselli et al. 2014; Tian et al. 2020). The working electrodes can be of various materials, gold, platinum, glassy carbon, carbon paint, and other conductive materials (Heard and Lennox 2020). The use of three planar electrodes screening-printed on solid substrates such as alumina, plastics, and other materials has been a current trend to miniaturize and to enable the development of point-of-care sensors. The material of a counter electrode must be electrochemically inert, since chemical adsorption or desorption of ion species during the measurements can cause additional currents masking the results. Additionally, it is important that the counter electrode has a physical area much larger than the working electrode. This is due to the capacitance increases with the size of the area, so considering that the working and auxiliary electrodes are positioned electrically in series, the resulting current AC component from the auxiliary electrode will be minimal or negligible, as desirable (Ianeselli et al. 2014; Zhang et al. 2014). In general, auxiliary electrodes are made of platinum, gold, or other carbon materials like carbon inks, composites with graphene carbon nanotube, among others, guaranteeing the electrochemical stability on the responses.

In the search for point-of-care sensors, where the detection of low concentrations of antigens or antibodies in biological samples is desirable, some characteristics must be taken into account, among which we can highlight the ability to miniaturize, possibility for mass production, easy scale-up and low cost, besides reliable and reproducible measurements. In the faradaic EIS, label immunosensors developed using screen-printed electrodes (SPE) seem to be quite reasonable, it is one of the reasons that SPE have been extensively explored in the last decades (Asav and Sezgintürk 2014; Mihailescu et al. 2015). Non-faradaic EIS has demonstrated that the use of capacitive and interdigitated electrodes (IDEs) can also be manufactured using low-cost and scale-up technologies like the screen-printed electrode (Khan et al. 2016). Thus, choice of the most suitable sensing method—faradaic or non-faradaic methods—is no easy task and depends on electrode configuration, material of the sensor platform, desirable sensitivity, and features of bioreceptor immobilizations and so on. Despite the fact that non-faradaic presents the advantages of absence of redox couple, the faradaic sensors have been more explored, probably attributed to easier conditions for more sensitivity achieved by resistance of charge transfer (Brosel-Oliu et al. 2019). Nevertheless, this choice is not so easy, especially if the production of point-of-care immunosensors is aimed. Interdigitated and screen printed electrodes have both shown act similarly as non-faradaic and faradaic mode respectively with relative advantages.

The use of SPEs in impedimetric systems is well established in the reports of immunosensors, reinforcing that, although cheaper, the use does not impact the sensitivity of the systems. Khan et al. related the construction of an immunosensor for myoglobin, reaching the linear detection range of 0.1–90 ng/mL, with sensibility of 0.74 k$\Omega$ ng/mL and LOD 0.08 ng/mL (Khan et al. 2016).

In addition to surface changes, to develop more sensitive impedimetric systems, it is also necessary to be attentive to the electrode used. Changes in the material and architecture of the electrodes may influence the levels of sensitivity achieved (Prodromidis 2010). Thus, interdigitated electrodes (IDEs) were developed to favor increased sensitivity by modifying the electrode architecture to promote increased sensitivity with use of the impedance technique. IDEs are microelectrodes deposited on a substrate by photolithography, arranged in two typically symmetrical combs, of the same size and material, similar to interlaced fingers separated by a small distance, in the order of micrometers, which corroborates the increase in sensitivity in electrochemical techniques, and particularly electrical impedance spectroscopy (Dizon and Orazem 2019). This production method allows the electrodes to be manufactured by depositing thin films, with competitive values for the purpose of disposal such as SPEs. This electrode can be used without the need for a reference electrode; the current response is measured between the combs as a function of the applied potential. The geometry of this sensor optimizes the penetration of the field applied over the system, making the current flow close to the sensor surface, as the penetration is approximately the sum of the width of the digits plus the distance between them. Sum of the impedances of each comb is the impedance of the electrode (Brosel-Oliu et al. 2019). Works such as Le et al. reported that the application of IDEs allows the production of very sensitive immunosensors, being able to reach detection limits in the order of pg/mL (Ngoc Le et al. 2019). In this work, a faradic system was developed based on the increase of the Rct for detection of peptide a$\beta$ 1–42, a biomarker for Alzheimer's, with low detection limits for both sera (100 pg/mL) as LCR (~500 pg/mL). Other works such as Ding et al. reached lower limits (0.24 pg/mL) and linear detection ranges smaller than the immunosorbent assays (ELISA) for the same protein (Ding et al. 2018).

## 5     EIS Immunosensors Based on Nanomaterials

In the last few decades, the best understanding and development of processes involving nanotechnology have resulted in new strategies for the introduction of nanomaterials on the sensing surfaces, causing a great impact on the sensitivity of the immunosensors, being mainly attributed to a more abundant amount of biomolecules immobilized on the sensing surface and the ease with which these they may be non-randomly and expose their binding sites or antigenic regions to their target analytes. Among the most used nanomaterials are carbon nanotubes, graphene, and nanoparticles. Physical–chemical processes allow these nanomaterials to be easily functionalized, in order to introduce reactive groups that enable stable protein coupling, including covalent bonds (Suresh et al. 2018). Due to the new properties acquired that are not present when in bulk, nanomaterials have promoted gains in sensitivity in their different presentations and in the different processes of electrical transduction, thus the particular features exhibited by each nanomaterial must be explored in line with the type of transduction (Dolatabadi and de la Guardia 2014; Jiang et al. 2013).

Carbon nanostructures (CN) have unique properties, due to the arrangement of atoms that form a hexagonal structure with $sp^2$ bonding, which facilitate the charge transfer and increases electrical conductivity. This is ideal for impedimetric immunosensors, biosensors, and other biomedical applications. The study of these nanomaterials began in the mid-1980s with the discovery of fullerenes by H. Kroto et al. (1985). The fullerenes are zero-dimension molecules and their structure is a spheroid, the most popular of which is C60 fullerene (Fig. 4a). A carbon molecule with 7A of diameter detected in stellar dust through mass spectrum. The C60 was quickly placed as one of the nanomaterials for optical applications, bone destruction treatment, delivery of drugs, lubricants, among other things (Holzinger et al. 2014). Nevertheless, due to its difficult synthesis its applications have been limited.

Years later, in 1991, Ijima discovered the carbon nanotube (CNT), a one-dimensional tubular structure (1D) (Fig. 4b) that stood out for having two different shapes: single-walled nanotube (SWCNT) and multiple-walled nanotube (MWCNT). The CNT has various applications in the mechanical, biomedical, optical, environmental, and chemical and many other areas (Eatemadi et al. 2014). However, the control of size, shape, and number of layers has been a challenge that has not yet been overcome. In 2004 Novoselov et al. manufactured a carbon sheet for the first time, this is a two-dimensional structure (2D) known as graphene (Geim and Novoselov 2007) (Fig. 4c), a unique material that, for a long time, was thought that it was thermodynamically impossible to produce. It has been the most outstanding for its great versatility and possesses unmatched properties, they are easy to functionalize allowing antibody or antigen bindings, and amplifying the electrochemical signals by increasing the charge transfer rate in Faradaic or non-Faradaic EIS (Yáñez-Sedeño et al. 2016). Proprieties of carbon nanomaterials, in general, lead to an increase of immunosensor sensitivity by characteristics shown in Fig. 4.

Carbon nanotubes (CNTs) can be categorized according to the number of walls single (SWCNTs) and in multiple wall (MWCNTs). The SWCNT diameter change

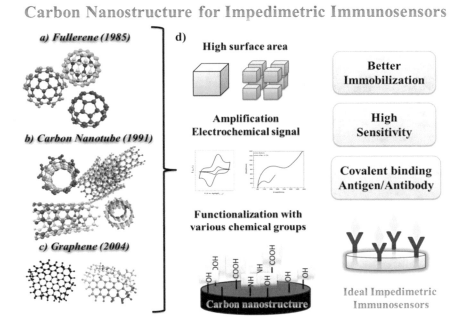

**Fig. 4** Carbon nanostructures (**a**) C60 fullerene, (**b**) MWCNT and SWCNT, (**c**) Graphene (**d**) Some of the most used properties for impedimetric immunosensors

approximately from 0.5 to 4 nm, while the MWCNT from 10 to 100 nm. CNTs have lengths from several hundred microns up to centimeters (100µm to 550 mm) (Zhang et al. 2013). The CNTs depend on their chirality, being classified as semiconductor or metallic. The metallic SWCNTs are more attractive for impedimetric immunosensors, for they produce an increase in the charge transfer allowing an increase in sensitivity. An impedimetric immunosensor for rapid detection of cancer biomarker (HER-3) using a SWCNTs in the working electrode was reported, improving sensitivity and linear range with a limit of detection (LOD) of 2 fg/mL HER-3 (Asav and Sezgintürk 2014). Nevertheless, SWCNTs are not easy to deposit in the working electrode due to strong electrostatic attraction between them, which makes it difficult for their dispersing in solvents and deposition by drop casting. The modification through functional groups is fundamental because this can provide great alternatives for covalent immobilization of a great number of biomolecules promoting an amplification of the signal in EIS. The LOD was lower, achieving 0.1 pg/mL and in comparison, to the one without CNT (LOD = 0.1 ng/mL) (Bourigua et al. 2010).

SWCNT-COOH polymeric composite formed by poly-L-lysine that has abundant amino groups, which is used for covalent binding of $\alpha$-fetoprotein antibodies, an important biomarker for the early diagnosis of cancer of liver. It has reached an LOD of 110 pg/mL measured by Rct in human blood plasma. This result shows that the

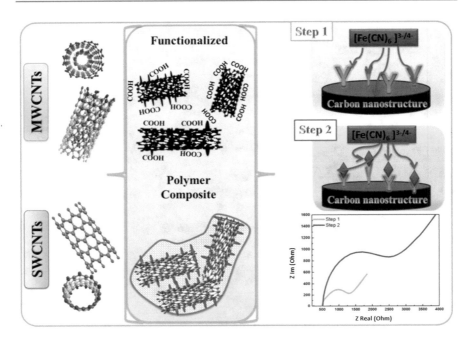

**Fig. 5** Scheme of CNT-based EIS immunosensors (**a**) SWCNT and MWCNT (**b**) Functional groups or polymer composite (**c**) Faradaic and Non-faradaic EIS approaches

impedimetric immunosensor has accuracy comparable to the results of the kit ELISA (Wang et al. 2016). SWCNTs are difficult to synthesize and the MWCNTs have been more widely used for many applications (Jacobs et al. 2010). As example, MWCNT-COOH previously modified with amino groups formed covalent binding with IgG with significant changes in faradaic EIS measurements. A LOD of 1.3 pg/mL was of lower value than the ELISA kit. A lower LOD is important for early detection of TGF-β1 (Sánchez-Tirado et al. 2016).

The use of polypyrrole conducting polymer associated with CNT has been studied as electrochemical supercapacitor (Valentini et al. 2013). Palomar et al. reported polypyrrole @ MWCNT-COOH in an impedimetric immunosensor for detection of the Dengue Virus NS1. This showed that the target (NS1 antibody) increases the capacitance and Rct (Palomar et al. 2018). Using a similar configuration, an ultra-sensitive impedimetric immunosensor for Cystatin C, a renal biomarker, detection had a LOD of 28 ng/mL appreciable for the clinical analysis of renal diseases in a non-faradaic EIS (Ferreira et al. 2020).

A schematic diagram using SWCNT and MWCNT in different approaches and the possibility of EIS immunosensor as faradaic and non-faradaic is showed in Fig. 5.

Graphene is a single layer of carbon atoms, which forms a two-dimensional hexagonal lattice array. It also has high electronic mobility, zero-gap semiconductor, and has a large surface area (of approx. 2630 $m^2$/g), among other properties

(Bonaccorso et al. 2015). The graphene synthesis is carried out by two processes from bottom-up or top-down. The first process consists of epitaxial growth through the chemical vapor deposition method. The method creates a pristine graphene with high quality and controlled number of layers (Loo et al. 2014). However, it is an expensive method as compared to the top-down process, which is cheaper since it is easier to be mass-produced. Top-down is a cost-effective process achieved by exfoliating the graphite oxide until obtaining a few layers: mechanical, salt-intercalation, chemical exfoliation, and more (Delle et al. 2015). The chemical exfoliation is more commonly used, although the graphene so obtained has many defects on lattice that hinder the charger transfer. Graphene oxide has a high density of oxygen-containing functional groups and is not very conductive due to disrupted conjugated $\pi$–$\pi$ bonds (Hammond et al. 2016). Despite this defect, they are important for obtaining functional groups, like the carboxyl, hydroxyl, carbonyl, and epoxy that facilitate the covalent immobilization of biomolecules (Sassolas et al. 2012; Baniukevic et al. 2012).

The reduced graphene oxide was employed in a faradaic EIS for the detection of histamine, obtaining concentrations of 0.1–1 mM by EIS measurements at low frequencies, and the results were compared with SPR assay, and both a good sensitivity and a linear concentration-dependent signal were achieved (Delle et al. 2015). Another EIS immunosensor useful for the detection of typhoid disease was developed using GO (Mutreja et al. 2016) and it was raised to a limit of detection of $1.07 \times 10^1$ CFU/mL of *Salmonella typhimurium* in water. Although better performance and a lower LOD were achieved using the rGO by an increase of electron charged transfer in the faradaic current, the use of GO has the advantage of easier immobilization by carboxyl groups present on the surface layer of GO sheets.

The rGO and GO were also investigated in the presence of conducting polymer by Gu et al. using the pyrrole-co-pyrrole-1-propionic acid and rGO (rGO-PPyPPA). They formed a thin film on the electrode, increased the charge transfer rate on the electrode due to the $\pi$–$\pi$ bonds on the rGO's lattice, and bonded with the aromatic ring of pyrrole. An important improvement in sensitivity was observed, probably due to the strong interaction of antibody with the thin film, which demonstrated a low LOD of 12 pmol/L that shows a high-sensitivity immunosensor platform for melamine additive detection (Gu et al. 2019).

Recently, a new nanostructure that has been used for the development of immunosensors is the graphene quantum dots (GQDs). This is the successor of graphene; it is a carbon nanostructure of zero dimensionality (0D) and it is a graphene fragment 2–30 nm in size. GQDs have properties of graphene and carbon dots, such as high surface area, good biocompatibility, quantum confinement, edge effects, and they can facilitate the bond of various functional groups (Mansuriya and Altintas 2019). It also can be grafted with several nanomaterials through the $\pi$–$\pi$ conjugate lattice, due to the structural defects of the heteroatoms of the edges of hexagonal carbon lattice (Mansuriya and Altintas 2019; Ganganboina and Doong 2019). The GQDs can enhance the LOD of targeted biomolecules due to their properties; many of these GQD nanomaterials have been used for impedimetric sensors as reported by Tuteja et al. achieving a sensitive impedimetric

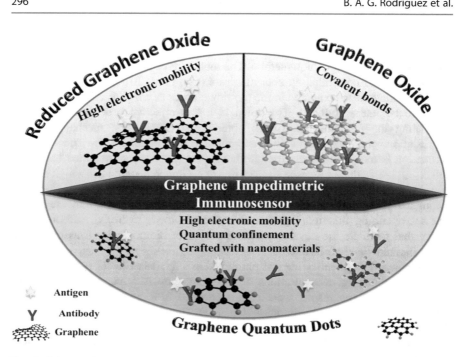

**Fig. 6** Scheme of graphene nanostructures and their properties for impedimetric immunosensor applications

immunosensor using GQDs for the detection of cardiac biomarker myoglobin (cMyo), observed the values of charge transfer resistance (Rct), which are monitored as a function of varying antigen concentration. These values of Rct were modeled and fitted to the impedance data in the Randles equivalent circuit, which display a LOD of 0.01 ng/mL, showing an improvement compared to the ELISA tests (Tuteja et al. 2016). On the other hand, also Ganganboina and Doong, demonstrate a new and efficient label-free impedimetric immunosensor based on GQDs@Au-polyaniline nanocomposite for detection of carcinoembryonic antigen (CEA). Rct increased proportionally upon increasing CEA concentration with LOD of 0.01 ng/mL (Ganganboina and Doong 2019).

Graphene-based carbon nanostructures, when compared to CNT, have the advantage of having a higher surface area/volume, which can be modified to reduce the amount of oxygen and increase $sp^2$ bonds, improving electronic mobility. GO has the advantage of providing functional groups for easy covalent immobilization of antibodies, serving both for faradaic and non-faradaic EIS. The combination of these nanostructures with conductive polymers increases the synergistic effects, also the electrochemical capacitance being attractive to develop electrochemical capacitors. Since GQD is the smallest of the nanostructures, they have a larger surface area by volume and are more interesting for the development of independent EIS immunosensors that can be easily reduced (Fig. 6).

## 6       Immunosensor Nanoparticles

The greatest use and performance of nanoparticles (NP) has been well established in recent decades. NP applications can serve as a strategy for label reagent molecules, immobilization of biomolecules, catalysis of chemical reactions, separation biomolecules, and optimization of electron transfer (Fig. 7). These properties can be controlled through their size and forms, which can be configured by adjusting the method chosen for the synthesis (Rodriguez et al. 2015). In general, synthesis processes are physical, chemical, and biological; the most used is frequently the chemical due to simplicity and easiness to perform since the size and shape of the nanoparticle is more controllable. In addition, the NPs can be classified by their dimensionality, (a) two-dimensional composed of nanoplates, such as stars, pentagons, squares, rectangles, or others. (b) One-dimensional composed of nanorods, nanowires, nanotubes, nanobelts, and nanospheroids (Elahi et al. 2018).

The gold nanoparticles (AuNPs) are the most explored and have been largely used in several fields, as electrochemical biosensors. These have an absorption in the visible region and the near-infrared region (NIR). This property has been used for label in biosensors; however, absorption depends on the shape and size of the AuNPs.

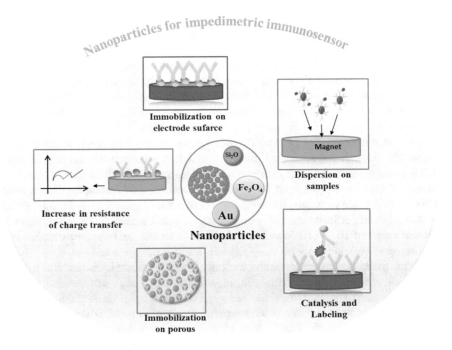

**Fig. 7** Nanoparticle for EIS immunosensors: advantages and characteristics

AuNPs can be functionalized to facilitate the covalent immobilization of biomolecules for immunosensor applications. AuNP was functionalized with thiol for an impedimetric immunosensor. This immunosensor showed significant changes in the Rct, due to the interference of biomolecule with the redox probe. However, capacitive (CdL) changes were not possible to detect, because electrode modifications were not sensitive to this change. This was caused by the functional groups that hinder the transfer of charges, causing the capacitance to decrease. Nevertheless, Rct was an adequate signal to detect the interfacial properties of the immunosensor, achieving a limited detection (LOD) of 5 ng/mL. This value is lower than that of many amperometric immunosensors (Liu et al. 2011).

On the other hand, the use of gold nanoparticles in immunosensors is also supported due to their good biocompatibility and easy of functionalization by covalent bonds with cysteine and amine groups. These characteristics allow to label and bind antibodies in a stable manner, as well to functionalize nanoparticles with other organic molecules. Labeled by AuNPs has favored an increase in sensitivity of sensors several studies such as that described by Lima et al. that developed an immunosensor for human chorionic gonadotrophin (hCG) using AuNP-secondary antibody for the formation of the immunocomplex to be revealed by the electrodeposition of gold (Lim et al. 2014). In this work, the use of antibodies labeled with AuNPs allowed a LOD of 5 pg/mL.

However, in the last few decades, there is an increasing trend toward the development of label-free immunosensors whose response is based on the direct interaction of the immobilized antibody with the antigen. This reaction model has shown good synergy with its application to impedimetric systems because it allows the measurement of field changes resulting from this interaction.

In this configuration, the optimization of the immobilization of antibodies process is of paramount importance, as it differently affects the availability of ligands for the important analyte so that lower detection limits are reached, and also requires stable connections between the biomolecule and the surface (Filik and Avan 2019). Interaction characteristics of AuNPs with ligands may allow them to be used as immobilization on the surface. Lima et al. used an oxidized carbon surface with AuNPs functionalized with arginine to immobilize anti-DHEAS (dehydroepiandrosterone sulfate) for diagnosis of the pediatric adrenocortical carcinoma (Lima et al. 2019). This strategy enabled the construction of a the immunosensor sensitive with good analytical performance and provided stability and precision. The study showed a linear range of 10.0–110.0µg/dL and LOD de 7.4µg/dL and the performance was confirmed in comparison with conventional tests for the same analyte.

Metal oxide nanoparticles are commonly used in immunosensors due to their properties: optical, magnetic, catalytic, chemical stability, high surface area, and fast electron transfer. These are ideal for the use of transducer platforms and biomolecule binders. For example, the iron oxide magnetic nanoparticles (MNPs) can be used for immobilization, separation, and purification of molecules. Nourani et al. proposed a system that uses MNPs for capture antibody against hepatitis B virus antigens (HBsAb) that were dispersed in sample (Nourani et al. 2013). After conjugation with the antigen, a magnet was activated on the sensing surface attracting the MNPs and allowing washing procedures and measurement of the generated current to be carried out.

The $Fe_3O_4$ NPs are used in immunosensors, due to the increase in mass transfer. For these reasons, they are used in sensors that enable to reach the detection limits and high signal-to-noise ratio, improving the sensitivity of the system. When using $Fe_3O_4$ Huang et al. observed an increase in the semicircle generated in the Nyquist graph and consequently charge transfer resistance (Rct), indicating that the nanoparticles obstruct the transfer occurred in the probe (Huang et al. 2010).

On the other hand, nanoparticles of silica ($SiO_2$) can also be used for this purpose. $SiO_2$ NPs accumulate characteristics such as thermal stability, controllable morphology and size, easy staining, biocompatibility, and large surface area; being considered a mesoporous material (Bagheri et al. 2020). Due to the high porosity, immobilization of biomolecules is an easy step. Zhang et al. used this porous structure covered with poly-brushes' carbonyl groups were modified with HRP enzyme. This complex was used with a label on their sensor for tetrabromobisphenol A and confirmed by EIS. The number of immobilized proteins blocked the electron transport on the electrode surface, increasing the diameter of the semicircle and the Rct (Zhang et al. 2018b).

Recent studies have evaluated the use of hybrid nanoparticles that combine materials in order to generate synergistic effects. A combination of AuNPs with $SiO_2$ microspheres reveal that the combination favors catalytic activities of AuNPs and inhibits the leak pore loads of $SiO_2$ (Bagheri et al. 2020). The studies show that the use of $Fe_3O_4$ along with AuNPs promotes an easier immobilization and an amplification of the signal (Butmee et al. 2020).

# 7 Conclusions and Outlook

Briefly, Table 1 summarizes some of the most important nanomaterials used in the construction of EIS-immunosensors. Important features of these immunosensors are the choice of nanomaterials, linear range, and limit of detection. It is concluded that there are still great challenges for the understanding of impedimetric electrochemical processes on different approaches.

Continued studies with new nanomaterials, such as GQD, fullerenes, metallic and oxide nanoparticles, or in combination with conducting polymers show that they offer improvements in sensitivity and in clinical range, so there is the possibility for them to be used in point-of-care testing, which is of utmost interest. Miniaturized SPE and IDE can be candidates for a new era of more reliable and reproductive immunosensors. The knowledge of faradaic and non-faradaic phenomena, taken from observation of electrical models and simulation with curve fittings, is very important for understanding the sensor components and the electrical vision. The charges transfer associated with the presence of new nanocomposites based on conductive polymers such as polypyrrole, polyaniline, and thiophene and their doping mechanisms may result in new strategies for approaches comparable to conventional laboratory analytical methods such as enzyme immunoassay

**Table 1** Nanomaterials for impedimetric immunosensors

| Nanomaterial | Detection | Application of nanomaterial | Linear range | LOD | Ref. |
|---|---|---|---|---|---|
| SWCNTs | HER-3 | Improved sensitivity | 2–14 fg/mL | 2 fg/mL | (Asav and Sezgintürk 2014) |
| SWCNT-COOH | Deep venous thrombosis | EIS signal amplification and improved sensitivity | 0.1 pg/mL to 2 μm/mL | 01 pg/mL | (Bourigua et al. 2010) |
| SWCNT/COOH_ Poly-L-lysine (nanocomposite) | α-Fetoprotein (anti-AFP) | Antibodies immobilization and improved sensitivity | 0.05–50 ng/mL | 110 pg/mL | (Wang et al. 2016) |
| MWCNT-COOH | Transforming growth factor β1 | Antibodies immobilization and improved sensitivity | 5 and 200 pg mL$^{-1}$ | 13 pg/mL | (Sánchez-Tirado et al. 2016) |
| MWCNT-COOH/polypyrrole (nanocomposite) | Dengue virus NS1 | EIS signal amplification, and improved sensitivity | 10$^{-12}$ to 10$^{-5}$ g/mL | 10$^{-12}$ g/mL | (Palomar et al. 2018) |
| MWCNT-COOH/polypyrrole (nanocomposite) | Anti-cystatin C | EIS signal amplification | 0–300 ng/mL | 28 ng/mL | (Ferreira et al. 2020) |
| rGO | Histamine | Improved sensitivity | 0.1–1 mM | --- | (Delle et al. 2015) |
| rGO/GO | *Salmonella typhimurium* | Antibodies immobilization and good electron transfer | 10$^{1}$ to 10$^{5}$ CFU/mL | 107 × 10$^{1}$ CFU/mL | (Mutreja et al. 2016) |
| rGO-PPyPPA (nanocomposite) | *Melamine* | Increased charge transfer | 10$^{-11}$ to 10$^{-2}$ mol/L | 12 pmol /L | (Gu et al. 2019) |
| GQDs@Au-polyaniline (nanocomposite) | Carcinoembryonic antigen | Improved sensitivity | 0.5–1000 ng/mL | 0.01 ng/mL | (Ganganboina and Doong 2019) |
| GQDs | Myoglobin (cMyo) | Improved sensitivity | 0.01–100 ng/mL | 0.01 ng/mL | (Mansuriya and Altintas 2019) |
| AuNPs | Anti-biotin IgG | Antibodies immobilization | 5–500 ng/mL | 5 ng/mL | (Liu et al. 2011) |
| AuNPs | hCG | Labelling of secondary antibodies | 0–500 pg/mL | 5 pg/mL | (Lim et al. 2014) |
| AuNPs | DHEAS | Antibodies immobilization | 10.0–110.0 μg/dL | 74 μg/dL | (Lima et al. 2019) |

| Fe$_3$O$_4$ NPs | HBsAb | Dispersion of antibodies in liquid samples | 1–15 pg/mL | 0.9 pg/mL | (Nourani et al. 2013) |
|---|---|---|---|---|---|
| Fe$_3$O$_4$ NPs | *Campylobacter jejuni* | Antibodies immobilization | $1.0 \times 10^7$ CFU/mL | $1.0 \times 10^3$ CFU/mL | (Huang et al. 2010) |
| SiO$_2$ | TBBPA-DHEE and TBBPA- MHEE | Targeted secondary antibodies are immobilized | 0.21–111.31 ng/mL | 0.08 ng/mL | (Zhang et al. 2018b) |

(ELISA) and eletrochemiluminescent assay (ECLIA). These studies will be important in the search for new nanomaterials and electrode configurations for analytical methods that can be safely used in clinical practice.

**Acknowledgments**  The authors thank the INCTBioanalytic Institute and support by CNPq Brazil agency. CAPES Brazil agency is also acknowledged by scholarships conceded to P.A.B. Ferreira and B.A.G. Rodriguez.

# References

Adegoke O, Morita M, Kato T, Ito M, Suzuki T, Park EY (2017) Biosens Bioelectron 94:513–522
Asav E, Sezgintürk MK (2014) Int J Biol Macromol 66:273–280
Bagheri E, Ansari L, Sameiyan E, Abnous K, Taghdisi SM, Ramezani M, Alibolandi M (2020) Biosens Bioelectron 153:112054
Baniukevic J, Kirlyte J, Ramanavicius A, Ramanaviciene A (2012) Procedia Eng 47:837–840
P. M. Biesheuvel, J. E. Dykstra, The difference between faradaic and nonfaradaic processes in electrochemistry, (2018). http://arxiv.org/abs/1809.02930
Bonaccorso F, Colombo L, Yu G, Stoller M, Tozzini V, Ferrari AC, Ruoff RS, Pellegrini V (2015) Science 347:1246501
Bourigua S, Hnaien M, Bessueille F, Lagarde F, Dzyadevych S, Maaref A, Bausells J, Errachid A, Renault NJ (2010) Biosens Bioelectron 26:1278–1282
Brosel-Oliu S, Abramova N, Uria N, Bratov A (2019) Anal Chim Acta 1088:1–19
Butmee P, Tumcharern G, Thouand G, Kalcher K, Samphao A (2020) Bioelectrochemistry 132: 107452
Camara AR, Dias ACMS, Gouvêa PMP, Braga AMB, Dutra RF, De Araujo RE, Carvalho ICS (2013) Fiber optic sensor with au nanoparticles for dengue immunoassay. In: Workshop on specialty optical fibers and their applications. https://doi.org/10.1364/WSOF.2013.W3.32
Cecchetto J, Carvalho FC, Santos A, Fernandes FCB, Bueno PR (2015) Sens Actuators B Chem 213:150–154
Chu Z, Peng J, Jin W (2017) Sens Actuators B Chem 243:919–926
Delle LE, Huck C, Bäcker M, Müller F, Grandthyll S, Jacobs K, Lilischkis R, Vu XT, Schöning MJ, Wagner P, Thoelen R, Weil M, Ingebrandt S (2015) Phys Status Solidi A Appl Mater Sci 212: 1327–1334
Ding S, Das SR, Brownlee BJ, Parate K, Davis TM, Stromberg LR, Chan EKL, Katz J, Iverson BD, Claussen JC (2018) Biosens Bioelectron 117:68–74
Dizon A, Orazem ME (2019) Electrochim Acta 327:135000
Dolatabadi JEN, de la Guardia M (2014) Anal Methods 6:3891–3900
Duffy GF, Moore EJ (2017) Anal Lett 50:1–32
Eatemadi A, Daraee H, Karimkhanloo H, Kouhi M, Zarghami N, Akbarzadeh A, Abasi M, Hanifehpour Y, Joo SW (2014) Nanoscale Res Lett 9:1–13
Elahi N, Kamali M, Baghersad MH (2018) Talanta 184:537–556
Ertekin Ö, Öztürk S, Öztürk ZZ (2016) Sensors 16:1–12
Faria RAD, Heneine LGD, Matencio T, Messaddeq Y (2019) Int J Biosens Bioelectron 5:29–31
Ferreira PAB, de Araujo MCM, Prado CM, de Lima RA, Rodríguez BAG, Dutra RF (2020) Colloids Surf B Biointerfaces 189:110834
Filik H, Avan AA (2019) Talanta 205:120153
Ganganboina AB, Doong RA (2019) Sci Rep 9:1–11
Geim AK, Novoselov KS (2007) Nat Mater 6:183–191
Gomes-Filho SLR, Dias ACMS, Silva MMS, Silva BVM, Dutra RF (2013) Microchem J 109:10–15
Grieshaber D, MacKenzie R, Vörös J, Reimhult E (2008) Sensors 8:1400–1458

Gu Y, Wang J, Pan M, Li S, Fang G, Wang S (2019) Sens Actuators B Chem 283:571–578

Hammond JL, Formisano N, Estrela P, Carrara S, Tkac J (2016) Essays Biochem 60:69–80

Heard DM, Lennox AJJ (2020) Angew Chem Int Ed 132:19026–19044

Heinze J, Bard AJ, Faulkner LF (1980) Electrochemical methods—fundamentals and applications. Wiley, New York

Holzinger M, Le Goff A, Cosnier S (2014) Front Chem 2:63

Huang J, Yang G, Meng W, Wu L, Zhu A, Jiao X (2010) Biosens Bioelectron 25:1204–1211

Ianeselli L, Grenci G, Callegari C, Tormen M, Casalis L (2014) Biosens Bioelectron 55:1–6

Jacobs CB, Peairs MJ, Venton BJ (2010) Anal Chim Acta 662:105–127

Jena SC, Shrivastava S, Saxena S, Kumar N, Maiti SK, Mishra BP, Singh RK (2019) Sci Rep 9:1–12

Jiang H, Lee PS, Li C, Tarascon JM (2013) Energ Environ Sci 6:41–53

Khan R, Pal M, Kuzikov AV, Bulko T, Suprun EV, Shumyantseva VV (2016) Mater Sci Eng C 68:52–58

Kokkinos C, Economou A, Prodromidis MI (2016) Trends Anal Chem 79:88–105

Kroto HW, Heath JR, O'Brien SC, Curl RF, Smalley RE (1985) Nature 318:162–163

Kumagai I, Tsumoto K (2002) Antigen-antibody binding. In: Encyclopedia of life sciences. Wiley

Lim SA, Yoshikawa H, Tamiya E, Yasin HM, Ahmed MU (2014) RSC Adv 4:58460–58466

Lima D, Inaba J, Lopes LC, Calaça GN, Weinert PL, Fogaça RL, de Moura JF, Alvarenga LM, de Figueiredo BC, Wohnrath K, Pessôa CA (2019) Biosens Bioelectron 133:86–93

Lindholm-Sethson B, Nyström J, Malmsten M, Ringstad L, Nelson A, Geladi P (2010) Anal Bioanal Chem 398:2341–2349

Lisdat F, Schäfer D (2008) Anal Bioanal Chem 391:1555–1567

Liu G, Liu J, Davis TP, Gooding JJ (2011) Biosens Bioelectron 26:3660–3665

Loo AH, Ambrosi A, Bonanni A, Pumera M (2014) RSC Adv 4:23952–23956

Mansuriya BD, Altintas Z (2019) Materials 13:1–30

Mihailescu C-M, Stan D, Iosub R, Moldovan C, Savin M (2015) Talanta 132:37–43

Mundinamani SP, Rabinal MK (2014) J Appl Chem 7:45–52

Mutreja R, Jariyal M, Pathania P, Sharma A, Sahoo DK, Suri CR (2016) Biosens Bioelectron 85:707–713

Ngoc Le HT, Park J, Chinnadayyala SR, Cho S (2019) Biosens Bioelectron 144:111694

Nourani S, Ghourchian H, Boutorabi SM (2013) Anal Biochem 441:1–7

Palomar Q, Gondran C, Marks R, Cosnier S, Holzinger M (2018) Electrochim Acta 274:84–90

Prodromidis MI (2010) Electrochim Acta 55:4227–4233

Randles JEB (1947) Kinetics of rapid electrode reactions. Discuss Faraday Soc 1:11–19

Rodriguez BAG, Trindade EKG, Cabral DGA, Soares ECL, Menezes CEL, Ferreira DCM, Mendes RK, Dutra RF (2015) Nanomaterials for advancing the health immunosensor. In: Biosensors—micro and nanoscale applications. InTech. https://doi.org/10.5772/61149

Sánchez-Tirado E, González-Cortés A, Yáñez-Sedeño P, Pingarrón JM (2016) Analyst 141:5730–5737

Sassolas A, Blum LJ, Leca-Bouvier BD (2012) Biotechnol Adv 30:489–511

Silva BVM, Cavalcanti IT, Mattos AB, Moura P, Sotomayor MDPT, Dutra RF (2010) Biosens Bioelectron 26:1062–1067

Silva BVM, Rodríguez BAG, Sales GF, Sotomayor MDPT, Dutra RF (2016) Biosens Bioelectron 77:978–985

Suresh L, Brahman PK, Reddy KR, Bondili JS (2018) Enzyme Microb Technol 112:43–51

Tian Y, Zhang P, Zhao K, Du Z, Zhao T (2020) Sensors 5:1394

Tuteja SK, Chen R, Kukkar M, Song CK, Mutreja R, Singh S, Paul AK, Lee H, Kim KH, Deep A, Suri CR (2016) Biosens Bioelectron 86:548–556

Valentini F, Fernàndez LG, Tamburri E, Palleschi G (2013) Biosens Bioelectron 43:75–78

Wang Y, Qu Y, Ye XX, Wu K, Li C (2016) J Solid State Electrochem 20:2217–2222

West K, Careem MA, Skaarup S (1993) Solid State Ion 60:153–159

Yáñez-Sedeño P, González-Cortés A, Agüí L, Pingarrón JM (2016) Electroanalysis 28:1679–1691

Yao C-Y, Fu W-L (2014) World J Gastroenterol 20:12485–12492
Zhang R, Zhang Y, Zhang Q, Xie H, Qian W, Wei F (2013) ACS Nano 7:6156–6161
Zhang F, Liu J, Ivanov I, Hatzell MC, Yang W, Ahn Y, Logan BE (2014) Biotechnol Bioeng 111: 1931–1939
Zhang Y, Wang J, Yu J, Wen D, Kahkoska AR, Lu Y, Zhang X, Buse JB, Gu Z (2018a) Small 14: 1704181
Zhang Z, Dong S, Ge D, Zhu N, Wang K, Zhu G, Xu W, Xu H (2018b) Biosens Bioelectron 105: 77–80

# Organic Electrochemical Transistors in Bioanalytical Chemistry

Ana Cristina Honorato de Castro, Suchismita Guha, and
Wendel Andrade Alves

## 1    Introduction

The synergy between organic electronics and biology is an emerging field in science. The main reason is to meet the needs of the medical community for the detection of analytes and pathogens in low concentrations, in addition to conditioning a favorable biological environment. Currently, the use of electrochemical biosensors has advantageous characteristics because they act as miniaturized devices and are cheaper and faster than current ones. They can meet the need for diagnosis in environments outside of specialized laboratories and are based on redox reactions, changes in potential or local impedance. However, biological signals are challenging and require amplification so that they can be detected, requiring more sensitive, specific, and biocompatible devices. Organic electrochemical transistors are important technological alternative (Strakosas et al. 2015).

Transistors were invented in 1947 by Shockley, Brattain, and Bardeen, and have been used to amplify or alter electronic signals since then. The basic principle is to measure the amount of the electric current that flows between the source electrode and the drain electrode proportional to an electric field applied to a third one, known as gate electrode. It has been almost 60 years since the invention of the metal–oxide–silicon field-effect transistor (MOSFET) and complementary metal-oxide semiconductor (CMOS), which now governs all technology. The discovery of conductive conjugated polymers in the 1970s has opened up an alternate path to Si electronics (Bai et al. 2019). Organic thin-film transistors, which may be classified into organic field-effect transistors (OFETs) and organic electrochemical transistors (OECTs),

A. C. H. de Castro · W. A. Alves (✉)
Centro de Ciências Naturais e Humanas, Universidade Federal do ABC, Santo André, SP, Brazil
e-mail: wendel.alves@ufabc.edu.br

S. Guha
Department of Physics and Astronomy, University of Missouri, Columbia, MO, USA

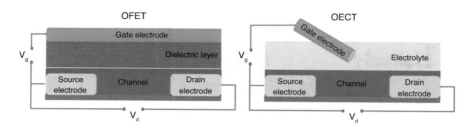

**Fig. 1** Representative scheme of electrochemical transistors, with $V_g$ gate voltage and $V_d$ drain voltage

are the building block elements of organic electronic circuits. Both use small organic molecules and conjugated macromolecules as semiconductors.

The typical structures of thin-film transistors are shown in Fig. 1. An organic semiconductor polymer channel connects the drain to the source electrodes. In OFETs, the gate electrode is separated from the other ones by a dielectric layer, and in OECTs, the separation occurs through an electrolyte solution (Gualandi et al. 2018).

The use of thin nanotechnological membranes can modify and adjust the operation of such devices with a wide range of applications. In the case of biological sensors, in particular, the matrices allow the formation of specific binding sites with biomolecules and thus assist in obtaining more reproducible responses (Kisner et al. 2012).

## 2    Organic Field Effect Transistors

In OFETs, the charge carriers are injected at the semiconductor/dielectric layer when a voltage is applied between the gate and source terminals, forming an accumulation layer that creates a conductive channel between the source and the drain. When a voltage is applied between the source and the drain, these charges generate a current flow, not only varying the electric field through the dielectric layer but also modifying the conductivity of the channel. The current flows in the longitudinal source-drain direction, creating a region called channel. The source and drain terminals are connected to the semiconductor, while the port is isolated from the semiconductor by the dielectric layer. Thus, the device can change between "on" and "off" states. At a basic level, an OFET can be considered as a charge sensor; any charge variation on the gate will induce a current variation in the channel. The ions do not penetrate the channel, but accumulate close to the surface of the dielectric layer (deposited in the channel), inducing the accumulation of charge inside the channel, which is investigated by variations in the electric field near and across the surface. Hence, these devices are chemically sensitive, and can measure the chemical changes based on variations in the electric field, when placed in a chemically reactive environment (Garrote et al. 2019).

OFETs are being actively developed for applications in electronics with flexible supports. Therefore, both the active semiconductor layer and the dielectric layer must be processed in solution. Peptide nanostructures pave the way for a new class of dielectrics in OFET architectures with applications in biosensors (Cipriano et al. 2014).

## 3 Organic Electrochemical Transistors

Unlike OFETs, where the source-drain current is modulated by the field-effect doping, the source-drain current in OECTs is modulated by electrochemical doping or de-doping. The electrolyte ions can penetrate the entire volume of the polymeric channel exhibiting high amplification characteristics in under-voltage operating regimes, preventing electrolysis, and extending the operating times, thus allowing for in vitro and in vivo applications.

OECTs are chemically sensitive where the electric drain current is controlled by the injection of ions from an electrolyte into the channel, changing the electronic charge density in all its volume (Garrote et al. 2019). They generate high capacitance values, which allow them to operate at low working voltages ($<1$ V) and provide impressive signal amplification properties, guaranteeing their electrochemical stability for working in aqueous solutions (Yeung et al. 2019). In addition, they are low-cost devices that can be built on flexible substrates, using simple manufacturing techniques. They present high sensitivity (Fan et al. 2019), biocompatibility, low toxicity (Hai et al. 2018), and the possibility to miniaturize and integrate into portable devices make them suitable for chemical and biological detection (Wang et al. 2017).

White and colleagues reported the first OECT in 1984, and it has been widely applied to biological interfaces and monitoring events, such as measuring the integrity of barrier tissues and cell layers, monitoring neural networks in vivo and in vitro, chemical and biological sensors, printed circuits, and clinical or biomedical research (Faria et al. 2017). The operational mechanism depends on the electrochemical doping and de-doping processes of the conducting polymer. Due to the absence of insulating layers, ions present in the electrolyte can both enter and exit the channel, which results in a change in their conductivity (Hai et al. 2018). The configuration of the transistor allows the modulation of the channel current between the source and the drain electrodes through electrochemical reactions in the gate electrode, resulting in very sensitive response, essential in bioanalytical monitoring.

The channel consists of species that have redox activity, such as organic conjugated semiconductor polymers with chemically tunability, and can be designed according to the needs of each application. Their use allows the conversion of the ionic current into an electric current and are thus good candidates for bioelectronics applications in which a biological event alters the conductivity of the polymer and translates it into an electrical charge (Fan et al. 2019). The organic conjugated semiconductor polymers have excellent biocompatibility for various types of cells, enabling better adhesion and ionic interactions with biomolecules.

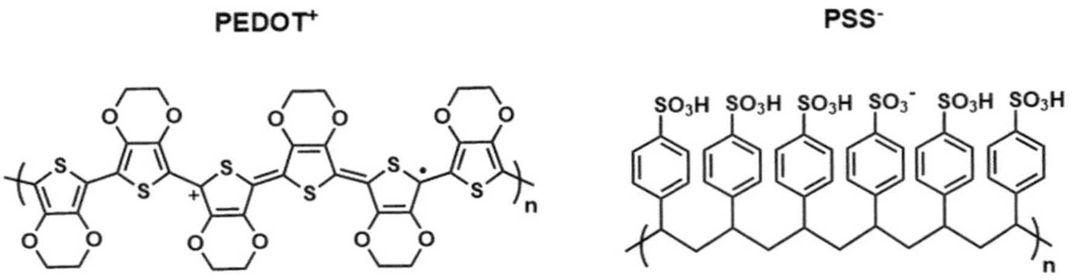

PEDOT⁺                                                          PSS⁻

**Fig. 2** The poly (3,4-ethylenedioxythiophene) is positively doped (PEDOT⁺) and balanced by the counter anion poly (styrene sulfonate) (PSS⁻)

For the construction of OECTs, several types of doped conjugated semiconductor polymers of type $p$ or type $n$ can be used. The most widely used is poly (3,4-ethylenedioxythiophene) doped with poly (styrene sulfonate), known as PEDOT:PSS. It is commercially available and can be readily standardized by photolithography, inkjet printing, and screen printing. Also, it features excellent thermal stability, ionic mobilities for small ions and good stability over a wide pH range (Contat-Rodrigo et al. 2017). PEDOT: PSS is a p-type polymer with high electronic conductivity. To improve its adhesion to substrates, it is often cross-linked with silanes, which may compromise its conductivity, but it still remains suitable for use with biological molecules. The PEDOT⁺ and PSS⁻ structures are shown in Fig. 2.

PEDOT is positively doped by removing electrons from the main polymer structure. Through this oxidative doping process, the charge carriers (polarons and bipolarons) and the positive dopants carried by PEDOT are balanced by the counter anion (PSS⁻), i.e., the sulfonate anions of PSS are added to stabilize the oxidized polymer and compensate the deficit of negative charges.

This doping is reversible and PEDOT:PSS not only undergoes oxidation/reduction reactions but also experiences change between its conductive state (Oxidized–PEDOT⁺) and its semiconductor form (Neutral–PEDOT⁰) (Faria et al. 2017). The widespread use of PEDOT:PSS as the active layer in OECTs is its high electronic and ionic mobility.

Some small organic molecules, such as hydroxyquinoline, pentacene, rubrene, and fullerene, can also be used as semiconductor channels. Additionally, conjugated polymers are also used since they allow the production of thin films using more straightforward and accessible methods at relatively low costs.

## 4    OECT Biological Sensors

The use of OECTs in electrochemical sensors depends on the type of biomolecules studied. Within the biological system, there may be redox active or inactive species. Redox active species can be directly reduced or oxidized at the electrode interface, while redox inactive ones can only be analyzed by the use of electroanalytical techniques for molecular biorecognition (Bai et al. 2019). All the devices treated below are low cost, fast, point of care, and sensitive alternatives for diagnosing various diseases.

## 5 OECTs for Redox Active Species

Dopamine (DO) is a neurotransmitter that plays an essential role in the nervous, renal, hormonal, and cardiovascular systems. Clinical studies show that neurodegenerative diseases are related to DO levels in the human body (Wang et al. 2019). However, the clinical concentration is very low, which makes it challenging to determine in biological samples. It is, therefore, necessary to develop alternative methods for rapid and sensitive detection. DO can be directly oxidized or reduced to a specific voltage on the electrodes. In this context, the construction of a high-performance electrode is crucial for high sensitivity detection.

Tang et al. investigated PEDOT:PSS-based OECTs and different types of gate electrodes for the determination of DO in biological samples. They found that the sensitivity of the device is related to the material of the electrode and the operating voltage. A platinum gate electrode characterized at +0.6 V showed the highest sensitivity, with a detection limit of less than 5 nM, which is an order of magnitude better than conventional electrochemical detection with the same electrode (Tang et al. 2011).

Contat-Rodrigo and collaborators used OECTs for fast, selective, cheap, and simple determination of ascorbic acid (AA) present in commercial orange juice. In this assay, the PEDOT:PSS polymer was employed to construct of printed OECTs and a detection limit of 80 mM was observed (Contat-Rodrigo et al. 2017).

Zhang et al. developed an OCET using a molecularly imprinted polymer (MIP) as the electrode. The combination of the amplification function of an OECT and the selective specificity of the MIPs provided a highly sensitive and selective sensor, reaching a detection limit of 10 nM for AA detection. The sensor demonstrated excellent specific recognition capacity, which prevented the interference of other compounds in the analysis and efficiency in working with real samples (Zhang et al. 2018).

## 6 OECTs for Redox Inactive Species

For specific species detection, the OECT channel or the gate electrode must be functionalized with a biorecognition element, such as enzymes, proteins, or nucleotides.

Nishizawa et al. developed one of the first biosensors based on OECTs with an enzymatic interface (Nishizawa et al. 1992). They used polypyrrole based OECTs containing the immobilized penicillinase enzyme, which oxidized penicillin to penicillic acid. This oxidation process promoted a change in the local pH and increased the conductivity of the polypyrrole. Thus, the operational process was based on the change in pH allowing the determination of penicillin. However, this system was not viable because the conductivity of the polypyrrole was impaired through the biological conditions.

Enzymes that are immobilized on the electrode gate convert their substrates into products. The electrons involved in the enzymatic reactions can be either obtained or lost, and the electrical signals are transferred to the gate electrode, leading to current changes in the channel.

The real-time measurement of glucose in serum is essential for the treatment of diabetes, the prevention of gestational diabetes, and other related diseases. Both high sensitivity and selectivity are sought. Wang et al. developed a glucose biosensor using a composite constructed from polypyrrole nanowires and reduced graphene oxide (rGO), where the presence of rGO improved the electrical performance of fiber transistors. They exhibited a response time of 0.5 s, a linear range from 1 nM to 5μM, and excellent repeatability and reversibility. They also featured a lower detection limit when compared to conventional methods and successful validation in real samples. The results obtained indicate that a nanotechnological architecture has a promising future for applications in bioanalytical techniques (Wang et al. 2017).

Due to the pioneering studies of Kohler and Milstein (1975), immunoassays could be used more widely in clinical analyses. The body in response to the presence of exogenous substances, called antigens, produces antibodies or they can be produced in autoimmune diseases or cancers. Both antibodies and antigens can act as biorecognition molecules and can be immobilized on a portal electrode. Based on this principle, Kim et al. demonstrated a sensitive immunosensor capable of determining the prostate-specific antigen/1-antichymotrypsin complex (PSA-ACT), an essential biomolecule for the diagnosis of prostate cancer. The OECT was constructed using PEDOT:PSS with AuNPs conjugated to specific secondary antibodies, and the reached detection limit was 1 pg mL$^{-1}$ (Kim et al. 2010).

Another developed immunosensor using PEDOT polymer, Macchia et al. have reported PSS-based OECTs for the detection of Immunoglobulin G (IgG). The IgG is the most abundant serum immunoglobulin in the immune system and protects against bacterial, viral, and fungal infections. Anti-immunoglobulin G (anti-IgG) was used as a capture protein and immobilized on the surface of the portal electrode so that IgG could be selectively detected, with a detection limit of 6 fM. This work shows that OECT devices, which are low cost, have ultra-sensitive properties and provide an excellent platform for applying immunoassay technology (Macchia et al. 2018).

Highly sensitive and low-cost DNA detection is a lofty aspiration due to the significant role of DNA molecules in genetic damage detection and mutations in the diagnosis of various diseases. Biosensors based on genetic biomolecules are considered promising bioanalytical tools (Heli et al. 2016). Li et al. explored an electrochemical graphene transistor for DNA detection. With graphene as the transistor's active element, the genetic probe was immobilized on the gold gate electrode to detect DNA molecules with a detection limit in the range of 1 fM to 5μM, demonstrating a highly sensitive and selective platform that can be used as a promising tool in genetic analysis (Li et al. 2019).

Aptamers are small artificial molecules selected from a set of random sequences and screened for specific binding to a target protein. The search for fast, sensitive, and specific diagnosis methods has directed researchers to explore genetic aptamers. They are selected to be complementary to the molecules of interest, achieving high specificity (Moosaviana and Sahebkar 2019).

Epinephrine is an excitatory neurotransmitter whose abnormal levels are directly related to the symptoms of some diseases such as myocardial infarction, arrhythmias, and other conditions associated to the heart. With that in mind, Saraf et al. developed a highly sensitive, specific, fast, and cheap biological sensor. They used OECTs, where the source and drain electrodes are connected by a layer of PEDOT:PSS and specific epinephrine aptamers were immobilized on the gate electrode by thiol connections. The sensor operation addressed the selective interaction of aptamers with epinephrine with high affinity and specificity. The reached detection limit was 90 pM, which is comparable to the normal physiological level (Saraf et al. 2018).

The above examples show that OECTs are a promising technology in the development of new devices for diagnosis and prognosis of diseases. The development of OECTs is multidisciplinary, which interfaces with bioelectronics and bioanalytical tools to obtain innovative, sensitive, selective, inexpensive, stable, miniaturized, and accessible technologies. They are easily integrated into biological systems and are compatible with photolithography techniques, which can facilitate the process of building useful portable platforms to monitor cell activities in vitro or in vivo. It is proven that their performance surpasses most traditional electrochemical methods and can detect both the presence of very low concentration metabolites and monitor physiological signals more precisely and efficiently.

# References

Bai L, Elósegui CG, Li W, Yu P, Fei L, Mao L (2019) Front Chem 17:1–16
Cipriano T, Knotts G, Laudari A, Bianchi RC, Alves WA, Guha S (2014) ACS Appl Mater Interfaces 6:21408–21415
Contat-Rodrigo L, Perez-Fuster C, Lidon-Roger JV, Bonfiglio A, García-Breijo E (2017) Org Electron 45:89–96
Fan J, Montemagno C, Gupta M (2019) Org Electron 73:122–129
Fan J, Rezaie SS, Facchini-Rakovich M, Gudi D, Montemagno C, Gupta M (2019) Org Electron 66:148–155
Faria GC, Duong CT, Salleo A (2017) Org Electron 45:215–221
Garrote BL, Fernandes FCB, Cilli EM, Bueno PR (2019) Biosens Bioelectron 127:215–220
Gualandi I, Scavetta E, Mariani F, Tonelli D, Tessarolo M, Fraboni B (2018) Electrochim Acta 268:476–483
Hai W, Goda T, Takeuchi H, Yamaoka S, Horiguchi Y, Matsumoto A, Miyahara Y (2018) Sensor Actuat B Chem 260:635–641
Heli H, Sattarahmady N, Hatam GR, Reisi F, Vais RD (2016) Talanta 156–157:172–179
Kim D-J, Lee N-E, Park J-S, Park I-J, Kim J-G, Cho HJ (2010) Biosens Bioelectron 25:2477–2482
Kisner A, Heggen M, Fischer W, Tillmann K, Offenhausser A, Kubota LT, Mourzina Y (2012) Nanotechnology 23:485301
Kohler G, Milstein C (1975) Nature 256:495–497

Li S, Huang K, Fan Q, Yang S, Shen T, Mei T, Wang J, Wang X, Chang G, Li J (2019) Biosens Bioelectron 136:91–96

Macchia E, Romele P, Manoli K, Ghittorelli M, Magliulo M, Kovàcs-Vajna Z-M (2018) Flex Print Electron 3:034002

Moosaviana AS, Sahebkar A (2019) Cancer Lett 448:144–154

Nishizawa M, Matsue T, Uchida I (1992) Anal Chem 64:2642–2644

Saraf N, Woods ER, Peppler M, Seal S (2018) Biosens Bioelectron 117:40–46

Strakosas X, Bongo M, Owens RM (2015) J Appl Polym Sci 41735:1–14

Tang H, Lin P, Chan HL, Yan F (2011) Biosens Bioelectron 26:4559–4563

Wang M, Cui M, Liu W, Liu X (2019) J Electroanal Chem 832:174–181

Wang Y, Qing X, Zhou Q, Zhang Y, Liu Q, Liu K, Wang W, Li M, Lu Z, Chen Y, Wang D (2017) Biosens Bioelectron 95:138–145

Yeung SY, Gu X, Tsang SM, Tsao SWT, Hsing I-M (2019) Sensors Actuators B Chem 297:126761

Zhang L, Wang G, Wu D, Xiong C, Zheng L, Ding Y, Lu H, Zhang G, Qiu L (2018) Biosens Bioelectron 100:235–241

# Quartz Crystal Microbalance in Bioanalysis

Zeki Naal and Rose Mary Zumstein Georgetto Naal

## 1 Introduction

Biosensors are analytical devices that have a biological recognition element attached to a physical transducer, which can be especially useful for biomarker analysis. Different types of transducers have been used for the development of biosensors, including optical, electrochemical, and piezoelectric transducers. Coupling electrical and mechanical energy makes piezoelectric materials useful in a wide range of applications. In general, biosensors use labels to transduce the signal. In many cases, these labels are responsible for false-positive results, opening a search for a technique without labels.

The QCM device represents a straightforward and valuable tool with which to analyze the formation or degradation of biopolymeric films or film composites at a surface–solution interface within which single or multiple assembly steps are required. Either the kinetics ($k_{assoc}$ or $k_{dissoc}$) of the process steps can be followed, or simply, the equilibrium states. A crucial advantage of the QCM technique is that these studies may be performed where optical techniques are impossible. For example, there are many experimental conditions such as complex systems involving surfactants (Lane et al. 2008; Afzal et al. 2017), self-assembling molecules, or biopolymers, where optically transparent solutions cannot be studied. In addition, another advantage of QCM is that not only the mass of the film but also the viscoelastic or energy dissipating behavior of the film can be measured simultaneously during its formation or at equilibrium, in this case called QCM-D (Dixon 2008; Normile 2013).

Finally, one of the unique features of QCM is its ability to measure the transport of ions in charged films, including polymers, during changes in the film's

Z. Naal (✉) · R. M. Z. G. Naal
Department of BioMolecular Sciences, School of Pharmaceutical Sciences of Ribeirão Preto, University of São Paulo, Ribeirão Preto, SP, Brazil
e-mail: zekinaal@usp.br

313

environment or oxidation state. The latter is conducted at a provided electrochemical potential applied at the underlying EQCM electrode. Because there is a broad range of systems under study that involve polymeric substrates and biological macromolecules (applied to biosensors in many cases), we discuss here systems and studies dominated by the biological component or biological analyte detection. Investigators have applied the QCM technique to a range of systems, some of which we describe below. These include electrochemical polymerization (Kushwaha et al. 2019; Yarman et al. 2018; Takeda et al. 2003; Marx 2003; Takada et al. 2002), conducting polymers (Hai et al. 2017), electroactive film redox properties (Marx 2003), micelle formation (Tabaei et al. 2015), self-assembling monolayers (Cornelio et al. 2015; Zhang et al. 2004), molecularly imprinted polymer films (Kartal et al. 2019; Saylan et al. 2019; Lieberzeit et al. 2016; Liu et al. 2018), nanoarchitecture (Peng et al. 2012), and layer-by-layer films (Chen et al. 2009).

Since piezoelectric devices can be used as transducers sensitive to change in mass and viscoelastic properties, which are intrinsic molecular interaction properties, they can be used for detection of unlabeled biological species.

Figure 1 demonstrates the growth in citation per year involving QCM or QCM-D or EQCM and bioanalysis.

Piezoelectric crystals, more commonly called quartz crystal microbalances (QCM), are mass-sensitive devices that transform a mass change on the surface of the gold electrode sensor into a change in the QCM's intrinsic resonance frequency.

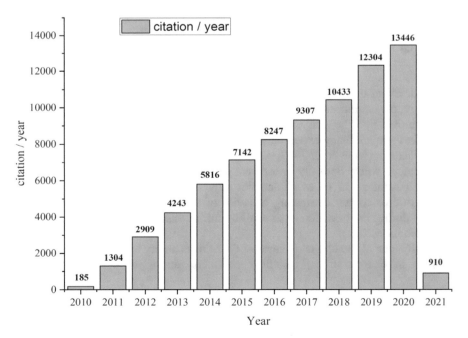

**Fig. 1** Analysis of citation per year from Web of Science from keywords in topic: QCM or QCM-D or EQCM and Bioanalysis in January 26, 2021

When a low-voltage AC signal is applied to thin-layer electrodes that sandwich a quartz crystal, the crystal's piezoelectric nature produces a shear (tangential) deformation, causing both surfaces to move in parallel, but in opposite directions, producing acoustic waves that propagate perpendicularly to the crystal surface. The frequency with which this happens is the resonant frequency of the crystal. As the material is added to the top of the sensor, the wave propagates to this layer and the resonance frequency changes. This is the basis of measurement. QCM can provide useful information about the amount of mass deposited and the rate of deposition (or removal) of such films, monitoring the change in frequency in real time. In the liquid environment, however, molecular adsorption includes contributions of liquid molecules associated with direct hydration/solvation and/or incorporation of species within the adsorbed film. Adsorption can produce soft or viscoelastic films, and the resulting layer may not fully connect to the oscillating crystal. This can cause damping or loss of energy from the oscillation.

The mass of such films cannot be accurately determined by measuring just the change in frequency. Both the change in frequency ($\Delta f$) and the loss of energy (measured as change in dissipation, $\Delta D$) must be measured to accurately determine the change in mass of a viscoelastic film. In addition, by monitoring both $\Delta f$ and $\Delta D$, some additional structural information (for example, conformational changes within the film, crosslinking, and swelling) can be obtained. This method is called Quartz Microbalance with Dissipation (QCM-D) (Teramura and Takai 2018; Thorn and Greenman 2012).

The basic principle of QCM is piezoelectricity, which is based on the application of a voltage to a certain material with consequent mechanical deformation, causing them to resonate at a certain frequency. When some substance is deposited on the surface of the material, the frequency decreases as can be seen in Fig. 2. Sauerbrey (Sauerbrey 1959) mathematically derived the relationship between frequency changes and mass loads, as shown in Eq. (1):

$$\frac{\Delta f}{f_0} = -\frac{\Delta m}{m} \tag{1}$$

On this equation, $\Delta f$ is the shift in frequency on loading of mass $\Delta m$, while $f_0$ is the fundamental resonating frequency and m is the mass of the unloaded resonator.

Change in frequency depends on several factors besides the change in mass, and that includes the physical constants of the material, such as density and shear modulus. Since one of the faces of the QCM device is in contact with the liquid phase, the viscosity of the liquid medium also needs to be considered to understand sensitivity, noise, and resolution issues (Rodriguez-Pardo et al. 2005). As a result, the relationship is changed as proposed by Kanazawa and Gordon (1985) in Eq. (2):

$$\Delta f = f_0^{3/2} \left(\frac{\eta \rho}{\pi \mu_q \rho_q}\right)^{1/2} \tag{2}$$

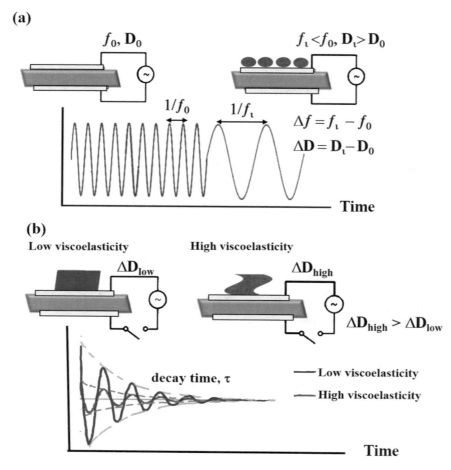

**Fig. 2** Schematic diagrams of QCM-D measurement. (**a**) Changes in the frequency of the oscillating sensor when the mass change occurs. (**b**) Difference in dissipations signal between different viscous layers. Adapted from Teramura and Takai (2018)

In this equation, $\eta$ and $\rho$ are viscosity and density of the liquid in contact with the probe, respectively, while $\rho_q$ represents the density of quartz, i.e., 2.648 g cm$^{-3}$ and $\mu_q$ is the shear modulus of quartz having a value $2.947 \times 10^{11}$ g cm$^{-2}$.

The quartz production process can influence both the temperature coefficient and the type of oscillation of the quartz. The deciding factor is the cutting angle at which the quartz disc is removed from the crystal. One of the most common angular cuts is the AT cut, which is used in the production of all crystals. This cut is made at 35.25° from the Z-axis of the quartz crystal, being one of the most selected cutting angles as well as having a good temperature coefficient.

Dixon, M. C. (2008). has written a review about QCM-D characterizing biological materials and their interactions. In this review, he has mentioned the conditions to be fulfilled for the Sauerbrey relationship to hold, and pointed out that the mass on the surface must be rigidly adsorbed.

The two main methods to address dissipation (D) due to viscoelastic film adsorption are to monitor the decay of a crystal's oscillation after rapid excitation near the resonant frequency (since the rate of decay is proportional to the oscillator's energy dissipation) or using electrochemical impedance spectroscopy analysis. QCM with dissipation monitoring (QCM-D) adjusts to the oscillatory decay voltage after a driving force is turned off to ensure that the quartz frequency decays close to the series resonant mode (Fig. 2b). The decay of the frequency amplitude over time depends on the properties of the oscillator and the medium that the oscillator is in contact with. The amplitude of the output voltage as a function of time, with a frequency given by the resonant frequency of the quartz crystal ($f_0$), is mixed with a reference frequency ($f_R$) and filtered with a low-band filter. This results in an output frequency ($f$) based on the difference between $f_R$ and $f_0$. This frequency difference is adjusted to an exponential sinusoidal decay according to Eq. (3) where $f = f_R - f_0$ and $\tau$ is the decay time, $A_0$ is the amplitude in $t = 0$ and $\alpha$ is the phase:

$$A(t) = A_0 e^{1/\tau} \sin(2\pi f t + \alpha) \tag{3}$$

The dissipation parameter is given by Eq. (4)

$$D = \frac{1}{\pi f \tau} \tag{4}$$

and is dimensionless, defined as the ratio between dissipated energy and stored energy following Eq. (5):

$$D = \frac{1}{Q} = \frac{E_{dissipated}}{2\pi E_{stored}} \tag{5}$$

The inverse of dissipation is the quality factor $Q$, $E_{\text{dissipated}}$ being the energy dissipated during one oscillatory cycle and $E_{\text{stored}}$ being the energy stored in the oscillating system.

In liquids, the resolution of the frequency is in the order of 0.1 Hz and the resolution of dissipation is $1 \times 10^{-7}$, whereas in the air or vacuum the resolution is an order of magnitude better. For measurements involving adsorption of proteins, vesicles, or cells, the resolution of the responses of ($f$) and ($D$) is in the range of 10–100 Hz and from 1 to $10 \times 10^{-6}$ of dissipation units, respectively. If the viscoelastic films are more than 100 nm thick, these measurements are an order of magnitude higher. The viscoelastic data allow a characterization where the system is outside the scope of the Sauerbrey relationship between $\Delta f$ and $\Delta m$ as can be seen in Fig. 3d that will be described below. Figure 3 describes the principal components of the QCM-D theory used with the appropriate instrumentation.

As can be seen in Fig. 3, the change in frequency in step (a) does not have a corresponding change in dissipation, suggesting the formation of a rigid film of

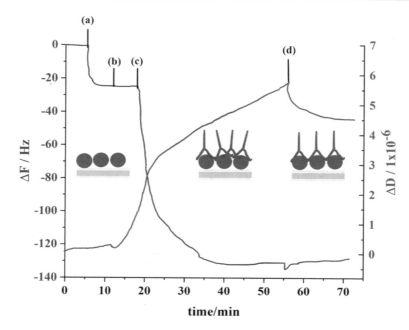

**Fig. 3** Presentation of data from a QCM-D experiment. Left side of the graph shows the frequency variation data ($\Delta f$) and the right side shows the dissipation variations ($\Delta D$). The additions of the substances are indicated by the letters (**a**) human serum albumin (HSA), (**c**) antibody to human serum albumin, (**b, d**) correspond to rinses with buffer. Adapted from Dixon (2008)

HSA. On the other hand, the adsorption of the antibody in step (c) produces a large variation in frequency and dissipation, indicating an increase in mass in the film as well as an increase in viscoelastic characteristics due to the incorporation of water. In addition, the rinse step (d) shows a change in dissipation due to the rearrangement of the antibody molecules after the elimination of physically adsorbed antibodies.

The film formed by the adsorption of a small protein, if it is rigid, can be monitored by the frequency variation according to the Sauerbrey equation. However, if water is incorporated, this change in frequency may be overestimated. It can be seen in the graph of Fig. 3 that the addition of HSA (a) produces a very small variation in dissipation, but with the addition of anti-HSA the film grows and incorporates a quantity of water (c), increasing the dissipation D. Thus, the combination of the $\Delta f$ and $\Delta D$ measurements makes it possible to quantify and separate viscoelastic variables in relation to shear viscosity and the packing process of the adsorbed materials. With this technique, specific surfaces can also be prepared and monitored step by step during preparation, in order to use them as sensor platforms to study other biomolecular interactions. This provides a wide scope for development of studies for biological material and bioanalytical systems.

## 2    QCM Applications: Diagnosis

The technology of analytical devices for diagnosis is in great demand in order to improve the quality of life and global health (Justino et al. 2016). Earlier and more reliable diagnosis leads to prevention and treatment of diseases with longer life expectancy (Ranjan et al. 2017; Jayanthi et al. 2017). Quartz crystal microbalance sensors can be applied to the quantification of small molecules, macromolecular compounds, and microorganisms (Skládal 2016). One of the advantages of this technique is the ability to detect the target analyte without any additional labeling molecule. According to Mujahid et al. (2018), the labeling process can change the dynamics of the target recognition since non-specific binding of labeling molecules may lead to false results during a simple bioanalytical test. Modern diagnostic devices advocate direct recognition of the analyte without using any labeling process. In the last decade, nanomaterials have been incorporated into bioanalytical devices to increase biocompatibility, binding sites, and sensitivity of the detection (Fenzl et al. 2016). In addition, nanostructured materials can help in the miniaturization of bioanalytical devices, with reduced sample volume, analysis time, and experimental costs. QCM-based devices are often constructed with different surfaces coating to gain sensing specificity and sensitivity for desired targets (Jackman et al. 2018; Farka et al. 2017; Latif et al. 2014; Speight and Cooper 2012). In most cases, the construction of the sensing surfaces on the QCM electrode is conducted by conventional chemical methods, or nanofabrication, in order to have polymer-modified surfaces (such as molecular imprinting polymers (MIPs) (Shandilya et al. 2019), self-assembled monolayers (including biological membrane model) (Damiati et al. 2018), DNA surfaces (Ma et al. 2020; Park and Lee 2019), metal nanoparticle (Park and Lee 2018), nanorods (Park and Lee 2019), among others (Bayramoglu et al. 2019; Mahmoodi et al. 2019; Hwang et al. 2019). Therefore, it is crucial to have a well-planned nanostructured sensor material on the electrode surface to bind desired bioreceptors (enzymes, antibodies, aptamers, DNA, etc.) and achieve a sensitive and specific responses in target analyte recognition. Despite having nanotechnology as a tool for increasing the sensitivity of piezoelectric sensors, the QCM-based assays often require additional strategies to improve the biorecognition signal. As examples, we can mention gold nanoparticles (Park and Lee 2019; Dong et al. 2018), nanobeads (Asai et al. 2018), and others (Justino et al. 2016). Based on the above context, we comment here on the bioanalytical applications (covering mainly the last 5 years) with emphasis on diagnosis.

## 3    Cancer

Cancer is considered a leading cause of mortality around the world (W. H. Organization 2018). One of the primary reasons for such high mortality level is the lack of an effective early diagnosis and disease monitoring (Shandilya et al. 2019). Unfortunately, it is very challenging to detect cancer in its early stages, and, in most cases, the detection occurs in advanced stages of the disease, making survival difficult. Until now, the major diagnostic modalities for cancer depend on the conventional methodologies that consist of tomography, magnetic resonance imaging, X-ray, and ultrasound. However, these technologies do not detect the early stage of the disease and do not differentiate the benign and malignant cells (Shandilya et al. 2019). For this reason, the development of low-cost and easy-to-use devices is urgent and requires mutual efforts from interdisciplinary areas such as chemistry, biology, and engineering (Farka et al. 2017; Shandilya et al. 2019; Roointan et al. 2019). Despite the difficulty of detecting cancer in its early stage, there are several biomarkers through which it is possible to develop bioanalytical detection methods with different techniques (Justino et al. 2016; Roointan et al. 2019). Several cancer biomarkers, such as nucleic acids (Dong et al. 2018), proteins (Park and Lee 2019), metabolites (Qiu et al. 2018), and even whole cells (Damiati et al. 2018; Ma et al. 2020) can be used in the diagnosis of tumors (Ranjan et al. 2017; Chung and Christianson 2014) and some of them will be commented here.

Expression of specific miRNAs arrangements can be used as biomarkers for several human diseases such as immune disorders, vascular diseases, infectious diseases, and cancer (Park and Lee 2019). Despite their pivotal role in several diseases, especially cancer, the reliable and sensitive detection of miRNAs, including miR-21, is still challenging, since these proteins are small, sparse, and easily degradable. Park and Lee (2019) have demonstrated the quantitative detection of miR-21 through the formation of miR-21–DNA hybrid duplexes and non-specific intercalation of surface-modified pyrene molecules. Selectivity was obtained through the gold nanoparticle probe conjugated to the complementary miRNA oligonucleotide. Sensitivity came from the association of intercalated complexes on the sensor surface, followed by gold staining signal amplification. The biosensor device was able to detect miR-21 concentrations as low as 3.6 pM. In addition, miR-21 could be detected in total RNA extracted from A549 cell line. Later, Ma et al. (2020) have used, for the first time, a nanoscale metal-organic framework (MOF) to improve the detection of miR-221 by QCM. Initially, a biotin-modified DNA was used as a probe to interact with miRNA. After, a streptavidin@MOF (MOF = UiO-66-NH$_2$) complex was introduced in the QCM system to work as both a signal amplifier and a specific recognition component via biotin–streptavidin interaction. This strategy reached the miR-221 detection limit of 6.9 fM, rising to 0.79 aM through coupling with isothermal exponential amplification reaction (EXPAR). This approach can detect traces of miR-221 that are significant for the early detection of cancer.

The most fatal cancer in women is cervical cancer that is caused by the human papillomavirus (HPV) (Mahmoodi et al. 2019). HPV types 16 and 18 are high risk because they invade epithelial cells and produce proteins E6 and E7 that interfere with the normal function of cells, converting them into malignant cells (Tomaic 2016). The Pap smear (Papanicolaou) is the oldest method for monitoring cervical cancer, but this method can result in false positives (from 13 to 70%) and false negatives (from 0 to 14%). A reliable assay was obtained through a QCM biosensor integrated with loop-mediated isothermal amplification (LAMP-QCM) to detect HPV16 DNA in cervical samples. The detection limit of LAMP-QCM was 10-fold more sensitive than TaqMan-qPCR with 100% specificity, 7.6% imprecision, and high correlation with qPCR. The assay can be promising for the sensitive detection of HPV16-DNA in a routine examination, facilitating clinical follow-up and therapeutic conduct (Jearanaikoon et al. 2016).

Most QCM devices detect only one analyte, which has not been enough for early diagnosis of cancer since the transformation of malignant cells is a complex process that involves the expression and alteration of numerous biological markers. Thus, the development of analytical devices for real-time detection and quantification of multiple bio-analytes is essential for cancer diagnosis and other diseases. Zhou et al. (2018) developed, for the first time, a multi-analyte QCM device based on the monitoring of glycoproteins whose expression and dysfunction are altered in tumor cells. The assay was managed through the folate receptor (FR), CD44 protein, and epidermal growth factor receptor (EGFR), all of them overexpressed on the cell membrane. Folate receptors (FR) mediate cellular uptake of folate and specifically bind to folate. CD44 is essential for metastasis and participates in many cellular events such as motility, differentiation, and growth. EGFR is involved in the starting and evolution of different types of cancers and also works as a biomarker. The real-time QCM cytosensor was engineered for in situ and continuous monitoring of multiple cell membrane glycoproteins, based on a signal recovery strategy. Gold nanoparticles (nanoprobes) were linked with the ligands (FR, CD44, and EGFR) in order to gain signal amplification. Mercaptosuccinic acid and poly-L-lysine were immobilized on the surface of the gold electrode, in sequence, for capturing MDA-MB-231 cells. Later, the three mass nanoprobes were injected into the QCM chamber, in sequence, to recognize the captured cells. Each mass nanoprobe was consecutively eluted using glycine-hydrochloric acid buffer, resulting in a quick recovery of the QCM signal. The quantification of MDA-MB-231 cells showed a range of linearity of $3.0 \times 10^4$ to $1.0 \times 10^6$ cells and a LOD of $5.0 \times 10^3$ cells. In addition, the multianalyte cytosensor provided information about the average numbers of glycoproteins per cell ($0.5 \times 10^6$, $0.2 \times 10^6$, and $1.4 \times 10^5$ for FR, CD44, and EGFR, respectively). Compared to the other multianalyte device, such as monolithic multichannel QCM (Jaruwongrungsee et al. 2015), the multianalyte QCM cytosensor, based on a single microbalance, can eliminate acoustic interference and reduce instrumental costs. The possibility of eluting the nanoprobe for the next use of the sensor fulfills one of the challenges to be overcome in the development of bioanalytical devices, which is the reusability. Despite the advantages of this approach, the measurement accuracy can be affected by the non-specific binding of proteins, another challenge to be solved in these kinds of QCM devices.

DNA biomarkers associated with diseases (such as cancer) have been identified, opening new perspectives for the development of an effective genosensor. Rasheed and Sandhyarani (Abdul Rasheed and Sandhyarani 2016) used a QCM genosensor to detect breast cancer gene (BRCA1 gene) with LOD of 10 aM. Gold nanoparticle clusters conjugated with a DNA-reporter probe were used for the hybridization with the DNA target, reaching a considerable increase of mass on the QCM surface, even at a very low concentration of the analyte. The high sensitivity of the QCM genosensor allowed the application in the early diagnosis of breast cancer and other types of cancer. In addition, this effective DNA detection protocol can be extended to the direct analysis of any non-amplified genomic DNA. Li et al. (2018) constructed a DNA nanostructure-based platform that enables a recyclable biointerface for ultrasensitive detection of nucleic acid. A chemically cross-linked branched DNA nanostructure was constructed as a probe DNA to obtain the biointerfaces for detection. This strategy increased considerably the exposition of the DNA probe to the solution enhancing the QCM signal dramatically. DNA functionalized $Fe_3O_4$ nanoparticles were used for further signal amplification, reaching a limit of detection of 500 fM.

Damiati et al. (2018) have explored a bacterial surface layer protein (Sbpa) as a basic platform to bind nanomolecules. A gold surface functionalized with folate modified Sbpa was used as a biosensor platform for breast cancer cells with a limited detection range of $1 \times 10^5$ cells $mL^{-1}$ and higher efficiency than traditional biosensors. Folate was used to recognize folate receptors highly expressed on cell membranes of breast cancer cells. The developed biosensor was able to distinguish between MCF-7 cells (breast cancer) and HepG2 cells (liver cancer) since the latter do not express folate receptors. In addition, this technique allows monitoring of the cell viability, opening future perspectives to investigate cellular response to chemo-therapeutic drugs (Damiati et al. 2018).

Nanoporous quartz crystal microbalance was developed by Park and Lee (2018) to detect epidermal growth factor receptor (EGFR) exon 19 mutations, an important indicator for treating lung cancer. Patients treated with EGRF inhibitor drugs have a longer life expectancy; however, the drug performance decreases with activating mutations on EGFR. The most common activating mutations are reported as deletions in exon 19. So, the monitoring of mutations in EGFR can be used to evaluate drug efficiency and treatment of lung cancer. The nanoporous structure was built on the QCM electrode through simple electrochemical deposition and printing processes, followed by the immobilization of the DNA probe on the nanoporous QCM electrode. The QCM signal was detected after DNA hybridization between probe DNA and EGFR mutant DNA. The recognition of EGFR mutation occurred within 30 mins with a limit detection of 1 nmol $L^{-1}$ (Park et al. 2018).

Yang et al. (2019) used nanorods for QCM signal amplification during the detection of poly(ADP-ribose) polymerase-1 (PARP-1), a potential biomarker and chemotherapeutic target for cancer. PARP-1, activated by the specific DNA, cleaves nicotinamide dinucleotide (NAD+) into nicotinamide and ADP-ribose in order to synthesize a ramified poly(ADP-ribose) polymer (nanorods) (Park and Lee 2019). Since microbalance is not very sensitive to PAR detection, positively charged cetyltrimethylammonium bromide (CTAB)-coated gold nanorods (GNRs) were added to the QCM chamber to improve the signal. The strong electrostatic interaction between CTAB nanorods and the negatively charged PAR resulted in high sensitivity with PARP-1 detection ranging from 0.06 to 3 nM and LOD of 0.04 nM. Methodology is promising for clinical diagnosis and drug screening since it can be used to study PARP-1 inhibitors and to detect PARP-1 activity in real cancer cells lysate (Park and Lee 2019).

Many biosensors have been built using molecular impression polymers (MIP) that can substitute antibody, antigen, and microorganisms as bio-recognition systems (Bayramoglu et al. 2019; Chen et al. 2016; Dejous et al. 2016). High selectivity of binding, similarity with natural antibodies, high thermal and chemical stability, compatibility with different carriers and ease and speed in their synthesis are the main advantages of MIPs (Mujahid et al. 2018; Cao et al. 2019). MIPs are designed to contain cavities formed by polymeric reactions that will interact specifically with target molecules (Emir Diltemiz et al. 2017; Poller et al. 2017). The combination of MIPs with QCM has become an extremely favorable detection system for the unlabeled recognition of various molecules of biological interest of different sizes, such as small molecules (volatile organic compounds), proteins, nucleic acid, antibodies, among others. MIPs for small molecules have been well-developed, although preparations for protein-MIPs remain a challenge (Chen et al. 2016) Qiu et al. (2018) have developed a QCM MIP for specific recognition of sialic acid in human urine. Sialic acid is a biomarker generally found in blood serum and used as an early-stage indicator of some cancers or cardiovascular diseases. The gold surface of QCM was modified by self-assembling allylmercaptan to introduce polymerizable double bonds on the gold surface. Next, sialic acid MIP nanofilm was attached to the modified QCM sensor to detect sialic acid in urine samples. The results have shown a linear response for sialic acid in urine in the range of 0.025–0.50 $\mu$mol L$^{-1}$ and a 1.0 nmol L$^{-1}$ limit of detection (LOD) (Qiu et al. 2018).

Several diseases, including lung cancer, can be related to the emitted volatile organic compounds (VOCs) from the human body, such as aldehydes, carboxylic acids, ketones, among others (Thorn and Greenman 2012). Odors coming from skin, breath, and sweat are related to the metabolic profile of healthy or sick people (Liu et al. 2018; D'Amico et al. 2008; Jha and Hayashi 2015). Therefore, the air around human beings contains significant information about the internal chemistry of the body, providing a very good tool to identify diseases. The potential of using odor in clinical diagnosis has encouraged the development of biosensor devices to be applied as an electronic nose for disease detection in a noninvasive way. Nowadays, gas chromatography with a mass spectrometry (GC/MS) is a commonly used analysis method to measure body odors (Pandey and Kim 2011). However, this technique is high cost, complex for operation, and limited to the laboratory level, which highlights the demand to build devices useful for real-time detection of body odors. Liu et al. (2018) have developed a MIP-nanobead-coated QCM array to detect carboxylic acid vapors in human body odor. MIP nanobeads (150–200 nm) were prepared by precipitation polymerization with methacrylic acid (MAA) as a functional monomer, trimethylolpropane trimethacrylate (TRIM) as a crosslinker, and carboxylic acids (Propionic acid, Hexanoic acid, and Octanoic acid the template molecules. Compared to previous work based on MIP films (Jha and Hayashi 2015), the MIP nanobead sensor showed improved sensitivity and selectivity for carboxylic acid vapors with concentrations as low as the ppm level (7.64 ppm for PA, 5.38 ppm for HA, and 4.11 ppm for OA) (Liu et al. 2018).

## 4      Diabetes

Diabetes is a chronic disease that occurs when the pancreas does not produce enough insulin or when the body cannot efficiently use the insulin produced. Millions of people around the world are affected by this disease and need an easy diagnosis and efficient disease control. Insulin is one of the main biomarkers of Diabetes, and its detection is crucial in the clinical diagnosis and monitoring therapy. Some analytical methods (radioimmunoassay, ELISA, and chromatography) are classic to detect insulin in the blood, but they are expensive, time-consuming, and require a professional operator. Kartal et al. (2019) fabricated a QCM-MIP device for real-time detection of insulin in artificial plasma. The QCM sensor surface was coated by a Poly(hydroxyethyl methacrylate)-N-methacryloyl-(L)-histidine methyl ester-based film to allow insulin detection. The insulin imprinted polymer-based QCM device displayed high sensitivity and stability, good selectivity, and a very low detection limit of 0.00158 ng mL$^{-1}$ at physiological pH (Kartal et al. 2019).

A multichannel quartz crystal microbalance array (MQCM) with three pairs of gold electrodes was engineered for the detection of two gaseous biomarkers, acetone and nitric oxide (NO), which are respectively related to diabetes and pulmonary diseases (Tao et al. 2016). The gold electrodes were deposited symmetrically on the sensor surface using photolithography, sputtering, and lift-off technologies. Two types of nanocomposites, titanium dioxide multiwalled carbon nanotubes and cobalt (II)phthalocyanine/silica, were synthesized as sensing materials to coordinate the target gases on the sensor surface. Acetone coordinates on multiwalled carbon nanotubes and NO coordinates on cobalt (II)phthalocyanine to give, respectively, a detection range of 4.33–129.75 ppmv for acetone and 5.75–103.45 ppbv for NO. The interference coming from resonators can be negligible if the resonators are constructed in an appropriate position and larger distances from each other (Tao et al. 2016).

A QCM-based biosensor for enzymatic detection of hemoglobin A1c (HbA1c) in whole blood was also developed for diabetes diagnosis (Park and Lee 2018). HbA1c is generated in the blood through the binding of glucose to N-terminal valine residue of one or both hemoglobin β-chains. One of the ways to quantify HbA1C in clinical laboratories is through the enzymatic assay using fructosyl amino acid oxidase (FAOD), which catalyzes the reaction of fructosyl amino acid with $H_2O$ and $O_2$ to generate $H_2O_2$. The concentration of $H_2O_2$ is proportional to that of HbA1c, allowing the development of a QCM biosensor for enzymatic detection of HbA1c. Mass change was measured by the size enlargement of the gold nanoparticles conjugated with thiol terminated Self Assembled Monolayers (SAMs) on the sensor surface. The $H_2O_2$ generated by the enzymatic reaction catalyzes the reduction of $HAuCl_4$ to Au (0), which is stacked on the surface of the colloidal gold nanoparticle, increasing the mass on the sensor surface. The proposed QCM biosensor was able to quantitatively analyze HbA1c with a detection limit of 0.147% HbA1c with respect to hemoglobin (Park and Lee 2018).

## 5 Neglected Diseases

Neglected diseases such as Dengue, Malaria, Tuberculosis, Leprosy, and others are a diverse group of communicable diseases that prevail in tropical and subtropical countries. They are caused by a variety of pathogens such as bacteria, viruses, helminths, and protozoa, and affect populations living in poverty, without adequate sanitation, and in contact with infectious vectors and domestic animals.

Dengue is caused by dengue virus (DENV), a member of the *Flaviviridae* family, currently classified in four serotypes called DENV-1, DENV-2, DENV-3, and DENV-4. The four serotypes differ in the nucleotide sequence by 25–35 base pairs, and each one can cause dengue (Mustafa et al. 2015; Parkash and Shueb 2015). The newest serotype, DENV-5, was recently identified through the screening of viral samples collected from an infected patient (Normile 2013). As clinical features of dengue are indistinguishable from other infectious diseases such as malaria, Zika, and chikungunya, the developed devices for dengue may also be applied for those diseases. QCM has been widely employed to detect viruses and other microorganisms (Saylan et al. 2019), using different sensing strategies such as bacterial cellulose nanocrystal (dengue) (Pirich et al. 2017), oligonucleotide-functionalized gold nanoparticles (dengue) (Chen et al. 2009), probe immobilized on silver QCM (malaria) (Wangmaung et al. 2014), self nanowell-based aptasensor (influenza virus) (Wang et al. 2017), trisaccharide-grafted conducting polymers (influenza virus) (Hai et al. 2017); among others (Park and Lee 2019; Afzal et al. 2017; de Almeida and Carabineiro 2013; Zhang et al. 2017).

A QCM biosensor was developed with thin films of bacterial cellulose (CN) nanocrystals to improve the sensitivity and the binding of the monoclonal immunoglobulin G (IgGNS1) for specific detection of the nonstructural protein 1 (NS1) of dengue. The system could detect NS1 serum protein, by QCM and QCM-D, with the lowest blood dilution reported so far (10-fold dilution). The limit detections were 0.1 $\mu$g mL$^{-1}$ for QCM-D and 0.32 $\mu$g mL$^{-1}$ for QCM showing that both microbalances are sensitive to detect dengue (Pirich et al. 2017). Chen et al. (2009) have engineered a quartz crystal microbalance (QCM) genosensor combining DNA-functionalized gold nanoparticles (AuNP probes) to detect DENV-2. Two types of specific Au-NPs (probes) were linked by the target DNA sequences onto the sensor surface of the QCM to work as signal amplifiers and target recognition systems. The genosensor was able to detect up to 2 PFU/mL of analyte in serum infected with dengue. The sensitivity and specificity of the method were comparable to the fluorescent real-time PCR assay (Chen et al. 2009).

Molecular imprinting polymers (MIPs) have been successfully used to detect dengue (Afzal et al. 2017; Li et al. 2019; Jia et al. 2018; Darwish et al. 2018). Qiu et al. (2018) were the first to demonstrate the detection of dengue virus using QCM and molecularly imprinted polymer (MIP). MIP was constructed on gold sensor to bind the NS-1 specific protein was constructed on the gold sensor of the QCM through the cross-linking of polymeric monomers in the presence of the epitope of NS-1. The QCM device was able to detect dengue in serum samples collected from infected patients with a detection limit in the range of 1–10 $\mu$g mL$^{-1}$. (Qiu et al.

2018) Later, Lieberzeit et al. (2016) used deactivated DENV molecularly imprinted polymer to coat the QCM surface and detect dengue. The QCM-MIP showed a considerable sensor effect between 100 and 400 Hz, while the non-imprinted platform was not sensitive to the DENV (Lieberzeit et al. 2016).

Malaria is caused by the *Plasmodium* spp., which is transmitted by mosquitoes of *Anopheles* spp. The more common plasmodia for humans are *P. vivax* and *P. falciparum* that infect the red blood cells after multiplication in the liver. Despite the advantages of the QCM technique, few biosensor devices have been developed in the last years using this technique to detect malaria. A silver fabricated QCM crystal was constructed through the biotinylated malaria probe (amplified DNA fragment of 18s rRNA gene) immobilized on the silver surface of the QCM. The new device was able to detect single and mixed plasmodium (*P. falciparum* and *P. vivax*) as well as identifying false positives and misdiagnosis in blood samples from patients suspected of infection by malaria (Wangmaung et al. 2014). A similar approach based on gold QCM was used by the same group to detect *P. falciparum* at the sub-nanogram concentration (Potipitak et al. 2011). Besides dengue and malaria, other neglected diseases have been detected by QCM devices (de Santana et al. 2018).

Leprosy and tuberculosis are both caused by specific mycobacteria, *Mycobacterium leprae* and *Mycobacterium tuberculosis,* respectively. These diseases are steadily increasing around the world and need reliable biosensors to detect them at an early stage of the disease. Santana et al. (2018) have used a mimotope-derived synthetic peptide (Ag85B peptide) to bind antibodies in leprosy diagnosis. QCM surface was coated by a protein previously conjugated to a phage-displayed mimotope specific to the *Mycobacterium leprae* antigen. The new device was sensitive enough to detect, for the first time, the anti-*M. leprae* antibodies in human serum, even in the sera of paucibacillary (low bacillary load) patients (de Santana et al. 2018).

More recently, polymeric nanoparticles printed with the epitope of the *Mycobacterium leprae* bacteria were electropolymerized on the electrochemical quartz crystal microbalance (EQCM) gold sensor to detect leprosy. The EQCM sensor was able to recognize the epitope of *Mycobacterium leprae* in infected blood samples without the interference of similar peptides or plasma proteins. The sensor fulfills all requirements for a reliable analytical tool with a detection limit of 0.161 nM and a quantification limit of 0.536 nM. The very LOD allows the sensor to be used to detect the early stages of leprosy (Kushwaha et al. 2019). Recently, Zhou et al. (2019) engineered a QCM device with silver nanoparticles (Ag-NPs) to recognize biomarkers for latent tuberculosis infection such as interferon-gamma (IFN-$\gamma$), tumor necrosis factor-alpha (TNF-$\alpha$), and interleukin-2 (IL-2). Silver nanoparticles (Ag-NPs) were linked to the specific antibodies in order to act as a signal amplifier and mass nanoprobe on the QCM surface. After each biomarker detection, the mass nanoprobe was dissolved by a chemical oxidation reaction with hydrogen peroxide to prevent steric hindrance and allow the continuous monitoring of the other biomarkers. This new methodology avoids acoustic interference compared to the multichannel QCM sensing and allows detection limits for TNF-α, IFN-γ, IL-2 of 7.3, 6.3, and 7.8 fg mL$^{-1}$, respectively (Ma et al. 2020).

## 6 Other Diseases: Influenza, HIV, Infections

Among the diseases already mentioned, others such as influenza, Alzheimer, HIV, and bacterial infections deserve comments about the detection strategies using QCM. Influenza, for example, is a disease easily transmitted by viruses whose characteristics change quickly, making diagnosis difficult. Avian influenza virus was successfully detected by the QCM technique using a conducting polymer covalently bound to saccharides with an affinity for the H1N1 virus. Specific interaction of 2,6-sialylactose with hemagglutinin in the envelope of human influenza H1N1 was detected by QCM with a detection limit of 0.12 HAU, two orders of magnitude higher than the commercially available detection kit (Hai et al. 2017). Another strategy was adopted by Wang et al. to detect the Avian influenza virus using a nanowell-based QCM aptasensor. The detection limit was 2–4 HAUs/50 µL with an analysis time of 10 min (Wang et al. 2017). Abdul Rasheed and Sandhyarani (2016) have detected HIV-1 antigen using QCM and gold nanoparticles to amplify the signal detection. The obtained limit of detection was 1 ng mL$^{-1}$, which was able to detect the HIV infection in the window phase.

Dysentery, syphilis, cholera, meningitis, pneumonia, etc. are all diseases caused by bacteria. *Salmonella*, for example, a leading cause of foodborne illness, was detected by Makhneva et al. (2018) using the antibody immobilization by the glutaraldehyde activation on cyclopropylamine polymers film. The detection limit reached by the immunosensor was 105 CFU mL$^{-1}$, with the advantage that it can be regenerated several times using 10 mM NaOH, enabling new measurements. ssDNA aptamers were selected for the first time by a whole-bacterium SELEX technique against *E. coli* O157:H7 and applied to the development of a QCM-aptasensor able to detect live *E. coli* O157:H7. The limit of detection was as low as $1.46 \times 10^3$ CFU mL$^{-1}$ with a response time of 50 min (Li et al. 2018). Other bacteria were also detected successfully by quartz crystal microbalance, such as *Listeria monocytogenes* (Wachiralurpan et al. 2020) *Bacillus anthracis* (Hao et al. 2011), and *Campylobacter jejuni* (Masdor et al. 2016).

## 7 Future Directions

In this part, we have presented the principle of the piezoelectric biosensor based on the QCM technique. We have summarized various examples and processes used for bioanalysis emphasizing diagnosis. It is clear that there are ever so many demanding situations, however, we might also consider some current problems making the repeatability of those devices more difficult: (1) heterogeneity existing at the nano-scale level, (2) stable form, size, and surface modification of nanomaterials, (3) high variability of such produced biosensors often depending on not optimized preparation techniques and not completely described properties of the components, and (4) general issues with immobilization of biological components on solid-state surfaces. In this sense, studies of immobilization and modification of the surface and biological components are very important in order to produce a more robust and reliable device.

The optimization of molecular recognition by the rational design of proteins in combination with the development of QCM sensor matrix substrates can lead to the innovation of a QCM platform for rapid analysis with high precision and sensitivity at a reduced cost. This type of label-free sensor can be used for various applications, including biotechnology, clinical diagnosis, and detection of infectious or pathogenic species, and even in forensic applications in addition to other related bioassays.

# References

Abdul Rasheed P, Sandhyarani N (2016) Anal Chim Acta 905:134–139

Afzal A, Mujahid A, Schirhagl R, Bajwa S, Latif U, Feroz S (2017) Chemosensors 5:7–31

Asai N, Shimizu T, Shingubara S, Ito T (2018) Sens Actuators B Chem 276:534–539

Bayramoglu G, Ozalp VC, Oztekin M, Arica MY (2019) Talanta 200:263–271

Cao YR, Feng TY, Xu J, Xue CH (2019) Biosens Bioelectron 141:111447–111464

Chen SH, Chuang YC, Lu YC, Lin HC, Yang YL, Lin CS (2009) Nanotechnology 20:215501–215510

Chen L, Wang X, Lu W, Wu X, Li J (2016) Chem Soc Rev 45:2137–2211

Chung C, Christianson M (2014) J Oncol Pharm Pract 20:11–28

Cornelio VE, Pedroso MM, Afonso AS, Fernandes JB, da Silva MF, Faria RC, Vieira PC (2015) Anal Chim Acta 862:86–93

Damiati S, Peacock M, Mhanna R, Søpstad S, Sleytr UB, Schuster B (2018) Sens Actuators B Chem 267:224–230

D'Amico A, Di Natale C, Paolesse R, Macagnano A, Martinelli E, Pennazza G, Santonico A, Bernabei M, Roscioni C, Galluccio G, Bono R, Agro EF, Rullo S (2008) Sens Actuators B Chem 130:458–465

Darwish NT, Sekaran SD, Khor SM (2018) Sens Actuators B Chem 255:3316–3331

de Almeida MP, Carabineiro SAC (2013) Gold Bull 46:65–79

de Santana JF, da Silva MRB, Picheth GF, Yamanaka IB, Fogaca RL, Thomaz-Soccol V, Machado-de-Avila RA, Chavez-Olortegui C, Sierakowski MR, de Freitas RA, Alvarenga LM, de Moura J (2018) Talanta 187:165–171

Dejous C, Hallil H, Raimbault V, Lachaud JL, Plano B, Delepee R, Favetta P, Agrofoglio L, Rebiere D (2016) Sensors 16:915–930

Dixon MC (2008) J Biomol Tech 19:151–158

Dong ZM, Jin X, Zhao GC (2018) Biosens Bioelectron 106:111–116

Emir Diltemiz S, Kecili R, Ersoz A, Say R (2017) Sensors 17:454–472

Farka Z, Jurik T, Kovar D, Trnkova L, Skladal P (2017) Chem Rev 117:9973–10042

Fenzl C, Hirsch T, Baeumner AJ (2016) Trends Anal Chem 79:306–316

Hai W, Goda T, Takeuchi H, Yamaoka S, Horiguchi Y, Matsumoto A, Miyahara Y (2017) ACS Appl Mater Interfaces 9:14162–14170

Hao RZ, Song HB, Zuo GM, Yang RF, Wei HP, Wang DB, Cui ZQ, Zhang Z, Cheng ZX, Zhang XE (2011) Biosens Bioelectron 26:3398–3404

Hwang SS, Chan H, Sorci M, Van Deventer J, Wittrup D, Belfort G, Walt D (2019) Anal Biochem 566:40–45

Jackman JA, Cho NJ, Nishikawa M, Yoshikawa G, Mori T, Shrestha LK, Ariga K (2018) Chem Asian J 13:3366–3377
Jaruwongrungsee K, Waiwijit U, Wisitsoraat A, Sangworasil M, Pintavirooj C, Tuantranont A (2015) Biosens Bioelectron 67:576–581
Jayanthi VSPKSA, Das AB, Saxena U (2017) Biosens Bioelectron 91:15–23
Jearanaikoon P, Prakrankamanant P, Leelayuwat C, Wanram S, Limpaiboon T, Promptmas C (2016) J Virol Methods 229:8–11
Jha SK, Hayashi K (2015) Talanta 134:105–119
Jia MF, Zhang Z, Li JH, Ma X, Chen LX, Yang XB (2018) Trends Anal Chem 106:190–201
Justino CIL, Duarte AC, Rocha-Santos TAP (2016) Trends Anal Chem 85:36–60
Kanazawa KK, Gordon JG (1985) Anal Chim Acta 175:99–105
Kartal F, Cimen D, Bereli N, Denizli A (2019) Korean J Couns Psychother 97:730–737
Kushwaha A, Srivastava J, Singh AK, Anand R, Raghuwanshi R, Rai T, Singh M (2019) Biosens Bioelectron 145:111698–111706
Lane TJ, Cheng CYH, Dixon MC, Oom A, Johal MS (2008) Anal Chem 80:7840–7845
Latif U, Qian J, Can S, Dickert FL (2014) Sensors (Basel) 14:23419–23438
Li F, Dong Y, Zhang Z, Lv M, Wang Z, Ruan X, Yang D (2018) Biosens Bioelectron 117:562–566
Li R, Feng Y, Pan G, Liu L (2019) Sensors 19:177–210
Lieberzeit PA, Chunta S, Navakul K, Sangma C, Jungmann C (2016) Procedia Eng 168:101–104
Liu C, Shang L, Yoshioka HT, Chen B, Hayashi K (2018) Anal Chim Acta 1010:1–10
Ma D, Zheng SR, Fan J, Cai SL, Dai Z, Zou XY, Teng SH, Zhang WG (2020) Anal Chim Acta 1095:212–218
Mahmoodi P, Fani M, Rezayi M, Avan A, Pasdar Z, Karimi E, Amiri IS, Ghayour-Mobarhan M (2019) Biofactors 45:101–117
Makhneva E, Farka Z, Skládal P, Zajíčková L (2018) Sens Actuators B Chem 276:447–455
Marx KA (2003) Biomacromolecules 4:1099–1120
Masdor NA, Altintas Z, Tothill IE (2016) Biosens Bioelectron 78:328–336
Mujahid A, Mustafa G, Dickert FL (2018) Biosensors (Basel) 8:52–72
Mustafa MS, Rasotgi V, Jain S, Gupta V (2015) Med J Armed Forces India 71:67–70
Normile D (2013) Science 342:415–415
Pandey SK, Kim K-H (2011) Trends Anal Chem 30:784–796
Park HJ, Lee SS (2018) Sens Actuators B Chem 258:836–840
Park HJ, Lee SS (2019) Analyst 144:6936–6943
Park H, You J, Park C, Jang K, Na S (2018) J Mech Sci Technol 32:1927–1932
Parkash O, Shueb RH (2015) Viruses 7:5410–5427
Peng S, Derrien TL, Cui J, Xu C, Luo D (2012) Mater Today 15:190–194
Pirich CL, de Freitas RA, Torresi RM, Picheth GF, Sierakowski MR (2017) Biosens Bioelectron 92:47–53
Poller AM, Spieker E, Lieberzeit PA, Preininger C (2017) ACS Appl Mater Interfaces 9:1129–1135
Potipitak T, Ngrenngarmlert W, Promptmas C, Chomean S, Ittarat W (2011) Clin Chem Lab Med 49:1367–1373
Qiu X, Xu XY, Chen X, Wu Y, Guo H (2018) Anal Bioanal Chem 410:4387–4395
Ranjan R, Esimbekova EN, Kratasyuk VA (2017) Biosens Bioelectron 87:918–930
Rodriguez-Pardo L, Rodriguez JF, Gabrielli C, Perrot H, Brendel R (2005) IEEE Sens J 5:1251–1257
Roointan A, Ahmad Mir T, Ibrahim Wani S, Mati Ur R, Hussain KK, Ahmed B, Abrahim S, Savardashtaki A, Gandomani G, Gandomani M, Chinnappan R, Akhtar MH (2019) J Pharm Biomed Anal 164:93–103
Sauerbrey G (1959) Zeitschrift Fur Physik 155:206–222
Saylan Y, Akgonullu S, Yavuz H, Unal S, Denizli A (2019) Sensors 19:1279–1297
Shandilya R, Bhargava A, Bunkar N, Tiwari R, Goryacheva IY, Mishra PK (2019) Biosens Bioelectron 130:147–165
Skládal P (2016) Trends Anal Chem 79:127–133

Speight RE, Cooper MA (2012) J Mol Recognit 25:451–473

Tabaei SR, Jackman JA, Kim SO, Zhdanov VP, Cho NJ (2015) Langmuir 31:3125–3134

Takada K, Naal Z, Park JH, Shapleigh JP, Bernhard S, Batt CA, Abruna HD (2002) Langmuir 18:4892–4897

Takeda K, Naal Z, Abruna HD (2003) Langmuir 19:5402–5406

Tao W, Lin P, Ai Y, Wang H, Ke S, Zeng X (2016) Anal Biochem 494:85–92

Teramura Y, Takai M (2018) In: The Surface Science Society of Japan. Compendium of surface and interface analysis. Springer, Singapore, pp 509–520

Thorn RM, Greenman J (2012) J Breath Res 6:024001–024025

Tomaic V (2016) Cancers 8:95–116

W. H. Organization (2018). https://www.who.int/news-room/fact-sheets/detail/cancer

Wachiralurpan S, Chansiri K, Lieberzeit PA (2020) Sens Actuators B Chem 308:127678–127686

Wang L, Wang R, Chen F, Jiang T, Wang H, Slavik M, Wei H, Li Y (2017) Food Chem 221:776–782

Wangmaung N, Chomean S, Promptmas C, Mas-oodi S, Tanyong D, Ittarat W (2014) Biosens Bioelectron 62:295–301

Yang H, Li P, Wang D, Liu Y, Wei W, Zhang Y, Liu S (2019) Anal Chem 91:11038–11044

Yarman A, Kurbanoglu S, Jetzschmann KJ, Ozkan SA, Wollenberger U, Scheller FW (2018) Curr Med Chem 25:4007–4019

Zhang B, Mao Q, Zhang X, Jiang T, Chen M, Yu F, Fu W (2004) Biosens Bioelectron 19:711–720

Zhang X, Feng Y, Yao Q, He F (2017) Biosens Bioelectron 98:261–266

Zhou B, Hao Y, Long D, Yang P (2018) Biosens Bioelectron 111:90–96

Zhou B, Hao Y, Chen S, Yang P (2019) Mikrochim Acta 186:212–219

# Scanning Electrochemical Microscopy (SECM): Fundamentals and Applications

Mauro Bertotti

## 1 Introduction

Scanning electrochemical microscopy (SECM) was first introduced in 1989 by two independent research groups (Bard and Engstrom) as a scanning probe technique used to probe surfaces and surface reactions based on electron transfer processes that take place through a microelectrode tip (Bard and Mirkin 2012; Engstrom and Pharr 1989). The technique belongs to the broader class of scanning probe microscopy and differs from other well-known scanning probe methods such as scanning tunneling (STM) and atomic force microscopy (AFM) because the information from the investigated substrate concerns to its electrochemical reactivity in liquid environments and can be quantitatively obtained through well-developed and rigorous theory (Bard and Mirkin 2012). Because SECM employs microelectrodes or, more recently, nanoelectrodes as a tip, spatially resolved phenomena localized at interfaces with micrometric resolution can be investigated, as opposed to more general electrochemical methods. It should be pointed out that the term SECM is employed to designate both the technique and the corresponding microscope.

The SECM signal is founded on the concept of diffusion-limited current that flows through an electrode with micrometric dimensions (radial diffusion) (Forster 1994). In a typical SECM experiment, the tip is positioned in the $x,y,z$ planes in close distance towards a sample surface with the application of piezoelectric elements and is then scanned stepwise at a certain scan rate according to a user-defined grid. The steady-state current is recorded at the microelectrode continuously during its movement. Because of the unique features of microelectrodes, the localized hemispherical diffusion zone of the investigated species will be influenced by the topography and/or electrochemical reactivity of a nearby surface. Information on the topography

M. Bertotti (✉)
Department of Fundamental Chemistry, Institute of Chemistry, University of São Paulo (USP), São Paulo, SP, Brazil
e-mail: mbertott@iq.usp.br

© The Author(s), under exclusive license to Springer Nature Switzerland AG 2022
L. T. Kubota et al. (eds.), *Tools and Trends in Bioanalytical Chemistry*,
https://doi.org/10.1007/978-3-030-82381-8_18

of a surface can be assessed by monitoring the extent by which the transport of an electroactive probe dissolved in solution is hindered, as such a current decrease should be noted when the diffusion layer is disrupted for an interrogated insulating surface very close to the tip. The electrochemical reactivity of a conducting substrate is usually evaluated by taking into account the extent of feedback associated with the regeneration of a probe species by the sample surface. By using suitable approaches, the effect of both topography and surface reactivity on the tip current can be properly deconvoluted.

The resolution in SECM depends mainly on the dimension of the tip and high-resolution images can be obtained by decreasing the size of the electrode. Even though the resolution of SECM images has been greatly improved, the technique cannot compete with that of AFM and STM (Engstrom and Pharr 1989). However, the electrochemical nature of the SECM tip makes it advantageous over AFM and STM for some studies. For instance, information on the electrochemical reactivity of a substrate, corrosion mechanisms, fuel cell catalyst evaluation, activity of biorecognition elements immobilized on a surface, flux of ions through pores of semi-permeable membranes, and the monitoring of the activity of living biological cells can be achieved with SECM. The continuous development of different approaches to fabricate tips with nanometric dimensions has greatly increased the ability of SECM to bring information on the analysis and characterization of single cells, molecules, and particles (Kim et al. 2016c; Amemiya 2015; Yi-Lun et al. 2017). In this chapter, the principles of SECM will be presented and recent advances in the field will be discussed. *Readers* are also referred to *a number of general reviews* that have been published in recent years covering the whole range of applications of SECM (Wittstock et al. 2007; Sun et al. 2007; Li et al. 2016; Polcari et al. 2016; Zoski 2016; Huang et al. 2018; Izquierdo et al. 2018; Kai et al. 2018).

## 2  SECM Instrumentation, Probes and Modes of Operation

### 2.1  SECM Instrumentation and Probes

A typical SECM equipment is constituted by four main parts: an electrochemical cell, a potentiostat (or bipotentiostat), a high-resolution positioning system, and a computer. The electrochemical cell is a conventional one and includes working (microelectrode), reference, and auxiliary electrodes (Bard et al. 1990). The interrogated sample is fixed to the cell bottom. The potentiostat is used to control the tip potential and a bipotentiostat is required in experiments where the substrate has to be biased. The equipment should be able to measure a large range of current values, from nA to fA for the tip, and higher currents for substrates with large area. Since very low currents should be measured at the tip and high precision tip positioning is needed, electronic noises and mechanical vibrations should be avoided. This can be accomplished by keeping the electrochemical cell inside a Faraday cage mounted on an anti-vibration table.

The SECM probe is generally a disk microelectrode sharpened to allow a closer approach to the substrate surface. Disk microelectrodes are the most used because they can be easily fabricated and the theoretical formulation is straightforward, even though microelectrodes with other geometries such as cones, hemispheres, and spheres can also be used as SECM tips (Lee et al. 2001). Disk microelectrodes are produced by sealing a microwire (typically platinum or gold) into pulled glass capillaries under vacuum using a laser pipette puller. Optimization of pulling parameters is very important to control the shape and size of the tip and relevant features of the tip geometry are the radius of the conductive fiber (a) and the total tip radius (r, the thickness of the insulating sheath plus the fiber). The ratio r/a is defined as the RG and such tip feature is relevant because of the extent of the hindered diffusion. For instance, for a microelectrode with RG = 10, the current is almost half of that of a microelectrode with RG = 2 over an insulating substrate and at L = 0.1, where L is the normalized distance (L is the ratio between the tip–substrate distance and the radius of the tip).

Potentiometric probes are less common and have been used especially for pH measurements over surfaces in corrosion processes and biological studies. Such probes have some disadvantages such as slow response time and short lifetime. Moreover, the positioning is not easy, even though there are some strategies to overcome such difficulty. These are mainly based on the use of dual-function SECM tips, i.e., the tip can operate independently for potentiometric pH measurements and in amperometric mode to evaluate the tip–substrate distance (Santos et al. 2016; Zhu et al. 2018a).

As the current response in SECM is highly dependent on tip-to-substrate distance, approaches have been developed to get electrochemical information at the tip with negligible contribution from the substrate topography. These include the combination of SECM with other techniques such as AFM, shear force, and others (Macpherson and Unwin 2000; Hengstenberg et al. 2000; Kurulugama et al. 2005; Fan and Bard 1999; Comstock et al. 2010; Lai et al. 2011).

## 2.2    Feedback Mode

Consider a disk-shaped micro or nanoelectrode immersed in a solution containing an electroactive species. For relatively low scan rates, radial rather semi-infinite linear diffusion will be the dominant mode of mass-transport and the voltammogram will have a sigmoidal shape. In this case, the transport to the electrode is much more effective owing to the expanding hemispherical layer which extends out into the solution, in comparison to a plane projecting into the solution for a diffusion layer that is much smaller than the size of the electrode. As a result, at potentials that exceed the redox potential of the electroactive species such special diffusion feature leads to a rapid establishment of a steady-state diffusion-limited current.

If the probe is positioned at the bulk solution (the separation between tip and substrate is greater than 10 times the radius of the conducting fiber), current will be equal to 4nFDCr, where n is the number of electrons exchanged in the redox

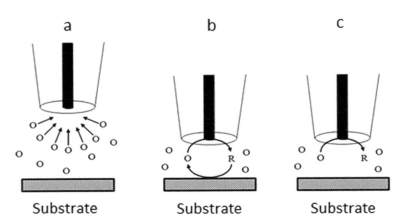

**Fig. 1** Electrochemical reduction of an oxidized species (O) at a microelectrode positioned far from a substrate (**a**) and close to a conductive (**b**) and an insulating (**c**) substrate

reaction, F is the Faraday constant, D is the diffusion coefficient of the electroactive species, C is the concentration of the electroactive species, and r is the radius of the microelectrode (Zoski 2002). The sensing zone of a microelectrode at steady-state conditions is localized within a delimited volume at the electrode/solution interface. Hence, the response of such microelectrode in a homogeneous solution containing an electroactive species does not depend on its actual position if it is located in the bulk solution. However, when the microelectrode is attached to a precise positioning device and carefully brought into close vicinity of an insulating substrate (a distance of 2–5 times that of the tip diameter), the hemispherical diffusion layer can interact with this surface. As a consequence, the presence of a nearby surface can hinder or increase the microelectrode response, i.e., the measured current at the tip changes from a bulk response to a local response, and this depends on the nature of the substrate.

When the probe is moved towards an insulating surface, the diffusion of the electroactive species is hindered because of the presence of a physical barrier (substrate) and current decreases in comparison to the bulk value. This hindered diffusion phenomenon (often called "negative feedback") is strongly influenced by the RG value. For instance, tips with larger RG values will experience a much more pronounced hindered diffusion because the walls of the insulating sheath will block more efficiently the transport of the electroactive species from the bulk to the conductive fiber and its corresponding diffusion layer. On the other hand, if the tip is approached to a conductive substrate the electroactive species that were previously reduced (or oxidized) at the tip can be reconverted into its original form. Such redox recycling leads to an increase in the tip response relative to the one measured at the bulk, resulting in the well-known SECM "positive feedback" effect (Fig. 1).

Depending on the nature of the substrate (insulating or conductive), the measurement of the current response as the tip is approached to the surface will lead to a current vs. distance plot named as "approach curve" and such behavior has been

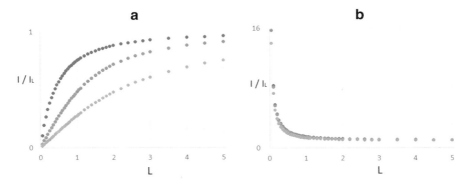

**Fig. 2** Theoretical approach curves (normalized current as a function of the ratio (L) between the tip–substrate distance and the radius of the tip) for hindered diffusion (negative feedback) (**a**) and positive feedback (**b**) for disk microelectrodes with different RG values (1000 (blue circles)), 10 (orange circles) and 1.5 (gray circles) (Amphlett and Denuault 1998)

described using analytical approximations (Amphlett and Denuault 1998; Lefrou and Cornut 2010). Figure 2 shows examples of approach curves for both negative and positive feedback and it is important to notice that RG values have a much stronger influence on approach curves for negative feedback because of the hindering effect. Approach curves are very important to get information on the tip position above a substrate and to assess the electrochemical reactivity of a conductive surface.

## 2.3 Generation and Collection Modes

Generator-Collector mode experiments are similar to those performed with a Rotating-Ring Disk Electrode (RRDE). Accordingly, at the "generator" electrode a redox reaction takes place and the electroactive product subsequently reaches the other electrode ("collector"), where it undergoes the reverse redox reaction. At the SECM apparatus these reactions can take place at the tip or substrate, hence one can have two different set-ups by simply changing polarization levels which define the origin of the redox reaction: Sample Generation–Tip Collection (SG–TC) and Tip Generation–Sample Collection (TG–SC) (Zhou et al. 1992). The SG–TC mode is used for measurements of concentration profiles or chemical flux from a substrate and currents are measured at both the tip and the substrate. As the area of the substrate is usually much larger than the one of the tip, the collection efficiency is very low, but the technique has found applications for both corrosion and enzymatic studies. In the TG–SC mode, the current at both electrodes is also measured and for relatively small tip–substrate distances all materials generated at the tip are collected at the substrate, hence the collection efficiency reaches 100%. Such approach is useful for investigations of homogeneous reaction kinetics (Martin and Unwin 1998a), determination of diffusion coefficients (Martin and Unwin 1998b), and evaluation of catalytic activity of modified surfaces (Fernandez and Bard 2003; Zhao and Wittstock 2005).

## 2.4 Redox Competition Mode

This mode is especially attractive to locally investigate catalytic activity on modified surfaces (Eckhard et al. 2006; Santana et al. 2010; Reisa et al. 2014), to study the activity of immobilized enzymes (Nogala et al. 2010a), or to get information on respiratory activity in single cells (Nebel et al. 2013a; Santos et al. 2017). When scanning at a constant tip–substrate distance in the vicinity of the substrate, the reduction current measured at the tip remains constant above the inactive area. As the SECM tip competes with the sample for the same analyte, a current decrease at the tip is noted when it is in close proximity to an active catalyst site on the surface. As the current is measured at the SECM tip and not the substrate, increased sensitivity is expected in comparison to the SG–TC mode because the background current contribution is significantly decreased.

## 2.5 Potentiometric Mode

As the concentration and oxidation state of the investigated electroactive species are unchanged during studies involving this SECM mode, the tip–substrate distance does not influence the measurements. Moreover, as the target compound is usually recognized by an ion-selective membrane or coating film, advantages such as high selectivity and the possibility of detecting non-electroactive species are expected when potentiometric SECM probes are employed (Horrocks et al. 1993). However, as no feedback effect appears in the potentiometric mode, a different approach is required to control the tip–substrate distance such as optical microscopes or the use of dual-function devices where one of the compartments operates in the amperometric mode (Santos et al. 2016; Izquierdo et al. 2011; Zhu et al. 2018b).

## 3 SECM Applications

A complete discussion on the various applications of SECM is beyond our present scope and *several reviews have appeared* recently in this field seeking to summarize past as well as current work (Wittstock et al. 2007; Sun et al. 2007; Li et al. 2016; Polcari et al. 2016; Zoski 2016; Huang et al. 2018; Izquierdo et al. 2018; Kai et al. 2018). In view of these, the extent of the present review has been restricted to a few groups of papers concerned with central themes involving relevant SECM applications which are of general interest.

## 3.1    Corrosion

SECM can be considered as an ideal tool for the investigation of local corrosion because passivated and active regions of surfaces as well as the formation of relevant species ($Fe^{3+}$, $O_2$, $H^+$, etc.) can be imaged with high resolution. Examples of the use of the technique for studies on precursor regions for pitting corrosion have been carried out on steel (Luong et al. 2003), Ti (Casillas et al. 1993), Al (Davoodi et al. 2007), Ni (Paik and Alkire 2001), Fe (Still and Wipf 1997), Mg (Pereda et al. 2010a), and Cu (Mansikkamaki et al. 2005). To gain information into the localized generation of certain ions during corrosion processes, SECM has been largely used such as in the monitoring of $Fe^{2+}$ at a Pt microelectrode (Volker et al. 2006) or in the potentiometric determination of $Zn^{2+}$ ions (Izquierdo et al. 2012). The SECM tip can also be used to initiate pitting corrosion by local generation of aggressive species such as chloride (Wipf 1994) and bromine (Casillas et al. 1993). A considerable number of studies focused on the development of strategies to prevent corrosion by using coatings, inhibitors, and surface treatments, and SECM have been employed to follow such treatments (Gonzalez-Garcia et al. 2010; Sidane et al. 2011; Dufek and Buttry 2009; Qian et al. 2017; Souto et al. 2010a; Singh et al. 2015). A recent thorough review on this topic can be found elsewhere (Payne et al. 2017).

## 3.2    Energy and Electrocatalysis

SECM has been continuously applied to investigations of heterogeneous chemical processes and chemical reactivity in materials involved in electrochemical energy conversion systems, such as batteries (Szot et al. 2009; Zampardi et al. 2013; Liu et al. 2019), fuel cells (Kucernak et al. 2000; Black et al. 2005; Kishi et al. 2010), biofuel cells (Scodeller et al. 2010; Karnicka et al. 2007), solar cells (Huang et al. 2011; Harati et al. 2010), and supercapacitors (Sumboja et al. 2012; Dey et al. 2017). Two electrocatalytic reactions, the hydrogen oxidation reaction (HOR) and the oxygen reduction reaction (ORR), have received special attention owing to their relevance for fuel cells and other energy applications (Bertoncello 2010). The rate of both reactions is strongly dependent on the nature of the substrate surface and great attention has been devoted to devising materials with superior catalytic activity. As current is influenced by the heterogeneous kinetics, it is not surprising that SECM has been an additional tool to investigate such systems (Amemiya et al. 2016; Park et al. 2017). This is particularly true when the material consists of conducting particles or mediators dispersed in an insulating binder, hence localized information on the active sites can be obtained with high resolution. With the advances in microfabrication technology, researchers are now able to investigate the properties of nanoparticles (NPs). Examples include the study of the geometric property and catalytic activity of individual Pt NPs in the hydrogen oxidation reaction (HOR) (Kim et al. 2016a), the relationship between catalysis and morphology of Pt NPs (Kim et al. 2016c), reactivity of hybrid materials between single or multi-walled carbon nanotubes and biomimetic compounds such as metalloporphyrins and

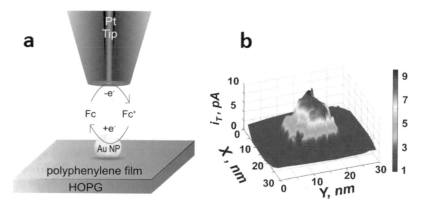

**Fig. 3** Schematic representation of the feedback mode (**a**) and constant-height SECM image of a 10 nm AuNP recorded with a 3 nm-radius tip ($E = 0.4$ V vs. Ag/AgCl in a 1 mM Fc and 0.1 M KCl solution (**b**). Adapted from Ref. (Sun et al. 2014)

metallophthalocyanines (Zagal et al. 2009), application of NiCu coatings for HER (hydrogen evolution reaction) (Solmaz et al. 2008), and the use of metallic NiPS$_3$@NiOOH structures as efficient and durable electrocatalysts for the oxygen evolution reaction (Konkena et al. 2017). Such an example of an SECM image with very high resolution is shown in Fig. 3, where the catalytic activity of a 10 nm AuNP was investigated with a 3 nm-radius Pt microelectrode (Sun et al. 2014).

## 3.3 Measurements of Heterogeneous Electron Transfer (ET) Kinetics

The study of charge transfer processes at solid/liquid interfaces is essential in understanding different chemical and biological processes. In order to get information on ET kinetics under steady-state conditions, the mass transfer rate has to be sufficiently high and the uncompensated resistive potential drop in solution should be sufficiently small so as not to influence the measurements. SECM is a useful tool to determine heterogeneous ET because very high mass transfer rates can be achieved by decreasing the tip–substrate distance (d) (Wipf and Bard 1991) Recent examples include the evaluation of the catalytic activity of surfaces modified with Au nanoparticles (Wain 2013), bimetallic materials (Nagaiah et al. 2009; Sanchez-Sanchez and Bard 2009; Li et al. 2012), carbon nanotube composites (Ma et al. 2015), platinum nanoparticles (Sanchez-Sanchez et al. 2010), and nanoporous Cu (Bae et al. 2016; Kumar et al. 2019). The technique has also been employed to correlate the heterogeneous ET rate to the defect density and defect distance of single-layer graphene (Zhang et al. 2019). It should be pointed out that for accurate measurements, the mass transfer rate (proportional to D/d) must be large in comparison to $k^o$, justifying efforts towards the design of nanogap-SECM devices where nanometer-scale SECM experiments can be performed (Kim et al. 2016b; Ma et al. 2017). For example, a $k^\circ$ value of 17 cm/s was reported for the electrochemical reduction of $Ru(NH_3)_6^{3+}$ at a Pt nanotip above a Au substrate (Martin and Unwin 1998b). Taking into account that interfaces are compositionally and/or structurally heterogeneous, the translation of SECM to the nanoscale has received increased attention because the identification and characterization of the structural properties of a particular surface can be related at the scale of the surface heterogeneities (Bentley et al. 2019a, b).

## 3.4    Studies on Rapid Homogeneous Reactions and Short-Lived Intermediates

The feedback mode of SECM consists of a suitable technique to measure kinetics in homogeneous reactions coupled to electrode reactions because of the high rate of mass transfer attainable at close tip–substrate separations (Unwin and Bard 1991; Cannan et al. 2011). The kinetics of the chemical step is probed by changing the distance between the tip and a conductive substrate in such a way that the diffusion time ($d^2/D$) of a species (Ox) generated at the tip to the substrate can be precisely controlled. In the presence of a homogeneous reaction, little Ox will survive the transit from the tip to the substrate, hence affecting the current measured at the tip owing to regeneration of the reactant. Examples include the determination of the homogeneous rate constant for the dimerization of 4-nitrophenolate in acetonitrile (Treichel et al. 1994), the reactivity of 1,4-dihydropyridines with superoxide (Bollo et al. 2005), the dimerization rate of trans-anethole (Demaille and Bard 1999), studies on electrogenerated nitro radical anions (Bollo et al. 2004), and more complicated systems such as ECE (electrochemical-chemical-electrochemical) and disproportionation reactions (Demaille et al. 1996).

## 3.5    Bio SECM or Electrochemical Imaging of Cells, Multicellular Structures and Tissues

Because of the possibility of positioning the sensor tip in close proximity to an investigated sample, SECM has been increasingly applied in the analysis of biological samples (living cells and immobilized biomacromolecules) to examine and visualize cellular activities with high spatial and temporal resolution (Bauermann et al. 2004). SECM has been used to study the activity of enzymes immobilized on different surfaces, which act as a conducting surface. The magnitude of the positive feedback signal produced has a correlation with the activity of the enzyme and several examples have been reported in the literature (Pierce et al. 1992; Ciobanu et al. 2008; Oyamatsu et al. 2003; Evans et al. 2005; Zhao et al. 2004). The characterization of proteins and corresponding building blocks has been an area of intensive research (Polcari et al. 2017; Alizadeh et al. 2011; Song et al. 2012; Abdelhamid et al. 2015; She et al. 2015), as well as studies on the release of

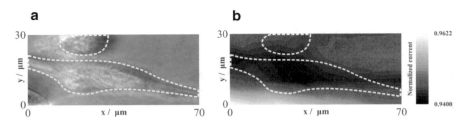

**Fig. 4** Optical image of two cultured HS578T cells (**a**) and normalized current map of the scanned area (500 nm-Pt microelectrode, $E = -0.4$ V vs. Ag/AgCl) (**b**). Adapted from Ref. (Santos et al. 2017)

metabolites of biological relevance from single cells, which depend on very small tips to get localized information in micrometric dimension (Borgmann et al. 2006; Liu et al. 2017; Zhang et al. 2011). SECM has also been a powerful tool to investigate DNA behavior (Yamashita et al. 2001; Wang and Zhou 2002; Turcu et al. 2004; Palchetti et al. 2007; Roberts et al. 2009; Shamsi and Kraatz 2010), to image fluxes of oxygen at living cells (Nebel et al. 2013a; Yasukawa et al. 2000; Nishizawa et al. 2002; Nebel et al. 2013b; Shiku et al. 2001; Date et al. 2011), and to study pH changes in metabolic cycles (Joshi et al. 2017; Barroso et al. 2019) and in the bioelectrochemical activity of yeast cells (Ramanavicius et al. 2017). For instance, Fig. 4 shows an SECM image corresponding to the oxygen consumption over a cell microenvironment, which demonstrates that the technique provides the advantage of high spatial resolution. In conclusion, SECM has been proven to be a powerful technique to investigate biochemical events related to cellular activity and several comprehensive review articles have been devoted to this field (Huang et al. 2018; Bauermann et al. 2004; Bard et al. 2006; Schulte and Schuhmann 2007; Beaulieu et al. 2011; Bergner et al. 2013; Conzuelo et al. 2018).

# 4    Perspectives

The integration of SECM to fluorescence microscopy systems, spectrometers, and hyphenated techniques (Polcari et al. 2016; Lai et al. 2011; Bentley et al. 2019a) is a new trend towards the development of analytical platforms with multifunctional applications, allowing investigations in more complex environments and materials to be performed. It should be pointed out, however, that SECM is still a high-technology technique whose application is limited to few laboratories with sophisticated facilities and very well-trained personnel. Hence, the transition of SECM from a technique with restricted access to a robust standard methodology would be very welcomed to continuously expand its usage as a tool to provide deeper insight and comprehensive aspects of the microscopic world.

# References

Abdelhamid ME, Piantavigna S, Bond AM, Graham B, Spiccia L, Martin LL, O'Mullane AP (2015) Electrochem Commun 51:11–14

Alizadeh V, Mousavi MF, Mehrgardi MA, Kazemi SH, Sharghi H (2011) Electrochim Acta 56: 6224–6229

Amemiya S (2015) In: Mirkin MV, Amemiya S (eds) Nanoelectrochemistry. Taylor and Francis, New York, pp 621–653

Amemiya S, Chen R, Nioradze N, Kim J (2016) Acc Chem Res 499:2007–2014

Amphlett JL, Denuault G (1998) J Phys Chem B 102:9946–9951

Bae JH, Yu Y, Mirkin MV (2016) J Phys Chem C 120:20651–20658

Bard AJ, Denuault G, Lee C, Mandler D, Wipf DO (1990) Acc Chem Res 23:357–363

Bard AJ, Li X, Zhan W (2006) Biosens Bioelectron 22:461–472

Bard AJ, Mirkin MV (2012) Scanning electrochemical microscopy, 2nd edn. Taylor and Francis, Boca Raton

Barroso IG, Santos CS, Bertotti M, Ferreira C, Terra WR (2019) Comp Biochem Physiol A 237: 110535

Bauermann LP, Schuhmann W, Schulte A (2004) Phys Chem Chem Phys 6:4003–4008

Beaulieu I, Kuss S, Mauzeroll J, Geissler M (2011) Anal Chem 83:1485–1492

Bentley CL, Edmondson J, Meloni GN, Perry D, Shkirskiy V, Unwin PR (2019a) Anal Chem 91: 84–108

Bentley CL, Kang M, Unwin PR (2019b) J Am Chem Soc 141:2179–2193

Bergner S, Vatsyayan P, Matysik FM (2013) Anal Chim Acta 775:1–13

Bertoncello P (2010) Energ Environ Sci 11:1620–1633

Black M, Cooper J, McGinn P (2005) Meas Sci Technol 16:174–182

Bollo S, Núñez-Vergara LJ, Squella JA (2004) J Electrochem Soc 151:E322–E325

Bollo S, Ulloa PJ, Finger S, Núñez-Vergara LJ, Squella JA (2005) J Electroanal Chem 577:235–242

Borgmann S, Radtke I, Erichsen T, Blöchl A, Heumann R, Schuhmann W (2006) ChemBioChem 7: 662–668

Cannan S, Cervera J, Steliaros RJ, Bitziou E, Whitworth AL, Unwin PR (2011) Phys Chem Chem Phys 13:5403–5412

Casillas N, Charlebois SJ, Smyrl WH, White HS (1993) J Electrochem Soc 140:142–144

Ciobanu M, Taylor DE, Wilburn JP, Cliffel DE (2008) Anal Chem 80:2717–2727

Comstock DJ, Elam JW, Pellin MJ, Hersam MC (2010) Anal Chem 82:1270–1276

Conzuelo F, Schulte A, Schuhmann W (2018) Proc R Soc A 474:0409

Date Y, Takano S, Shiku H, Ino K, Sasaki TI, Yokoo M, Abe H, Matsue T (2011) Biosens Bioelectron 30:100–106

Davoodi A, Pan J, Leygraf C, Norgren S (2007) Electrochim Acta 52:7697–7705

Demaille C, Bard AJ (1999) Acta Chem Scand 53:842–848

Demaille C, Unwin PR, Bard AJ (1996) J Phys Chem 100:14137–14143

Dey MK, Sahoo PK, Satpati AK (2017) J Electroanal Chem 788:175–183

Dufek EJ, Buttry DA (2009) J Electrochem Soc 156:C322–C330

Eckhard K, Chen X, Turcu F, Schuhmann W (2006) Phys Chem Chem Phys 8:5359–5365

Engstrom RC, Pharr CM (1989) Anal Chem 61:1099A–1104A

Evans SAG, Brakha K, Billon M, Mailley P, Denuault G (2005) Electrochem Commun 7:135–140

Fan FRF, Bard AJ (1999) Proc Natl Acad Sci 96:14222–14227

Fernandez JL, Bard AJ (2003) Anal Chem 75:2967–2974

Forster RJ (1994) Chem Soc Rev 23:289–297

Gonzalez-Garcia Y, Santana JJ, Gonzalez-Guzman J, Izquierdo J, Gonzalez S, Souto RM (2010) Prog Org Coat 69:110–117

Harati M, Jia J, Giffard K, Pellarin K, Hewson C, Love DA, Lau WM, Ding Z (2010) Phys Chem Chem Phys 12:15282–15290

Hengstenberg A, Kranz C, Schuhmann W (2000) Chem A Eur J 6:1547–1554

Horrocks BR, Mirkin MV, Pierce DT, Bard AJ, Nagy G, Toth K (1993) Anal Chem 65:1213–1224
Huang KC, Huang JH, Wu CH, Liu CY, Chen HW, Chu CW, Lin JT, Lin CL, Ho KC (2011) J Mater Chem 21:10384–10389
Huang L, Li Z, Lou Y, Cao F, Zhang D, Li X (2018) Materials 11:1389
Izquierdo J, Knittel P, Kranz C (2018) Anal Bioanal Chem 410:307–324
Izquierdo J, Nagy L, Varga A, Bitter I, Nagy G, Souto RM (2012) Electrochim Acta 59:398–403
Izquierdo J, Nagy L, Varga A, Santana JJ, Nagy G, Souto RM (2011) Electrochim Acta 56:8846–8850
Joshi VS, Sheet PS, Cullin N, Kreth J, Koley D (2017) Anal Chem 89:11044–11052
Kai T, Zoski CG, Bard AJ (2018) Chem Commun 54:1934–1947
Karnicka K, Eckhard K, Guschin DA, Stoica L, Kulesza PJ, Schuhmann W (2007) Electrochem Commun 9:1998–2002
Kim J, Dick JE, Bard AJ (2016c) Acc Chem Res 49:2587–2595
Kim J, Renault C, Nioradze N, Arroyo-Currás N, Leonard KC, Bard AJ (2016a) J Am Chem Soc 138:8560–8568
Kim J, Renault C, Nioradze N, Arroyo-Curras N, Leonard KC, Bard AJ (2016b) Anal Chem 88: 10284–10289
Kishi A, Inoue M, Umeda M (2010) J Phys Chem C 114:1110–1116
Konkena B, Masa J, Botz AJR, Sinev I, Xia W, Koßmann J, Drautz R, Muhler M, Schuhmann W (2017) ACS Catal 7:229–237
Kucernak AR, Chowdhury PB, Wilde CP, Kelsall GH, Zhu YY (2000) Electrochim Acta 45:4483–4491
Kumar A, Selva JSG, Gonçalves JM, Araki K, Bertotti M (2019) Electrochim Acta 322:134772
Kurulugama RT, Wipf DO, Takacs SA, Pongmayteegul S, Garris PA, Baur JE (2005) Anal Chem 77:1111–1117
Lai SCS, Dudin PV, Macpherson JV, Unwin PR (2011) J Am Chem Soc 133:10744–10747
Lee Y, Amemiya S, Bard AJ (2001) Anal Chem 73:2261–2267
Lefrou C, Cornut R (2010) ChemPhysChem 11:547–556
Li W, Fan FRF, Bard AJ (2012) J Solid State Electrochem 16:2563–2568
Li Y, Ning X, Ma Q, Qin D, Lu X (2016) Trends Anal Chem 80:242–254
Liu S, Liu D, Wang S, Cai X, Qian K, Kang F, Li B (2019) J Mater Chem A 7:12993–12996
Liu L, Zhang L, Dai Z, Tian Y (2017) Analyst 142:1452–1458
Luong BT, Nishikata A, Tsuru T (2003) Electrochemistry 71:555–561
Ma W, Hu K, Chen Q, Zhou M, Mirkin MV, Bard AJ (2017) Nano Lett 17:4354–4358
Ma L, Zhou H, Xin S, Xiao C, Li F, Ding S (2015) Electrochim Acta 178:767–777
Macpherson JV, Unwin PR (2000) Anal Chem 72:276–285
Mansikkamaki K, Ahonen P, Fabricius G, Murtomäki L, Kontturi K (2005) J Electrochem Soc 152: B12–B16
Martin RD, Unwin PR (1998a) J Chem Soc Faraday Trans 94:753–759
Martin RD, Unwin PR (1998b) Anal Chem 70:276–284
Nagaiah TC, Maljusch A, Chen X, Bron M, Schuhmann W (2009) ChemPhysChem 10:2711–2718
Nebel M, Gruetzke S, Diab N, Schulte A, Schuhmann W (2013a) Faraday Discuss 164:19–32
Nebel M, Grützke S, Diab N, Schulte A, Schuhmann W (2013b) Angew Chem Int Ed 52:6335–6338
Nishizawa M, Takoh K, Matsue T (2002) Langmuir 18:3645–3649
Nogala W, Szot K, Burchardt M, Jönsson-Niedziolka M, Rogalski J, Wittstock G, Opallo M (2010a) Bioelectrochemistry 79:101–107
Oyamatsu D, Hirano Y, Kanaya N, Mase Y, Nishizawa M, Matsue T (2003) Bioelectrochemistry 60:115–121
Paik CH, Alkire RC (2001) J Electrochem Soc 148:B276–B281
Palchetti I, Laschi S, Marrazza G, Mascini M (2007) Anal Chem 79:7206–7213
Park J, Kumar V, Wang X, Lee PS, Kim W (2017) ACS Appl Mater Interfaces 9:33728–33734
Payne NA, Stephens LI, Mauzeroll J (2017) Corrosion 73:759–780

Pereda MD, Alonso C, Burgos-Asperilla L, del Valle JA, Ruano OA, Perez P, de Mele MAFL (2010a) Acta Biomater 6:1772–1782

Pierce DT, Unwin PR, Bard AJ (1992) Anal Chem 64:1795–1804

Polcari D, Dauphin-Ducharme P, Mauzeroll J (2016) Chem Rev 116:13234–13278

Polcari D, Perry SC, Pollegioni L, Geissler M, Mauzeroll J (2017) ChemElectroChem 4:920–926

Qian H, Xu D, Du C, Zhang D, Li X, Huang L, Deng L, Tu Y, Mol JMC, Terryn HA (2017) J Mater Chem A 5:2355–2364

Ramanavicius A, Vilkonciene IM, Kisieliute A, Petroniene J, Ramanaviciene A (2017) Colloids Surf B Biointerfaces 149:1–6

Reisa RM, Valima RB, Rocha RS, Lima AS, Castro PŠ, Bertotti M, Lanza MRV (2014) Electrochim Acta 139:1–6

Roberts WS, Davis F, Higson SPJ (2009) Analyst 134:1302–1308

Sanchez-Sanchez CM, Bard AJ (2009) Anal Chem 81:8094–8100

Sanchez-Sanchez CM, Solla-Gullon J, Vidal-Iglesias FJ, Aldaz A, Montiel V, Herrero E (2010) J Am Chem Soc 132:5622–5624

Santana JJ, González-Guzmán J, Fernández-Mérida L, González S, Souto RM (2010) Electrochim Acta 55:4488–4494

Santos CS, Kowaltowski AJ, Bertotti M (2017) Sci Rep 7:11428–11434

Santos CS, Lima AS, Battistel D, Daniele S, Bertotti M (2016) Electroanalysis 28:1441–1447

Schulte A, Schuhmann W (2007) Angew Chem Int Ed 46:8760–8777

Scodeller P, Carballo R, Szamocki R, Levin L, Forchiassin F, Calvo EJ (2010) J Am Chem Soc 132: 11132–11140

Shamsi MH, Kraatz H-B (2010) Analyst 135:2280–2285

She Z, Topping K, Shamsi MH, Wang N, Chan NWC, Kraatz H-B (2015) Anal Chem 87:4218–4224

Shiku H, Shiraishi T, Ohya H, Matsue T, Abe H, Hoshi H, Kobayashi M (2001) Anal Chem 73: 3751–3758

Sidane D, Devos O, Puiggali M, Touzet M, Tribollet B, Vivier V (2011) Electrochem Commun 13: 1361–1364

Singh A, Lin Y, Obot IB, Ebenso EE, Ansari KR, Quraishi MA (2015) Appl Surf Sci 356:341–347

Solmaz R, Doner A, Kardas G (2008) Electrochem Commun 10:1909–1911

Song W, Yan Z, Hu K (2012) Biosens Bioelectron 38:425–429

Souto RM, González-García Y, Izquierdo J, González S (2010a) Corros Sci 52:748–753

Still JW, Wipf DO (1997) J Electrochem Soc 144:2657–2665

Sumboja A, Tefashe UM, Wittstock G, Lee PS (2012) J Power Sources 207:205–211

Sun P, Laforge FO, Mirkin MV (2007) Phys Chem Chem Phys 9:802–823

Sun T, Yu Y, Zacher BJ, Mirkin MV (2014) Angew Chem Int Ed 53:14120–14123

Szot K, Nogala W, Niedziolka-Jonsson J, Jonsson-Niedziolka M, Marken F, Rogalski J, Kirchner CN, Wittstock G, Opallo M (2009) Electrochim Acta 54:4620–4625

Treichel DA, Mirkin MV, Bard AJ (1994) J Phys Chem 98:5751–5757

Turcu F, Schulte A, Hartwich G, Schuhmann W (2004) Angew Chem Int Ed 43:3482–3485

Unwin PR, Bard AJ (1991) J Phys Chem 95:7814–7824

Volker E, Inchauspe CG, Calvo EJ (2006) Electrochem Commun 8:179–183

Wain AJ (2013) Electrochim Acta 92:383–391

Wang J, Zhou FM (2002) J Electroanal Chem 537:95–102

Wipf DO (1994) Colloids Surf A Physicochem Eng Asp 93:251–261

Wipf DO, Bard AJ (1991) J Electrochem Soc 138:469–474

Wittstock G, Burchardt M, Pust SE, Shen Y, Zhao C (2007) Angew Chem Int Ed 46:1584–1617

Yamashita K, Takagi M, Takenaka S, Uchida K, Kondo H (2001) Analyst 126:1210–1211

Yasukawa T, Kaya T, Matsue T (2000) Electroanalysis 12:653–659

Yi-Lun Y, Ding Z, Zhan D, Long Y-T (2017) Chem Sci 8:3338–3348

Zagal JH, Griveau S, Ozoemena KI, Nyokong T, Bedioui F (2009) J Nanosci Nanotechnol 9:2201–2214

Zampardi G, Ventosa E, La Mantia F, Schuhmann W (2013) Chem Commun 49:9347–9349
Zhang J, Guo J, Chen D, Zhong J-H, Liu J-Y, Zhan D (2019) Appl Surf Sci 491:553–559
Zhang MMN, Long YT, Ding Z (2011) J Inorg Biochem 108:115–122
Zhao C, Sinha JK, Wijayawardhana CA, Wittstock G (2004) J Electroanal Chem 561:83–91
Zhao C, Wittstock G (2005) Biosens Bioelectron 20:1277–1284
Zhou F, Unwin PR, Bard AJ (1992) J Phys Chem 96:4917–4924
Zhu Z, Liu X, Ye Z, Zhang J, Cao F, Zhang J (2018a) Sens Actuators B Chem 255:1974–1982
Zhu Z, Ye Z, Zhang Q, Zhang J, Cao F (2018b) Electrochem Commun 88:47–51
Zoski CG (2002) Electroanalysis 14:1041–1051
Zoski CG (2016) J Electrochem Soc 163:H3088–H3100

# Modified Electrodes Surface with Inorganic Oxides and Conducting Polymers

Luan Pereira Camargo, Bruna M. Hryniewicz, Marcio Vidotti, and Luiz Henrique Dall'Antonia

## 1    Introduction

The development of techniques for highly sensitive, fast, and precise bioanalysis is of considerable significance for several subjects such as disease diagnosis, medicinal research, food safety, and environmental monitoring (Venn 2013). In this way, the electroanalytical methods are an excellent way to provide rapid and accurate quantification of molecules. However, some electrodes used in electrochemistry do not present sensitivity and selectivity for some molecules. The modification of electrode surfaces ensures that electroanalytical analyzes can have a significant gain in terms of sensitivity and selectivity. Since they provide platforms where reactions at the electrode/solution interface can be controlled from modifications with compounds in order to develop responses suitable for different applications.

The modification usually occurs on a conductive electrode surface, like gold (Bianchi et al. 2014; Xiang et al. 2009; Carvalhal et al. 2005), platinum (Merz and Bard 1978; Ciszewski and Milczarek 1999), glassy carbon (Chen and McCreery 1996; Chen et al. 1995), doped-boron diamond (Hutton et al. 2011), or indium tin oxide glass (ITO); (Aydın and Sezgintürk 2017), once the bare surface cannot present selectivity and/or sensitivity for some molecules. The choice and design of a modifier depend on the interactions and properties of both the modifier and the molecule of interest to be quantified. The modification can take place by different compounds, such as different types of oxides, polymers, composites, or even molecules in specific arrangements (Ribeiro and Kubota 2006). However, in this

L. P. Camargo · L. H. Dall'Antonia (✉)
Departamento de Química, Universidade Estadual de Londrina, Londrina, PR, Brazil
e-mail: luizh@uel.br

B. M. Hryniewicz · M. Vidotti
Departamento de Química, Universidade Federal do Paraná, Curitiba, PR, Brazil

chapter only electrodes modified with inorganic oxides and conductive polymers will be reviewed.

## 2    Inorganic Oxides Modified Electrodes

Oxides can be chemically synthesized in several ways: precipitation (Sudha et al. 2019; Fazli et al. 2019), hydrothermal (Qian et al. 2020; Tan et al. 2019), solvothermal (El-Said et al. 2019; Kamble et al. 2019), sol-gel (Younus et al. 2019; Jia et al. 2005), combustion synthesis (Afonso et al. 2016; Avinash et al. 2019), sonochemical (Vidotti et al. 2008; Chen et al. 2019), electrosynthesis (Hutton et al. 2011; Pelissari et al. 2019) among many others. However, each synthesis leads to morphology controls, surface and structural properties, among others that modify the entire interaction between the modifiers, electrodes, and molecules of interest. However, aiming at applications of these electrodes some factors are very important such as sensitivity, selectivity, detection limit, simple preparation, cost, among others. Thus, the variables are many, which makes these modified electrodes liable for numerous researches around the world. This can be evidenced by the extremely high number of works published over the past 30 years (Brazaca et al. 2020; Sousa et al. 2019; Parnianchi et al. 2018; Freitas et al. 2018).

Easy methods to deposit the oxide synthesized previously are accessible: drop-casting (Yang et al. 2015), electrospinning (Chen et al. 2019), and electrophoresis (Maaoui et al. 2017; Amin et al. 2018). Each technique can lead to thin films with different properties and characteristics, from nanostructures up to thick films (Seshan 2012).

Regarding simple techniques for deposition of oxide, the layer-by-layer deposition can provide nanostructure films with good reproducibility. Afonso et al. (2016) synthesized $BiVO_4$–$Bi_2O_3$ composite particles by combustion synthesis (SCS) and immobilized on the ITO surface by layer-by-layer (LbL) technique.

The $BiVO_4$–$Bi_2O_3$ composite electrode was applied in the amperometric determination of atenolol (ATN) in pharmaceutical formulations and in urine using 0.25 mol $L^{-1}$ $NaNO_3$ as a supporting electrolyte. A well-defined oxidation peak of ATN has been observed in 1.12 V vs. Ag/AgCl (3.0 mol $L^{-1}$ KCl) by cyclic voltammetry. Analytical curve presented good linearity with a correlation coefficient of 0.997 and it was obtained in the concentration range of 50.0–800 μmol $L^{-1}$ ATN with a slope of 2.73 × $10^5$ μA mol$^{-1}$ L. The detection limit is found to be 0.459 μmol $L^{-1}$. The proposed method was used for ATN determination in pharmaceutical formulations and urine samples and the recoveries range of 94.0–104%. $BiVO_4$–$Bi_2O_3$ electrode exhibited good selectivity and repeatability, besides simplicity and low-cost construction as well as stability without the need for surface renewal before each measure.

Among the different techniques that an electrode can be modified, the direct electrochemical modification on the electrode surface stands out. The advantages of this method lie in the fact that the interaction of the formed oxide film is closely linked to the electrode, which facilitates the transfer of charge from the film to the electrode, increasing the sensitivity and detection limit of the electrodes, as well as the control of the growth and morphology of the film. Recently, Pelissari et al. (2019) describe the use of electrodeposited $Co(OH)_2$ on transparent conducting fluorine-doped tin oxide electrode (FTO) as a functional material for ascorbic acid electrooxidation in different electrolyte solutions. The $Co(OH)_2$ film presented electrocatalytic activity in ascorbic acid electrooxidation, being the sensitivity values found were 182.3 and 119.4 mA L $mol^{-1}$ $cm^{-2}$ in KOH and KCl solutions, respectively.

Moreover, Vidotti et al. (2008) describe electrodes for urea determination by using nickel hydroxide and compared electrodes modified by bulk electroformed films and by nanoparticles; the influence of cobalt atoms was also investigated in order to avoid the oxygen evolution reaction. Besides, the utilization of less positive operating potentials is very interesting due to reduction of possible interference signal in future studies for sensor applications. Sensitivities were found to be $3 \times 10^8$ and $34.4 \times 10^8$ mA $cm^2$ $C_{Urea}^{-1}$ $mol_{M(OH)_2}^{-1}$ for bulk film and nanostructured modified electrodes, respectively, when normalized by the amount of deposited material. It is clearly noticed the great superficial area achieved by nanostructured hydroxide, enhancing the sensitivity for urea oxidation increasing by 10 times the sensitivity.

Recently, the semiconductor inorganic oxide has been used to modify electrodes surface with the aim to develop new photoelectrochemical (PEC) sensors. Throughout the photoelectrochemical process, the photogenerated electron–hole pairs of the semiconductor material are separated and transferred to the interface between the electrode surface and the electrolyte solution to generate an oxidation–reduction reaction, thus generating a photocurrent. PEC sensors are a modern sensing technology based on the development of electrochemical sensors, in which the photoelectric conversion ability of semiconductor materials is commonly utilized to quantify the target (Zhao et al. 2015; Ge et al. 2019; Xu et al. 2019).

A new photoelectrochemical (PEC) immunosensor for prion diseases has been developed based on hemin-induced switching of photocurrent direction (Yang et al. 2019). Herein, taking the normal cellular form of prion ($PrP^C$) as a model owing to a high risk of pathogenicity of $PrP^{Sc}$. In the presence of $PrP^C$, nitrogen-doped porous carbon-hemin polyhedra labeled with secondary antibody were introduced onto the CdS-chitosan (CS) nanoparticles-modified indium–tin oxide (ITO) electrode via the antigen–antibody specific recognition. Because of the matched energy level between CdS and hemin, the high-efficiency switch of photocurrent direction of the ITO/CdS-CS photoelectrode from anodic to cathodic photocurrent was observed even at very low concentration (0.4 amol $L^{-1}$) of $PrP^C$ (Yang et al. 2019).

Tang and collaborators had developed a versatile photoelectrochemical bioanalysis platform for sensitive and specific screening of low-abundance kanamycin antibiotics. The platform was originally designed using rGO-$Bi_2WO_6$-Au as a photoactive matrix and target-induced branched hybridization chain reaction (t-bHCR) for efficient signal amplification. The addition of reduced graphene oxide (rGO) and Au nanoparticles (Au NPs) significantly enhanced the electron transfer and augmented photoactivity. Under the optimized testing conditions, the t-bHCR-based photoelectrochemical bioanalysis platform exhibited superior analytical performance with a linear range of 1 pmol $L^{-1}$ to 5 nmol $L^{-1}$ target kanamycin and limit of detection down to 0.78 pmol $L^{-1}$. Additionally, favorable stability, great anti-interference ability, and adequate precision for the analysis of samples were obtained (Zeng et al. 2019).

Also, regarding the photoelectrochemical application of $Bi_2WO_6$ semiconductor, a sandwich-type based on flower-like $Bi_2WO_6/Ag_2S$ nanoparticles (F-$Bi_2WO_6$/ $Ag_2S$) photoelectrochemical immunosensing platform was designed for detection of amino-terminal pro-B-type natriuretic peptide (NT-pro BNP) (Qian et al. 2019). As NT-pro BNP is one of the most important biochemical markers for the diagnostics and prognostics of heart failure. Specially, a cascade-like band-edge level between F-$Bi_2WO_6$ and $Ag_2S$ effectively improved the photocurrent conversion efficiency and improved the photocurrent response. Under optimal experimental conditions, the proposed sandwich-type PEC immunosensor presented a desirable linear relationship ranging from 0.1 to 100 pg $mL^{-1}$ for NT-pro BNP with the detection limit of 0.03 pg $mL^{-1}$ (S/N = 3).

Troponin I (cTnI) is also a biomarker of myocardial cell damage even for acute coronary syndrome or acute myocardial infarction. Based on manganese doped CdS (CdS:Mn) sensitized graphene (G)/$Cu_2MoS_4$ composite a label-free photoelectrochemical immunosensor was developed for the detection of cardiac troponin I. Graphene as an excellent 2D conductive material, combined with $Cu_2MoS_4$ could improve its charge transfer efficiency. CdS:Mn nanoparticles loaded on G/$Cu_2MoS_4$ further enlarged the light absorption range of $Cu_2MoS_4$ and restricted the electron–hole pairs recombination. Under optimal conditions, the proposed PEC immunosensor responded sensitively to cTnI with a low detection limit of 0.18 pg $mL^{-1}$ and a wide linear range (0.005–1000 ng $mL^{-1}$). Additionally, the immunosensor also exhibited high sensitivity, excellent selectivity, and good stability (Chi et al. 2019).

Some living tumor cancer cells can be detected by a small amount of hydrogen peroxide released (Szatrowski and Nathan 1991). Zhao and collaborators had developed a two-dimensional (2-D) hybrid material produced by the immobilization of ultrasmall gold nanoparticles (AuNPs, ∼3 nm) onto sandwich-like periodic mesoporous silica (PMS) coated with reduced graphene oxide (RGO) (Maji et al. 2014). The hybrid-based electrode sensor showed attractive electrochemical performance for sensitive and selective nonenzymatic detection of hydrogen peroxide ($H_2O_2$) in 0.1 mol $L^{-1}$ phosphate buffered saline, with wide linear detection range (0.5 μmol $L^{-1}$ to 50 mmol $L^{-1}$), low detection limit (60 nmol $L^{-1}$), and good sensitivity (39.2 μA $mmol^{-1}$ L $cm^{-2}$), and without any interference by common interfering agents. In addition, the sensor exhibited a high capability for glucose sensing and $H_2O_2$ detection in human urine. More remarkably, the hybrid was found to be nontoxic, and the electrode sensor could sensitively detect a trace amount of $H_2O_2$ in a nanomolar level released from living tumor cells (human embryonic kidney cells (HEK 293, a normal cell line), human cervical cancer cells (HeLa), and human hepatoma cancer cells (HepG2).

Ultrasensitive and high-selective detection of low amounts of nucleic acids has attracted significant attention due to its crucial role in preventive detection of genetic or pathogenic diseases and prognosis of cancer (Liu et al. 2016; Deng et al. 2020; Sadighbayan et al. 2019). In this way, an ultrasensitive sandwich-type electrochemical biosensor for DNA detection was developed based on spherical silicon dioxide/molybdenum selenide ($SiO_2@MoSe_2$) and graphene oxide–gold nanoparticles (GO–AuNPs) hybrids as carrier triggered Hybridization Chain Reaction (HCR) coupling with multi-signal amplification (Shuai et al. 2017). The proposed sensor assay utilized a spherical $SiO_2@MoSe_2$/AuNPs as sensing platform and GO–AuNPs hybrids as carriers to supply vast binding sites. $H_2O_2$ + HQ system is used for DNA detection and HCR as the signal and selectivity enhancer. The sensor was designed in sandwich type to increase the specificity. As a result, the present biosensor exhibits a good dynamic range from 0.1 fmol $L^{-1}$ to 100 pmol $L^{-1}$ with a low detection limit of 0.068 fmol $L^{-1}$ (S/N = 3).

The detection of biomolecules in bioanalysis is related to the morphology and surface arrangement in the electrode (Dai and Ju 2012). In materials synthesis, many different morphologies can be obtained varying parameters such as templates, temperature, concentration, among many others. Su and collaborators reported a one-step and template-free on-electrode synthesis method for making shape-controlled hierarchical flower-like gold nanostructures (HFGNs) on indium tin oxide substrates, which provide an electrochemical sensing platform for ultrasensitive detection of nucleic acids (Su et al. 2016). It is found that the sensitivity for electrochemical DNA sensing is critically dependent on the morphology of HFGNs. A highly sensitive electrochemical biosensor is developed for the label-free detection of microRNA-21 (miRNA-21), a biomarker for lung cancers. Importantly, it is demonstrated that this biosensor can be employed to measure the miRNA-21 expression level from human lung cancer cell (A549) lysates and works well in 100% serum.

Regarding the gold nanostructures, a new method has been developed to improve sensing performances of electrochemically grown Au nanostructures (AuNSs) based on the pre-seeding of the electrode (Siampour et al. 2020). The pre-seeding modification is simply carried out by vacuum thermal deposition of 5 nm thin film of Au on the substrate followed by thermal annealing at 500 °C. For more positive potentials, the pre-seeding leads to the growth of porous and hole-possess networks of AuNSs on the surface. Using less positive potentials, AuNSs with carved stone ball shapes are obtained. The best electrode was succeeded from AuNSs developed at 0.1 V for 900 s with pre-seeding modification. The sensing properties of this electrode toward glucose detection show a high sensitivity of 184.9 µA $mmol^{-1}$ L $cm^{-2}$, with a remarkable detection limit of 0.32 µmol $L^{-1}$ and a wide range of linearity.

# 3      Conduction Polymers Modified Electrodes

Conducting polymers (CPs) are a distinguished class of materials firstly reported by the Nobel Prize winners Alan MacDiarmid, Alan Heeger, and Hideki Shirakawa relating the oxidation of acetylene, finding out metallic characteristics in a polymeric matrix (MacDiarmid 2001; Macdiarmid et al. 1987). Up to date, many different works in literature have been published to the unfolding of the physical-chemical properties of these materials where the doping level of each polymer (Cochet et al. 2000; Louarn et al. 1996) leads to differences in the optical (Takahashi et al. 2002; Yue et al. 1991), structural (Lapkowski et al. 1995; ĆiriĆ-MarjanoviĆ et al. 2008; Salaneck et al. 1999) and electrochemical (Marchesi et al. 2015; Quintanilha et al. 2014) features. Up to date, many different reviews in literature can be found dealing with the characterization and applications of conducting polymers (Furukawa 1996; Heeger 2010; Wolfart et al. 2017; Holze and Wu 2014).

Although the very first reported conducting polymer was acetylene, nowadays, the most reported polymers are poly(aniline), poly(pyrrole), and poly (3,4-ethylenedioxythiophene) and from now on named as PANI, PPy, and PEDOT. The synthesis of these polymers can be found in either chemical or electrochemical methodologies, each one presenting their advantages. Regarding the chemical synthesis, undoubtedly the main advantage relies on the possibility of large amounts of material formed (Sapurina and Stejskal 2008; Malinauskas 2001) besides the easy combination of foreign structures on the formation of hybrid materials, such as inorganic oxides (Leroux and Besse 2001; Herring 2006) or carbon nanotubes (Salvatierra et al. 2010). Besides, the tunable morphologies are also found in chemical synthesis, the different structures are achieved by the polymerization in the presence of surfactants (Zhang et al. 2006; Hryniewicz et al. 2018), gums (Cornelsen et al. 2015; Lucht et al. 2015), oil droplets (Liu et al. 2010; Nagao et al. 2012), and others (Gangopadhyay and De 2000; Long et al. 2011). Also, the coupling with other energy sources, such as microwave (García-Escobar et al. 2016) and ultrasound (Hostert et al. 2016) promotes the formation of new morphologies with different properties, in particular those related to the ionic diffusion through the solid material. In Fig. 1 are shown some of these different morphologies generated by conducting polymers. The characterization of the conducting polymers has gained a lot of attention due to the development of new electronic microscopy techniques where many details can be easily identified. In the context of structural characterization, the spectroscopic techniques are always employed although for sure the Raman spectroscopy plays an important role in this field as due to the resonant effect (Furukawa 1996), the signal intensity is boosted several times and a fine structural characterization is possible, even in the presence of other foreign materials (Cho et al. 2015).                          ·

Some techniques are frequently used for the electrode modification, using a dispersion of the CP material previously synthesized, the most common is the simple drop casting where a certain amount of material is dropped on the electrode and the solvent is let dry, generating a solid film (Lee et al. 2011; Nicolas et al. 2006). Another widespread methodology consists in alternating layers of positively and

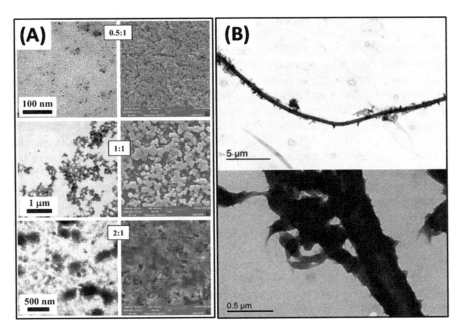

**Fig. 1** Images obtained from chemical synthesis of (**a**) PANI composites formed with different contents of Gum Arabic (GA) where it was possible to tune up different morphologies depending on the proportion of monomer: GA in the synthetic solution, reproduced with permission from Ref. Quintanilha et al. (2014). (**b**) PEDOT nanotubes chemically formed using Sodium bis(2-ethylhexyl) sulfosuccinate (AOT) as a template and $Fe^{3+}$ ions as an oxidizing agent, reproduced with permission from Ref. Hryniewicz et al. (2018)

negatively charged species, the so-called layer-by-layer deposition (LbL) (Decher 1997). In this methodology, the electrode is placed in different diluted solutions of the charged species and the thickness of the deposit is controlled by the simple repetition of these immersions (Schneider and Decher 2004; Marmisollé and Azzaroni 2016), due to its simplicity and the good covering of the surface, the LbL technique is especially employed in electrochromic electrodes (Kim and Jung 2005; DeLongchamp et al. 2003; Quintanilha et al. 2014). Another immobilization technique that takes advantage of the electrically charged species in solution is the electrophoretic deposition (EPD) (Vidotti and Córdoba de Torresi 2009; Augusto et al. 2013). In this methodology, electrodes are parallel placed, and an electric field is created by an external voltage. The charged particles move to the opposite charge electrode and parameters such as electric field and time of deposition can be easily tuned to create different morphologies on the electrode surface.

The electrochemical synthesis has some interesting advantages over the chemical route, in particular to the science of modified electrodes. In most electrochemical applications, the CP material must be placed on the surface of the electrode and depending on this step, many advantages generated in the chemical synthesis are lost as the morphology of the synthesized material can change drastically, especially due to the drastic solvent changes during the drying step. So, it is quite important the direct immobilization on the electrode surface, avoiding by this way an extra experimental step and using much lower material and energy and most importantly, avoiding any huge change of the morphologic features. In simple ways, the electrochemical deposition can be divided into the oxidation of the monomer at the electrode surface followed by the polymerization itself and precipitation of the polymer at the interface (Licona-Sánchez et al. 2010, 2011; Palomar-Pardavé et al. 2005). In this scenario, different electrochemical techniques have been employed: (1) Chronoamperometry, where the electrochemical potential is kept constant and the amount of electric charge passed through the electrode is monitored by the time (Biallozor and Kupniewska 2005); (2) Potentiometry, or Galvanostatic deposition by choosing a constant anodic current density, with less control of the electrochemical potential at the interface (Biallozor and Kupniewska 2005); and finally using (3) Cyclic voltammetry where it is possible to visualize both the growth of the polymer and the electrochemical properties of the fresh formed PC (Popov et al. 2019).

Regardless of the electrochemical way of synthesis, huge advances in the understanding of the mechanism of deposition and the role of the counterions during synthesis were obtained by the "in situ" quartz microbalance (MEQC) (Varela et al. 2001; Otero and Broschart 2006). The use of different counterions in the electrolyte also provides an easy experimental setup to tune up the morphology of the deposit. Due to the strong interactions between the carbonic chain and the PCs, surfactants have attracted much attention of the scientific community, presenting some other advantages such as strong adherence with the substrate, conductivity, and electroactivity, globular morphology and facilitates the ionic transport through in/out the PC matrix, increasing the durability of the modified electrode (Stejskal et al. 2003; Omastová et al. 2003). Another interesting route relies on the use of templates such as methyl orange. Depending on the pH of the electrolyte, MO molecules precipitate into a nanotubular morphology where the electrodeposition of PCs takes place covering those structures (Hryniewicz and Vidotti 2018; Hryniewicz et al. 2019). In Fig. 2 are shown some interesting morphologies of PCs electrosynthesized.

As commented above, there are many different applications based on CP-modified electrodes. Nowadays, with the increasing demand for energy, several reports about batteries (Hyder et al. 2011), solar cells (Grätzel 2001; Neophytou et al. 2017), and supercapacitors (Dubal et al. 2015; Wolfart et al. 2017; Bach-Toledo et al. 2020) have been shown up, leading to an irreversible trend in the CP technology. Important findings also can be found in the electrocatalysis and photocatalysis fields, where the degradation of organic pollutants seems to be an important perspective with increasing interest (Hryniewicz et al. 2020; Soares et al. 2019). Many efforts are also found in analytical sciences to the construction of platforms for extraction analysis (Augusto et al. 2010; Qi et al. 2015; de Lazzari et al. 2019), sensors (Ramanavičius et al. 2006; Hryniewicz et al. 2018; Naveen et al. 2017), and biosensors (Gonçales et al. 2011; Cosnier 2005).

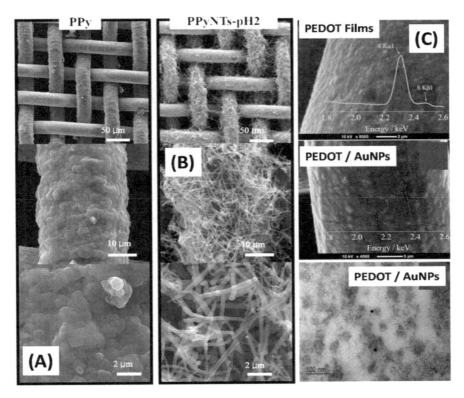

**Fig. 2** Images of CPs electrosynthesized under substrates: (**a**) PPy films using sodium dodecylbenzenesulfonate as counterion during synthesis, (**b**) PPy nanotubes formed using methyl orange as template, and (**c**) PEDOT thin films using gold nanoparticles (AuNPs) as dopant, both electrosynthesized on the surface of stainless steel mesh electrodes. Figures (**a**, **b**) reproduced with permission from Ref. Hryniewicz et al. (2019) and (**c**) from Ref. Soares et al. (2019)

Many studies have been focused on the employment of conducting polymers in bioanalytical applications, especially for biosensors development (Park et al. 2016). In this context, PANI has received attention because of easy synthesis, good environmental stability, high electrical conductivity, and tunable properties (Dhand et al. 2011). In addition, PANI has two redox couples in the right potential window, acting as a self-contained electron transfer mediator. Synthesis of PANI nanostructures, as nanofibers and nanotubes, leads to a higher surface area, which decreases the diffusional path for the biomolecules and makes the electroactive sites more exposed, improving the sensing performance (Zare et al. 2020). Although PANI is often used for biosensors construction, PPy and PEDOT are also very applied. They have a larger potential window when compared to PANI, being possible to detect a wide variety of electroactive species and, in the case of PEDOT, the presence of the sulfur atom in the structure can help to attach gold

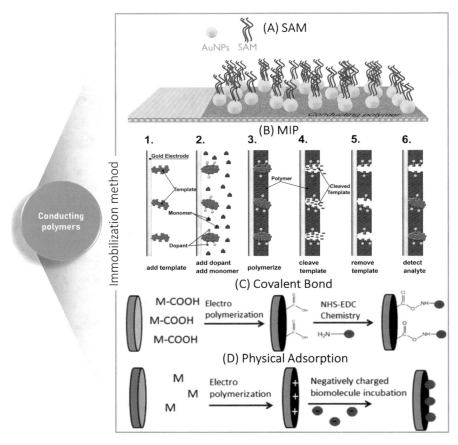

**Fig. 3** (**a**) Self-assembly monolayer on gold nanoparticles modified conducting polymer. (**b**) Scheme of MIP fabrication on gold electrodes, modified and reproduced with permission from Ref. Komarova et al. (2015). Biomolecules immobilization by (**c**) covalent bond and (**d**) physical adsorption, modified, and reproduced with permission from Ref. Aydemir et al. (2016)

nanoparticles on its surface, which can be important for further immobilization steps (Su et al. 2013; Kim et al. 2019).

One of the major steps to fabricate a biosensor is the immobilization of the recognition element, which can be made by different methodologies. Among them, the self-assembly monolayer (SAM) promotes an organized and dense system, controlling the orientation to bond the biomolecules (Samanta and Sarkar 2011). When the conducting polymer is decorated with gold nanoparticles, the strong sulfur–gold interaction can facilitate the formation of a SAM of molecules containing thiol, sulfides, or disulfides groups, which can be further modified with the element of recognition (Fig. 3a). Another methodology to fabricate a biosensor is the development of a molecularly imprinted polymer (MIP), that can be used as an alternative to antibodies or receptors (Komarova et al. 2015). To do so, the polymerization is performed in the presence of the analyte, producing a polymer with the molecular structure imprinted when the analyte is removed (Fig. 3b). Conducting polymers can be used to fabricate MIPs; they can be synthesized by chemical (Wang et al. 2018) or electrochemical polymerization, confining a specificity to the electrode for electrochemical sensing (Komarova et al. 2015).

Conducting polymers have also been modified with the element of biorecognition by a covalent bond and physical adsorption (Aydemir et al. 2016). The former usually employs the N-hydroxysuccinimide/1-ethyl-3-(3-dimethylaminopropyl) carbodiimide (NHS/EDC) (Xia et al. 2013) to bond the -COOH group on the -NH$_2$ groups of the conducting polymer and probe, respectively, providing a robust biding between the probe and the material (Fig. 3c). The latter is usually employed with the positively charged conducting polymer and anionic biomolecules (Fig. 3d), but other interactions can be present, as van der Waals and hydrophobic forces (Ahuja et al. 2007). This physical adsorption does not require any functionalization of the monomer, but since this force is relatively weak, the biomolecules can leach out of the surface.

The transduction step is also of great importance, and when it comes to conducting polymers, generally it can be based on cyclic voltammetry, amperometry, and electrochemical impedance spectroscopy (EIS) (Park et al. 2016). Nanostructured conducting polymers with high surface area have a huge capacitive current, making the detection of small amounts of the analyte by cyclic voltammetry or amperometry difficult. However, these nanostructures can be used in impedimetric biosensors, where the high surface area makes the electroactive sites of conducting polymers very exposed, facilitating the charge transfer process in the material/electrolyte the interface, which means that the charge transfer resistance (*Rct*) is small (Hryniewicz and Vidotti 2018). Once the analyte is bonded or adsorbed at the polymeric surface due to the interaction with the recognition element, some polymeric electroactive sites are blocked, increasing the *Rct* value (Soares et al. 2019). Thus, is possible to detect small amounts of the analyte by monitoring the changes on the *Rct* without the need of a redox probe, which is usually used in impedimetric sensors (Prodromidis 2010).

**Acknowledgments** The authors are thankful to the INCT in Bioanalytics (FAPESP grant no. 2014/50867-3 and CNPq grant no. 465389/2014-7).

# References

Afonso R, Eisele APP, Serafim JA, Lucilha AC, Duarte EH, Tarley CRT, Sartori ER, Dall'Antonia LH (2016) J Electroanal Chem 765:30–36
Ahuja T, Mir I, Kumar D, Rajesh (2007) Biomaterials 28:791–805
Amin HMA, El-Kady MF, Atta NF, Galal A (2018) Electroanalysis 30:1757–1766
Augusto F, Carasek E, Silva RGC, Rivellino SR, Batista AD, Martendal E (2010) J Chromatogr A 1217:2533–2542

Augusto T, Teixeira Neto É, Teixeira Neto ÂA, Vichessi R, Vidotti M, de Torresi SIC (2013) Sol Energy Mater Sol Cells 118:72–80

Avinash B, Ravikumar CR, Kumar MRA, Nagaswarupa HP, Santosh MS, Bhatt AS, Kuznetsov D (2019) J Phys Chem Solid 134:193–200

Aydemir N, Malmström J, Travas-Sejdic J (2016) Phys Chem Chem Phys 18:8264–8277

Aydın EB, Sezgintürk MK (2017) Trends Anal Chem 97:309–315

Bach-Toledo L, Hryniewicz BM, Marchesi LF, Dall'Antonia LH, Vidotti M, Wolfart F (2020) Mater Sci Energy Technol 3:78–90

Biallozor S, Kupniewska A (2005) Synth Met 155:443–449

Bianchi RC, da Silva ER, Dall'Antonia LH, Ferreira FF, Alves WA (2014) Langmuir 30:11464–11473

Brazaca LC, Sampaio I, Zucolotto V, Janegitz BC (2020) Talanta 210:120644

Carvalhal RF, Sanches Freire R, Kubota LT (2005) Electroanalysis 17:1251–1259

Chen K, Chou W, Liu L, Cui Y, Xue P, Jia M (2019) Sensors 19:1–19

Chen P, Fryling MA, McCreery RL (1995) Anal Chem 67:3115–3122

Chen P, McCreery RL (1996) Anal Chem 68:3958–3965

Chen T-W, Sivasamy Vasantha A, Chen S-M, Al Farraj DA, Soliman Elshikh M, Alkufeidy RM, Al Khulaifi MM (2019) Ultrason Sonochem 59:104718

Chi H, Han Q, Chi T, Xing B, Ma N, Wu D, Wei Q (2019) Biosens Bioelectron 132:1–7

Cho E-C, Li C-P, Huang J-H, Lee K-C, Huang J-H (2015) ACS Appl Mater Interfaces 7:11668–11676

Ćirić-Marjanović G, Trchová M, Stejskal J (2008) J Raman Spectrosc 39:1375–1387

Ciszewski A, Milczarek G (1999) Anal Chem 71:1055–1061

Cochet M, Louarn G, Quillard S, Buisson JP, Lefrant S (2000) J Raman Spectrosc 31:1041–1049

Cornelsen PA, Quintanilha RC, Vidotti M, Gorin PAJ, Simas-Tosin FF, Riegel-Vidotti IC (2015) Carbohydr Polym 119:35–43

Cosnier S (2005) Electroanalysis 17:1701–1715

Dai Z, Ju H (2012) Trends Anal Chem 39:149–162

de Lazzari AC, Soares DP, Sampaio NMFM, Silva BJG, Vidotti M (2019) Microchim Acta 186:398

Decher G (1997) Science 277:1232–1237

DeLongchamp DM, Kastantin M, Hammond PT (2003) Chem Mater 15:1575–1586

Deng H, Chai Y, Yuan R, Yuan Y (2020) Anal Chem 92:8364–8370

Dhand C, Das M, Datta M, Malhotra BD (2011) Biosens Bioelectron 26:2811–2821

Dubal DP, Ayyad O, Ruiz V, Gómez-Romero P (2015) Chem Soc Rev 44:1777–1790

El-Said WA, Abdel-Rahman MA, Sayed EM, Abdel-Wahab AA (2019) Electroanalysis 31:829–837

Fazli G, Esmaeilzadeh Bahabadi S, Adlnasab L, Ahmar H (2019) Microchim Acta 186:821

Freitas M, Nouws HPA, Delerue-Matos C (2018) Electroanalysis 30:1584–1603

Furukawa Y (1996) J Phys Chem 100:15644–15653

Gangopadhyay R, De A (2000) Chem Mater 12:608–622

García-Escobar CH, Nicho ME, Hu H, Alvarado-Tenorio G, Altuzar-Coello P, Cadenas-Pliego G, Hernández-Martínez D (2016) Int J Polym Sci 2016:1–9

Ge L, Liu Q, Hao N, Kun W (2019) J Mater Chem B 7:7283–7300

Gonçales VR, Matsubara EY, Rosolen JM, Córdoba de Torresi SI (2011) Carbon N Y 49:3039–3047

Grätzel M (2001) Nature 414:338–344

Heeger AJ (2010) Chem Soc Rev 39:2354–2371

Herring AM (2006) J Macromol Sci C Polymer Rev 46:245–296

Holze R, Wu YP (2014) Electrochim Acta 122:93–107

Hostert L, de Alvarenga G, Vidotti M, Marchesi LF (2016) J Electroanal Chem 774:31–35

Hryniewicz BM, Bach-Toledo L, Vidotti M (2020) Appl Mater Today 18:100538

Hryniewicz BM, Lima RV, Wolfart F, Vidotti M (2019) Electrochim Acta 293:447–457

Hryniewicz BM, Orth ES, Vidotti M (2018) Sens Actuators B: Chem 257:570–578

Hryniewicz BM, Vidotti M (2018) ACS Appl Nano Mater 1:3913–3924
Hryniewicz BM, Winnischofer H, Vidotti M (2018) J Electroanal Chem 823:573–579
Hutton LA, Vidotti M, Patel AN, Newton ME, Unwin PR, Macpherson JV (2011) J Phys Chem C 115:1649–1658
Hyder MN, Lee SW, Cebeci FÇ, Schmidt DJ, Shao-Horn Y, Hammond PT (2011) ACS Nano 5: 8552–8561
Jia N, Zhou Q, Liu L, Yan M, Jiang Z (2005) J Electroanal Chem 580:213–221
Kamble BB, Naikwade M, Garadkar KM, Mane RB, Sharma KKK, Ajalkar BD, Tayade SN (2019) J Mater Sci Mater Electron 30:13984–13993
Kim M, Iezzi R, Shim BS, Martin DC (2019) Front Chem 7:1–11
Kim E, Jung S (2005) Chem Mater 17:6381–6387
Komarova E, Aldissi M, Bogomolova A (2015) Analyst 140:1099–1106
Lapkowski M, Berrada K, Quillard S, Louarn G, Lefrant S, Pron A (1995) Macromolecules 28: 1233–1238
Lee BH, Park SH, Back H, Lee K (2011) Adv Funct Mater 21:487–493
Leroux F, Besse J-P (2001) Chem Mater 13:3507–3515
Licona-Sánchez T d J, Álvarez-Romero GA, Mendoza-Huizar LH, Galán-Vidal CA, Palomar-Pardavé M, Romero-Romo M, Herrera-Hernández H, Uruchurtu J, Juárez-García JM (2010) J Phys Chem B 114:9737–9743
Licona-Sánchez T d J, Álvarez-Romero GA, Palomar-Pardavé M, Galán-Vidal CA, Páez-Hernández ME, Silva MTR, Romero-Romo M (2011) Int J Electrochem Sci 6:1537–1549
Liu M, Nie F-Q, Wei Z, Song Y, Jiang L (2010) Langmuir 26:3993–3997
Liu X, Shuai HL, Liu YJ, Huang KJ (2016) Sens Actuators B Chem 235:603–613
Long Y-Z, Li M-M, Gu C, Wan M, Duvail J-L, Liu Z, Fan Z (2011) Prog Polym Sci 36:1415–1442
Louarn G, Lapkowski M, Quillard S, Pron A, Buisson JP, Lefrant S (1996) J Phys Chem 100:6998–7006
Lucht E, Rocha I, Orth ES, Riegel-Vidotti IC, Vidotti M (2015) Mater Lett 149:116–119
Maaoui H, Singh SK, Teodorescu F, Coffinier Y, Barras A, Chtourou R, Kurungot S, Szunerits S, Boukherroub R (2017) Electrochim Acta 224:346–354
MacDiarmid AG (2001) Angew Chem Int Ed 40:2581–2590
Macdiarmid AG, Chiang JC, Richter AF, Epstein AJ (1987) Synth Met 18:285–290
Maji SK, Sreejith S, Mandal AK, Ma X, Zhao Y (2014) ACS Appl Mater Interfaces 6:13648–13656
Malinauskas A (2001) Polymer 42:3957–3972
Marchesi LF, Jacumasso SC, Quintanilha RC, Winnischofer H, Vidotti M (2015) Electrochim Acta 174:864–870
Marmisollé WA, Azzaroni O (2016) Nanoscale 8:9890–9918
Merz A, Bard AJ (1978) J Am Chem Soc 100:3222–3223
Nagao D, Ohta T, Ishii H, Imhof A, Konno M (2012) Langmuir 28:17642–17646
Naveen MH, Gurudatt NG, Shim Y-B (2017) Appl Mater Today 9:419–433
Neophytou M, Griffiths J, Fraser J, Kirkus M, Chen H, Nielsen CB, McCulloch I (2017) J Mater Chem C 5:4940–4945
Nicolas M, Guittard F, Géribaldi S (2006) Langmuir 22:3081–3088
Omastová M, Trchová M, Kováŕová J, Stejskal J (2003) Synth Met 138:447–455
Otero TF, Broschart M (2006) J Appl Electrochem 36:205–214
Palomar-Pardavé M, Scharifker BR, Arce EM, Romero-Romo M (2005) Electrochim Acta 50: 4736–4745
Park C, Lee C, Kwon O (2016) Polymers 8:249
Parnianchi F, Nazari M, Maleki J, Mohebi M (2018) Int Nano Lett 8:229–239
Pelissari MRS, Archela E, Tarley CRT, Dall'Antonia LH (2019) Ionics 25:1911–1920
Popov A, Brasiunas B, Mikoliunaite L, Bagdziunas G, Ramanavicius A, Ramanaviciene A (2019) Polymer 172:133–141
Prodromidis MI (2010) Electrochim Acta 55:4227–4233
Qi F, Li X, Yang B, Rong F, Xu Q (2015) Talanta 144:129–135

Qian Y, Feng J, Fan D, Zhang Y, Kuang X, Wang H, Wei Q, Ju H (2019) Biosens Bioelectron 131: 299–306
Qian J, Wang Y, Pan J, Chen Z, Wang C, Chen J, Wu Z, Yangyue (2020) Mater Chem Phys 239: 122051
Quintanilha RC, Orth ES, Grein-Iankovski A, Riegel-Vidotti IC, Vidotti M (2014) J Colloid Interface Sci 434:18–27
Quintanilha RC, Rocha I, Vichessi RB, Lucht E, Naidek K, Winnischofer H, Vidotti M (2014) Quim Nova 37:677–688
Ramanavičius A, Ramanavičienė A, Malinauskas A (2006) Electrochim Acta 51:6025–6037
Ribeiro ES, Kubota LT (2006) Microchim Acta 154:303–308
Sadighbayan D, Sadighbayan K, Khosroushahi AY, Hasanzadeh M (2019) Trends Anal Chem 119: 115609
Salaneck WR, Friend RH, Brédas JL (1999) Phys Rep 319:231–251
Salvatierra RV, Oliveira MM, Zarbin AJG (2010) Chem Mater 22:5222–5234
Samanta D, Sarkar A (2011) Chem Soc Rev 40:2567
Sapurina I, Stejskal J (2008) Polym Int 57:1295–1325
Schneider G, Decher G (2004) Nano Lett 4:1833–1839
Seshan K (2012) Handbook of thin film deposition. Elsevier, Oxford
Shuai HL, Wu X, Huang KJ, Zhai ZB (2017) Biosens Bioelectron 94:616–625
Siampour H, Abbasian S, Moshaii A, Omidfar K, Sedghi M, Naderi-Manesh H (2020) Sci Rep 10: 1–11
Soares AL, Zamora ML, Marchesi LF, Vidotti M (2019) Electrochim Acta 322:134773
Sousa CP, Ribeiro FWP, Oliveira TMBF, Salazar-Banda GR, de Lima-Neto P, Morais S, Correia AN (2019) ChemElectroChem 6:2350–2378
Stejskal J, Omastová M, Fedorova S, Prokeš J, Trchová M (2003) Polymer 44:1353–1358
Su W, Cho M, Nam J-D, Choe W-S, Lee Y (2013) Electroanalysis 25:380–386
Su S, Wu Y, Zhu D, Chao J, Liu X, Wan Y, Su Y, Zuo X, Fan C, Wang L (2016) Small 12:3794–3801
Sudha V, Annadurai K, Kumar SMS, Thangamuthu R (2019) Ionics 25:5023–5034
Szatrowski TP, Nathan CF (1991) Cancer Res 51:794–798
Takahashi K, Nakamura K, Yamaguchi T, Komura T, Ito S, Aizawa R, Murata K (2002) Synth Met 128:27–33
Tan W, Zhu Z, Yang J, Li H, Li S, Wu D, Qin Y, Kong Y (2019) Synth Met 258:116193
Varela H, de Albuquerque Maranhão SL, Mello RMQ, Ticianelli EA, Torresi RM (2001) Synth Met 122:321–327
Venn RF (2013) Principles and practice of bioanalysis. CRC Press, Boca Raton
Vidotti M, Córdoba de Torresi SI (2009) Electrochim Acta 54:2800–2804
Vidotti M, Silva MR, Salvador RP, de Torresi SIC, Dall'Antonia LH (2008) Electrochim Acta 53: 4030–4034
Wang Q, Xue R, Guo H, Wei Y, Yang W (2018) J Electroanal Chem 817:184–194
Wolfart F, Hryniewicz BM, Góes MS, Corrêa CM, Torresi R, Minadeo MAOS, Córdoba de Torresi SI, Oliveira RD, Marchesi LF, Vidotti M (2017) J Solid State Electrochem 21:2489–2515
Wolfart F, Hryniewicz BM, Marchesi LF, Orth ES, Dubal DP, Gómez-Romero P, Vidotti M (2017) Electrochim Acta 243:260–269
Xia N, Xing Y, Wang G, Feng Q, Chen Q, Feng H, Sun X, Liu L (2013) Int J Electrochem Sci 8: 2459–2467
Xiang C, Zou Y, Sun L-X, Xu F (2009) Sens Actuators B Chem 136:158–162
Xu Y-T, Yu S-Y, Zhu Y-C, Fan G-C, Han D-M, Qu P, Zhao W-W (2019) Trends Anal Chem 114: 81–88
Yang C, Denno ME, Pyakurel P, Venton BJ (2015) Anal Chim Acta 887:17–37
Yang R, Zou K, Zhang X, Du C, Chen J (2019) Biosens Bioelectron 132:55–61
Younus AR, Iqbal J, Muhammad N, Rehman F, Tariq M, Niaz A, Badshah S, Saleh TA, Rahim A (2019) Microchim Acta 186:471

Yue J, Wang ZH, Cromack KR, Epstein AJ, MacDiarmid AG (1991) J Am Chem Soc 113:2665–2671

Zare EN, Makvandi P, Ashtari B, Rossi F, Motahari A, Perale G (2020) J Med Chem 63:1–22

Zeng R, Zhang L, Su L, Luo Z, Zhou Q, Tang D (2019) Biosens Bioelectron 133:100–106

Zhang X, Lee J-S, Lee GS, Cha D-K, Kim MJ, Yang DJ, Manohar SK (2006) Macromolecules 39:470–472

Zhao W-W, Xu J-J, Chen H-Y (2015) Chem Soc Rev 44:729–741

# The Role of Gas Chromatography in Bioanalysis

Nathália de Aguiar Porto and Leandro Wang Hantao

## 1    Introduction

Today, gas chromatography (GC) has become an important technique in metabolomic investigations due to the unparalleled chromatographic efficiency and improved peak capacity. The foundation of the chromatographic separation is the differential migration attained by the solute bands as they migrate through in the GC column (McNair and Miller 2008). The chromatographic resolution arises from this differential migration, which is dependent on the distribution constant of the analyte between gas phase (i.e., carrier gas) and stationary phase, while minimizing band broadening during such process (Poole and Poole 2009). So, the net retention depends on analyte solubility, which is a temperature-influenced property and it is affected by the intermolecular interactions between solute and stationary phase (Grob and Barry 2004). Furthermore, the carrier gas does not impact the partition constant of the analyte and its main role is to carry the solute bands through the capillary column. In this context, the carrier gas may be nitrogen ($N_2$), hydrogen ($H_2$), or helium (He). The most employed carrier gases are hydrogen and helium that exhibit optimum linear velocities at 30–60 cm s$^{-1}$ (Grob and Barry 2004). The latter (He) is most employed in gas chromatography coupled with mass spectrometry (GC-MS)-based experiments, because of improved sensitivity (Muñoz-Guerra et al. 2011; Impens et al. 2001). For instance, a 50-fold improvement in peak area was reported in the analysis of steroids by GC-MS using He compared to $H_2$ (Muñoz-Guerra et al. 2011). Furthermore, the durability of some metallic components of the MS is also higher when using He with respect to $H_2$, due to unwanted reactions between hydrogen and the component's surface (Muñoz-Guerra et al. 2011). The latter is particularly important for GC-MS experiments wherein a high volumetric

N. de Aguiar Porto · L. W. Hantao (✉)
Department of Analytical Chemistry, Institute of Chemistry, University of Campinas (UNICAMP),
Campinas, SP, Brazil
e-mail: wang@unicamp.br

flow of carrier gas is employed, like flow-modulated comprehensive two-dimensional gas chromatography (FM-GC × GC).

## 1.1    Sample Introduction

A broad range of neutral organic metabolites can be analyzed using GC-based methods, as long as the solutes exhibit significant vapor pressure (0.1 Torr) at the operating temperature (Grob and Barry 2004). Furthermore, analytes must be thermally stable at the operating temperature of the GC inlet and the chromatographic conditions must not favor undesirable surface adsorption of the solutes. In Table 1 is shown the impact of the inlet operating temperature to analyte degradation by pyrolysis (Grob and Barry 2004).

The standard split/splitless injector (SSL) is a vaporization-based module that operates at a constant temperature. The SSL is typically used when analytes are temperature resistant. However, thermally sensitive metabolites may still be analyzed by GC, if the method employs a mild sample introduction technique. Milder introduction techniques include the cold on-column (OC) injector and the programmable temperature vaporizing (PTV) inlet. For instance, GC-MS analyses of ethinylestradiol hormone and thalidomide, nifedipine, torcetrapib, and maraviroc drugs were successfully performed using an OC module, wherein all peaks were symmetric and showed no signs of degradation (Brunelli et al. 2010). Analyte adsorption to active sites of the inner surface of the gas chromatograph is a challenge when dealing with sub-ppm levels of solutes (Maštovská et al. 2005; Fujiyoshi et al. 2016). Analyte protectants may be spiked to the final sample solution to reduce the impact of such unwanted adsorption by competing with the analytes for the active sites (Maštovská et al. 2005).

The samples amenable to GC analyses may be solids, liquids, or gases. Sample preparation plays an important role in GC-based methods to provide metabolite derivatization, matrix cleanup, and/or analyte preconcentration. Modern sample preparation methods may also integrate sampling and sample preparation into a single step for in vivo investigations. In general, analyses of permanent gases ($N_2$, $O_2$, CO, $CO_2$, $CH_4$, etc.) do not provide useful information for metabolomic studies. The most common analytes are volatile (molecular weight up to 200 Da (Marriott et al. 2001)) and semi-volatile (weight above 200 Da) organic compounds found in

**Table 1**  Impact of inlet operating temperature to analyte degradation

| Temperature range | Description |
| --- | --- |
| 100–300 °C | Analyte vaporization. Some thermal degradation may occur depending on the chemical reactivity of the analyte |
| 300–500 °C | Mild pyrolysis. Cleavage of C–C bond is minimal |
| 500–1100 °C | Pyrolysis. Extensive cleavage of C–C bond producing smaller fragment molecules |

liquid and solid samples. Among the most used techniques of sample preparation, liquid–liquid extraction (LLE), solid-phase extraction (SPE), solid-phase microextraction (SPME), and purge-and-trap extraction (PandT) are highlighted in the following sections, alongside case studies of GC-MS-based metabolic profiling.

## 1.2    Column Selection

The retention of the analyte is readily affected by the nature of the stationary phase. In bioanalysis, liquid polymeric phases are typically employed for gas-liquid chromatography using wall-coated open-tubular (WCOT) columns. Most stationary phases are derived from poly(siloxanes) and poly(ethylene glycols) (PEG). The chemical composition of the phase determines the type and intensity of the intermolecular interactions that analytes experience during gas-liquid chromatography. For instance, PEG-based phases are hydrogen-bond basic and can engage in dipole-type interactions, which typically results in selective separations (Poole and Poole 2008). For example, germacrene D ($C_{15}H_{24}$) and alpha-farnesene ($C_{15}H_{24}$) coelute in GC analysis using a nonpolar OV1 column, but are baseline resolved using a PEG-20M-based column (Cagliero et al. 2017). Poly(cyanopropylmethylsiloxane) phases are also hydrogen-bond basic and can engage in dipole-type interactions. Such polar phases typically exhibit maximum allowable operating temperatures (MAOT) of ca. 250–260 °C (Poole and Poole 2008). Recent advances in column technology have enabled manufacturers to produce PEG-based phases with MAOT values of 280 °C/290 °C (DB-HeavyWAX from Agilent Technologies) and 300 °C (MEGA-WAX HT from MEGA S.r.l.). Alternatively, poly(diphenyldimethylsiloxanes) may be used for metabolic profiling of complex samples, wherein higher elution temperatures are required for GC analysis (Kouremenos et al. 2010; Gullberg et al. 2004). For example, 66 endogenous metabolites of *Arabidopsis thaliana* were derivatized into its trimethylsilyl (TMS) counterparts, including ribose, alpha-ketoglutaric acid, glucosamine, oxalic acid, proline, and stearic acid, and separated by GC using a nonpolar DB-5ms column (Gullberg et al. 2004). Noteworthy, an oximation reaction was used to protect the carbonyl-moieties of the metabolites. Poly(dimethylsiloxanes) (PDMS) are low selectivity phases which may engage in non-specific dispersive interactions. The polarity of poly(diphenyldimethylsiloxane) phases may be enhanced by increasing the content of diphenylsiloxane monomer of the copolymer (Poole and Poole 2008). So, a poly(diphenyldimethylsiloxane) with 50% (w/w) monomer incorporation becomes more hydrogen-bond basic and can engage in stronger dipole-type interactions compared to PDMS. An important characteristic of poly(diphenyldimethylsiloxanes) is its higher MAOT values (ca. 300–430 °C), which enables the execution of highly efficient separations at high temperatures (above 260 °C). However, poly(diphenyldimethylsiloxane) phases are typically less polar than poly(cyanopropylmethylsiloxanes) and PEG phases (Poole and Poole 2008).

Column selection is based on the physicochemical properties of the analytes. For instance, primary metabolites typically exhibit active hydrogens (e.g., polyols or organic acids), which require suitable sample preparation and higher operating temperatures. Hence, most GC-based methods for the analyses of primary metabolites use poly(diphenyldimethylsiloxanes) stationary phases. Secondary metabolites that are amenable to GC analysis consists of volatile and semi-volatile organic compounds, which comprises hydrocarbons (terpenes and sesquiterpenes), aldehydes, ketones, esters, and alcohols. The most popular phases used for the analyses of these metabolites are poly(diphenyldimethylsiloxanes) with 5% (w/w) diphenylsiloxane monomer incorporation (e.g., DB-5, Rti-5, and SLB-5), PEGs (e.g., DB-WAX, Stabilwax, and SUPELCOWAX 10), and poly (cyanopropylmethylsiloxanes)/poly(biscyanopropylsiloxanes)     (e.g.,     DB-624, Rt-2560, and SP-2560). The most selective separations are attained by using more polar GC columns such as SP-2560 and SUPELCOWAX 10. For instance, terpinene-4-ol ($C_{10}H_{18}O$) and menthol ($C_{10}H_{20}O$) were baselines resolved using a SUPELCOWAX 10 column compared to the overlapping peaks obtained using a SLB-5ms column (Ragonese et al. 2015). However, databases available for qualitative analyses are mainly compiled using retention data obtained with SLB-5-like columns.

The most challenging samples are mixtures with an overwhelming quantity of compounds to be separated, which simply exceeds the peak capacity of single-column-based GC separations, or mixtures that comprises isomers with subtle structural differences (Marriott et al. 2001). For instance, plant- and microbiome-based metabolomics frequently investigate secondary metabolites by GC-based separations (Belinato et al. 2018). These matrices are rich in terpenes, which result from the head-to-tail linkage of 2-methyl-1,3-butadiene (isoprene, $C_5H_8$) moieties. The terpenes are classified according to the number of isoprene units: monoterpenes ($C_{10}$, 2 isoprene units), sesquiterpenes ($C_{15}$, 3 isoprene units), diterpenes ($C_{20}$, 4 isoprene units), sesterterpenes ($C_{25}$, 5 isoprene units), triterpenes ($C_{30}$, 6 isoprene units), and tetraterpenes ($C_{40}$, 8 isoprene units) (Marriott et al. 2001). To illustrate the multifaceted sample complexity, the separation of 46 terpenes of *Mentha piperita* was attained by GC using an ionic liquid-based column (Ragonese et al. 2015), but a 29 mixture of terpenes was not baseline resolved using a highly selective ionic liquid-based column (Cagliero et al. 2017). Hence, single-column-based GC methods may not produce the best chromatographic separations for metabolic profiling depending on the sample complexity.

In this context, multidimensional chromatographic techniques were developed to effectively increase the peak capacity of the composite system and may be best suited to tackle complex samples, such as those found in metabolomics. Multidimensional gas chromatography (MDGC) comprises two large groups of techniques, namely, heart-cutting (GC-GC) and comprehensive two-dimensional gas chromatography (GC × GC). In both groups, two consecutive stages of separation are combined by using GC columns with different solvation properties. However, only selected analytes are submitted to both stages of separation in GC-GC, while the entire sample mixture is resolved using both dimensions in GC × GC. The interested reader is directed elsewhere for an extensive discussion on GC-GC (Tranchida et al. 2012) and GC × GC (Bahaghighat et al. 2019).

## 1.3    Comprehensive Two-Dimensional Gas Chromatography

In GC $\times$ GC, all the effluent from the primary column ($^1$D) is subjected to an additional separation on the secondary column ($^2$D) (Liu and Phillips 1991). In some GC $\times$ GC configurations, it is not possible to analyze all effluent from the $^1$D, but rather a representative fraction of solutes is used instead by low-duty cycle interfaces (Seeley et al. 2018). Even so, the qualitative ability of the method is maintained in such GC $\times$ GC instruments. A primary requisite of any GC $\times$ GC experiment is that both columns involved in the separation process must feature distinct solvation properties (i.e., selectivity). Also, the second requisite is that the experiment must preserve the chromatographic resolution attained by each stage. Consequently, unprecedented peak capacities have been achieved by GC $\times$ GC (Pollo et al. 2018) enabling exciting opportunities for fundamental and applied research in metabolomics.

The most important interface of the GC $\times$ GC instrument is the modulator. The modulator is responsible for the successful hyphenation of two GC-based dimensions and it contributes directly to the overall peak capacity of the composite system by generating sharp reinjection bands. Briefly, the modulator continuously samples the effluent from the $^1$D and transfers the bands to the $^2$D as sharp pulses. Each pulse is then subjected to a highly efficient and fast separation in the $^2$D, typically with durations of 2–12 s. For instance, a peak with 6 s width in the $^1$D is modulated to generate a reconstructed peak with 50–400 ms width in the $^2$D. The compression ratio is dependent on the principle of the modulator and compression values of up to 120 may be attained using cryogenic modulation. The modulation process is continuous throughout the analysis with a well-defined frequency to ensure reproducible analysis. The time interval between successive reinjections by the modulator is defined as the modulation period.

A general guideline for the selection of the modulation period is the modulation ratio. An average modulation ratio of 3 is recommended to maintain the resolution achieved in the primary separation step to preserve resolution attained by each dimension of the composite system (Seeley and Seeley 2013). The interested reader is directed elsewhere for further details on modulation ratio (Khummueng et al. 2006).

Thermal modulation is an important technique, which originated from multiplex GC (Phillips et al. 1985) and it was later extended to the seminal development of GC $\times$ GC (Liu and Phillips 1991). Modulation may be based on many principles, including thermal, valve-based, and flow interfaces (Bahaghighat et al. 2019; Seeley and Seeley 2013). Today, the two most popular modulation principles are thermal and flow modulation. The best performing thermal interfaces are the cryogenic modulators, which trap the $^1$D effluent in a cryogenically cooled region and reinject it into the $^2$D by thermal desorption (Bahaghighat et al. 2019). For example, the ZX-1 interface (Zoex Corporation) uses liquid nitrogen to achieve highly efficient modulation. However, the operational costs and safety precautions when handling liquid nitrogen may dampen the widespread use of cryogenic interfaces. Hence, consumable-free interfaces have been considered for efficient, robust, and cost-

**(A)**

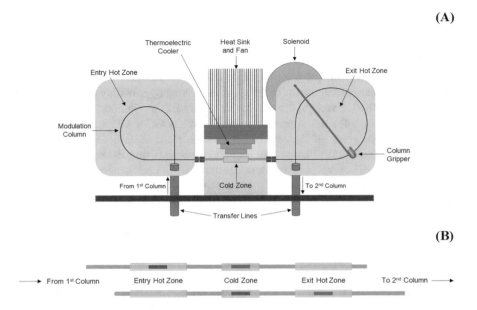

**(B)**

**Fig. 1** Schematic of the solid-state modulator (**a**) and diagram of the dual-stage modulation using a moving trapping capillary (**b**). Boswell, H.; Tarazona Carrillo, K.; Górecki, T. Evaluation of the Performance of Cryogen-Free Thermal Modulation-Based Comprehensive Two-Dimensional Gas Chromatography-Time-of-Flight Mass Spectrometry (GC × GC-TOFMS) for the Qualitative Analysis of a Complex Bitumen Sample. *Separations* 2020, 7, 13. Open access article distributed under the Creative Commons Attribution License

effective GC × GC analyses. The ZX-2 interface (Zoex Corporation) uses a multi-stage cooling system that bypasses the need for cryogenic fluids. Noteworthy, the ZX-1 and ZX-2 modulators use a two-stage loop assembly, wherein the interface is housed inside the GC oven (Harynuk and Górecki 2003, 2005). Recently, a solid-state modulator (SSM) (JandX Technologies) has gained attention by exhibiting a compact design placed outside the GC. This interface exhibits fixed heating and cooling zones and a moving sorbent capillary that is used to emulate thermal modulation, as shown in Fig. 1 (Boswell et al. 2020), exhibiting a modulation range of $C_8$ to $C_{40}$. The concept underlying the SSM was already described elsewhere (Marriott and Kinghorn 1997; Zoccali et al. 2019; Luong et al. 2016).

Differential flow modulators (FM) are interesting alternatives to thermal modulation. Although the reinjection of cryogenic interfaces outperforms that of flow modulators, the latter have provided similar qualitative results. Flow-based interfaces exhibit a simple design and employ differential pressure to actuate the $^1$D effluent into two stages: sample accumulation and reinjection (Seeley and Seeley 2013), as shown in Fig. 2. Current FM uses microfluidic components and it may be arranged in two configurations: forward fill/flush (FFF) (Seeley et al. 2008; Tranchida et al. 2014) or reverse fill/flush (RFF) (Griffith et al. 2012). The primary difference in its performance (RFF vs. FFF) is the analysis of complex mixtures

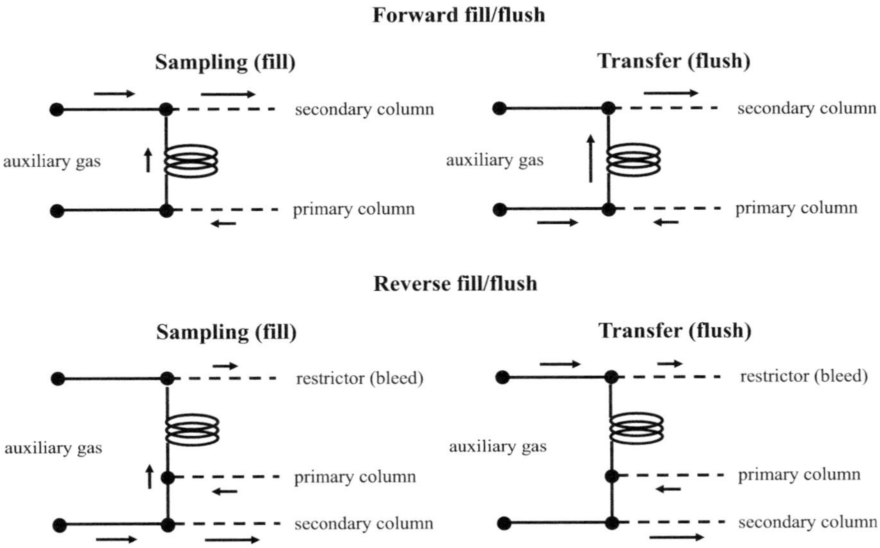

**Forward fill/flush**

**Sampling (fill)**          **Transfer (flush)**

secondary column                    secondary column

auxiliary gas                       auxiliary gas

primary column                      primary column

**Reverse fill/flush**

**Sampling (fill)**          **Transfer (flush)**

restrictor (bleed)                  restrictor (bleed)

auxiliary gas                       auxiliary gas

primary column                      primary column

secondary column                    secondary column

**Fig. 2** Schematic of the forward and reverse fill/flush flow modulators

where analyte overloading occurs (Griffith et al. 2012). Commercial FM interfaces are available in both configurations (FFM and RFF), such as INSIGHT (SepSolve Analytical) and Agilent's Capillary Flow Technology. Even though the first GC × GC report was dated back in 1991, modulation is still an active research topic in the GC community. The advances are mainly attributed to the rapid development of practical technologies and the growing interest of companies in such market opportunities for common practitioners.

An important consideration in GC × GC-MS analysis is the acquisition rate of the mass analyzer. A recommended sampling rate of 30 scans $s^{-1}$ is indicated for proper peak reconstruction in the GC × GC chromatograms for quantitative purposes, ensuring reproducible peak volumes (Mondello et al. 2008; van Stee and Brinkman 2016). In other words, seven data points across the chromatographic peak may be adequate for qualitative purposes. If much less than 7 data points are obtained, the quadrupole mass analyzer (QMS), a scanning instrument, is strongly affected by mass spectral skewing (Mondello 2011). However, non-scanning instruments do not exhibit spectral skewing, like the time-of-flight mass analyzer (TOFMS) and the Fourier transform Orbitrap mass analyzer (FT-Orbitrap) (Watson and Sparkman 2007).

The visualization of the chromatogram is attained by means of contour plots, wherein the peaks exhibit two retention coordinates ($^1t_R$ and $^2t_R$), and peak intensity is shown using a user-defined color scale (Seeley and Seeley 2013). A 3D surface plot may also be used but its applicability is limited due to visualization restrictions. Common software for GC × GC data processing includes ChromaTOF (LECO Corporation), GC Image (Zoex Corporation), ChromSpace (SepSolve Analytical), and Canvas (J&X Technologies). In many platforms, scripting functions are also available to extract meaningful information based on the mass spectral patterns (Reichenbach et al. 2005). Alternatively, Matlab (The MathWorks Inc.) is a powerful environment for matrix operations, which enables the development of algorithms and chemometric techniques (Berrier et al. 2020). Moreover, in-house scripts have been developed for visualization and advanced processing of GC×GC data, which are easily customizable and have provided unparalleled flexibility.

## 1.4    Metabolite Derivatization and Retention Indices

Derivatization protocols are critical for the analysis of hydrogen bond acid metabolites, like polyols and organic acids, to improve thermal stability, peak symmetry, and detection limits (Poole 2013). Noteworthy, short-chain free organic acids such as ethanoic, propanoic, and butanoic acids may be analyzed in its native form by using acid-modified PEG columns (e.g., Nukol) (Batista et al. 2018). In metabolomic investigations using GC-based methods, derivatization approaches may use alkylsilylating reagents. For instance, trimethyl-, alkyldimethyl-, and *t*-butyldimethyl-based reagents are frequently employed. Analytes protected with *t*-butyldimethylsilyl possesses the best hydrolytic stability compared to trimethylsilyl derivatives (Poole 2013). Polyols such as carbohydrates are incompletely derivatized with *t*-butyldimethyl-based reagents, like N-methyl-*N*-*t*-butyldimethylsilyltrifluoroacetamide (MTBSTFA) due to steric hindrance of such reagents (Fiehn 2016). In Ref. Poole (2013), the interested reader may find an extensive list of reactions for alkylsilyl derivatives, including the silyl-acceptor reactivity order: alcohols (more reactive) > phenols > carboxylic acids > amines > amides (less reactive).

Derivatization of metabolites with silylating reagents must be performed with great caution. Some compounds, such as carbonyl-containing metabolites, form additional unexpected derivatives (i.e., by-products, artifacts) generating multiple peaks for the same analyte. An extensive list of artifacts in TMS derivatization reactions and proposed ways to avoid it are reported elsewhere (Little 1999). Some carbonyl-containing metabolites are difficult to derivatize because by-products are formed during silylation. Carbonyl moieties engage in keto-enol tautomerization by movement of the alpha-hydrogen and reorganization of the bonding electrons. An important route to eliminate carbonyl-related artifacts is by first protecting the carbonyl moiety, followed by silylating the active hydrogen-containing group. Protection of the carbonyl is accomplished by using a methoximation (MeOX) reaction (Madala et al. 2014). Additionally, serine can be used to monitor the quality of the deactivation reaction (Fiehn 2016). Complete derivatization is attained when only the *N,O,O-tris*-TMS serine peak is detected, while the presence of the *O,O-bis*-TMS serine indicates partial derivatization.

In addition, other methods for analyte derivatization are available. Microwave-assisted derivatization is an interesting alternative to conventional indirect heating of reactional media. A broad set of metabolites, including amino acids, sugars, organic acids, and free fatty acids, were successfully converted into its TMS-analogs by using a microwave-assisted protocol (150 W power for 90 s) (Kouremenos et al. 2010). The use of greater volumes of TMS reagent ensured complete derivatization. Moreover, liquid matrices free of proteins and lipids may be derivatized using in-port GC derivatization. Estrogens, namely, estrone, estradiol, estriol, and ethynyl estradiol may be converted into their TMS-derivatives using a PTV injector (González et al. 2019). For solvent-free extraction techniques such as SPME (Belinato et al. 2018) on-fiber derivatization methods are available (Parkinson et al. 2010) for laboratory and on-field analyses using reusable standard reagent generating devices (Poole et al. 2016).

The analyses of lipids including fatty acids, glycerolipids, and glycerophospholipids, also require a derivatization step. The most common method converts the lipids into its respective fatty acid methyl esters (FAMEs). Lipids are important small molecule metabolites that have roles in a wide variety of physiological processes. The crucial role of lipids in cell, tissue, and organ physiology is demonstrated by many genetic studies and by many human diseases that involve the disruption of lipid metabolic enzymes and pathways (Wenk 2005). Base-catalyzed transesterification is applicable to glycerolipids and glycerophospholipids under mild conditions, but it does not react with free fatty acids (Bogusz et al. 2012; Batista et al. 2014). An acid-catalyzed reaction converts both glycerol-bound lipids and free fatty acids, but it required heating for complete derivatization (Mondello et al. 2003). An interesting feature of FAME analysis by GC $\times$ GC is the usefulness of the clear elution patterns (i.e., chromatographic structure) of the chromatograms. The GC $\times$ GC chromatograms are used for identification, because the retention time coordinates may be used to pinpoint the number of carbons of the FAME (e.g., $C_{16}$, $C_{18}$, $C_{20}$, $C_{22}$, and $C_{24}$), level of unsaturation, and position of the unsaturation (e.g., n-3, n-6, and n-9) (Mondello et al. 2003; de Geus et al. 2001). The improved signal-to-noise level, alongside enhanced peak capacity, provided by thermal-modulated GC $\times$ GC allowed the detection for the first time of odd carbon number FAMEs (Mondello et al. 2003).

The PEG phases are not suitable for bioanalytical methods that use some derivatizing reagents. Silylating agents are troublesome to PEG-based phases, because it reacts with the active hydroxyl groups of the coating. However, TMS reagents are much less damaging to poly(siloxanes). Other reagents are known to produce acid by-products such as perfluoro acid anhydride acylation reagents, such as trifluoroacetic anhydride (TFAA) (Sensue 2012a, b). For this reason, it is recommended to evaporate the sample extract to dryness to remove excess reagent.

Qualitative analyses that rely on mass spectral and retention index databases (e.g., NIST, Wiley, Adams (2007), Fiehnlib (Kind et al. 2009), Binbase (Skogerson et al. 2011)) require standardization of sample preparation. For instance, isoleucine, an alpha-amino acid, will exhibit distinct mass spectra depending on the silylation reagent (trimethylsilyl- or *tert*-butyldimethylsilyl-derivative) and will also yield different retention in GC analysis. A common metric to standardize reporting of retention times is the retention index. Two forms are available to determine the retention index: the Kovats retention (Eq. 1) and van Den Dool and Kratz (Eq. 2)

**Table 2** Seven golden rules for heuristic filtering of molecular formulas obtained by accurate mass spectrometry (Kind and Fiehn 2007)

| Seven golden rules |
| --- |
| (1) Apply heuristic restrictions for number of elements during formula generation |
| (2) Check if chemical bonding rules are satisfied by the assigned molecular formula (e.g., LEWIS and SENIOR) |
| (3) Perform isotopic pattern filter |
| (4) Perform hydrogen/carbon (H/C) ratio check (hydrogen/carbon ratio) |
| (5) Perform N-, O-, P-, S-element ratio check (N/C, O/C, P/C, S/C ratios) |
| (6) Perform heuristic H-, N-, O-, P-, S-element probability check (H/C, N/C, O/C, P/C, S/C high probability ratios) |
| (7) Perform TMS check (for GC-MS if a silylation step is involved) |

indexes are used for isothermal and linear temperature programmed analysis, respectively (van Den Dool and Kratz 1963). The retention index is the most reliable metric to express retention coordinates of an analyte, as it accounts for small deviations of the GC method, such as carrier gas flow and slight changes in the temperature settings of the oven (von Mühlen and Marriott 2011).

$$I = 100 \times \left( n + \frac{log(t_{R,\text{sol}}) - log(t_{R,n})}{log(t_{r,N}) - log(t_{R,n})} \right) \tag{1}$$

$$\text{LTPRI} = 100 \times \left( n + \frac{t_{R,\text{sol}} - t_{R,n}}{t_{r,N} - t_{R,n}} \right) \tag{2}$$

where $t_R$ is the retention time of the solute (sol) eluting between two adjacent $n$-alkanes ($n$ and $N$), $n$ is the carbon number of the less retained alkane and $N$ is the carbon number of the most retained $n$-alkane.

Tentative identification of the metabolites is readily achieved by combining mass spectral similarity searchers and retention index filtering. For instance, values for mass spectral similarity are above 85% and a tolerance of $\pm 20$ retention index units are accepted for tentative identification. The assigned identity may be confirmed by the use of authentic standards and/or by accurate mass measurements (if available). Accurate mass measurements enable the attribution of the elemental composition of the molecular ion or adduction, while the isotopic pattern of the cluster may be explored for confirmation purposes (Peterson et al. 2014). Examples of high-resolution/high-mass accuracy mass spectrometers for GC-MS coupling are the time of flight (Gómez-Pérez et al. 2014) and Orbitrap mass analyzers (Peterson et al. 2010, 2014). A set of seven heuristic rules for filtering candidate elemental compositions was proposed (Kind and Fiehn 2007) and apply a number of chemical rules, elemental ratios, and elemental probabilities, and evaluate the accuracy of the experimental mass isotopomer abundance distribution, as shown in Table 2. Together with a MS having high mass and isotopomer ratio accuracy, these rules allow assignment of the correct formula 98% of the time for compounds present in a database (Peterson et al. 2014).

## 1.5    Applications of GC×GC in Human Metabolomics

In this section, we focused on human-based metabolic profiling using GC-based methods. Considering the human metabolome, roughly 50% of exposome research uses urine samples, 25% blood (blood, serum, or plasma) samples, and the remainder matrices comprise other body fluids (e.g., breath, sweat, feces, and cells) (Weggler et al. 2020).

Biological fluids are challenging matrices in metabolomics due to the presence of proteins and salts from the sample matrix. Hence, protein removal by precipitation using trifluoroacetic acid or acetonitrile is advised prior to GC $\times$ GC analysis. In addition, protein removal is particularly important if a derivatization step is required to improve the peak shape and response factor of the analytes. Liquid-liquid extraction (LLE) is likely the most common method used for sample preparation, such as the Folch (Folch et al. 1957) and Bligh and Dyer (1959) protocols. Quenching of the metabolism may be required if dealing with in vitro sampling to obtain a representative snapshot of the metabolome (Villas-Bôas et al. 2005). Depending on the sample matrix, mechanical or chemical cell disruption is required to successfully extract the metabolite content (Iverson et al. 2001). However, the investigation of the exometabolome, i.e., extracellular metabolites, does not demand cell lyses (Belinato et al. 2018).

A major benefit of employing GC $\times$ GC for volatile organic compounds (VOCs) analyses is the enhanced peak detection. A recent investigation showcased that a twofold improvement in peak detection was obtained with GC $\times$ GC-TOFMS compared to 1D-GC-MS fitted with a FT-Orbitrap mass analyzer (Misra et al. 2018). The GC $\times$ GC-TOFMS platform permitted $\sim$380 properly identified metabolites to be assigned to 60 different Kyoto Encyclopedia of Genes and Genomes (KEGG)-based pathways.

Two recent applications of GC $\times$ GC-based metabolomics analyzed the metabolites in human serum samples. The first application studied the alterations in human serum metabolome induced by a marathon. Hence, analyses of the primary metabolites required solvent fractionation and analyte derivatization (Stander et al. 2018). Briefly, 3-phenylbutyric acid internal standard was spiked to the serum sample. The samples were cooled, and proteins were precipitated with acetonitrile. Samples were vortexed and centrifuged. Methoxamine hydrochloride (MOX) was used to protect the carbonyl groups. A solution of $N,O$-bis-(trimethylsilyl)-trifluoroacetamide (BSTFA) with 1% trimethylchlorosilane (TMCS) was used for TMS derivatization, followed by GC $\times$ GC-TOFMS analyses. Chemometric analysis of the metabolome highlighted that alteration to the levels of amino acids, fatty acids, ketones, and tricarboxylic acid cycle intermediates were related to metabolic shifts between fuel substrate systems. These findings indicated the strains on energy producing pathways of marathon athletes, leading to autophagy, inhibition of the rapamycin complex 1, oxidative stress, and protein degradation (Stander et al. 2018).

The second application investigated the effect of physical training on the respiratory infection risk and lipid peroxidation in older adults. Metabolic profiling was obtained by combining headspace SPME (HS-SPME) with GC × GC-TOFMS (Silva et al. 2019). This approach enables simultaneous analyte pre-concentration and sample cleanup in a single and reliable step (Arthur and Pawliszyn 1990; Jiang and Pawliszyn 2012; Piri-Moghadam et al. 2017). For the HS-SPME assay, aliquots of 500 μL of serum were used. The addition of 0.1 g of sodium chloride ensured the 'salting-out' effect. Vigorous stirring was required to improve mass transfer in the multiphasic system. Analyte isolation was attained using a mixed divinylbenzene/carboxen phase dispersed in poly(dimethylsiloxane) (DVB/CAR/PDMS). Sample equilibration was performed at 40 °C for 30 min. The metabolites were separated using a 30 m × 0.32 mm (0.25 μm film thickness) HP-5 primary column and a 0.79 m × 0.25 mm (0.25 μm film thickness) DB-FFAP secondary column (nitroterephthalic acid-modified PEG). In this study, 76 metabolites including alcohols, aldehydes, alkanes, and ketones were successfully identified. This information showed a relation between the metabolites and respiratory infections, wherein a pattern of lipid peroxidation was associated with the number of infections.

Analysis of metabolites in urine samples from newborns can be performed by combining SPE and GC × GC-MS (Bileck et al. 2019). SPE is an exhaustive sample preparation method, wherein the liquid sample mixture is loaded onto a sorbent-packed cartridge (or disk) (Hennion 1999). Next, matrix cleanup may be achieved by using carefully selected solvents, followed by analyte desorption using a strong organic solvent. Low detection limits are obtained by maintaining a high phase ratio between sample volume and final volume of the extract (Hennion 1999). Urine samples are complex matrices with highly congested chromatograms, which makes it difficult for proper peak integration and identification (Bileck et al. 2019). SPE enabled cleanup of interfering compounds that are incompatible with GC-based analysis, while 21 progesterone metabolites were successfully separated and identified by GC × GC-TOFMS. The steroids were derivatized with MOX and TMS-silylated. A 15 m × 0.25 mm (0.25 μm film thickness) Rxi-1ms primary column and a 2 m × 0.1 mm (0.1 μm film thickness) BPX-50 secondary column were used in the GC × GC experiments. Such methods are important to study steroids implicated in a high variety of physiological processes and are typically analyzed for the diagnosis of hormonal disorders, which affects growth patterns in early infancy (Bileck et al. 2019).

Aromatic amines are an important class of harmful components of cigarette smoke (Lamani et al. 2015). Nevertheless, only a few of them have been reported to occur in urine, which raises questions on the fate of these compounds in the human body. A recent investigation used SPME combined to GC × GC-QMS (QMS, mass spectrometer with a single transmission quadrupole mass analyzer) to determine such metabolites in complex matrices. More than 150 aromatic amines were successfully identified in the urine of a smoking person (Fig. 3), including

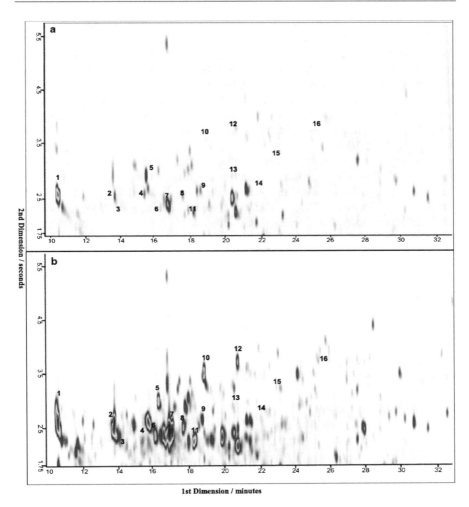

**Fig. 3** Total ion GC × GC chromatograms (m/z 70–460) showing aromatic iodine compounds proposed to be present in the urine of (**a**) a person without exposure to cigarette smoke and (**b**) a smoking person. The same scale was used on both chromatograms. The red color indicates the high intensity of the analyte followed by yellow, green, and blue as the intensity decreases. The samples were measured by SPME-GC × GC-QMS after hydrolysis, liquid–liquid extraction, and derivatization. Numbers in the figure refer to iodinated derivatives. Lamani, X., Horst, S., Zimmermann, T. et al. Determination of aromatic amines in human urine using comprehensive multi-dimensional gas chromatography-mass spectrometry (GCxGC-QMS). *Anal Bioanal Chem* 407, 241–252 (2015). Open access article distributed under the Creative Commons Attribution License

alkylated and halogenated amines as well as substituted naphthylamines, due to the improved resolving power of GC × GC (Lamani et al. 2015).

The analysis of exhaled breath is a noninvasive diagnostic tool and has gained increased attention in precision medicine particularly in molecular pathology. Such an approach requires PandT exhaustive sampling to achieve low detection limits and enable decoupling of sampling and sample analyses, since VOCs are stable inside the sorbent tube for a short time period (e.g., weeks). Briefly, the exhaled breath (gaseous mixture) is percolated through a sorbent-packed tube for analyte enrichment. Next, the sampling tube is sealed, followed by GC × GC analysis. Thermal desorption (TD) is an effective technique for sample introduction of volatile and semi-volatile organic compounds pre-sorbed to solid-phase adsorbents (Baltussen et al. 2002; Piotrowski et al. 2018). The online analyte introduction is attained by heating the TD tube under flow of carrier gas. Peak focusing using a temperature gradient may be used to improve the peak shape of the analytes, increasing the peak capacity of the GC-based method (Baltussen et al. 2002; Piotrowski et al. 2018). The VOCs in exhaled breath were analyzed by coupling TD and FM-GC × GC with parallel detection using flame ionization detection (FID) and QMS, revealing a wealth of metabolomic information (Wilde et al. 2019) A high-quality breath sample is defined by a good record of sample history as much as good sampling technique.

Good sampling technique requires ensuring a secure seal when capping the TD tubes; not leaving the tubes uncapped for any length of time except during sampling; once sampled keeping the tubes refrigerated in a low odor environment; conditioning face masks or silicon-based products before use; taking regular environmental samples (e.g., sampling the room air and air supply) (Wilde et al. 2019). GC × GC-MS for the investigation of breath gas and bronchoalveolar lavage analysis has been reported for mycobacterial infection (Beccaria et al. 2019), lung cancer (Pesesse et al. 2019), and asthmatic phenotypes (Schleich et al. 2019). These preliminary studies showed promise in differentiation between different states of the investigated matrices by combining GC × GC studies with sophisticated chemometric data treatment techniques (Weggler et al. 2020).

# References

Adams RP (2007) Identification of essential oil components by gas chromatography/mass spectrometry, 4th edn

Arthur CL, Pawliszyn J (1990) Anal Chem 62:2145–2148

Bahaghighat DH, Freye CE, Synovec RE (2019) Trends Anal Chem 113:379–391

Baltussen E, Cramers C, Sandra P (2002) Anal Bioanal Chem 373:3–22

Batista AG, Lenquiste SA, Cazarin CBB, Silva JK, Luiz-Ferreira A, Bogusz S Jr, Hantao LW, Souza RN, Augusto F, Prado MA, Maróstica MR Jr (2014) J Funct Foods 6:450–461

Batista AG, Silva-Maia JK, Mendonça MCP, Soares ES, Lima GC, Bogusz Junior S, Cruz-Höfling MA, Maróstica Júnior MR (2018) J Funct Foods 48:266–274

Beccaria M, Bobak C, Maitshotlo B, Mellors TR, Purcaro G, Franchina FA, Rees CA, Nasir M, Black A, Hill JE (2019) J Breath Res 13:016005

Belinato JR, Dias FFG, Caliman JD, Augusto F, Hantao LW (2018) Anal Chim Acta 1040:1–18

Berrier KL, Prebihalo SE, Synovec RE (2020) Sep Sci Technol 12:229–268

Bileck A, Fluck CE, Dhayat N, Groessl M (2019) J Steroid Biochem Mol Biol 186:74–78

Bligh EG, Dyer WJ (1959) Can J Biochem Physiol 37:911–917

Bogusz S Jr, Hantao LW, Braga SCGN, França VCRM, Costa MF, Hamer RD, Ventura DF, Augusto F (2012) J Sep Sci 35:2438–2444

Boswell H, Carrillo KT, Górecki T (2020) Separations 7:1–12

Brunelli C, Pereira A, Dunkle M, David F, Sandra P (2010) LCGC Eur 2010:396–405
Cagliero C, Bicchi C, Cordero C, Liberto E, Rubiolo P, Sgorbini B (2017) J Chromatogr A 1495: 64–75
de Geus H-J, Aidos I, de Boer J, Luten JB, Brinkman UAT (2001) J Chromatogr A 910:95–103
Fiehn O (2016) Curr Protoc Mol Biol 114:30.4.1–30.4.32
Folch J, Lees M, Stanley GHS (1957) J Biol Chem 226:497–509
Fujiyoshi T, Ikami T, Sato T, Kikukawa K, Kobayashi M, Ito H, Yamamoto A (2016) J Chromatogr A 1434:136–141
Gómez-Pérez ML, Plaza-Bolaños P, Romero-González R, Vidal JLM, Frenich AG (2014) J Am Soc Mass Spectrom 25:899–902
González A, Clavijo S, Cerdà V (2019) Talanta 194:852–858
Griffith JF, Winniford WL, Sun K, Edam R, Luong JC (2012) J Chromatogr A 1226:116–123
Grob RL, Barry EF (2004) Modern practice of gas chromatography, 4th edn
Gullberg J, Jonsson P, Nordström A, Sjöström M, Moritz T (2004) Anal Biochem 331:283–295
Harynuk J, Górecki T (2003) J Chromatogr A 1019:53–63
Harynuk J, Górecki T (2005) J Chromatogr A 1086:135–140
Hennion M-C (1999) J Chromatogr A 856:3–54
Impens S, De Wasch K, De Brabander H (2001) Rapid Commun Mass Spectrom 15:2409–2414
Iverson SJ, Lang SLC, Cooper MH (2001) Lipids 36:1283–1287
Jiang R, Pawliszyn J (2012) Trends Anal Chem 39:245–253
Khummueng W, Harynuk J, Marriott PJ (2006) Anal Chem 78:4578–4587
Kind T, Fiehn O (2007) BMC Informatics 8:105
Kind T, Wohlgemuth G, Lee DY, Lu Y, Palazoglu M, Shahbaz S, Fiehn O (2009) Anal Chem 81: 10038–10048
Kouremenos KA, Harynuk JJ, Winniford WL, Morrison PD, Marriott PJ (2010) J Chromatogr B 878:1761–1770
Lamani X, Horst S, Zimmermann T, Schimidt TC (2015) Anal Bioanal Chem 407:241–252
Little JL (1999) J Chromatogr A 844:1–22
Liu Z, Phillips JB (1991) J Chromatogr Sci 29:227–231
Luong J, Guan X, Xu S, Gras R, Shellie RA (2016) Anal Chem 88:8428–8432
Madala NE, Piater LA, Steenkamp PA, Dubery IA (2014) SpringerPlus 3:254
Marriott PJ, Kinghorn RM (1997) Anal Chem 69:2582–2588
Marriott PJ, Shellie R, Cornwell C (2001) J Chromatogr A 936:1–22
Maštovská K, Lehotay SJ, Anastassiades M (2005) Anal Chem 77:8129–8137
McNair HM, Miller JM (2008) Basic gas chromatography, 2nd edn
Misra BB, Bassey E, Bishop AC, Kusel DT, Cox LA, Olivier M (2018) Rapid Commun Mass Spectrom 32:1497–1506
Mondello L (2011) Comprehensive chromatography in combination with mass spectrometry. Wiley, New Jersey
Mondello L, Casilli A, Tranchida PQ, Dugo P, Dugo G (2003) J Chromatogr A 1019:187–196
Mondello L, Tranchida PQ, Dugo P, Dugo G (2008) Mass Spectrom Rev 27:101–124
Muñoz-Guerra JA, Prado P, García-Tenorio SV (2011) J Chromatogr A 1218:7365–7370
Parkinson D-R, Warren JJ, Pawliszyn J (2010) Anal Chim Acta 661:181–187
Pesesse R, Stefanuto PH, Schleich F, Louis R, Focant J-F (2019) J Chromatogr B 1114–1115:146–153
Peterson AC, Balloon AJ, Westphall MS, Coon JJ (2014) Anal Chem 86:10044–10051
Peterson AC, Hauschild J-P, Quarmby ST, Krumwiede D, Lange O, Lemke RAS, Grosse-Coosmann F, Horning S, Donohue TJ, Westphall MS, Coon JJ, Griep-Raming J (2014) Anal Chem 86:10036–10043
Peterson AC, McAlister GC, Quarmby ST, Griep-Raming J, Coon JJ (2010) Anal Chem 82:8618–8628
Phillips JB, Luu D, Pawliszyn JB, Carle GC (1985) Anal Chem 57:2779–2787
Piotrowski PK, Tasker TL, Burgos WD, Dorman FL (2018) J Chromatogr A 1579:99–105

Piri-Moghadam H, Alam MN, Pawliszyn J (2017) Anal Chim Acta 984:42–65
Pollo BJ, Alexandrino GL, Augusto F, Hantao LW (2018) Trends Anal Chem 105:202–217
Poole CF (2013) J Chromatogr A 1296:2–14
Poole JJ, Grandy JJ, Gómez-Ríos GA, Gionfriddo E, Pawliszyn J (2016) Anal Chem 88:6859–6866
Poole CF, Poole SK (2008) J Chromatogr A 1184:254–280
Poole CF, Poole SK (2009) J Chromatogr A 1216:1530–1550
Ragonese C, Sciarrone D, Grasso E, Dugo P, Mondello L (2015) J Sep Sci 39:537–544
Reichenbach SE, Kottapalli V, Ni M, Visvanathan A (2005) J Chromatogr A 1071:263–269
Schleich FN, Zanella D, Stefanuto P-H, Bessonov K, Smolinska A, Dallinga JW, Henket M, Paulus V, Guissard F, Graff S, Moermans C, Wouters EFM, Van Steen K, van Schooten F-J, Focant J-F, Louis R (2019) Am J Respir Crit Care Med 200:444–453
Seeley JV, Micyus NJ, McCurry JD, Seeley SK (2008) Am Lab 914:35744
Seeley JV, Schimmel NE, Seeley SK (2018) J Chromatogr A 1536:6–15
Seeley JV, Seeley SK (2013) Anal Chem 85:557–578
A. Sensue, in Capillary GC column killers – part 1 2012a. https://blog.restek.com/?p=5001. Accessed 30 Apr 2020
A. Sensue, in Capillary GC column killers – part 2 2012b. https://blog.restek.com/?p=5003. Accessed 30 Apr 2020
Silva D, Arend E, Rocha SM, Rudnitskaya A, Delgado L, Moreira A, Carvalho J (2019) Eur J Sport Sci 19:384–393
Skogerson K, Wohlgemuth G, Barupal DK, Fiehn O (2011) BMC Bioinformatics 12:321
Stander Z, Luies L, Mienie LJ, Keane KM, Howatson G, Clifford T, Stevenson EJ, Loots DT (2018) Metabolomics 14:150
Tranchida PQ, Franchina FA, Dugo P, Mondello L (2014) J Chromatogr A 1372:236–244
Tranchida PQ, Sciarrone D, Dugo P, Mondello L (2012) Anal Chim Acta 716:66–75
van Den Dool H, Kratz PD (1963) J Chromatogr A 11:463–471
van Stee LLPP, Brinkman UAT (2016) Trends Anal Chem 83:1–13
Villas-Bôas SG, Mas S, Åkesson M, Smedsgaard J, Nielsen J (2005) Mass Spectrom Rev 24:613–646
von Mühlen C, Marriott PJ (2011) Anal Bioanal Chem 401:2351–2360
Watson JT, Sparkman OD (2007) Introduction to mass spectrometry: instrumentation, applications and strategies for data interpretation. Wiley, New Jersey
Weggler BA, Gruber B, Focant J-F (2020) Curr Opin Environ Sci Health 15:16–25
Wenk MR (2005) Nat Rev Drug Discov 4:594–610
Wilde MJ, Cordell RL, Salman D, Zhao B, Ibrahim W, Bryant L, Ruszkiewicz D, Singapuri A, Free RC, Gaillard EA, Beardsmore C, Paul Thomas CL, Brightling CE, Siddiqui S, Monks PS (2019) J Chromatogr A 1594:160–172
Zoccali M, Giocastro B, Tranchida PQ, Mondello L (2019) J Sep Sci 42:691–697

# Liquid Chromatography in Bioanalysis

Mariana R. Gama, Gisláine C. da Silva, and Carla B. G. Bottoli

## 1 Introduction

New analytical techniques have been developed for several research areas; however, high-performance liquid chromatography (HPLC) remains as one of the most employed analytical tools for many purposes. These HPLC systems consist of a solvent delivery system, a sample injection valve, high-pressure columns, detectors, and a computer to control the system.

HPLC and related techniques are versatile separation approaches for diverse applications, such as environmental monitoring, food and pharmaceutical analysis, metabolomics and clinical screening. Liquid chromatography is important because many compounds are not sufficiently volatile nor thermally stable for gas chromatography. In bioanalytical analysis, the increasing complexity of (bio)samples requires continuous development of HPLC analytical instrumentation, column packings, and detection. Recent advances in analytical instrumentation and sample preparation methods have propelled biological analyses for the identification of these interesting analytes. Very low amounts of analytes in complex matrices, such as metabolites, pharmaceutical and biological compounds, and clinical targets demand analytical strategies with improved performance. Thus, the results are highly affected by the correct choice of the following steps: sample pretreatment, separation mode, and detection.

The direct injection of most biological samples is frequently not possible for most chromatographic techniques because of the complexity of the matrix to be analyzed

M. R. Gama
Department of Inorganic Chemistry, Institute of Chemistry, Federal University of Rio Grande do Sul (UFRGS), Porto Alegre, Rio Grande do Sul, Brazil

G. C. da Silva · C. B. G. Bottoli (✉)
Department of Analytical Chemistry, Institute of Chemistry, University of Campinas (UNICAMP), Campinas, SP, Brazil
e-mail: carlab@unicamp.br

© The Author(s), under exclusive license to Springer Nature Switzerland AG 2022
L. T. Kubota et al. (eds.), *Tools and Trends in Bioanalytical Chemistry*,
https://doi.org/10.1007/978-3-030-82381-8_21

and the presence of interfering compounds. For this reason, a sample pretreatment step is mandatory before any (bio)chromatographic analysis, and the development of more specific sorbents is also welcome for sample preparation. Separation performance is especially critical for bioanalytical applications. As the column is the core component of an HPLC system, and specific interactions established between analytes and stationary phase occur on the column packing, the properties of stationary phases can greatly affect the analyte separation. Different stationary phases are responsible for the development of different modes of HPLC separation, such as reversed or normal phase, ion exchange, and hydrophilic interaction liquid chromatography, among others.

Detection plays an important role in bioanalytical separations. In view of its widespread use, UV/Vis detection has dominated HPLC, often requiring pre- or post-column derivatization. Clinical, biomedical, or pharmaceutical applications require good detectability and, often a universal detection method, such as that provided by mass spectrometric (MS) detection.

This chapter presents the general aspects of separation modes and their characteristics for bioanalytical liquid chromatographic applications. Materials for specific separations, such as restricted-accessed materials and imprinted stationary phases, and packings for bio affinity and hydrophilic interaction chromatography will also be presented. Particularities for detection of bioanalytes are also included, as well as multidimensional approaches for resolution of typical complex matrices.

## 2 Separations Modes in Bioanalytical Liquid Chromatography

### 2.1 Ion pairing

The reversed-phase (RP) mechanism is the most popular of the liquid chromatographic separation modes, being applied for a wide range of analytes. Hydrophobic interactions provided by alkyl chains bonded to silica particles retain hydrophobic and less polar analytes. An aqueous mobile phase allows the separation of moderately polar molecules and expands this separation mode for several classes of compounds. The analyte retention can be modulated by changes in the organic solvent content and by the presence of mobile phase additives, which allow changes in selectivity based on mobile phase composition.

The weak interactions established by highly polar and/or ionized analytes and hydrophobic stationary phases promote poor retention and peak shapes in RP separations, limiting the use of this separation mode for polar targets.

To overcome the low retention of polar compounds in reversed-phase systems, ion-pair liquid chromatography was introduced in the 1970s. This approach is based on the separation of charged analytes using a specific mobile phase additive and a typical RP column containing C8 or C18 alkyl moieties (Snyder et al. 1997).

The ion-pair additive is oppositely charged to the analyte and presents a lipophilic moiety. This enables its simultaneous interaction with the alkyl group on the bonded

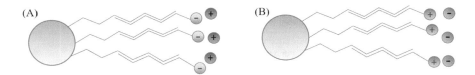

**Fig. 1** Schematic representation of ion-pair chromatography, showing the alkyl bonded group attached to the silica particle, the ion-pair reagent adsorbed to the bonded group, and the ionic analytes. Counter ions from ion-pair reagents are not shown. (**a**) Separation of a positively charged analyte with a cationic ion-pairing additive. (**b**) Separation of a negatively charged analyte with an anionic ion-pairing additive

stationary phase and the analyte (Cecchi 2015). Sulfate and sulfonate alkyl salts and tertiary and quaternary ammonium alkyl salts are usually employed as cationic and anionic ion-pair reagents, respectively. These are added to the mobile phase at concentrations up to 50 mmol $L^{-1}$. Acidic compounds are separated using alkyl quaternary ammonium salts with pH of the mobile phase adjusted to 7.5, while basic samples are usually separated by addition of alkyl sulfonic salts with the pH adjusted to 3.0–4.0 (Dolan 2008).

Figure 1 shows a simple schematic representation of ion-pairing chromatography. The ion-pair reagent is added to the mobile phase, enabling the strong interaction between its hydrophobic end, the alkyl group bonded to the stationary phase. Then, the charged functional group from the immobilized ion-pair reagent is available for interaction with the ionic analytes with opposite charge, providing chromatographic retention.

In addition to sulfonic salts and quaternary ammonium alkyl salts, ionic liquids (ILs) are also employed as ion-pair reagents. ILs are liquid salts at room temperature, which have been applied in several research areas, including catalysis, electrochemical techniques, and separation sciences. Theoretically, it is estimated there are more than $10^{18}$ combinations of ILs, due to the possibilities of interchanging cationic or anionic moieties (Carmichael and Seddon 2000). About 1000 of these compounds are described in the literature, and approximately 300 are commercially available. Advantages of ILs include low toxicity, good thermal stability, adjustable viscosity, and miscibility with both aqueous and organic solvents, as well as being potential green solvents (Hong et al. 2016).

Although not named as ion-pair chromatography, ILs have been used as additives in the mobile phase since the 2000s, aiming to minimize band broadening and to improve peak symmetry for basic analytes in HPLC and to increase the reproducibility in capillary electrophoresis (CE) (Xiaohua et al. 2004; He et al. 2003). However, the low volatility of ILs hinders detection by some techniques, such as evaporative light scattering and mass spectrometry (Berthod and Carda-Broch 2004; Han and Row 2010; García-Alvarez-Coque et al. 2015).

A combination of retention mechanisms improves the efficiency of the separation of ionic compounds using ion-pairing, showing better results than ion-exchange chromatography (Meyer 2010). Both hydrophobic and ion-exchange interactions

**Table 1** Effect of experimental conditions on the retention of ionized analytes in ion-pair chromatography. Adapted from Ref. (Ståhlberg 2000) with permission

| Experimental parameter | Retention of oppositely charged analyte | Retention of similarly charged analyte |
|---|---|---|
| Increase the charge of analyte | Increases | Increases |
| Increase the charge of ion-pair additive | Decreases | Decreases |
| Increase the concentration of pairing ion | Increases | Decreases |
| Increase the hydrophobicity of pairing ion | Increases | Decreases |
| Increase the concentration of organic modifier | Decreases | Decreases |
| Increase the polarity of the organic modifier | Decreases | Increases |
| Increase the ionic strength | Decreases | Increases |
| pH of eluent | Increases | Decreases |

mediate analyte retention in ion-pair LC, while ionic interactions predominate in ion-exchange mode. Evaluating the chromatographic behaviors of 8-hydroxyquinoline in different LC modes, such as reversed-phase, ion-pair, and ion-exchange, better separation performance was achieved by an ion-pairing method, with shorter retention times, better peak symmetry, and higher plate number than reversed-phase and ion-exchange separation methods (the efficiency was 30% higher for the ion-pairing method) (Ma et al. 2010).

According to Ståhlberg (2000), the retention factor for an oppositely charged analyte can be increased by a factor up to 20 with an increase in ion-pairing concentrations, when compared to the retention with no addition of the ion-pair reagent. Likewise, the decrease in the concentration of the ion-pair reagent promotes the reduction of the retention factor of a similarly charged analyte by a factor up to 20. However, the increase in the reversed-phase character of the column packing with the addition of an ion-pair reagent to the mobile phase may also decrease the retention time for a neutral compound (Dolan 2008).

Retention and selectivity in ion-pair LC are greatly influenced by experimental conditions; the most important parameters are related to the type and concentration of the ion-pair additive and organic modifier, and to the pH of the mobile phase, since the interactions between the analyte and ion-pair reagent are pH-dependent (Cecchi 2017). Table 1 summarizes the effect of the main experimental conditions in the retention of ordinary ionized analytes. Although having many advantages, ion-pair chromatography also presents several limitations. The column must be dedicated to ion-pair applications since the pairing additives cannot be completely removed from the column. This also decreases the column lifetime. Many ion-pair additives possess chromophore groups and have substantial UV-absorption, inducing the poor baseline stability and interferences in UV detection (Cecchi 2017). However, signal suppression can be exploited as an analytical tool in ion-pair chromatography, enabling indirect determination of analytes by the decreasing signal of the baseline.

Ion-pairing reagents are avoided in MS or MS/MS detection systems, since they are mostly non-volatiles, increase the normal background levels, and contribute to additional noise. Consequent signal loss, poor reproducibility, and reduced column lifetimes due to ion-pair residues in the instrumental system are also listed as ion-pair drawbacks and discourage the application of ion-pairing reagents for routine analysis (Holčapek et al. 2004; Annesley 2003). However, although discouraged, validated methods of ion-pair LC–MS/MS have been applied for routine application in clinical studies (Thuboy et al. 2018).

Due to several limitations, the applicability of ion-pairing LC for the analysis of polar compounds has been discontinued and replaced for more attractive chromatographic modes, such as separations by mixed-mode columns (generally exploiting embedded polar stationary phases) and hydrophilic interaction (liquid) chromatography (HILIC).

Commonly analyzed by ion-pairing LC methods using positively charged amines as mobile phase additives (Kaczmarkiewicz et al. 2019), a new methodology was fully validated for quantitation of oligonucleotides in plasma (MacNeill et al. 2019). According to the authors, ion-pairing reagents presented negative effects in both sample preparation and separation steps, which were avoided by a chromatographic method exploiting the HILIC mode prior to MS detection.

## 2.2    HILIC

Hydrophilic interaction (liquid) chromatography (HILIC) is an interesting alternative for the analysis of polar substances. HILIC is defined as a separation mode that combines a stationary phase usually used in the normal phase mode, and a mobile phase used in reversed-phase separations. Typical applications of HILIC involve highly hydrophilic stationary phases, such as chemically bonded silica or polar polymers. In most cases, the mobile phase is a polar organic solvent (most frequently acetonitrile, sometimes methanol) containing up to 30% of water (Gama et al. 2012). Figure 2 shows the most frequent stationary phases employed in HILIC separations in the last 5 years.

Although it was previously employed in another context, Alpert (1990) coined the term HILIC, also describing a complex mechanistic separation, in which the polar groups attached to the different types of stationary phase attract water molecules, forming an aqueous layer over the surface. Thus, a polar analyte that is dissolved in the mobile phase undergoes partitioning between two liquid phases: the semi-immobilized aqueous layer and the mobile phase, which also has some aqueous content. Polar solutes have higher affinities for the semi-immobilized aqueous layer than for the principal organic mobile phase since they are better solvated in the former. This preference leads to increased interaction of the solute with the aqueous layer, also increasing the retention. For this reason, the elution order in HILIC

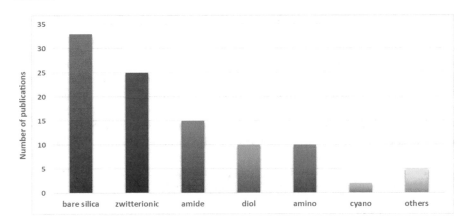

**Fig. 2** Stationary phases employed in HILIC mode separations in the last 5 years, according to the Web of Science. "Others" corresponds to polyethylene glycol, saccharides, or cyclodextrin-based phases

separations is the opposite from that observed in reversed-phase runs, since the polar analytes are more retained on a HILIC column (Buszewski and Noga 2012).

Combined mechanisms in HILIC offer an interesting alternative for separations that are unresolved with normal and reversed-phases, or even with ion-pairing or ion-exchange chromatography. Figure 3 describes the most common interactions in a typical HILIC separation of neutral and ionized polar analytes on a silica-based stationary phase.

Generally, polar interactions increase in the following sequence: cyanopropyl < diol < aminopropyl < silica. Thus, for the same analyte, bare silica columns usually have significantly higher retention than cyanopropyl type columns. However, the strong electrostatic attraction presented by bare silica can cause residual interactions and peak asymmetry in the HILIC mode, reducing the separation efficiency for some classes of analytes. The poor reproducibility on bare silica is reduced when using polar bonded stationary phases or hydride silica (Gama et al. 2012).

According to Kawachi et al. (2011) to select the proper stationary phases for a separation target, one must know the retention, selectivity, and separation efficiency of HILIC columns for that specific application. These authors suggest an inclusive test scheme for HILIC stationary phases using nucleosides, saccharides, xanthines, sodium p-toluenesulfonate, and trimethylphenylammonium chloride to describe some parameters, such as the degree of hydrophilicity, the selectivity for hydrophobic and hydrophilic groups, positional selectivity, the configuration of hydrophilic groups, anion and cation exchange properties, the local pH conditions on the stationary phases, and shape selectivity. The authors observed that strongly hydrophilic phases included amide-bonded phases and zwitterionic phases. Phases such as cyclodextrin-, diol-, trizol-, amine-bonded, and bare silica phases could be categorized as weakly hydrophilic phases.

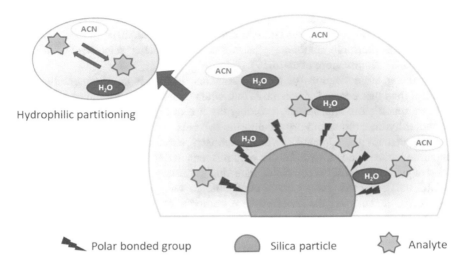

Fig. 3 Common interactions established between a hydrophilic stationary phase and polar analytes in a typical HILIC separation. Blue and water regions correspond to the water later and organic solvent (most frequently acetonitrile) in the mobile phase, respectively. The water layer formed on the particle surface retains most polar analytes by hydrogen bonding, while analyte partitioning simultaneously occurs between the same water layer and acetonitrile. Electrostatic interactions are mainly enabled by polar groups bonded to the support surface, which also helps to form a water layer

Since its inception, the coupling of HILIC phases and mass spectrometry (MS) detection has been reported as advantageous (Nguyen and Schug 2008). Before, the most used detectors were based on spectroscopy, such as absorbance and fluorescence, which, basically, depend on the nature of the analyte, sometimes requiring an additional step of derivatization. A high content of organic solvent in the mobile phase made easier the step of fragmentation in the MS system, and confirmed the successful coupling of HILIC-MS and HILIC-MS/MS for separation and detection of polar analytes, with a recent popularization of the latter (Tang et al. 2016).

Although advantageous in some aspects, HILIC has some limitations in comparison with reversed-phase separations. The separation mechanism of HILIC is not straightforward and is different from that of reversed-phase, which makes it difficult to predict the effect of changing separation conditions (McCalley 2010). In addition, HILIC is not suitable for hydrophobic analytes because they often do not present sufficient interactions to promote as good a separation as with reversed-phase columns, unless a mixed-mode column with both hydrophilic and hydrophobic interactions is employed (Sýkora et al. 2019).

The difficult separation of mixtures containing structurally similar analytes is another limitation of HILIC discussed in the literature (Bisceglia et al. 2010). Because of the similar hydrophilic interactions established with the stationary phase, analogous compounds could coelute. However, this statement has been contradicted by other experimental observations, and successful HILIC separations for similar bio compounds have been described (Fatima et al. 2019), especially those exploiting zwitterionic stationary phases that present better selectivity than other polar phases (Sonnenberg et al. 2019).

By using large volumes of organic solvents, HILIC is environmentally more harmful than the reversed-phase mode but, when compared to normal phase, HILIC appears much more appropriate, avoiding the use of highly toxic solvents, such as nonpolar hydrocarbons and organochlorine solvents.

In agreement with Green Analytical Chemistry, exploitation of U-shape retention of some polar bonded hydrophilic columns can reduce the quantity of organic solvent used in HILIC separations. The dual retention mechanism — where the HILIC and reversed-phase effects contribute to analyte retention to a different degree — depends on the mobile phase composition and the type of stationary phase. By modulation of the U-shape effect, it is possible to retain polar compounds with high contents of water in the mobile phase (Jandera and Hájek 2009).

HILIC has a special acceptance in omics sciences, which created new fields heavily dependent on faster and more reliable procedures capable of identifying and quantifying sets of compounds synthesized by a biological system. Omics also need comprehensive analytical techniques suitable to process complex biological samples, some even containing trace amounts of unstable or labile analytes. Based on the varied characteristics of analytes, HILIC appears as a complementary way to reversed-phase separations, respecting the versatility of several compounds from biological matrices. For this reason, HILIC was pointed out as the main analytical tool for coupling to mass spectrometry in omics research, such as metabolomics (Haselberg et al. 2019), lipidomics (Lange et al. 2019; Wolrab et al. 2019), and glycomics (Yang and Bartlett 2019).

## 2.3    Restricted-Access Materials

Restricted-access materials (RAMs) are biocompatible packings used for the direct extraction of analytes from biological fluids, such as plasma, serum, or blood.

RAM technology was introduced in 1991 (Desilets et al. 1991) and its separation mechanism is mainly based on the size exclusion of analytes, which limits their accessibility to the pores. Small analytes interact with binding sites within the pores, while large molecules are excluded from the pores and interact only with the outer surface of the particle sorbent. Additional interactions are promoted by hydrophilic groups on the particle surface, which also minimizes the adsorption of proteins from the matrix (Souverain et al. 2004).

Modifications of the external surfaces of conventional supports, such as silica-based and polymeric materials, are employed for the obtention of several RAMs, through variations of the nature of the pore size barrier and on the surface coatings (Abrão et al. 2019).

A unimodal RAM has similar ligands bonded to both external and internal surfaces, while a bimodal material has distinct bonded groups on its different

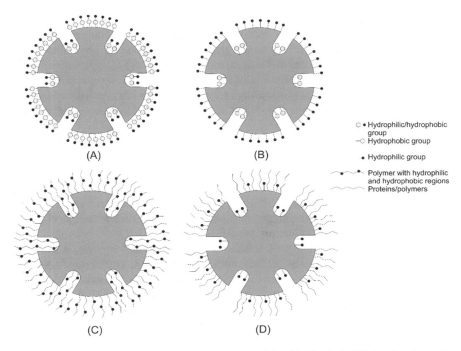

**Fig. 4** Configurations of restricted silica-based materials. (**a**) Physical diffusion barriers with unimodal phases. (**b**) Physical diffusion barriers with bimodal phases. (**c**) Chemical diffusion barriers with unimodal phases. (**d**) Chemical diffusion barriers with bimodal phases. Reprinted from Ref. (de Faria et al. 2017) with permission

surfaces (for example, a hydrophilic external surface and a hydrophobic internal surface). Additionally, the exclusion mechanism of RAM could be obtained by physical or chemical diffusion barriers, based on the pore size or the presence of external hydrophilic bonded groups, respectively (Fig. 4). The most popular RAM is the bimodal porous phase, characterized by an outer hydrophilic layer and an inner pore surface with a hydrophobic bonded phase. The combination of a hydrophilic surface—with minimal protein interaction—and small hydrophobic pores promotes the exclusion of large interferents, while small bioanalytes enter the pores and are retained by hydrophobic interactions with alkyl bonded phases (Boos and Rudolphi 1997; Rudolphi and Boos 1997).

Although described as non-clogging sorbents, RAM eventually presents fouling with the repeated injection of untreated biological fluids. Additionally, with no optimization of pH or organic solvent content of the mobile phase, protein precipitation and clogging will disable the RAM, which requires special attention by the chemical analyst (de Faria et al. 2017).

Single-column mode or multidimensional LC can exploit RAM as column packing. In the single-column mode, the exclusion of interferents is first performed, followed by a gradient elution for elution and separation of the targeted compounds.

Although this approach has suitable performances, there are some chances of clogging the RAM because of a lack of re-equilibration after gradient elution or because of inadequate selectivity of the hydrophobic stationary phase inside the pores (Sadílek et al. 2007). For this reason, multidimensional LC approaches tend to be more popular due to the inclusion of an additional reversed-phase column that allows the backflushing and regenerating of the RAM phase after each analysis. This arrangement prevents fouling of the RAM and increases its long-term stability even with repeated usage of the RAM column (Kataoka and Saito 2012).

More recently, RAMs have been applied in coupling with molecularly-imprinted polymeric materials. While RAM excludes the macromolecules, the imprinted material has specific recognition sites for small analytes, especially drug metabolites and pharmaceutical compounds from biological matrices (Abrão et al. 2019; Souza et al. 2016; Oliveira et al. 2016). This approach is especially useful for small analytes in real samples, since the advantages of both, such as the efficient matrix clean-up and convenient analyte isolation, are combined.

Expanding the applicability of RAMs beyond traditional ligands, a boronate affinity-RAM with an external surface built of hydrophilic bottlebrush polymers with excellent capability for protein exclusion was developed (Xu et al. 2010). This newly synthesized material was employed for the extraction of catecholamines from serum samples, with recoveries of 87–114%, and high selectivity for molecules containing cis-diol groups.

## 2.4  Imprinted Materials

Imprinting is a process by which selected functional monomers are self-assembled around a molecular or ionic template, and subsequently polymerized in the presence of a cross-linker. Once the template compound is removed from the polymeric structure, a cavity complementary in shape and chemical properties is present in the structure, and becomes available to bind ions or molecules identical to or closely related to the template (Fig. 5) (Gama and Bottoli 2017). Concisely, an imprinted material has a memory of the shape and functional groups of one or more templates. Thus, each material has high recognition properties, which is interesting for specific interactions, such as those required in bioanalysis. Non-covalent interactions, such as hydrogen bonds, dipole–dipole, and ion interactions established between the template and functional groups present in the material cavity drive the molecular recognition phenomena (Gama and Bottoli 2017; Maciel et al. 2019). Molecularly-(MIP) and ion-imprinted bead polymers (IIP) and macroporous molecularly-imprinted monoliths (MIM) are described as imprinted materials with application in sample preparation and separation techniques.

MIPs have been demonstrated as a promising sorbent for sample preparation due to its specific recognition that provides high enrichment factors and great extraction analyte recoveries (Ashley et al. 2017). The first report of MIP as a sorbent for solid-phase extraction was proposed in the 1990s (Sellergren 1994). A polyacrylate-based polymer selective for pentamidine (an antimicrobial agent) was synthesized and the

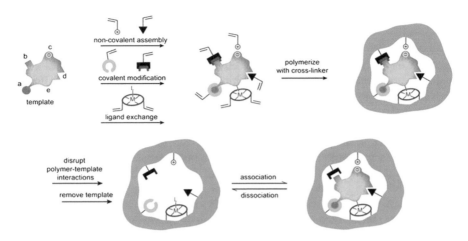

**Fig. 5** Schematic representation of the molecular imprinting process. The creation of affinity in the binding site for the template may involve one or more interactions: (**a**) reversible covalent interactions; (**b**) covalent binding groups activated by template cleavage; (**c**) electrostatic interactions; (**d**) hydrophobic or van der Waals interactions; (**e**) coordination with a metal center. Rebinding/release of the template from the final polymer site involves a facile condensation/hydrolysis reaction at site (**b**). Reproduced from Ref. (Bedwell and Whitcombe 2016) with permission

resulting material proved to be selective towards pentamidine for extractions from water and urine, with an enrichment factor of 54. Since then, applications of MIPs for sample preparation were widely reported (Moein et al. 2019; BelBruno 2019; Turiel and Martín-Esteban 2010), including for bioanalysis (Gama and Bottoli 2017).

MIMs combine the properties of MIP with those from ordinary monoliths. They have been used as stationary phases in chromatography and as sorbents for solid-phase extraction in several formats, such as fibers, disks, or columns. These materials stand out due to the highly selective extraction of analytes, encompassing applications from preconcentration to sample clean-up.

During their synthesis, MIM properties can be adjusted for application to complex matrices, enabling the extraction of either small or large molecules. The selection of the template and porogen solvent is critical because these materials can affect the final morphology and performance of the process (Tan et al. 2012). For example, binding properties of a MIM for SPE of ciprofloxacin were up to threefold higher for a given solvent composition and proportion of monomer (Vlakh et al. 2016).

For bioanalysis, the functionalization of MIM with a biostructure is the usual approach for molecular recognition. For example, by using a short fragment of the target peptide as a template, the resultant MIM was able to recognize the amino acid sequence of the related homologous peptides (Ji et al. 2015).

## 2.5    Affinity Chromatography

Antibody–antigen interactions are naturally highly specific and can be adopted as the basis for the development of highly selective materials. Therefore, immuno-based materials are created for specific recognition using natural antibodies and allow a high degree of molecular selectivity, that could extract and isolate the targeted analyte from complex matrices in a single step, thus avoiding the extraction of interfering compounds (Pichon 1999).

The first step in the preparation of an immunosorbent is the design of antibodies with molecular recognition ability for either one or a group of analytes. After the selection of the target analyte, the preparation of antibodies demands sensitization of a suitable laboratory animal, generally a rabbit, with the analyte coupled to a large protein, such as bovine serum albumin. This conjugate will induce the immune response and antibody production in the animal. Afterwards, the animal blood plasma is collected, followed by isolation and purification of the desired antibody from the IgG plasma fraction (Pichon 1999; Stevenson 2000).

After antibody obtention, they can be covalently immobilized onto a suitable support solid or gel having appropriate requirements, such as large sized-pores — due to also large size of antibodies, and high hydrophilicities — to avoid any non-specific interaction and possible deactivation of the antibodies. Common supports for immune ligands are silica gel and polysaccharides, such as Sepharose or cellulose; glass, alumina, and polymers are less used as supports (Buszewski and Szultka 2012). Limitations of immuno-based stationary phases include low stability under extreme pH values and high temperatures, and the high cost of preparation (Buszewski and Szultka 2012).

In a common immunochromatographic separation, a sample containing the targeted analyte is incubated with an excess amount of a labeled binding agent for this target, such as an intact antibody or its fragment. The non-bound labeled agent that remains after the incubation step is removed by passing this mixture through a column that contains an immobilized analog of the target.

A signal related to the analyte in the original sample is then obtained by measuring the amount of non-retained labeled binding agent that elutes from the column or that is captured by this column, as shown in Fig. 6 (Zhang et al. 2020). An on-line purification system was developed for extraction of human serum albumin from biological samples with further entrapping within a small column for high-performance affinity chromatography (Rodriguez et al. 2020). This separation technique combines the use of a support with the selective and reversible interactions that can occur between a given analyte and an immobilized biologically-related binding agent for this target. Columns for high-performance affinity chromatography are often prepared by covalent immobilization of a protein onto a support; however, it is well-known this immobilization can lead to an incorrect attachment or improper orientation of the protein, and consequent loss of its activity (Kim and Hage 2005). In the proposed method, the first immunoextraction column containing polyclonal anti-protein antibodies was developed to isolate albumin and was followed by a cation exchange column to recapture and focus the eluted protein, prior to UV

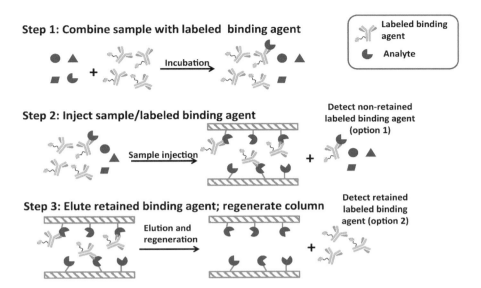

**Fig. 6** Schematic representation of an immunochromatographic separation. Reprinted from (Zhang et al. 2020) with permission

detection. Immune-based columns stable over several weeks were achieved, useful for about 50–60 sample injections in drug binding studies (Rodriguez et al. 2020).

In another branch of affinity chromatography, the aptamer-based separation method was described as an alternative of a highly specific bioanalytical tool.

Aptamers are oligonucleotide or peptide molecules capable of binding to a specific target molecule. They are short (up to 100 bp) and can fold in characteristic shapes that establish highly specific bindings; the recognition arises from hydrogen bonding, van der Waals forces, and dipole–dipole and stacking interactions (Luzi et al. 2003; Mascini 2009).

Specific aptamers towards a targeted analyte can be isolated through the SELEX cycle (Systematic Evolution of Ligands by Exponential Enrichment, Fig. 7), in which a random pool of DNA, RNA, or peptide is doped with the analyte, and the sequences that bind to the target are selected through iterative cycles of separation, isolation, and amplification (Yüce et al. 2018). This in vitro method of ligand screening may be automated, and reasonable amounts of highly specific aptamers for the targeted compound can be obtained and then immobilized onto a support.

The SELEX strategy has been successfully applied to proteins, which easily bind to nucleic acids and are initially supposed to be the main targets of the aptamers, as well as ions, small chemical moieties, nanomaterials, large biomolecules, cell surface proteins, growth factors, organelles, and even whole cells and viruses (Jayasena 1999).

Once obtained, the next step is the aptamer immobilization on a support. As a medium-size folded molecule composed of a few numbers of nucleotides with a

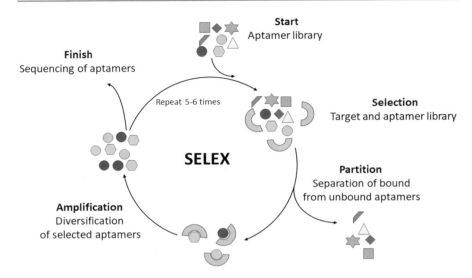

**Fig. 7** SELEX cycle employed for obtention of aptamers. The process starts with an aptamer library for initiation of the SELEX cycle. The library is incubated with the target, which is washed for removing the unbounded aptamers. Then, bound aptamers are eluted and amplified by PCR. The amplified sequences seed the next round of SELEX. Typically, several cycles are performed before sequencing and aptamer characterization. Reprinted from Ref. (Kinghorn et al. 2018) under a Creative Commons license

negative dominant charge from the phosphate groups, partially balanced by the positive charge of purine and pyrimidine bases, immobilization strategies are extremely dependent on the presence of binding sites on the support, where the aptamer will be grafted (Perret and Boschetti 2018). Hydroxyl, amine, or carboxylic acid groups are necessary for aptamer immobilization due to their easy coupling to the sorbent and to avoid the risk of tertiary structure denaturation and consequent loss of activity (Perret and Boschetti 2018).

The first chromatographic application of aptamers was exploited for the purification of a recombinant human L-selectin-Ig fusion protein from hamster ovary cells using columns packed with an aptamer-based sorbent (Romig et al. 1999). Since then, chromatographic applications of aptamers have been focused on aptamer affinity liquid chromatography, extraction and concentration using solid phases, and applications in microfluidic formats (Zhao et al. 2012).

In a forensic application, cocaine was isolated from human plasma before its HPLC determination, using a column packed with Sepharose modified with an anti-cocaine aptamer (Madru et al. 2009). The extraction recoveries were close to 90%, and the cocaine isolation was more selective on the aptamer column than on traditional C18 packing, which also eluted interfering compounds. The reusability of the synthesized sorbent for plasma samples was also investigated, and three consecutive extractions of plasma samples were performed, showing a decrease in extraction recovery (72, 43, and 41%), probably due to the progressive binding of proteins from the matrix (Madru et al. 2009).

Even as a very promising approach, there is still a limited number of reports using aptamer-modified materials, when compared to other selective molecular recognition packings, such as molecularly-imprinted and immunosorbents. Noticeable advantages for obtention of aptamer-based materials include demanding minimum amounts of the template (a limitation found in the preparation of imprinted materials, that spend high amounts of analyte template, critical for rare or expensive targets), being also a faster method when compared to immunosorbents. In addition, the synthesis of aptamer-based phases is easier than immunosorbents, which requires the use of laboratory animals to produce the antibodies, which may pose legal and ethical limitations on the research in many countries, and limits the obtention of these structures (Pichon et al. 2015). However, a broad knowledge in molecular biology required for production of aptamers, and the high costs of precursors and reagents hinder the widespread use of aptamer affinity chromatography.

## 3    Detection

LC bioanalyses make use of many of the detectors available. The choice of the most suitable one greatly depends on the analyte characteristics and its concentration, and on the matrix and on how much sample preparation efforts are needed for accurate determination. In summary, as long as a validated and reliable method is developed for the stated purpose, any detector can be appropriate, even though some of them might be more advantageous than others in certain aspects.

Given the fact that a great number of compounds absorb UV and visible radiation, classical UV-Vis based detectors such as the photodiode-array (PDA) detector are still currently used in bioanalytics, being a simple and cost-effective means of sensing compounds eluted from a chromatographic column. Dalvi et al. (2018) validated an HPLC-PDA method for the determination of rufinamide, an antiepileptic drug, in rat plasma and brain for pharmacokinetics studies. Limits of quantification (LOQ) of the drug were 13.84 ng mL$^{-1}$ in plasma and 105.24 ng g$^{-1}$ in brain tissue, both pre-treated by a simple protein precipitation protocol. Wang and co-workers (Wang et al. 2018) also developed a validated HPLC-PDA method for the determination of 10 drugs (adrenaline, salbutamol, timolol, carteolol, ephedrine, mexiletine, chlorprenaline, clenbuterol, propranolol, and carvedilol) in rat plasma. By coupling an on-line turbulent flow chromatography (TFC) extraction system prior to the separation step, the method provided LOQ values between 10 and 40 ng mL$^{-1}$ (Dalvi et al. 2018; Wang et al. 2018; Swartz 2010).

Absorbance detectors, however, exhibit lower detectivities for certain compounds and might require considerable sample preparation to improve it, aiming at the concentration of the analytes. For example, moxifloxacin, a broad-spectrum antibiotic, was quantified at a 40 ng mL$^{-1}$ level in human plasma (LOQ) after a labor-intensive liquid–liquid extraction procedure prior to HPLC-UV. A simple protein precipitation protocol might facilitate its analysis in such a matrix, but also might lead to analyte instability. The choice of a fluorescence detector (FL) circumvented these problems by providing a LOQ of 20 ng mL$^{-1}$ with a simple and time-saving deproteination sample preparation protocol, turning the monitoring of moxifloxacin easier for pharmacokinetic studies and clinical routines (Wichitnithad et al. 2018).

HPLC-FL can be also found in official compendia, such as the AOAC official method 2005.06 for the determination of saxitoxins in shellfish. Since saxitoxins are naturally accumulated by these organisms, and they are harmful to human health at certain levels, HPLC-FL of oxidized samples ascertains the safety for shellfish consumption and the quality control of such products (Wichitnithad et al. 2018; Turner et al. 2019).

The evaporative light scattering detector (ELSD) has also been used recently in HPLC analyses of biological samples. This type of detector works on the principle of nebulization of the mobile phase and the measurement of the light scattered by the dry analyte particles. No chromophores or fluorophores are required, broadening the spectra of analyzable molecules. HPLC-ELSD has been reported for the analysis of sugars in floral nectars and plant extracts (Swartz 2010; Lindqvist et al. 2018; Shao et al. 2018).

Mass spectrometry coupled to LC is the top-class technique for many routine analyses and in research laboratories. Considered almost as a universal-type detector, MS is highly sensitive, detecting most compounds at trace levels, and extremely selective when the acquisition is set to certain scan modes. Tandem mass spectrometry (MS/MS) combines two or more analyzers and enables the fragmentation of a compound, aiding in its identification. Drawbacks in the speed of data collection and the provision of friendly interfaces have been circumvented, and current LC–MS devices, depending on the matrix, allow a dilute-and-shoot sample preparation or reduce the need for extensive interference clean-up or analyte derivatization. MS also helped to lead to the miniaturization of LC for better coupling of the techniques and systems like nano-LC–MS can deliver sharp-peak chromatograms. Nevertheless, the LC–MS instrumentation is still rather more expensive than LC coupled to the classical spectroscopic detectors, and most bioanalytical challenges can be addressed with either one or other type of detection (Kočová Vlčková et al. 2018).

LC–MS is advantageous for targeted and non-targeted analyses. In targeted analyses, the compounds to be detected are known and previously defined. In these situations, selected reaction monitoring (SRM) or single ion monitoring (SIM) acquisition modes are generally used due to high sensitivity and extreme selectivity. An extracted ion chromatogram (EIC), in which only the $m/z$ of interest are read from a total ion count (TIC) chromatogram, is also a feasible strategy with the same aim, as used by Eckenrode et al. (2019) for the determination of mithramycin and its analogs in mouse plasma with high accuracies. The calibration curves were linear in the range of 5 to 100 ng mL$^{-1}$ for all the analytes (Kočová Vlčková et al. 2018; Eckenrode et al. 2019).

In non-targeted analysis, an exploratory and comprehensive analysis on the sample is done and all the analytes may not be known. "Omics" analyses generally go through this path and require high-resolution mass spectrometers (HRMS), containing mainly quadrupole time-of-flight (Q-TOF) or Orbitrap analyzers for the identification of new metabolites and biomarkers in biological samples. TIC and EIC are strategies for data acquisition and post-treatment in this analytical mode (Kočová Vlčková et al. 2018).

Proteins are large and complex structures to which two approaches can be applied, requiring specific understanding of LC–MS. The first and most traditional one is the analysis of enzyme-digested proteins. The analyte is turned into a mixture of smaller peptides and the quantification is based on one or a few of these molecules. This generally needs high chromatographic resolutions and systems such as UHPLC–MS or nano LC–MS have been used. Multidimensional liquid chromatography for a better peptide resolution has also been reported. A second approach is the analysis of intact proteins, which require large pore stationary phases and HRMS (van de Merbel 2019).

Detection in LC bioanalyses has been following the advances in chromatography coupled with multiple techniques for molecular identification, but classical tools and equipment are still valuable for solving the bioanalytical challenges. For routine tasks, speed and the cost-benefit relation are the elements to be considered. Low detectivity and specificity choices might require higher time consumption in sample preparation. For research purposes, depending on its nature, sophisticated tools may be needed, but classical detectors still find their space.

## 4 Multidimensional Liquid Chromatography

Even though much progress has been made in sample preparation and analytic separation, the extreme complexity of biological samples still poses a challenge in the resolution, identification, and quantification of their compounds. Multidimensional liquid chromatography comes towards overcoming these hurdles to make bioanalytics even more effective. In this chromatographic approach, samples are subjected to at least two complementary mechanisms of separation to attain higher resolutions. A sample or its fractions of interest are collected from a first separation step (first dimension—1D), and manually (off-line mode) or automatically (on-line mode) injected into a second column (second dimension—2D) for further separation. This scheme can be repeated with other serial dimensions, if necessary (Mogollón et al. 2014).

Multidimensional liquid chromatography is capable of reducing or eliminating coelutions existing in the classic one-dimension analyses of complex samples and, by providing higher selectivity for the different compounds, the detectivity of the analyses is improved as well (Chen et al. 2020).

The following sections discuss the principal concepts and some examples of multidimensional LC use for different purposes.

## 4.1    Basic Concepts of Multidimensional Liquid Chromatography

The first consideration when referring to multidimensional LC is the nomenclature used for the methodology's designation. When the whole eluate from one dimension is transferred to the following one, the chromatography is said to be comprehensive and an "×" symbol is used for abbreviation (e.g.: LC × LC). Otherwise, if only fractions are collected for the next step of separation, in a mode known as heart-cut, the dimensions are separated with a "-" (e.g.: LC-LC) (Li et al. 2015).

It is important to highlight the advantage of multidimensional separations over unidimensional ones. The main reason for the use of a multidimensional LC system is its increased peak capacity ($n$), which is defined as the number of peaks that can be resolved within a separation space. The peak capacity can be maximized by the coupling of chromatographic columns with different mechanisms of separation so that the highest resolution of the sample compounds can be achieved. In this sense, orthogonality among the dimensions is an important factor for the achievement of a high $n$ value (Holland et al. 2016).

A system can be considered as orthogonal when the separation mechanisms involved in each dimension are completely independent and exhibit different selectivities for the compounds in a sample. Thus, an orthogonal LC system could be composed of a normal phase and a reversed-phase column, but such a system is difficult to be achieved due to mobile phase incompatibilities between these dimensions and to the solute solubility in them. Therefore, a wide range of column combinations can be made and the choices are guided by the sample features and the strategies available for the best separation to be accomplished (Holland et al. 2016).

The orthogonality of a system can also be achieved by a combination of columns with the same separation mode, reducing the solvent incompatibility problems. This is the case of the coupling of two RP columns for the generation of high peak capacity. Among the reversed-phase columns, there are many types of stationary phases with different composition (silica, polymer, inorganic oxide) and functional groups attached to them. This gives RP columns a broad range of different selectivities for various compounds. As an example, the combinations of a C18 in $^1$D with cyano, amino, SB-phenyl, polyamine, carbon clad zirconia column, or C18 monolithic in $^2$D have been reported. On the other hand, when C18 was allocated in $^2$D, fluoroalkylsilyl, cyano, polyethylene glycol (PEG), zirconia carbon, porous graphitic carbon (PGC), human serum albumin (HSA), and immobilized liposome chromatography have been used in $^1$D (Gilar et al. 2005; Bassanese et al. 2015).

Despite the advantages of same selectivities in $^1$D and $^2$D, the combination of columns with different separation mechanisms is welcome in multidimensional separations. Such a variety of options commercially available and under development might make orthogonality checking an empirical and time-consuming task, and efforts for in silico predictions have been reported (Marlot et al. 2018).

As mentioned, the transference of fractions between the two dimensions can be off- or on-line. In an off-line based mode, the fractions of the sample are collected in flasks at the end of one dimension and each portion is then injected into the second dimension for a new round of separations. An off-line LC × LC method was

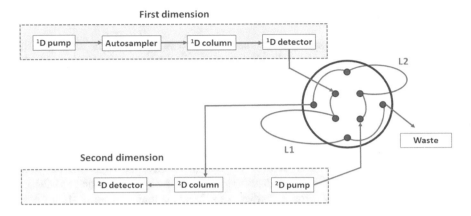

**Fig. 8** Common configuration of a 2D-LC instrument. An existing LC system composed of a pump, autosampler, column, and detector is connected by an interface with a second set up (pump, column, and detector). Here, the interface is an eight-port valve and two sample loop capillaries (L1 and L2) used to collect effluent from the 1D column and then inject it into the 2D column for further separation. Adapted from Ref. (Stoll et al. 2018)

developed by Kadivar et al. (2013) for the analysis of stereospecific triacylglycerols (TAGs) in cocoa butter equivalents. These are non-lauric fats made from other vegetable oils, but similar and compatible to cocoa butter. In the production of these TAGs, which are mainly symmetrical molecules, part of them is converted into non-symmetrical isomers which influence the products to which they are incorporated. Given the low polarity of the analytes and the existence of positional isomers to be resolved, a two-dimensional system comprising a C18 reversed-phase chromatography column was used as $^1$D for the separation of the TAGs based on their degree of unsaturation and number of carbons. The fractions of $^1$D were recovered by a fraction collector and they were subsequently separated in an Ag$^+$ column in $^2$D for the resolution of the isomers, which could not be resolved by unidimensional C18 (Kadivar et al. 2013).

On-line multidimensional LC dismisses a fraction collector and uses an interface instead. The interface is generally a high-pressure valve that guarantees the collection of samples from $^1$D, and their injection into $^2$D. For fast analyses, a loop interface optionally containing a stationary phase within it is the principal one used. Figure 8 shows an example of an on-line two-dimensional system comprising a first pair of pumps, which lead the mobile phase towards the $^1$D column, a multiple port valve, a second pair of pumps to guide another mobile phase through the interface and to inject the samples into the $^2$D column, a detector, and a computer. The main advantages of an on-line system are the lower time of analysis since both injections are automated and, therefore, less demanding, and lower sample loss, as the analysis circuit is closed, and the sampling loops have defined volumes (Stoll and Carr 2017). These might be the reasons why the greatest part of the scientific literature related to bioanalysis by multidimensional LC makes use of automated systems.

In such a scheme, sampling from one dimension to the other, the choice of the mobile phase compositions and the elution mode must be carefully chosen in order not to decrease the peak capacity. Ideally, for a good sampling, $^1$D should be operated with a low mobile phase flow to allow the collection of several fractions of a peak, while $^2$D should be operated with a faster and effective flow. The correct mobile phase composition choice may diminish peak widening effects that may occur in the system. By starting the separations with a weak solvent in $^1$D and using a strong one in $^2$D, the compounds of a sample are focalized at the beginning of the second dimension and eluted in narrower bands. Finally, the right elution mode guarantees the peak capacity and the orthogonality of the system (Pirok et al. 2018).

In two-dimensional systems, a gradient of solvents is mainly used in $^2$D for peak narrowing and higher resolutions. The gradients can be: (1) complete for each fraction collected from $^1$D; (2) fragmented, starting with weaker mobile phases for the first fractions and stronger ones for the final eluates from $^1$D; (3) parallel, with a constant increase in the stronger eluent concentration, or (4) shift gradient, in which the mobile phase constitution changes in bands (Li and Schmitz 2013). These programmed elution approaches can provide a significant improvement in peak capacity with narrow bandwidths, especially for strongly retained compounds.

The mobile phase strength can be also modulated by a temperature gradient (Stoll et al. 2006). This is especially beneficial when using aqueous solvents, which exhibit different chromatographic performances according to the temperature. In these situations, fewer organic solvents are required and, therefore, separations become more cost-effective and environmentally friendly. Temperature elevation also changes the chromatographic selectivity of ionizable analytes, for which ionization equilibria are displaced by heat. This speed up the separations carried out by organic solvents because of the decrease in the mobile phase viscosity and the elevation in the analyte diffusion without loss in resolution. However, not all the stationary phases present mechanical and chemical resilience to high temperatures and the use of this type of gradient is limited. Also, the re-equilibration time needed is not favorable for routine analyses. Even so, a carbon clad zirconia stationary phase exhibits good chemical and mechanical stabilities at high temperature and with high flow rate mobile phases, being also suitable for operation with a temperature gradient (Holland et al. 2016; Stoll et al. 2006).

Still concerning ionizable analytes, the mobile phase pH is another important factor. When the analytes display different ionization constants, a pH change can increase orthogonality even if similar stationary phases are used in the different dimensions of the system. Therefore, the choice of mobile phase additives may be able to increase the performance of a multidimensional separation. And when it comes to organic solvents, the classical methanol, acetonitrile, isopropanol, and THF are the most used in multidimensional liquid chromatography due to their compatibility with water and the existing detectors (Li et al. 2015).

After the separations, the peaks are detected and the chromatograms are recorded. On-line multidimensional liquid chromatography delivers a matrix of data from

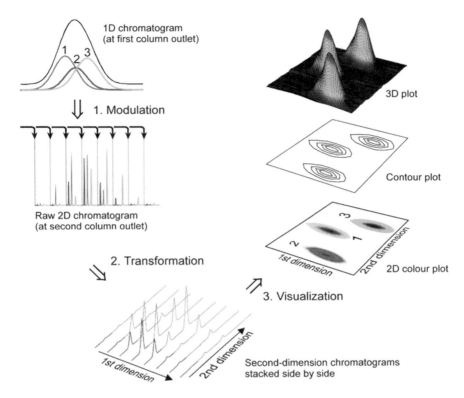

**Fig. 9** The steps for a two-dimensional chromatogram construction. (**a**) Unidimensional chromatogram. (**b**) Rough two-dimensional chromatogram. (**c**) Two-dimensional chromatogram. (**d**) 3D plot. (**e**) Contour plot. (**f**) Color plot. Adapted from Ref. (Dallüge et al. 2003) with permission

several measured parameters as a result. Each signal is reported against its retention time in each dimension of the system, and the final chromatogram contains an elevated quantity of points from which information such as the area under a peak and its spectral data, depending on the type of the detector used in the analysis, can be gathered. Figure 9 shows a scheme of a multidimensional chromatogram construction. Manually obtaining the area under each chromatographic peak is a non-precise and ineffective task since each peak often contains data from various fractions and the areas must be combined for the quantification of a compound. Another issue is that generally the peaks are not baseline resolved in multidimensional separations, especially in $^2$D: so, deconvolution techniques are required for the accurate calculation of the areas under overlapping peaks, and chemometric tools are needed for the best interpretation of such chromatograms. Algorithms in multidimensional chromatography for background and retention time alignment corrections are used, and multiway tools such as PARAFAC (Parallel Factor Analysis) and MPCA (Multiway Principal Component Analysis) aid data interpretation (Dallüge et al. 2003; Stevenson and Guiochon 2013).

## 4.2    Multiple Uses of Multidimensional Liquid Chromatography

In most of the reported cases for bioanalytical purposes, multidimensional LC is restricted to two separation dimensions and much work has been done for the evaluation of amino acids and proteins by this technique. When coupled with MS/MS, 2D-HPLC allowed the determination of d-amino acids in mammalian fluid samples. Samples eluted from a C18 column with an aqueous ACN gradient in the first dimension (360 min run), followed by elutions of different fractions from the first dimension with mixtures of MeOH and ACN in a chiral Pirkle-type column (35 min run), allowed quantification at the femto and picomol levels (Ishii et al. 2018). A similar approach for the analysis of the citrulline (Cit) and ornithine (Orn) enantiomers in healthy and d-amino acid oxidase deficient mice urine was used, but using a faster configuration composed by a capillary monolithic ODS column in $^1$D and a narrow-bore chiral Pirkle-type column in $^2$D. Cit and Orn were determined at the trace levels and the elution times for $^1$D and $^2$D were 18 and 30–40 min, respectively (Koga et al. 2016).

In fact, the long time required for multidimensional separations has been an issue to be faced and efforts have been applied to overcome this limitation. For faster analyses, one can either use faster columns, such as monolithic or core-shell stationary phases, or split the $^1$D flow into various columns for simultaneous separations in $^2$D. For an E. coli protein lysate analysis, an array of three RP-phase $^2$D columns for simultaneous resolution of multiple fractions from an ion-exchange $^1$D capillary column has been reported. The parallel $^2$D columns captured samples from $^1$D each 45 min for separation (Zhu et al. 2018).

Lower solvent consumption and higher sensitivity systems have also been demanded. In this sense, miniaturized systems have also been reported in the literature. The separation of trypsin digested casein fractions was accomplished by two C18 capillary columns, connected through two 2-position 6-port nano switching valves, eluted with gradients of a high-pH mobile phase in $^1$D and a low-pH one in $^2$D. The method exhibited good retention time repeatability and precision, and even though similar stationary phases were used in both dimensions, the mobile phases made the system highly orthogonal (Sommella et al. 2012). Another system, having one monolithic ion-exchange capillary column in $^1$D, and 12 monolithic RP capillary columns in $^2$D operated in two batches of six each time, tackled simultaneously the requirements for lower time of analysis, lower solvent consumption, and higher detectivity, resolving over 900 protein fractions from an E. coli lysate (Ren et al. 2018).

Multidimensional separations and chemometric tools have also been useful for untargeted analyses of components from biological matrices. In an untargeted analysis, the overall measurement of a certain group of compounds is done independent of the pathways from which they are originated. No specific list of compounds is searched for. The effects of arsenic exposure on the lipidomics of rice were evaluated by an RP-HILIC 2D-HPLC system with a triple quadrupole mass detector. A set of chemometric tools including multivariate curve resolution by an alternating least squares method (MCR-ALS), which gives the pure spectra and elution profiles of the components in a mixture, and principal component analysis (PCA) for the clustering of results led to the conclusion that high exposure of rice to As implies a lipidomic change in this plant (Navarro-Reig et al. 2018).

Still, in the field of metabolism products, pharmacological and toxicological studies have also benefited from the high peak capacity of multidimensional liquid chromatography. Using a 2D-HPLC system consisting of a strong cation exchange column in $^1$D and a RP column in 2D, the herbicide atrazine and its 11 metabolites and hydrolysis products were resolved in samples of 500µL of urine or less, with limits of detection on the order of a few µL L$^{-1}$, allowing the assessment of atrazine in occupational and non-occupational studies (Kuklenyik et al. 2012).

For plasma samples, there are some reports of $^1$D as a sample preparation column, which separates proteins from the analytes of interest without the need for a prior pre-purification step. For this, a RAM column in $^1$D was used to exclude plasma proteins from samples comprising modafinil enantiomers and their metabolites. These were further eluted into a chiral column for separation (Cass and Galatti 2008). A thermo-responsive polymeric stationary phase for plasma pretreatment in $^1$D has also been reported (Mikuma et al. 2017). By control of the temperature and the ionic strength of the mobile phase, plasma proteins from samples containing barbiturates and benzodiazepines were successfully excluded and the analytes were retained for further elution into $^2$D (Mikuma et al. 2017).

Plant metabolites in natural products have also been separated by the technique for different purposes. To improve the analysis of phenolics, a HILIC-RP 2D-LC method was developed and applied to rooibos tea, wine, and grape samples by utilizing the dilution of $^1$D flow and the injection of a large volume in $^2$D to overcome incompatibility problems (Muller et al. 2019). The $^2$D retention times, the UV-Vis, and the MS data made it possible to identify 72 compounds of different classes in grape seeds, 32 in rooibos tea, and 45 in wine and grapes.

Multidimensional chromatography shows applicability also in the isolation of natural molecules. A 2D semi-preparative system was used for the purification of paclitaxel from a *Taxus baccata* cell culture using a combination of C8 and C18 columns (Ghassempour et al. 2007). A yield and a purity of 89% was obtained by this process. Aside from these applications, 2D-HPLC with chemometric data treatment is also a potential tool for industrial quality control. For example, an industrial opiate production from *Papaver somniferum* could be monitored by analyzing 2D fingerprints from samples of different stages of the process by PCA (Stevenson et al. 2017). Although greater resolutions are achieved by a 2D system in comparison with 1D chromatography, the complexity of some biological matrices does not allow the full resolution of their compounds. Systems with higher numbers of dimensions have been proposed for the analyses of proteomes and proteins. 3D-LC analyses began with columns packed with three phases and evolved to off-line, partial on-line, and totally on-line systems. As well as 2D separations, 3D on-line platforms require orthogonality among the phases and mobile phase compatibility, thus needing skill to permit the set ups. In its beginnings, huge sample loss throughout the system by non-specific adhesion to tubes and connections was an issue overcome by higher sample loadings, but the advances in micro and nano-scale chromatography might make 3D separations more feasible. However, it seems that on-line 3D-LC peak capacity is not greatly improved in relation to 2D-LC separations for protein analyses, and the difficulties in system design and method transfer make off-line and mixed-mode 3D-LC more attractive, at least for proteomic studies (Duong et al. 2020).

Despite the observed drawbacks of 3D on-line systems, there is a recent report on a 4D-LC platform for the analysis of chemical modifications in therapeutic antibodies, in which an off-line multistep and time-consuming routine was transformed into an on-line one. The system comprised an ionic exchange column in $^1$D for the separation and the fractionation of antibody charged species. The fractions of $^1$D were further resolved in an RP column ($^2$D), followed by a digestion step in a TPCK-Trypsin column ($^3$D). The fourth dimension, a UHPLC C18 column, was dedicated to peptide trapping, washing, and elution to an MS-MS detector (Gstöttner et al. 2018).

## 5    Conclusions and Perspectives

Bioanalysis requires an efficient analyte isolation step, which must provide complete removal of interferents from the matrix and unequivocal analyte identification and quantification. This isolation should preferably be carried out in a few steps, preventing the analytes from degrading or being subject to the formation of artifacts during the sample clean-up. Clean-up systems performed as a prior step in the same separation column are welcome, since sample treatment can take place in the same system in which the chromatographic separation will be performed, saving time and treatment steps. The pretreatment also aims to increase the concentration of the analytes enabling its detection.

Mostly, detection of bioanalytes is done by mass spectrometry, and LC–MS and LC–MS/MS couplings are consolidated. This detection technique allows the quantification of analytes in very low concentrations.

The exploitation of miniaturized (portable) systems enables *in loco* analyses, facilitating (bio)analyte quantification in places of difficult access or with limited availability of HPLC instruments.

Finally, the overall view from recent reports on multidimensional liquid chromatography shows this emerging technique as a promising tool for the assessment of the chemical compositions of biological matrices for the next years, although further development of the technique itself is needed to facilitate its incorporation into research and analytical routines. Better resolutions can be achieved when the most appropriate set of columns, mobile phases, and separation times are elected, but the task of finding these "best conditions" is not trivial and, sometimes, it finds technical and chemical compatibility bottlenecks to be overcome. Data treatment also demands special tools for the correct interpretation, but multidimensional separations can reveal data never assessed before. While in theory more than two dimensions seem to provide even higher peak capacities to a multidimensional system, there are some hurdles for the achievement of such platforms, but there are chances that such designs will arise as automated versions of laborious and manual analytical tasks.

# References

Abrão LCC, de Faria HD, Santos MG, Barbosa AF, Figueiredo EC (2019) Handbook of smart material analytical chemistry. Wiley, Chichester, pp 411–438

Alpert AJ (1990) J Chromatogr A 499:177–196

Annesley TM (2003) Clin Chem 49:1041–1044

Ashley J, Shahbazi MA, Kant K, Chidambara VA, Wolff A, Bang DD, Sun Y (2017) Biosens Bioelectron 91:606–615

Bassanese DN, Holland BJ, Conlan XA, Francis PS, Barnett NW, Stevenson PG (2015) Talanta 134:402–408

Bedwell TS, Whitcombe MJ (2016) Anal Bioanal Chem 408:1735–1751

BelBruno JJ (2019) Chem Rev 119:94–119

Berthod A, Carda-Broch S (2004) Anal Bioanal Chem 380:168–177

Bisceglia KJ, Roberts AL, Schantz MM, Lippa KA (2010) Anal Bioanal Chem 398:2701–2712

Boos KS, Rudolphi A (1997) LGGC N Am 15:602–611

Buszewski B, Noga S (2012) Anal Bioanal Chem 402:231–247

Buszewski B, Szultka M (2012) Crit Rev Anal Chem 42:198–213

Carmichael AJ, Seddon KR (2000) J Phys Org Chem 13:591–595

Cass QB, Galatti TF (2008) J Pharm Biomed Anal 46:937–944

Cecchi T (2015) J Liq Chromatogr Relat Technol 38:404–414

Cecchi T (2017) Ion-pair chromatography and related techniques. Taylor and Francis Group

Chen Z, Gao Y, Zhong D (2020) Biomed Chromatogr 34:e4798

Dallüge J, Beens J, Brinkman UAT (2003) J Chromatogr A 1000:69–108

Dalvi AV, Uppuluri CT, Bommireddy EP, Ravi PR (2018) J Chromatogr B 1102–1103:74–82

de Faria HD, Abrão LCC, Santos MG, Barbosa AF, Figueiredo EC (2017) Anal Chim Acta 959:43–65

Desilets CP, Rounds MA, Regnier FE (1991) J Chromatogr A 544:25–39

Dolan JW (2008) LCGC Eur 21:258–263

Duong VA, Park JM, Lee H (2020) Int J Mol Sci 21:1524

Eckenrode JM, Mitra P, Rohr J, Leggas M (2019) Biomed Chromatogr 33:e4544

Fatima Z, Jin X, Zou Y, Kaw HY, Quinto M, Li D (2019) J Chromatogr A 1606:360245

Gama MR, Bottoli CBG (2017) J Chromatogr B 1043:107–121

Gama MR, Silva RGC, Collins CH, Bottoli CBG (2012) Trends Anal Chem 37:48–60

García-Alvarez-Coque MC, Ruiz-Angel MJ, Berthod A, Carda-Broch S (2015) Anal Chim Acta 883:1–21

Ghassempour A, Noruzi M, Zandehzaban M, Talebpour Z, Yari Khosroshahi A, Najafi NM, Valizadeh M, Poursaberi T, Hekmati H, Naghdibadi H, Aboul-Enein HY (2007) J Liq Chrom Relat Technol 31:382–394

Gilar M, Olivova P, Daly AE, Gebler JC (2005) Anal Chem 77:6426–6434

Gstöttner C, Klemm D, Haberger M, Bathke A, Wegele H, Bell C, Kopf R (2018) Anal Chem 90:2119–2125

Han D, Row KH (2010) Molecules 15:2405–2426

Haselberg R, Pirok BWJ, Gargano AFG, Kohler I (2019) LCGC Eur 32:465–480

He L, Zhang W, Zhao L, Liu X, Jiang S (2003) J Chromatogr A 1007:39–45

Holčapek M, Volná K, Jandera P, Kolářová L, Lemr K, Exner M, Církva A (2004) J Mass Spectrom 39:43–50

Holland BJ, Conlan XA, Francis PS, Barnett NW, Stevenson PG (2016) Anal Methods 8:1293–1298

Hong T, Yang X, Xu Y, Ji Y (2016) Anal Chim Acta 931:1–24
Ishii C, Akita T, Mita M, Ide T, Hamase K (2018) J Chromatogr A 1570:91–98
Jandera P, Hájek T (2009) J Sep Sci 32:3603–3619
Jayasena SD (1999) Clin Chem 45:1628–1650
Ji X, Li D, Li H (2015) Biomed Chromatogr 29:1280–1289
Kaczmarkiewicz A, Nuckowski Ł, Studzińska S, Buszewski B (2019) Crit Rev Anal Chem 49:256–
    270
Kadivar S, De Clercq N, Nusantoro BP, Le TT, Dewettinck K (2013) J Agric Food Chem 61:7896–
    7903
Kataoka H, Saito K (2012) Bioanalysis 4:809–832
Kawachi Y, Ikegami T, Takubo H, Ikegami Y, Miyamoto M, Tanaka N (2011) J Chromatogr A
    1218:5903–5919
Kim H, Hage D (2005) Handbook affinity chromatography, 2nd edn. Taylor and Francis, Boca
    Raton, pp 36–68
Kinghorn A, Fraser L, Liang S, Shiu S, Tanner J (2018) Int J Mol Sci 18:2516
Kočová Vlčková H, Pilařová V, Svobodová P, Plíšek J, Švec F, Nováková L (2018) Analyst 143:
    1305–1325
Koga R, Miyoshi Y, Sato Y, Mita M, Konno R, Lindner W, Hamase K (2016) J Chromatogr A
    1467:312–317
Kuklenyik Z, Panuwet P, Jayatilaka NK, Pirkle JL, Calafat AM (2012) J Chromatogr B 901:1–8
Lange M, Ni Z, Criscuolo A, Fedorova M (2019) Chromatographia 82:77–100
Li D, Schmitz OJ (2013) Anal Bioanal Chem 405:6511–6517
Li D, Jakob C, Schmitz O (2015) Anal Bioanal Chem 407:153–167
Lindqvist DN, Pedersen HÆ, Rasmussen LH (2018) J Chromatogr B 1081–1082:126–130
Luzi E, Minunni M, Tombelli S, Mascini M (2003) Trends Anal Chem 22:810–818
Ma Q, Wang C, Bai H, Ma W, Zhou X, Dong Y, Zhang Q (2010) Chromatographia 71:469–474
Maciel EVS, de Toffoli AL, Neto ES, Nazario CED, Lanças FM (2019) Trends Anal Chem 119:
    115633
MacNeill R, Hutchinson T, Acharya V, Stromeyer R, Ohorodnik S (2019) Bioanalysis 11:1155–
    1167
Madru B, Chapuis-Hugon F, Peyrin E, Pichon V (2009) Anal Chem 81:7081–7086
Marlot L, Batteau M, De Beer D, Faure K (2018) Anal Chem 90:14279–14286
Mascini M (2009) Aptamers in bioanalysis. Wiley
McCalley DV (2010) J Chromatogr A 1217:858–880
Meyer VR (2010) Practical high-performance liquid chromatography, 5th edn. Wiley, Chichester
Mikuma T, Uchida R, Kajiya M, Hiruta Y, Kanazawa H (2017) Anal Bioanal Chem 409:1059–
    1065
Moein MM, Abdel-Rehim A, Abdel-Rehim M (2019) Molecules 24:2889
Mogollón NGS, de Lima PF, Gama MR, Furlan MF, Braga SCGN, Prata PS, Jardim ICSF, Augusto
    F (2014) Quim Nova 37:1680–1691
Muller M, Tredoux AGJ, de Villiers A (2019) Chromatographia 82:181–196
Navarro-Reig M, Jaumot J, Tauler R (2018) J Chromatogr A 1568:80–90
Nguyen HP, Schug KA (2008) J Sep Sci 31:1465–1480
Oliveira FM, Segatelli MG, Tarley CRT (2016) Anal Methods 8:656–665
Perret G, Boschetti E (2018) Biochimie 145:98–112
Pichon V (1999) Trends Anal Chem 18:219–235
Pichon V, Brothier F, Combès A (2015) Anal Bioanal Chem 407:681–698
Pirok BWJ, Gargano AFG, Schoenmakers PJ (2018) J Sep Sci 41:68–98
Ren J, Beckner MA, Lynch KB, Chen H, Zhu Z, Yang Y, Chen A, Qiao Z, Liu S, Lu JJ (2018)
    Talanta 182:225–229
Rodriguez EL, Poddar S, Choksi M, Hage DS (2020) J Chromatogr A 1136:121812
Romig TS, Bell C, Drolet DW (1999) J Chromatogr B 731:275–284
Rudolphi A, Boos KS (1997) LGGC N Am 15:814–823

Sadílek P, Šatínský D, Solich P (2007) Trends Anal Chem 26:375–384

Sellergren B (1994) Anal Chem 66:1578–1582

Shao J, Cao W, Qu H, Pan J, Gong X (2018) PLoS One 13:e0198515

Snyder LR, Kirkland JJ, Glajch JL (1997) Practical HPLC method development, 2nd edn. Wiley, Hoboken

Sommella E, Cacciola F, Donato P, Dugo P, Campiglia P, Mondello L (2012) J Sep Sci 35:530–533

Sonnenberg RA, Naz S, Cougnaud L, Vuckovic D (2019) J Chromatogr A 1608:60419

Souverain S, Rudaz S, Veuthey JL (2004) J Chromatogr B 801:141–156

Souza ID, Melo LP, Jardim ICSF, Monteiro JCS, Nakano AMS, Queiroz MEC (2016) Anal Chim Acta 932:49–59

Ståhlberg J (2000) Encyclopedia of separation sciences. Elsevier, pp 676–684

Stevenson D (2000) J Chromatogr B 745:39–48

Stevenson PG, Guiochon G (2013) J Chromatogr A 1308:79–85

Stevenson PG, Burns NK, Purcell SD, Francis PS, Barnett NW, Fry F, Conlan XA (2017) Talanta 166:119–125

Stoll DR, Carr PW (2017) Anal Chem 89:519–531

Stoll DR, Cohen JD, Carr PW (2006) J Chromatogr A 1122:123–137

Stoll DR, Venkatramani CJ, Rutan SC (2018) LCGC N Am 36:356–361

Swartz M (2010) J Liq Chromatogr Relat Technol 33:1130–1150

Sýkora D, Řezanka P, Záruba K, Král V (2019) J Sep Sci 42:89–129

Tan J, Jiang ZT, Li R, Yan XP (2012) Trends Anal Chem 39:207–217

Tang DQ, Zou L, Yin XX, Ong CN (2016) Mass Spectrom Rev 35:574–600

Thuboy B, Kellermann T, Castel S, Norman J, Joubert A, Garcia-Prats AJ, Hesseling AC, Wiesner L (2018) Biomed Chromatogr 32:e4269

Turiel E, Martín-Esteban A (2010) Anal Chim Acta 668:87–99

Turner AD, Hatfield RG, Maskrey BH, Algoet M, Lawrence JF (2019) Trends Anal Chem 113: 124–139

van de Merbel NC (2019) Bioanalysis 11:629–644

Vlakh EG, Stepanova MA, Korneeva YM, Tennikova TB (2016) J Chromatogr B 1029–1030:198–204

Wang M, Liu L, Yin Z, Lu Y (2018) RSC Adv 8:5816–5821

Wichitnithad W, Kiatkumjorn T, Jitavech P, Thanawattanawanich P, Ratnatilaka Na Bhuket P, Rojsitthisak P (2018) Pharmazie 73:625–629

Wolrab D, Jirásko R, Chocholoušková M, Peterka O, Holčapek M (2019) Trends Anal Chem 120: 115480

Xiaohua X, Liang Z, Xia L, Shengxiang J (2004) Anal Chim Acta 519:207–211

Xu W, Su S, Jiang P, Wang H, Dong X, Zhang M (2010) J Chromatogr A 1217:7198–7207

Yang X, Bartlett MG (2019) J Chromatogr B 1120:29–40

Yüce M, Kurt H, Hussain B, Budak H (2018) Biomedical applications functionality nanomaterials. Elsevier, pp 211–243

Zhang C, Lott S, Clarke W, Hage DS (2020) J Chromatogr A 1610:460558

Zhao Q, Wu M, Chris Le X, Li XF (2012) Trends Anal Chem 41:46–57

Zhu Z, Chen H, Ren J, Lu JJ, Gu C, Lynch KB, Wu S, Wang Z, Cao C, Liu S (2018) Talanta 179: 588–593

# The Role of Capillary Electromigration Separation Techniques in Bioanalysis

José Alberto Fracassi da Silva, Alexandre Zatkovskis Carvalho, Emanuel Carrilho, Dosil Pereira de Jesus, and Marcone Augusto Leal de Oliveira

## 1 Capillary Electromigration Separations Techniques

In electromigration techniques, the application of an electric field promotes the separation of charged species. The first attempts to develop analytical separation using electric field date back to the 1930s when Arne Tiselius (Nobel prize in Chemistry in 1948) applied a voltage in a U-shaped tube to separate serum proteins in a technique called the "moving-boundary method." Although the electrokinetic phenomena were already known by the 1800s Tiselius was the first to establish a methodology for their use as an analytical separation tool. The technique was developed to include sieving media that is still useful today in slab gel electrophoresis. A good view of the electrophoresis's historical development stands out in a paper published by Pier Giorgio Righetti (Righetti 2005).

J. A. F. da Silva (✉) · D. P. de Jesus
Department of Analytical Chemistry, Institute of Chemistry, University of Campinas (UNICAMP), Campinas, SP, Brazil

National Institute of Science and Technology in Bioanalytics (INCTBio), Campinas, SP, Brazil
e-mail: fracassi@unicamp.br

A. Z. Carvalho
Center for Natural Sciences and Humanities (CCNH), Federal University of ABC (UFABC), Santo André, SP, Brazil

E. Carrilho
National Institute of Science and Technology in Bioanalytics (INCTBio), Campinas, SP, Brazil

Instituto de Química de São Carlos, Universidade de São Paulo, São Carlos, SP, Brazil

M. A. L. de Oliveira
National Institute of Science and Technology in Bioanalytics (INCTBio), Campinas, SP, Brazil

Departamento de Química Analítica, Universidade Federal de Juiz de Fora, Instituto de Ciências Exatas, Juiz de Fora, Brazil

© The Author(s), under exclusive license to Springer Nature Switzerland AG 2022
L. T. Kubota et al. (eds.), *Tools and Trends in Bioanalytical Chemistry*,
https://doi.org/10.1007/978-3-030-82381-8_22

Jumping several decades of development, the electromigration techniques in the capillary format encompass several methodologies that differ on the separation mechanism (Riekkola et al. 2004). So, one can find capillary zone electrophoresis (CZE), micellar electrokinetic chromatography (MEKC), capillary isotachophoresis (CITP), capillary isoelectric focusing (CIEF), capillary sieving electrophoresis (CSE), capillary affinity electrophoresis (CAE), and capillary electrochromatography (CEC). What makes the technique powerful is that these separation modes can be switched only by replacing the background electrolyte (BGE) filling the capillary and using the same instrumentation. The geometry and dimensions of capillary tubes effectively dissipate the heat generated when an electric current passes through a resistive medium (Joule heating), circumventing the problems associated with sample thermal degradation and convection. This latter issue relates to band spreading. In this way, the use of high voltages leads to fast and high-resolution separations.

## 2    Instrument Setup

The instrumentation associated with electromigration techniques is simple when compared to liquid and gas chromatography, and completely automated instruments exist in the market. An electromigration instrument comprises a high voltage power supply capable of delivering $\pm30$ kV, a sample injection system, a detection system, a temperature control, and an acquisition and control module. The heart of the separation system is the capillary filled with the separation media (background electrolyte, BGE). Typically, fused-silica capillaries with an internal diameter ranging from 20 to 100 μm are the choice. The separation media is usually an aqueous buffer solution to avoid pH changes due to the electrolysis on the cathode (negative electrode) and anode (positive electrode). However, there are also separations using non-aqueous media (NACE, but this separation mode is CZE as well).

The injection performs by applying positive pressure to the capillary inlet or negative pressure to the outlet (hydrodynamic injection). Gravity can also promote a pressure gradient on the capillary for sample injection. As a recommendation, only 10% of the total capillary volume must be filled with the sample to limit dispersion due to the injection process. Alternatively, the electric field can inject samples directly (electrokinetic injection), but this strategy introduces bias to the injection process because ions with higher velocity preferentially migrate into the capillary. However, if it is imperative to use electrokinetic injection, an internal standard is mandatory to keep the accuracy of the methodology. Table 1 shows the critical parameters for the injection process.

Detection performs similarly as in HPLC. Typically, UV-visible absorption (including diode array detection), fluorescence and laser-induced fluorescence (LIF), mass spectrometry (MS), and capacitively coupled contactless conductivity ($C^4D$) detectors are commercially available. Non-commercial detectors include electrochemical (contact conductivity, potentiometry, amperometry, voltammetry), spectrometric (atomic emission and absorption), spectroscopic (infrared, thermal

**Table 1** Injection modes, physical principles, and quantitative equations for sample introduction in capillary electromigration methods

| Injection type | Actuation | Equation |
|---|---|---|
| Hydrodynamic | Differential pressure ($\Delta P$) | $V = \frac{\Delta P r^A \pi t}{8 \eta L}$ (1) |
| | Gravity | $V = \frac{\rho g \Delta h r^A \pi t}{8 \eta L}$ (2) |
| Electrokinetic | Electric field | $Q = \frac{V C t r^2 (\mu_{ef} + \mu_{eo})}{L}$ (3) |

$V$ Injected volume, $r$ capillary radius, $t$ injection time, $\eta$ solution viscosity, $L$ total capillary length, $\rho$ solution density, $g$ gravity acceleration, $\Delta h$ sample vial elevation, $Q$ Injected quantity (mol), $C$ concentration, $\mu_{ef}$ effective mobility, $\mu_{eo}$ electroosmotic flow mobility

lens, nuclear magnetic resonance, surface plasmon resonance), chemiluminescence, and electrochemiluminescence, among other.

Usually, the limits of detection in electromigration techniques are typically in the range of $\mu mol\ L^{-1}$ to sub- $\mu mol\ L^{-1}$; however, laser-induced fluorescence and mass spectrometry achieve outstanding limits of detection. In this way, the techniques are promptly suitable for quality control of drugs and many metabolites of clinical interest. For compounds at trace levels, preconcentration is necessary. There are several strategies for on-capillary preconcentration, such as field-amplified sample injection (FASI), large volume sample injection, and sweeping. Isotachophoresis is a technique that intrinsically promotes the concentration of the sample ions into narrow zones and, in the same direction, does the capillary isoelectric focusing. It is possible to integrate solid-phase extraction (SPE) packs to the separation capillary to perform online SPE.

Electromigration techniques are fully compatible with miniaturization. Pan and colleagues (Pan et al. 2018) presented a handheld instrument equipped with a laser-induced fluorescence detector. Look for more details on electromigration techniques in the microchip format in Chap. 24.

## 3 Electroosmotic Flow

The electroosmotic flow (EOF) plays a crucial role in electromigration techniques. This electrokinetic phenomenon manifests in capillary systems filled with electrolyte solutions under the action of an electric field with the characteristic movement of the bulk solvent toward the cathode in bare fused-silica capillaries. The EOF occurs due to the unbalancing of ions on the region of the electrical double layer (EDL). In its turn, the unbalancing at the solution side is due to the surface charge of the capillary since the excess charge of the surface must be counteracted by the same number of charges of opposite signal giving rise to the compact and diffuse regions of the EDL. The model of the EOF (Oldham and Myland 1994) predicts that the mobility of the EOF ($\mu_{eo}$) is proportional to the zeta potential ($\zeta$) of the surface:

$$\mu_{eo} = \frac{v_{eo}}{E} = -\frac{\varepsilon\zeta}{\eta} \tag{4}$$

where $v_{eo}$ is the velocity of the EOF, $E$ is the electric field, $\varepsilon$ and $\eta$ are the permittivity and viscosity of the solution, respectively.

By inspecting Eq. 4, it is possible to conclude that negatively charged surfaces (as in the case of bare fused-silica capillaries) promote the transport of solvent to the cathode. The surface of the capillary can be modified to attach positive charges dynamically or permanently. In this case, the EOF flows toward the anode. The viscosity of the solution can be adjusted to control the EOF, and in this regard, the control of the temperature is a must to avoid fluctuations (usually, the viscosity of the background electrolyte changes 2% per degree Celsius). Moreover, the pH and ionic strength of the BGE affect the zeta potential, and consequently the EOF magnitude. The implications of the EOF in electromigration are of utmost importance since all mobilities will be affected.

# 4    Modes of Electromigration Techniques and Applications

In this section, we will present the most used electromigration separation techniques applied to bioanalysis. We deliberated skip the technique called capillary isotachophoresis (CITP) since it is not extensively used in bioanalysis, comparatively. Nevertheless, it is worthwhile to mention that CITP offers an astonishing resolution power, and it is an intrinsic online preconcentration technique. Indeed, transient ITP can be applied to compact zones and promote exciting preconcentration factors, but this is out of the scope of this chapter.

## 4.1    Capillary Zone Electrophoresis (CZE)

Capillary zone electrophoresis is by far the most popular electromigration technique. The easiness of the instrumental setup and parameter optimization makes the CZE the first-choice technique—simulation tools provided by the group of Professor Bohuslav Gas (https://web.natur.cuni.cz/gas/) demonstrate the power of CZE (Jaros et al. 2005) effortlessly. Basically, in CZE, one has to maximize the differences between the apparent mobilities ($\mu_a$) of the target analytes. The apparent mobility is the vector sum of the effective ($\mu_{ef}$) and EOF mobilities. Figure 1 summarizes the possibilities considering the EOF in the same direction of the cations (cathodic EOF). Cations and anions can be separated and detected in the same run, given that the resultant vector of apparent mobility might have the same direction for the ions (Fig. 1).

Operationally, CZE is very simple. A small volume of the sample introduced at the inlet of a capillary filled with BGE sets the separation stage. After injection, the

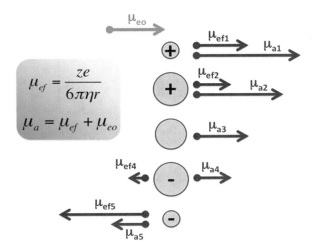

**Fig. 1** The separation in CZE relies on the differences in apparent mobilities of the analytes. Neutral species will move with the same velocity of the EOF ( $\mu_{a3} = \mu_{eo}$ ). By inspecting the figure, one can note that out of the two negatively charged species, the species 4 presents apparent mobility with the same direction of cations because its effective mobility ($\mu_{ef4}$) is lower than that of the EOF. On the equation, $e$, $z$, $r$, and $\eta$ stand for the electron charge, the effective charge of the ion, the solvated radius of the ion, and the viscosity of the media, respectively. Effective mobilities are positive for cations and negative for anions

**Fig. 2** Mechanism of separation in CZE. The sample enters the capillary filled with BGE. Note that the same BGE surrounds the sample zone, representing a homogenous system. After the application of voltage, the differences in their apparent mobilities separate the compounds. The downside blue triangle represents the position of the detection system

capillary returns to the BGE vial, and the application of the high voltage drives the separation (Fig. 2).

CZE has been applied extensively in the bioanalysis of complex biofluid samples with less pre-treatment compared to HPLC. This mode is particularly suitable for separating small charged molecules, such as amino acids and peptides, organic acids, vitamins, carbohydrates, inorganic ions, drugs and their metabolites, neurotransmitters, among many others. Although not usual, macromolecules such as proteins can also be separated in CZE as long as there are differences in effective mobilities.

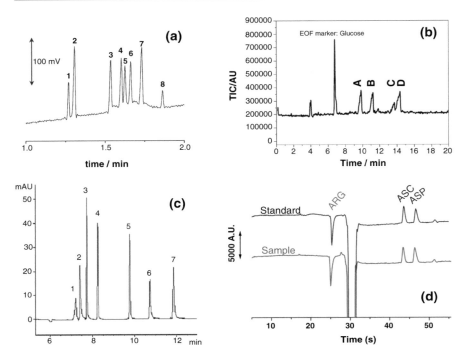

**Fig. 3** Examples of CZE separation. (**a**) Electropherogram of a sample containing eight metal cations (50 μmol L$^{-1}$ each); BGE: 20 mmol L$^{-1}$ lactic acid and 20 mmol L$^{-1}$ histidine, pH 4.9; Voltage: 20 kV; Gravity injection at 10 cm for 30 s; Peaks: (1) Cs$^+$, (2) K$^+$, (3) Ba$^{2+}$, (4) Sr$^{2+}$, (5) Na$^+$, (6) Ca$^{2+}$, (7) Mg$^{2+}$, and (8) Li$^+$; C$^4$D detection; reproduced from reference (da Silva et al. 2003). (**b**) Total Ion Count (TIC) electropherogram of 4.61 mmol L$^{-1}$ S-nitrosoglutathione stored in the solid-state for 6 months, then dissolved in BGE just before analysis; BGE: 20 mmol L$^{-1}$ ammonium carbonate buffer, pH 8.5; Injection: 50 mbar, 3 s; Voltage: 20 kV; MS in positive mode; Peaks A: S-nitrosoglutathione, B: oxidized glutathione, C: glutathione sulfinic acid, and D: glutathione sulfonic acid; Neutral marker glucose at 5 mmol L$^{-1}$; reproduced from reference (Ismail et al. 2015). (**c**) Separation of seven antimicrobials at 100 μg mL$^{-1}$; Peaks: (1) chloramphenicol, (2) danofloxacin, (3) ciprofloxacin, (4) enrofloxacin, (5) sulfamethazine, (6) sulfaquinoxaline, and (7) sulfamethoxazole; BGE: 60 mmol L$^{-1}$ sodium phosphate, and 20 mmol L$^{-1}$ sodium tetraborate, pH 8.5; Voltage: 24 kV; UV detection at 270 nm for chloramphenicol and fluoroquinolones and 203 nm for sulfonamides; reproduced from reference (Mamani et al. 2008). (**d**) Single run separation of 290 μmol L$^{-1}$ arginine, 500 μmol L$^{-1}$ ascorbic acid, and 290 μmol L$^{-1}$ aspartic acid; BGE: 20 mmol L$^{-1}$ N-tris(hydroxymethyl)-methyl]-3-aminopropanesulfonic acid (TAPS) and 10 mmol L$^{-1}$ of NaOH, pH 8.7; Voltage: 25 kV; C$^4$D detection at 1.1 MHz and 4 V$_{pp}$; reproduced from reference (Costa et al. 2019). All figures reproduced with permission

By including chiral selectors in the BGE, it is also possible to perform chiral separation in CZE. Remarkably, CZE achieves a higher resolution of the enantiomer pair with short separation time and reduced costs than HPLC. Figure 3 shows some examples of CZE separations using bare fused-silica capillaries.

## 4.2    Micellar Electrokinetic Chromatography (MEKC)

Micellar electrokinetic chromatography is an electromigration technique introduced by Shigeru Terabe in 1984 that allows the separation of neutral compounds (Terabe 2009). The separation mechanism in MEKC relies on the interaction of the analytes with micelles included in the BGE. For example, by using sodium dodecyl sulfate (SDS), the net charge of the micelle will be negative, yielding effective mobility toward the anode. Commonly MEKC is performed under the condition of strong cathodic EOF, which results in apparent mobility ($\mu_{mc}$) of the micelle toward the cathode. If the analyte is attached to the micelle, its mobility will be equal to $\mu_{mc}$, while on the contrary, the mobility will be equal to the mobility of the EOF, as shown in Fig. 4. Just like happens in chromatography, time spent by the analyte on the separation capillary will depend on the time spent on each media, i.e., the aqueous solution, or the micelle. In MEKC, the retention factor is the ratio of the analyte in the micelle ($n_{mc}$) and the aqueous media ($n_{aq}$) (Terabe 2009):

$$k = \frac{n_{mc}}{n_{aq}} \tag{5}$$

Considering a neutral compound and cathodic EOF, the maximum velocity obtained by the analyte will be the velocity of the EOF ($k = 0$), while the minimum will be the velocity of the micelle ($k \to \infty$). For charged compounds, the mechanism can be more complicated since one has to consider the electrostatic attraction or repulsion from the micelle and the effective mobility of the analyte. The representative electrokinetic chromatogram is portraited in Fig. 5. Note that methanol elutes with the EOF-marker since it does not interact with the micelles. Instead, the capacity factor for Sudan III approaches infinity, and it marks the velocity of micelles (Terabe et al. 1985).

BGE additives are also common in MEKC. For example, organic solvents increase the solubility of non-polar species, but at higher concentrations can have

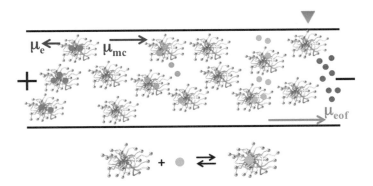

**Fig. 4** Mechanism of separation in Micellar Electrokinetic Chromatography (MEKC). The blue downside triangle illustrates the position of the detection system

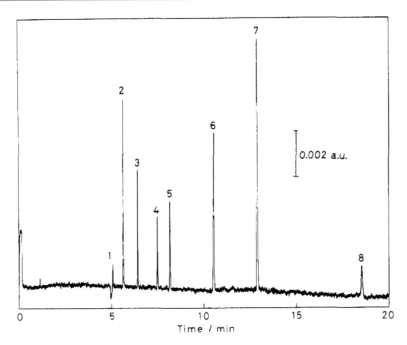

**Fig. 5** Electrokinetic chromatogram of (1) methanol, (2) resorcinol, (3) phenol, (4) p-nitroaniline, (5) nitrobenzene, (6) toluene, (7) 2-naphthol, (8) Sudan III; BGE: 50 mmol $L^{-1}$ SDS in 100 mmol $L^{-1}$ borate and 50 mmol $L^{-1}$ phosphate buffer, pH 7.0; Capillary: 50 μm i.d., 65 cm total length and 50 cm to the detector; Voltage: 15 kV; Current: 26 μA; UV detection at 210 nm; temperature 35 °C. Reproduced from reference (Terabe et al. 1985) with permission. Copyright, 1985, American Chemical Society

deleterious effects due to the disruption of the micellar structure. Chiral selectors can also be incorporated into the BGE to promote the resolution of enantiomeric pairs.

## 4.3    Capillary Isoelectric Focusing (CIEF)

Separation of biomolecules often means that we are talking about the separation of macromolecules, spotlight to DNA, RNA, and proteins. Size separation is the most used approach applicable to the separation of macromolecules due to the versatility of this parameter, since size may indicate the identity of some molecules, may infer the geometrical conformation of a macromolecule, and even can identify interactions between macromolecules by a shift in the migration time (or distance). Pure size separation is the mechanism of CSE, while mobility shifts regulate the selectivity in CAE. However, some points require a more careful look, as follows: first, due to the high complexity of biological samples, CSE is not enough to resolve completely several analytical challenges; second, the CAE strategy is exact and selective, but it is highly demanding of a good understanding of the target under study and is more demanding on data analysis. Polymerase chain reaction (PCR) is the technique of choice to study nucleic acids, with or without the aid of separation techniques, but PCR does not apply to proteins.

In common, DNA, RNA, and proteins are polymeric molecules; however, RNA and DNA are repetitions of a few (only four) different units that maintain some similarity in chemical characteristics. For large molecules such as RNA or DNA, the variation of isoelectric point (pI) is not useful from the analytical point of view. On the other hand, proteins are a combination of 20 different amino acids, seven of them with ionizing side chains, viz. cysteine, tyrosine, lysine, arginine, histidine, aspartic acid, and glutamic acid, reason why proteins vary widely in terms of pI.

Proteins may be separated according to their pI using a technique named isoelectric focusing electrophoresis (IEF), performed in general in a cylindrical gel or into a capillary (CIEF). Before discussing the technique in more detail, it is important to stress that orthogonal separations are an essential approach to resolve complex matrices, including a protein pool. Two-dimensional separation is characterized by orthogonality, i.e., when the separation mechanisms of the first and the second dimensions are different. Combining a pI-based separation (IEF) as a first dimension and a size-based separation (sodium dodecyl sulfate-polyacrylamide gel electrophoresis, SDS-PAGE) as a second dimension represents an essential tool in proteomics, known as 2D-gel electrophoresis.

CIEF separation mechanism relies on forming a pH gradient along the separation capillary, which is an EOF-canceled column. To establish the gradient, a heterogeneous BGE, composed of an acidic solution, fills the anodic reservoir of the capillary, while an alkaline solution fills the cathodic end. In this mode of separation, a mix of the sample with a pool of ampholytes that must cover the range of the expected pIs of the analytes fills the whole capillary. Once separation voltage is turned on, a titration process occurs, and all zwitterions start to migrate according to their charges. After some time of this described process, it is created a pH gradient along the capillary, where the ampholytes help to make this gradient stable and buffered. The sample zwitterions behave similarly to the ampholytes, ceasing migration in the region of the column where the pH matches the pI of the molecule during the focusing process. This step requires monitoring the current profile, starting from around 10–20 µA, and waiting until dropping to a residual 0.5–1 µA. At this point, the point of minimal current establishes the formation of the gradient and the focusing of all substances.

There are several commercial mixtures of ampholytes that can cover a wide or narrow range of pH with higher or lower resolution, depending, respectively, on the range of the pI and the number of different substances (with diverse pIs as well). Although there are many options and brands, all ampholytes mixtures are composed of relatively small molecules (less than 1 kDa) with sulfonate, carboxylic, free primary and secondary amino groups, amino acids, and polyamines.

To improve the stability of the focusing it is necessary to prevent gradient drift. To achieve stability, an anodic and a cathodic substance are added to the procedure and will act as the ampholytes closest to the capillary ends. It is common to find iminodiacetic acid (pI 2.2) and arginine (pI 10.7) as anodic and cathodic spacers, respectively. The quantity of the spacer has to be large enough to cover the detection window. Finally, to acquire the CIEF electropherogram, the focused molecules must be mobilized, what might be achieved hydrodynamically by pressurizing the capillary to drive the substances toward the detector or, preferably, employing chemical mobilization. The mobilization chemically promoted can be accomplished by changing the cathodic alkaline BGE using acetic acid, which will migrate into the capillary and ionize the mobilized zwitterions from the substance with higher to the lower pI. Chemical mobilization leads to sharper peaks than pressure mobilization. CIEF is useful for quantification; however, the primary purpose of this technique is to assess the pI of the analytes. A pI calibration curve requires spiking the sample with substances of known pIs, generally small peptides (3–6 amino acids). Figure 6 illustrates the process discussed herein.

The quality of the focusing step requires monitoring a marker. Poor focusing leads to a bad shaped/split peak. On the other hand, excessive focusing may lead to protein precipitation due to the formation of a region of high concentration of proteins on their isoelectric point. The solubility of proteins is generally low at their isoelectric pH. Solving this problem without diluting the sample and losing detectability typically involves adding a moderated denaturant that improves the solubility of the protein. Urea represents the most used reagent for this purpose, added in concentrations from 1 to 6 mol $L^{-1}$.

CIEF works isolating from small peptides to intact microorganisms (Salplachta et al. 2012), ranging from viruses to larger cells like yeast. The separation of microorganisms through CIEF is possible due to the chargeable molecules present on the external surface of the particles. Besides CIEF, CZE also separates microorganisms. The pI of intact or empty viruses (without the genetic material) is the same, which is an advantage to CIEF over CZE since the electrophoretic mobility of intact or empty viruses is different. Thus, CIEF can be used as a safer strategy compared to CZE to investigate hazardous bioparticles.

CIEF is a versatile tool not only for protein separation, but also for pI identification or use in proteomics, as discussed before. Just to mention a few CIEF applications, it is a useful tool to identify proteins mutation-driven changes, separation of isoforms, verification of the number of variants among a pool of antibodies, and identification and quantification of chemical modifications in proteins like disulfide bonding, deamination, and glycation.

## 4.4 Capillary Sieving Electrophoresis (CSE)

Capillary Sieving electrophoresis is the electromigration mode that is closest to slab gel electrophoresis. CSE selectivity is a function of the size of porous structures generated in capillary filled by a sieving media or gel. Longer chain macromolecules tend to twine on the polymeric network with a consequent decrease in mobility. Therefore, CSE is a separation technique that sorts molecules by size and is of particular interest in protein, DNA, and RNA analysis, but shows useful for other

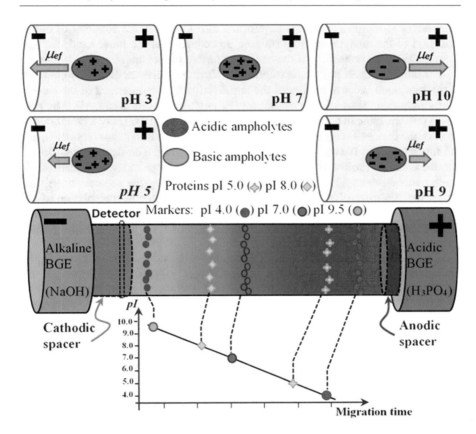

**Fig. 6** Schematic illustration of the CIEF method. The upper part of the figure shows the electrophoretic behavior of a zwitterion at different pH values, indicating that if the buffer pH matches the substance pI, the effective electrophoretic mobility ($\mu_{ef}$) is null. The core of the figure represents a capillary with the separation ampholytes and sampled proteins already focused according to their respective pIs. The illustration depicts the anodic (acidic) BGE, the cathodic (alkaline) BGE, and the anodic and cathodic spacers (usually iminodiacetic acid and arginine, respectively). Although the mobilization process is not pictured, the graph at the bottom represents the expected plot of the pI in the function of the migration times as the analyte bands pass through the detector

applications such as nanoparticles, whole cells, synthetic polymers, and microorganism separations.

In the early days of the technique, a rigid (chemical) gel was applied, such as agarose and polyacrylamide. The degree of cross-linking between the polymer chains defines the pore size of the gel. The preparation of the capillary is of utmost importance since the gel needs to be attached to the capillary walls, which turns such preparation of capillaries laborious. The high surface-to-volume ratio in the capillary system makes possible using high electric fields without deleterious heat effects, promoting faster separations than slab gel electrophoresis, reducing the analysis time from hours to minutes. The capillary format also facilitates the coupling of online detection of the analytes without staining procedures. On the other hand, the slab format allows the collection of the bands for further processing.

The introduction of polymeric solutions as sieving media was of great importance to the technique because it allowed the introduction and replacement of the sieving media by pressure, which is essential for the automation of the analysis. The pore size is customizable in polymeric solutions by tailoring the type (polyacrylamide and others), the concentration (usually in the range from 2 to 10% (mass/volume)), and the molecular size (from 0.4 to 4 MDa) of the polymer. The development of the Human Genome Project skipped years by introducing large capacity sequencers, in some cases containing an array of 96 capillaries running the separation on CSE mode. In CSE, astonishing separation power is typical, reaching a resolution of one base pair over a thousand bases and separation efficiency on the millions of plates level. Figure 7 shows examples of the application of CSE for both rigid gel and polymeric solution sieving media.

In a broad view, the mechanism of separation relies on the sieving promoted by porous media inserted in the capillary, but different migration behaviors occur for the separation of long-chain molecules such as nucleic acids (Slater et al. 1996; Quesada 1997). For small molecules, there is a lack of selectivity since the porous are too big comparatively. This scenario makes the separation based on charge-to-size ratio, just as in CZE, but this ratio is almost constant for nucleic acids. Conversely, for huge fragments, the nucleic acid is trapped on the polymeric network, and mobility drops close to zero. On DNA of intermediate sizes, the mobility is highly affected by the size of the fragments: in Ogston sieving (I) the mobility decreases exponentially with the nucleic acid size; in the entropic trapping (II) the coil nucleic acid jumps from pore to pore, in reptation regime (III) the mobility assumes a linear (log-log) relationship with size followed by a decrease on mobility due to reptation trapping (IV), and in oriented reptation (V) the molecules are aligned on the electric field yielding a constant mobility in this size range. So, the useful range for sieving falls on the regions I-III. Refer to the work of Slater (Slater et al. 1996) and Quesada (Quesada 1997) for an in-depth discussion on the mobility behavior of nucleic acids in CSE.

## 4.5    Capillary Electrochromatography (CEC)

Capillary Electrochromatography (CEC) is a term used to describe an electromigration separation mode, which takes into account the merge of capillary electrophoresis (CE) and high-performance liquid chromatography (HPLC). This hybrid technique involves the use of fused-silica capillary columns with an internal diameter of 50–100 μm, containing a stationary phase (SP), through which percolates the mobile phase (MP) pumped by EOF under an electric field (Dittmann and Rozing 1996; Svec et al. 2000; Majors 2009; Hilder et al. 2004; Gusev et al. 1999). This association results in potentialities for separation of neutral compound classes, where MEKC does not get satisfactory results due to high hydrophobic

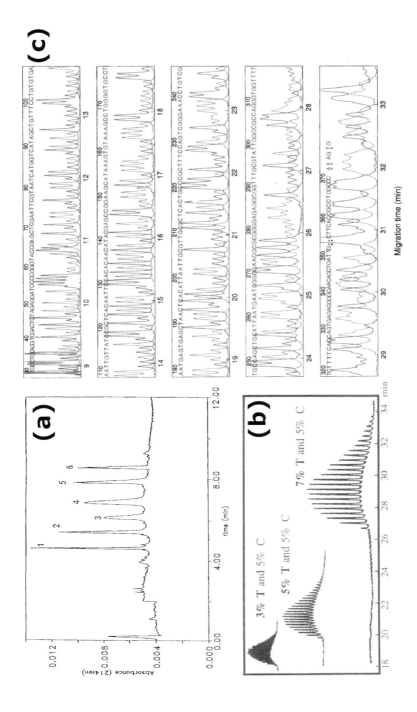

**Fig. 7** Example of CSE applications. (**a**) Separation of SDS–protein complexes in 10% w/v dextran (MW 2,000,000) sieving media containing 0.1% SDS; Electric field: 400 V/cm; (1) myoglobin, (2) carbonic anhydrase, (3) ovalbumin, (4), bovine serum albumin (5) β-galactosidase, and (6) myosin; UV detection at 214 nm, reproduced from reference (Ganzler et al. 1992), (**b**) Effect of the polyacrylamide gel concentration on the separation of poly(deoxyadenylic acid)$_{40-60}$;

characteristic, limited peak capacity (time interval between the unretained solute and the solute totally retained by the micelle), and the use of surfactants, which can compromise mass spectrometry detection systems. The implementation of CEC is simple, once the CE commercial equipment is promptly adaptable to CEC mode through the external pressure device activation. Figure 8 depicts the schematic illustration of a traditional CEC mode with Diode Array Detection (DAD). However, it is essential to highlight that CEC can be performed in one- or two-dimensional separations or through a microscale system (Zhang and El Rassi 2006; Gottschlich et al. 2001).

There are three main types of chromatographic SP in CEC. The first type consists of a column filled with a spherical and porous material, or spherical particulate stationary phase (SPSP), filling the capillary up to the detection window (Dittmann and Rozing 1996; Schmeer et al. 1995; Norton et al. 2003). The second refers to the open tubular stationary phase (OTSP), where the chromatographic material consisting of a porous polymeric film remains linked to the inner surface of the capillary (Xie et al. 2001). Moreover, there is also the monolithic columns or monolithic stationary phases (MSP), which present a three-dimensional filling network attached to the inner surface of the capillary with a porous polymeric material. Figure 9 illustrates the structural details of the three main modes of capillary columns used in CEC.

The bioanalysis applications involving blood, plasma, saliva, serum, urine, and other biological fluids evoke the necessity of sophisticated analytical tools to achieve appropriate information about the system under investigation, owing to the high complexity of the samples. In this sense, when molecular information is mandatory, electromigration separation techniques hyphenated to mass spectrometry become necessary. A critical point to consider is the composition of the MP, which in the present case, is the background electrolyte (BGE) and must be compatible with the electrospray ionization (ESI) process. CEC becomes a promising strategy within this context because different approaches considering selective SP preparations for specific biological compounds can be optimized. Hence, knowing that for CEC mode, the separation mechanism also involves competition among the analyte molecules and those of the MP for the active sites of the SP, the MP can contain a very low concentration of ions, minimizing adverse effects in the mass spectrometer detector. However, it is essential to emphasize full compatibility with other types of detectors, mentioned a priori, in harmony with CEC.

Sample preparation prior to CEC analysis is an important issue. It is essential to keep in mind that due to the inherent complexity of biosamples, sample preparation procedures should often exist in the method to mitigate the presence of matrix

---

**Fig. 7** (continued) T and C are the total concentration of monomer and the cross-linking agent, respectively; Electric field: 200 V/cm; UV detection at 260 nm; reproduced from reference (Baba et al. 1992). (**c**) Sequencing of M13mp18 phage using fluorescent dye-labeled primers FAM and JOE in a modified Sanger–Coulson strategy; Sieving medium: replaceable 6% polyacrylamide; Electric field: 250 V/cm; reproduced from reference (Ruiz-Martinez et al. 1993). Copyright 1992 and 1993, American Chemical Society

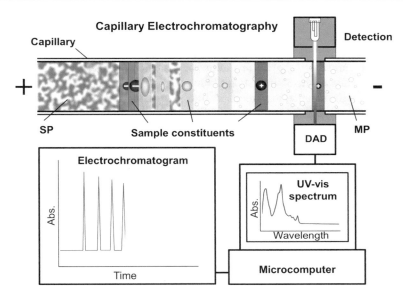

**Fig. 8** Illustration of a CEC run, showing some instrumental components, SP and MP inside the capillary, and the constituents of the introduced sample

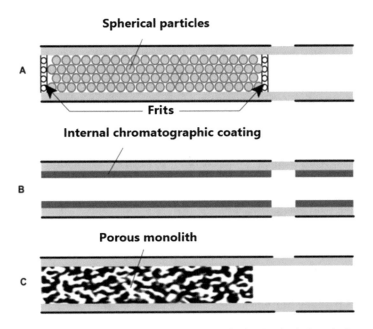

**Fig. 9** Representation of three types of columns used in CEC: (**a**) spherical particulate stationary phase; (**b**) open tubular stationary phase; and (**c**) monolithic stationary phase

artifacts that can compromise the quality of the analyzes. Not less important is to preserve the useful life of the stationary phase. The assembly of analytical apparatus by means of a valve system and multiple switching channels are useful for online sample preparation or to perform multidimensional separation. This multidimensional system allows the optimization of bioanalyses. The first dimension carries out the cleanup of the sample, the preconcentration of the analyte, or the pre-separation of the mixture through an orthogonal CE mode, and in the second dimension, the CEC separation. This multidimensional approach is possible in a traditional or on a microscale system (Zhang and El Rassi 2006; Gottschlich et al. 2001).

There is an extensive list of CEC-related articles published since the 1990s. However, for those interested in developing research in this area, there is still a lot of space to be explored, given that the commercial CEC columns available are far from meeting the huge and specific demand for the existing and the rapidly growing analytical challenges.

## 4.6    Capillary Affinity Electrophoresis (CAE)

Every separation technique of choice depends upon the physical and chemical characteristics of the analytes. As a general and unpretentious overview, gas chromatography explores the volatility and polarity, liquid chromatography explores the polarity, charge, and size, and capillary electrophoresis relies on the charge-to-size ratio, polarity, size, and isoelectric point of the analytes. Of course, intermediate or mixed separation mechanisms exist if one closely observes each analytical approach, especially the so-called affinity separations.

One of the most exciting ways to separate biomolecules is employing affinity interactions, which work on the impressive recognition and attachment between molecules. Not only the strong binding forces are something to emphasize here, but also the high specificity of this interaction. It is important to stress that we are talking about non-covalent bonds, which forms due to the sum of several non-covalent forces, such as hydrogen bonds, electrostatic attractions, hydrophobic interactions, van der Waals forces (dipole, induced dipole, and London dispersion), pi-interactions (pi-stacking, pi-anion, and pi-polar), and hydrophobic effects. Although all these forces are often individually observed among ions and relatively small molecules, they occur in combination and with the addition of steric effects when dealing with macromolecules.

Affinity capillary electrophoresis (CAE) is the CE mode that exploits affinity interactions between molecules. CAE mechanism of separation makes use of the ability of molecules to bind to other molecules, with high applicability to biomolecules. Once an analyte binds to another molecule, there is a modulation on the electrophoretic mobility of the analyte, which depends on the size, charge, and complex equilibrium constant. The CAE mechanism, however, is also present in CZE when using a complexing agent additive to modulate the charge of an analyte, causing a shift in the migration time of a given analyte. For example, adding 18-crown-6 in the BGE allows the separation of $K^+$ and $NH_4^+$ cations, which present the same electrophoretic mobility in an acidic medium. Due to the selectivity of the crown ether to complex $K^+$, the mobility of this alkaline ion reduces due to the complexation. Thus, CAE initially developed as a dynamic shift technique, and for this reason, CAE is also known as mobility shift CAE (ms CAE). Later, step-profile elution and vacancy CE phenomena appeared as options to develop affinity studies. There are several affinity interactions which might be studied by means of CAE, as displayed in Table 2. In this discussion, we will name the substances as binding agents and targets; however, one can consider the binding agent as a target and vice-versa or use other denominations, like ligands. Separations based on immobilized molecules are also possible in CAE, especially for the study of enzymatic reactions and in immunoassays. In this chapter section, however, we addressed only free solution CAE.

There are several variations of CE available to assess affinity interactions. As just discussed, the mobility shift of a binding agent due to the interaction with a target comes from observing the electropherogram and registering the migration time of the molecule in the study with and without the presence of a target molecule in the BGE. This mechanism is basically the situation described for 18-crown-6 ether and $K^+$ interaction. The higher the concentration of the ether on BGE, the higher the shift in the migration time of $K^+$, up to a maximum level, observed as asymptotic mobility as a function of the ether concentration. The stronger the interaction between a binding agent to a target molecule, the larger the migration time change. This approach is known merely as CAE or as mobility shift CAE (ms CAE).

Due to the presence in the BGE of a responding target molecule to the detection system, detectability in ms CAE is challenging. A method variation that can overcome this problem is the plug-plug kinetic capillary electrophoresis (ppKCE), also known as partial filling CAE (PF-CAE). The separation in ppKCE is best because it allows the passage of the analytes by the detector within a low signal BGE. When binding agents and targets are an enzyme, and the respective subtract this setup may be used to study the kinetics of the enzymatic reaction, being then be referred to as an enzymatically mediated microanalysis (EMMA) (Zhang et al. 2020).

An exciting feature of electromigration techniques exploited in affinity studies is the vacancy phenomenon. When a sample plug enters the CE column, not only the physically present molecules will migrate through the capillary, but also the vacancies from the BGE ions that are absent in the sample plug. For example, cetyltrimethylammonium bromide (CTAB) is a typical cationic surfactant used to determine fast anions due to its capability to reverse the EOF direction toward the anode. Approximately 200 $\mu$mol $L^{-1}$ of CTAB added to BGE is enough to promote an anodic EOF; however, if using a conductometric detector, a negative peak with the same migration time of $Br^-$ appears in the electropherogram. This negative peak is the vacancy peak. A vacancy peak may appear if a binding agent depletes the local concentration of the target in the injection plug. If the complex has a high formation constant, fast kinetics, and electrophoretic mobility sufficiently different from the binding agent and/or the target, one may observe the vacancy of the target or the binding agent, depending upon the separation setup. Vacancy peak method

**Table 2** Common binding interactions that can be assessed by Capillary Electrophoresis

| Binding agent | Targets | Description, relevance, and application |
|---|---|---|
| Aptamers | Carbohydrates, proteins, peptides, small molecules | Aptamers are sequences of single-stranded DNA or RNA (10–100 nucleotides), naturally occurring as riboswitches or oligopeptides that binds to a specific target. Aptamers might be selected or synthesized in vitro and can be used alternatively to antibodies |
| Carbohydrates | Proteins, lectins, metal ions, small molecules | Mono- and polysaccharides, linked or not to proteins (glycoproteins), are essential in molecular recognition, as predominantly observed among hormonal function, cell signalization, antigen–antibody interaction |
| Metal ions | Proteins, carbohydrates, DNA and RNA | Metal ions play an essential role in enzymatic function, often interacting in the catalytic sites as complexation and redox agents. Specialized proteins usually make transport and storage of metal ions |
| Antibodies | Antigens | Antibodies are macromolecules produced by the immune system responsible for recognizing and helping to neutralize foreign agents. Antibodies have a highly conserved chain and a randomized chain, which is the region responsible for binding the antigen |
| Lectins | Carbohydrates, glycoprotein, glycolipids | Lectins are a particular class of proteins and glycoproteins, not related to the immune system that binds to specific carbohydrates. Lectins are ubiquitous, found from bacteria to mammals. Carbohydrate recognition is particularly important in cell signalizing, immune response, and hormonal regulation |
| Enzymes | Carbohydrates, RNA, DNA, proteins, peptides, small molecules | Enzymes are very specialized biocatalytic agents that can increase the velocity of a reaction in orders up to $10^{17}$ (for example, urease increases decomposition of urea to $NH_3$ $10^{14}$-fold faster than a non-catalyzed reaction). Enzymes are proteins, although RNA can present catalytic activity through ribozymes |
| Drugs | Carbohydrates, proteins, DNA, RNA | Drugs category is a comprehensive collection of substances that can interfere with the metabolism. Drugs may interact with DNA improving or inhibiting replication and or expression, might interact with enzymes changing the catalysis of a reaction, interact with membrane glycoprotein receptors interfering hormonal signaling, etc. A particularly important topic of study is the interaction of drugs with plasmatic proteins like albumin |

(VP CAE), vacancy CAE (VCAE), and Hummel–Dreyer CAE (HD CAE) are methods based on vacancies that are more detailed in this chapter.

Step methods are also crucial to the study of affinity interactions. In this kind of approach, there is not a formation of typical peaks but plateaus, or steps. By forming a trending of discrete regions, the heights of the plateaus serve to infer the formation constants of the affinity complex in a very reproducible and exact way. Frontal analysis CAE (FA CAE) and continuous non-equilibrium CE of equilibrium mixtures (cNECEEM) are important methods based on this phenomenon.

Other methods that are not fully addressed but mentioned in this text are non-equilibrium CE of equilibrium mixtures (NECEEM), equilibrium CE of equilibrium mixtures (ECEEM), and the methods based on sweeping, just like the approach of the same name used as stacking strategy in MEKC. Namely, these methods are the sweeping CE (SweepCE), short SweepCE (sSweepCE), and short SweepCE of equilibrium mixtures (sSweepCEEM).

CAE methods are also named kinetics capillary electrophoresis (KCE) by some authors. Most important information achieved employing CAE (KCE) methods are equilibrium parameters, including dissociation and binding constant, the stoichiometry of the interaction, and even the determination of the concentration of a target molecule without external calibration. The ms CAE, FA CAE, and ppKCE are the most cited CAE variants in recent articles (Busch et al. 1997; Olabi et al. 2018) Although some data analysis demands a hard-mathematical work, semi-quantitative and preliminary screening to infer good and bad ligands, competitive interactions, etc., can be easily obtained using CAE. Table 3 displays a compilation of the methods and respective experimental parameters, including other methods not addressed in detail here, namely non-equilibrium CE of equilibrium mixtures (NECEEM), equilibrium CE of equilibrium mixtures (ECEEM), and the methods based on sweeping.

Considering only one binding site, fundamental kinetics write as:

$$B + T \underset{k_{off}}{\overset{k_{on}}{\rightleftarrows}} C \text{ (formation of the affinity complex)} \tag{6}$$

$$K_d = \frac{k_{off}}{k_{on}} \text{ (equilibrium constant)} \tag{7}$$

where B is a binding agent, T is a target molecule, C is the complex BT, $k_{on}$ and $k_{off}$ are the formation and dissociation rate constants, respectively, and $K_d$ is the equilibrium dissociation constant. For a detailed discussion see (Galievsky et al. 2015).

### 4.6.1   Mobility Shift CAE

In ms CAE, the BGE that contains the target molecule fills the capillary and the reservoirs. A small plug of the binding agent dissolved in target-free BGE is introduced hydrodynamically to the capillary inlet. During separation, a steady-state between the binding agent and target forms and the migration time of the complex will depend on the concentration and mobility of the complex,

**Table 3** Experimental setup and data analyzing the most frequently applied ACE methods

| | Experimental setup | | | | Data reading | |
|---|---|---|---|---|---|---|
| Method name | Inlet reservoir | Outlet reservoir | Capillary | Injection plug(s) | Typical signal[a] | Assessed (measured) parameter |
| ms CAE | B | B | B | T | | $K_d$ (migration times) $K_d$, $K_{on}$, and $K_{off}$ (peak heights and widths, mobilities) |
| VP CAE and VCAE | EM | EM | EM | BGE | | $K_d$ (VP: Peak areas or heights) $K_d$ (VCAE: mobilities) |
| FA CAE | BGE | BGE | BGE | EM large | | $K_d$ (plateau heights) |
| HD CAE | T | T | T | EM | | $K_d$ (peak areas or heights) |
| NECEEM | BGE | BGE | BGE | EM | | $K_d$ and $K_{off}$ (peak areas and migration times) |
| ppKCE (PF-CAE) | BGE | BGE | BGE | B\|T | | $K_{on}$ and $K_{off}$ (ppKCE: Peak areas and migration times) $K_d$ (PF-ACE: mobilities) |
| ECEEM | B | B | B | EM | | $K_d$ (mobilities) $K_d$, $K_{on}$, and $K_{off}$ (peak heights and widths, migration times) |
| cNECEEM | EM | BGE | BGE | None (BGE) | | $K_d$ (plateau heights) |

Abbreviations used: *B* binding agent, *T* target molecule, *BGE* background electrolyte (running buffer), *EM* equilibrium mixture of B and T, *B\|T* two plugs injected in sequence, in the capillary, first T and after B

[a]Typical signal expected with a detectable target molecule (T). Adapted from reference (Galievsky et al. 2015) with permission. Copyright 2015, American Chemical Society

concentration and mobility of the free binding agent, and the binding constant. For this approach to work, the equilibrium reaction must show very fast kinetics.

It is also desirable to detect the binding agent at low concentrations. Since migration time is the analytical data to be acquired, peak shape must be taken into account since the maximum peak height might be no longer the peak center due to electromigration dispersion (EMD). To correct this issue, the application of the Haarhoff-Van der Linde function provides the apparent mobility of a distorted peak. It is also crucial to use an internal migration marker to correct for fluctuations in the separation parameters. A neutral compound (EOF-marker) or a substance of a very established electrophoretic mobility permits this correction. Figure 10 brings a schematic explanation of a mobility shift-based method.

By plotting the observed mobility of an analyte (target) as a function of the concentration of a binding agent added to the separation BGE, it is possible to observe an increasing or decreasing asymptotic curve, depending on if the complex has higher or lower electrophoretic mobility compared to the free target, respectively. It is essential to choose the substance with its electrophoretic mobility more affected by the complex formation to be the injected one. For example, considering the combination of a large protein (P) with a small ligand molecule (T), it is expected that the electrophoretic mobilities of P and the complex P-T are practically the same. In this case, the protein should compose the BGE, and the small molecule injected as the sample plug.

### 4.6.2   Hummel–Dreyer CAE (HD CAE)

In the HD CAE, the peak area or height of one compound is the measured data. For a given situation when the complex B-T forms, it is highly likely to have one of the ligands with practically the same electrophoretic mobility of the complex. In this case, this is the substance that is better to be injected (differently from ms CAE). There are two approaches for method calibration, the internal and the external calibration. In the external calibration, the capillary, filled with BGE plus some amount of one compound of the complex, let us adopt T and assume that the mobilities of B and B-T are the same. With the capillary filled this way, two injections suffice, one with neat BGE and the other with BGE plus some amount of B. In both injections, the formation of a trough peak (a vacancy peak with the electrophoretic mobility of T) appears, being the differences between the negative areas of the peaks the amount of T bonded to B. Due to the formation of a complex B-T, the injection of B will generate a larger peak than the trough peak formed with neat BGE.

In the HD CAE with internal calibration, the experimental setup is almost the same as described for the external calibration. However, more injections of B are necessary, adding a known quantity of T in the samples $[T_s]$. From samples with B only to samples in which the concentration of T is higher than its concentration in the BGE ($[T_{BGE}]$), the T peak goes from a negative peak to a positive one. By plotting the T peak areas as a function of the concentration of this compound in the sample, it is possible to obtain a curve that crosses the abscissa in the concentration of

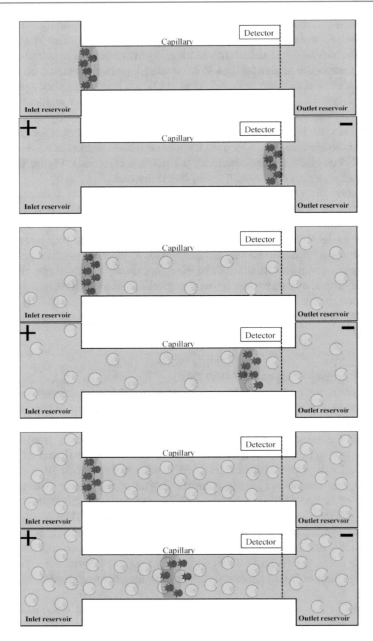

**Fig. 10** Schematic representation of an ms CAE method. Each consecutive couple of diagrams represents the capillary at injection time and after a fixed equal separation time. Three separations conditions are schematically represented. From top to bottom, a binding agent is added in increasing concentrations, and the mobility of its respective target is slowed proportionally to this parameter. Adapted from reference (Zhang et al. 2020) with permission

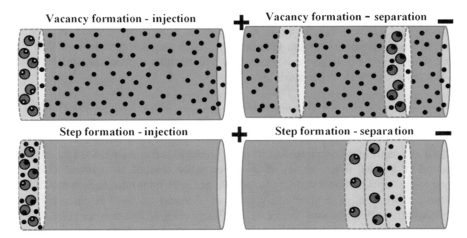

**Fig. 11** On top, vacancy formation by sequestration of a target ligand (black circle) due to the injection of a binding agent (large red circle). The amplitude of the trough peak is proportional to the concentration of the formed complex, on bottom steps (plateaus) formation typical of the frontal analysis. Adapted from reference (Zhang et al. 2020) with permission

equilibrium, where the concentration of the free T in the sample is equal to the $[T_{BGE}]$. Thus, for the crossing point we have $[T_s] - [T_{BGE}] = [B\text{-}T]$.

### 4.6.3 Vacancy Peak Method (VP CAE), Vacancy CAE (VCAE)

In addition to HD CAE, VP CAE, and VCAE are also methods based on vacancies events observed in CE. In VP CAE, the capillary contains the BGE with some amount of both compounds of the complex, B and T. The concentration of one of the compounds is constant, for instance, [B], and the concentration of the other ([T]) varies. For each condition, we inject a small plug of the plain BGE, and two negative peaks appear, one due to the B + B-T vacancy, assuming again that the mobilities of B and B-T are the same. The second negative peak originates from the vacancy of T. This peak area is used for the calculations similarly as described for HD CAE.

VCAE uses the same setup as VP CAE; however, the calculations are made on the mobility shift of T. Thus, VP CAE uses the area of the trough peak, and VCAE uses the migration time of the peak, taking the same care against variations on the separation conditions as described to ms ACE. Figure 11 shows a vacancy formation, typical of the HD CAE, VP CAE, and VCAE schematically.

### 4.6.4 Frontal Analysis CAE (FA CAE)

In FA CAE, the capillary filled with the BGE receives a large sample plug containing binding agent (B) and target (T) in equilibrium, i.e., B + T ⇆ B-T. As assumed in previous discussions, the B-T complex and B have approximately the same electrophoretic mobility, and the mobility of T differs largely from the mobility of the complex. When the separation voltage rises, T will develop higher electrophoretic velocity and will detach from the injection plug, forming a step-shaped separation. Through an external calibration method, the system concentration of free T in the sample is assessed through the height of the plateau, also allowing determining the dissociation equilibrium constant.

## 5    Conclusions and Future Trends

Capillary electromigration techniques are well-established and have solid theoretical ground. Together, these modes of operation when applied to a broad class of chemical compounds, in a variety of matrices, and with minimum sample treatment, show the potential of a great analytical tool. Because of these features and the compatibility with aqueous solutions, capillary electromigration techniques are especially suitable and powerful for bioanalysis, involving from small molecules to macromolecules, particularly in biofluids such as blood, plasma, serum, and urine. With the popularization of these techniques, particularly CZE, CIEF, and CSE, an increasing number of standard methods using capillary electromigration techniques should emerge. Portable and autonomous instrumentation and the microdevices depicted in Chap. 24 allow applying these electromigration techniques in the field and point-of-care.

## References

Baba Y, Matsuura T, Wakamoto K, Morita Y, Nishitsu Y, Tsuhako M (1992) Anal Chem 64:
    1221–1225
Busch MHA, Carels LB, Boelens HFM, Kraak JC, Poppe H (1997) J Chromatogr A 777:311–328
Costa BMC, Prado AA, Oliveira TC, Bressan LP, Muñoz RAA, Batista AD, da Silva JAF, Richter
    EM (2019) Talanta 204:353–358
da Silva JAF, Ricelli NL, Carvalho AZ, do Lago CL (2003) J Braz Chem Soc 14:265–268
Dittmann MM, Rozing GP (1996) J Chromatogr A 744:63–74
Galievsky VA, Stasheuski AS, Krylov SN (2015) Anal Chem 87:157–171
Ganzler K, Greve KS, Cohen AS, Karger BL (1992) Anal Chem 64:2665–2671
Gottschlich N, Jacobson SC, Culbertson CT, Ramsey JM (2001) Anal Chem 73:2669–2674
Gusev I, Huang X, Horvath C (1999) J Chromatogr A 855:273–290
Hilder EF, Svec F, Frechet JM (2004) J Chromatogr A 1044:3–22
Ismail A, d'Orlyé F, Griveau S, da Silva JAF, Bedioui F, Varenne A (2015) Anal Bioanal Chem
    407:6221–6226
Jaros M, Soga T, van de Goor T, Gas B (2005) Electrophoresis 26:1948–1954
Majors RE (2009) LC GC North America 27:1032–1039
Mamani MCV, Amaya-Farfan J, Reyes FGR, da Silva JAF, Rath S (2008) Talanta 76:1006–1014
Norton D, Zheng J, Shamsi SA (2003) J Chromatogr A 1008:205–215
Olabi M, Stein M, Watzig H (2018) Methods 146:76–92
Oldham KB, Myland JC (1994) Fundamentals of electrochemical science. Academic Press, San
    Diego, CA
Pan J-Z, Fang P, Fang X-X, Hu T-T, Fang J, Fang Q (2018) Sci Rep 8:1791
Quesada MA (1997) Curr Opin Biotechnol 8:82–93
Riekkola ML, Jonsson JA, Smith RM (2004) Pure Appl Chem 76:443–451

Righetti PG (2005) J Chromatogr A 1079:24–40
Ruiz-Martinez MC, Berka J, Belenkii A, Foret F, Miller AW, Karger BL (1993) Anal Chem 65: 2851–2858
Salplachta J, Kubesová A, Horká M (2012) Proteomics 12:2927–2936
Schmeer K, Behnke B, Bayer E (1995) Anal Chem 67:3656–3658
Slater GW, Mayer P, Drouin G (1996) Methods Enzymol 270:272–295
Svec F, Peters EC, Sykora D, Frechet JMJ (2000) J Chromatogr A 887:3–29
Terabe S (2009) Annu Rev Anal Chem 2:99–120
Terabe S, Otsuka K, Ando T (1985) Anal Chem 57:834–841
Xie MJ, Feng YQ, Da SL, Meng DY, Ren LW (2001) Anal Chim Acta 428:255–263
Zhang MQ, El Rassi Z (2006) J Proteome Res 5:2001–2008
Zhang CH, Woolfork AG, Suh K, Ovbude S, Bi C, Elzoeiry M, Hage DS (2020) J Pharm Biomed Anal 177:112882

# Introduction to Bioanalytical Mass Spectrometry

Daniel Nunes Martins, Pedro Henrique Vendramini,
Ana Valéria Colnaghi Simionato, and Alessandra Sussulini

## 1    Introduction

Mass spectrometry (MS) is a powerful analytical technique that has the ability to characterize molecular structures and quantify organic molecules or elements based on the measurement of the mass-to-charge ratio ($m/z$) of ions formed in the gas phase and their respective relative abundances (Murray et al. 2013). Since the experiment that measured the mass and charge of the electron, carried out in 1897 by the physicist J. J. Thomson (with the help of his laboratory assistant E. Everett in building the apparatus for magnetic deflection of cathode rays) and the study with natural stable isotopes performed by his student Francis Aston (Griffiths 2008; Paré and Yaylayan 1997; Finehout and Lee 2004), MS has experienced an impressive evolution in terms of accuracy, sensitivity, and selectivity. This progress involves the participation and effort of several scientists and, nowadays, MS is considered a reference technique in different science fields.

In analytical and bioanalytical chemistry, MS is the state-of-the-art technique for the identification and quantification of a wide range of compounds, such as sugars, proteins, oligonucleotides, and lipids, as well as elements (metals, metalloids, and non-metals) in complex biological matrices. An important milestone for the application of MS in bioanalysis occurred in the 1940s with the study of biological

D. N. Martins · A. Sussulini (✉)
Laboratory of Bioanalytics and Integrated Omics (LaBIOmics), Department of Analytical Chemistry, Institute of Chemistry, University of Campinas (UNICAMP), Campinas, SP, Brazil
e-mail: sussulini@unicamp.br

P. H. Vendramini
Laboratory of Neuroproteomics, Institute of Biology, University of Campinas (UNICAMP), Campinas, SP, Brazil

A. V. C. Simionato
Laboratory of Analysis of Biomolecules Tiselius (LABi Tiselius), Department of Analytical Chemistry, Institute of Chemistry, University of Campinas (UNICAMP), Campinas, SP, Brazil

© The Author(s), under exclusive license to Springer Nature Switzerland AG 2022
L. T. Kubota et al. (eds.), *Tools and Trends in Bioanalytical Chemistry*,
https://doi.org/10.1007/978-3-030-82381-8_23

processes, such as the production of $CO_2$ in animals, using heavy stable isotopes as tracers (Finehout and Lee 2004). Alfred Nier is considered the father of modern MS and supported many biologists by preparing [13]C enriched carbon compounds (Finehout and Lee 2004; Nier and Gulbransen 1939; De Laeter and Kurz 2006), making important contributions to the construction and development of mass spectrometers, especially the magnetic sector field mass spectrometer in 1940 (De Laeter and Kurz 2006). Furthermore, in partnership with Johnson, an instrument named Nier-Johnson mass spectrometer was developed, in which a symmetrical electrostatic mass analyzer and an asymmetric magnetic one resulted in a large angular divergence for the ion beam, optimizing the separation of the produced ions (De Laeter and Kurz 2006; Daolio et al. 1987). The works carried out by the American scientists Fred McLafferty, Klaus Biemann and Carl Djerassi (Finehout and Lee 2004) in the mid-1950s were extremely relevant for the use of MS in elucidating the fragmentation mechanisms of different classes of organic molecules, allowing the determination of unknown molecular structures, which also advanced the technique, especially in biological research (Griffiths 2008). The studies performed by these scientists also increased the industrial interest in MS, which until then was limited to qualitative analyses, due to the high cost of instrumentation and difficulty in using the technique. Biemann presented valuable works in which MS was applied to the elucidation of unknown structures in natural products, establishing rules for the comprehension of the fragmentation process of alkaloids and peptides, which was fundamental for later analysis of proteins and identification of peptide sequencing by MS (Griffiths 2008). Nevertheless, until the 1980s, MS was mainly applied to the analysis of small molecules. Only between the years 1980 and 1981, with the research conducted by Barber et al. (Gaskell and Gale 2015) with fast atom bombardment (FAB), an ionization technique that allowed the analysis of large molecules by bombarding thermally labile and intact polar compounds dissolved in a liquid matrix was established. Although this process promoted a rapid decomposition of the compounds of interest, the possibility of applying MS to the analysis of large molecules, such as proteins, moved scientists towards the development of new ionization techniques (Gaskell and Gale 2015). In 1984, Masamichi Yamashita and John Fenn (1984) presented the electrospray ionization (ESI) technique applied to the study of proteins. In 1985, Hillenkamp and Karas developed the matrix-assisted laser desorption/ionization (MALDI) technique, publishing a manuscript in 1988 reporting the application of MALDI for the analysis of proteins with molecular masses above 10,000 Da (Karas et al. 1985; Karas and Hillenkamp 1988). A few months before this publication, Koichi Tanaka also published an important manuscript reporting a similar method for the successful analysis of proteins and polymers (Tanaka et al. 1988). Then, new paths for MS have been opened, attracting the interest of biologists and the medical research community. The possibility of expanding the range of molecules to be analyzed with ESI and the advantages of producing isolated charged species of peptides and proteins with MALDI brought new developments focused on refined applications, such as the coupling of MS with separation techniques, like liquid chromatography (LC), gas chromatography (GC) and capillary electrophoresis (CE) (El-Aneed et al. 2009).

Mass analyzers, which are considered the heart of an MS system for promoting the separation of the ions according to their $m/z$ (El-Aneed et al. 2009), underwent important developments that allowed the improvement of obtained information about the species of interest. From the development in the 1950s of the quadrupole (Q) and the time-of-flight (TOF) (El-Aneed et al. 2009; Boesl 2017) to the development of the Orbitrap (OT) in the late 1990s (Eliuk and Makarov 2015), mass analyzers have become increasingly efficient. The expansion of the mass range coverage, the improvements in the scanning speed, transmission (the fraction of produced ions that reach the detector) and resolution, as well as the reduction of the mass error (Paré and Yaylayan 1997; Eliuk and Makarov 2015; Alaee et al. 2001; Solliec et al. 2015) are characteristics that enlarged the applications of MS, especially in bioanalytical chemistry. This fascinating history of scientific developments in MS has overlapped with important milestones in science, such as the Human Genome Project, which started in 1990 (Fieggen and Ntusi 2019), opening a new horizon for applications of MS in bioanalysis. From the determination of the genome (i.e., the sequencing of the DNA of an organism), it became necessary to expand the understanding of structurally complex biological systems phenotypes and their interactions with the environment (Fieggen and Ntusi 2019; Goodwin et al. 2016). In the 1990s and early 2000s, the fields of proteomics, metabolomics, and metallomics (Shah and Misra 2011; Klassen et al. 2017; Haraguchi 2004) emerged with the aim of elucidating the sets of proteins, metabolites, and elements, respectively, in matrices like cell cultures, tissues, and biofluids, in order to fully comprehend biological systems. The complexity of the interactions of organisms with the environment and the variety of biological processes made interdisciplinarity increasingly necessary, bringing analytical chemistry and biology closer. The possibility of understanding the molecular bases of pathologies, impacts of environmental degradation and pollution on human and animal health, the relationship between food and changes in cellular metabolism, as well as the search for biomarkers and new therapeutic targets in the context of personalized medicine, made omic sciences stand out as the main application of MS in bioanalytical chemistry.

Furthermore, mass spectrometry imaging (MSI), which emerged in the 2000s, revealed an immense potential for application in bioanalysis due to its ability to acquire molecular/elemental specific images in complex matrices such as human, animal, and plant tissues, as well as cell cultures (Wu et al. 2013; Becker et al. 2012; Prideaux and Stoeckli 2012; Weaver and Hummon 2013). This technique is extremely efficient for investigating the spatial distribution of biomolecules, such as peptides, proteins, lipids, and drugs, in addition to elements in different types of biological matrices. In the following sections, MS instrumentation and application in bioanalysis to omic sciences and imaging approaches will be discussed.

## 2 Instrumentation

The mass spectrometer consists basically of a sample introduction system, an ion source, mass analyzer(s), and a detector connected to a computing system for data acquisition and analysis (Klassen et al. 2017; Haag 2016) (Fig. 1). Bioanalytical

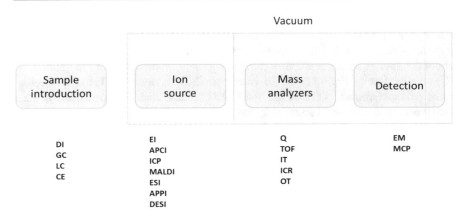

**Fig. 1** Basic components of a mass spectrometer with examples for each component. The dashed line shows the instrument parts possibly operating at low pressure: the white background area indicates operation at ambient pressure or vacuum, and the gray background area indicates operation at vacuum

applications of MS often require hyphenation of these systems with separation techniques, such as LC, GC, and CE, due to the complexity of biological samples (Maurer 2006; Scheffer et al. 2008). Ion sources that perform direct extraction of the analytes from the samples, such as MALDI, desorption electrospray ionization (DESI), and inductively coupled plasma combined with laser ablation (LA-ICP), can be used for mass spectrometry imaging, as further discussed in Section MASS SPECTROMETRY IMAGING (MSI). Direct infusion (DI) is a sample introduction option when fingerprinting approaches, which aim at obtaining general characteristic mass spectra for a sample type or condition, are employed (Kirwan et al. 2014; Lin et al. 2010; Rigano et al. 2019; Wilschefski and Baxter 2019). The most important component of the system is the mass analyzer, where the separation of ions occurs according to their $m/z$ (Daas 2007; Glish and Burinsky 2008). There are several mass analyzers currently available and suitable for different analyses according to the characteristics of the samples and analytes. Amongst the mass analyzers, Q, TOF, ion trap (IT), and Fourier transform mass analyzers (ion cyclotron resonance—ICR, and OT) (Haag 2016) are the most widely used in bioanalyses. In the following sections, mass spectrometer components will be discussed.

## 2.1    Ion Sources

MS analysis requires that the analytes of interest are converted to free ions into the gas phase. Different ionization techniques allow the transference of sufficient energy to promote the ionization of analytes (Rigano et al. 2019). The ionization process and the amount of applied energy define whether the ionization is soft or energetic, generating scarce analyte fragmentation (spectrum containing mainly molecular

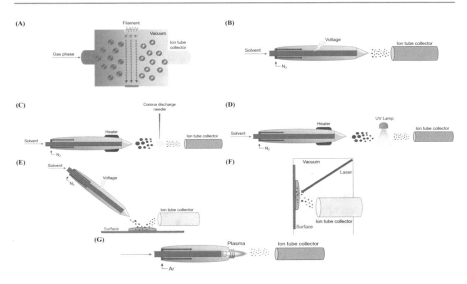

**Fig. 2** Simplified illustration of the main ion sources used in bioanalysis: (**a**) EI, (**b**) ESI, (**c**) APCI, (**d**) APPI, (**e**) DESI, (**f**) MALDI, and (**g**) ICP

mass information) or intensive analyte fragmentation (spectrum containing structural information), respectively (Rigano et al. 2019). Ionization techniques used in MS may be classified according to the ion sources pressure. For example, EI and MALDI operate under vacuum, while other sources, which do not operate under vacuum, are sub-classified into atmospheric pressure ion sources (e.g., ESI, APCI, APPI, and ICP) and ambient ion sources (e.g., DESI). The main ionization techniques applied to bioanalysis are presented in Fig. 2 as a simplified scheme and will be further discussed.

### 2.1.1 Electron Ionization (EI)

EI is the most common ionization technique in gas chromatography coupled to mass spectrometry (GC-MS). This technique was used in 1918 by Arthur Dempster and, later, optimized by Bleakney and Nier when it was named electron impact ionization (Traeger 2016). The principle of the technique that operates under high vacuum and produces extensive analyte fragmentation, providing structural characterization, is based on the production of a high-energy electron beam by an incandescent tungsten filament (McNair and Miller 1997). The beam is accelerated towards an anode, generally resulting in an energy of 70 eV, and collides with gas phase analytes previously separated by GC. The beam energy is generally higher than that from analytes chemical bonds, thus promoting the fragmentation of molecules, with consequent production of positively or negatively charged ions and neutral species in an excited electronic state (Traeger 2016; McNair and Miller 1997). Afterwards, ions are forced out of the ionization chamber by applying a positive charge to the repeller and focused into the mass analyzer, while negative ions are attracted by a positive cathode, and neutral fragments are ejected out of the system (McMaster 2008) (Fig. 2a).

Most EI applications in bioanalysis involve the use of GC-MS to the determination of metabolites extracted from plant, human, and animal tissues and diagnosis of inborn errors of metabolism (Kanani et al. 2008; Jiye et al. 2005; Hiller et al. 2009; Mastrangelo et al. 2015; Civallero et al. 2018). Commonly, EI associated with GC-MS has been considered as a complement to the use of LC-MS and CE-MS techniques to obtain a wider coverage of the global metabolic profile in biological samples (Mastrangelo et al. 2015).

## 2.1.2   Electrospray Ionization (ESI)

Since its first use reported in 1968 by M. Dole and the application to the study of proteins by Masamichi Yamashita and John Fenn in 1984 (Yamashita and Fenn 1984), ESI has allowed the analysis of polar, ionic, thermolabile, and high molecular mass compounds, including the determination of three-dimensional protein structures, non-covalent interactions, amino acid sequencing and post-translational modifications (Banerjee and Mazumdar 2012). In recent years, this technique has been commonly used in the analysis of biological molecules, especially for providing a significant amount of information about their structure and the straight-forward coupling to LC and CE (Pitt 2009).

ESI is a soft ionization technique based on the formation of an electrostatic spray from the analytes present in a liquid phase, which is pumped under constant flow through a heated capillary made of stainless steel or silica quartz (Banerjee and Mazumdar 2012; Pitt 2009; Kavadiya and Biswas 2018). In this ionization process, a high voltage source (1–7 kV) is connected at the end of the capillary, with concomitant nitrogen flow around it, causing the dispersion of the solution and consequent formation of an aerosol with charged droplets. As the solvent droplets evaporate, droplets radii decrease, with consequent charge density augment, resulting in droplet deformation and the release of ions into the gas phase—the Taylor cone process. Ions are directed to the skimmer cone and accelerated into the mass analyzer (Fig.2b) (Pitt 2009; Wilm 2011). ESI may form molecular ions, quasi-molecular ions, and anionized/cationized molecules, due to redox reaction, acid/base reactions, or coordination with cations or anions (Juraschek et al. 1999; Chen et al. 2019a), so that analyses may be performed in positive or negative ion mode. Adduct formation is a common effect in ESI, both in positive and negative modes. Although it is not a completely understood process, the effect is defined as the formation of an ion that contains all the atoms or molecules from the interaction of a precursor ion with other atoms or molecules (Murray et al. 2013). Many studies mention the possibility of controlling these adducts using additives such as sodium and lithium salts (Banerjee and Mazumdar 2012; Hu et al. 2013; Kruve and Kaupmees 2017; Ghosh and Jones 2015). In order to prevent ionization impairment or a difficult mass spectrum interpretation, ion suppression must be avoided by minimizing the salt amount in the sample preparation step, as well as in the mobile phase and background electrolyte (BGE) (for LC and CE, respectively) (Ghosh and Jones 2015). The use of detergents such as Triton X-100 and Tween 20 for protein extraction should also be avoided to prevent ion suppression (Šustáček et al. 1991; Annesley 2003).

In general, analytes that are readily ionizable are more appropriate for ESI. However, the analysis of compounds of varying nature may diverge in terms of ionization efficiency, resulting in different sensitivities according to the analytes that compose the sample. Hence, the prediction of ionization efficiency may be based on the pKa and molecular structure of the analytes (Annesley 2003; Oss et al. 2010). Samples in low concentrations (pmol $L^{-1}$) can be analyzed upon ESI and are often prepared with the addition of organic solvents to favor droplets formation and solvent evaporation. The addition of small amounts of weak acids (such as acetic acid and formic acid) or bases (such as ammonium hydroxide) aids analytes ionization (Oss et al. 2010). ESI has also been applied to the analysis of high molecular mass analytes since multi-charged ions formation is favored. Therefore, a wide range of biological molecules, such as intact proteins, DNA, hormones, organic polymers, and metabolites such as sugars, amino acids, carboxylic acids, and lipids, have been analyzed using ESI (Loo 1997; Beck et al. 2001; Immler et al. 1998; Chowdhury and Guengerich 2011).

### 2.1.3 Atmospheric Pressure Chemical Ionization (APCI)

APCI was developed in 1973 by Stillwell and coworkers (Horning et al. 1973), initiating a new generation of ion sources under atmospheric pressure. The mechanism of ionization occurs within a probe containing a heater block, which assists solvent droplet evaporation until analytes are in the gas phase. A corona discharge needle produces ions from air components, such as $N_2^+$, $N_4^+$ and $H_2O^+$, which transfer charge to the analytes through chemical reactions in the gas phase (Fig. 2c). Therefore, APCI generates low analyte fragmentation, resulting in mass spectra mostly composed of protonated ions, and is mainly applied for the analysis of small molecules (up to 1500 Da), which are easily protonated. Several applications of LC-MS using APCI to the determination of vitamins, drugs, and metabolites in biological matrices, such as human plasma, have been reported (Nannapaneni et al. 2017; Sangster et al. 2004).

### 2.1.4 Atmospheric Pressure Photoionization (APPI)

In 2000, Bruins and coworkers introduced the APPI source, which may be considered an APCI modification, since it works with the same solvent flow rate (200–2000 $\mu L$ $min^{-1}$) (Revelskii et al. 2019). APPI mechanism involves the incidence of photons generally produced by a Krypton lamp, which emits enough energy to ionize analytes, but not enough to ionize air components, promoting a mass spectrum with less interference. Photoionization may be performed directly in the analytes or indirectly, through dopants or solvents, with subsequent charge transfer to the analytes (Fig. 2d). Hence, protonated and radical cations are produced, as well as fragment ions (Kauppila et al. 2017). Nowadays, APPI is considered one of the three main ion sources at atmospheric pressure, in conjunction with ESI and APCI (Robb et al. 2000), and has been mainly applied for the analysis of natural products and drugs in human plasma (Xing et al. 2007; van den Broek et al. 2006; Coe et al. 2006).

### 2.1.5 Desorption Electrospray Ionization (DESI)

DESI was developed by Cook's group in 2004, constituting a pioneer work on ambient ion sources (Takáts et al. 2004). The ionization occurs via the incidence of microfilm of charged solvent on the sample, which promotes the desorption of analytes and subsequent transfer to the gas phase (Fig. 2e). The transference of ions from the liquid to the gas phase occurs under the same mechanism as for ESI (Badu-Tawiah et al. 2013). Since analyte desorption occurs by a charged solvent flush on the sample surface, the analytes must be compatible with the extracting solvent. Different organic solvents may be used, such as methanol (MeOH), acetonitrile (ACN), N,N-dimethylformamide (DMF), and others, combined or not with water (Wu et al. 2013). A derivatization step or addition of a dopant to the sample may be performed when analytes do not ionize properly, e.g., steroids (Wang et al. 2019; Wu et al. 2009).

DESI source parameters optimization (such as geometric parameters and capillary protrusion) is crucial to obtain the maximum signal-to-noise (S/N) ratio. On the other hand, the solvent flow rate is directly related to the type of support where the sample is displaced. Indeed, supports with high porosity and solvent adsorption, such as paper and polytetrafluoroethylene (PTFE), requires a high solvent flow (*ca.* 10 $\mu$L min$^{-1}$) to form the desorption microfilm, while supports like glass requires a low solvent flow (*ca.* 1 $\mu$L min$^{-1}$) (Schwab et al. 2014; Bodzon-Kulakowska et al. 2014). DESI is the most used ion source for ambient ionization and has been currently applied for imaging assays in life sciences. However, the main limitations of this source are performing quantitative and reproducible analysis, sensitivity, molecular coverage, and ion suppression (Kuo et al. 2020). Several references mention the use of DESI for the imaging of animal and human tissue, as well as drugs in biological fluids, thus constituting a useful technique for clinical research, such as in cancer studies (Abbassi-Ghadi et al. 2015; Wiseman et al. 2006, 2010).

### 2.1.6 Matrix-Assisted Laser Desorption/Ionization (MALDI)

MALDI is considered a soft ionization technique, which has been crucial in the analysis of biomolecules in several biological matrices, especially proteins. The principle of operation of this technique is based on the ionization of the compounds of interest dissolved in a matrix (usually organic acids, such as $\alpha$-cyano-4-hydroxycinnamic acid—$\alpha$-CHCA—and sinapinic acid) (Singhal et al. 2015), which transfers protons to the analytes when irradiated by a pulsed laser beam. The prepared samples are deposited on a plate suitable for MALDI, where matrix crystallization occurs upon solvent evaporation, co-crystallizing the samples. A pulsed laser beam, such as nitrogen laser or an Er:YAG (2.94 $\mu$m) operating in the ultraviolet or in the infrared range, respectively (Diehn et al. 2018; Costello 1999; Dreisewerd 2003), is applied on the sample spots promoting matrix and analytes desorption, with consequent volatilization. Subsequently, protons from the matrix are transferred to the analytes resulting in ionization and transition to the gaseous state (Costello 1999) (Fig. 2f). MALDI has a low analytical capacity for compounds with masses lower than 800 Da, due to the effect of matrix self-dissociation, in addition to reproducibility problems related to non-homogeneous crystallization (He et al. 2019).

The ions produced by MALDI are usually directed to a TOF mass analyzer, although energy propagation between ions results in wide peaks with consequent limitation of the resolution and precision of mass assignment. However, a delayed extraction system is frequently applied to minimize flight time dispersion, circumventing those issues and resulting in high sensitivity, high resolution, and low analyte fragmentation (Costello 1999). MALDI-TOF-MS systems have been applied to a variety of bioanalyses, from pathology studies involving the characterization of high molecular masses analytes (such as intact proteins and lipids in human tissues) to the characterization and differentiation of the microbial proteome (Singhal et al. 2015; Diehn et al. 2018).

### 2.1.7 Inductively Coupled Plasma (ICP)

Houk and coworkers (Houk et al. 1980; Houk 2010) published the pioneer application of ICP as an ion source for MS in 1980. In 1983, ICP-MS instrumentation was commercially available (Meermann and Nischwitz 2018), thus consolidating it as one of the main techniques for multi-elemental and isotopic analysis in a variety of matrices (from geological materials to biological samples) (Meermann and Nischwitz 2018; Liang et al. 2000). The principle of the technique consists of a source with high energy, where ions are produced by an argon plasma at atmospheric pressure, which promotes the decomposition of analytes into atoms (C-Celis and Encinar 2019) (Fig. 2g). Both liquid and solid samples may be analyzed by ICP-MS, using nebulization or LA for sample introduction into the plasma, respectively (Wilschefski and Baxter 2019). The nebulizer converts solutions to aerosols, which then pass through a spray chamber, removing most of the largest droplets, limiting the total charge of the solvent entering the ICP (Olesik 2014), and converting the analytes to free ions or atoms through the plasma. Most atoms are excited by electron collision, according to the corresponding concentration and speed, while gas temperature controls samples desolvation and vaporization (Olesik 1996). In LA-ICP-MS, a laser beam is focused on the solid, desorbing the material and producing an aerosol that is transferred to the ICP with the aid of an argon gas stream for elemental analysis (Sylvester and Jackson 2016). ICP ion source has been widely used in metallomics and quantification of metals, non-metals, and metalloids in plant and human tissues or fluids (Morton et al. 2017; de Vlieger et al. 2011; Limbeck et al. 2015).

## 2.2 Sample Introduction and Hyphenated Separation Techniques

There are several advantages in hyphenating separation techniques to the MS system, especially when complex matrices, such as biological samples, are manipulated. The main advantages of such hyphenation are the previous separation of the analytes present in the biological matrix before introduction into the ion source, higher selectivity, increased accuracy for analytes identification, interfering compounds removal (with consequent improvement of mass spectra data quality), reduction of ion suppression, among others (Daas 2007).

GC-MS is applied to the analysis of samples composed of volatile and thermally stable molecules, as they are vaporized in the injector port and carried by a gaseous mobile phase (generally helium) into a chromatographic column, where analytes are separated and then introduced into an ionization chamber, promoting both analytes ionization and possible fragmentation. GC-MS has several advantages, such as low cost (compared to MS hyphenations with LC and CE), high compatibility with EI source (since analytes separation occurs in the gas phase), excellent chromatographic resolution and fragmentation pattern, as well as the use of complete spectral libraries, such as the one from National Institute of Standards and Technology (NIST) (Kopka 2006; De Leenheer et al. 1986).

Despite several advantages, GC-MS also has a limitation for the analysis of low volatile/polar metabolites and other thermally unstable compounds, which may decompose at elevated temperatures, requiring a derivatization procedure that usually occurs in two stages, namely oximation and silylation (Miyagawa and Bamba 2019). This process may present some drawbacks such as incomplete derivatization, manipulation of toxic reagents—such as pyridine, trimethylsilyl (TMS), N,O-bis-(trimethylsilyl)-trifluoroacetamide (BSTFA), and N-methyltrimethylsilyltrifluoroacetamide (MSTFA), in addition to reproducibility problems related to the extensive sample preparation step. Besides, the higher the sample preparation complexity, the more prone to error propagation becomes the method. Coupling of MS to comprehensive two-dimensional gas chromatography (GC×GC-MS) presents superior resolution power due to the use of two columns with different dimensions and stationary phase compositions connected by a modulator, allowing separation of specific fractions transferred from the first column to the second one (Purcaro et al. 2017; Rosso et al. 2020; Bahaghighat et al. 2019). Even with all the advantages presented by this technique, its use is still limited by data analysis/interpretation complexity.

In comparison to GC-MS, the hyphenation of LC with MS has the advantage of identifying a wider range of compounds with extremely different molecular masses, without restrictions of volatility and thermal stability (Rigano et al. 2019; Perez et al. 2016). Other advantages reside in the possibility to perform proteomics analysis and no requirement of prior sample derivatization, resulting in little sample handling, thus making it the method of choice in most bioanalytical chemistry studies. Since LC-MS employs a liquid mobile phase, the use of interfaces between the separation system and the ion source is necessary in order to remove solvent excess while transferring the maximum number of eluted compounds, promoting an efficient ionization process with concomitant maintenance of the MS vacuum system (Daas 2007). Recently, improvement of LC systems resulted in ultrahigh-performance liquid chromatography (UHPLC), where analytes are separated at higher pressures than in HPLC, using shorter columns containing smaller diameters of stationary phase particles ($<1.7$ µm), yielding excellent resolution and system stability.

NanoLC, where columns with capillary dimensions are used, has also emerged as a new technique, and it is widely applied to proteomics (Ishihama 2005).

CE-MS has also been used in bioanalysis. It is a simple system consisting of a silica capillary (usually covered with polyimide), an energy source for potential application, electrodes to connect BGE vials to the energy source and conduct the required potential, a cooling system to reduce the temperature gradient inside the capillary caused by Joule effect, a sample introduction system and a detector system (Godzien et al. 2019). The coupling of CE to MS is based on the ion sources used for LC-MS. However, due to CE particularities, a sheath liquid or a sheathless interface for CE-MS coupling is required (Hirayama et al. 2018). The advantages of CE-MS reside in the simple sample preparation step, the use of small sample volumes (in the order of nL) and BGE (in the order of µL) per analysis, the low consumption of organic solvents, and the high speed of analysis. Moreover, it is a complementary analytical technique to GC-MS and LC-MS regarding the classes of analytes that are possible to be analyzed, ensuring a higher analyte coverage in untargeted metabolomics investigations, for example. However, CE-MS has not been used as often as LC-MS and GC-MS (Stolz et al. 2019; Gahoual et al. 2019) due to its lower reproducibility, robustness, and more recent development. Therefore, with CE-MS system consolidation and companies' commercialization, the technique will certainly be widespread used.

## 2.3   Mass Analyzers and Detectors

Ions formed in the ion source are separated according to their respective $m/z$ in mass analyzers (Haag 2016), which may be classified according to ions introduction (continuous or pulsed modes), ions transmission (ion beam or ion trapping), among others (Hoffman and Stroobant 2007; Niessen and Falck 2015).

In tandem mass spectrometry (MS/MS), fragment ions from the analytes are generated, thus enriching mass spectra information and performing qualitative analysis. Mass analyzers may be used sequentially (for example, QqQ) or different types can be combined (hybrid instruments), such as quadrupole time-of-flight (QTOF). Tandem mass spectrometry allows the selection of a precursor ion in the first analyzer, followed by the dissociation in a collision cell, and the transmission and separation of the product ions into a subsequent analyzer. This process occurs in spatially separated analyzers—MS/MS in space. Alternatively, the fragmentation steps may occur within a single mass analyzer (trapping instruments, such as IT, FT-ICR, and OT) in multiple stages ($MS^n$, with $n \geq 2$), where the product ions are analyzed at different time intervals—MS/MS in time (Glish and Burinsky 2008; Niessen and Falck 2015). Methods used to obtain precursor and fragment ions mass information are based on data-dependent acquisition (DDA) or data-independent acquisition (DIA). In DDA methods, MS/MS fragmentation spectra are acquired automatically only for the selected precursor ions, whereas in DIA, MS/MS fragmentation spectra are acquired for all precursor ions (Rutledge et al. 2016). Table 1

**Table 1** Characteristics of mass analyzers and most common applications

| Mass analyzer | Resolution $(m/z/\Delta m)^a$ | Tandem MS | Most common application |
|---|---|---|---|
| Quadrupole (Q) | 1 | $MS^2$ | Quantitative and qualitative analysis |
| Time-of-flight (TOF) | 70,000 | $MS^2$ | Qualitative analysis |
| Fourier transform-ion cyclotron resonance (FT-ICR) | 10,000,000 | $MS^n$ | Qualitative analysis |
| Ion trap (IT) | 1 | $MS^n$ | Quantitative and qualitative analysis |
| Orbitrap (OT) | 1,000,000 | $MS^n$ | Quantitative and qualitative analysis |

$^a$Resolution at $m/z$ 200

summarizes some important characteristics of selected mass analyzers, as well as their main applications.

The Q mass analyzer may be considered a mass filter due to the ability to discriminate and filter ions of different $m/z$ (Haag 2016; Niessen and Falck 2015). It consists of four rods in cylindrical or hyperbolic shape, arranged in two pairs of parallels. A potential of direct current (DC) (El-Aneed et al. 2009; Haag 2016; Daas 2007) is applied to a pair of parallel rods, while a radiofrequency (Rf) potential is applied to the other pair. The ions formed in the ion source are accelerated towards the Q by an electric field in the range of 5 kV so that an ion that is positively charged is moved towards a negatively charged rod (El-Aneed et al. 2009). However, the repetitive alteration of rods polarities alters the ion path before it reaches the rod, making it go through a spiral trajectory towards the detector. Ions whose trajectory is unstable collide with one of the rods and do not pass through the filter (El-Aneed et al. 2009; Niessen and Falck 2015). Therefore, the selection of ions $m/z$ to be monitored before the analysis will be converted to the corresponding rods potentials in order to avoid such ions collision, with the consequent resonance of the selected ions until they reach the detector.

TOF mass analyzers consist of ions undergoing high-speed acceleration at the ion source launched into a flight tube (Daas 2007; Niessen and Falck 2015). If two ions with different $m/z$ are accelerated at the source with the same kinetic energy, both will pass through the whole flight tube, reaching the detector at different times. Indeed, all ions inserted into the flight tube will reach the detector, constituting one of the advantages of this mass analyzer (Niessen and Falck 2015). Equation (1) correlates $m/z$ with total flight time $(t_f)$:(El-Aneed et al. 2009)

$$m/z = t_f^2\, 2Es/(2s + x) \qquad (1)$$

where $E$ is the applied voltage, $s$ is the length of the ion acceleration region, and $x$ is the length of the free-flight region.

IT mass analyzer is a modification of Q (Niessen and Falck 2015), since the separation of ions occurs according to time instead of space, in which two hyperbolic electrode plates are facing each other, and a hyperbolic ring is placed between them.

An oscillating Rf field superimposed by an electric field produced by a DC (El-Aneed et al. 2009) traps the produced ions between the electrodes. The variation in the Rf ejects ions of different $m/z$ from the trap towards the detector (El-Aneed et al. 2009). For decades, 3-D IT has been the most used IT mass analyzer. However, 2-D IT, also known as linear trap, has currently become more popular. In linear traps, ions are confined to an axial dimension by applying an electrical potential at the end of the trap and radially by an Rf field through a quadrupolar field (Niessen and Falck 2015). OT and FT-ICR are both mass analyzers that use the Fourier transform so that a signal produced by ions that oscillate within a trap is converted from the time domain to the frequency domain (El-Aneed et al. 2009). However, in OT, there is no application of a magnetic field, and the detection of the image current produced by ions that perform an axial oscillation is induced by the application of an electric field while they spin around a cylindrical internal electrode (Niessen and Falck 2015; Du and Douglas 1999). On the other hand, FT-ICR, developed by Alan Marshall and Melvin Comisarow (Griffiths 2008), is a mass analyzer based on the detection of image current from the movement of ion cyclotron coherently excited and on the effects of the electric and magnetic fields on the trajectories of charged particles (Marshall and Hendrickson 2002; Abboud et al. 1999).

High-resolution mass spectrometry (HRMS) enables the measurement of the exact masses of ions, including those originated from analytes present in complex matrices, such as biological samples. It is also a valuable tool for the determination of a synthesized or purified compound's exact mass (Xian et al. 2012). Some examples of HRMS analyzers are QTOF, OT, and FT-ICR.

The separation and identification of molecules can still be improved by the use of ion mobility (IM), a technique that is based on differences in the mobility of ions under an electric field capable of promoting ions separation in the gas phase (Cumeras et al. 2015). IM separates the ionized molecules according to size, shape, charge, and mass, which result in different ion mobilities when submitted to low or high electric fields (Leaptrot et al. 2019). This technique allows the separation of molecules due to differences in two-dimensional collision cross-section (CCS), defined as an average measure of the cross-sectional area of the ion in the gas phase, which allows differentiation of isomers, for example (Leaptrot et al. 2019; Hu and Zhang 2018). Currently, IM is divided into traveling-wave IM, drift-time IM, and field asymmetric IM (Hu and Zhang 2018).

After the separation in mass analyzers, ions are directed to the detector, where measurement of ions relative abundances occurs, and a mass spectrum is generated by an acquisition system (software). Currently, electron multiplier (EM) and microchannel plate (MCP) are the most used detectors (Koppenaal et al. 2005). The EM is made up of dynodes that are bombarded by positive or negative ions, causing the emission of secondary particles containing secondary ions, electrons, and neutral particles. Secondary particles are accelerated in the EM dynodes with enough energy to dislodge electrons that collide with the subsequent dynodes, producing more electrons (an electron cascade) (Koppenaal et al. 2005). The MCP consists of an array of continuous channels organized in parallel that act as electron multipliers when these are generated by the incident ions (Koppenaal et al. 2005;

Gilmore and Seah 2000). When an ion strikes the emissive surface near the entrance of the microchannel, an electron burst begins, and the cloud of ions exiting the microchannel is directed to an anode so that the generated electric current is measured (Koppenaal et al. 2005).

## 3    Mass Spectrometry-Based Omics

Omic sciences constitute an important field of study composed of several approaches that provide valuable information about biological processes derived from cellular activity and their interaction with the environment in an organism. The four main pillars of molecular biology are based on the study of genes, transcripts, proteins, and metabolites. Therefore, a comprehensive analysis of all species would be impossible using a single strategy, either by limitation of the selected analytical technique or by challenges imposed by the biological matrix (Klassen et al. 2017).

There are five main omic strategies, named genomics, transcriptomics, proteomics, metabolomics, and metallomics, as well as some sub-areas of them. Downward the omics cascade (Fig. 3), the system complexity increases since it is directly associated with the phenotype, making the metabolome the most complex omics area (since it is composed of a wide variety of biological molecules functions in a diverse concentration range), while the genome is the least complex one (since it is basically composed by four nucleotides) (Peedicayil 2019; Biz et al. 2019; Lemberger and Babu 2019). After the success of the Human Genome Project (Fieggen and Ntusi 2019; Goodwin et al. 2016; Shah and Misra 2011), studies in the field of systems biology were rapidly and significantly developed (Goodwin et al. 2016), using holistic and systematic research aiming to understand the molecular mechanisms involved in several biological processes, and how external agent factors can promote changes in the biological system. The following sessions will describe the main MS-based omic approaches.

## 3.1    Metabolomics

Metabolomics aims to study low molecular mass biomolecules (up to 1500 Da) that constitute the metabolome, which is defined as the set of intermediate or final compounds produced by cellular processes and their interactions with the environment, involved in biochemical reactions catalyzed by enzymes, and are found in cells, tissues, and biological fluids (Khan et al. 2019).

The necessity of understanding the role of the metabolome in biological systems, their altered levels and the metabolic pathways involved in these changes in the intracellular (fingerprinting) and extracellular (footprinting) media (Klassen et al. 2017) conducted metabolomics to a significant increase in scientific production and MS became a reference technique for metabolomics research. Currently, metabolomics has provided information applied to several areas, such as clinical (Montaner et al. 2020; Badhwar et al. 2020; Eicher et al. 2020; Chen et al. 2019b;

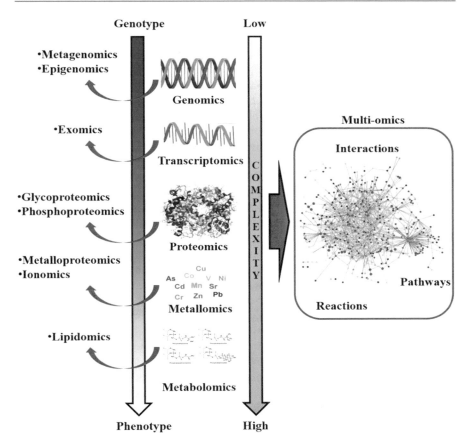

**Fig. 3** Omic sciences: genomics, transcriptomics, proteomics, metabolomics, metallomics, and sub-areas. Omics-interactions and pathways constitute multi-omics investigation

Peña-Bautista et al. 2019), environmental (Wei et al. 2020; Liu et al. 2018), nutrition and food (Rizo et al. 2020; Li et al. 2020; Suárez et al. 2017), sports (Bongiovanni et al. 2019; Harshman et al. 2019), plant physiology (van Der Hooft et al. 2020; Feng et al. 2020), parasitology and microbiology (Li et al. 2020; Dunham et al. 2017; Xu 2017; Wang et al. 2018), and forensic toxicology (Steuer et al. 2019; Dinis-Oliveira 2019; Nielsen et al. 2016).

In the 1940s, the scientist Roger Williams proposed that individuals had a metabolic profile characterized by their body fluids, and throughout the 1950s and 1960s, several studies attempting to identify metabolic pathways were carried out. In 1955, Donald Nicholson joined several metabolic reactions in a single panel, displaying for the first time a compilation that presented a comprehensive cellular metabolome (Zamboni et al. 2015; Dettmer et al. 2007). In 1964, Ernest Boris Chain was awarded a Nobel Prize for his achievements in the elucidation of biochemical pathways (Zamboni et al. 2015).

Only in the late 1990s and early 2000s, with works presented by Oliver, Nicholson, and Fiehn, the term metabolome was established, and studies on alterations in the levels of metabolites caused by different biological processes rapidly increased (Klassen et al. 2017; Fiehn 2002; Oliver et al. 1998). Despite all the achieved advances in recent years about development and optimization of analytical platforms applied to metabolomic studies, it still presents several challenges related to the comprehension of the entire metabolome of a biological system due to sample complexity, as well as the chemical diversity and different concentrations of metabolites (Kuehnbaum and Britz-McKibbin 2013; Fiehn et al. 2007).

In the last years, several metabolomics investigations have employed MS tools, highlighting the hyphenation with separation techniques (GC-MS, LC-MS, CE-MS), and, to a lesser extent, the direct sample introduction (DI) into the ion source. Generally, ESI is the most used ion source for LC and CE coupling with MS, and EI is the most used one for GC-MS, while QqQ, OT, IT, and QTOF are within the most used mass analyzers (Arevalo Jr et al. 2020; Begou et al. 2017).

Metabolomics investigations may be divided into untargeted and targeted approaches. Untargeted metabolomics aims to carry out an exploratory and qualitative analysis on a biological sample to obtain as much information as possible about the set of metabolites. For this purpose, sample preparation involves minimum steps to avoid analytes loss, usually precipitating/removing proteins with cold solvents, such as MeOH. Mass spectrometers monitor the ions on the scan mode, and mass analyzers must have high resolving power, wide mass range, and high mass accuracy (Klassen et al. 2017). Untargeted metabolomics analyses result on several molecules (molecular features, characterized by their $m/z$ and retention time), which endeavors the researcher towards metabolites identification by a thorough analysis before searching in public databases. Afterward, metabolomics data is subjected to a careful statistical analysis to determine which metabolites are biologically significant within the investigated cohort (Rutledge et al. 2016; Arevalo et al. 2020; Begou et al. 2017; Manier et al. 2019; Vinayavekhin and Saghatelian 2010). The use of hyphenated separation techniques allows adopting other strategies that maximize the capacity of metabolites identification, ensuring that they are ionized at different times after the previous separation, and also considering their polarity characteristics with the use of appropriate stationary phases. For instance, columns with the octadecylsilane stationary phase (C18) used in reversed-phase liquid chromatography (RPLC) are applied for the analyses of low to medium polarity analytes, while columns with stationary phases for hydrophilic interaction liquid chromatography (HILIC) are applied to the analysis of polar metabolites. Analyses performed by GC-MS or GC×GC-MS also allow the selection of columns with different stationary phases (Rosso et al. 2020; Bahaghighat et al. 2019; Liu et al. 2018). Targeted metabolomics is intended for quantitative and semi-quantitative analysis using internal standards to assist the quantification of specific metabolites or even a set of metabolites, i.e., the investigation initially proposes the search for pre-defined targets. Comprehension about enzyme activities, metabolic pathways, and their final products, and data obtained in untargeted studies allows the definition of specific targets to be monitored in a biological matrix (Rutledge et al. 2016; Zamboni et al. 2015; Manier et al. 2019; Vinayavekhin and Saghatelian 2010; Roberts et al. 2012; Gowda and Djukovic 2014). The characteristics of mass analyzers, such as average efficiency, time, and speed of scanning cycles, are sufficient to perform targeted metabolomics analyses. A significant part of targeted studies uses the hyphenation of MS with separation techniques, especially GC-MS and LC-MS.

Generally, analyses using Q mass analyzers are performed in the selected ion monitoring (SIM) mode, which allows an increase in sensitivity by selecting and filtering only one or a few ions. On the other hand, the QqQ mass analyzer allows the performance of selected reaction monitoring (SRM) experiments, where the selection of a specific precursor ion in the first quadrupole is performed, followed by fragmentation in the second quadrupole (acting as a collision cell using an inert gas), and monitoring of specific fragments in the third one (Vidova and Spacil 2017; Liu et al. 2019).

As previously mentioned, MS-based metabolomics strategy has been used to develop studies in different fields, from health to plant biology, exploiting the full potential of MS tools and hyphenated techniques, from the extensive use of chromatography to columns that allow the separation of compounds focused on their physicochemical characteristics.

In the clinical area, there are countless applications, especially with a focus on the elucidation of the etiology of several pathologies, the search for biomarkers aiming the development of more accurate clinical exams, the mechanisms of action of drugs, the exposure of an organism to environmental factors, pathogen-host relationships, and the correlation between metabolism and immunity (Klassen et al. 2017; Khan et al. 2019; Montaner et al. 2020; Badhwar et al. 2020; Eicher et al. 2020; Chen et al. 2019b; Peña-Bautista et al. 2019). A review article published in 2019 by Badhwar et al. focused on a multi-omics approach applied to the discovery of multimodal biomarkers in biological fluids to evaluate the treatment and development of new drugs for Alzheimer's disease (Badhwar et al. 2020). MS was used to build metabolite panels from the discriminating metabolites found in studies, using LC-MS, GC-MS, and LC-MS/MS, identifying compounds such as arachidonic acid, *N,N*-dimethylglycine, and thymine associated with the metabolism of lipids, amino acids and nucleic acids, suggesting a metabolic dysregulation at the beginning of Alzheimer's disease (Badhwar et al. 2020). Chen et al. evaluated the potential of metabolomics and its workflow charts to assess possible changes related to breast cancer, which still presents many challenges for obtaining an early diagnosis, treatments, and monitoring the prognosis of the disease (Chen et al. 2019b). In this article, metabolites such as carnitine, lysophosphatidylcholine, proline, alanine, and 2-octanedioic acid were evaluated in urine samples using LC-MS and GC-MS to compare groups of breast cancer patients and healthy individuals, as well as comparing strains of less aggressive cancer cells with more aggressive ones.

Metabolomics also stands out in the environmental field for evaluating the response of organisms to the environment in which they are inserted or in experiments that simulate such environments, or the exposure to pollutants and pathogens under controlled conditions. In 2020, Wei et al. evaluated the health risks caused by contamination by hydroxylated polybrominated diphenyl ethers associated with problems such as thyroid hormone disorders, estrogen effects and neurotoxicity (Wei et al. 2020). Hydroxylated polybrominated diphenyl ethers, widely used as flame retardants in textile industries, electronic circuit boards and plastics, are found in several samples such as human, animal, and environmental matrices. The investigation carried out in a review article (Wei et al. 2020), highlighted the importance of MS, mainly in applications for GC-MS and LC-MS for the identification of metabolites of polybrominated diphenyl ethers and hydroxylated polybrominated diphenyl ethers.

In 2018, Liu et al. evaluated samples of marine medaka fish for the hepatotoxicity of the drug sulfamethazine, which is widely used in agriculture and has persistence in the marine environment, using GC×GC-TOF-MS (Liu et al. 2018). The increased level of amino acids, such as alanine, threonine, lysine, tyrosine, among others, revealed an important biological role of amino acids during fish exposure in altering energy, oxidative stress and disorders of immune function (Liu et al. 2018).

In addition to the examples previously mentioned, metabolomics has several applications in the most varied research topics within bioanalysis, some of which are listed in Table 2. Another important area that emerged in 2003 is lipidomics (Yang and Han 2016), a term coined initially by Han and Gross (Hsu 2018), which is applied to study the lipidome (the set of lipids in a biological system), and its performance in several cellular and metabolic processes. Lipids can be divided into several classes and subclasses (Hsu 2018) and lipidome comprehension requires the use of techniques that allow its identification and quantification in biological matrices.

**Table 2** Examples of MS bioanalytical applications to metabolomics

| MS techniques | Application | Metabolites | References |
| --- | --- | --- | --- |
| LC-QTOF-MS/MS, LC-Orbitrap-MS/MS, MALDI-TOF-MS, GC-MS, and CE-MS | Nutrition and food | Amino acids, fatty acids, carbohydrates, organic acids, and phenolic compounds | Rizo et al. (2020), Li et al. (2020), and Suárez et al. (2017) |
| GC-TOF-MS, LC-TOF-MS, and HILIC-MS | Sports | Lipids, amino acids, acylcarnitines, and organic acids | Bongiovanni et al. (2019) and Harshman et al. (2019) |
| LC-MS, GC-MS, CE-MS, GC-TOF-MS, and MALDI-TOF-MS | Plant physiology and natural products | Flavonoids, terpenes, saccharides, and vitamins | van Der Hooft et al. (2020) and Feng et al. 2020 |
| MSI, LC-MS, GC-MS, and CE-MS | Parasitology and microbiology | Alkyl quinolines, amino acids, and fatty acids | Li et al. (2020), Dunham et al. (2017), Xu (2017), and Wang et al. (2018) |
| LC-MS, GC-MS, HILIC-MS, GC-MS/ MS, and LC-MS/MS | Forensic toxicology | Psychoactive substances, carbohydrates, lipids, amino acids, steroids, fatty acids, and acylcarnitines | Steuer et al. (2019), Dinis-Oliveira (2019), and Nielsen et al. (2016) |

MS-based strategies are employed in lipidomics and are the most efficient for identification and determination of lipid profiles (Hsu 2018), especially the coupling with LC, the direct infusion shotgun MS often combined with ion sources such as ESI and HRMS such as QTOF, OT, and FT-ICR, in addition to MSI for the analysis of human, animal, and plant tissues. Due to the high complexity and structural variety of lipids, IM has been increasingly used in lipidomic applications, allowing efficient separation of isobaric/isomeric lipid species (Leaptrot et al. 2019; Hu and Zhang 2018).

## 3.2    Proteomics

Proteomics is a complex and important approach within the omic sciences that investigates the proteome, defined as the set of proteins encoded by the genome and other proteins produced by post-transcriptional and post-translational processes (Ahrens et al. 2010). The human genome can synthesize *ca.* 1.8 million different protein species (Jensen 2004). The study of the proteome is extremely complex, presenting numerous challenges and applications to understand the interaction between proteins that have different functions and how they act within biological systems. Proteomic approaches have numerous applications in different fields, making MS an extremely relevant technique in its development and the ability to provide answers to detailed scientific investigations.

The term proteomics was defined for the first time in 1994 by Marc Wilkins (Parker et al. 2010) and has achieved many advances in recent years, especially due to the development of improved mass analyzers (Yamashita and Fenn 1984; Karas et al. 1985; Karas and Hillenkamp 1988; Tanaka et al. 1988). Nowadays, MS has been used as a reference technique for protein sequencing and identification, determination of molecular mass, and intact protein complexes structures, using either ESI or MALDI, which may be considered as complementary ionization techniques. However, ESI may hamper the analysis of intact proteins due to mass range limitation, while MALDI has a practical mass limit of 250 kDa (Parker et al. 2010). Such ion sources are combined with mass analyzers like IT, QTOF, FT-ICR, and OT (Lange et al. 2008; Piehowski et al. 2013). Some studies report the combination of IM with ESI-MS as an appropriate method to study conformational changes in protein complexes (Young et al. 2016; Woodall et al. 2019).

Several MS/MS methodologies are available for peptide sequencing, allowing the analysis of small peptides obtained by the previous cleavage of proteins by proteases (bottom-up strategy) or the analysis of intact proteins to be fragmented within the mass spectrometer (top-down strategy) (Kelleher et al. 1999; Gregorich et al. 2014). The bottom-up strategy (Amunugama et al. 2013) is used to perform untargeted analysis and identify as many proteins as possible (Saraswathy and Ramalingam 2011; Zhang et al. 2019). It is based on the cleavage of proteins that generates a mixture of peptides, preceded by protein extraction from the biological matrix and purification/fractionation, which are usually performed by fast protein liquid chromatography (FPLC) (Schuster et al. 2012) and two-dimensional gel electrophoresis (2-DE) (Strader et al. 2006; David et al. 2007; Lee and Lee 2004).

Proteins extracted from biological matrices are generally digested by trypsin (Strader et al. 2006) with cleavage occurring at C-terminal regions of lysine and arginine residues, and the resulting peptide mixture is analyzed by MALDI-MS or ESI-MS (in this case, a previous chromatographic separation—usually by nanoLC—is employed before the peptides are directed to the ion source) (Young et al. 2016; Aburaya et al. 2017). Protein identification can be performed by peptide mass fingerprinting (PMF), where the exact masses of small peptides are determined, and a comparison of mass spectrum peaks obtained experimentally with those listed in databases is performed, limiting the analyses by PMF to those proteins previously inserted in databases (Saraswathy and Ramalingam 2011). This strategy is usually employed when MALDI-TOF-MS is used. Shotgun proteomics is carried out by nanoLC-ESI-MS/MS for the determination of peptides amino acid sequences and masses. Tandem mass spectra are compared with those predicted from a sequence database, and the combination of multiple peptide identifications allows protein identification.

The top-down approach is used for the analysis of intact proteins by MS, providing structural and post-translational modification (PTM) information, as well as evaluation of amino acids polymorphisms (Armirotti and Damonte 2010). In this case, the mass spectrometer fragments the proteins to obtain the respective peptides. Typically, ESI combined with IT, FT-ICR, or OT are employed in this strategy, since it provides enough energy to break the chemical bonds of proteins.

The fragmentation of peptides and intact proteins can be accomplished by different mechanisms (Tabb et al. 2004; Pingitore et al. 2004), such as collision-induced dissociation (CID), high-energy collision dissociation (HCD) (Tabb et al. 2004; Pingitore et al. 2004; Jedrychowski et al. 2011), electron transfer dissociation (ETD)(Lee and Lee 2004) and electron capture dissociation (ECD) (Kim and Pandey 2012). CID and HCD are the most used processes for the analysis of peptides, while ETD and ECD are mainly applied to the analysis of intact proteins (although HCD can also be applied) (Parker et al. 2010). CID is a low-energy process where the ion is excited by the conversion of kinetic energy into internal energy due to the collision of ions with inert gas (Zhurov et al. 2013; Gstaiger and Aebersold 2009; Ngoka and Gross 1999), generating fragment ions. This process results in the cleavage of the amide bond, forming type "b" fragments ions (containing terminal N) and "y" (containing terminal C from the acylium-ion site) (Parker et al. 2010; Gstaiger and Aebersold 2009; Kim et al. 2010; Tabb et al. 2003; Aebersold and Mann 2003). HCD is a fragmentation process similar to CID; however, it uses higher energies. ETD and ECD are fragmentation processes that involve the transfer of electrons from an anion to molecules with multiple protonations, with consequent release of energy and reduction of the ion charge, forming an odd-electron ion (Kim and Pandey 2012).

Quantitative proteomics may be absolute when the exact concentration of a protein is determined, or relative, when the levels of a specific protein are compared in different conditions to determine whether there are alterations in their relative abundances (Wasinger et al. 2013). There are many challenges in using quantitative methods in MS-based proteomics to understand systems biology and personalized medicine, but this methodology potentially provides valuable data from the proteome, improving information about the biochemical state of a biological system (Schubert et al. 2017). In 1999, Oda et al. performed a proteomics analysis for the identification and simultaneous relative quantification of individual proteins using whole cell stable isotope labeling for the evaluation of modifications levels at proteins specific sites (Oda et al. 1999; Tang et al. 2020). Since then, several articles have been published using relative quantification methods, where proteins or peptides abundances are compared in different samples of biological fluids, tissues, and plants (Tang et al. 2020; Liu et al. 2020a, b; El-Khateeb et al. 2019; Xia et al. 2020). Stable isotope labeling by/with amino acids in cell culture (SILAC) (Gstaiger and Aebersold 2009; El-Khateeb et al. 2019), isotope-coded affinity (ICAT) (Gstaiger and Aebersold 2009), isotope-coded protein label (ICPL) (Gstaiger and Aebersold 2009), and isobaric tags for relative and absolute quantification (iTRAQ) (Gstaiger and Aebersold 2009; Tang et al. 2020; El-Khateeb et al. 2019) are the most used relative quantification methods. Targeted (absolute) methods, although limited to some proteins, present sensitive and quantitative capacity to provide answers based on hypotheses and clinical studies aimed at investigating possible biomarkers (Schubert et al. 2017). There is a variety of applications of proteomics in bioanalysis. Some examples include the study of the pathogen-host relationship (Tang et al. 2020) and the pathogenesis of diseases in animals with economic importance (Liu et al. 2020a).

El-Khateeb et al. evaluated quantitative proteomic methods applied to translational pharmacology, an area that requires the extrapolation of in vitro observations to predict the results of in vivo therapies to assist the development of model-based drugs (El-Khateeb et al. 2019). The authors report that LC-MS has been fundamental in quantifying individual protein levels in heterogeneous biological matrices, especially when the lack of specific substrates for enzymes, receptors, and transporters makes the use of abundance data an approach for in vitro-in vivo extrapolation. In addition, the sensitivity and selectivity of MS are essential for enzymes and pharmacologically active transporters, which are found at low levels in tissue membranes and cellular systems to be detected and quantified. Still, in the clinical and pharmacological area, Xia et al. used MS-based proteomics to try to elucidate the molecular mechanism by which the drug metformin inhibits the proliferation and invasion of cervical cancer cells, identifying 53 proteins that were differentially expressed in cervical cancer cells after treatment with the drug. The study was conducted using SiHA and HeLa cell lines from human cervical cancer and proteomic quantification was performed by LC-MS (Xia et al. 2020).

Resistance to trastuzumab, a drug used for cancer treatment, has also been the subject of an investigation by Wen-Hu Liu et al. in 2020 (Liu et al. 2020b). The research was conducted using gastric cancer cells to investigate the signaling pathways regulated by GATA6 that are related to drug resistance. The results obtained from the quantification of 5792 proteins by LC-MS/MS were used to analyze the involved pathways and demonstrated the influence of GATA6 and

**Table 3** Examples of different bioanalytical proteomics applications using MS

| MS techniques | Application | Proteins | References |
|---|---|---|---|
| LC-MS/MS, nanoLC-MS/MS | Clinical | Hexosaminidase-A, aspartate aminotransferase, alpha-1-acid glycoprotein 2, and zinc transporter Zip14 | Iwan et al. (2021), Brady et al. (2018), and Yao et al. (2021) |
| MALDI-TOF-MS, nanoLC-MS/MS, LC-MS/MS | Plant biology | LEA Proteins, AT3G45980 proteins, glutathione S-transferase, and superoxide dismutase | Sghaier-Hammami et al. (2021), Salvato et al. (2019), and Timperio et al. (2008) |
| LC-MS/MS, 2-DE, MALDI-TOF-MS | Food | 33-mer α-2-gliadin, keratin, kinase, FGG protein, and AMBP protein | Marzano et al. (2020), Ogilvie et al. (2020), and Mouzo et al. (2020) |
| LC-MS/MS, MALDI-TOF-MS | Forensic toxicology | Myosin, ovalbumin, salivary annexin A1, fibronectin, and tropomyosin | Pieri et al. (2019), Procopio et al. (2018), and Handke et al. (2017) |

significant changes in mitochondrial transport, apoptosis, DNA damage, glucose metabolism, pyruvate metabolism, TCA cycle and degradation pathways of Wnt/β-catenin (Liu et al. 2020b). Table 3 presents more examples of proteomics applications in different areas in the scope of bioanalysis.

## 3.3   Metallomics

The term metallomics was used for the first time in 2004 to describe a new scientific field that aimed to promote studies related to biometals (Haraguchi 2004, 2017). Since then, metallomics has received attention as an emerging area in omic sciences (Haraguchi 2017). Within this context, the entire set of metal ions and other inorganic elements (metallome) is comprehensively evaluated to understand their role in biological systems better (Banci and Bertini 2013; Vogiatzis and Zachariadis 2014). Metal ions are involved in several cellular and physiological processes, such as the synthesis and metabolic function of genes and proteins. They are also incorporated into the three-dimensional structure of proteins, and some studies suggest that alterations in their concentration levels are related to neuropsychiatric pathologies (Scassellati et al. 2020; Cruz et al. 2020), plant diseases, or metabolism changes, among others (Gómez-Ariza et al. 2004; Li and Maret 2008; Lee et al. 2020; Arruda and Azevedo 2009).

ICP-MS is the technique of choice in metallomics, metalloproteomics (the study of metals bound to proteins), and metallometabolomics (the study of metals associated with metabolites) (Shi and Chance 2008; Montes-Bayón et al. 2018), allowing speciation analysis in biology with high selectivity and sensitivity, since analytes in the trace and ultra-trace concentration range may be measured (Vogiatzis and Zachariadis 2014; Szpunar 2004). For this purpose, ICP is used as an ion source and is usually coupled to Q and double-focusing sector field (SF) mass analyzers, presenting a suitable multi-elemental capacity, optimum accuracy, and fast isotopic analysis (Beauchemin 2010; Yasuda et al. 2013; Sun and Chai 2010; Albarède et al. 2017).

Speciation analysis, which aims at identifying and quantifying the different physicochemical forms of an element, is widely used in studies of the metallome, and metalloproteome/metallometabolome, where a combination of elemental (ICP-MS) and molecular (ESI, MALDI) MS is required. With this combination, the organic portion of a metallobiomolecule may be identified, allowing a better characterization of the biological system under investigation (Montes-Bayón et al. 2018).

The applications of metallomics, as well as other approaches of omic sciences in bioanalytical chemistry, include several areas. Scassellati and collaborators discussed the importance of metallomics in identifying potential biomarkers for neurodevelopmental diseases with a focus on schizophrenia, attention-deficit hyper-activity disorder, autism, and epilepsy, reporting the role of cobalt, copper, iron, manganese, selenium, and zinc, and how they can influence diagnosis and treatment (Scassellati et al. 2020). Santa Cruz et al. published a study focusing on schizophre-nia and bipolar disorder, investigating the association between trace elements in blood serum samples considering the pharmacological treatments applied to the treatment of these pathologies (Cruz et al. 2020). In the exploratory study, ICP-MS was used to quantify the serum concentrations of the trace elements selenium, zinc, iron, potassium, calcium, magnesium, phosphorus, aluminum, cop-per, manganese, and nickel. The authors verified that the concentrations of selenium and zinc were significantly lower for patients with schizophrenia and bipolar disor-der when compared to the group of healthy controls. On the other hand, serum iron concentrations were significantly higher for schizophrenia and bipolar disorder patients who did not use the drug lithium in their treatments. In addition, the results also showed a negative correlation between selenium and iron in patients with bipolar disorder treated with lithium and a significantly higher ratio between copper and zinc in the group of schizophrenia patients in comparison to individuals in the healthy control group (Cruz et al. 2020). Gómez-Ariza and collaborators performed an interesting discussion about the use of MS techniques for the characterization of protein-bound metals in biological systems and the possibilities of using a metallomics approach in the environmental, food, and clinical areas (Gómez-Ariza et al. 2004).

It is necessary to carry out the determination of individual chemical species to understand their impacts and different behaviors regarding toxicity, mobility, or bioavailability. Due to the instability of many chemical species and low concentrations in different matrices, such determination is still an analytical chal-lenge. ICP-MS technique is widely used for the determination of endogenous and exogenous metal species in biological systems, including those bound to biomolecules like amino acids, polysaccharides, and proteins, due to its advantages of sensitivity, selectivity, and ability to characterize ions, thus replacing classic techniques such as atomic absorption spectrometry in conducting studies in the field of metallomics (Gómez-Ariza et al. 2004).

**Table 4** Examples of bioanalytical applications of metallomics

| MS techniques | Application | Metal ions | References |
|---|---|---|---|
| ICP-MS | Clinical | Zinc, manganese, selenium, iron, molybdenum, barium, copper, calcium, lithium, and aluminum | Jia et al. (2021), Zhao et al. (2018), and Fiore et al. (2020) |
| ICP-MS | Environmental | Mercury, arsenic, cadmium, and chromium | Chen et al. (2020), Gomes et al. (2017), and González-Domínguez et al. (2016) |
| ICP-MS and LA-ICP-MS | Pharmaceutical | Magnesium, iron, copper, zinc, calcium, and gallium | Theiner et al. (2021), Ossipov et al. (2013), and Wenzel and Casini (2017) |
| ICP-MS and ICP-MS/MS | Foods | Selenium, mercury, cadmium, chromium, copper, and cerium | Palomo et al. (2014), Nong et al. (2021), and Wang et al. (2012) |

Yuan Li and Wolfgang Maret presented an interesting discussion about the challenges faced by researchers on developing research on metallothionein and attempts to relate the structure of a protein to its function (Li and Maret 2008). The integration of metallomics and proteomics information is possible when combining the ability of ICP-MS to measure the concentration of elements in proteins with molecular characterization by ESI- or MALDI-MS (Li and Maret 2008). Metallomics has expanded to several areas of knowledge in the last years and became a necessary approach for numerous investigations in the environmental, food, and pharmaceutical areas, among others. Table 4 describes applications of metallomics in different areas.

# 4    Mass Spectrometry Imaging (MSI)

Mass spectrometry imaging (MSI) is a regiospecific technique, which is based on the principle of desorption and ionization of molecules, and other chemical species, directly from surfaces, allowing their responses in a spatial context (Van Hove et al. 2010; McDonnell and Heeren 2007). Theoretically, any molecular or element species that can be desorbed from a surface and converted to the gas phase as an ion is eligible to be analyzed by MSI. MSI has undergone expressive growth in recent decades, including its application in the field of biomedical and life sciences, where it has been used for the analysis of relevant biological samples, such as histological sections. Several studies on the diversity and scope of the technique have been recently reported (Xue et al. 2019; Takats et al. 2017; Linscheid 2019).

MSI is a powerful tool to investigate the distribution of biomolecules, such as proteins, peptides, lipids, secondary metabolites, in addition to elements found in complex matrices such as human, animal, and plant tissues, and cells. The main advantages of MSI include (a) analyses of a wide range of analytes, (b) ability to monitor multiple analytes from different chemical classes in a single analysis without using chemical labeling or a derivatization step, and (c) preservation of biologically relevant spatial information in the studied samples. Therefore, MSI is a comprehensive technique and can be potentially applied to any solid, flat tissue, or substrate sample (Bagga et al. 2019).

Despite the advantages of the MSI technique, quantitative analysis is still a major challenge, since ion suppression in a heterogeneous matrix can promote different analytes ionization responses. Another important factor is the difficulty of desorption of the analytes by some ion sources (mainly ambient ionization sources, like DESI). However, great efforts are underway to solve this problem. The two most applied methodologies consist in the use of standard isotope labeling internal standard (SILIS) or free-SILIS, which can be a virtual calibration or with tissue extinction coefficient (Zhao and Cai 2020). Another essential point to be highlighted is the adequate preparation of the sample to obtain a reliable image. Ideally, a very regular surface of the solid tissue (plant or animal) or also microorganisms should be obtained. When the surface is not regular, it can promote different ions extraction, generating a non-corresponding result to the real chemical composition of the analyzed region (Bagga et al. 2019).

The collection and storage of the samples must also be considered since any biological sample degrades quickly under environmental conditions. Therefore, as soon as biological tissues are removed from the plant or animal, they must be quickly and properly stored. Fresh tissues storage at $-80\ ^{\circ}C$ is the most suitable for MSI analysis, especially when it comes to lipidomics. One example of sample pre-processing is paraffinization and deparaffinization, which unlike proteomics analysis, in which proteins maintain their location, lipids are leached, making the lipidomic analyses unfeasible (Dill et al. 2011a). The defrost and analysis processes can be a critical point for MSI. Recently, Dannhorn and colleagues presented a new method that minimizes analytical response variation due to different thawing conditions, which may occur both in different batches of samples and in the same sample, for hours-long analysis. The methodology is based on the embedding at the low temperature of fresh samples frozen in a hydrogel matrix composed of hydroxypropylmethylcellulose (HPMC) and polyvinylpyrrolidone (PVP), followed by a cut in a cryomicrotome, and finally taken to the ion source for imaging (Dannhorn et al. 2020).

Although several ionization techniques can be used in MSI, the principle of operation is the same: a platform or laser moves across the surface of interest guided by Cartesian coordinates (x and y), while a mass spectrum is recorded for each position (x, y), herein called pixel. Each dot or pixel in the two-dimensional image represents a mass spectrum acquired for a different part of the surface. Data analysis programs combined with two-dimensional information with an optical image of the surface generate heatmaps showing the spatial location (x, y) of any $m/z$ value measured by the mass spectrometer, as presented in Fig. 4 (Bagga et al. 2019; Walch et al. 2008).

Ion sources like MALDI, LA-ICP, and DESI have been used for MSI, mainly for biological investigations related to diseases like cancer, neurological disorders, and others. Ion source selection must also consider the image spatial resolution since

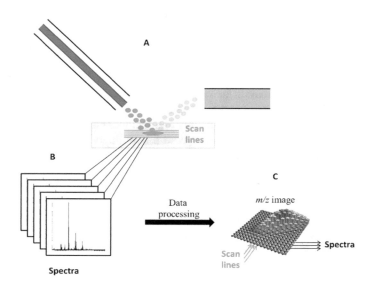

**Fig. 4** Overview of the 2-D imaging system. The pixel is composed of the information of the detected analytes—scanning line (**a**) and mass spectrum acquired for each pixel (**b**). 2-dimensional image reconstructions (**c**) and the relative abundance of each ion are observed according to the color difference

each of them allows obtaining images with different pixel sizes. In general, the location of chemical species on the sample surfaces varies with spacing up to 7 µm (MALDI) and 35 µm (DESI) (Abbassi-Ghadi et al. 2015; Xu and Li 2019).

Theoretically, MSI responds to any ionizable species within a selected mass range while the spatial information is maintained. A single pixel in an MSI image contains all the ionizable species detected for the specific mass range, while an MSI image may contain hundreds of thousands of pixels. Thus, the images produced by MSI have a wealth of information in the form of signals from thousands of analytes in each image. The images produced by MSI can be compared with images produced by other techniques, such as immunohistochemistry, conventional cytogenetics, and/or molecular genetic markers, providing useful and complementary information. Therefore, MSI results can be correlated with histological staining techniques, such as hematoxylin-eosin (H&E) staining, providing chemical information in a spatially defined context in clinical studies, such as oncological cases (Walch et al. 2008; Dill et al. 2011b). The choice of the MSI technique depends on several factors, such as mass range, spatial resolution, and classes of ionizable compounds, among others (Vaysse et al. 2017; Chen et al. 2019c). Table 5 presents relevant parameters to support the MSI technique selection.

Two of these parameters (analyte class and spatial resolution) are the most relevant for the ionization technique selection. Figure 5 demonstrates the correlation between these parameters for different ionization techniques (secondary ion mass spectrometry—SIMS, MALDI, LA-ICP, and DESI), with emphasis on the

**Table 5** Characteristics of ionization techniques used for MSI studies

| MSI technique | Ionization type | Ionization method | Sample preservation | Spatial resolution | Mass range | Analyte classes |
|---|---|---|---|---|---|---|
| DESI | Ambient | Solvent | Non-destructive | ~35 μm | <2000 Da | Peptides, lipids, and small molecules |
| MALDI | Vacuum | Laser | Destructive | ~7 μm | >400 Da | Proteins, peptides, and lipids |
| SIMS | Vacuum | Ion beam | Destructive | ~50 nm | <1000 Da | Lipids and elements |
| LA-ICP | Ambient | Plasma | Destructive | ~1 μm | <150 Da | Elements |

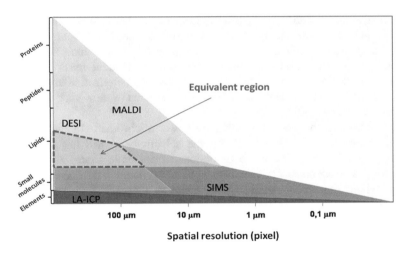

**Fig. 5** Comparison of MSI ionization techniques and their potential to identify analytes and respective spatial resolutions, with emphasis on the region of biomolecule ionization exposure (Abbassi-Ghadi et al. 2015)

equivalent ionization region between the three methodologies (lipids). It is noteworthy that DESI is a technique with the lowest spatial resolution amongst all, but that requires little or no sample preparation, making it more adequate and straightforward for MSI than the other ones. On the other hand, the LA-ICP technique presents elemental quantitative information that complements the molecular data obtained by the other strategies (Passarelli and Winograd 1811). Although there are more pronounced challenges, such as spatial resolution for some techniques that depend on solvent-assisted desorption, techniques that perform desorption and ionization assisted by laser or ion beam already perform single-cell level analyses. On the other hand, when the resolution is very small, the amount of desorbed material may represent a limitation in detectability (Ščupáková et al. 2019; De Samber et al. 2020).

The field of action of MSI techniques is vast in bioanalysis. Currently, we have long established the analysis of lipids and small molecules by DESI, lipids, peptides, and proteins by MALDI, elemental analyses by LA-ICP, and small molecules and elements by SIMS. Despite the clinical area presents great prominence, especially for oncology, other areas have also applied MSI, such as microbiology, forensics, environment, food, pharmacology, among others (Bagga et al. 2019).

# 5 Conclusions and Perspectives

MS has experienced a significant advance since the beginning of its development, counting on the efforts of several scientists until it became a powerful technique for bioanalysis, aiding in the elucidation of complex questions about biological systems. Since the establishment of ESI and MALDI ion sources in the 1980s, which allowed the qualitative and quantitative analyses of large molecules such as proteins, MS has become one of the most widely employed analytical techniques in a multiplicity of fields, for example, clinical, toxicology, environmental, and food sciences. The development of high-resolution mass analyzers and tandem mass spectrometry improved the abilities of this technique in terms of characterization and quantification, respectively, for the analysis of biomolecules even at ultra-trace levels and in complex matrices as biological samples.

The combination between the separating power of chromatographic and electrophoretic techniques with the sensitivity and selectivity of MS has enhanced the amount of possible information to be obtained from omic strategies, particularly metabolomics, proteomics, and metallomics that are crucial in the context of systems biology, which aims at characterizing organisms as a whole.

MSI has most recently emerged as an interesting approach since it can provide molecular and/or elemental information on specific regions of a biological tissue with high precision, which is especially relevant in the clinical area for diagnosis. MSI still presents several challenges for quantitative analysis purposes. However, scientists have been working on the minimization/elimination of such issues by using standard isotope labeling internal standard (SILIS), for instance.

Efforts on the improvement of MS instrumentation, especially on novel mass analyzers and ion sources, have continuously being made with the aim of increasing resolution power, mass accuracy, and sensitivity, which are relevant to the bioanalysis of compounds and elements at very low concentrations in more challenging and rare samples, including single cells.

# References

Abbassi-Ghadi N, Jones EA, Veselkov KA, Huang J, Kumar S, Strittmatter N, Golf O, Kudo H, Goldin RD, Hanna GB, Takats Z (2015) Anal Methods 7:71–80

Abboud J-LM, Notario R (1999) FT ICR. Basic principles and some representative applications. In: da Piedade MEM (ed) Energetic of stable molecules and reactive intermediates, vol 535. Springer Science, Berlin, pp 281–302

Aburaya S, Aoki W, Minakuchi H, Ueda M (2017) Biosci Biotechnol Biochem 81:2237–2243

Aebersold R, Mann M (2003) Nature 422:198–207

Ahrens CH, Brunner E, Qeli E, Basler K, Aebersold R (2010) Nat Rev Mol Cell Biol 11:789–801

Alaee M, Sergeant DB, Ikonomou MG, Luross JM (2001) Chemosphere 44:1489–1495

Albarède F, Télouk P, Balter V (2017) Rev Mineral Geochem 82:851–885

Amunugama R, Jones R, Ford M, Allen D (2013) Adv Wound Care 2:549–557

Annesley TM (2003) Clin Chem 49:1041–1044

Arevalo Jr R, Ni Z, Danell RM (2020) J Mass Spectrom 55:e4454

Armirotti A, Damonte G (2010) Proteomics 10:3566–3576

Arruda MAZ, Azevedo RA (2009) Ann Appl Biol 155:301–307

Badhwar A, McFall GP, Sapkota S, Black SE, Chertkow H, Duchesne S, Masellis M, Li L, Dixon RA, Bellec P (2020) Brain Commun 143:1315–1331

Badu-Tawiah AK, Eberlin LS, Ouyang Z, Cooks RG (2013) Annu Rev Phys Chem 64:481–505

Bahaghighat HD, Freye CE, Synovec RE (2019) Trends Anal Chem 113:379–391

Banci L, Bertini I (2013) Metallomics and the cell: some definitions and general comments. In: Banci L (ed) Metallomics and the cell. Springer, Berlin, pp 1–9

Banerjee S, Mazumdar S (2012) Int J Anal Chem 2012:1–40

Beauchemin D (2010) Anal Chem 82:4786–4810

Beck JL, Colgrave ML, Ralph SF, Sheil MM (2001) Mass Spectrom Rev 20:61–87

Becker JS, Kumtabtim U, Wu B, Steinacker P, Otto M, Matusch A (2012) Metallomics 4:284–288

Begou O, Gika HG, Wilson ID, Theodorids G (2017) Analyst 142:3079–3100

Biz A, Proulx S, Xu Z, Siddartha K, Indrayanti AM, Mahadevan R (2019) Biotechnol Adv 37:107379

Bodzon-Kulakowska A, Drabik A, Ner J, Kotlinska JH, Suder P (2014) Rapid Commun Mass Spectrom 28:1–9

Boesl U (2017) Mass Spectrom Rev 36:86–109

Bongiovanni T, Pintus R, Dessi A, Noto A, Sardo S, Finco G, Corsello G, Fanos V (2019) Eur Rev Med Pharmacol Sci 23:11011–11019

Brady JV, Troyer RM, Ramsey SA, Leeper H, Yang L, Maier CS, Goodall CP, Ruby CE, Albarqi HAM, Taratula O, Bracha S (2018) Transl Oncol 11:1137–1146

C-Celis F, Encinar JR (2019) J Proteome 198:11–17

Chen T, Yao Q, Nasaruddin RR, Xie J (2019a) Angew Chem Int Ed 58:11967–11977

Chen Z, Li Z, Li H, Jiang Y (2019b) Onco Targets Therapy 12:6797–6811

Chen K, Baluya D, Tosun M, Li F, Maletic-Savatic M (2019c) Metabolites 9:135

Chen B, Hu L, He B, Luan T, Jiang G (2020) Trends Anal Chem 126:115875

Chichi JF, David O, Villers F, Schaeffer B, Lutomski D, Huet S (2007) J Chromatogr B 849:261–272

Chowdhury G, Guengerich FP (2011) Curr Protoc Nucleic Acid Chem 47:7.16.1-7.16.11

Civallero G, de Kremer R, Giugliani R (2018) J Inborn Errors Metab Screen 6:1–6

Coe RA, Rathe JO, Lee JW (2006) J Pharm Biomed Anal 42:573–580

Costello CE (1999) Curr Opin Biotechnol 10:22–28

Cruz ECS, Madrid KC, Arruda MAZ, Sussulini A (2020) J Trace Elem Med Biol 59:126467

Cumeras R, Figueras E, Davis CE, Baumbach JI, Gràcia I (2015) Analyst 140:1376–1390

Daas C (2007) Fundamentals of contemporary mass spectrometry. Wiley-Interscience, Chichester, pp 10–12

Dannhorn A, Kazanc E, Ling S, Nikula C, Karali E, Serra MP, Vorng J-L, Inglese P, Maglennon G, Hamm G, Swales J, Strittmatter N, Barry ST, Sansom OJ, Poulogiannis G, Bunch J, Goodwin RJA, Takats Z (2020) Anal Chem 92:11080–11088

Daolio S, Facchin B, Pagura C (1987) Int J Mass Spectrom Ion Process 76:277–286

De Laeter J, Kurz MD (2006) J Mass Spectrom 41:847–854

De Leenheer AP, Lefevere MF, Thienpont LMR (1986) J Pharm Biomed Anal 4:735–745

De Samber B, De Rycke R, De Bruyne M, Kienhuis M, Sandblad L, Bohic S, Cloetens P, Urban C, Polerecky L, Vincze L (2020) Anal Chim Acta 1106:22–32

de Vlieger JSB, Giezen MJN, Falck D, Tump C, van Heuveln F, Giera M, Kool J, Lingeman H, Wieling J, Honing M, Irth H, Niessen WMA (2011) Anal Chim Acta 698:69–76

Dettmer K, Aronov PA, Hammock BD (2007) Mass spectrometry-based metabolomics. Mass Spectrom Rev 26:51–78

Diehn S, Zimmermann B, Bağcıoğlu M, Seifert S, Kohler A, Ohlson M, Fjellheim S, Weidner S, Kneipp J (2018) Sci Rep 8:16591

Dill AL, Eberlin LS, Costa AB, Ifa DR, Cooks RG (2011a) Anal Bioanal Chem 401:1949–1961

Dill AL, Eberlin LS, Ifa DR, Cooks RG (2011b) Chem Commun 47:2741–2746

Dinis-Oliveira RJ (2019) J Forensic Legal Med 61:128–140

Dreisewerd K (2003) Chem Rev 103:395–426

Du Z, Douglas DJ (1999) J Am Soc Mass Spectrom 10:1053–1066

Dunham SJB, Ellis JF, Li B, Sweedler JV (2017) Acc Chem Res 50:96–104

Eicher T, Kinnebrew G, Patt A, Spencer K, Ying K, Ma Q, Machiraju R, Mathé EA (2020) Metabolites 10:202

El-Aneed A, Cohen A, Banoub J (2009) J Appl Spectrosc Rev 44:210–230

Eliuk S, Makarov A (2015) Annu Rev Anal Chem 8:61–80

El-Khateeb E, Vasilogianni A-M, Alrubia S, Al-Majdoub ZM, Couto N, Howard M, Barber J, Rostami-Hodjegan A, Achour B (2019) Pharmacol Ther 203:107397

Feng Z, Ding C, Li W, Wang D, Cui D (2020) Food Chem 310:125914

Fieggen KJ, Ntusi NAB (2019) S Afr Med J 109:204–206

Fiehn O (2002) Metabolomics—the link between genotypes and phenotypes. In: Town C (ed) Functional genomics. Springer, Berlin, pp 155–171

Fiehn O, Robertson D, Griffin J, van der Werf M, Nikolau B, Morrison N, Sumner LW, Goodacre R, Hardy NW, Taylor C, Fostel J, Kristal B, Kaddurah-Daouk R, Mendes P, van Ommen B, Lindon JC, Sansone S-A (2007) Metabolomics 3:175–178

Finehout EJ, Lee KH (2004) Biochem Mol Biol Educ 32:93–100

Fiore M, Barone R, Copat C, Grasso A, Cristaldi A, Rizzo R, Ferrante M (2020) J Trace Elem Med Biol 57:126409

Gahoual R, Leize-Wagner E, Houzé P, François Y-N (2019) Rapid Commun Mass Spectrom 33:11–19

Gaskell SJ, Gale PJ (2015) Michael Barber. In: Gross ML, Caprioli RM (eds) The encyclopedia of mass spectrometry—historical perspectives, part B: notable people in mass spectrometry. Elsevier Science, Amsterdam, pp 11–12

Ghosh B, Jones AD (2015) Analyst 140:6522–6531

Gilmore IS, Seah MP (2000) Int J Mass Spectrom 202:217–229

Glish GL, Burinsky DJ (2008) J Am Soc Mass Spectrom 19:161–172

Godzien J, López-Gonzálvez A, García A, Barbas C (2019) Metabolic phenotyping using capillary electrophoresis mass spectrometry. In: Lindon JC, Nicholson JK, Holmes E (eds) The handbook of metabolic phenotyping. Elsevier Science, Amsterdam, pp 171–204

Gomes MAC, Hauser-Davis RA, Suzuki MS, Vitória AP (2017) Ecotoxicol Environ Safety 140:55–64

Gómez-Ariza JL, García-Barrera T, Lorenzo F, Bernal V, Villegas MJ, Oliveira V (2004) Anal Chim Acta 524:15–22

González-Domínguez R, Santos HM, Bebianno MJ, García-Barrera T, Gómez-Ariza JL, Capelo JL (2016) Mar Pollut Bull 113:117–124

Goodwin S, McPherson JD, McCombie WR (2016) Nat Rev Genet 17:333–351

Gowda GAN, Djukovic D (2014) Overview of mass spectrometry-based metabolomics: opportunities and challenges. In: Raftery D (ed) Mass spectrometry in metabolomics. Springer Protocols, Berlin, pp 3–12

Gregorich ZR, Chang Y-H, Ge Y (2014) Pflugers Archiv 466:1199–1209

Griffiths J (2008) Anal Chem 80:5678–5683

Gstaiger M, Aebersold R (2009) Nat Rev Genet 10:617–627

Haag AM (2016) Mass analyzers and mass spectrometers. In: Mirzaei H, Carrasco M (eds) Modern proteomics – Sample preparation, analysis and practical applications. Springer, Berlin, pp 157–169

Handke J, Procopio N, Buckley M, van der Meer D, Williams G, Carr M, Williams A (2017) Forensic Sci Int 281:1–8

Haraguchi H (2004) J Anal At Spectrom 19:5–14

Haraguchi H (2017) Metallomics 9:1001–1013

Harshman SW, Pitsch RL, Schaeublin NM, Smith ZK, Strayer KE, Phelps MS, Qualley AV, Cowan DW, Rose SD, O'Connor ML, Eckerle JJ, Das T, Barbey AK, Strang AJ, Martin JA (2019) J Chromatogr B 1126–1127:121763

He H, Guo Z, Wen Y, Xu S, Liu Z (2019) Anal Chim Acta 1090:1–22

Hiller K, Hangebrauk J, Jäger C, Spura J, Schreiber K, Schomburg D (2009) Anal Chem 81:3429–3439

Hirayama A, Abe H, Yamaguchi N, Tabata S, Tomita M, Soga T (2018) Electrophoresis 39:1382–1389

Hoffman E, Stroobant V (2007) Mass spectrometry: principles and applications. Wiley-Interscience, Chichester, pp 85–175

Horning EC, Horning MG, Carroll DI, Dzidic I, Stillwell RN (1973) Anal Chem 45:936–943

Houk RS (2010) J Anal At Spectrom 25:1801–1802

Houk RS, Fassel VA, Flesch GD, Svec HJ, Gray AL, Taylor CE (1980) Anal Chem 52:2283–2289

Hsu F-F (2018) Anal Bioanal Chem 410:6387–6409

Hu T, Zhang J-L (2018) J Sep Sci 41:351–372

Hu B, So P-K, Yao Z-P (2013) J Am Soc Mass Spectrom 24:57–65

Immler D, Gremm D, Kirsh D, Spengler B, Presek P, Meyer HE (1998) Electrophoresis 19:1015–1023

Ishihama Y (2005) J Chromatogr A 1067:73–83

Iwan K, Clayton R, Mills P, Csanyi B, Gissen P, Mole SE, Palmer DN, Mills K, Heywood WE (2021) iScience 24:102020

Jedrychowski MP, Huttlin EL, Haas W, Sowa ME, Rad R, Gygi SP (2011) Mol Cell Proteomics 10: M111.009910

Jensen ON (2004) Curr Opin Cell Biol 8:33–41

Jia H, Yuan X, Liu S, Feng Q, Zhao J, Zhao L, Xiong Z (2021) J Pharm Biomed Anal 193:113705

Jiye A, Trygg J, Gullberg J, Johansson AI, Jonsson P, Antti H, Marklung SL, Moritz T (2005) Anal Chem 77:8086–8094

Juraschek R, Dülcks T, Karas M (1999) J Am Soc Mass Spectrom 10:300–308

Kanani H, Chrysanthopoulos PK, Klapa MI (2008) J Chromatogr B 871:191–201

Karas M, Hillenkamp F (1988) Anal Chem 60:2299–2301

Karas M, Bachmann D, Hillenkamp F (1985) Anal Chem 57:2935–2939

Kauppila TJ, Syage JA, Benter T (2017) Mass Spectrom Rev 36:423–449

Kavadiya S, Biswas P (2018) J Aerosol Sci 125:182–207

Kelleher NL, Lin HY, Valaskovic GA, Aaserud DJ, Fridriksson EK, McLafferty FW (1999) J Am Chem Soc 121:806–812

Khan MM, Ernst O, Manes NP, Oyler BL, Fraser IDC, Goodlett DR, Nita-Lazar A (2019) ACS Infect Dis 5:493–505

Kim M-S, Pandey A (2012) Proteomics 12:530–542

Kim M-S, Kandasamy K, Chaerkady R, Pandey A (2010) J Am Soc Mass Spectrom 21:1606–1611

Kirwan JA, Weber RJM, Broadhurst DI, Viant MR (2014) Sci Data 1:140012

Klassen A, Faccio AT, Canuto GAB, Cruz PLR, Ribeiro HC, Tavares MFM, Sussulini A (2017) Metabolomic strategies involving mass spectrometry combined with liquid and gas chromatography. In: Sussulini A (ed) Metabolomics: from fundamentals to clinical applications. Springer, Berlin, pp 3–99

Kopka J (2006) J Biotechnol 124:312–322
Koppenaal DW, Barinaga CJ, Denton MB, Sperline RP, Hieftje GM, Schilling GD, Andrade FJ, Barnes JH (2005) Anal Chem 77:418–427
Kruve A, Kaupmees K (2017) J Am Soc Mass Spectrom 28:887–894
Kuehnbaum NL, Britz-McKibbin P (2013) Chem Rev 113:2437–2468
Kuo T-H, Dutkiewicz EP, Pei J, Hsu C-C (2020) Anal Chem 92:2353–2363
Lange E, Tautenhahn R, Neumann S, Gröpl C (2008) BMC Bioinf 9:375
Leaptrot KL, May JC, Dodds JN, McLean JA (2019) Nat Commun 10:985
Lee W-C, Lee KH (2004) Anal Biochem 324:1–10
Lee M-H, Gao Y-T, Huang Y-H, McGee EE, Lam T, Wang B, Shen M-C, Rashid A, Pfeiffer RM, Hsing AW, Koshiol J (2020) Hepatology 71:917–928
Lemberger T, Babu MM (2019) Mol Syst Biol 15:e9376
Li Y, Maret W (2008) J Anal At Spectrom 23:1055–1062
Li S, Tian Y, Jiang P, Lin Y, Liu X, Yang H (2020) Crit Rev Food Sci Nutr 61:1448–1469
Liang Q, Jing H, Gregoire DC (2000) Talanta 51:507–513
Limbeck A, Galler P, Bonta M, Bauer G, Nischkauer W, Vanhaecke F (2015) Anal Bional Chem 407:6593–6617
Lin L, Yu Q, Yan X, Hang W, Zheng J, Xing J, Huang B (2010) Analyst 135:2970–2978
Linscheid MW (2019) Mass Spectrom Rev 38:169–186
Liu Y, Wang X, Li Y, Chen X (2018) Aquat Toxicol 198:269–275
Liu C, Gu C, Huang W, Sheng X, Du J, Li Y (2019) J Chromatogr B 1113:98–106
Liu P-F, Xia Y, Hua X-T, Fan K, Li X, Zhang Z, Liu Y (2020a) Fish Shellfish Immunol 104:213–221
Liu W-H, Yuan J-B, Chang J-X (2020b) Chin J Anal Chem 48:187–196
Loo JA (1997) Mass Spectrom Rev 16:1–23
Manier SK, Keller A, Schäper J, Meyer MR (2019) Sci Rep 9:2741
Marshall AG, Hendrickson CL (2002) Int J Mass Spectrom 215:59–75
Marzano V, Tilocca B, Fiocchi AG, Vernocchi P, Levi Mortera S, Urbani A, Roncada P, Putignani L (2020) J Proteomics 215:103636
Mastrangelo A, Ferrarini A, Rey-Stolle F, García A, Barbas C (2015) Anal Chim Acta 900:21–35
Maurer HH (2006) J Mass Spectrom 41:1399–1413
McDonnell LA, Heeren RMA (2007) Mass Spectrom Rev 26:606–643
McMaster MC (2008) GC-MS - A practical user's guide. Wiley-Interscience, Chichester, pp 4–17
McNair HM, Miller JM (1997) Basic gas chromatography. Wiley-Interscience Publication, Chichester, pp 153–166
Meermann B, Nischwitz V (2018) J Anal At Spectrom 33:1432–1468
Miyagawa H, Bamba T (2019) J Biosci Bioeng 127:160–168
Montaner J, Ramiro L, Simats A, Tiedt S, Makris K, Jickling GC, Debette S, Sanchez J-C, Bustamante A (2020) Nat Rev Neurol 16:247–264
Montes-Bayón M, Sharar M, Corte-Rodriguez M (2018) Trends Anal Chem 104:4–10
Morton J, Tan E, Suvarna SK (2017) J Trace Elem Med Biol 43:63–71
Mouzo D, Rodríguez-Vázquez R, Lorenzo JM, Franco D, Zapata C, López-Pedrouso M (2020) Trends Food Sci Technol 99:520–530
Murray KK, Boyd RK, Eberlin MN, Langley GJ, Li L, Naito Y (2013) Pure Appl Chem 85:1515–1609
Nannapaneni NK, Jalalpure SS, Muppavarapu R, Sirigiri SK (2017) Talanta 164:233–243
Ngoka LCM, Gross ML (1999) J Am Soc Mass Spectrom 10:360–363
Nielsen KL, Telving R, Andreasen MF, Hasselstrøm JB, Johannsen M (2016) J Proteome Res 15:619–627
Nier AO, Gulbransen EA (1939) J Am Chem Soc 61:697–698
Niessen WMA, Falck D (2015) Introduction to mass spectrometry, a tutorial. In: Kool J, Niessen WMA (eds) Analyzing biomolecular interactions by mass spectrometry. Wiley-Interscience, Chichester, pp 1–43

Nong Q, Dong H, Liu Y, Liu L, He B, Huang Y, Jiang J, Luan T, Chen B, Hu L (2021) Chemosphere 263:128110

Oda Y, Huang K, Cross FR, Cowburn D, Chait BT (1999) Proc Natl Acad Sci 96:6591–6596

Ogilvie O, Roberts S, Sutton K, Domigan L, Larsen N, Gerrard J, Demarais N (2020) Food Chem 333:127466

Olesik JW (1996) Anal Chem 68:469–474

Olesik JW (2014) Inductively coupled plasma mass spectrometers. In: Reference module in earth systems and environmental sciences treatise on geochemistry, vol 15, pp 309–336

Oliver SG, Winson MK, Kell DB, Baganz F (1998) Trends Biotechnol 16:373–378

Oss M, Kruve A, Herodes K, Leito I (2010) Anal Chem 82:2865–2872

Ossipov K, Foteeva LS, Seregina IF, Perevalov SA, Timerbaev AR, Bolshov MA (2013) Anal Chim Acta 785:22–26

Palomo M, Gutiérrez AM, Pérez-Conde MC, Cámara C, Madrid Y (2014) Food Chem 164:371–379

Paré JRJ, Yaylayan V (1997) Mass spectrometry: principles and applications. In: Paré JRJ, Bélanger JMR (eds) Instrumental methods in food analysis. Elsevier Science, Amsterdam, pp 239–266

Parker CE, Warren MR, Mocanu V (2010) Mass spectrometry for proteomics. In: Alzate O (ed) Neuroproteomics. CRC Press/Taylor & Francis, Boca Raton

Passarelli MK, Winograd N (2011) Biochim Biophys Acta 1811:976–990

Peedicayil J (2019) Front Genet 10:985

Peña-Bautista C, Roca M, Hervás D, Cuevas A, López-Cuevas R, Vento M, Baquero M, García-Blanco A, Cháfer-Pericás C (2019) J Proteomics 200:144–152

Perez ER, Knapp JA, Horn CK, Stillman SL, Evans JE, Arfsten DP (2016) J Anal Toxicol 40:201–207

Perez CJ, Bagga AK, Prova SS, Taemeh MY, Ifa DR (2019) Rapid Commun Mass Spectrom 33:27–53

Piehowski PD, Petyuk VA, Orton DJ, Xie F, Moore RJ, Ramirez-Restrepo M, Engel A, Lieberman AP, Albin RL, Camp DG, Smith RD, Myers AJ (2013) J Proteome Res 12:2128–2137

Pieri M, Silvestre A, De Cicco M, Mamone G, Capasso E, Addeo F, Picariello G (2019) J Proteomics 209:103524

Pingitore F, Polce MJ, Wang P, Wesdemiotis C, Paizs B (2004) J Am Soc Mass Spectrom 15:1025–1038

Pitt JJ (2009) Clin Biochem Rev 30:19–34

Prideaux B, Stoeckli M (2012) J Proteomics 75:4999–5013

Procopio N, Williams A, Chamberlain AT, Buckley M (2018) J Proteomics 177:21–30

Purcaro G, Tranchida PQ, Mondello L (2017) Comprehensive gas chromatography methodologies for the analysis of lipds. In: Holčapek M, Byrdwell WC (eds) Handbook of advanced chromatography/Mass spectrometry techniques. Elsevier Science, Amsterdam, pp 407–444

Revel'skii IA, Yashin YS, Revelskii AI (2019) J Anal Chem 74:192–197

Rigano F, Tranchida PQ, Dugo P, Mondello L (2019) Trends Anal Chem 118:112–122

Rizo J, Guillén D, Farrés A, Díaz-Ruiz G, Sánchez S, Wacher C, Rodríguez-Sanoja R (2020) Crit Rev Food Sci Nutr 60:791–809

Robb DB, Covey TR, Bruins AP (2000) Anal Chem 72:3653–3659

Roberts LD, Souza AL, Gerszten RE, Clish CB (2012) Curr Protoc Mol Biol 98:1–24

Rosso MC, Mazzucotelli M, Bicchi C, Charron M, Manini F, Menta R, Fontana M, Reichenbach SE, Cordero C (2020) J Chromatogr A 1614:460739

Rutledge ACS, Codreanu SG, Sherrod SD, McLean JA (2016) J Am Soc Mass Spectrom 27:1897–1905

Salvato F, Loziuk P, Kiyota E, Daneluzzi GS, Araújo P, Muddiman DC, Mazzafera P (2019) Label-free quantitative proteomics of enriched nuclei from sugarcane (Saccharum ssp) stems in response to drought stress. Proteomics Syst Biol 19:190004

Sangster T, Spence M, Sinclair P, Payne R, Smith C (2004) Rapid Commun Mass Spectrom 18: 1361–1364

Saraswathy N, Ramalingam P (2011) Concepts and techniques in genomics and proteomics. Woodhead Publishing, Cambridge, pp 185–192

Scassellati C, Bonvicini C, Benussi L, Ghidoni R, Squitti R (2020) J Trace Elem Med Biol 60: 126499

Scheffer A, Engelhard C, Sperling M, Buscher W (2008) Anal Bioanal Chem 390:249–252

Schubert OT, Röst HL, Collins BC, Rosenberger G, Aebersold R (2017) Nat Protoc 12:1289–1294

Schuster SA, Boyes BE, Wagner BM, Kirkland JJ (2012) J Chromatogr A 1228:232–241

Schwab NV, Ore MO, Eberlin MN, Morin S, Ifa DR (2014) Anal Chem 86:11722–11726

Ščupáková K, Balluff B, Tressler C, Adelaja T, Heeren RMA, Glunde K, Ertaylan G (2019) Clin Chem Lab Med 58:914–929

Sghaier-Hammami B, Castillejo MA, Baazaoui N, Jorrín-Novo JV, Escandón M (2021) J Proteomics 233:104087

Shah TR, Misra A (2011) Proteomics. In: Misra A (ed) Challenges in delivery therapeutic genomics and proteomics. Elsevier Science, Amsterdam, pp 387–427

Shi W, Chance MR (2008) Cell Mol Life Sci 65:3040–3048

Singhal N, Kumar M, Kanaujia PK, Virdi JS (2015) Front Microbiol 6:791

Solliec M, Roy-Lachapelle A, Sauvé S (2015) Anal Chim Acta 853:415–424

Steuer AE, Brockbals L, Kraemer T (2019) Front Chem 7:319

Stolz A, Jooß K, Höcker O, Römer J, Schlecht J, Neusüß C (2019) Electrophoresis 40:79–112

Strader MB, Tabb DL, Hervey WJ, Pan C, Hurst GB (2006) Anal Chem 78:125–134

Suárez M, Caimari A, del Bas JM, Arola L (2017) Trends Anal Chem 96:79–88

Sun H, Chai Z-F (2010) Ann Rep Progress Chem 106:20–38

Šustáček V, Foret F, Boček P (1991) J Chromatogr A 545:239–248

Sylvester PJ, Jackson SE (2016) Elements 12:307–310

Szpunar J (2004) Anal Bioanal Chem 378:54–56

Tabb DL, Smith LL, Breci LA, Wysocki VH, Lin D, Yates JR (2003) Anal Chem 75:1155–1163

Tabb DL, Huang Y, Wysocki VH, Yates JR (2004) Anal Chem 76:1243–1248

Takáts Z, Wiseman JM, Gologan B, Cooks RG (2004) Science 306:471–473

Takáts Z, Strittmatter N, McKenzie JS (2017) Appl Mass Spectr Imaging Cancer 134:231–256

Tanaka K, Waki H, Ido Y, Akita S, Yoshida Y, Yoshida T, Matsuo T (1988) Rapid Commun Mass Spectrom 2:151–153

Tang X, Zhang Y, Zhou Y, Liu R, Shen Z (2020) J Invertebr Pathol 172:107355

Theiner S, Schoeberl A, Schweikert A, Keppler BK, Koellensperger G (2021) Curr Opin Chem Biol 61:123–134

Timperio AM, Egidi MG, Zolla L (2008) J Proteomics 71:391–411

Traeger JC (2016) The development of electron ionization. In: Gross ML, Caprioli RM (eds) The encyclopedia of mass spectrometry -historical perspectives, part a: the development of mass spectrometry. Elsevier Science, Amsterdam, pp 77–82

van den Broek I, Sparidans RW, Huitema ADR, Schellens JHM, Beijnen JH (2006) J Chromatogr B 837:49–58

van Der Hooft JJJ, Mohimani H, Bauermeister A, Dorrestein PC, Duncan KR, Medema MH (2020) Chem Soc Rev 49:3297–3314

Van Hove ERA, Smith DF, Heeren RMA (2010) J Chromatogr A 1217:3946–3954

Vaysse P-M, Heeren RMA, Porta T, Balluff B (2017) Analyst 142:2690–2712

Vidova V, Spacil Z (2017) Anal Chim Acta 964:7–23

Vinayavekhin N, Saghatelian A (2010) Curr Protoc Mol Biol 90:1–24

Vogiatzis CG, Zachariadis GA (2014) Anal Chim Acta 819:1–14

Walch A, Rauser S, Deininger S-O, Höfler H (2008) Histochem Cell Biol 130:421–434

Wang Q, Ma X, Zhang W, Pei H, Chen Y (2012) Metallomics 4:1105–1112

Wang J, Wang C, Liu H, Qi H, Chen H, Wen J (2018) Crit Rev Biotechnol 38:1106–1120

Wang X, Hou Y, Hou Z, Xiong W, Huang G (2019) Anal Chem 91:2719–2726

Wasinger VC, Zeng M, Yau Y (2013) Int J Proteomics 2013:180605

Weaver EM, Hummon AB (2013) Adv Drug Deliv Rev 65:1039–1055

Wei J, Xiang L, Cai Z (2020) Mass Spectrom Rev 40:255–279

Wenzel M, Casini A (2017) Coord Chem Rev 352:432–460

Wilm M (2011) Mol Cell Proteomics 10:M111.009407

Wilschefski SC, Baxter MR (2019) Clin Biochem Rev 40:115–133

Wiseman JM, Ifa DR, Song Q, Cooks RG (2006) Angew Chem Int Ed 45:7188–7192

Wiseman JM, Evans CA, Bowen CL, Kennedy JH (2010) Analyst 135:720–725

Woodall DW, El-Baba TJ, Fuller DR, Liu W, Brown CJ, Laganowsky A, Russel DH, Clemmer DE (2019) Anal Chem 91:6808–6814

Wu C, Ifa DR, Manicke NE, Cooks RG (2009) Anal Chem 81:7618–7624

Wu C, Dill AL, Eberlin LS, Cooks RG, Ifa DR (2013) Mass Spectrom Rev 32:218–243

Xia C, Yang F, He Z, Cai Y (2020) Biomed Pharmacother 123:109762

Xian F, Hendrickson CL, Marshall AG (2012) Anal Chem 84:708–719

Xing J, Xie C, Lou H (2007) J Pharm Biomed Anal 44:368–378

Xu Y-J (2017) Trends Anal Chem 96:14–21

Xu G, Li J (2019) J Comp Neurol 527:2158–2169

Xue J, Bai Y, Liu H (2019) Trends Anal Chem 120:115659

Yamashita M, Fenn JB (1984) J Phys Chem 88:4451–4459

Yang K, Han X (2016) Trends Biochem Sci 41:954–969

Yao Z, Zhang Y, Yan J, Yan L (2021) J Gynecol Obstetr Human Reprod 50:102043

Yasuda H, Yasuda Y, Tsutsui T (2013) Sci Rep 3:1199

Young LM, Saunders JC, Mahood RA, Revill CH, Foster RJ, Ashcroft AE, Radford SE (2016) Methods 95:62–69

Zamboni N, Saghatelian A, Patti GJ (2015) Mol Cell 58:699–706

Zhang B, Whiteaker JR, Hoofnagle AN, Baird GS, Rodland KD, Paulovich AG (2019) Nat Rev Clin Oncol 16:256–268

Zhao C, Cai Z (2020) Mass Spectr Rev:1–19. https://doi.org/10.1002/mas.21674

Zhao S, Cao S, Luo L, Zhang Z, Yuan G, Zhang Y, Yang Y, Guo W, Wang L, Chen F, Wu Q, Li L (2018) Clin Chim Acta 485:323–332

Zhurov KO, Fornelli L, Wodrich MD, Laskay ÜA, Tsybin YO (2013) Chem Soc Rev 42:5014–5030

# Microchip-Based Devices for Bioanalytical Applications

Kemilly M. P. Pinheiro, Thaisa A. Baldo, Lucas P. Bressan, José A. F. da Silva, and Wendell K. T. Coltro

## 1 Introduction to Microfabricated Devices

Miniaturization of systems has been a trend in many fields of knowledge and technological development. Probably, the electronic industry is the best example of how miniaturization can be beneficial. The substitution of vacuum tubes for solid-state transistors caused a considerable impact on equipment size and power consumption. Today's microprocessors can integrate an astonishing number of 54 billion metal oxide semiconductor field-effect transistors (MOSFET). And as occurred in electronics, analytical chemistry instrumentation and methods also took advantage of miniaturization. Year by year, the sample volume or amount has been decreasing, and it is common to find methodologies that use $10^{-12}$ g or even $10^{-15}$ g of samples and describe the handling of $10^{-9}$ to $10^{-12}$ L (Cousino et al. 1997).

In this direction, microfluidic devices (or simply microchips) are the natural evolution of analytical systems. In 1979, Terry and colleagues described a device dedicated to gas chromatography built on a 5-cm-diameter silicon wafer containing an injection port, a 1.5-m-long separating capillary column, and a thermal conductivity detector (Terry et al. 1979). This pioneering work anticipated the upcoming

K. M. P. Pinheiro · T. A. Baldo
Instituto de Química, Universidade Federal de Goiás, Goiânia, GO, Brazil

L. P. Bressan
Instituto de Química, Unicamp, Campinas, SP, Brazil

J. A. F. da Silva
Instituto de Química, Unicamp, Campinas, SP, Brazil

Instituto Nacional de Ciência e Tecnologia de Bioanalítica, Campinas, SP, Brazil

W. K. T. Coltro (✉)
Instituto de Química, Universidade Federal de Goiás, Goiânia, GO, Brazil

Instituto Nacional de Ciência e Tecnologia de Bioanalítica, Campinas, SP, Brazil
e-mail: wendell@ufg.br

© The Author(s), under exclusive license to Springer Nature Switzerland AG 2022
L. T. Kubota et al. (eds.), *Tools and Trends in Bioanalytical Chemistry*,
https://doi.org/10.1007/978-3-030-82381-8_24

revolution of the miniaturized analytical devices. The concept of integrating many analytical steps into a single substrate to process analysis was introduced by Andreas Manz in the early 1990s and became known as Micro Total Analysis Systems (µTAS) (Manz et al. 1990). The general idea is the transfer of bench procedures, such as sample treatment, injection, target analyte isolation (separation included), detection, and data processing to a compact device.

Microfluidic devices present several advantages compared to standard-sized equipment, such as low power and reagents consumption, reduced fabrication and operation cost, disposability, enhanced sensitivity, mechanical stability, extended dynamic range, reduced analysis time, and parallel processing, among others (Whitesides 2006). Microfluidics is a well-established science and technology field, and one can find thousands of applications of microfluidic devices and several companies that commercialize microdevices and associated instrumentation (Whitesides 2006).

In this chapter, we cover some aspects of microfluidic devices in bioanalysis, emphasizing 3D printing microfabrication, microchip electrophoresis, paper-based devices, and droplet-based and digital microfluidics.

## 2     Microfabrication Strategies

The most suitable microfabrication technique is highly dependent on the type of substrate material or combination of materials, and several variations can be found in the literature. In the beginning, the microfabrication of devices used the well-established photolithographic protocols derived from the electronic industry since the preferred materials were silicon, glass, and quartz. In 1998, Duffy et al. introduced the rapid prototyping of microchips in poly(dimethylsiloxane) (PDMS), which popularized the field and allowed devices' production in few hours (Duffy et al. 1998). The technique is based on the replication of high relief features previously prepared (mold or template). The use of laser ablation, hot embossing, and injection molding techniques is suitable for polymeric substrates. Direct lamination protocol using polyester sheets and toner is a very low-cost protocol for fabricating microfluidic devices with a resolution of around 100 µm (do Lago et al. 2003). Recently, 3D printing has been successfully used in the production of microdevices. The advantage of 3D printing resides in the easiness of the production of complex microfluidic networks in a single step. Considering the 3D printing techniques as promising to produce microfluidic devices, we detail their different types and applications.

## 2.1     3D Printing Microfabrication

Creating complex 3D structures for microchips is cumbersome since it involves multiple processes requiring assembly and a careful approach that can take several days to construct based on established lithographic techniques (Waldbaur et al.

2011). Instead of this, 3D printing arises as a tool that can significantly change the microfluidic fabrication due to its inherently layer-by-layer additive construction process (Au et al. 2016). It also facilitates the design of microfluidic chips and their change and optimization. The printing process involves creating a 3D model in a computer-aided design (CAD) software that allows easy edition. This allows for the iteration of designing, printing, and testing to get the final device to be done in a matter of hours, not days or weeks (Gross et al. 2014).

Due to these characteristics, 3D printing is emerging as a fabrication tool, hence the drawbacks it still presents in creating devices with high spatial resolution, low surface roughness, transparent optical windows, and composed of multi-materials (Gross et al. 2017). These drawbacks are being addressed by several research groups that are focused on 3D printed-based microfluidic devices, expanding the application of the 3D printers especially by increasing the printer resolution to achieve results similar to those of PDMS microfluidic devices (Nielsen et al. 2020).

There are different 3D printers based on the materials used and mode of construction. Each 3D printer has its limitations and advantages, as explained below. However, the cost of acquiring both the printer and their materials causes some of them to be more widespread than others either in science or everyday use.

## 2.2 Fused Deposition Modelling

Fused deposition modeling (FDM) or fused filament fabrication consists of a polymer that is presented in the form of a filament (usually 1.75 mm or 3.00 mm diameter) that is fed through a hot nozzle to molten the plastic and allow its delivery to a specific location which is controlled by three motors in the $X$-, $Y$- and $Z$-planes, thus creating the 3D printed object. The materials used come in the form of a roll of filament and are usually poly(lactic acid) (PLA), polyethylene terephthalate glycol-modified (PETG), composite materials (PLA-carbon nanotubes), acrylonitrile butadiene styrene (ABS), and flexible materials such as thermoplastic polyurethane (TPU), among others (Salentijn et al. 2017).

It is the lowest cost of all the 3D printers due to the acquisition cost of the printer (starting ~250 USD) and the materials used (starting ~20 USD $kg^{-1}$) to create the microfluidic device. As listed above, the polymeric materials are diverse and find different applications based on their characteristics: PLA is biocompatible but has low heat resistance to temperatures higher than 60 °C; ABS is heat resistant but is attacked by organic solvents; TPU is a thermoplastic that is heat resistant and offers good transparency, but the devices made are usually flexible.

FDM-based 3D printed microfluidic devices are usually large and lack resolution to print smooth surfaces while also presenting drawbacks such as low bridging tolerance and expansion/shrinkage after printing. The resolution of the printed object is directly tied to the nozzle diameter used. Although several groups managed to print channels with dimensions sub-100 μm with this technique, it is more suited to microfluidic devices with channel dimensions ranging from 200 to 500 μm. Another drawback of the technique is that it lacks inherent transparency due to air trapping inside the extruded material. Hence, several research groups focused on developing transparent devices either by careful selection of printing parameters or including an optical window into the microfluidic device (Bressan et al. 2019; Nelson et al. 2019).

## 2.3    Digital Light Processing/Stereolithography

Stereolithography (SLA) is the process of curing a resin presenting monomers that are polymerized using an ultraviolet light source with a movable platform to create a 3D printed object (Amin et al. 2016). This technique has evolved to a faster and higher resolution process where the whole layer is projected into the resin in a single step, where Digital Light Processing (DLP) equipment is used, such as a projector.

One of the drawbacks of DLP/SLA printers is that they present a higher cost of acquisition for either the equipment (starting at ~2000 USD) and the resins are usually expensive (starting ~200 USD $L^{-1}$), and most of them have proprietary composition. This technique also requires post-processing where the 3D printed object is washed with organic solvents and exposed to UV light to finish the curing of the resin, increasing the time from printing start to actually using the 3D printed device.

Despite its cost, currently, DLP/SLA 3D printers are the most cost-effective to print high-resolution microfluidic devices. The channel widths are usually sub-200 μm, and with the custom resin and apparatus, it can achieve sub-20 μm resolution (Beauchamp et al. 2017). This allows for interesting competition with established techniques such as soft lithography. Regarding the major drawback FDM-based microfluidic devices present, DLP-based 3D printed objects are usually transparent and present low-surface roughness compared to them, meaning they can be used after a single-step printing process (Dixit et al. 2018).

## 2.4    Other 3D Printing Techniques

There are several other 3D printers available, but they usually present higher costs and are not very broad in the materials they use to print. Inkjet- or polyjet-based 3D printers are similar to 2D printers as they deliver the material in the form of droplets into a surface (Walczak and Adamski 2015). After one layer is finished, another layer of material is deposited on top to create the 3D printed object. This printer has a higher acquisition cost (starting at ~25,000 USD) but can achieve high resolution and allow the use of different polymeric materials in the same model natively. However, they require the use of a sacrificial material that must be removed in order for the printed object to be used, increasing the post-processing time and requiring the use of solvents.

Laser sintering is another form of 3D printing based on the fusing of a powder by a high-energy laser that creates the complex 3D structure. It is usually applied to metal powders that can create structures that are further annealed to the desired applications. Once again, its resolution is not optimal for microfluidics structures, and they require extensive post-processing (Yap et al. 2015).

## 2.5    Applications of 3D Printed Devices in Bioanalytical Sciences

Fluidic control tools are one of the most common applications of 3D printed microfluidic devices since they allow the manipulation of fluids with higher accuracy than the reference batch methods and usually are built in less time and with fewer materials than the reference lithography construction (Bhattacharjee et al. 2016).

Another area that has an extensive application of 3D printing with bioanalytical applications is sample preparation by means of microfluidic devices. 3D printed microfluidic devices can act to concentrate, dilute, and extract the analyte from the sample by directing using the polymeric material or inclusion of another class of materials such as monoliths and filters (Kalsoom et al. 2018; Calderilla et al. 2018).

A wide scope of bioanalytical sensors uses 3D printed microfluidic devices since they allow rapid prototyping. They can be made either from single 3D printed microfluidic devices or a combination with other techniques, such as paper- and thread-based devices (Palenzuela and Pumera 2018; Sharafeldin et al. 2018).

## 3    Microchip Electrophoresis

Electrophoresis is a separation technique based on the migration time of the species under the influence of an electrical field. Electrophoresis on microsystems was first reported in the 1990s (Manz et al. 1990; Harrison et al. 1992). Since then, they are being used for several applications (Martin et al. 2012; Saylor and Lunte 2015), demonstrating to be faster than conventional electrophoresis and also allowing low consumption of reagents and samples, automation, and miniaturization. Regarding bioanalytical analyzes, microchip electrophoresis has been applied for DNA, protein, peptides, and neurotransmitters detection. Some recent reviews report clinical biomarkers (Pagaduan et al. 2015) and biomedical analysis (Nuchtavorn et al. 2015), and also protein and peptides analysis in the human body (Dawod et al. 2017; Stepanova and Kasicka 2019).

Jabasini et al. developed a methodology for DNA polymorphisms analysis on the human Y-chromosome using an Agilent 2100 Bioanalyzer with epifluorescence detection. It was possible to analyze genomic polymorphisms on the Y-chromosome (Y *Alu* polymorphism, 47z/*Stu*I, and 12f2) in approximately 100 s with high resolution and precision (Jabasini et al. 2002). DNA traces have also been analyzed using microchip gel electrophoresis coupled to electrochemical detection (ED). Buffer containing hydroxypropyl cellulose (HPC) and the electrode was modified with gold nanoparticles to improve pre-concentration, separation, and detection of 13 fragments present in a 100-bp DNA ladder (Shiddiky and Shim 2007). Polymeric microchip electrophoresis with fluorescent detection was used for end-labeled DNA products after PCR for genetic diagnostics (Hall et al. 2016).

Some studies demonstrate methodologies for DNA analysis for screening the occurrence of cancer (Zhang et al. 2019a) and cardiac diseases (Fujihara et al. 2019).

Microchip electrophoresis has also been applied for cell and cell lysis products determination. A methodology using a polymeric (SU-8) microchip-electrospray ionization mass spectrometry was developed for cytochrome C, b-lactoglobulin, ovalbumin, and BSA identification originated by human muscle cell lysates, demonstrating to be effective to detect human proteins from the complex cellular matrix (Nordman et al. 2010). Liu's group developed a microfluidic platform coupled to a nano-electrospray emitter for ionization in mass spectrometry, allowing nano-electrode cell lysis for automated single-cell analysis. PC 12 neuronal cells were analyzed, being able to detect and separate dopamine and glutamic acid at the cellular levels (Li et al. 2016a). This same group also reported the real-time controlling release of neurotransmitters in the single nerve cell. Monitoring chiral compounds (D-serine to L-serine) was important to associate them with the nature of neuronal communications (Li et al. 2016b). Siegel et al. described an electrophoresis microchip coupled with electrochemical detection for nitrite detection in cell lysates and its separation from electroactive cell constituents and interferences. Nitrite was determined as an indicator of nitric oxide production after cell lysis, which, when in excess, has been associated with neurodegenerative diseases and cardiovascular disorders (Siegel et al. 2019).

A hybrid device composed of polystyrene (PS) and PDMS coupled to electrochemical detection was developed to monitor neurotransmitter release after PC 12 cells culture and immobilization. The device includes continuous sampling from a cell culture dish, on-chip pump and valving, microchip electrophoresis, and detection to enable separation and measurement of norepinephrine and dopamine (Johnson et al. 2015). Neurotransmitters were also analyzed in human urine samples (Zhang et al. 2016). A microchip electrophoresis system coupled with laser-induced fluorescence (LIF) detection was used for the separation of dopamine, norepinephrine, and serotonin based on the field-amplified stacking (FASS) and reversed-field stacking (RFS) methods. Adenosine, inosine, hypoxanthine, and guanosine were separated in less than 85 s using microdialysis coupled to microchip electrophoresis with amperometric detection. It was effective to monitor the in vitro enzymatic conversion of adenosine to inosine, once they are biomarkers for cerebral ischemia (Gunawardhana and Lunte 2018).

Peptides and proteins study is also relevant for bioanalytical applications, and these compounds are widely analyzed using electrophoresis microdevices. Isomers peptides were described by Ollikainen et al. using a microchip electrophoresis-electrospray ionization mass spectrometry. It was possible to separate within 1 min two monophosphorylated positional isomers of insulin receptor peptide, a triply phosphorylated insulin receptor peptide, and a non-phosphorylated peptide (Ollikainen et al. 2016). A study of sample derivatization was reported to enable comprehensive glycoprotein characterization. For this purpose, mass spectrometry detection was coupled to microfluidic electrophoresis to analyze the release of glycans, glycopeptides, and monosaccharides (Fig. 1) (Khatri et al. 2017). Fresta et al. used microchip electrophoresis with laser-induced fluorescence detection for

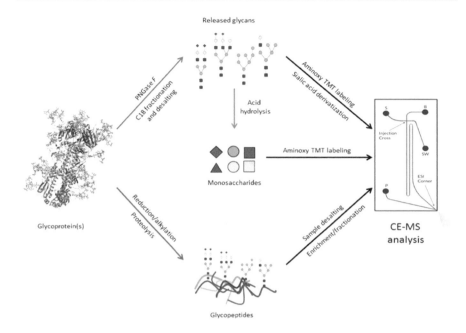

**Fig. 1** Overview of released glycans, glycopeptides, and monosaccharides analysis for glycoprotein characterization. Reprinted from reference (Khatri et al. 2017) with permission

carnosine determination in murine RAW 264.7 macrophage cells. Carnosine levels increased under pro-inflammatory conditions, suggesting that macrophages use carnosine as a defense mechanism (Fresta et al. 2017). An online multi-stacking preconcentration technique involving field enhanced sample injection and micelle-to-solvent stacking was reported for vancomycin in human plasma analysis using microchip electrophoresis coupled to a contactless conductivity detector ($C^4D$). The LOD obtained for this antibiotic was 2.5 g $L^{-1}$, with recoveries in spiked human plasma between 99.0 and 99.2% (Chong et al. 2017).

Microchip electrophoresis with a LED-induced fluorescence detection was used to detect creatinine in urine samples. Determination of creatinine levels in biological fluids is an important marker to evaluate renal function and muscular dysfunctions. Creatinine detection was obtained in approximately 30 s with LOD of 2.87 µmol $L^{-1}$ and recovery ranging from 96.0 to 107% (Wang et al. 2012). Analysis of alpha-fetoprotein, which is an oncofetal glycoprotein, was performed using a microchip electrophoresis with chemiluminescence detection. Its detection was performed within 1 min, and the methodology was successfully applied in human serum for healthy people and for cancer patients, demonstrating to be effective as a tumor marker (Liu et al. 2017). Tumor markers for cancer diagnostics were also evaluated by Gan et al. They demonstrated the use of microfluidic chip electrophoresis-based multiplex immunoassay for alpha-fetoprotein, carcinoembryonic antigen, and carbohydrate antigen 199 detection in human serum, demonstrating that a DNA platform can be used to detect any target based on endonuclease labels (Gan et al. 2018). A microchip gel electrophoresis was used to analyze perchloric acid-soluble serum proteins in healthy people and in patients with systemic inflammatory diseases. Even though the protein composition of perchloric acid-soluble serum fraction was not explored, the marked differences in the protein patterns present in the patients with inflammatory diseases demonstrated to be a rapid screening of inflammation (Makszin et al. 2019).

# 4    Droplet-Based Microfluidics

The use of droplet-based microfluidics systems has increased in the last decade. They are based on tiny amounts of fluids generating and manipulating droplets in small channels. These systems are able to carry cells, DNA, drugs, and proteins, allowing reactions and analysis for many bioanalytical applications (Shembekar et al. 2016; Huang et al. 2017; Feng et al. 2019).

Pekin et al. described a method based on a droplet-based microfluidic system to perform digital PCR in droplets to detect and quantify the ratio of mutant and wild-type genes in DNA. It was possible to determinate genomic DNA by confining it inside droplets with on TaqMan specific for mutant genomic which generates the green fluorescent signal, and other for wild-type DNA, which generates a red fluorescent signal, demonstrating to be effective for clinical oncology applications (Pekin et al. 2011). A droplet-based system was also used for sequencing DNA molecules with a FRET-based assay. This system demonstrated efficiency for genotyping analysis since it discriminates probes that complement a target molecule from the ones which do not (Abate et al. 2013). Islam et al. fabricated a low-cost chip for droplet generation and in-line polymerization. The droplets were polymerized into DNA-coated hydrogel microparticles. Acrylic materials demonstrated to be adequate with a more stable surface than PDMS, being a good option for microfluidics approaches (Islam et al. 2018).

Droplets-based microfluidics systems have also been applied for cell analysis. El Debs et al. developed a microfluidic platform that allows the screening of up to 300,000 individual hybridoma cell clones in less than a day. Besides that, this methodology can be applied to non-immortalized primary B-cells. For this purpose, each cell is encapsulated in a droplet and assayed directly for the release of antibodies, and a fluorescence signal demonstrated the drug target inhibition (El Debs et al. 2012). A microfluidic platform was also developed for cytokine (IL-2, IFN-$\gamma$, TNF-$\alpha$) secretion of single and activated T-cells detection. A PDMS chip was used to encapsulate cells in agarose gel droplets with functionalized cytokine-capture beads for secreted cytokines detection (Fig. 2) (Chokkalingam et al. 2013).

Circulating tumor cells as cancer markers were also analyzed by droplets-based systems. Droplets were used for individual cell analysis measuring the secretion of acid from tumor cells. It was possible to detect tumor cells among white cells in the blood of metastatic patients (Del Ben et al. 2016). A system based on microchip SERS-active silver nanoparticles spectra of cell lysates was demonstrated by

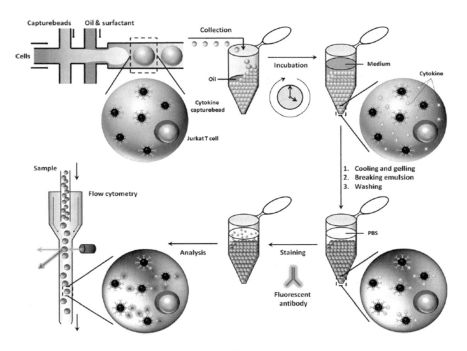

**Fig. 2** Overview of the droplet-based microfluidic system for individual cell encapsulation and detection of secreted cytokines. Reprinted from reference (Chokkalingam et al. 2013) with permission

Hassoun et al. A droplet-based quartz chip was used for cell lysate identification using surface-enhanced Raman scattering, and it was possible to classify three leukemia cell lines, Jurkat, THP-1, and MONO-MAC-6 (Hassoun et al. 2018).

An automated microfluidic platform was developed employing microvalves controlled by a computer to generate laminar co-flows of samples and reagents. In this system was possible to control the number, composition, and size of droplets. Fluorescent and colorimetric signals were produced after enzymatic reaction involving target analyte and assay reagents inside the droplets. They analyzed LDH, glucose, and bile acid in the same small volume of media. To demonstrate the bioanalytical application, the platform was connected to a cell culture system containing hepatocyte spheroid, being possible to detect the increase in LDH and the decrease in bile acid synthesis, markers of cytotoxicity, and hepatic function, respectively (Cedillo-Alcantar et al. 2019).

A droplet-based chip using the surface-enhanced Raman scattering (SERS) technique was developed for detecting 6-thioguanine. This anticancer drug was successfully detected in human serum, demonstrating to be a promising tool for therapeutic drug analysis (Zhang et al. 2019b). Buryska et al. report a droplet-based microfluidic system to characterize enzymes that convert hydrophobic substrates by controlling oil/water partitioning. They demonstrated that hydrophobic compounds can be distributed from an oil phase into aqueous droplets, allowing reactions of compounds with limited water solubility (Buryska et al. 2019).

## 5 Paper-Based Devices

The use of paper as a substrate for chip fabrication has become attractive, once this substrate attaches advantages as cost-effective, simplified analytical systems and portable platforms for point-of-care assays (Kaneta et al. 2018; Yang et al. 2017; Noviana et al. 2020). Due to the hydrophilic characteristic, the paper allows that aqueous solutions flow through the porous structure by capillarity and consequently does not require added mechanical constructions for pumping (Channon et al. 2019; Schaumburg et al. 2020). Therefore, microfluidic paper-based analytical devices (µPADs) have become increasingly recognized with potential applications to identify a range of analytes in health diagnosis (Teengam et al. 2017; Campbell et al. 2018), pathogens detection, biomolecules monitoring (Murdock et al. 2013), immunoassays (Vashist and Luong 2018), and others allowing multiple bioassays with only a droplet of sample (Santhiago et al. 2014). The µPADs are usually formed by a hydrophobic barrier on a paper platform that delimits channels to confine sample solutions and restricts fluid traveling (Martinez et al. 2007).

Teengam et al. described a novel electrochemical biosensor for human papillomavirus (HPV) using a screen-printed carbon electrode modified with an anthraquinone-labeled pyrrolidinyl peptide nucleic acid (acpcPNA) probe (AQ-PNA) and graphene-polyaniline (G-PANI). The DNA biosensor was used to detect a synthetic 14-base oligonucleotide target with a sequence corresponding to human papillomavirus (HPV) type 16 DNA by measuring the square-wave voltammetry signal response of the AQ label using before and after hybridization. The proposed biosensor was applied to DNA detection extracted from SiHa cell line, HPV type 16 and 393 amplified using PCR, and showed successful results indicating potential device to detect HPV type 16 DNA in clinical samples (Teengam et al. 2017). A nucleic acid paper-based detection chip was used to *Listeria monocytogenes* diagnostics as described by Fu et al., where the chip was integrated with DNA extraction and signal amplification using magnetic silica beads and, a helical continuous-flow polymerase chain reaction (PCR) performed the denaturing, annealing, and extension phases of the PCR (Fu et al. 2018).

The µPADs have also been explored in the cancer field to detect proteins and cells related to the disease. The transmembrane glycoprotein, epithelial cell adhesion molecule (EpCAM), was evaluated by fluorescence resonance energy transfer (FRET) modifying the paper matrix with aptamer-linked quantum dots (QDs-Apt) and Cy3 labeled complementary DNA (cDNA). The cDNA binds to EpCAM, resulting in competitive bioassay and the reduction of FRET in a reaction time of no more than 60 min (Das and Krull 2017). Simultaneous detection of carcinoembryonic antigen (CEA) and neuron-specific enolase (NSE) was studied by Wang et al. with an electrochemical paper device fabricated through wax printing and screen printing techniques. The working electrodes were modified with amino

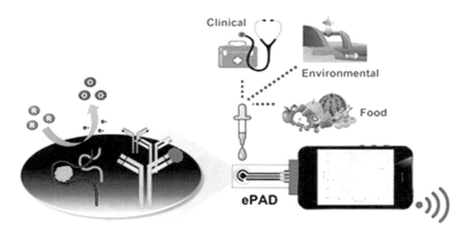

**Fig. 3** Overview of released electrochemical paper-based devices analysis for clinical, environmental, and food analysis. Reprinted from reference (Noviana et al. 2020) with permission

functional graphene (NG)-Thionin (THI)-gold nanoparticles (AuNPs) and Prussian blue (PB)- poly (3,4- ethylenedioxythiophene) (PEDOT)-AuNPs nanocomposites to promote immobilization of the CEA and NSE aptamers. The proposed aptasensor shows the limit of detection (LOD) of 2.0 pg mL$^{-1}$ for CEA and 10 pg mL$^{-1}$ for NSE. The device also was evaluated in clinical samples and showed a good correlation with large electrochemical luminescence (ECL) for both biomarkers indicating a promising platform for early cancer diagnostics (Das and Krull 2017). A colorimetric paper sensor for citrate detection was reported by Abarghoei et al. aiming at early-stage detection of prostate cancer (Abarghoei et al. 2019). The biomimetic sensor was performed using a paper assay on a Y-shaped microfluidic device modified with cysteine-capped gold nanoclusters (Cys-AuNCs) that in the presence of $H_2O_2$ catalyze the oxidation of 3,3′,5,5′-tetramethylbenzidine (TMB) with high efficiency to produce a blue dye (with an absorbance maximum at 650 nm). This sensor showed advantages as short response time, very low reagent volume required, low fabrication cost, etc. A detection limit of 0.4 μmol L$^{-1}$ was achieved through the paper test, and a good linear range was observed between 1.0 μmol L$^{-1}$ and 10 mmol L$^{-1}$. The bioanalytical application for disease diagnosis has been attracting interesting with paper devices due to facility to surface modification, simple to accomplish to portable platforms, and different techniques to analyte detection (Fig. 3) (Noviana et al. 2020).

Paper-based Enzyme Linked Immunosorbent Assay (ELISA) has also been applied to bioanalysis combining the convenience and low-cost of PADs platforms with sensitivity and specificity of the immunoassay (Murdock et al. 2013). Ortega et al. developed a magnetic paper-based ELISA for IgM-dengue detection (Ortega et al. 2017). The core-shell magnetite@polydopamine nanoparticles were deposited onto cellulose surface and conjugated with antibodies isotype IgM detection. This magnetic assay provided analytical response two orders more sensitive and lower limit of detection (LOD) about 700 times than traditional ELISA or using magnetic beads without depositing, showing appropriate accuracy for real sample detection, low cost, and effortless and easy handling. Neuropeptide Y detection was performed with silver enhancement of AuNPs conjugated with IgG antibodies, enzyme-free system, reported by Murdock et al. (Murdock et al. 2013) The determination of antigen concentrations was able to be performed on Windows- and Android-based mobile platforms.

## 6     Digital Microfluidics

Digital microfluidics (DMF) consists of a droplet-based microfluidic system that can be fabricated by photolithography (Wang and Fatoyinbo 2014; Fair 2007) and screen printing technique (Abadian et al. 2017). The DMF consists of a pathway composed of a grid of hydrophobic dielectric coated electrodes, which delimits a pathway where the droplets will move due to the sequential application of electrical potential. The droplets may be added independently and in parallel and do not require external modules or complicated geometries such as pumps or valves (Kwon et al. 2003; Pollack et al. 2000). This technique allows manipulating nanoliter to microliters (de Campos et al. 2019; Rackus et al. 2015; Dryden et al. 2013). The DMF chips show primary advantages as automation and reduction of consumed samples and reagents, lower reaction time, and high controllability in sample manipulation. Besides that, they can be mass-produced and be merged as a device of choice for point-of-care applications (Rackus et al. 2015). Also, DMF electrodes can be fabricated with different designs, and as a result, the device can be applied for multiple applications (Wang and Fatoyinbo 2014; Fair 2007; Abdulwahab et al. 2017).

   Campos et al. reported DMF embedded in electrochemical sensors for the detection of biomolecules, a "plug-n-play DMF sensing" (PnP-DMF) (de Campos et al. 2019). The PnP-DMF was assembled with commercial biosensors for glucose and β-ketone, a custom paper-based electrochemical sensor for lactate, and a generic screen-printed electroanalytical cell. Figure 4 illustrates the assembled steps. The sensor was suitable to combine an electrochemical detection of glucose with a chemiluminescent magnetic bead-based sandwich immunoassay for insulin.

## 7     Conclusions

This chapter has summarized an overview of conventional and emerging microfluidic platforms with great potential for bioanalytical applications. Outstanding examples involving 3D printing, microchip electrophoresis, paper-based microfluidic devices, droplet microfluidics, and digital microfluidic platforms were successfully presented and discussed, making clear the huge potentiality for studies in the subject.

   As highlighted in this chapter, 3D printing, droplet, and digital microfluidics have appeared as powerful tools to be explored in bioanalytical chemistry. Among them,

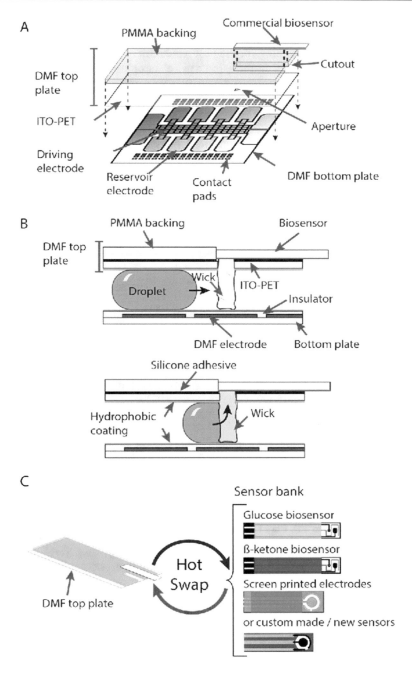

**Fig. 4** Overview of "Plug-n-play" digital microfluidics (PnP-DMF). (**a**) Cartoon (not to scale) illustrating how a PnP-DMF top plate is assembled and interfaced with a bottom plate. (**b**) Cartoon side view (not to scale) illustrating the composition of the DMF device and how droplets are wicked into the electroanalytical cell. (**c**) Cartoon illustrating "hot swapping" sensors into the DMF top plate. A sensor bank provides options for a variety of applications that can be chosen on-the-fly. Reprinted from reference (de Campos et al. 2019) with permission

3D printing is currently globally known, and its use for the development of microfluidic devices has received great attention in the field. The optimization of the fabrication protocols aiming for the best resolution in terms of dimension is the main drawback to be solved. In the same way, electrophoresis chips have become increasingly popular for bioanalysis. A few advantages, including the low required sample volume, the short analysis time, and the sample-in-answer-out capability, make electrophoresis chips a current trend for commercialization, which a few instruments are already available in the market. Lastly, it is important to emphasize the current impact of paper-based platforms for bioanalytical chemistry. Due to the inherent low cost, the global affordability, and the biocompatibility of this platform, many examples of point-of-care diagnostics have been demonstrated in the literature. We are sure that this new emerging platform can be helpful for direct diagnostic testing to detect active SARS-CoV-2 infections.

# References

Abadian A, Manesh SS, Ashtiani SJ (2017) Microfluid Nanofluid 21:65

Abarghoei S, Fakhri N, Borghei YS, Hosseini M, Ganjali MR (2019) Spectrochim Acta A Mol Biomol Spectrosc 210:251–259

Abate AR, Hung T, Sperling RA, Mary P, Rotem A, Agresti JJ, Weiner MA, Weitz DA (2013) Lab Chip 13:4864–4869

Abdulwahab S, Ng AHC, Chamberlain MD, Ahmado H, Behan LA, Gomaa H, Casper RF, Wheeler AR (2017) Lab Chip 17:1594–1602

Amin R, Knowlton S, Hart A, Yenilmez B, Ghaderinezhad F, Katebifar S, Messina M, Khademhosseini A, Tasoglu S (2016) Biofabrication 8:022001

Au AK, Huynh W, Horowitz LF, Folch A (2016) Angew Chem Int Ed 55:3862–3881

Beauchamp MJ, Nordin GP, Woolley AT (2017) Anal Bioanal Chem 409:4311–4319

Bhattacharjee N, Urrios A, Kang S, Folch A (2016) Lab Chip 16:1720–1742

Bressan LP, Adamo CB, Quero RF, de Jesus DP, da Silva JAF (2019) Anal Methods 11:1014–1020

Buryska T, Vasina M, Gielen F, Vanacek P, van Vliet L, Jezek J, Pilat Z, Zemanek P, Damborsky J, Hollfelder F, Prokop Z (2019) Anal Chem 91:10008–10015

Calderilla C, Maya F, Leal LO, Cerdá V (2018) Trends Anal Chem 108:370–380

Campbell JM, Balhoff JB, Landwehr GM, Rahman SM, Vaithiyanathan M, Melvin AT (2018) Int J Mol Sci 19:2731

Cedillo-Alcantar DF, Han YD, Choi J, Garcia-Cordero JL, Revzin A (2019) Anal Chem 91:5133–5141

Channon RB, Nguyen MP, Henry CS, Dandy DS (2019) Anal Chem 91:8966–8672

Chokkalingam V, Tel J, Wimmers F, Liu X, Semenov S, Thiele J, Figdor CG, Huck WT (2013) Lab Chip 13:4740–4744

Chong KC, Thang LY, Quirino JP, See HH (2017) J Chromatogr A 1485:142–146

Cousino MA, Jarbawi TB, Halsall HB, Heineman WR (1997) Anal Chem 69:A544–A549

Das P, Krull UJ (2017) Analyst 142:3132–3135

Dawod M, Arvin NE, Kennedy RT (2017) Analyst 142:1847–1866

de Campos RPS, Rackus DG, Shih R, Zhao C, Liu X, Wheeler AR (2019) Anal Chem 91:2506–2515

Del Ben F, Turetta M, Celetti G, Piruska A, Bulfoni M, Cesselli D, Huck WT, Scoles G (2016) Angew Chem Int Ed 55:8581–8584

Dixit CK, Kadimisetty K, Rusling J (2018) Trends Anal Chem 106:37–52 do Lago CL, da Silva HDT, Neves CA, Brito-Neto JGA, da Silva JAF (2003) Anal Chem 75:3853–3858

Dryden MD, Rackus DD, Shamsi MH, Wheeler AR (2013) Anal Chem 85:8809–8816

Duffy DC, McDonald JC, Schueller OJA, Whitesides GM (1998) Anal Chem 70:4974–4984

El Debs B, Utharala R, Balyasnikova IV, Griffiths AD, Merten CA (2012) Proc Natl Acad Sci USA 109:11570–11575

Fair RB (2007) Microfluid Nanofluid 3:245–281

Feng H, Zheng T, Li M, Wu J, Ji H, Zhang J, Zhao W, Guo J (2019) Electrophoresis 40:1580–1590

Fresta CG, Hogard ML, Caruso G, Costa EEM, Lazzarino G, Lunte SM (2017) Anal Methods 9:402–408

Fu Y, Zhou X, Xing D (2018) Sensors Actuators B Chem 261:288–296

Fujihara J, Takinami Y, Ueki M, Kimura-Kataoka K, Yasuda T, Takeshita H (2019) Clin Chim Acta 497:61–66

Gan N, Xie L, Zhang K, Cao Y, Hu F, Li T (2018) Sensors Actuators B Chem 272:526–533

Gross BC, Erkal JL, Lockwood SY, Chen C, Spence DM (2014) Anal Chem 86:3240–3253

Gross B, Lockwood SY, Spence DM (2017) Anal Chem 89:57–70

Gunawardhana SM, Lunte SM (2018) Anal Methods 10:3737–3744

Hall GH, Glerum DM, Backhouse CJ (2016) Electrophoresis 37:406–413

Harrison DJ, Manz A, Fan Z, Ludi H, Widmer HM (1992) Anal Chem 64:1928–1932

Hassoun M, Ruger J, Kirchberger-Tolstik T, Schie IW, Henkel T, Weber K, Cialla-May D, Krafft C, Popp J (2018) Anal Bioanal Chem 410:999–1006

Huang H, Yu Y, Hu Y, He X, Berk Usta O, Yarmush ML (2017) Lab Chip 17:1913–1932

Islam MM, Loewen A, Allen PB (2018) Sci Rep 8:8763

Jabasini M, Zhang L, Dang F, Xu F, Almofli MR, Ewis AA, Lee J, Nakahori Y, Baba Y (2002) Electrophoresis 23:1537–1542

Johnson AS, Mehl BT, Martin RS (2015) Anal Methods 7:884–893

Kalsoom U, Nesterenko PN, Paull B (2018) Trends Anal Chem 105:492–502

Kaneta T, Alahmad W, Varanusupakul P (2018) Appl Spectrosc Rev 54:117–141

Khatri K, Klein JA, Haserick JR, Leon DR, Costello CE, McComb ME, Zaia J (2017) Anal Chem 89:6645–6655

Kwon CS, Hyejin M, Chang-Jin K (2003) J Microelectromech Syst 12:70–80

Li X, Zhao S, Hu H, Liu Y-M (2016a) J Chromatogr A 1451:156–163

Li X, Zhao S, Liu Y-M (2016b) Chem Select Commun 1:5554–5560

Liu J, Zhao J, Li S, Zhang L, Huang Y, Zhao S (2017) Talanta 165:107–111

Makszin L, Kustan P, Szirmay B, Pager C, Mezo E, Kalacs KI, Paszthy V, Gyorgyi E, Kilar F, Ludany A, Koszegi T (2019) Electrophoresis 40:447–454

Manz A, Graber N, Widmer HM (1990) Sensors Actuators B B1:244–248

Martin A, Vilela D, Escarpa A (2012) Electrophoresis 33:2212–2227

Martinez AW, Phillips ST, Butte MJ, Whitesides GM (2007) Angew Chem Int Ed 46:1318–1320

Murdock RC, Shen L, Griffin DK, Kelley-Loughnane N, Papautsky I, Hagen JA (2013) Anal Chem 85:11634–11642

Nelson MD, Ramkumar N, Gale BK (2019) J Micromech Microeng 29:095010

Nielsen AV, Beauchamp MJ, Nordin GP, Woolley AT (2020) Annu Rev Anal Chem 13:1.1–1.21

Nordman N, Sikanen T, Aura S, Tuomikoski S, Vuorensola K, Kotiaho T, Franssila S, Kostiainen R (2010) Electrophoresis 31:3745–3753

Noviana E, McCord CP, Clark KM, Jang I, Henry CS (2020) Lab Chip 20:9–34

Nuchtavorn N, Suntornsuk W, Lunte SM, Suntornsuk L (2015) J Pharm Biomed Anal 113:72–96

Ollikainen E, Bonabi A, Nordman N, Jokinen V, Kotiaho T, Kostiainen R, Sikanen T (2016) J Chromatogr A 1440:249–254

Ortega GA, Pérez-Rodríguez S, Reguera E (2017) RSC Adv 7:4921–4932

Pagaduan JV, Sahore V, Woolley AT (2015) Anal Bioanal Chem 407:6911–6922

Palenzuela CLM, Pumera M (2018) Trends Anal Chem 103:110–118

Pekin D, Skhiri Y, Baret JC, Le Corre D, Mazutis L, Salem CB, Millot F, El Harrak A, Hutchison JB, Larson JW, Link DR, Laurent-Puig P, Griffiths AD, Taly V (2011) Lab Chip 11:2156–2166
Pollack MG, Fair RB, Shenderov AD (2000) Appl Phys Lett 77:1725–1726
Rackus DG, Dryden MD, Lamanna J, Zaragoza A, Lam B, Kelley SO, Wheeler AR (2015) Lab Chip 15:3776–3784
Salentijn GIJ, Oomen PE, Grajewski M, Verpoorte E (2017) Anal Chem 89:7053–7061
Santhiago M, Nery EW, Santos GP, Kubota LT (2014) Bioanalysis 6:89–106
Saylor RA, Lunte SM (2015) J Chromatogr A 1382:48–64
Schaumburg F, Kler PA, Carrell CS, Berli CLA, Henry CS (2020) Electrophoresis 41:562–569
Sharafeldin M, Jones A, Rusling JF (2018) Micromachines 9:394
Shembekar N, Chaipan C, Utharala R, Merten CA (2016) Lab Chip 16:1314–1331
Shiddiky MJA, Shim Y-B (2007) Anal Chem 79:3724–3733
Siegel JM, Schilly KM, Wijesinghe MB, Caruso G, Fresta CG, Lunte SM (2019) Anal Methods 11: 148–156
Stepanova S, Kasicka V (2019) J Sep Sci 42:398–414
Teengam P, Siangproh W, Tuantranont A, Henry CS, Vilaivan T, Chailapakul O (2017) Anal Chim Acta 952:32–40
Terry SC, Jerman JH, Angell JB (1979) IEEE Trans Electron Devices 26:1880–1886
Vashist SK, Luong JHT (2018) Chapter 15. Lab-on-a-chip (LOC) immunoassays. In: Vashist SK, Luong JHT (eds) Handbook of immunoassay technologies, pp 415–431, Academic Press
Walczak R, Adamski K (2015) J Micromech Microeng 25:085013
Waldbaur A, Rapp H, Länge K, Rapp BE (2011) Anal Methods 3:2681–2716
Wang K, Fatoyinbo HO (2014) Chapter 4. Digital microfluidics. In: RSC detection science series, vol 5, pp 84–135
Wang S, Li X, Yang J, Yang X, Hou F, Chen Z (2012) Chromatographia 75:1287–1293
Whitesides GM (2006) Nature 442:368–373
Yang Y, Noviana E, Nguyen MP, Geiss BJ, Dandy DS, Henry CS (2017) Anal Chem 89:71–91
Yap CY, Chua CK, Dong ZL, Liu ZH, Zhang DQ, Loh LE, Sing SL (2015) Appl Phys Rev 2: 041101
Zhang Y, Zhang Y, Wang G, Chen W, Li Y, Zhang Y, He P, Wang Q (2016) J Chromatogr B 1025: 33–39
Zhang Y, Zhang Y, Zhu L, He P, Wang Q (2019a) Electrophoresis 40:425–430
Zhang W-S, Wang Y-N, Wang Y, Xu Z-R (2019b) Sensors Actuators B Chem 283:532–537

# Microarrays Application in Life Sciences: The Beginning of the Revolution

Regiane Fátima de Travensolo, Vinícius Guimarães Ferreira, Maria Teresa Federici, Eliana Gertrudes Macedo de Lemos, and Emanuel Carrilho

## 1    Microarrays: What Are and How It Works?

As the reader may know, every significant progress in science starts with the sum of several small steps. After all, the Wright brothers could never develop in 1890s a modern Airbus A320, neither could Airbus build their aircraft without the Wright brothers and Santos Dumont pioneering projects (Alsina 2020). Nowadays, it is common to enter a pharmacy store and find different kits for DNA analysis, from paternity tests to ancestry check. Indeed, the point we are now is certainly less impressive than the point we will be in 20 years from now, and we may say the same about the past. Today, we can sequence our DNA and check for a few diseases and learn where our ancestors lived; however, only 20 years ago, science was only crawling to these features. In this somehow nostalgic chapter, the reader will travel over time and learn how the genomic revolution began.

Before the development of Next Generation DNA Sequencing (NGS), scientists needed to perform a laborious process for DNA sequencing, called Sanger sequencing, that was extremely time and resource consuming. For instance, the Human Genome Project spent a total cost of about 3 billion dollars on human DNA sequencing (Gibbs 2020; Carrasco-Ramiro et al. 2017). Indeed, only a really few magnates hold this amount of money to spend on DNA analysis, correct? On the other hand, after Next Generation Sequencing, the reader may find in pharmacy

R. F. de Travensolo · V. G. Ferreira · E. Carrilho (✉)
Instituto de Química de São Carlos, Universidade de São Paulo, São Carlos, SP, Brazil
e-mail: emanuel@iqsc.usp.br

M. T. Federici
Instituto Nacional de Investigación Agropecuaria, Parque Tecnológico del LATU, Montevideo, Uruguay

E. G. M. de Lemos
Faculdade de Ciências Agrárias e Veterinárias, Universidade Estadual Paulista, Via de Acesso Prof. Paulo Donato Castellane, Jaboticabal, SP, Brazil

© The Author(s), under exclusive license to Springer Nature Switzerland AG 2022
L. T. Kubota et al. (eds.), *Tools and Trends in Bioanalytical Chemistry*,
https://doi.org/10.1007/978-3-030-82381-8_25

stores DNA tests costing less than a couple hundred dollars. Although even more expressive than the gigantic difference in cost is the idea of NGS evaluating especially the expressed genes, analogously, considering DNA as the book of our body, NGS highlights only the most important chapters, letting behind the not so interesting ones.

However, NGS was not developed in just a couple of weeks, and the idea of analyzing the expressed DNA was inherited from another DNA analysis technology, the microarrays (Wöhrle et al. 2020). Straightforwardly, DNA microarrays, also known as DNA chips, gene chips, or biochips, are a collection of high-density single-stranded DNA molecules attached to a solid surface, as summarized in Fig. 1. We could suppose a case where scientists want to analyze the different gene expressions between normal and specific tumor cells for a better understanding of the molecular mechanisms. Firstly, the scientists need to prepare the microarray chip by attaching several known DNA sequences, or genes, on different spots on the chip. Later, mRNA is extracted from the tissue samples, and a reverse transcription-polymerase chain reaction (RT-PCR) is performed to obtain the complementary DNA chains (cDNA), that are later labeled with a fluorescent tag, for example, green for normal cell cDNAs and red for cancer cell cDNAs (Lashkari et al. 1997; Schena et al. 1995; Van Hal et al. 2000).

The cDNA samples from both tissues are mixed and added to the microarray chip, where cDNA will hybridize with their matching pair, attached in the microarray. The microarray chip is then taken to a microarray scanner to read the fluorescent tags on each (micrometer size) spot. The basic principle here is that if the scanner reads a green fluorescence in a particular spot, it means that the normal cell cDNA hybridized with the DNA single strand in the microarray, meaning that this particular gene is expressed in the normal cell but not expressed in the cancer cell. In parallel, if the scanner reads a red tag in a different spot, it means that this particular gene is expressed only in the cancer cell. If the reading is yellow, it means that this particular gene is expressed in both cells (Fig. 1).

Therefore, microarrays were an efficient manner to evaluate differential gene expression in different conditions or organisms (Diehn and Relman 2001; Bhatia and Dahiya 2015). This technology was vastly employed in the late 1990s and early 2000s, when sets of non-redundant cDNAs became widely available, enabling the researchers to analyze several genes in a single microarray device. At this period, even before the complete sequencing of the human genome, different organisms had their complete genome sequenced, allowing the production of sets of microarrays encompassing the whole genome of those organisms (Lashkari et al. 1997; Richmond et al. 1999).

Beyond genetic research advances, several breakthroughs in high-precision robotic deposition processes were responsible for microarray technology dissemination, allowing the preparation of more specific and cheap microarrays, employing different surfaces, such as nylon, glass slides, and silicon chips (DeRisi et al. 1997; Kaushik et al. 2018). Besides, microarray applications evolved to the application of oligomers arrays, providing an increase in specificity for the intended binding target, as oligos could be designed to target specific regions of the genes or genome

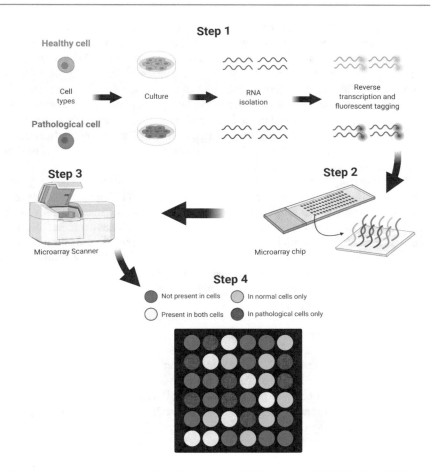

**Fig. 1** Microarray analysis process. Step 1) cell culture, RNA extraction, RT-PCR of the RNA, and cDNA fluorescence tagging. Step 2) Adding the cDNAs into the microarray chip. The cDNA will hybridize in the matching single-strand DNA attached to the chip. Step 3) The microarray chip is taken into a microarray scanner in order to read the fluorescent tags in each spot. Step 4) Data analysis. Each color in the array corresponds to the cDNA from a specific organism; in this example, green is for normal cells, red for pathological cells, and when the attached gene is expressed in both cells, the red color is yellow. Created with BioRender (https://biorender.com/)

(Bumgarner 2013). To date, microarray platforms rely on short oligonucleotides synthesized in situ (25 bases) or longer oligonucleotides (50–80 bases) chemically adsorbed to the chip (Raghavendra and Pullaiah 2018).

Indeed, microarray technology has advanced to such a point that researchers now demand microarrays that are cost-effective, flexible, and have a high-quality assurance. Some authors describe that the simplicity of the oligonucleotide approach makes it the most attractive option for gene expression research (Deyholos and Galbraith 2001). The use of microarrays to study gene expression profiles is based on two principles. The first is that for many genes, changes in expression mean changes in the quantity of mRNA, and the second is that only the DNA strands can hybridize with complementary sequences forming a stable double-stranded molecule. The microarray exploits this property by immobilizing millions of single strands of DNA on a solid support. These molecules are hybridized with single strands of complementary labeled DNA that contain, proportionally, all the genes being expressed in a given organism or tissue. Another set of important information given by microarrays is the quantitative measurement of the expressed genes, once the fluorescence intensity is directly proportional to the amount of transcription of each gene (Deyholos and Galbraith 2001).

The microarray technology was responsible for an important advance in genomic studies to discover new drugs and toxicological research. Unlike other forms of hybridization (such as microplates or dot blot), glass slide microarrays allow significant miniaturization of thousands of DNA probes and/or clones arranged on a single slide. As a result, this technology was ideal for the extensive identification of nucleic acids and gene expression analysis. The simultaneous analysis of multiple genes makes it possible to determine a complete genetic profile of a single microorganism or distinguish a microorganism from a single experiment collection.

## 2    Microarray Construction

As mentioned, microarray technology's main gear is attaching a known DNA on a solid support, which will serve to identify another DNA chain through hybridization. The attached DNA is called a probe, while the cDNA, from the sample, is usually called the target (Phimister 1999). Nowadays, besides the commercially available chips, researchers may develop different microarray chips for their specific application. The production of DNA microarrays can be divided into three stages: (1) probe determination; (2) probe amplification, purification, and concentration; and (3) robotic probe deposition. In the first stage, the probes are generally PCR amplification products with an approximate 400–2000 bases, as denoted in Fig. 2 (Schena et al. 1995; Welford et al. 1998). In the second stage, these amplified products must be purified, quantified, and aliquoted to the same concentration, presenting a single fragment in electrophoretic analysis (see Fig. 2). After the probe's entire production process, the material will be deposited by robots on a solid support, usually on glass sheets covered with certain substances to increase the probe's adhesion. The so-called robots are equipment developed specifically for microarray fabrication and work as a very precise arrangement of pipettes, adding the reagents needed for linking the exact DNA single strand to the exact desired position (see Fig. 2).

DNA binding to glass slides is commonly performed with poly-lysine, which binds to glass through non-covalent bonds involving the lysine residues' primary amines. These amines also electrostatically adsorb themselves to the DNA phosphate groups, immobilizing these molecules on the slides; in addition, other hydrophobic forces and hydrogen bonds can contribute to DNA binding to poly-lysine.

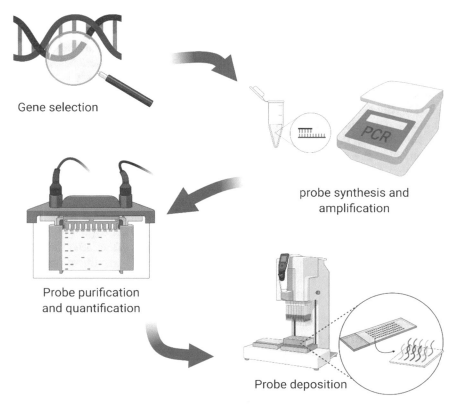

**Fig. 2** Microarray construction process. The process starts with the gene, or genes, selection for the analysis. The chosen DNA that will be used as the microarray probe is later synthesized and amplified to generate enough material. The amplified DNA is purified and quantified by gel electrophoresis, and the pure DNA strands are bonded into glass slides, in a process called probe deposition. Created with BioRender (https://biorender.com/)

Another type of treatment that can be applied on glass slides is based on amino silane, which has positively charged amine groups that contribute to DNA binding, but unlike poly-lysine, it is covalently bound to glass (Doug Chung and Le Roch 2013). Other types of slide treatments provide covalent DNA bonds, thus decreasing their loss during the hybridization process. However, it has already been demonstrated that the non-covalent binding of DNA to the slides is sufficient for the production of the microarrays (Lashkari et al. 1997; Welford et al. 1998).

Various equipment types can do an automatic deposition of the material, but generally, they only require a volume of 0.3– 1 nL of the solution with a spacing of approximately 150μm between the spots. In this process, the control of the temperature and humidity is essential to avoid evaporation of the samples (Deyholos and Galbraith 2001). Different types of negative controls to check for possible cross-hybridization or non-specific hybridization should be included as probes, such as, gene sequences from genetically distant species that are non-coded genomic sequences containing only poly (dA) or poly (dT).

The cDNA molecules are normally marked with the fluorochromes Cyanine 3 (Cy3) and Cyanine 5 (Cy5), which have different spectral characteristics, allowing two different samples to be analyzed simultaneously co-hybridized in the solid matrix, under conditions similar to those used in other hybridization techniques (Van De Rijn and Gilks 2004). Saline solution with citrate (SSC, 300 mM NaCl and 30 mM sodium citrate) is used to reduce electrostatic repulsion between the probe and the labeled cDNA. Agents such as yeast tRNA, salmon sperm DNA, and poly (dA) are used as non-specific hybridization blockers. Hybridization must be carried out in a humid chamber with temperature control, and buffers with different degrees of stringency must be used to wash the material (Cheung et al. 1999). After the entire hybridization and washing process, the amount of hybridized fluorescent material will be visualized through the excitation of the fluorophores by a laser. The hybridization signal produced in each probe corresponds to the expression level of the respective gene in a given sample at the time of the study. Thus, after the marked targets' competitive hybridization, the signals are detected, quantified, integrated, and normalized with specific software and reflect the gene transcription profile for each biological sample (Van De Rijn and Gilks 2004).

## 3    Microarray Data Analysis

As with any analytical method, data analysis is a crucial step in microarray analysis. This step is essential to perform a data normalization to decrease the effect of random and systematic variation in the experiments. The term normalization refers to the mathematical process of removing these systematic variations observed for further evaluation of gene expression. The use of an adequate set of controls is essential for normalization (Bassett et al. 1999).

Normally, in microarray data analysis, two main sources of errors can be observed: biological and operational. They may be present in different steps of the microarray analysis process, including RNA isolation, cDNA labeling, hybridization, and microarray imaging. The efficiency of each step may vary accordingly to the equipment and the expertise of the analyst. Therein, normalization and use of adequate controls are fundamental. The simplest and most employed normalization approach is to normalize each spot's intensity with the total image intensity of all detected genes in the microarray chip. Normalization minimizes differences in intensity resulting from nonbiological variables, such as the amount of cDNA applied to the probe and the efficiency of incorporating and detecting the fluorescent signal (Bassett et al. 1999).

The analysts must choose a proper test for each of their datasets, and the best approach may be to examine the intersection of genes identified by different statistical tests or by the more conservative rank-sum test and non-parametric $t$-test (Chang and Chen 2009). Methods based on the conventional $t$-test allow the evaluation of the randomness of a gene modification. However, in the case of large-scale gene expression analysis, even a $p = 0.01$, which in the context of the analysis of a few genes is significant, is not sufficient here. For example, in an array with 10,000 genes, a conventional analysis considering $p = 0.01$ would indicate that there is a probability that 100 genes are being misplaced as differentially expressed. Problems of this type led the professor of the biostatistics group, Robert Tibshirani, from Stanford University, to develop a statistical method called Significance Analysis of Microarrays (SAM) (Tusher et al. 2001). The data analysis by SAM relies on a series of specific $t$-tests for each gene, adapted for the large-scale detection of differentially expressed genes. Based on the observation that random fluctuations are specific to each gene, the SAM test is based on the ratio between the difference between the means of the control and treated situations and the standard deviation of each gene, calculated from experimental repetition.

Many different forms of visualizing the information may be applied to microarrays, being clustering analysis is one of the most widely employed. Clustering is a method that organizes data and groups of genes according to their similarity in gene expression patterns; thence, samples with similar expression patterns are grouped. Different algorithms provide clustering, such as using centroid-based ($k$-means), density-based clustering, and the most applied hierarchical clustering. Particularly in microarrays, the average linkage is well established to create hierarchical clusters, generating a cluster dendrogram that reflects the similarity between the samples by their similar gene expression (Doug Chung and Le Roch 2013).

## 4    Microarray Applications

As mentioned before, the development of the microarrays revolutionized molecular biology since it was the first time scientists were able to perform a global analysis on the expression of tens of thousands of genes (Wöhrle et al. 2020; Brewster et al. 2004; Choudhuri 2004). The technology allowed the analyzes of biomarker determination, disease classification, and studies on gene regulation, for example (Chang et al. 2006). Besides qualitative analysis, microarrays allowed a quantitative comparison between samples through the use of two differently labeled mRNA samples. Compared to previous existing methods, the microarray technique's biggest advantage was the simultaneous analysis of a large number of genes and samples (Van Hal et al. 2000).

However, besides many advances and great characteristics attributed to microarrays, because their production remained "lab-made" for a while, its use remained narrowly adopted. Nonetheless, due to the technique's potential, commercial platforms began to be produced, driving its use to the next level. Manufacturers as Affymetrix, Agilent, Nimblegen, Ilumina, among others, were indispensable for the spread of microarray applications. The Affymetrix company was the first to launch oligonucleotide microarrays through a combination of micro photolithography and combinatorial chemistry. Their successful industrial method for producing high-quality DNA microarrays was based on each plate holding a GeneChip®, which consists of arrays of oligonucleotides directly attached to a surface, instead of attaching the cDNA probes in specific positions on a quartz plate. Therefore, GeneChips® is considered more accurate and more versatile than printed cDNA arrays (Miklos et al. 2005).

In earlier years, scientists have employed microarrays for gene expression monitoring, mutation detection, polymorphic analyses, gene mapping, drug discovery, and organism evolution studies (Schena et al. 1995; Kaushik et al. 2018; Miller and Tang 2009; Schena 1999). However, one of the most interesting applications was the analysis of expression patterns for disease and control conditions simultaneously, an innovative approach feasible by hybridizing cDNAs labeled with different fluorochromes.

Regarding DNA mutations, microarrays have been used for single nucleotide polymorphism (SNP) genotyping, which consists of a mutation of a single nucleotide on a gene sequence. Between the most common microarray technologies applied to SNPs, we may highlight the allele discrimination by hybridization, the allele-specific extension followed by an oligo hybridization, and approaches in which the evaluated DNA is extended the SNP through a single nucleotide extension reaction (Bumgarner 2013).

Comprehensive gene expression (CGE) analysis by DNA microarray has been applied in various human diseases, such as cancer, immunological health issues, and pathophysiological diseases (Kotani et al. 2015). CGE can also be performed focusing on different objectives, such as evaluating normal tissue taxonomy, disease diagnosis, disease prognosis, and biological mechanisms studies (Yee and Ramaswamy 2010). However, besides the disease or the goal, the basic principle remains the same, detect the genes or the gene expressions to access the features of the target condition, which is exactly the focus of microarray technology (Yee and Ramaswamy 2010).

Many studies have been conducted using expression profiling for diagnosis, especially in cancer. Oligonucleotides arrays were used to examine the expression of 6817 genes in 72 patients related the acute myelogenous leukemia (AML) and acute lymphoblastic leukemia (ALL) (Golub et al. 1999). Studies involving different cancer types found an expression signature that can serve as a biomarker for patient prognosis (Yee and Ramaswamy 2010).

A somehow more analytical approach for gene expression monitoring is quantitative gene expression. This approach was initially developed using the plant *Arabidopsis thaliana* as a model due to its small genome, enabling evaluation of the whole-genome expression (Schena et al. 1995). The major application of DNA chips has been to measure gene expression in miniaturized arrays, allowing a higher sensitivity, besides enabling the parallel screening of a large number of genes, employing smaller amounts of starting material (Van Hal et al. 2000). Normally, the microarrays are lab-made for these applications, initiating with RNA extraction from biological samples, followed by conversion to a labeled cDNA, as shown previously. The two most frequently employed quantitative gene expression methods are the incorporation of fluorescently labeled nucleotides in the cDNA synthesis step and the biotin-labeled nucleotide in the cRNA synthesis step (Bumgarner 2013).

This technology has shown great potential for use in microbial diagnosis. Pathogenic microorganisms are commonly identified by biochemical and immunological markers. An alternative is the use of ribosomal DNA for the identification of the microorganism (Chizhikov et al. 2001). The conventional methods are well established and effective; however, they are time consuming, and do not directly characterize the identified organisms' virulence factors. Researchers had made great progress in the study of expression in *Arabidopsis thaliana* (Schena et al. 1995), humans (DeRisi et al. 1997), *Escherichia coli* (Blattner et al. 1997), yeast (Lashkari et al. 1997), and *Mycobacterium tuberculosis* (Barry and Schroeder 2000).

Microarrays have also been applied in Functional Genomics, a field of research that attempts to connect each gene to their function and interactions. In this field, the most often used microarray technique is the high-density DNA microarrays, which consist of microarrays with lower spacings between the bases than usual, allowing the study of the structure, organization, and function of thousands of genes and their simultaneous expression in a single experiment. Undoubtedly, the expansion of DNA sequences deposited in gene banks allowed the improvement of new strategies to elucidate gene expression patterns in different organisms. All fields apart, microarrays have caused a high impact on the analysis of pathogen–host interaction once the pattern of gene expression differs when the host is infected (Diehn and Relman 2001). This information may be used not only to identify the pathogen but also to dictate better treatments.

An example of microarrays applied to bacteria research is the work developed by Richmond et al. (1999) where the authors used the microarray technique to analyze the expression of *Escherichia coli* (*E. coli*) genes. In their paper, 4290 oligonucleotides were drawn from Open Readings Frame (ORFs) of the complete genome to analyze the difference in gene expression when the bacteria were exposed to heat shock and treatment with isopropyl β-D-thiogalactopyranoside (IPTG). Likewise, Wei et al. (2001) compared the mRNA levels of *E. coli* cells cultured in rich and minimal media. Through the microarray technique, the expression of the bacteria genes grown in each condition was detected, and the results compared with what is already known about the physiology of the bacteria.

Investigating global changes in gene expression in *Bacillus subtilis*, Ye et al. (2000) built an arrangement containing 4020 ORFs, used to investigate differences in the expression of *B. subtilis* genes under aerobic and anaerobic conditions in the exponential growth phase. The authors reported that thousands of genes involved in various cellular functions, such as carbon and iron metabolism, electron transport, antibiotic production, and stress response, are differently regulated under anaerobic conditions.

A comparison between gene expression methods was made by Loos et al. (2001) through the study with *Corynebacterium glutamicum*. Comparing the changes in gene expression that occur between the fermentation phases (exponential growth and lysine production), the authors stated that microarray results for most genes correspond to those obtained through other methodologies as well.

Zheng et al. (2001) used a microarray arrangement composed of 4169 ORFs to examine a cDNA of *E. coli* cells previously treated with hydrogen peroxide. By measuring gene expression between isogenic and mutant strains, the authors confirmed that the OxyR regulator activates most of the genes induced by hydrogen peroxide. The DNA arrangements allowed the identification of several genes activated by the OxyR regulator, such as hemH (heme group biosynthesis) and operon suf (assembly or repair of Fe-S redox centers).

Strains of *Klebsiella* spp. were studied by different researchers since it is an endophyte commonly found in corn roots (Fisher et al. 1992; Chelius and Triplett 2000), a multimillionaire commodity, and the high pathogenicity of some *Klebsiella* species to humans. *Klebsiella* spp. is considered diazotrophic (Palus et al. 1996), i.e., it is able to fixate nitrogen on the soil due production of dinitrogenase reductase within the corn roots. On public health *Klebsiella pneumoniae* is the most important *Klebsiella* spp., and since its similarity with *E. coli*, Dong et al. (2001) hybridized the DNA of *K. pneumoniae* strain 342 (Chelius and Triplett 2000), with an array containing 4098 ORFs of *E. coli* strain K-12, completely sequenced by Blattner et al. (1997) The results obtained showed that the two bacteria have approximately 76% of common genes involved in energy metabolism, amino acids and fatty acids, cofactor synthesis, transport, cell division, DNA replication, transcription and translation, and regulatory protein synthesis. Dong et al. (2001) also mentioned that these results are in agreement with the physiological characteristics of the two organisms and that the comparison of interspecies genomes, through hybridizations, can quickly identify thousands of common genes among related organisms, even if one of them has not been completely sequenced. In other cases, for example, on the context of food and microbiological analysis, Spiro et al. (2000) applied microarrays to establish microbial profiles and quantification of changes in soil, sediment, and water microbiota as indicators of bioremediation.

Nowadays, newer techniques have been developed, and a great example of this movement is the direct sequencing of transcripts by high-performance technologies, called high throughput RNA-Seq, which has become an additional alternative to microorganisms and begun to replace the techniques of Serial Analysis of Gene Expression (SAGE) and Massive Parallel Signature Sequencing (MPSS) (Busch and Lohmann 2007). The most powerful ability of RNA-Seq is its independence on previous genomic annotation, meaning not requiring previous probes selection and avoiding the bias of microarray hybridization. On the other hand, RNA-Seq employs newer algorithms and logistical challenges, besides its lengthy procedures. Therefore, RNA-Seq is the method of choice in projects that use non-model organisms and discover new transcripts and genomic annotation. Microarrays, due to the intense processes for data processing and sample analysis, are still preferred for projects involving a large number of samples and for the analysis of transcriptomes in model organisms with well-characterized and annotated genomes (Baginsky et al. 2010).

# 5 Microarrays in Brazil

It may be noted that Brazilian scientists also recognized the innovation capacity of microarray technology and started in the early 2000s to apply microarrays to evaluate genetic properties and expression of their studied material, resulting in more than 2000 papers published up to nowadays. Several Brazilian institutions published papers applying microarrays, but Universidade de São Paulo was distinguished as the major Brazilian institution on microarray publications, leading São Paulo state as the responsible for most Brazilian papers using microarrays, as shown in Fig. 3. Led mainly by medicine and biochemistry research programs, Brazil has been active in microarray publications every year, even after NGS and RNA-Seq technologies arrived. In this section, the reader will be guided to a quick view of Brazilian achievements applying microarrays.

Nunes et al. (Nunes et al. 2003; de Souza et al. 2003) presented evidence of a coordinated control on the transcription of horizontal transfer elements by comparing 12 isolates of *X. fastidiosa,* in two different culture media, by microarray technique. The results showed most genes being transcriptionally active, and their expressions have been influenced by stimulating the environmental conditions to which the bacteria are cultivated.

Still in the early 2000s, Marques (2003) were pioneers in analyzing the expression of genes related to sucrose regulation in plants. To this purpose, 1370 genes were analyzed by comparing two commercial hybrid varieties of sugarcane,

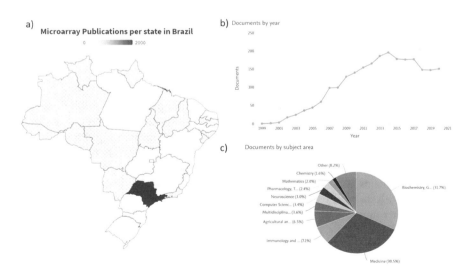

**Fig. 3** Brazilian scenario on microarray publications. (**a**) Graphical representation on publication distribution through Brazilian states. (**b**) Number of Brazilian publications on microarrays per year. (**c**) Most important research areas in Brazil applying microarrays. Source: Scopus database, search for "microarray" within article name, abstract or keywords, and "Brazil" in filiation country. The search was performed on January 13, 2021

SP87–396, and SP83–2847, presenting opposite characteristics regarding efficiency and speed in sucrose accumulation.

Souza et al. (2003) analyzed gene expression through microarrays in two stages of growth in *X. fastidiosa* and its relationship with pathogenicity. It was observed that most of the genes induced after a passage were associated with adhesion and probably adaptation to the environment. Thus, the authors also analyzed the pattern of gene expression of *X. fastidiosa* during the formation of biofilms through microarrays, noting that the gene expression in the cells that are developing the biofilm is similar to the other systems already characterized.

Koide et al. (2004) compared pathogenic and non-pathogenic *X. fastidiosa* isolates, observing that isolates J1 to 12, when inoculated in tobacco and citrus, did not induce symptoms. Once compared with isolate 9a5c, considered highly pathogenic, it did not present genes related to adhesion and pathogenicity. Sequentially, the authors analyzed the global gene expression of the response to high temperatures (heat shock) through microarrays, revealing a complex network of genes that act together in response to the temperature.

Travensolo et al. (2009a, b) analyzed the expression of the bacterium *X. fastidiosa* in different culture media using microarray. The expressions of genes related to the metabolism of amino acids, proteins, nucleotides, energy, among others, have been observed, suggesting that this bacterium is able of synthesizing substances according to its needs and according to the shortage of nutrients in the culture medium.

Zaini et al. (2008) studied the gene expression profile of *X. fastidiosa* in response to iron limitation through microarrays and observed the induction of genes that encode pilis type IV, as well as bactericin-type Colicine V.

## 6 Conclusions and Perspectives

Through this chapter the reader travelled back to the past and could comprehend more about how the amazing DNA analysis scenario from nowadays was slowly been created, from human genome project and Sanger sequencing, to microarray analysis, culminating on the Next Generation Sequencing. Although microarray technology is not as powerful for DNA analysis as NGS, we could see that the technique remains useful over the past years, allowing quick and reliable analysis when the analysts need data from specific genes. Furthermore, microarrays entered on its own evolution scenario, with the emergence of commercial microarrays and automatization of the protocols, supporting even more applications.

From the authors' point of view, although NGS has become cheaper and faster over the years, microarrays technology will remain useful for a long time, since its simple application does not require highly trained staff, and can be easily applied for clinical diagnosis, besides its use in research. For the future, as genomic research discovers newer genetic markers, new microarrays may be developed to detect and analyze the expression of the biomarkers, helping to achieve the everyday closer dream of personalized medicine.

# References

Alsina MJ (2020) Technol Cult 61:976–978
Baginsky S, Hennig L, Zimmermann P, Gruissem W (2010) Plant Physiol 152:402–410
Barry CE, Schroeder BG (2000) Trends Microbiol 8:209–210
Bassett DE, Eisen MB, Boguski MS (1999) Nat Genet 21:51–55
Bhatia S, Dahiya R (2015) Tissue culture science. Elsevier, Amsterdam
Blattner FR, Plunkett G, Bloch CA, Perna NT, Burland V, Riley M, Collado-Vides J, Glasner JD, Rode CK, Mayhew GF, Gregor J, Davis NW, Kirkpatrick HA, Goeden MA, Rose DJ, Mau B, Shao Y (1997) Science 277:1453–1462
Brewster JL, Beason KB, Eckdahl TT, Evans IM (2004) Biochem Mol Biol Educ 32:217–227
Bumgarner R (2013) Curr Protoc Mol Biol 101:22.1.1–22.1.11
Busch W, Lohmann JU (2007) Curr Opin Plant Biol 10:136–141
Carrasco-Ramiro F, Peiró-Pastor R, Aguado B (2017) Gene Ther 24:551–561
Chang HY, Chen X (2009) Microarray analysis of stem cells and differentiation, 2nd edn. Elsevier, Amsterdam
Chang HY, Thomson JA, Chen X (2006) Methods Enzymol 420:225–254
Chelius MK, Triplett EW (2000) Appl Environ Microbiol 66:783–787
Cheung VG, Morley M, Aguilar F, Massimi A, Kucherlapati R, Childs G (1999) Nat Genet 21: 15–19
Chizhikov V, Rasooly A, Chumakov K, Levy DD (2001) Appl Environ Microbiol 67:3258–3263
Choudhuri S (2004) J Biochem Mol Toxicol 18:171–179
de Souza AA, Takita MA, Coletta-Filho HD, Caldana C, Goldman GH, Yanai GM, Muto NH, de Oliveira RC, Nunes LR, Machado MA (2003) Mol Plant-Microbe Interact 16:867–875
DeRisi JL, Iyer VR, Brown PO (1997) Science 278:680–686
Deyholos MK, Galbraith DW (2001) Cytometry 43:229–238
Diehn M, Relman DA (2001) Curr Opin Microbiol 4:95–101
Dong Y, Glasner JD, Blattner FR, Triplett EW (2001) Appl Environ Microbiol 67:1911–1921
Doug Chung DW, Le Roch KG (2013) Genome-wide analysis of gene expression, 2nd edn. Elsevier, Amsterdam
Fisher P, Petrini O, Scott H (1992) New Phytol 122:299–305
Gibbs RA (2020) Nat Rev Genet 21:575–576
Golub T, Slonim D, Tamayo P, Huard C, Gassenbeck M, Mesirov J, Coller H, Loh M, Downing J, Caligiuri M, Bloomfield CD, Lander ES (1999) Science 286:531–537
Kaushik S, Kaushik S, Sharma D (2018) Encyclopedia of bioinformatics and computational biology: ABC of bioinformatics, vol 1–3. Elsevier, Amsterdam, pp 118–133
Koide T, Zaini PA, Moreira LM, Vêncio RZN, Matsukuma AY, Durham AM, Teixeira DC, El-Dorry H, Monteiro PB, da Silva ACR, Verjovski-Almeida S, da Silva AM, Gomes SL (2004) J Bacteriol 186:5442–5449
Kotani K, Kawabe J, Morikawa H, Akahoshi T, Hashizume M, Shiomi S (2015) Mediat Inflamm 2015:349215
Lashkari DA, Derisi JL, Mccusker JH, Namath AF, Gentile C, Hwang SY, Brown PO, Davis RW (1997) Proc Natl Acad Sci U S A 94:13057–13062
Loos A, Glanemann C, Willis LB, O'Brien XM, Lessard PA, Gerstmeir R, Guillouet S, Sinskey AJ (2001) Appl Environ Microbiol 67:2310–2318
Marques JO (2003) Expressed Sequence Tags de Cana-de-Açúcar Diferencialmente Expressos em Folha e Colmo Maduros. Universidade Estadual Paulista, Jaboticabal
Miklos DA, Freyer GA, Crotty D (2005) A Ciência do DNA, 2nd edn. Artmed, Porto Alegre
Miller MB, Tang YW (2009) Clin Microbiol Rev 22:611–633
Nunes LR, Rosato YB, Muto NH, Yanai GM, da Silva VS, Leite DB, Gonçalves ER, de Souza AA, Coletta-Filho HD, Machado MA, Lopes SA, de Oliveira RC (2003) Genome Res 13:570–578
Palus JA, Borneman J, Ludden PW, Triplett EW (1996) A diazotrophic bacterial endophyte isolated from stems of zea mays l. and Zea luxurians iltis and doebley. Plant soil. Kluwer Academic, Boston, pp 135–142

Phimister B (1999) Nat Genet 21:1

Raghavendra P, Pullaiah T (2018) Advances in cell and molecular diagnostics. Academic Press, Amsterdam, pp 57–84

Richmond CS, Glasner JD, Mau R, Jin H, Blattner FR (1999) Nucleic Acids Res 27:3821–3835

Schena M (1999) DNA microarrays: a practical approach. Oxford University Press, Oxford

Schena M, Shalon D, Davis RW, Brown PO (1995) Science 270:467–470

Spiro A, Lowe M, Brown D (2000) Appl Environ Microbiol 66:4258–4265

Travensolo RF, Carareto-Alves LM, Costa MVCG, Lopes TJS, Carrilho E, Lemos EGM (2009a) Genet Mol Biol 32:340–353

Travensolo RF, Costa MVCG, Carareto-Alves LM, Carrilho E, Lemos EGM (2009b) Braz Arch Biol Technol 52:555–566

Tusher VG, Tibshirani R, Chu G (2001) Proc Natl Acad Sci U S A 98:5116–5121

Van De Rijn M, Gilks CB (2004) Histopathology 44:97–108

Van Hal NLW, Vorst O, Van Houwelingen AMML, Kok EJ, Peijnenburg A, Aharoni A, Van Tunen AJ, Keijer J (2000) J Biotechnol 78:271–280

Wei Y, Lee JM, Richmond C, Blattner FR, Rafalski JA, Larossa RA (2001) J Bacteriol 183: 545–556

Welford SM, Gregg J, Chen E, Garrison D, Sorensen PH, Denny CT, Nelson SF (1998) Nucleic Acids Res 26:3059–3065

Wöhrle J, Krämer SD, Meyer PA, Rath C, Hügle M, Urban GA, Roth G (2020) Sci Rep 10:1–9

Ye RW, Tao W, Bedzyk L, Young T, Chen M, Li L (2000) J Bacteriol 182:4458–4465

Yee AJ, Ramaswamy S (2010) DNA microarrays in biological discovery and patient care, 1st edn. Elsevier, Amsterdam

Zaini PA, Fogaça AC, Lupo FGN, Nakaya HI, Vêncio RZN, da Silva AM (2008) J Bacteriol 190: 2368–2378

Zheng M, Wang X, Templeton LJ, Smulski DR, LaRossa RA, Storz G (2001) J Bacteriol 183: 4562–4570

# Chemometrics in Bioanalytical Chemistry

Marcelo Martins Sena, Jez Willian Batista Braga,
Márcia Cristina Breitkreitz, Marco Flores Ferrão, and
Carolina Santos Silva

## 1 Introduction

Chemometrics can be broadly defined as the research area that applies multivariate statistics to chemical data (Otto 1999; Sena et al. 2017). This definition can also include data from similar areas, such as biology, pharmacy, environmental sciences, etc. The discipline and the word chemometrics appeared at the beginning of the 1970s aligned with the spread of microprocessors in chemical laboratories at that time. Jointly with the emergence of microcomputers, this technological advancement drastically increased the capacity of chemists and biochemists to generate data, thus demanding new methods to process them and extract relevant information. In the early days of chemometrics, two researchers were the main pioneers, Bruce R. Kowalski, in the University of Washington at Seattle, USA, and Svante Wold, in the Umea University, Sweden. Since the beginning, chemometric applications in analytical chemistry have stood out. As quoted by one of the pioneers, analytical

M. M. Sena (✉)
Chemistry Department, Institute of Exact Sciences, Universidade Federal de Minas Gerais, Belo Horizonte, MG, Brazil
e-mail: marcsen@ufmg.br

J. W. B. Braga
Chemistry Institute, Universidade de Brasília, Brasília, DF, Brazil

M. C. Breitkreitz
Analytical Chemistry Department, Chemistry Institute, Universidade Estadual de Campinas, Campinas, SP, Brazil

M. F. Ferrão
Inorganic Chemistry Department, Chemistry Institute, Universidade Federal do Rio Grande do Sul, Porto Alegre, RS, Brazil

C. S. Silva
Faculty of Health Sciences, Department of Food Sciences and Nutrition, University of Malta, Msida, Malta

© The Author(s), under exclusive license to Springer Nature Switzerland AG 2022
L. T. Kubota et al. (eds.), *Tools and Trends in Bioanalytical Chemistry*,
https://doi.org/10.1007/978-3-030-82381-8_26

chemistry had the data and had the need (Geladi and Esbensen 1990). An updated definition of chemometrics prescribed by the International Union of Pure and Applied Chemistry (IUPAC) is "the science of relating measurements made on a chemical system or process to the state of the system via application of mathematical or statistical methods" (Hibbert 2016).

Bioanalytical data are often multivariate in nature, and bioanalytical samples are in most cases more complex than pharmaceutical formulations, food, and environmental samples. One of the most typical goals of chemometric methods is the modeling of latent multivariate data patterns as principal components. In many types of applications, such as frequently found in pharmaceutical quality control analysis, the estimated factors are associated with singular chemical molecules. However, in bioanalytical analysis involving more complex biological or pharmaceutical samples, e.g., in in vivo analysis, each factor or principal component is more often related to the cumulative response of several molecules (Kumar and Cava 2017).

Bioanalytical data are remarkably diverse, originating from different fields, such as plant science, toxicology, clinical analysis, disease diagnosis, personalized medicine, pharmaceutical and environmental research, etc. An area of particular interest is metabolomics, which can be considered an amalgam of traditional fields that include bioanalytical development, metabolite analysis, analytical separations, and chemometrics (Khoo and Al-Rubeai 2007). Typically, metabolomics has the goal of distinguishing subpopulations within an overall sample population and to detect and identify the respective biomarkers (Sandusky 2017). In this context, a key step in the search for new biomarkers is related to the interpretation of multivariate data obtained after perturbations in biological systems, caused by diseases or controlled treatments (Kotlowska 2014; Madsen et al. 2010). Perturbed systems are compared to unperturbed reference ones and for this aim univariate statistics is very limited. Nevertheless, until some years ago, chemometric methods were largely overlooked by biologists, who preferred the traditional univariate statistical tools (Trygg et al. 2007). This situation has changed in the last years due to the steady trend in the capacity of biologists and bioanalytical chemists to generate higher amounts of data, which are used to characterize samples, objects, experiments, or time points. For instance, in clinical analysis assays based on only traditional known analytes have been continuously replaced by metabolomic approaches, in which multivariate statistics plays a key role in the discovery of new disease biomarkers (Madsen et al. 2010).

In general, chemometric models have been developed mainly based on data generated by molecular spectroscopy. The most employed analytical techniques have been vibrational, UV-visible, mass spectrometry (MS), and nuclear magnetic resonance (NMR) spectroscopies (Brereton 2003). Particularly in the development of supervised qualitative (classification) and quantitative models, the combination of chemometrics and vibrational spectroscopy has provided non-destructive (or minimally destructive), rapid, and low-cost analytical methods based on the direct determination of the analytes or some intrinsic properties. Multivariate modeling eliminates the requisite of signal resolution, and the selectivity of the analytical technique is no longer a requirement. The physical separation of the interferences can be substituted by the chemometric separation of the analytical signals, what is sometimes figuratively called "mathematical chromatography" (Bro et al. 2010). Other advantages of this kind of methodology are that analysis minimizes or even does not consume reagents or solvents nor generates chemical waste. In addition, the trends of miniaturization and portability in spectroscopic equipment (Crocombe 2018), more pronounced in vibrational spectrophotometers (Santos et al. 2013), have great potential to increasingly simplify and democratize analytical and bioanalytical chemistry sciences. We can imagine soon the incorporation of these technologies as a part of the Internet of Things (Crocombe 2018).

In metabolomic studies, chromatographic data have also been commonly combined with chemometrics, and vibrational spectroscopy has been less widely utilized. Vibrational techniques, such as near and mid infrared (NIR and MIR), and Raman spectroscopies have been largely employed in other fields such as food analysis for the quantification of macroconstituents, detection of adulterants, or origin assignment (Sena et al. 2017). However, in bioanalytical chemistry applications, the use of these techniques is less common due to their lower sensitivity, which can be demanded for determinations in some types of biological samples.

A limitation in the use of chemometrics as applied to bioanalytical chemistry is the lack of deep knowledge of many users about this discipline. There are typical misunderstandings commonly present in the literature in general, such as the belief that a good principal component analysis (PCA) model provides necessarily a good separation of the sample clusters, the common absence of model (spectral) interpretation, and the lack of validation of supervised classification and calibration models (Kjeldahl and Bro 2010). Specifically, in bioanalytical applications, some authors have drawn attention to the lack of validation and model interpretation in most of the metabolomic literature. Regardless of the chosen method, statistical/chemometric validation is so crucial as biological validation. However, in practice, the absence of model validation seems tempting, since it provides results that might look better for a non-specialist at a first glance (Madsen et al. 2010). Model interpretation is also crucial, since metabolomic papers should not be limited to the question "what is there?", but rather should intend to answer other points, such as "what is its/their relation to?" and "what is the difference between?" (Trygg et al. 2007).

One of the aims of this chapter is to warn bioanalytical chemists about these types of misunderstandings in chemometrics, which may be present in the literature. Our main goal is to describe the most important chemometric methods that will be divided according to their traditional objectives. In this sense, the following sections will present and discuss data preprocessing, design of experiments, exploratory analysis (unsupervised classification), supervised classification, multivariate calibration, and curve resolution. The standard algebraic notation will be used to describe equations, in which scalars are represented by italic lowercase letters, vectors by boldface lowercase letters, and matrices by boldface uppercase letters. The discussion will be restricted to linear multivariate methods considering the principle of parsimony, the high collinearity often found in metabolomic and bioanalytical data, and the drawback caused by limitations in the interpretation of non-linear multivariate models (Madsen et al. 2010).

## 2    Data Preprocessing

Preprocessing is a key step in multivariate analysis aiming to remove unwanted variance from the data (Engel et al. 2013; Mishra et al. 2020; Rinnan et al. 2013). These unwanted sources of variance are often called artifacts, which can be originated from different factors, such as instrumental drifts, noise, measurement modes, and other external variations. Bioanalytical chemists should have knowledge about all the chemical, physical, and environmental sources of variation in the data in order to make the appropriate choices. Inappropriate preprocessing choices can undermine chemometric models by removing systematic variance crucial for the prediction performance of the method. The purely empirical testing of preprocessing options available in data analysis softwares should be strongly discouraged.

The two most basic preprocessing methods are mean centering and autoscaling (Bro and Smilde 2003; van den Berg et al. 2006). In practice, almost all of the datasets should be submitted to one of these two preprocessing methods. These two methods are applied to the columns of the data matrix. Mean centering consists of subtracting the mean of each column/variable from each value. This corresponds to the translation of the natural origin of the data to the multivariate mean. Spectral data is usually mean centered. Autoscaling consists of two steps. Firstly, data are mean centered, and then the resultant values are divided by the standard deviation of the respective column. This preprocessing provides equal weights for all the variables and is recommended when modeling discrete non-spectral variables. When data is only mean centered, the subsequent analysis is performed based on covariances, while when autoscaling is applied, this is based on correlations (van den Berg et al. 2006). Autoscaling of spectral data should be avoided in the absence of a particular reason. Variables related to the top of spectral bands usually contribute with more relevant systematic variance to the predictive model than variables associated with low intense signals. In a negative way, autoscaling improves the contribution of noisy spectral regions. Other alternatives sometimes used in metabolomic studies are Pareto scaling, range scaling, and vast scaling (van den Berg et al. 2006). The last two types of scaling are less common, and Pareto scaling is like autoscaling with each mean centered value being divided by the square root of the standard deviation, instead of the raw standard deviation itself. All the previously mentioned scaling methods are column normalizations. Other preprocessing methods are based on row normalizations. Thus, authors should avoid informing that data was normalized without specification of which type (Sena et al. 2017).

Important sources of unwanted variance in bioanalytical data, mainly in molecular spectra, are noise, baseline deviations, light scattering, and misalignments (Engel et al. 2013). To deal with these artifacts, several row preprocessing methods are available. Noisy data/spectra demand the increase of signal-to-noise ratio. This is performed by applying digital filters with the aim to smooth the data. The most used method has been the Savitzky–Golay smoothing (Savitzky and Golay 1964), which is a moving-average filter that fits a polynomial of low degree to neighboring data point windows by least squares. The size of the window should be adjusted with care, since small windows may not sufficiently filter noise while large ones may distort the analytical signal causing loss of systematic information. Typical window sizes are between 7 and 15 points. Another alternative is the Norris derivative filter (Pan et al. 2020), in which three parameters are fitted to the data, the derivative order, the numbers of smoothing points and of differential gaps.

Baseline deviations are artifacts commonly observed in spectroscopic data. These deviations can be linear or non-linear. Linear baseline deviations are associated with the need to correct a constant offset and may be caused by problems in the blank compensation. Specific methods of offset correction and first derivatives can be used to overcome this kind of artifact (Barak 1995). Second derivative is used to correct non-linear baseline deviations. However, the use of derivatives has some drawbacks, such as the improvement of the noise and the difficulty in the (spectral) interpretation of the informative vectors obtained from chemometric models. Aiming to avoid the decrease in the signal-to-noise ratio, the simultaneous use of smoothing jointly with derivatives is recommended. The use of higher-order derivatives should be avoided.

Spectral drifts are non-linear baseline deviations caused by light scattering. These artifacts are typically observed in vibrational spectra (NIR, MIR, Raman) of solids and suspensions obtained in reflectance modes. The preprocessing methods most used to eliminate the variance of drifts are multiplicative scatter correction (MSC) and standard normal variate (SNV) (Rinnan et al. 2013). On the other hand, misalignments (shifts in the abscissa axis) are typically observed in NMR spectra and chromatographic data. For NMR data, preprocessing methods specifically developed for correcting the occurrence of shifts between peaks across different spectra have been proposed in omics studies (Vu and Laukens 2013), but the most used is binning or bucketing (Smolinska et al. 2012). In this method, NMR spectra are divided into small bins for which the respective integrals of peak areas will replace the original variables. A side effect is the decrease in the spectral resolution. For aligning chromatographic data, the most known preprocessing methods are interval correlation shifting (icoshift) and correlation optimized warping (COW) (Kumar 2018). While icoshift is based on an insertion and deletion approach, COW acts expanding and compressing variable intervals. The COW algorithm is composed of three steps. First, a reference sample is selected, then two warping parameters are optimized, the length and the slack of the segment, and finally the analytical signal is rebuilt. Both reference and unaligned samples are split into an equal number of segments. The alignment is based on the maximization of the correlation coefficients of the variables and the signals are rebuilt through linear interpolation.

Finally, it is interesting to mention the orthogonal signal correction (OSC) (Westerhuis et al. 2001). This is a filter that eliminates variance in the independent variables (spectral or other analytical signals, $\mathbf{X}$ matrix) uncorrelated with the dependent variables (analyte concentration or property to be quantified, or dummy variables in classification, $\mathbf{Y}$ matrix). The use of OSC is recommended for models in which the first latent variables (LV) explain much more variance in $\mathbf{X}$ than in $\mathbf{Y}$. The number of components to be removed should not be higher than two, otherwise models tend to overfitting. The first removed component represents a linear baseline correction while the second OSC component is equivalent to correcting multiplicative effects. The combination of OSC with partial least squares (PLS) and PLS discriminant analysis (PLS-DA) originated orthogonal PLS (OPLS) and PLS-DA (OPLS-DA) (Bylesjö et al. 2006), specific methods most common in metabolomic applications and that will be discussed in the next sections.

## 3    Design of Experiments

Design of experiments (DOE) is a broad term that encompasses different multivariate methods aiming at setting up experiments and extracting information from data in many areas of knowledge. The pioneering work of Ronald A. Fisher, in the 1920s–30s, introduced statistical thinking and principles into designing experimental investigations, including the factorial design concept and the analysis of variance (ANOVA). The work of George E. P. Box and K. B. Wilson in the 1950s set the foundations of the field of response surface methodology (RSM). The introduction of structured experimentation to the industrial scenario was accomplished by Taguchi's work in the 1970s, who used fractional factorial designs and orthogonal arrays to help decreasing the variability of industrial processes and developed the concept of "robust engineering" (Montgomery 2001).

DOE methods can be classified according to the nature of the involved factors (statistically independent/non-independent variables), the goal of the experiment, and the number of variables involved. Figure 1 is a schematic representation of the most important DOE methods.

The objective of any experimental design is to establish experiments in a statistically appropriate manner aiming to study variables that are of interest to the researcher. These variables are represented in a multidimensional space, with the number of dimensions equals to the number of variables studied. Levels of the factors should be selected, determining an experimental region to be investigated, and experiments are carried out using combinations of these factor levels.

If the number of variables is high (typically higher than 5), screening designs can be used to select the most influential factors. After identifying the most important factors, complete designs can be set and the results used to build models by multiple linear regression (MLR), describing the behavior of the response variables ($y$) along with the variations of experimental parameters ($x_i$). In addition to the organization of the experimental work in a systematic way, covering the whole experimental domain, this set-up brings several advantages as compared to the univariate one-factor-at-a-time (OFAT) approach. One major advantage of this methodology is the identification of possible interactions between the variables, which could not be achieved by the univariate method. Interactions occur when the effect observed by varying variable A depends on the level of variable B. In this case, it is said that A and B interact. Another significant advantage is the possibility of modeling

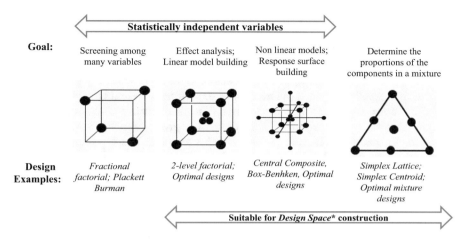

**Fig. 1** A representative illustration of the most important DOE methods. (*) The term Design Space will be explained further on in this section

responses as a function of the experimental parameters and, therefore, building response surface graphs that are highly informative for the researcher. It should be highlighted that fractional factorial designs are not suitable for model building, except those of resolution higher than V, due to the aliases (confounding) of the effects.

The models can be linear or of a higher order, according to the type of design used. Two-level factorial designs allow the development of linear models and center points in the middle of experimental domain allow evaluating the curvature of the response surface. On the other hand, Box–Behnken or central composite designs allow the estimation of quadratic coefficients of non-linear models, besides the linear and interaction coefficients. The coefficients (b) and their variances can be calculated according to Eqs. (1) and (2), respectively,

$$\mathbf{b} = \left(\mathbf{X}^T\mathbf{X}\right)^{-1}\mathbf{X}^T\mathbf{y} \tag{1}$$

$$s^2{}_\mathbf{b} = \left(\mathbf{X}^T\mathbf{X}\right)^{-1}\mathbf{X}^T\mathrm{MSr} \tag{2}$$

where $\mathbf{X}$ is the experimental matrix containing the structured $x_i$ experiments, $\mathbf{y}$ is the response vector containing the property of interest, the superscript "T" stands for the transpose of a matrix; $s^2{}_\mathbf{b}$ represents the variance of the coefficients, and MSr refers to the mean square of residuals from an ANOVA (analysis of variance), which could be alternatively replaced by the mean square of pure error ($\mathrm{MS_{pe}}$). Aiming to obtain reliable response surfaces and predictions, models should be carefully evaluated using ANOVA and residual diagnostic plots (Box et al. 2005). After evaluating their validity, models can be used to define a region that meets the desired requirements of the investigator.

There are many situations in which a standard experimental design does not fit the research requirements of the problem, for example, when constraints need to be imposed on the design region, when a nonstandard model is necessary to adequately explain the response or when the number of experiments should be constrained. In these situations, optimal designs can be used. They are versatile computer-generated designs that set the experiments according to a given predefined mathematical criterion, for example, D-, G-, A-, I-optimality criteria. The D-optimal criterion aims at maximizing the determinant of the $\mathbf{X}^T\mathbf{X}$ (information matrix), i.e., experiments are chosen within a candidate set in a way that this criterion is achieved. The determinant maximization aims at covering the largest possible volume in the experimental region. Maximizing the determinant of the information matrix $(\mathbf{X}^T\mathbf{X})$ is equivalent to minimizing the determinant of the dispersion matrix $(\mathbf{X}^T\mathbf{X})^{-1}$, minimizing the average variance of the regression coefficients. Similarly, to D-optimal, A-optimal designs aim at minimizing the average variance of the regression coefficients, but in this case by minimizing the trace of the $(\mathbf{X}^T\mathbf{X})^{-1}$ matrix. On the other hand, G and I criteria seek the minimization of the average variance of the response predicted values (Johnson et al. 2011).

## 3.1 DOE in the Quality by Design Framework for Pharmaceutics and Biopharmaceutics

The use of DOE methods in the pharmaceutical and biopharmaceutical fields has significantly increased in the last decades. These industries are currently facing a paradigm shift from the less effective OFAT approach to the multivariate way of thinking, especially driven by the recommendations of regulatory agencies. The scenario started changing in 2004 when the FDA (Food and Drug Administration) launched the guidelines for good manufacturing practices (GMP) for the twenty-first century, motivating the use of scientific knowledge and multivariate tools (Department of Health and Human Services, Food and Drug Administration 2004). Following the same trend, in 2005 International Conference on Harmonization (ICH), a group created to harmonize drug registration requirements worldwide, launched the "*ICH Q8 Pharmaceutical Development*" guide, which became a milestone in pharmaceutical development because it introduced the concept of "quality by design" (QbD) for pharmaceutical development (International Conference of Harmonisation (ICH) 2008). QbD is a framework whose major goal is to understand how raw materials, formulation, and process factors affect product performance. Therefore, there is a change of focus to building the quality during the process, instead of only testing it at the end of the production chain ("quality by testing"). In the QbD framework, the critical quality attributes (CQA) of the product or process should be described as functions of the critical material attributes (CMA) and critical process parameters (CPP), using multivariate statistical methods, mainly

DOE-based. CMA and CPP are identified by a previous risk analysis step and are the input $x_i$ variables of the design, whereas the CQA are the responses (y variables).

A very important step of QbD is the definition of the "design space" (DS), which is "the multidimensional combination and interaction of input variables (e.g., CMA and CPP) that have been demonstrated to provide assurance of quality" (International Conference of Harmonisation (ICH) 2008). According to this definition, the DS is considered a robust region of work. Therefore, within this region experimental variations do not cause changes in CQA. Since DS is normally built as a compromise of different CQA, multi-criteria tools are needed, such as the study of the coefficients in the fitted equations and overlaying contour maps, and desirability functions (Derringer and Suich 1980). It should be emphasized that a combination of acceptable ranges based on univariate optimization does not constitute a DS; the acceptable ranges must be based on multivariate optimization taking into account the main effects, as well as variables interactions. After establishing the DS, a control space should be determined encompassing the operating conditions of the process, and in this step other chemometric methods may be used, such as PCA and PLS.

The current status, challenges, and future perspectives of QbD for pharmaceutical development were recently summarized (Grangeia et al. 2020). ICH Q8 concepts of QbD and DS, initially proposed for pharmaceutical development of synthetic drugs, were readily extended to biotechnological processes, products, and even clinical results (Rathore and Winkle 2009; Rathore 2009). The processes are typically complex, composed of many unit operations and present a high raw material variability to be accounted for. The establishment of the DS not only brings extensive knowledge of the process and guarantees its consistency even in the presence of small changes, but also provides regulatory flexibility for the manufacturer (International Conference of Harmonisation (ICH) 2008).

Harms et al. have presented a stepwise approach for defining the process DS for a biologic product involving *P. pastoris* fermentation (Harms et al. 2008). First, risk analysis via failure modes and effects analysis (FMEA) was performed to identify parameters for process characterization, which were studied by a fractional factorial design. The results indicated that the fermentation unit operation was very robust with a wide DS and no critical operating parameters. A case study has been presented to illustrate the step-by-step development of a purification process to produce a biosimilar product. In this paper, advantages of QbD-based process development over traditional approaches, such as higher yield and increased product quality, were highlighted and discussed (Rathore 2016). Rathore and Winkle have performed a two-phase study to describe a chromatographic step of a biotech process (Rathore and Winkle 2009). First, a screening was performed to identify the process parameters presenting the higher impact on product quality (% purity) and process consistency (% recovery). Then, a DOE was carried out to study temperature, buffer A pH, buffer B pH, flow rate, product loading, and bed height. The results allowed establishing both the operating and the acceptable ranges, as well as categorizing the variables as key or critical.

The concepts and applications of different DOE methods in recombinant protein biotechnology have recently been detailed reviewed (Papaneophytou 2019), highlighting the advantages of multivariate methods over the traditional OFAT. The literature on QbD for biopharmaceutics, including perspectives and case studies, has been organized in a book. This book presents examples of the definition of process DS for a microbial fermentation step, purification processes, tangential flow filtration operations, raw materials assessment, formulation DS, among other topics (Kozlowski and Swann 2009).

## 3.2    Analytical Quality by Design (AQbD)

Subsequently, the concept of QbD was extended to the development of analytical methods in laboratory scale, originating the branch of analytical QbD (AQbD) (Vogt and Kord 2011; Orlandini et al. 2013). In this branch, the objective is to use DOE tools for establishing multivariate relationships between the experimental variables and the quality parameters of the analytical method. The ultimate goal is to build the quality and gather knowledge about method performance under different conditions during its development, instead of only evaluating it in the final validation step (Hubert et al. 2015). The development of methods using univariate strategies, besides being very time-consuming, does not provide satisfactory solutions in the presence of interactions between variables. In addition, results are obtained only at the points in which the experiments were carried out, not allowing an overview of the whole experimental domain, as already discussed.

High performance and ultra-high performance liquid chromatography (HPLC and UPLC, respectively) combined with different detectors have been the most used techniques for the analysis of drugs (impurities, degradation products) and medicines (content, uniformity, dissolution), and different types of bioanalysis. In liquid chromatography, some examples of experimental variables affecting a separation are chromatographic column (type of stationary phase, particle size and length), type and percentage of organic modifier(s), pH of the mobile phase, gradient variables (time or slope), column temperature, mobile phase flow, injection volume, etc. The most influential variables can be identified by risk analysis and are called critical procedure parameters (CPP) or critical method parameters (CMP). Some examples of typical CQA of chromatographic methods are resolution, separation factor, asymmetry factor, number of theoretical plates, analysis time, etc.

If the number of variables is high, a two-step approach can also be used in AQbD, i.e., first a screening design is employed to identify the most important variables, and then an optimization step is carried out using, for example, central composite design (CCD), Box–Behnken (BB), or optimal designs. The result of the second step is the determination of DS, which is also called method operable design region (MODR) (Rozet et al. 2013; Bhutani et al. 2014). In order to ensure that quality parameters are achieved for all compounds in the chromatogram, a procedure for the simultaneous optimization of various responses should be employed. With this aim, Candioti et al. have reviewed the use of desirability functions (Candioti et al. 2014). A comprehensive review of the DOE methods for HPLC method development has recently been published, including theory and applications (Sahu et al. 2018). For analytical method development, mixture designs can be used for the optimization of the mobile phase composition.

The DS definition allows the selection of working conditions in a safe manner during the method development. As the entire experimental domain is known, it is possible to select a region in which the answers have higher desirability values and simultaneously are robust to small variations. Thus, it is possible to define the limits of failure (edges of failure) from which the method performance becomes unacceptable, allowing risks upon transfer. Likewise, as in QbD applications for product/process, careful statistical analysis of models used to build DS should be performed. Another important step of increasing interest is the evaluation of DS uncertainty limits. Monte-Carlo simulations, Bayesian modeling, and bootstrapping are useful tools to accomplish this task (Deidda et al. 2018; Dispas et al. 2018). Estimate of confidence intervals for the predicted responses also represents a way to take into account the uncertainty of DS.

Recently, a complete description of the development of an HPLC reversed-phase method for quantitative analysis of ferulic acid in human plasma according to AQbD has been published (Saini et al. 2020). An Ishikawa fishbone diagram was constructed to evaluate several potential variables influencing the analytical target profile (peak area, theoretical plates count, retention time, and peak tailing as the critical analytical attributes). Risk assessment using a risk estimation matrix and factor screening studies employing Taguchi design helped to identify two critical method parameters (ratio of acetonitrile:water in the mobile phase and flow rate). Subsequently, the optimum operational conditions of the method were delineated using face-centered CCD. Monte-Carlo simulations were used to check the robustness of the MODR. The method was then validated and applied to biological samples. This study encompassed all important elements of the AQbD strategy, with DOE as an important tool of a broader framework. DOE is sometimes wrongly mentioned in the literature as a synonym of QbD or AQbD, which is not accurate.

An integrated AQbD approach for the development and validation of an HPLC method to determine valsartan in rat plasma has been recently published (Bandopadhyay et al. 2020). Studies were carried out in two steps, involving the optimization of chromatographic conditions with a Box–Behnken design (CMP: mobile phase pH; flow rate and % of acetonitrile in the mobile phase), followed by the optimization of the extraction process using a face-centered CCD for achieving maximum recovery. Nevertheless, neither for the extraction nor for the analytical method the uncertainties of the MODR were estimated. Sample preparation is a key step in the development of chromatographic methods, especially when applied to complex bioanalytical samples. Experimental designs for solid-phase microextraction (SPME) method development in bioanalysis have been reviewed (Marrubini et al. 2020). Different DOE methods published to study/optimize extraction by SPME during the period of 2009–2019 were described by the authors.

# 4    Unsupervised Pattern Recognition Methods

One of the first types of chemometric methods that had its development and application in bioanalytics was unsupervised pattern recognition. Much of chemistry involves using multivariate data to determine patterns. For example, in metabolomics studies, these methods have been used to detect and characterize metabolic profiles based on MS, NMR, and other spectroscopic techniques. The application of chemometrics can help answer questions such as: (1) can analytical instrumentation assist in the discovery of new diagnostics and therapies for diseases? (2) is it possible to optimize the production of industrial biotechnology? (3) how can we evaluate and improve food quality?

These and other questions are certainly of interest not only to bioanalytical chemists, but also to those in the biological and medical sciences. And unsupervised pattern recognition tools, such as principal component analysis (PCA) and hierarchical cluster analysis (HCA), can guide researchers in interpreting instrument data. Unsupervised classification methods allow us to discover groups or trends in the data. The word "unsupervised" here implies the analyzed data are not labeled as members of predefined classes (Ren et al. 2015). In Fig. 2, the main areas of bioanalytical chemistry to which the most common unsupervised methods (PCA and HCA) have been applied are shown.

Among the most widespread applications of exploratory analysis, metabolomics stands out. Commonly defined as the study of the metabolic profiles of organisms and changes in these profiles, whether due to natural fluctuations or induced by external disturbances. Metabolomics seeks new insights into biochemical processes and pathways, aiming to relate them to characteristics at lower levels (DNA, RNA, and protein) and at higher organizational levels (sensory, physiological, and phenotype). These characteristics range from stages of the disease, production of

**Fig. 2** Main omics sciences employing chemometric tools to interpret data

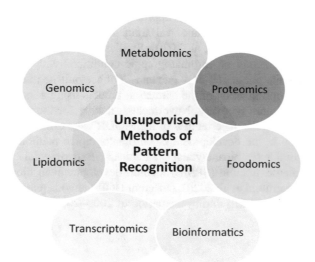

metabolites, as well as sensory characteristics to health status (Hendriks et al. 2011). On the other hand, the advances in MS instrumentation in the last decade, together with the establishment of standardized libraries of chemical fragmentation, greater computing power and availability, new data analysis softwares (including free programs), new scientific applications and commercial perspectives, have made mass spectrometry-based metabolomics the latest technology in demand. This methodology allows, for instance, cataloging and quantifying the amounts of femtomol from cellular metabolites. In this context, the application of exploratory analysis tools must be followed by supervised methods, thus expanding the applicability and validation of chemometrics (Mishur and Rea 2012).

Chemometric exploratory methods are valuable tools to extract information in fields such as proteomics and lipidomics. A recent review has highlighted the importance of determining proteins in human urine, since urinary protein biomarkers can be important for diagnosing many diseases (Aitekenov et al. 2021). Particularly, vibrational spectroscopic techniques combined with pattern recognition are presented as promising tools for the development of new, rapid, and low-cost diagnostic methods. As an example of this type of application, a recent paper has discriminated urine samples from patients with end-stage kidney disease and healthy individuals based on Raman spectra. PCA was used as an important unsupervised tool to interpret these data, being complemented by supervised discriminant analysis (Senger et al. 2020). This last study has utilized Rametrix, a Raman chemometrics toolbox for Matlab specifically developed aiming to simplify and spread the application of pattern recognition methods to biological samples (Fisher et al. 2018). In lipidomic applications, PCA has been used in combination with chromatographic data to discriminate peanut, corn, soy, sunflower, olive, and sesame oils based on their free fatty acid profiles (Qu et al. 2015).

An important emerging field is foodomics, which involves studies in the domains of food and nutrition through the application and integration of other advanced omics technologies. This integration together with chemometric tools, particularly pattern recognition, has enabled the ability to extract information from food biological systems. Exploratory multivariate methods, such as PCA, have great potential in the search for bioactive food compounds acting as biomarkers, helping to interpret the association between food exposure and health (Valdés et al. 2017). Aiming to expand the multidisciplinary studies in foodomics, applications of unsupervised methods in food microbiology have also been discussed, related to the evaluation of food safety, quality control, food nutrition, and health studies (Xu 2017).

Omics data include several types of analytical signals related to the factors controlling biological systems. Unsupervised methods of pattern recognition can extract and visualize latent low-dimensional structures from large datasets, and their consequent interpretation allows to discover and model relevant biological interactions. A recent review has discussed different multivariate exploratory tools, including PCA and more advanced methods. When applied to the same datasets, they may model different sets of factors. Thus, since a specific biological question must employ a specific tool tailored to the problem at hand, bioanalytical chemists should have an in-depth knowledge of chemometric tools useful to enable genomic data integration (Stein-O'Brien et al. 2018).

## 4.1    Principal Component Analysis (PCA)

PCA is probably the most widespread multivariate chemometric method and, due to the increasing importance of multivariate measurements in chemistry, it is considered the tool that has most significantly changed the view of analytical and bioanalytical chemists on the analysis of multivariate data. The knowledge about PCA is also a prerequisite to understand most of the other chemometric methods. PCA was introduced by Karl Pearson in 1901 (Pearson 1901), but the formal treatment of the method is due to the work of Harold Hotelling (Hotelling 1933), popularized in the 1930s and which caused a revolution in the use of multivariate methods initially in the field of psychology.

PCA is a method used to project multivariate collinear data in a smaller space, while retaining most of the variance. Thus, there is a reduction in the dimensionality of the original space of the dataset, without affecting the relationships among samples. Through this methodology it is possible to visualize and interpret possible clusters of analyzed samples and their contrasts, as well as observing how the variables contribute to each sample cluster. PCA also allows to detect samples exhibiting a distinct (atypical) behavior, since after data projection they tend to become evident (Brereton 2003; Ferreira 2015).

The PCA decomposition of a data matrix $\mathbf{X}$ $(n, m)$ is represented by the product of two matrices, one of scores $\mathbf{T}$ $(n, A)$ and one of loadings $\mathbf{P}$ $(m, A)$, as shown in Eq. (3) (Sena et al. 2017; Massart et al. 1997):

$$\mathbf{X} = \mathbf{t}_1\mathbf{p}_1^{\mathrm{T}} + \mathbf{t}_2\mathbf{p}_2^{\mathrm{T}} + \ldots + \mathbf{t}_A\mathbf{p}_A^{\mathrm{T}} + \mathbf{E} = \mathbf{T}\mathbf{P}^{\mathrm{T}} + \mathbf{E} \tag{3}$$

in which $A$ is the selected number of principal components (PCs), $\mathbf{t}_1$ and $\mathbf{p}_1$ are the score and loading vectors of the first PC (and so on), respectively. $\mathbf{E}$ contains the residuals, the variance not accounted for by the $A$ modeled PC. Each PC is calculated aiming to account for the maximum data variance and constrained to be orthogonal. Thus, the first PC represents the direction in the multivariate space that explains most of the variance, the second PC most of the remaining variance, and so on. The constraint of orthogonality means that each PC is independent from each other. Consequently, PC will always model the amount of variance in decreasing order.

To illustrate an application of PCA in bioanalytical chemistry, consider the data presented in Table 1, which consists of clinical parameters for medical diagnosis obtained from five patients (samples 1–5), plus the extreme reference values of Brazilian legislation (samples 6–7). The following blood parameters were measured: glucose, uric acid, urea, creatinine, cholesterol, HDL, and iron. Please, note that variables were measured at different scales and have different distribution ranges. Therefore, $\mathbf{X}$ must be preprocessed by autoscaling, providing a matrix with zero mean and unit variance.

**Table 1** Clinical analysis data used to illustrate a PCA application

| Samples | Glucose mg/dL | Uric acid mg/dL | Urea mg/dL | Creatinine mg/dL | Cholesterol mg/dL | HDL mg/dL | Iron μg/dL |
|---------|---------------|-----------------|------------|------------------|-------------------|-----------|------------|
| S01 | 144 | 5.3 | 22 | 0.74 | 209 | 62 | 137 |
| S02 | 155 | 4.0 | 30 | 0.80 | 280 | 70 | 140 |
| S03 | 80 | 5.0 | 33 | 0.75 | 89 | 80 | 120 |
| S04 | 87 | 5.2 | 28 | 0.85 | 101 | 78 | 112 |
| S05 | 94 | 4.0 | 30 | 0.90 | 96 | 82 | 89 |
| S06[a] | 75 | 3.4 | 10 | 0.70 | 80 | 40 | 65 |
| S07[a] | 99 | 7.0 | 50 | 1.30 | 190 | 100 | 175 |

[a]Extreme reference values of Brazilian legislation

**Table 2** PCA explained and cumulative variance for raw and autoscaled data

| PC | Raw data | | Autoscaled data | |
|----|------------|-------------------|------------|-------------------|
|    | % Variance | % Var. Cumulative | % Variance | % Var. Cumulative |
| 1 | 85.9 | 85.9 | 62.0 | 62.0 |
| 2 | 12.0 | 97.9 | 28.7 | 90.7 |
| 3 | 1.2 | 99.1 | 5.0 | 95.7 |
| 4 | 0.9 | 100.0 | 3.4 | 99.1 |
| 5 | 0.0 | 100.0 | 0.8 | 99.9 |
| 6 | 0.0 | 100.0 | 0.1 | 100.0 |
| 7 | 0.0 | 100.0 | 0.0 | 100.0 |

Why should data be autoscaled? Observe the amount of variance explained by each PC in Table 2, which shows the results for PCA models built with raw and autoscaled data. When raw data are modeled, the parameters with the highest natural scales (glucose, cholesterol, and iron) tend to concentrate the variance in a few PC. Reciprocally, the parameters with the lowest scales (uric acid and creatinine) are practically neglected when decomposing the data matrix. This is particularly critical for spectroscopic data in which samples have similar characteristics, resulting in models with the first PC accounting for more than 99% of the total data variance. In these cases, it is recommended to mean center the data matrix, so the average spectrum is disregarded from the model and the differences between samples become more evident and interpretable.

In the PCA with autoscaled data, the first three PC explain more than 95% of the variance. This indicates that the seven original variables can be represented by only 3 PC, facilitating the interpretation of the contribution of each of the parameters. Figure 3a, b shows the scores plots for PC1xPC2 and PC1xPC3, respectively. The starting point to interpret a PCA model is to know that the scores, which represent the projections of each sample in the reduced PC space, show the relations among samples, while the loadings show the influence of each variable on each PC. It is observed in Fig. 3 that samples S06 and S07 are on opposite sides of PC1. These

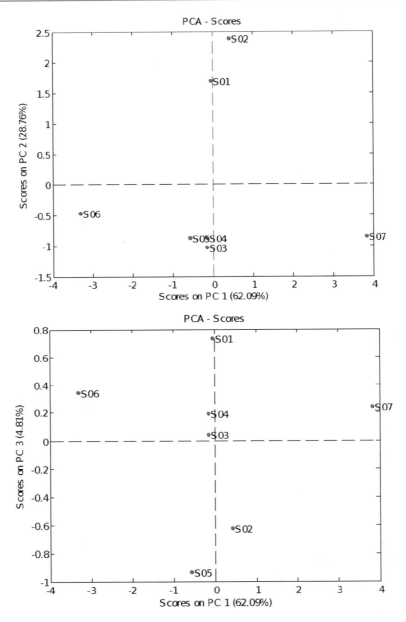

**Fig. 3** Scores plots for the autoscaled data. (**a**) PC1xPC2; (**b**) PC1xPC3

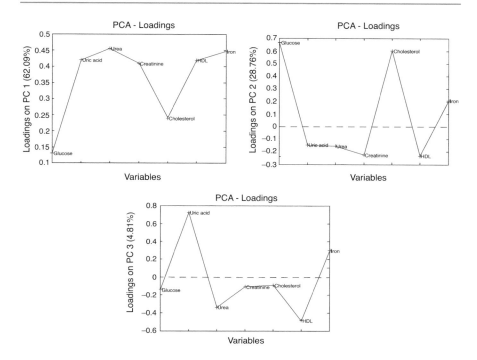

**Fig. 4** Loadings plots for the autoscaled data. (**a**) PC1; (**b**) PC2; (**c**) PC3

samples correspond to the lower (S06) and upper (S07) limits recommended by the Brazilian legislation.

As a consequence, the contributions of the loadings of PC1 are all positive (see Fig. 4a), and sample S07 presents the greatest contribution of all parameters in PC1, while S06 presents the respective smallest contribution. Thus, PC1 can be interpreted as a contrast between the two extremes of the established limits for these parameters. On the other hand, PC2 is responsible for separating two groups of samples, the first formed by samples S01 and S02 on the positive side, and the second one by samples S03, S04, and S05, on the negative side. Examining the PC2 loadings plot (Fig. 4b), we can see that the parameters glucose, cholesterol, and iron are responsible for this separation. These three parameters, mainly glucose, show the most positive loadings on PC2. Thus, these parameters are directly related to the cluster of S01–02, and inversely related to the other cluster (S03–5). The bioanalytical interpretation of this model indicates that the group of patients S01–2 has developed or is developing a disease such as diabetes and must have medical monitoring. The group of samples S03, S04, and S05 are healthy, with the parameter values within the appropriate ranges. In brief, PC2 is a component that discriminates against healthy patients (in the negative part) from patients showing some signs of problems related to diabetes (in the positive part). This is the second most important PC in terms of variance, but the most important in terms of bioanalytical interpretation. PC3 can be viewed as a contrast between patient S01 (Fig. 3b), showing the highest content of uric acid (Fig. 4c), and patients S02 and S05. However, this trend in the data is less important, accounting for only 5.0% (Table 2). Other PC explained non-significant variance and therefore did not need to be interpreted. In general, PCs accounting for small amounts of variance represent only noise.

One important aspect to realize is that the positive or negative values of scores and loadings are arbitrary, since PCA solutions present rotation ambiguity (Sena et al. 2017). This means that each PC can be inverted from positive to negative values, without affecting the model solution. Thus, the interpretation of each PC should be performed relating positive scores directly with positive loadings and inversely with negative loadings, and so on. Another key aspect is related to the number of PC to be included in the model. As already mentioned, PC associated with variance below the noise level should not be considered. Some authors have proposed specific rules to choose the number of PC before slight variations in the amount of variance for each PC are observed. Nevertheless, the key aspect is the bioanalytical interpretation of each significant PC, which always should be clearly performed. PCA has been considered a first-pass method used to visualize (bio)-chemical differences in the data, mainly in fields such as metabolomics, quality control, and process monitoring. Since no class/group membership information is provided to the model, it is the least biased method for discrimination that should be used previously to the building of supervised classification models (Worley and Powers 2016).

Some softwares allow displaying the projections of samples (scores) and variables (loadings) in a single graph, which is called biplot. In this type of representation, the interpretation of sample clusters and the respective contribution of variables are easier visualized. As an example, in Fig. 5, the PC1xPC2 biplot is shown. In this plot, the conclusions previously discussed can be more easily realized, with emphasis on samples cluster formed by S01 and S02 with the main contribution of the levels of glucose and cholesterol is easily noted. Biplots are useful options to observe the results of a PCA built with discrete data, as in the previous example. However, biplots should not be used for spectral or continuous data. In this case, plots of the individual loadings as a function of the spectral or time variables should be preferred (Sena et al. 2017).

## 4.2   Hierarchical Cluster Analysis (HCA)

Numerical taxonomy was developed mainly by biologists to study the similarity between organisms of different species, genus, family, etc. The analysis of clusters by hierarchical methods (HCA) had its origin in numerical taxonomy. This is yet another unsupervised pattern recognition method, suitable for discovering natural patterns of behavior among individuals (samples) based on their multivariate profile. This is a well-known tool, and there are already several algorithms for HCA analysis available, many considered classic. HCA is useful to reduce the dimensionality of

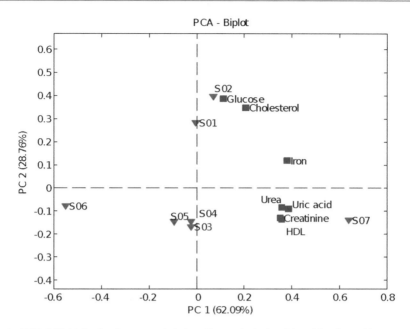

**Fig. 5** PC1xPC2 biplot for the autoscaled data. Scores (red triangle) and loadings (blue squares)

large datasets, for example, it allows dozens of large genes to be represented by a few groups of genes of similar behavior or, also, for the detection of differentiated behavior (anomalous) in a set of data (Brereton 2003; Ferreira 2015).

The main advantage of HCA, among the unsupervised methods, is the possibility of viewing multivariate data in an integrated two-dimensional tree diagram, called dendrogram, in which the hierarchical structure of the data can be observed. In this method, all the information contained in the data matrix is used to build the dendrogram, and, consequently, to calculate the similarity between samples. Can protein sequences from different animals be related and does this tell something about the molecular basis of evolution? Can chemical fingerprints of wines be related and does this tell something about the origins and taste of a particular wine? Unsupervised cluster analysis can be used to answer this type of question by clustering different samples/objects based on biochemical measurements (Brereton 2003). HCA is based on similarity measurements between the samples (commonly metric distances), enabling all of them to be compared with each other. In the next step, a linkage criterion is employed to group samples and form clusters. This linkage criterion aims at maximizing both the homogeneity within the clusters and the heterogeneity between the clusters (Sena et al. 2017).

The most used similarity measurement is the Euclidean distance. It represents the distance between the positions of the samples (data points) in the $m$ dimensional space defined by the variables, as shown in Eq. (4),

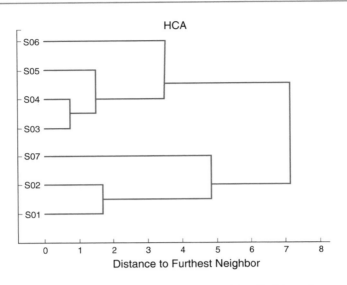

**Fig. 6** Dendrogram obtained for the autoscaled data applying Euclidean distance and complete linkage method

$$d_{AB} = \sqrt{(x_{a1} - x_{b1})^2 + (x_{a2} - x_{b2})^2 + \ldots + (x_{aj} - x_{bj})^2} \qquad (4)$$

in which $x_{a1}$ and $x_{b1}$ are the numerical values of the $j$th coordinates of $A$ and $B$, respectively. Another alternative is the Manhattan distance, also known as "taxi distance." It measures the sum of the absolute values of the differences between the coordinates of the samples for all the variables. Less often used, Mahalanobis distance has the advantages of being dimensionless and independent on data scale. In addition, when using it the relevance of some variables might be greater than others due to the utilization of the variance–covariance matrix.

Once chosen the similarity measurement and formed the first cluster including the two closest samples, it is necessary to calculate the distance between this cluster and all other samples. For this aim, several linkage criteria are available and can be tested, such as the $K$th nearest neighbor, furthest neighbor, pair-group average, centroid, median, and Ward's method. In the application example discussed in PCA subsection, complete linkage clustering, also known as furthest neighbor, was used. In this criterion, the distance between two clusters is determined by the largest distance between the objects of these two groups. As a result, the dendrogram shown in Fig. 6 is obtained. Based on this dendrogram it is possible to clearly observe the formation of two clusters. One formed by samples S01, S02 e S07 (with the highest values for the parameters in Table 1), and other by samples S03, S04, S05 e S06. These results agree with the PCA model previously presented. However, the simple inspection of this dendrogram does not permit to know any information on how the studied variables contributed to the formation of the clusters. This is the main limitation of HCA in comparison to PCA.

## 5 Supervised Classification

PCA is one of the most applied multivariate methods in general, and particularly in metabolomics. The use of PCA can be considered a starting point in many chemometric applications aiming to perform an exploratory analysis. However, due to its unsupervised nature, multivariate analysis should not be limited to PCA, often demanding the use of supervised classification methods (Trygg et al. 2007). In a more general context of analytical and bioanalytical chemistry, this type of methods is under the scope of qualitative analysis. In the last years, applications of multivariate supervised classification have grown steadily to the point that some authors defined the relationship between qualitative analysis and chemometrics as vibrant (Szymanska et al. 2015). Clearly, chemometrics has contributed to the extension of qualitative analysis far beyond the traditional compound identification. Its applications should be highlighted mainly in fields such as food analysis (Sena et al. 2017; Jimenez-Carvelo et al. 2019), forensics (Sena et al. 2020a), and metabolomics (Kotlowska 2014; Madsen et al. 2010; Trygg et al. 2007; Bylesjö et al. 2006). Moreover, these methods have gained widespread use in the authentication of food and beverages, recognition of evidences in forensic analysis, and diseases classification associated to the search for biomarkers in medical diagnosis (Lee et al. 2018). Particularly, multivariate classification is important in metabolomics, foodomics, and other omics sciences regarding the challenges posed by the recent trend in moving from targeted to non-targeted analysis (McGrath et al. 2018).

In supervised classification, the information about class assignment is previously supplied during the model building. A representative set of samples is used in the step of calibration/training, while other set of independent samples should be employed to validate/test the model. The limits for each class are defined in the multivariate space of the variables during the training step (Sena et al. 2017; Brereton 2003; Ferreira 2015; Massart et al. 1997). Cross-validation should be used to define the dimensionality of the model, namely the number of PC or LV. Since the definition of the classes and the classification problem in general could be flexible, authors should have an in-deep knowledge of the available methods to make the appropriate choices. The simplest situation involves binary classification models, in which only two predictions are possible: belongs or not belongs. As an example, we can imagine a model to detect the presence or absence of a disease based on spectra of biological samples, such as urine, plasma, etc. On the other hand, when there are more than two classes, such as in a study to classify several subtypes of a disease, a multiclass model is needed. In this last situation, the model can be constrained to assign samples to only one class, to permit classification simultaneously to more than one class, or to assign samples to none of the classes, depending on the problem at hand.

Multivariate supervised classification methods may be divided in several manners (Sena et al. 2017; Szymanska et al. 2015). These methods can be linear or non-linear. However, the discussion in this chapter will be limited to the linear methods due to the reasons already mentioned in the end of the Introduction. Classification methods can also be divided into parametric and non-parametric. This chapter will be limited to present the most used parametric methods, which are based on the establishment of probabilistic distributions for the assignment of samples to classes. In contrast to probabilistic methods, non-parametric methods are commonly based on the estimate of distances between samples in the multivariate space. The most used non-parametric method is K-nearest neighbors (KNN) (Wang et al. 2007), which is a supervised extension of HCA. Examples of relevant applications of KNN in bioanalytical chemistry have been related among other applications to the detection of biochemical alterations in lung cancer based on Raman spectra of tissue samples (Zheng et al. 2020), and the detection of wound biomarkers in a plant by UPLC–MS data in a metabolomic study (Boccard et al. 2010). Lastly, parametric classification methods can be divided into discriminant and class modeling methods. Discriminant methods have been most common in the literature, what has given rise to a controversy about which type of method should be preferred (Rodionova et al. 2016). This controversy will be presented in a subsection in the following, jointly with new class modeling methods originated from this discussion. Before that, the most used supervised classification methods, soft independent modeling of class analogy (SIMCA), PLS-DA and its variant OPLS-DA, will be presented in specific sections. A last subsection will discuss complementary aspects, including analytical validation and common mistakes found for classification methods.

## 5.1 Soft Independent Modeling of Class Analogy (SIMCA)

Class modeling methods are supervised classification tools that build models considering exclusively the samples of each class at a time. In contrast to discriminant methods, the information about samples of other classes is not employed to define the limits of a class. Thus, these methods can be used to build one-class models based on only one target class, or two class and multiclass models. The most used class modeling method is SIMCA, which was developed during the 1970s by S. Wold, one of the pioneers in chemometrics (Wold 1976). SIMCA can be considered a supervised extension of PCA and aims at handling within-class variance. For each class, a separate PCA model is constructed and its dimensionality, the number of PC, is defined by cross-validation (Brereton 2003; Ferreira 2015; Massart et al. 1997).

There are small variations in the use of SIMCA, but currently the most used version defines the limits of each class based on two parameters: Hotelling $T^2$ values and Q residues (Sena et al. 2017; Brereton 2003; Ferreira 2015; Massart et al. 1997). The first parameter is related to the variance used to model each class, while the second one is a measure of the residual variance left out of the model. In short, these two parameters measure the variances in and out of each class PCA model, respectively. In fact, these parameters are typically used to detect outliers in PCA models, and their use is historically related to applications in the field of multivariate statistical process control (Nomikos and MacGregor 1995).

Some authors have pointed out the difficulties in interpreting SIMCA models as its main disadvantage in comparison to discriminant models, such as PLS-DA (Bylesjö et al. 2006). The spectral (or other type of analytical signal) interpretation is complicated, because information related to differences between classes is not easily available. In truth, parameters such as modeling power and discriminatory power can be used to estimate the contribution of each variable to SIMCA models (Sena et al. 2017; Brereton 2003). The first parameter measures the contribution to each class model, while the second one evaluates the discrimination between two specific classes. Nevertheless, we agree that the spectral interpretation of PLS-DA models is much easier than SIMCA models. It should be stressed the importance of both the goals in supervised classification, the discrimination between classes, and the interpretation of this discrimination in terms of the analytical signals, aiming to understand the underlying bioanalytical phenomena.

It is interesting to cite some recent bioanalytical applications employing SIMCA. By combining variables of different origins, obtained from interviews, cognitive assessments, and clinical analysis (blood plasma), it has been possible to differentiate healthy volunteers and patients with deficit and nondeficit schizophrenia. These results reinforced the conclusion that schizophrenia is not a unitary disease (Kanchanatawan et al. 2018). The combination of SIMCA and support vector machine (SVM) classification models has allowed to differentiate the geographical origin of hazelnuts based on UPLC–MS data. In this foodomics study, 20 key metabolites were selected and identified (Klockmann et al. 2016). SIMCA has also played an important role in a study that developed a new data mining methodology for metabolite identification based on gas chromatography–MS (GC–MS) data (Tsugawa et al. 2011). In another article, the development of classification models based on PCA and SIMCA to discriminate nasal aspirates from patients infected or not with influenza virus based only on Vis-NIR spectra (600–1100 nm) was considered surprising (Sakudo et al. 2012). Our surprise is related to the well-known sensitivity limitations of this spectroscopic technique (limits of detection optimistically above 0.1% w/w), which has been applied to determinations of some analytes in quantities far below reasonable values (Sena et al. 2020a; Pasquini 2003, 2018).

## 5.2 Partial Least Squares Discriminant Analysis (PLS-DA)

Discriminant methods estimate delimiters between classes in the multivariate space. For binary models, only one delimiter is estimated, while for multiclass models more delimiters are needed according to the number of classes. The most used discriminant method in chemometrics is PLS-DA (Brereton and Lloyd 2014), which has also been the most used supervised classification method in general in the last 10 years. It can be considered an extension of linear discriminant analysis (LDA), the most used classification method in other areas. It is also an extension of the most popular multivariate calibration method, PLS, adapted for classification. The first idea of using PLS for classification was published in 1987 (Stahle and Wold 1987), but PLS-DA was only formalized in 2003 (Barker and Rayens 2003).

PLS-DA works in a similar manner to PLS, establishing a correlation between an **X** block of independent variables and a **Y** block of dependent variables. Instead of quantitative y values, PLS-DA will correlate the analyte signals with the so-called dummy variables, which typically assume the value 1.0 when a sample belongs to the modeled class, and 0.0 when it does not belong. **X** and **Y** data are concomitantly decomposed aiming to account for most of the variance in both the blocks and simultaneously maximize the discrimination between classes. The number of LV is selected by cross-validation (venetian blinds cross-validation is recommended), according to the lowest cross-validation classification error (CVCE). This is a measure of the number of samples incorrectly assigned. It should be stressed that root mean square error of cross-validation (RMSECV), the parameter used to the choice of the LV number in quantitative PLS models, is not useful for PLS-DA models (Kjeldahl and Bro 2010). Considering that predicted y values will rarely be exactly 1.0 or 0.0, it is necessary to estimate a threshold for sample classification. Currently, Bayesian thresholds are adopted. They simultaneously minimize the number of false negative and false positive predictions in the training set (Botelho et al. 2015; Pulido et al. 2003). Other aspects of PLS regression will be discussed in the next section of this chapter, about multivariate calibration.

Two variants of PLS-DA can be used (Brereton and Lloyd 2014). PLS1-DA models each class/**y** vector at a time while PLS2-DA models all the classes simultaneously (**Y** is a matrix with the number of columns equal to number of classes). For binary models, PLS1 should mandatorily be used. For multiclass models, PLS2-DA presents the disadvantage of assuming interactions among the **Y** columns, which could not be realistic depending on the problem at hand. In these cases, the alternative is to build PLS1-DA models for each class. Specific constraints may be adopted as a function of each situation. For instance, when the objective of the model is to assign samples to only one class, they are constrained to be classified in the class with the highest probability, even if some of them had been predicted above the thresholds of more than one class. A final remark about the need of PLS-DA model interpretation should be stressed. Latent bioanalytical patterns can be investigated through the interpretation of spectra or another instrumental signal contained in the **X** block. This interpretation relies on the so-called informative vectors (Sena et al. 2017; Teófilo et al. 2009). The most important are variable importance in projection (VIP) scores (Chong and Jun 2005) and regression vectors, which should be jointly analyzed.

One of the most common types of bioanalytical applications using PLS-DA has involved analysis of human plasma by different analytical techniques. Very recently published PLS-DA models have been able to discriminate mid infrared spectra of plasma samples spiked with Aspergillus species (Elkadi et al. 2021), and non-fatigued cancer patients from control groups based on UPLC–MS data. This last article was a complete metabolomic study that identified metabolites associated with cancer but not fatigue, and proposed specific pathways (Feng et al. 2021). Nine different LC–MS platforms have been compared in an untargeted lipidomics study with biochemical profiles of plasma samples obtained from patients after postprandial meal. PLS-DA allowed discrimination between instrumental platforms and the detection of the most discriminant lipids based on the interpretation of VIP scores (Cajka et al. 2017). PLS-DA models have also permitted the discrimination between patients that had chronic ischemic stroke and a control group as well as among three subtypes of strokes based on the plasma contents of 16 amino acids determined by GC–MS. (Goulart et al. 2019)

## 5.3 Orthogonal Partial Least Squares Discriminant Analysis (OPLS-DA)

The use of OSC was already mentioned in this chapter, in the section Data Preprocessing (Section 2). It is a filter to remove orthogonal information from the **X** block uncorrelated to the predicted dependent variables, which was originally proposed in 1998 by S. Wold (Wold et al. 1998). It must be used with caution to avoid overfitting, since no more than two factors should be removed (Westerhuis et al. 2001). In the following years, OSC was incorporated to PLS and PLS-DA resulting in the new variants OPLS (Trygg and Wold 2002) and OPLS-DA (Bylesjö et al. 2006).

It is interesting to note that, in the last years, the use of OPLS-DA became widespread in metabolomics, but not in fields such as food authenticity, forensic analysis, and analytical chemistry in general. In other fields, PLS-DA has been much more utilized. While it is not generally agreed, the main advantages of OPLS-DA over PLS-DA would be better prediction ability and easier interpretation of the model. The information from within-class variance would be better modeled by separating the part of the variation linearly related to the **Y** block from the orthogonal (independent) variation in **X** (Trygg et al. 2007). OPLS-DA would bring together the good qualities of SIMCA for handling within-class variance and PLS-DA for maximum class separation, especially for multiclass models (Bylesjö et al. 2006). OPLS-DA would "disentangle" the class predictive variance from the unrelated class variance, providing more parsimonious models (Worley and Powers 2016). By quantifying how much of the **X** block variance is actually relevant for classification, variables with the most discriminant power will be easier detected and may be assigned to biomarkers. Variants of this method have continued to improve discriminant ability, such as OPLS-effect projections (OPLS-EP). This new variant has provided less complex models with higher predictive ability for discriminating patients before and after prostatectomy based on chromatographic data of plasma samples (Jonsson et al. 2015).

Some recent metabolomic studies successfully applying OPLS-DA models included the discrimination between autistic and healthy children based on urinary volatile organic compounds determined by GC–MS after SPME (Cozzolino et al. 2014), and the monitoring of metabolites of anlotinib, an inhibitor of multiple receptor tyrosine kinases used in the treatment of patients with advanced lung cancer, in rat plasma by UPLC–MS/MS (Du et al. 2020). Nevertheless, both of these articles lacked robust validation by an independent test set. They were limited to internal cross-validation and permutation tests.

## 5.4    New One-Class Modeling Methods

Since more than 10 years ago, there is a clear predominance in the chemometric literature of discriminant methods over class modeling methods. The predictive ability of PLS-DA has outperformed SIMCA in most of the cases, provided that within-class variance is low (Bylesjö et al. 2006). Nevertheless, some important research groups have criticized the predominance of discriminant methods. This discussion is mainly focused on food authentication methods, and the core argument is about the difficulties in obtaining a set of samples representative of all possible types of adulterations. Therefore, these authors have considered discriminant analysis as an inappropriate tool for authentication in many situations and recommended the preferential use of one-class modeling methods (Rodionova et al. 2016; Oliveri and Downey 2012; Oliveri 2017). Since discriminant methods split the multivariate space variance between classes, they would not be able to produce robust models in the presence of new sources of variance such as new types of adulteration. On the other hand, one-class methods model only the authentic class without considering variation from the other classes. This is a controversial discussion and alternatives to deal with new sources of variance based on robust outlier detection have been proposed (Martins et al. 2017).

A positive side effect of this controversy has been the development of new one-class classification methods, such as PLS density modeling (PLS-DM) (Oliveri et al. 2014), one-class PLS (OCPLS) (Xu et al. 2013), and data-driven SIMCA (DD-SIMCA) (Zontov et al. 2017). These new methods have predominantly been applied to food authentication, and rarely found use in bioanalytical chemistry. However, it is possible to imagine a good potential for them in one-class problems. For instance, in a metabolomic study a target class can be defined for a specific disease, and control samples are too heterogeneous due to factors such as gender, genes, diet, age, etc. (Trygg et al. 2007). In fact, only one article was found applying DD-SIMCA to the geographical classification of rice samples based on metabolomic fingerprinting of volatile organic compounds (Ch et al. 2021).

## 5.5    Final Remarks on Validation and Good Practices

All the steps in the model building should be carried out with caution in relation to good practices, and the literature might be critically evaluated. Sample selection should homogeneously span the representative experimental domain in a systematic way and in sufficient number for each class (Trygg et al. 2007). Unbalanced class sample sets should be avoided or minimized whenever possible (Brereton and Lloyd 2014). Replicates must not be used as independent samples to artificially enlarge the training and test sets, introducing an unacceptable optimistic bias in the model. As already mentioned in PLS-DA section, classification models need interpretation through informative vectors aiming to extract biologically meaningful results, such as the detection of the metabolites responsible for class separation. Particularly, (O)PLS-DA scores are not a reliable way of detecting groups in the data, especially in the absence of proper model validation (Madsen et al. 2010; Kjeldahl and Bro 2010). The interpretation of these scores is also misleading if the number of variables far exceeds the number of samples (Brereton and Lloyd 2014). It is hard to understand why many metabolomic papers have chosen to show the scores of a discriminant model instead of a plot with the predictions for dummy variables.

It is of utmost importance to validate classification models with a test set of independent samples. Only internal cross-validation or permutation tests are not enough to validate models. Unfortunately, a large number of papers have lacked proper validation, especially in metabolomic studies aiming to develop diagnostic methods (Kotlowska 2014; Madsen et al. 2010). The lack of validation is more dangerous because overfitted (O)PLS-DA models forced separation between classes by modeling weaker sources of variance, regardless whether signal or noise. This has spawned a pattern of misuse in metabolomics and related fields, leading to a quite frequent publication of statistically unreliable results (Worley and Powers 2016).

A final remark is needed on the figures of merit used for analytical validation in qualitative models (Botelho et al. 2015; Isabel López et al. 2015). The most used are sensitivity and specificity rates, corresponding to true positive and true negative results, respectively. A more global figure, reliability rate, embeds both the rates of false positive and false negative results.

# 6 Multivariate Calibration

When young analytical chemists or other researchers start to learn about chemical analysis, they usually dream of putting their raw sample into the equipment and, in a matter of seconds, get the results with all the analytes identified and quantified. Unfortunately, even with all the advances in instrumentation and available chemical knowledge, getting the answers for a new analytical problem always requires hard work in method development. If the analysis at hand involves the determination of analytes in biological samples, the situation might be even more challenging, since these samples usually present a high number of compounds with overlapped signals, low concentration levels, and matrix effects. The application of separation techniques, such as HPLC, GC, or capillary electrophoresis (CE) using MS detection, to ensure the selectivity of all target analytes jointly with simple univariate regression has been the most common approach for these samples. However, multivariate calibration methods have been effectively applied to many sources of instrumental data, mostly spectroscopic but also chromatographic, to develop models to estimate the concentration of selected analytes in bioanalytical samples. Unlike univariate regression, multivariate models enable the quantification even from rather unselective signals.

**Table 3** Classification of the types of data, calibration set, and main models used for quantitative chemical analysis (Escandar et al. 2006; Escandar and Olivieri 2017, 2019)

| Classification | Data array order | Order of the calibration data set | Main models | Second-order advantage |
|---|---|---|---|---|
| Univariate | Zeroth-order | One-way/vector | Linear regression | No |
| Multivariate | First-order | Two-way/matrix | MLR PCR PLS | No |
| | Second-order unfolded to first-order | Two-way/matrix | MCR-ALS BLLS/RBL U-PLS/RBL | Yes |
| | Second-order | Three-way/data cube | PARAFAC PARAFAC2 PARALIND ATLD related methods N-PLS/RBL N-PLS/RTL | Yes |
| | Third-order | Four-way | PARAFAC TLLS/RTL N-PLS/RTL | Yes |

*MLR* multiple linear regression, *PCR* principal component regression, *PLSR* partial least squares, *MCR-ALS* Multivariate curve resolution alternating least squares, *BLLS/RBL* Bilinear least squares with residual bilinearization, *U-PLS/RBL* unfolded PLS with residual bilinearization, *PARAFAC* parallel factor analysis, *PARALIND* parallel profiles with linear dependencies, *ATLD* alternating trilinear decomposition, *N-PLS/RBL* N-way partial least squares with residual bilinearization, *N-PLS/RTL* N-way partial least squares with residual trilinearization, *TLLS* trilinear least squares (TLLS) with residual trilinearization

Table 3 presents the standard notation for data of various orders, as well as the nomenclature of the corresponding training sets and the main multivariate calibration models applied to bioanalytical samples. It is important to realize that the order of the calibration model means the order of the data array obtained per sample. Unfortunately, the detailed description of all these models is out of the scope of this chapter. Therefore, only the ones presenting the most general and wide applications in bioanalytical samples will be briefly discussed.

Zero-order calibration models present as main advantages the simplest requirements for data measurement and statistical data analysis. However, the signal must be fully selective for the analyte of interest or, if other sources contribute to the measured signal, the contribution of these sources should be constant, in order to be eliminated by background correction or modeled by an intercept (Escandar et al. 2006). Therefore, its application to biological matrices always requires laborious sample clean-up and the use of separation techniques. In zero-order calibration, only a single measurement (zero-order tensor or a scalar) is used for each sample, resulting in a vector (first-order tensor) when data of all calibration samples are gathered together. The statistical analysis, in this case, consists mostly of linear univariate regression (Olivieri 2014, 2015).

First-order models involve the acquisition of a first-order array (vector) of data for each sample (e.g., NIR, UV-Vis, fluorescence spectra, chromatogram, electrophero-gram, etc.), which nowadays is easily obtained by most of the instrumental techniques available in a chemical laboratory. The multivariate analysis of the signals measured for different variables now opens new possibilities in many different areas: (1) simultaneous determination of more than one analyte, even with overlapped signals; (2) the lack of selectivity due to the presence of interferences can be modeled in the calibration set; and (3) outlier samples can be efficiently identified using the parameters of the model (Olivieri 2014; Valderrama et al. 2007).

As a result of these three aspects, analytical procedures usually can be carried out with no or simpler sample clean-up or physical separation, in a short time and significantly reducing the chemical waste, showing a high adherence to the principles of the green chemistry (Koel 2016). In bioanalytical applications, the whole composition of the samples is usually unknown, which restricts the regression analysis to inverse calibration models. Although there are many methods and regression approaches in the literature, MLR, principal component regression (PCR), and PLS are the most applied models. The latter is the most prevalent due to the performance of its results, the easiness of its implementation, and the avail-ability of software. Despite the structural differences of these models, considering a single analyte problem, the estimate of concentration ($y_i$) for a prediction sample $i$ can be represented by the same equation:

$$y_i = x_i b_{\text{reg}} \tag{5}$$

where $x_i$ (1, m) is the vector containing the m instrumental measurements and $b_{\text{reg}}$ (m, 1) is the vector of regression coefficients obtained in each model. In MLR, $b_{\text{reg}}$ is obtained by least squares fit of the instrumental measurements ($\mathbf{X}$ (n, m)) against the reference concentrations (y (n, 1)) for the n calibration samples ($b_{\text{MLR}} = (X^T X)^{-1} X^T y$). One of the main drawbacks of MLR is that the number of samples in the calibration set (n) should be larger than the number of variables (m), which can be crucial in most bioanalytical applications. In addition, MLR is not robust in the presence of collinearity in the $\mathbf{X}$ matrix, which is always present in spectroscopic (and other types of instrumental signals) data. This data collinearity leads to an ill-conditioned problem, a mathematical term that is related to the lack of robustness of MLR in these conditions.

PCR regression coefficients ($b_{\text{PCR}}$) are obtained analogously to MLR, but instead of the instrumental data, this model makes use of the scores of the PCA decomposi-tion (Eq. (3), Sect. 4.1) of the calibration set ($b_{\text{PCR}} = (T^T T)^{-1} T^T y$). This apparent simple distinction between MLR and PCR has a great practical impact, since PCA can compress the relevant information in $\mathbf{X}$ into a few number PC, thus removing the main disadvantages observed in MLR. Regression coefficients for PLS models are expressed by $b_{\text{PLSR}} = W(P^T W)^{-1} q^T$, where $\mathbf{W}$ and $\mathbf{P}$ are the weight and loadings matrices of $\mathbf{X}$, and $\mathbf{q}$ contains the loadings of $\mathbf{y}$. The deduction of this expression is beyond the scope of this text, but it can be found in important references describing PLS (Dayal and MacGregor 1997; ASTM 2012). Unlike PCR, which maximizes the explained variance in $\mathbf{X}$ and then performs the regression against $\mathbf{y}$, PLS maximizes the covariance between $\mathbf{X}$ and $\mathbf{y}$ while both matrices are simultaneously decomposed by PCA. For this reason, in PLS models the new variables are called latent variables (LV).

The number of PC/LV is a crucial parameter in PCR/PLS models. The key point is to select a PC/LV number that models as much of the complexity of the system without causing overfit, which is always accomplished by cross-validation procedures. Shortly, the model is calculated with a specific number of PC/LV on $n - j$ calibration samples, and the model developed with this calibration subset is used to predict the concentration of the $j$ samples not included in the previous modeling. This process is repeated a total of $n/j$ times until each sample has been left out once. RMSECV is determined using the prediction errors stored for each model. This procedure is performed with an increased number of PC/VL and, ideally, the model dimensionality is determined when no significant decrease is observed between two consecutive RMSECV values, being selected the lower number of factors (Haaland and Thomas 1988).

Another important aspect in first-order calibration methods is the possibility to identify outliers, which can be defined as any sample that cannot be part of the statistical population that characterizes the remaining of the calibration data. Based on this definition, before the outlier identification, we must be sure that the data used for calibration is representative of the bioanalytical system intended to be described by the model. All sources of variation expected in future samples should be included in the calibration data, otherwise biased predictions for the analyte concentration and a significant detection of outliers may be observed. The main parameters used for outlier detection in PCR/PLS models are based on extreme leverages and large modeling residuals of the experimental measurements ($\mathbf{X}$) or in the estimation of the concentration values of the analyte ($\mathbf{y}$) (ASTM 2012; Fernández-Pierna et al. 2002). Valderrama et al. have presented a detailed procedure for the identification and exclusion of outliers during the model optimization and validation for the direct determination of quality parameters in sugar cane juice by NIR, in a study performed in the alcohol industry (Valderrama et al. 2007). Also, this paper highlighted the importance of a representative sampling for both calibration and validation sets for predictions in complex systems containing high external variation in the samples. To properly model the variations regarding the different sugarcane species, harvest season, harvest methods, and interferences, the model required approximately 900 and 350 samples for calibration and validation, respectively. The high number of samples used in this study may look discouraging. However, we should bear in mind that, different from many bioanalytical problems, in this application a high number of samples were available, each NIR spectrum was recorded in about 1 min., and almost no sample pretreatment was performed. Therefore, after the model development, the method can provide efficient and reliable determinations with high analytical throughput in the laboratory.

In contrast with the previous example, some challenging bioanalytical problems may be accomplished by PLS modeling using a relatively small set of samples. The quantitative analysis of host cell protein (HCP) in cell culture supernatants using MIR spectroscopy has been proposed using a calibration set composed of only 35 samples. Although the validation results obtained with only five samples are little questionable and, other figures of merit such as precision and robustness have not been included in this paper, these results looked promising (Capito et al. 2013).

Hyperspectral data from image analysis is another increasing application area for first-order calibration models. Hyperspectral instruments enable to obtain simultaneously spatial and spectral information from a heterogeneous sample, which can be used to determine chemical composition, physicochemical properties, or other quality parameters for each pixel in the image. Most of the procedures for model development (e.g., preprocessing, variable selection, model dimensionality, etc.) remain the same as in conventional spectroscopy applications. However, before the actual quantitative modeling, steps such as image compression, removal of background, spiked points, and dead pixels should be observed by the researcher (Vidal and Amigo 2012; Amigo et al. 2015). Some of these aspects have been highlighted in the ripeness monitoring of two cultivars of nectarine using Vis-NIR and PLS (Munera et al. 2017).

The ability to enable quantitative analysis even in the presence of interferences is a great advantage in first-order models. However, as mentioned before, to model these interferences in complex systems sometimes a considerable high number of samples are required in the calibration set, which can be a serious limitation in certain bioanalytical applications. Furthermore, the composition and the interferences may vary between different biological samples. Therefore, even the increase of the number of samples in the calibration set could not be effective to allow good predictions in new samples. In this context, second-order calibration models can offer a new perspective, as they allow the determination of calibrated analytes in new samples presenting uncalibrated constituents. This is known as "second-order advantage" (Escandar et al. 2006). This revolutionary characteristic means that calibration sets no longer require a diverse and large number of samples to cover all possible interferences appearing in unknown samples. Providing that no matrix effects are present, the calibration set can be formed by a small set of pure samples only containing the analytes of interest, analogous to univariate calibration, and the determination can be performed in complex bioanalytical samples even in the presence of high overlapped signals.

Second-order calibration models require to measure a matrix of signals for each sample, which can currently be obtained by many different instrumental techniques, such as excitation–emission spectrofluorimetry, spectroscopy measurements varying some chemical property as a function of time, chromatography or electrophoresis with multiple detection systems (e.g., diode-array detection, DAD, MS, or infrared), etc. Considering the mathematical structure of the models presented in Table 3, they can be classified into two categories, based on bilinear or trilinear decomposition. In the first category, the matrix of signals from each sample is unfolded into a vector, and the calibration samples are arranged in a matrix for data analysis. Multivariate curve resolution alternating least squares (MCR-ALS) can be considered the most applied model in this category and will be discussed in more detail in the following section of this chapter. Bilinear least squares (BLLS) is the natural generalization of classical least squares (CLS) in the second-order scenario. In this sense, it can provide qualitative information about the calibrated analytes but requires that all the compounds in the calibration samples be known a priori. To achieve the second-order advantage, BLLS makes use of the residual bilinearization (RBL) procedure when the residuals of a prediction sample indicate the presence of interferences. This model has shown good results for the determination of the antibiotic ciprofloxacin in human urine by excitation–emission fluorescence measurements, presenting average errors of 5%, and 5 mg $L^{-1}$ as detection limit (Damiani et al. 2004). Also, BLLS has successfully been used for the quantification of the propranolol enantiomers in pharmaceutical formulations (Valderrama and Poppi 2008).

On the other hand, unfolded PLS (U-PLS) was one of the first calibration models proposed to handle second-order data. It is in fact the application of the first-order PLS model to the unfolded sample data (Wold et al. 1987). However, it was only after its association with RBL that the model obtained the second-order advantage and was efficiently applied to bioanalytical samples presenting unexpected interferences (Olivieri 2005). The first application that proved the efficiency of this model to biological samples has utilized spectrofluorimetric measurements to determine the antibiotic tetracycline and the anti-inflammatory salicylate in human serum (Marsili et al. 2005). The comparison between the results of U-PLS/RBL and parallel factor analysis (PARAFAC) shows that the first method provided better results due to its ability to model some interactions among the sample components of the system.

Models in Table 3 that perform the analysis of the dataset in its original matrix structure can be considered the true or native second-order calibration models. PARAFAC is the most important model in this category, which can be expressed as (Bro 1997):

$$x_{i,j,k} = \sum_{f=1}^{F} a_{i,F} b_{j,F} c_{k,F} + e_{i,j,k} \tag{6}$$

where, $x\,(i, j, k)$ is an element of the three-way data ($\underline{\mathbf{X}}$) for the $i$th sample, $j$th and $k$th variables for the second and third modes, respectively; the values of $a$, $b$, and $c$ are the parameters describing the importance of the sample/variables for each factor $f$, $F$ is the total number of factors, and $e\,(i, j, k)$ is an element of the residual array ($\underline{\mathbf{E}}$) containing the variation not captured by the model. The values of $a$, $b$, and $c$ are usually collected into the loading matrices $\mathbf{A}(I, F)$, $\mathbf{B}(J, F)$, and $\mathbf{C}(K, F)$, respectively. Figure 7 shows a schematic representation of this model, in which it can be observed one of the data arrangements usually applied to the analysis of bioanalytical samples. All $I$ calibration standards and an unknown sample are stacked on top of each other and a three-way array $\underline{\mathbf{X}}$, with dimensions ($I + 1$, $J$, $K$) is obtained. In this disposition, each unknown sample is analyzed individually, which may turn the modeling easier, since there are several different standard

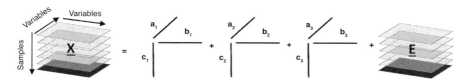

**Fig. 7** Representation of the trilinear decomposition of a data cube into a three-factor PARAFAC model. Vectors **a**, **b**, and **c** represent the profiles of each dimension. (gray) calibration samples, (black) validation sample

samples and just one unknown. The decomposition is accomplished through the ALS algorithm (Braga et al. 2007).

The number of factors of the model can be accessed by the prior knowledge of the system, or by monitoring a PARAFAC internal parameter known as core consistency. Usually, core consistency is computed for a few trial factors, and $F$ is chosen as the number for which its value drops from a high value ($\sim$100%) to less than 50%. The initialization of the algorithm is performed based on initial estimates for the profiles related to the variables (usually **B** and **C**), which can be obtained by several procedures (e.g., direct trilinear decomposition, DTLD, using random values, or initializing the algorithm from several different starting points) (Sanchez and Kowalski 1990). Besides, depending on the system under study, some constraints, such as non-negativity, unimodality, and orthogonality, could be applied to the matrices **A**, **B**, and/or **C** to obtain physically meaningful information or better decomposition. After the convergence, the loadings matrices related to the variables (**B** and **C**) should provide good estimates of the normalized pure chemical profiles of each model factor. In quantitative applications, the factors associated with the analytes are generally identified by comparing the respective estimated profiles with reference spectral profiles for the analytes. The regression coefficients for a given analyte are then obtained by the linear regression fit of the first I elements of its specific column in the **A** loadings matrix against the reference concentrations (**y**) of the calibration samples. Finally, the estimated concentration of the analyte in the unknown sample is obtained by applying these regression coefficients to the $I + 1$ element of the analyte column of the **A** matrix. Note that the regression step is performed after the PARAFAC model was fit, meaning that PARAFAC works as a curve resolution model that intents to estimate pure chemical profiles for each analyte and interferences present in the sample, while the calibration is performed afterwards following a univariate procedure.

The most common application of the PARAFAC model is for the analysis of excitation–emission fluorescence data, since the physicochemical description of the fluorescence phenomena agrees perfectly with Eq. (6). A very good example of this type of application has combined a detailed description of the experimental conditions and the model building for the determination of flufenamic and meclofenamic acids in human urine samples (Muñoz de la Peña et al. 2006). This article also used micellar-enhanced excitation–emission fluorescence to enhancement the analyte signal in relation to the biological background.

In the occurrence of matrix effects unfortunately not even second-order calibration models can enable external calibration protocols. In this case, we can say that a chemical problem should be solved using the chemical way, which is the standard addition calibration method. However, different papers have shown that the combination of PARAFAC and the standard addition method can be effective by avoiding clean-up procedures or the application of separation techniques. This chemometric strategy has been illustrated through the quantification of ibuprofen enantiomers in biological fluids using fluorescence data (Valderrama and Poppi 2011).

Once PARAFAC requires that all components should present a unique characteristic profile in each dimension, the application of this model in some instrumental data, such as chromatographic, may be challenging, since the time shifts observed in these separation techniques cause lack of data trilinearity. This difficulty may be overcome using time alignment algorithms (Braga et al. 2007; Bloemberg et al. 2013), or the PARAFAC2 model (Bro et al. 1999). Both of these approaches have been applied to the detection of prostate carcinoma using GC–MS data (Amante et al. 2019). Other type of lack of trilinearity not modeled by PARAFAC is observed in closed chemical systems with analytes involved in an equilibrium. However, a multi-way model called parallel profiles with linear dependencies (PARALIND) has been proposed to specifically handle those systems (Bro et al. 2009).

The advance of the instrumental methods is making possible the acquisition of an increasing amount of data. Nowadays, a three or even four-way array of data can be obtained for a single sample, giving rise to fourth- and fifth-order calibration models, respectively. Fourth-order models have been applied to the simultaneous analysis of 15 Environmental Protection Agency-polycyclic aromatic hydrocarbons (EPA-PAHs) in soil samples using time-resolved excitation–emission fluorescence data cubes measured at the liquid helium temperature (4.2 K) (Goicoechea et al. 2012). In this application, PARAFAC and U-PLS/RTL results showed sensitivity values allowing determinations at concentration levels varying from ng $g^{-1}$ to pg $g^{-1}$, depending on the analyte, without sample pre-concentration. The synergy of the models with the high-resolution information present in the data eliminated the need of sample clean-up steps and chromatographic separation. Three-dimensional GC–MS data is another example to produce four-way data for each sample (Watson et al. 2017). Escandar and Olivieri have reported an increase of sensitivity typically observed in fourth- and fifth-order calibration models in comparison with second-order calibration models (Escandar and Olivieri 2017, 2019). Considering the complexity and low concentration values commonly found in bioanalytical samples, as well as the gradual increase in the availability of instrumental techniques that enable the acquisition of high-order data, it is also expected a steady increase on the application of these models in bioanalytical samples.

After the careful development of the quantitative method, independent on which type of calibration model, it must still be properly validated before its application in routine analysis. This validation can be performed by estimating figures of merit for the method in accordance with the requirements established previously to ensure that it will fulfill its purpose. For quantitative methods, the most common figures of merit are trueness, precision, prediction uncertainty, robustness, linearity, analytical range, sensitivity, analytical sensitivity, and detection capabilities (limits of detection and quantification). The estimate of figures of merit in multivariate calibration has been an active research topic in chemometrics and extensively reviewed in the literature (Olivieri 2014, 2015; ASTM 2012; Brereton et al. 2018; Allegrini and Olivieri 2014). We encourage the readers to look for these references to find the details of the estimate of each figure of merit. The extension of validation to be performed, or the specific figures of merit to be determined, will also depend on the objectives of the method, following the principle of "fitness for purpose" (Magnusson and Örnemark 2014). Additionally, as mentioned before for the calibration set, the validation should also be carried out with a representative dataset covering the significant variations that can be observed in future samples.

## 7    Multivariate Curve Resolution

Curve resolution methods consist of powerful tools to handle mixture analysis problems. Those types of problems are particularly common in bioanalytical chemistry, due to the variety of analytes, interferences, complex sample matrices, and overlapping signals. The spectral profiles acquired by instrumental analyses are the result of a combination of contributions from the various pure compounds present in the sample. In certain cases, retrieving the profile of a single compound and its respective contribution in a sample may be a challenging task that can be circumvented by curve resolution methods. Among the different methods available to accomplish this task, MCR-ALS (de Juan and Tauler 2021; Ruckebusch and Blanchet 2013) is widely accepted due its simplicity, flexibility, and the wide range of possible applications (Gómez and Callao 2008; Felten et al. 2015). Therefore, MCR-ALS will be the focus of the present section. For further details on other methods and free packages, see the following literature (de Juan et al. 2009; Liang 2009; Camp 2019; Jaumot et al. 2015; Wehrens et al. 2015).

The general mathematical concept behind MCR-ALS is intimately related to the Beer–Lambert law, which makes this technique particularly suitable for spectroscopic datasets (de Juan et al. 2014). In both cases a bilinear model is followed, and the structural basis of MCR-ALS model is described in Eq. (7). The $\mathbf{X}$ matrix, containing $n$ spectral observations (e.g., different samples, spectra acquired at different times, etc.) and $m$ variables, can be decomposed into two matrices, $\mathbf{S}^{\mathbf{T}}$ and $\mathbf{C}$. In MCR-ALS, the $\mathbf{S}^{\mathbf{T}}$ matrix represents the profiles of the $k$ pure components present in the sample containing $m$ variables (e.g., spectral channels), while $\mathbf{C}$ contains the relative intensities of each $k$ component in the $n$ observations. Ideally, the product $\mathbf{CS}^{\mathbf{T}}$ must explain the main variance sources of $\mathbf{X}$ to be considered a reliable model. Therefore, it is expected that the residual matrix $\mathbf{E}$ should not contain relevant structured or systematic information.

$$\mathbf{X}_{(n,m)} = \mathbf{C}_{(n,k)}\mathbf{S}^{\mathbf{T}}_{(k,m)} + \mathbf{E}_{(n,m)} \tag{7}$$

To solve Eq. (7), there are two initial information required: the number of components $k$ (i.e., independent sources of variability) and the initial estimates for either $\mathbf{C}$ or $\mathbf{S^T}$. If the matrix $\mathbf{S^T}$ is initially provided, $\mathbf{C}$ can be calculated using Eq. (8). The new $\mathbf{C}$ matrix obtained can be employed according to Eq. (9) to obtain a new matrix $\mathbf{S^T}$. This iterative process will be carried out in an alternating manner to optimize both $\mathbf{C}$ and $\mathbf{S^T}$ using a least squares approach until the outer product $\mathbf{CS^T}$ is a good representation of $\mathbf{X}$ and a convergence criterion is achieved.

$$\mathbf{C} = \mathbf{XS}\left(\mathbf{S^T S}\right)^{-1} \tag{8}$$

$$\mathbf{S^T} = \left(\mathbf{C^T C}\right)^{-1}\mathbf{C^T X} \tag{9}$$

Although MCR-ALS Eq. (7) is rather similar to Eq. (3) (Sect. 4.1), that is the structural basis of PCA, in the latter, the scores ($\mathbf{T}$ matrix) are constrained to be orthogonal, whereas for MCR-ALS models the constraints are set to provide a physicochemical meaning to the outputs. This means that MCR-ALS dyads ($\mathbf{C}$ and $\mathbf{S^T}$) can be interpreted as actual relative concentrations and spectral profiles. One advantage of employing MCR-ALS is that its model decomposition can handle multiset analysis, providing one result simultaneously for several datasets. This strategy can be performed by concatenating matrices in both spectra and concentration modes. In this case, the $\mathbf{X}$ matrix can be arranged in three different ways: by concatenating data from several experiments performed with the same spectroscopic technique (e.g., $i$ different chromatographic runs, batches, chemical images, etc.), as described in Eq. (10); by concatenating data from the same chemical system evaluated by $j$ different spectroscopic techniques (Eq. 11); and by the combination of both approaches, in which several experiments are performed with more than one spectroscopic technique (Eq. 12) (Ferré et al. 2020).

$$\begin{bmatrix} X^1 \\ \dots \\ X^i \end{bmatrix} = \begin{bmatrix} C^1 \\ \dots \\ C^i \end{bmatrix} S^T + \begin{bmatrix} E^1 \\ \dots \\ E^i \end{bmatrix} \tag{10}$$

$$\begin{bmatrix} X_1 \dots X_j \end{bmatrix} = C\begin{bmatrix} S_1^T \dots S_j^T \end{bmatrix} + \begin{bmatrix} E_1 \dots E_j \end{bmatrix} \tag{11}$$

$$\begin{bmatrix} X_1^1 & \dots & X_j^1 \\ \dots & \dots & \dots \\ X_1^i & \dots & X_j^1 \end{bmatrix} = \begin{bmatrix} C^1 \\ \dots \\ C^i \end{bmatrix}\begin{bmatrix} S_1^T & \dots & S_j^T \end{bmatrix} + \begin{bmatrix} E_1^1 & \dots & E_j^1 \\ \dots & \dots & \dots \\ E_1^i & & E_j^1 \end{bmatrix} \tag{12}$$

As already mentioned, the determination of the number of components, or the data rank, is a crucial step in MCR-ALS, and will directly affect model performance. If the number of components is lower than the actual sources of variation, the data variability will not be well explained by the model, resulting in a high percentage of lack of fit, with structured and high residuals. If the number of components is higher than the expected, noisy profiles without a chemical meaning will be obtained as results. In some cases, when the system is well-known or controlled, the determination of the number of components is an easy task. However, this is not always the case (especially in bioanalytical analysis), and different techniques such as PCA and SVD (singular value decomposition) can be employed to assess the proper number of components. This approach has been employed to analyze datasets from a $2^3$ full-factorial design and assess the number of insulin forms in each experiment performed (Martí-Aluja and Larrechi 2013). This MCR-ALS application was proposed to evaluate the processes of aggregation and association of human insulin. Through the DoE approach, the authors have identified the interaction among all three variables analyzed (temperature, pH, and ionic strength). The MCR-ALS analyses performed provided additional information regarding the association process, enabling to identify specific structural changes in the insulin molecule through its infrared spectra. In some complex scenarios such as omics, a data-driven approach using unsupervised methods has been proposed to assess reliable MCR-ALS models regardless the number of components initially used (Motegi et al. 2015).

Once the number of components is set, some initial guesses for $\mathbf{C}$ or $\mathbf{S}^T$ are needed to start the least squares iterations and solve Eq. (7) (de Juan and Tauler 2003). In situations in which the analyst knows the pure compounds present in the system (e.g., pharmacological processes), spectra can be directly acquired from the pure compounds and used as initial guesses for $\mathbf{S}^T$ to build the MCR-ALS model. If the spectra are not available, approaches based on purest variable selection can be employed to obtain more accurate initial estimates. One of the main methods designed to achieve this goal is the simple-to-use iterative self-modeling analysis (SIMPLISMA) (Windig and Guilment 1991). SIMPLISMA is a versatile tool that can be used for all types of datasets to provide initial estimates for both $\mathbf{C}$ and $\mathbf{S}^T$ matrices. Employing simple statistics such as mean and standard deviation, the method searches for purest spectral channels or observations through the calculation and ranking of a purity index.

In the particular case of process control applications, other approaches can be employed to obtain the number of components and the initial estimates. Those types of data have a structured direction in which the mixture evolves in a sequential way, and are commonly found in bioanalytical chemistry. One typical example comes from spectra recorded at different time intervals of a reaction or process in which intermediate compounds are formed. In this case, it is possible to acquire spectra from the initial reagents and final products; however, the intermediate compounds are not always well-known or stable enough for allowing spectra acquisition. In these cases, it is difficult to generate the initial estimates for $\mathbf{S}^T$. Other common process control datasets commonly found in bioanalytical problems are obtained by the hyphenation of chromatographic systems and spectroscopic methods for detection. In many biological systems, co-eluted compounds cannot be completely separated by chromatography even after optimizing the experimental parameters. Therefore, mathematical approaches such as MCR-ALS can be used to resolve the overlapped peaks and estimate the concentrations in the samples (Amigo et al. 2010;

Gemperline 1984). In those cases, the estimate of $\mathbf{S}^{\mathrm{T}}$ can also be challenging, and other mathematical strategies may be more appropriate.

Although SIMPLISMA can be employed in the above mentioned cases, evolving factor analysis (EFA) is a powerful alternative for those types of datasets (de Juan and Tauler 2003; Maeder 1987). EFA will perform sequential PCA analysis in forward and backward directions of the $\mathbf{X}$ matrix to estimate not only the number of components, but also to provide initial estimates for the $\mathbf{C}$ matrix. If one cannot assume, or does not know, the sequential behavior of the dataset, then EFA should not be employed. In this case, SIMPLISMA presents an advantage since it can be applied to any type of dataset, including chemical images. Theoretically, random guesses can also be used as initial estimates, although they should not be preferred.

Considering that the acceptable solutions for Eq. (7) must have a physicochemical meaning, it is natural to conclude that not all mathematical solutions are suitable for the problem. This is due to the fact that a perfectly acceptable solution math-wise can provide, for instance, negative values for the $\mathbf{C}$ matrix, even though concentrations cannot be negative in real-life. In that sense, the knowledge one has about the analyzed system can be mathematically imposed as constraints, restricting the possibility of acceptable solutions so that they are consistent with the reality of the analyzed system. Those constraints can be categorized in natural, mathematical, or model-related, and can be applied individually for each factor/component (Ferré et al. 2020).

Natural constraints are related to the physicochemical properties inherent to the system. The most commonly applied constraints are: non-negativity, unimodality, and closure. Naturally, non-negativity in concentration and spectral modes are appropriate to spectroscopic data. However, if a preprocessing method used provides negative values in any mode (e.g., derivatives), non-negativity can no longer be applied. Unimodality is a constraint that can be set to obtain solutions in which the profiles must have a single maximum, what is expected from elution profiles of pure components, for instance. Other common constraint suitable to certain systems is closure, which can be applied to the concentration mode and is related to reactions' mass-balance. Besides the three well-established constraints previously mentioned, it is also possible to use known spectral or concentration profiles as constraints to improve model performance.

Mathematical and model-related constraints can also be imposed to solve Eq. (7). In certain scenarios, it is possible to provide information regarding the presence and absence of one or more components in specific observations of $\mathbf{X}$. This information can be inputted as local rank (absence of one or more components) or selectivity (only one component is present) constraint in the concentration mode. Additionally, information related to specific parametric models can be used to produce interpretable solutions for the studied system. Correlation constraint can be used to directly establish a mathematical relationship between the relative concentration profiles obtained by the MCR-ALS and real concentration values through a univariate calibration model. It is possible to notice that there is a variety of available constraints to be used in different scenarios, and the advances observed in the MCR-ALS field, regarding data complexity and new technologies, promote the development of new constraints (de Juan and Tauler 2021).

Ideally, the constraints are used to solve or minimize the ambiguity problems inherent to MCR-ALS. The use of augmented $\mathbf{X}$ matrix can also help in the search for a stable and meaningful solution of Eq. (7). However, even by using those approaches, the ambiguity issues related to permutation, intensity, and rotation can still hamper the analysis of complex systems. The permutation ambiguity refers to the fact that different mathematical solutions can be obtained by permuting matrices $\mathbf{C}$ and $\mathbf{S}^{\mathbf{T}}$. The adequate use of constraints can avoid this issue. Intensity ambiguities are related to different mathematical solutions in which $\mathbf{C}$ and $\mathbf{S}^{\mathbf{T}}$ have the same profiles, but vary in scale. This issue can be overcome by normalizing the profiles. While permutation and intensity ambiguities are easier to deal with, rotational ambiguity poses a major challenge for MCR-ALS models. Different approaches have been developed to assess the extension of this ambiguity in the model. Some of those are focused in the search for the extreme values possible for $\mathbf{C}$ and $\mathbf{S}^{\mathbf{T}}$ (Gemperline 1999; Tauler 2001), while others provide an assessment of all feasible solutions available (Borgen and Kowalski 1985).

One great advantage of MCR-ALS promoting its widespread use in different bioanalytical applications is related to calibration models. MCR-ALS is a soft-modeling method that can be employed for quantitative analysis to both first- and second-order data. First-order datasets are obtained when the analytical measurement of one sample provides a single vector (e.g., a spectrum per sample). Common first-order calibration techniques, such as PLS and PCR, usually provide abstract loadings and regression coefficients. Although, those informative vectors can be used to identify the analytes, this is not a straightforward task. In contrast, MCR-ALS recovers estimates of the pure spectral profiles of analytes and interferences, which is an advantage from a characterization standpoint (Ahmadi et al. 2015).

As previously mentioned, some measurements can provide one whole matrix per sample (e.g., spectra over time per sample), generating what is known as second-order data. Those types of data are being increasingly applied to quantify, among other analytes, antibiotics and anti-inflammatories in biological samples (Gómez and Callao 2008). This multivariate approach is particularly interesting for complex systems, such as body fluids in general, due to the second-order advantage, which refers to the fact that reliable calibration models can be obtained even in the presence of unknown interferences. In fact, MCR-ALS performance is often compared with PARAFAC to handle three-way array datasets. While PARAFAC has the advantage of providing a unique solution without ambiguity issues, MCR-ALS is more flexible and often shows a better performance if the data's inner-structure does not follow a trilinear behavior (de Juan and Tauler 2001).

A MCR-ALS approach has been proposed to quantify fluoroquinolones in porcine blood by capillary electrophoresis with diode-array detection (CE-DAD) (Teglia et al. 2017). The electropherograms of three out of nine analyzed fluoroquinolones shown overlapping peaks due to unknown interferences present in real samples, hampering quantitative determination of the analytes by univariate methods. HPLC–DAD has also been employed for quantitative purposes (Vosough et al. 2014). Phenobarbital and carbamazepine were quantified in human serum samples of patients before and after surgery. In both studies, second-order calibration was performed using MCR-ALS to achieve an accurate methodology in the presence of unknown interferences from biological matrices.

Finally, MCR-ALS is such a versatile method, that several different types of its applications can be found in the literature. In contrast with the previously mentioned applications, chemical images of biological matrices provide a wide field for MCR-ALS (Felten et al. 2015). Some authors have reported the use of MCR-ALS jointly with NIR chemical images for presumptive identification of biofluids stains on fabrics with forensic purposes (Silva et al. 2017). Additional constraints for the MCR-ALS model have been further proposed to extract texture components of the images and improve their detection (Vitale et al. 2020; Ahmad et al. 2020). MCR-ALS has also been employed in photoacoustic imaging data to locate gold nanorods (GNR) contrast agent in biological tissues (Maturi et al. 2021). Authors reported a significant improvement of MCR-ALS in the localization of the regions containing GNR, regardless the presence of blood interference.

## 8    Conclusions

The continuous development of the instrumental techniques applied to chemical analysis, as well as the data acquisition, storing and processing capabilities, have enabled the generation of an increasing amount of data. Chemometric methods have demonstrated to be an effective way of extracting meaningful information from this large amount of data even in the presence of overlapped signals, significant noise, unknown interferences, external factors, etc. Consequently, nowadays the number of multivariate models available in equipment's software is rapidly expanding, as well as the application of these models to many different fields of science, including bioanalytical chemistry. Despite the positive aspects, it is important to highlight the danger of applying chemometrics as a "black box" using automatic softwares without the knowledge of the main features of the models. The excessive pursuit for good and rapid results may lead to non-robust overfitted models and wrong conclusions. Therefore, before chasing for results, all researchers and users should really take care in obtaining in-depth knowledge about the applied chemometric model and the (bio)chemical system under study. Furthermore, biased models and non-reliable estimates for figures of merit may be obtained if the model was built from an unrealistic dataset. Reliable models will always require a representative sampling in both training/calibration and test/validation datasets. This chapter presented and discussed relevant aspects of some of the most important chemometric models applied to bioanalytical science. It also cited many important and representative references to help the readers to look for further details and information. The application of chemometric models might be a way to solve many problems and make life easier in the laboratory. However, everyone should realize that the way to obtain these earnings always requires hard work and the deep knowledge in chemometrics and (bio)analytical chemistry. We finished this chapter by quoting and paying a tribute to Professor R. J. Poppi, a pioneer of chemometrics in South

America and the mentor of many Brazilian chemometricians: "we need to teach students how to think analytically (or chemometrically) and not simply how to push buttons" (Sena et al. 2020b).

# References

Ahmad M, Vitale R, Silva CS, Ruckebusch C, Cocchi M (2020) J Chemom 34:e3295
Ahmadi G, Tauler R, Abdollahi H (2015) Chemom Intell Lab Syst 142:143–150
Aitekenov S, Gaipov A, Bukasov R (2021) Talanta 223:121718
Allegrini F, Olivieri AC (2014) Anal Chem 86:7858–7866
Amante E, Salomone A, Alladio E, Vincenti M, Porpiglia F, Bro R (2019) Molecules 24:1–10
Amigo JM, Skov T, Bro R (2010) Chem Rev 110:4582–4605
Amigo JM, Babamoradi H, Elcoroaristizabal S (2015) Anal Chim Acta 896:34–51
ASTM (2012) Standard practices for infrared multivariate quantitative Analysis, E1655–05. ASTM International, West Conshohocken
Bandopadhyay S, Beg S, Katare OP, Sharma T, Singh B (2020) J Chromatogr Sci 58:606–621
Barak P (1995) Anal Chem 67:166–173
Barker M, Rayens W (2003) J Chemom 17:166–173
Bhutani H, Kurmi M, Singh S, Beg S, Singh B (2014) Pharma Times 46:71–75
Bloemberg TG, Gerretzen J, Lunshof A, Wehrens R, Buydens LMC (2013) Anal Chim Acta 781: 14–32
Boccard J, Kalousis A, Hilario M, Lantéri P, Hanafi M, Mazerolles G, Wolfender J-L, Carrupt P-A, Rudaz S (2010) Chemom Intell Lab Syst 104:20–27
Borgen OS, Kowalski BR (1985) Anal Chim Acta 174:1–26
Botelho BG, Reis N, Oliveira LS, Sena MM (2015) Food Chem 181:31–37
Box GEP, Hunter WG, Hunter JS (2005) Statistics for experimenters: design, innovation and discovery, 2nd edn. Wiley, New Jersey
Braga JWB, Bottoli CBG, Jardim ICSF, Goicoechea HC, Olivieri AC, Poppi RJ (2007) J Chromatogr A 1148:200–210
Brereton RG (2003) Chemometrics: data analysis for the laboratory and chemical plant. Wiley, Chichester
Brereton RG, Lloyd GR (2014) J Chemom 28:213–225
Brereton RG, Jansen J, Lopes J, Marini F, Pomerantsev A, Rodionova O (2018) Anal Bioanal Chem 410:6691–6704
Bro R (1997) Chemom Intell Lab Syst 38:149–171
Bro R, Smilde AK (2003) J Chemom 17:62–71
Bro R, Andersson CA, Kiers HAL (1999) J Chemom 309:295–309
Bro R, Harshman RA, Sidiropoulos ND, Lundy ME (2009) J Chemom 23:324–340
Bro R, Viereck N, Toft M, Toft H, Hansen PI, Engelsen SB (2010) Trends Anal Chem 29:281–284
Bylesjö M, Rantalainen M, Cloarec O, Nicholson JK, Holmes E, Trygg J (2006) J Chemom 20: 341–351
Cajka T, Smilowitz JT, Fiehn O (2017) Anal Chem 89:12360–12368
Camp CH Jr (2019) J Res Natl Inst Stand Technol 124:1–10
Candioti LV, de Zan MM, Cámara MS, Goicoechea HC (2014) Talanta 124:123–138
Capito F, Skudas R, Kolmar H, Stanislawski B (2013) Biotechnol Bioeng 110:252–259
Ch R, Chevallier O, McCarron P, McGrath TF, Wu D, Duy LND, Kapile AP, McBride M, Elliott CT (2021) Food Chem 334:127553
Chong I-G, Jun C-H (2005) Chemom Intell Lab Syst 78:103–112
Cozzolino R, de Magistris L, Saggese P, Stocchero M, Martignetti A, di Stasio M, Malorni A, Marotta R, Boscaino F, Malorni L (2014) Anal Bioanal Chem 406:4649–4662
Crocombe RA (2018) Appl Spectrosc 72:1701–1751

Damiani PC, Nepote AJ, Bearzotti M, Olivieri AC (2004) Anal Chem 76:2798–2806
Dayal BS, MacGregor JF (1997) J Chemom 11:73–85
de Juan A, Tauler R (2001) J Chemom 15:749–771
de Juan A, Tauler R (2003) Anal Chim Acta 500:195–210
de Juan A, Tauler R (2021) Anal Chim Acta 1145:59–78
de Juan A, Rutan SC, Tauler R (2009) In: Brown S, Tauler R, Walczak B (eds) Comprehensive chemometrics, vol 2. Elsevier, Oxford, pp 325–344
de Juan A, Jaumot J, Tauler R (2014) Anal Methods 6:4964–4976
Deidda R, Orlandini S, Hubert P, Hubert C (2018) J Pharm Biomed Anal 161:110–121
Department of Health and Human Services, Food and Drug Administration (2004) PAT guidance for industry: a framework for innovative pharmaceutical development, manufacturing and quality assurance. FDA, Rockville
Derringer G, Suich R (1980) J Qual Technol 12:214–219
Dispas A, Avohou HT, Lebrun P, Hubert P, Hubert C (2018) Trends Anal Chem 101:24–33
Du P, Hu T, An Z, Li P, Liu L (2020) Analyst 145:4972–4981
Elkadi OA, Hassan R, Elanany M, Byrne HJ, Ramadan MA (2021) Spectrochim Acta A 248: 119259
Engel J, Gerretzen J, Szymanska E, Jansen JJ, Downey G, Blanchet L, Buydens LMC (2013) TrAC Trends Anal Chem 50:96–106
Escandar GM, Olivieri AC (2017) Analyst 142:2862–2873
Escandar GM, Olivieri AC (2019) J Chromatogr A 1587:2–13
Escandar GM, Damiani PC, Goicoechea HC, Olivieri AC (2006) Microchem J 82:29–42
Felten J, Hall H, Jaumot J, Tauler R, de Juan A, Gorzsás A (2015) Nat Protoc 10:217–240
Feng LR, Barb JJ, Regan J, Saligan LN (2021) Cancer Med. https://doi.org/10.1002/cam4.3749
Fernández-Pierna JA, Wahl F, de Noord OE, Massart DL (2002) Chemom Intell Lab Syst 63:27–39
Ferré J, Boqué R, Faber NM (2020) In: Brown S, Tauler R, Walczak B (eds) Comprehensive chemometrics, vol 2. Elsevier, Oxford, pp 233–266
Ferreira MMC (2015) Quimiometria: Conceitos, Métodos e Aplicações. Unicamp, Campinas
Fisher AK, Carswell WF, Athamneh AIM, Sullivan MC, Robertson JL, Bevan DR, Senger RS (2018) J Raman Spectr 49:885–896
Geladi P, Esbensen K (1990) J Chemom 4:337–354
Gemperline PJ (1984) J Chem Inf Comput Sci 24:206–212
Gemperline PJ (1999) Anal Chem 71:5398–5404
Goicoechea HC, Yu S, Moore AFT, Campiglia AD (2012) Talanta 101:330–336
Gómez V, Callao MP (2008) Anal Chim Acta 627(2):169–183
Goulart VAM, Sena MM, Mendes TO, Menezes HC, Cardeal ZL, Paiva MJN, Sandrim VC, Pinto MCX, Resende RR (2019) Biomed Res Int 2019:8480468
Grangeia HB, Silva C, Simões SP, Reis MS (2020) Eur J Pharm Biopharm 147:19–37
Haaland DM, Thomas EV (1988) Anal Chem 60:1193–1202
Harms J, Wang X, Kim T, Yang X, Rathore A (2008) Biotechnol Prog 24:655–662
Hendriks MMWB, van Eeuwijk FA, Jellema RH, Westerhuis JA, Reijmers TH, Hoefsloot HCJ, Smilde AK (2011) Trends Anal Chem 30:1685–1698
Hibbert DB (2016) Pure Appl Chem 88:407–443
Hotelling H (1933) J Educ Psychol 24:498–520
Hubert C, Houari S, Rozet E, Lebrun P, Hubert P (2015) J Chromatogr A 1395:88–98
International Conference of Harmonisation (ICH)(2008) Harmonised tripartite guideline: Q8 (R1) pharmaceutical development
Isabel López M, Pilar Callao M, Ruisánchez I (2015) Anal Chim Acta 891:62–72
Jaumot J, de Juan A, Tauler R (2015) Chemom Intell Lab Syst 140:1–12
Jimenez-Carvelo AM, Gonzalez-Casado A, Bagur-Gonzalez MG, Cuadros-Rodriguez L (2019) Food Res Int 122:25–39
Johnson RT, Montgomery DC, Jones BA (2011) Qual Eng 23:287–301

Jonsson P, Wuolikainen A, Thysell E, Chorell E, Stattin P, Wikström P, Antti H (2015) Metabolomics 11:1667–1678

Kanchanatawan B, Sriswasdi S, Thika S, Sirivichayakul S, Carvalho AF, Geffard M, Kubera M, Maes M (2018) Metab Brain Dis 33:1053–1067

Khoo SHG, Al-Rubeai M (2007) Biotechnol Appl Biochem 47:71–84

Kjeldahl K, Bro R (2010) J Chemom 24:558–564

Klockmann S, Reiner E, Bachmann R, Hackl T, Fischer M (2016) J Agric Food Chem 64:9253–9262

Koel M (2016) Green Chem 18:923–931

Kotlowska A (2014) Drug Dev Res 75:283–290

Kozlowski S, Swann P (2009) Considerations for biotechnology product quality by design. In: Rathore AS, Mhatre R (eds) Quality by design for biopharmaceuticals: perspectives and case studies. Wiley, New Jersey

Kumar K (2018) Anal Methods 10:190–203

Kumar K, Cava F (2017) Analyst 142:1916–1928

Lee LC, Liong CY, Jemain AA (2018) Analyst 143:3526–3539

Liang Y (2009) In: Brown S, Tauler R, Walczak B (eds) Comprehensive chemometrics, vol 2. Elsevier, Oxford, pp 309–324

Madsen R, Lundstedt T, Trygg J (2010) Anal Chim Acta 659:23–33

Maeder M (1987) Anal Chem 59:527–530

Magnusson B, Örnemark U (eds) (2014) Eurachem guide: the fitness for purpose of analytical methods—a laboratory guide to method validation and related topics, 2nd edn. isbn:978-91-87461-59-0. Available at www.eurachem.org. Assessed Mar 2021

Marrubini G, Dugheri S, Cappelli G, Arcangeli G, Mucci N, Appelblad P, Melzi C, Speltini A (2020) Anal Chim Acta 1119:77–100

Marsili NR, Lista A, Fernandez Band BS, Goicoechea HC, Olivieri AC (2005) Analyst 130:1291–1298

Martí-Aluja I, Larrechi MS (2013) Chemom Intell Lab Syst 127:49–54

Martins AR, Talhavini M, Vieira ML, Zacca JJ, Braga JWB (2017) Food Chem 229:142–151

Massart DL, Vandeginste BGM, Buydens LMC, de Jong S, Lewi PJ, Smeyers-Verbeke J (1997) Handbook of chemometrics and qualimetrics. Part B, 1st edn. Elsevier, Amsterdam

Maturi M, Armanetti P, Menichetti L, Franchini MC (2021) Nano 11:142

McGrath TF, Haughey SA, Patterson J, Fauhl-Hassek C, Donarski J, Alewijn M, van Ruth S, Elliott CT (2018) Trends Food Sci Technol 76:38–55

Mishra P, Biancolillo A, Roger JM, Marini F, Rutledge DN (2020) Trends Anal Chem 132:116045

Mishur RJ, Rea SL (2012) Mass Spectrom Rev 31:70–95

Montgomery DC (2001) Design and analysis of experiments, 5th edn. Wiley, New Jersey

Motegi H, Tsuboi Y, Saga A, Kagami T, Inoue M, Toki H, Minowa O, Noda T, Kikuchi J (2015) Sci Rep 5:1–12

Munera S, Amigo JM, Blasco J, Cubero S, Talens P, Aleixos N (2017) J Food Eng 214:29–39

Muñoz de la Peña A, Mora Díez N, Bohoyo Gil D, Olivieri AC, Escandar GM (2006) Anal Chim Acta 569:250–259

Nomikos P, MacGregor JF (1995) Technometrics 37:41–59

Oliveri P (2017) Anal Chim Acta 982:9–19

Oliveri P, Downey G (2012) Trends Anal Chem 35:74–86

Oliveri P, Isabel López M, Casolino MC, Ruisánchez I, Pilar Callao M, Medini L, Lanteri S (2014) Anal Chim Acta 851:30–36

Olivieri AC (2005) J Chemom 19:253–265

Olivieri AC (2014) Chem Rev 114:5358–5378

Olivieri AC (2015) Anal Chim Acta 868:10–22

Orlandini S, Pinzauti S, Furlanetto S (2013) Anal Bioanal Chem 405:443–450

Otto M (1999) Chemometrics—statistics and computer application in analytical chemistry. Wiley-VCH, Weinheim

Pan T, Zhang J, Shi XW (2020) NIR News 31:24–27
Papaneophytou C (2019) Mol Biotechnol 61:873–891
Pasquini C (2003) J Braz Chem Soc 14:198–219
Pasquini C (2018) Anal Chim Acta 1026:8–36
Pearson K (1901) Philos Mag 2:559–572
Pulido A, Ruisanchez I, Boqué R, Rius XF (2003) Trends Anal Chem 22:647–654
Qu S, Du Z, Zhang Y (2015) Food Chem 170:463–469
Rathore AS (2009) Trends Biotechnol 27:546–553
Rathore AS (2016) Trends Biotechnol 34:358–370
Rathore AS, Winkle H (2009) Nat Biotechnol 27:26–34
Ren S, Hinzman AA, Kang EL, Szczesniak RD, Lu LJ (2015) Metabolomics 11:1492–1513
Rinnan A, van den Berg S, Engelsen SB (2013) Trends Anal Chem 28:1201–1222
Rodionova OY, Titova AV, Pomerantsev AL (2016) Trends Anal Chem 78:17–22
Rozet E, Lebrun P, Debrus B, Boulanger B, Hubert P (2013) Trends Anal Chem 42:157–167
Ruckebusch C, Blanchet L (2013) Anal Chim Acta 765:28–36
Sahu PK, Ramisetti NR, Cecchi T, Swain S, Patro CS, Panda J (2018) J Pharm Biomed Anal 147:
    590–611
Saini S, Sharma T, Patel A, Kaur R, Tripathi SK, Katare OP, Singh B (2020) J Chromatogr B 1155:
    122300
Sakudo A, Baba K, Ikuta K (2012) Clin Chim Acta 414:130–134
Sanchez E, Kowalski BR (1990) J Chemom 4:29–45
Sandusky PO (2017) J Chem Educ 94:1324–1328
Santos CAT, Lopo M, Páscoa RNMJ, Lopes JA (2013) Appl Spectrosc 67:1215–1233
Savitzky A, Golay M (1964) Anal Chem 36:1627–1639
Sena MM, Almeida MR, Braga JWB, Poppi RJ (2017) Multivariate statistical analysis and
    chemometrics. In: Franca AS, Nollet LML (eds) Spectroscopic methods in food analysis.
    CRC Press (Taylor & Francis Group), Boca Raton, pp 273–314
Sena MM, Rocha WFC, Braga JWB, Silva CS, Urbas A (2020a) Chemometrics in forensics. In:
    Brown S, Tauler R, Walczak B (eds) Comprehensive chemometrics, 2nd edn. Elsevier,
    Amsterdam
Sena MM, Braga JWB, Scarminio IS, Bruns RE (2020b) J Chemom 34:e3284
Senger RS, Sullivan M, Gouldin A, Lundgren S, Merrifield K, Steen C, Baker E, Vu T, Agnor B,
    Martinez G, Coogan H, Carswell W, Kavuru V, Karageorge L, Dev D, Du P, Sklar A, Pirkle J Jr,
    Guelich S, Orlando JL (2020) PLoS One 15:e0227281
Silva CS, Pimentel MF, Amigo JM, Honorato RS, Pasquini C (2017) Trends Anal Chem 95:23–35
Smolinska A, Blanchet L, Buydens LMC, Wijmenga SS (2012) Anal Chim Acta 750:82–97
Stahle L, Wold S (1987) J Chemom 1:185–196
Stein-O'Brien GL, Arora R, Culhane AC, Favorov AV, Garmire LX, Greene CS, Goff LA, Li Y,
    Ngom A, Ochs MF, Xu Y, Fertig EJ (2018) Trends Genet 34:790–805
Szymanska E, Gerretaen J, Engel J, Geurts B, Blanchet L, Buydens LMC (2015) Trends Anal Chem
    69:34–51
Tauler R (2001) J Chemom 15:627–646
Teglia CM, Cámara MS, Vera-Candioti L (2017) Electrophoresis 38:1122–1129
Teófilo RF, Martins JPA, Ferreira MMC (2009) J Chemom 23:31–48
Trygg J, Wold S (2002) J Chemom 16:119–128
Trygg J, Holmes E, Lundstedt T (2007) J Proteome Res 6:469–479
Tsugawa H, Tsujimoto Y, Arita M, Bamba T, Fukusaki E (2011) BMC Bioinf 12:131
Valderrama P, Poppi RJ (2008) Anal Chim Acta 623:38–45
Valderrama P, Poppi RJ (2011) Chemom Intell Lab Syst 106:160–165
Valderrama P, Braga JWB, Poppi RJ (2007) J Agric Food Chem 55(21):8331–8338
Valdés A, Cifuentes A, León C (2017) Trends Anal Chem 96:2–13
van den Berg RA, Hoefsloot HCJ, Westerhuis JA, Smilde AK, van der Werf MJ (2006) BMC
    Genom 7:142

Vidal M, Amigo JM (2012) Chemom Intell Lab Syst 117:138–148
Vitale R, Hugelier S, Cevoli D, Ruckebusch C (2020) Anal Chim Acta 1095:30–37
Vogt FG, Kord AS (2011) J Pharm Sci 100:797–812
Vosough M, Ghafghazi S, Sabetkasaei M (2014) Talanta 119:17–23
Vu TN, Laukens K (2013) Metabolites 3:259–276
Wang J, Neskovic P, Cooper LN (2007) Pattern Recogn Lett 28:207–213
Watson NE, Prebihalo SE, Synovec RE (2017) Anal Chim Acta 983:67–75
Wehrens R, Carvalho E, Fraser PD (2015) Metabolomics 11:143–154
Westerhuis JA, de Jong S, Smilde AK (2001) Chemom Intell Lab Syst 56:13–25
Windig W, Guilment J (1991) Anal Chem 63:1425–1432
Wold S (1976) Pattern Recogn 8:127–139
Wold S, Geladi P, Esbensen K, Öhman J (1987) J Chemom 1:41–56
Wold S, Antti H, Lindgren F, Ohman J (1998) Chemom Intell Lab Syst 44:175–185
Worley B, Powers R (2016) Curr Metabol 4:97–103
Xu Y-J (2017) Trends Anal Chem 96:14–21
Xu L, Yan S-M, Cai C-B, Yu X-P (2013) Chemom Intell Lab Syst 126:1–5
Zheng Q, Li J, Yang L, Zheng B, Wang J, Lv N, Luo J, Martin FL, Liu D, He J (2020) Analyst 145:
    385–392
Zontov Y, Rodionova OY, Kucheryavskiy SV, Pomerantsev A (2017) Chemom Intell Lab Syst
    167:23–28

# Recent Advances and Future Trends in Bioanalytical Chemistry

Patricia Batista Deroco, Dagwin Wachholz Junior, and Lauro Tatsuo Kubota

## 1    Introduction

The bioanalytical science represents a combining of many research areas such as materials science, chemistry, biotechnology, electrical engineering, software engineering, among others, being a very important topic of science for application in medicine, security, forensic science, environmental monitoring and so on (Keçili et al. 2019).

Currently, biomolecule analysis is mainly focused on new materials and strategies to improve miniaturization and portability in order to supply the demand for point-of-care diagnoses, that is, devices that have the potential to perform analysis directly in the field, without needing expensive equipment or high levels of expertise. These devices can perform measurements and transfer detection information via wireless communication in a totally integrated system. It is in this vision that the lab-on-chip, wearable sensors, home-testing systems and continuous monitoring technologies became important (Vashist et al. 2015; Mejía-Salazar et al. 2020; Li et al. 2020; Solanki et al. 2020).

Therefore, modern analytical chemistry aims at developing and producing portable, simple, low cost and easy-to-use analytical devices, so that they are accessible to anyone (trained people or not). It offers the advantages of widening accessibility to diagnoses, minimal sample volumes, reduced costs, and rapid analysis time (Coluccio et al. 2020).

Today, the most popular commercial devices are lateral flow pregnancy tests and electrochemical glucose biosensors, which have been marketed for about 40 years (Portable meter to aid diabetics, Pittsburgh Press, November 6 1981; Chard 1992). Since then, the research in this field and the search for similar devices that allow to monitor different analytes have not stopped.

P. B. Deroco · D. Wachholz Junior · L. T. Kubota (✉)
Institute of Chemistry, University of Campinas – UNICAMP, Campinas, SP, Brazil
e-mail: kubota@unicamp.br

© The Author(s), under exclusive license to Springer Nature Switzerland AG 2022          543
L. T. Kubota et al. (eds.), *Tools and Trends in Bioanalytical Chemistry*,
https://doi.org/10.1007/978-3-030-82381-8_27

These devices may be used in the monitoring of relevant biomarkers with interest for diagnostics, safety/security, and other fields. Certainly, in the healthcare field, bioanalysis plays an important role, and it has not been only a way to identify a specific disease in a patient, but also to provide diagnoses to support clinical development of drugs, predict disease before symptoms appear, forecast the progress of a disorder, and identify patients who are most likely to respond (or not respond) to a specific treatment. This segment supports the emerging personalized medicine, a medical approach that uses molecular information to identify, prevent, diagnose, and treat diseases (Sengupta et al. 2019; Kim et al. 2019; Noah and Ndangili 2019).

This tendency has made bioanalytical chemistry an important sector of the economy, with the introduction and preparation of several types of devices (Sharma 2018; Otero and Magner 2020; Kundu et al. 2019). Biological molecules-based devices, called biosensors, are part of a market that was estimated at \$21.2 billion in 2019, and it is expected to raise \$31.5 billion by 2024 (https://www. marketsandmarkets.com/Market-Reports/biosensors-market-798.html? gclid=CjwKCAjw797P795BRBQEiwAGflV796a793kLaEnUsqiXHiSRwyF-cP792LEKbz795ZXQgsGmiSeLZ791kr-klht_eKxoCeVcQAvD_BwE)

Thus, possibilities, opportunities, and prospects are numerous in the bioanalytical fields, as well as the challenges that still need to be overcome before routine sensing is revolutionized, especially in transferring bioanalytical technological innovation from academic scientific research to the industry and the market.

The trends in the field of bioanalytical have been driven by recent advances in emerging technologies such as microfabrication techniques, materials, and electronics, which facilitate the production of smaller, portable, and cheaper bioanalytical devices. The hot topics in bioanalytical science are summarized in Fig. 1.

This chapter provides a view of the main advances currently achieved within the bioanalytical research field and discusses the remaining challenges and future prospects that could make an important impact on the market.

## 2 Emerging Technologies for Bioanalytical Devices Development

### 2.1 Smartphone-Based Technologies

It is common sense that smartphones are replacing computers in their numerous tasks. The main features such as operating system, internal memory, hardware, data processing and analysis are similar to those from computers, and they are still more accessible and cheaper than the computers are (Quesada-González and Merkoçi 2017). Thus, smartphones are being used more and more as potential sensitive platforms (sensors) to different matrices, allowing a rapid application and on-site analysis for preliminary and meaningful information extraction in the so-called mobile diagnostics (Rateni et al. 2017).

The smartphones are equipped with numerous components that can be employed for application in bioanalytical, such as a fast multicore processor, high-quality

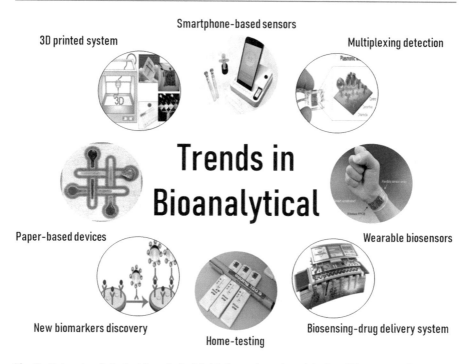

**Fig. 1** Future trends in the bioanalytical field: Smartphone-based devices [Figure reproduced with permission from reference (Min et al. 2018) (Copyright© 2018, American Chemical Society); Multiplexing detection, [Figure reproduced with permission from reference (Soler et al. 2017) (Copyright© 2017, Elsevier); Wearable biosensor [Figure reproduced with permission from reference (Gao et al. 2016) (Copyright© 2016, Springer Nature); Biosensing-drug delivery system [Figure reproduced with permission from reference (Lee et al. 2018) (Copyright© 2018, Elsevier); Home-testing [Figure reproduced with permission from reference (Burki 2020) (Copyright© 2018, Elsevier); New biomarkers discovery [Figure reproduced with permission from reference (Yang et al. 2019) (Copyright© 2019, Elsevier); Paper-based devices [Figure reproduced with permission from reference (Schilling et al. 2013) (Copyright© 2001, Royal Society of Chemistry); 3D printed system [Figure reproduced with permission from reference (Khosravani and Reinicke 2020) (Copyright© 2020, Elsevier)

camera lenses, battery, visual display, and intuitive user interface. Smartphones also possess several wireless data transfer features (e.g., USB, NFC, Wi-Fi, Bluetooth), allowing test results to be displayed immediately to the user and/or transmitted to cloud databases (Rateni et al. 2017).

The direct measurement of physical quantities using the integrated sensors, the facility of a wireless connection and an easy operation offer the opportunity to advance analytical sensing systems and their applications, resulting in more researchers focusing on smartphones to design portable biosensing devices (Roda et al. 2016). These mobile diagnostic devices are cheaper and much easier for application than the routine laboratory analysis, which are generally performed by skilled analysts, expensive equipment and materials, and complex techniques—

improving the diagnosis, quantification, and monitoring, particularly in low-resource countries, leading to a democratization in measurement science, thanks to the massive volume of mobile phone users spread globally, including in remote and underdeveloped areas (Ozcan 2014).

Recent literature reports suggest that smartphone-based sensing applications have gradually been expanded in all the sensing fields, such as healthcare, food safety, environmental monitoring, drug analysis and biosecurity, with the most diverse types of detectors, organized in four main groups: color (absorbance and reflectance); luminescence (bio, electro and photo-chemiluminescence and fluorescence), SPR (surface plasmon resonance) and electrochemical-based detection (Quesada-González and Merkoçi 2017).

Considering the smartphones-based sensors, two types of devices can be reported: (1) smartphone as a detector and (2) smartphone as an interface for an instrument (Roda et al. 2016). In the first case, the smartphone camera is used as a detector to identify the signal output, used coupled with other instruments (simple apparatus) or as a main detector (Kanchi et al. 2018). For the second use, an external device performs the measurement, and the smartphone is used to control the experimental setups and to displays the test result on the screen. This approach is less frequently reported in the literature, but some examples are found for commercial applications, as glucometers, ethylometers, and abuse drugs detection strips (Roda et al. 2016).

Studies show that this novel field of research represents a promising area that has high scientific and commercial impact (Rateni et al. 2017). However, as it is a developing field, some challenges and obstacles must be overcome. The main concerns are the still overestimated technology and wide use of smartphones, once the results obtained through it are still not compared to other miniaturized conventional devices, such as portable photometers or luminometers. Moreover, these smartphone-based biosensors have relatively low sensitivity and are therefore applicable for analytes present in high concentrations in biological fluids, environmental, and food samples (Roda et al. 2016). The repeatability of the measure, intrinsically guaranteed by a laboratory apparatus, also becomes a delicate condition to be met in the case of portable modules for on-site analysis. Suitable optimizations must be evaluated at the design stage for the physicality of the instrument to exclude or minimize any matrix or sample interference and external errors (Rateni et al. 2017). The applicability in real samples is also a problem to overcome for the application of these smartphone-based sensors in diagnosis. One of the greatest challenges for smartphone-based biosensors is the complete integration with other technologies and detectors in a fully integrated biorecognition process, providing a standalone biosensor that can be easily operated by everyone (Kanchi et al. 2018). Today, the use of these sensors by unskilled personnel is made difficult by the complexity of the various device accessories (smartphone's physical and electronic components), the need for laboratory instruments (such as pipettes), the difficulty of storing reagents, and the operation times (Roda et al. 2016). For the future, these standalone sensors should be simple and easy to operate by the users and should meet the commercial and technological standards in order to ease functional expansions and upgrades by the developers (Zhang and Liu 2016).

Some perspectives in the development of sensing devices include the use of smartphones for a continuously monitoring human health program integrated with other gadgets, like smartwatches and contact lenses (Liu et al. 2019). This complete automatized system should monitor real-time important health biomarkers, such as small molecules, e.g., glucose, oxygen, $Na^+$ and $Li^+$ ions and also more complex molecules such as proteins and nucleic acids in different biological fluids, in order to detect physiological deregulation (or an early stage diagnosis of a disease). Due to simple wireless connection, these systems could be directly linked to health care services, like hospitals, doctors, health professionals, emergency services, for a rapid and automatic response in urgent cases.

In the environmental and food area, there are some challenges, including the ability to detect multiple analytes simultaneously, especially for food and water safety, because the selection of only one parameter as a safety indicator is not effective enough to ensure the quality of food or drinking water. In the future, food and industrialized products may have internal sensors, such as pHmeter, redox potential and contaminants sensors, to assure the quality and spare date. The results may be easily checked with a smartphone platform with Bluetooth or NFC connection, and QR code reader, in a simple integrated system (Rateni et al. 2017; Vashist et al. 2014). As seen, there are an infinity of applications and possibilities with smartphone-based sensing devices, each one related to the real need. Therefore, these devices will be powerful point-of-care and point-of-need tools for the next-generation technology for detecting and monitoring countless analytes.

## 2.2    3D Printed System

Three-dimensional (3D) printing (or additive manufacturing) is an emerging eco-friendly technology with significant potential for several developments in different areas of science. This technology is able to manufacture geometrically complex, highly flexible objects in a short period of time, with minimum waste compared to the traditional manufacturing processes (O'Donnell et al. 2016; Ngo et al. 2018).

3D printing is advancing in the manufacture of smart and functional devices, making it possible to reduce the costs and time of manufacturing. In addition, the iterative resources allowed by 3D printing are becoming indispensable for researchers (Ngo et al. 2018).

In the area of bioanalytical, this innovative form of manufacture allows to manufacture devices with complex geometries and impossible before. For instance, researchers at Washington State University in the United States (2018), using 3D printing technology, developed the first 3D printed flexible electrochemical biosensor to monitor glucose with improved stability and sensitivity over those manufactured by traditional methods. The device is based on a wearable, non-invasive, needle-free biosensor, which facilitates glucose monitoring in children, for example. In addition, with this interesting technology, the biosensor can be manufactured customized specifically for each patient, adjusting to individual needs (Nesaei et al. 2018).

In a more recent study (2020), researchers at the University of Minnesota developed a 3D printing system able to print biosensors on living human organs. This system estimates the movement and deformation of the target surface to adapt the path in real time. The researchers demonstrated the applicability of the system by printing a hydrogel-based sensor to the pig's lung, under deformation induced by breathing. This sensor was able to provide the continuous spatial mapping of the deformation via electrical impedance tomography. Looking ahead, this 3D printing approach may allow flexible and compatible biomedical devices and sensors to be manufactured on the skin and inside the human body. This means a potential breakthrough in portable monitoring, which will facilitate diagnosis, disease monitoring and patient treatment (Zhu et al. 2020).

Despite these interesting studies, the impact of existing 3D printing technologies in bioanalytical still remains at an early stage. As the case of the majority of new developments, there are many challenges and opportunities facing this relatively new field, both in fundamental and applied aspects.

These challenges include (1) low resolution and surface smoothness, which makes it difficult to manufacture nanochannels; (2) void formation between subsequent layers of materials resulting in additional porosity during the manufacturing process, which can lead to problems of biological immobilization and substrate transport; (3) the electrochemical behavior of 3D printed sensors is still inferior to that from the sensors manufactured by consolidated methods and (4) limited biocompatibility and sensitivity of printable materials (Otero and Magner 2020; Manzanares Palenzuela and Pumera 2018; Salentijn et al. 2017). Thus, currently and in upcoming years, the research trend will be considerably focused on the development of new printable materials or on updating existing ones with the intended properties. Also, the development of printers capable of achieving higher resolution, to overcome some of these challenges.

Further research is required to manufacture 3D-printed electrodes at a large scale and with the performance for relevant analytes to make it competitive with traditional methods in mass production. Due to the exponential growth of the number of publications in analytical chemistry involving this technology, it is likely to happen in the foreseeable future (Gross et al. 2017).

## 2.3  Paper as Smart Substrate Material

Paper-based substrates have been used as a platform to develop analytical devices for over a decade (Martinez et al. 2007), and it has a prominent position in the development of devices for bioanalytical. That is because the paper has interesting properties such as strong capillary action and biocompatibility, but mainly due to its abundance, low cost and easy disposal, which makes it an environmentally friendly platform (Noviana et al. 2019; Deroco et al. 2020; Arduini et al. 2020).

The global concern for sustainable development is increasingly consolidated. The trend of sustainable development requires the use of cleaner/greener materials (Tobiszewski et al. 2010). In this way, paper as a substrate material for applications in bioanalytical will continue to attract more attention as a sustainable and disposable alternative to the more conventionally used materials such as glass or plastic.

Although the field of bioanalytical microdevices on paper continues to develop at an exponential rate with notable impacts on academic and industrial communities, it is an open field for several applications (Gutiérrez-Capitán et al. 2020; Suntornsuk and Suntornsuk 2020; Ratajczak and Stobiecka 2020; Carrell et al. 2019).

Nowadays, the main opportunities related to the use of paper for applications in bioanalytical is the replacement of the initial classic formats for a variety of innovative architectures, maintaining the simplicity of operation and the low cost of manufacturing (Fava et al. 2019; Sun et al. 2018; Chinnadayyala et al. 2019; Wang et al. 2018).

Therefore, the future trends will be driven by research and development of (1) microdevices with new, more complex 3D paper formats, which may be the future of multi-analyte analysis using a combination of different techniques; (2) innovative paper devices with various functionalities serving, for example, as a substrate for cleaning, chromatographic separations, sample pre-treatment integrated system, and so on; (3) in addition, more types of papers with exclusive properties should be explored and applied to perform more specific functions. These future developments will provide comprehensive results, real portability and a more environmentally friendly approach, opening a new horizon in the discovery and diagnosis of drugs and diseases.

# 3 Trends Platforms and Applications

## 3.1 Wearable Biosensor

Regarding portable monitoring devices, wearable biosensors are certainly an important tendency in bioanalytical for the near future. Wearable devices still represent a minimal share in the biosensor market; however, the growing demand for real-time analysis and the continuous monitoring of relevant biomolecules has spurred the development in the field of wearable technology aiming at non-invasive data collection at a low cost. As a result, the wearable market had a historic growth in 2019, which was more than double compared to 2014 (https://www.idtechex.com/en/research-report/wearable-technology-forecasts-2019-2029/2680). It is expected that wearable biosensors continue as a fast-growing product due to the need to improve the current healthcare infrastructure for the global population.

In this scenario, the trends related to wearable sensors appear in the development of devices capable of collecting highly accurate health data and connect remotely to the personal health professional, allowing the vital sign monitoring or the health progress; and still, check if the patient is using the medications correctly.

The challenges in the implementation of these devices on the market involve the reliability of detection for over extended periods (days or even months), stability of the bio-recognizer, body fluid transport through the device, sensor biofouling, and long-term self-powered operation. As a consequence, future efforts will be focused on facing these problems, seeking for more efficient materials (Lou et al. 2020).

Furthermore, a new tendency in wearable devices is the combination of monitoring and therapy delivery, namely biosensing-drug delivery systems. This system combines biosensors and drug delivery technology; thus, it is possible to monitor the disease's biomarker and, subsequently, release the required therapeutic dose (Meng and Sheybani 2014).

It can pave the way for more effective treatment strategies for biomedical applications, as in the case of chronic diseases (e.g., hypertension, diabetes, heart and respiratory problems) that require continuous monitoring and regular medication administration (Bernell and Howard 2016). An example of a biosensing-drug delivery system that has been developed is an intelligent patch that easily mimics the regulatory function of the pancreas. The device is composed of microneedles preloaded with insulin, which release the drug quickly when it detects high levels of blood sugar, and the release of insulin decreases when blood glucose returns to normal (Yu et al. 2020a). This device has shown promising results when used in mice and minipigs, and according to the researchers, this smart patch, if demonstrated to be safe and effective in tests with humans, would revolutionize the patient's experience in the treatment of diabetes (Yu et al. 2020a).

However, this complex structure is at an early stage, still remains several challenges and opportunities to research, mainly in their in vivo application.

Therefore, in the future, it is hoped that wearable devices sophisticated and fully functional, including those that are capable of wireless data transfer, will be available for several applications. Even though the promising efforts and applications of wearable sensors are more focused on personalized medicine and sports health, there is interest in other relevant and emerging areas, for example, in the defense and food safety (Mishra et al. 2017, 2018), and animal health management (Neethirajan 2017).

## 3.2    Home-/Self-testing Kit

Another emerging trend for bioanalysis are the home-testing kits. These tests, also called self-tests or home-use tests, are typically sold over the counter (OTC) and allow users to test self-collected specimens and interpret the results on their own without the help of skilled health professionals. These types of tests differ from home-collection tests, which require users to collect samples at home, mail them to a laboratory or clinic for analysis, and obtain the results a few days later. These tests unify analytical platforms and medical diagnostic concepts in a simplified way to create highly autonomous devices. Currently, few examples of commercial home-tests are available, and almost all of them are related to the detection of sexually transmitted diseases, such as HIV (Ibitoye et al. 2014). Other examples of home-tests available on the market are pregnancy tests and glucometers. Despite the lack of commercial tests, some recent studies have developed new kits to detect or facilitate the diagnosis of a variety of diseases, such as the detection of vaginal infection (Geva et al. 2006), cardiovascular diseases (Leddy et al. 2019), Influenza (Rumbold et al. 2017), and Malaria disease (Jelinek et al. 1999). Other tests for cholesterol and oral anticoagulation monitoring (Ryan et al. 2010) and allergies and cancer diagnoses are also being developed for further application in home-test kits (Ryan et al. 2006).

These home tests are a major revolution on how health care will be regulated in the future. Many clinical tests will be done directly at home, saving time and hospital resources. However, the use of these tests and the interpretation of results by untrained or unskilled people can lead to controversial results and diagnosis, often leading to the patient's negligence of some disease (Ickenroth et al. 2010). The ease of use of the tests, mainly related to the analyzed sample, is a challenge that will have to be faced in the future. Studies have shown that blood tests were the most difficult to perform, and participants typically struggled with obtaining adequate blood samples or applied too little blood to the test strips (Ibitoye et al. 2014). For instance, in a study of blood glucose testing, it was observed at baseline that only 17% of participants were able to perform the test without errors (Müller et al. 2006). For tests using saliva or sweat as samples, patients had no problem performing the analysis, showing that in the future, these samples should be the main ones used. Another key point is the interpretation of the results that must be obtained directly in the tests, without the need for conversions or additional calculations, making the user's interference zero, minimizing the possibility of errors in the results. For instance, in a blood-based cholesterol test that required users to convert the reading obtained on a thermometer-like display using a results chart, over half (51.9%) of the patients were unable to obtain any result at all (McNamara et al. 1996).

Regarding the transparency of the tests, it is essential that the developers add clear information about the operation of the device, as well as possible explanations on how the interferences may lead to an unexpected result and especially make it clear that these tests do not replace the clinical and medical monitoring available in hospitals or laboratory tests. The regulation of these tests will be vital to guarantee that the analytical parameters, such as reproducibility, sensitivity, and accuracy, may be respected and meet the standard analytical techniques and detection limits for each method. Currently, one of the greatest challenges of these tests is to ensure good reliability in the results, mainly in eliminating false positive and false-negative results. Few tests are now approved by regulatory agencies, and most are still in the initial stages of testing and implementation (Grispen et al. 2014; Huppert et al. 2012).

Developing countries facing public health problems are unlikely to have access to reliable home-tests kits, further increasing inequality and dependency on rich countries' technologies (An et al. 2013). Thus, one of the main challenges is to develop inexpensive self-test kits that are affordable to all, especially for more vulnerable populations, in order to even serve as monitoring strategies for epidemiological surveillance, such as outbreaks of infectious diseases (Mahase 2020), like Zika virus and Ebola. In the future, home-tests are also expected to detect other tropical infectious diseases, such as Dengue fever, Yellow fever and Chagas disease. In addition, it hopes that these self-tests can also be used for early diagnosis of heart and neurodegenerative diseases and even the most diverse types of cancer.

Home-tests have also proved to be a good alternative for exceptional situations such as the Coronavirus Disease 2019 (COVID-19) pandemic, declared by the World Health Organization on March 11, 2020 (World Health Organization 2020). The use of home-tests, even on an emergency basis, to detect cases of the new coronavirus, or other viral diseases, can mean greater control over the information on the number of cases, and together with smartphone technologies (GPS, Wi-fi, Bluetooth) register and map areas of risk and with high potential for contamination, serving as very important platforms for health authorities in taking initiatives in epidemiological controls, such as lockdown (Mahase 2020; Torjesen 2020). Some home-tests for the COVID-19 were made available on the market and served as a complement to the gold tests (RT-PCR, Reverse Transcription-Polymerase Chain Reaction). It is important to note, however, that these tests are not a substitute for laboratory tests and do still not have the same degree of reliability.

Regarding the environment, few studies of the applicability of these tests are found. Therefore, it is expected that in the future, these home-tests will not be directed only at the health area, but at food and environmental safety, with the development of simple home-tests to detect, for example, some contaminant in water or food (pesticides) and for general use of the population in daily life.

Finally, we have to bear in mind that these home-tests can only be used on a large scale, when the regulatory agencies act rigorously in evaluating the performance of the devices and when the necessary information on the performance of the tests is available to all users (Ibitoye et al. 2014).

### 3.3   New Biomarkers and Sensing Approaches for Diagnostic

The discovery of novel biomarkers that enables earlier or more accurate diagnosis of the disease is a major challenge of biomedical research. The analytical and clinical factors can make the test validation of these biomarkers challenging, once sample collection, storage, and handling can influence test results and lead to errors in diagnostic studies (Byrnes and Weigl 2018). The complexity of developing and evaluating can also influence the designing of diagnostic systems and test implementation. Additionally, the novel biomarkers must be both disease-specific and universal in a diseased population for reliable and accurate measurement across a range of patients (Strimbu and Tavel 2010). When a new assay or device is designed, the selection of the suitable type of biomarker for a specific disease plays an important role, reducing development time and cost. Often, there is more than one biomarker for a disease and selecting the right biomarker(s) usually affects aspects of a final diagnostic test such as sensitivity, specificity, cost, and logistics (Vashist et al. 2015). An effective biomarker is also able to identify asymptomatic and early disease states, which would, in turn, decrease the time between diagnosis and treatment, reducing adverse outcomes and disease transmission (DiClemente et al.

2004). The expansion of access for new devices and less complex assays should be coupled with the continuous need for basic research to identify new biomarkers. It is important to consider how and where a test will be used and its cost-effectiveness, in order to select an appropriate biomarker.

Diagnostic tests are designed for ultra-sensitivity and quantification that can identify single target molecules: these types of tests are often referred to as digital assays, where a positive signal indicates the presence of one target molecule and a negative indicates no target molecule. In the future, multiplexed or panel-based tests will help identify biomarkers indicative of specific treatment pathways. Thus, biomarkers will be referred to a set of molecules (multiplexing) that together describe the molecular signature of a disease state. As sensing techniques become more readily available, multiple molecules can simultaneously be probed to understand a disease state more completely. This makes the multiplexing approach one of the most challenging goals that the new biomarkers must overcome (Simon and Ezan 2017). The parallel detection of multiple biomolecular tags associated with cellular biomarkers (i.e., multiplexing) is critical to the utility of flow cytometry in high-throughput biomedical research and clinical diagnosis. For this purpose, researchers have exploited the brightness, photostability, size-dependent optoelectronic properties, nanoscale interface and superior multiplexing capability of quantum dots (QDs) for many applications (Petryayeva et al. 2013). Some of the more prominent applications include in vitro diagnostics, energy transfer-based sensing, cellular and in vivo imaging, and drug delivery. QDs are also well suited to multiplexed staining of tissue biopsy specimens. In general, more insight can be obtained from the detection of more biomarkers in parallel, suggesting a key role for the multiplexing advantages of QDs in diagnose pathologies. This technical capability is ideal for pairing with the multiplexing advantages of QDs to facilitate the detection and visualization of multiple biomarkers in complex biological specimens. Although QD materials have been used for research and development purposes, one of the challenges will be the transference from bench-top studies into commercialized sensing devices (Petryayeva et al. 2013; Giri et al. 2011; Xing et al. 2007).

Another future interesting platform for biomarkers is the use of magnetic particles (nano and micro) in diagnostics and sensing devices. The magnetic particles can combine two very selective processes in bioanalysis: the specific binding of analytes to the particle surface based on molecular recognition and the specific isolation of magnetic objects from complex sample mixtures. The surface of such particles can be decorated with functional groups of biomolecules using well-established surface linking strategies followed by the manipulation of these particles with an external magnetic field, which has no contact with the analyte solution. This provides an excellent strategy for the development of high-sensitive biomarkers for diagnostic application. The manipulation of sized droplets generated within microfluidic devices and utilized as nanolitre or picolitre reaction vessels provides an elegant pathway for bioassays, such as DNA extraction, once these droplets can be split or moved to certain locations. One droplet contained magnetic particles that could be moved by switching the embedded electromagnets to achieve merging and splitting from the other fixed droplets.(Zhang et al. 2011; Lehmann et al. 2006)

The new arising technology for flexible electrochemical biosensors enables an in situ detection of a plenty of analytes present in biological fluids and is a great alternative for personalized health monitoring. Some characteristics like intrinsic wearability, high sensitivity and selectivity, and low cost are some of the great advantages of these devices, representing an improvement to detection of biomarkers compared to other conventional analytical techniques. Recent studies have used these new technologies for the detection of biomarkers of interest (ranging from metabolites, electrolytes, heavy metals, drugs, to neurotransmitters, proteins, and hormones) in a variety of body fluids, including blood, sweat, saliva, tears, cerebrospinal, and interstitial fluid (Yu et al. 2020b). Although some important progress on flexible electrochemical devices was achieved in the last years, some challenges still need to be overcome. Today, the most common flexible biosensors are focused on the detection of a very limited number of biomarkers, for example, metabolites and electrolytes. However, our body contains other abundant biomarkers (e.g., proteins, hormones, and peptides) in trace amounts, which are closely related to various health conditions and could be used for the health monitoring of patients. It occurs that the quantification of these biomarkers is hard to regenerate in vivo with current technologies once they are based on affinity recognition. Other challenges for flexible electrochemical biosensors include the selectivity, sensitivity, stability and repeatability of the method. Mechanical reliability and system robustness of the device also needs to be addressed when used continuously in body fluids, once factors such as fluid pH, volume, and flow rate can affect the accuracy of the detection (Yu et al. 2020b).

Despite numerous advances in sensing approaches and platforms with high-resolution techniques, the main challenge lies in the development of specific biomarkers for a specific disease. The diagnostics and management of localized cancer, for example, is complicated because of cancer heterogeneity and differentiated progression in various subgroups of patients. There is currently a significant challenge in identifying and validating novel biomarkers for any type of cancer. Novel biomarkers should allow in the future not only detecting cancer in primary biopsies, but also disease staging and choosing therapeutic options, such as performing or not a repeated biopsy, which can help prevent overtreatment of patients. The target in the search for novel biomarkers is mainly focused on breast, prostate, lung, and cervical cancer, but it has also expanded to other types of disease (Sharma 2018; Chistiakov et al. 2018). Future studies should be driven to the collection of new biomarkers for neurodegenerative—like Alzheimer and Parkinson—and cardiovascular diseases. The development of biomarkers for different psychiatric illnesses is also essential, given the need for specific diagnosis and differentiation between depression and bipolar disorder; and the diagnosis of different types of schizophrenia based on specific biomarkers for each one. This collection of novel biomarkers can be useful for solving various diagnostic and prognostic tasks such as primary screening, risk stratification, prediction of adverse endpoints, decision-making for treatment and selection of candidates for intensive therapy or active surveillance.

Recent advances in gene expression assessment ("*omics*" technology) combined with an improved understanding of the roles of different genes in malignant transformation have opened an intriguing possibility of customizing the novel biomarkers for future diagnostic and therapeutic tools (Lorente et al. 2014). As a result, we are getting closer to enjoying personalized medicine, which uses systematically detailed biological information about an individual patient to make decisions about the best treatment. From molecules such as proteins, DNA, RNA, or lipids, new biomarkers have been recognized as indicators to identify, prevent, and treat disease according to the particularities of each organism. Despite the fact that genetic testing is widely available, and this is an area of intense research, the genetic map is still slightly used in the clinical environment for patient care in real life; however, it is expected that this is going to change in the coming years (Ely 2009). In the future this technology will be used for the specific development of individual drugs, with the goal of treatment and eradicating the disease without side effects. And still to know if the individual is predisposed to any disease before it actually happens so that doctors will be able to act before people get sick. Aside from great benefits to patients' health, this approach holds the promise to reduce healthcare costs (Drucker and Krapfenbauer 2013).

It is also worth mentioning that even though a lot of research and billions of dollars are spent to develop new diagnostic tests based on biomarkers, most are still only classified as "promising" (Lumbreras et al. 2009). The distance between benchtop research and clinical research may hinder the path from discovery to implementation. Some primary research fails to answer questions relevant for clinicians and patients, and there are a great number of biomedical discoveries without effective translation in healthcare (Ioannidis et al. 2014). To avoid (or at least reduce) knowledge waste, some authors have called for the prompt identification of scientific discoveries that have the ability to affect health and have emphasized the importance of adding value to existing evidence to prioritize research gaps before starting a new line of research. Furthermore, to minimize error and stimulate progress from biomarker discovery to clinical application, a formal validation strategy based on available evidence is essential (Ioannidis 2016).

## 4    Final Remarks

Bioanalytical is being developed in an extraordinary way. Advances in fields such as nanotechnology, molecular biology and medicine have boosted the bioanalysis devices capacity to extract quantitative and innovative information, revolutionizing the analysis in living systems, the diagnostic biomarkers detection and the therapeutic effects evaluation in model systems. Furthermore, advances in bioanalytical are also extended to fields such as environmental monitoring, forensic science, food and water quality control.

It is hoped that in the future commercial devices will be smaller than palm size and will be capable of disease diagnosis such as cancer and degenerative diseases, and even rare diseases. These devices will be miniaturized systems integrated with sample collection and pre-treatment, multiplex analysis, and wireless telecommunication capacity, and it will generate data easily interpreted by users. In this way, consumers will be able to use real-time in situ testing at home, supermarkets, restaurants, fields, industries for health, quality and safety assessments, even using their smartphones or/and other smart handheld devices.

Although a few bioanalytical devices are already available on the market for some specific applications, they are still insufficient for the high demand of existing and growing problems in medicine, agri-food field and the environment. There are still several opportunities to improve the sensitivity, selectivity, long-term stability and reliability of devices and multiplexing capabilities, which will make it possible to manufacture/commercialize them on a large scale for an increasing number of applications. The enhancing multidisciplinary synergism of chemistry, biology, and engineering will play a central role in creating innovative portable and wearable microdevices with higher performance.

## References

An Q, Prejean J, McDavid Harrison K, Fang X (2013) Am J Public Health 103:120–126
Arduini F, Micheli L, Scognamiglio V, Mazzaracchio V, Moscone D (2020) TrAC Trends Anal Chem 128:115909
Bernell S, Howard SW (2016) Front Public Health 4:159–159
Burki TK (2020) Lancet Respir Med 8:e63–e64
Byrnes SA, Weigl BH (2018) Expert Rev Mol Diagn 18:19–26
Carrell C, Kava A, Nguyen M, Menger R, Munshi Z, Call Z, Nussbaum M, Henry C (2019) Microelectron Eng 206:45–54
Chard T (1992) Hum Reprod 7:701–710
Chinnadayyala SR, Park J, Le HTN, Santhosh M, Kadam AN, Cho S (2019) Biosens Bioelectron 126:68–81
Chistiakov DA, Myasoedova VA, Grechko AV, Melnichenko AA, Orekhov AN (2018) Semin Cancer Biol 52:9–16
Coluccio ML, Pullano SA, Vismara MFM, Coppedè N, Perozziello G, Candeloro P, Gentile F, Malara N (2020) Micromachines 11:123
Deroco PB, Giarola JDF, Wachholz Júnior D, Lorga GA, Kubota LT (2020) Chapter 4. Paper-based electrochemical sensing devices. In: Merkoçi A (ed) Comprehensive analytical chemistry, vol 89. Elsevier, Amsterdam, pp 91–137
DiClemente RJ, Wingood GM, Harrington KF, Lang DL, Davies SL, Hook EW III, Oh MK, Crosby RA, Hertzberg VS, Gordon AB, Hardin JW, Parker S, Robillard A (2004) JAMA 292:171–179
Drucker E, Krapfenbauer K (2013) EPMA J 4:7
Ely S (2009) Transl Res 154:303–308
Fava EL, Silva TA, Prado TMD, Moraes FCD, Faria RC, Fatibello-Filho O (2019) Talanta 203:280–286
Gao W, Emaminejad S, Nyein HYY, Challa S, Chen K, Peck A, Fahad HM, Ota H, Shiraki H, Kiriya D, Lien D-H, Brooks GA, Davis RW, Javey A (2016) Nature 529:509–514
Geva A, Bornstein J, Dan M, Shoham HK, Sobel JD (2006) Am J Obstet Gynecol 195:1351–1356

Giri S, Sykes EA, Jennings TL, Chan WCW (2011) ACS Nano 5:1580–1587
Grispen JE, Ickenroth MH, de Vries NK, van der Weijden T, Ronda G (2014) Health Expect 17: 741–752
Gross B, Lockwood SY, Spence DM (2017) Anal Chem 89:57–70
Gutiérrez-Capitán M, Baldi A, Fernández-Sánchez C (2020) Sensors 20:967
https://www.marketsandmarkets.com/Market-Reports/biosensors-market-798.html? gclid=CjwKCAjw797P795BRBQEiwAGflV796a793kLaEnUsqiXHiSRwyF-cP792LEKbz795ZXQgsGmiSeLZ791kr-klht_eKxoCeVcQAvD_BwE. Accessed 5 June 2020
https://www.idtechex.com/en/research-report/wearable-technology-forecasts-2019-2029/2680. Accessed. 15 July 2020
Huppert JS, Hesse EA, Bernard MC, Bates JR, Gaydos CA, Kahn JA (2012) J Adolesc Health 51: 400–405
Ibitoye M, Frasca T, Giguere R, Carballo-Diéguez A (2014) AIDS Behav 18:933–949
Ickenroth MHP, Ronda G, Grispen JEJ, Dinant G-J, de Vries NK, van der Weijden T (2010) BMC Fam Pract 11:77
Ioannidis JPA (2016) PLoS Med 13:e1002049
Ioannidis JPA, Greenland S, Hlatky MA, Khoury MJ, Macleod MR, Moher D, Schulz KF, Tibshirani R (2014) Lancet 383:166–175
Jelinek T, Amsler L, Grobusch MP, Nothdurft HD (1999) Lancet (London, England) 354:1609
Kanchi S, Sabela MI, Mdluli PS, Inamuddin BK (2018) Biosens Bioelectron 102:136–149
Keçili R, Büyüktiryaki S, Hussain CM (2019) TrAC Trends Anal Chem 110:259–276
Khosravani MR, Reinicke T (2020) Sensors Actuators A Phys 305:111916
Kim J, Campbell AS, de Ávila BE-F, Wang J (2019) Nat Biotechnol 37:389–406
Kundu M, Krishnan P, Kotnala RK, Sumana G (2019) Trends Food Sci Technol 88:157–178
Leddy J, Green JA, Yule C, Molecavage J, Coresh J, Chang AR (2019) BMC Nephrol 20:132
Lee H, Song C, Baik S, Kim D, Hyeon T, Kim D-H (2018) Adv Drug Deliv Rev 127:35–45
Lehmann U, Vandevyver C, Parashar VK, Gijs MA (2006) Angew Chem Int Ed Eng 45:3062–3067
Li F, You M, Li S, Hu J, Liu C, Gong Y, Yang H, Xu F (2020) Biotechnol Adv 39:107442
Liu J, Geng Z, Fan Z, Liu J, Chen H (2019) Biosens Bioelectron 132:17–37
Lorente D, Mateo J, de Bono JS (2014) Molecular characterization and clinical utility of circulating tumor cells in the treatment of prostate cancer. American Society of Clinical Oncology educational book. American Society of Clinical Oncology. Annual Meeting, pp e197–e203
Lou Z, Wang L, Jiang K, Wei Z, Shen G (2020) Mater Sci Eng R Rep 140:100523
Lumbreras B, Parker LA, Porta M, Pollán M, Ioannidis JP, Hernández-Aguado I (2009) Clin Chem 55:786–794
Mahase E (2020) Br Med J 368:m621
Manzanares Palenzuela CL, Pumera M (2018) TrAC Trends Anal Chem 103:110–118
Martinez AW, Phillips ST, Butte MJ, Whitesides GM (2007) Angew Chem 119:1340–1342
McNamara JR, Warnick GR, Leary ET, Wittels E, Nelson FE, Pearl MF, Schaefer EJ (1996) Prev Med 25:583–592
Mejía-Salazar JR, Rodrigues Cruz K, Materón Vásques EM, Novais de Oliveira O Jr (2020) Sensors 20:1951
Meng E, Sheybani R (2014) Lab Chip 14:3233–3240
Min J, Nothing M, Coble B, Zheng H, Park J, Im H, Weber GF, Castro CM, Swirski FK, Weissleder R, Lee H (2018) ACS Nano 12:3378–3384
Mishra RK, Hubble LJ, Martín A, Kumar R, Barfidokht A, Kim J, Musameh MM, Kyratzis IL, Wang J (2017) ACS Sens 2:553–561
Mishra RK, Martín A, Nakagawa T, Barfidokht A, Lu X, Sempionatto JR, Lyu KM, Karajic A, Musameh MM, Kyratzis IL, Wang J (2018) Biosens Bioelectron 101:227–234
Müller U, Hämmerlein A, Casper A, Schulz M (2006) J Pharma Pract (Granada) 4:195–203
Neethirajan S (2017) Sens Bio-Sens Res 12:15–29
Nesaei S, Song Y, Wang Y, Ruan X, Du D, Gozen A, Lin Y (2018) Anal Chim Acta 1043:142–149
Ngo TD, Kashani A, Imbalzano G, Nguyen KTQ, Hui D (2018) Compos Part B 143:172–196

Noah NM, Ndangili PM (2019) J Anal Methods Chem 2019:2179718
Noviana E, McCord CP, Clark KM, Jang I, Henry CS (2019) Lab Chip 20:9–34
O'Donnell J, Kim M, Yoon H-S (2016) J Manuf Sci Eng 139:10801
Otero F, Magner E (2020) Sensors 20:3561
Ozcan A (2014) Lab Chip 14:3187–3194
Petryayeva E, Algar WR, Medintz IL (2013) Appl Spectrosc 67:215–252
Portable meter to aid diabetics. Pittsburgh Press, November 6, 1981, pA-1986
Quesada-González D, Merkoçi A (2017) Biosens Bioelectron 92:549–562
Ratajczak K, Stobiecka M (2020) Carbohydr Polym 229:115463
Rateni G, Dario P, Cavallo F (2017) Sensors 17:1453
Roda A, Michelini E, Zangheri M, Di Fusco M, Calabria D, Simoni P (2016) TrAC Trends Anal Chem 79:317–325
Rumbold B, Wenham C, Wilson J (2017) BMC Med Ethics 18:33
Ryan A, Wilson S, Greenfield S, Clifford S, McManus RJ, Pattison HM (2006) J Public Health 28:370–374
Ryan F, O'Shea S, Byrne S (2010) Int J Lab Hematol 32:e26–e33
Salentijn GIJ, Oomen PE, Grajewski M, Verpoorte E (2017) Anal Chem 89:7053–7061
Schilling KM, Jauregui D, Martinez AW (2013) Lab Chip 13:628–631
Sengupta P, Khanra K, Chowdhury AR, Datta P (2019) Lab-on-a-chip sensing devices for biomedical applications. In: Pal K, Kraatz H-B, Khasnobish A, Bag S, Banerjee I, Kuruganti U (eds) Bioelectronics and medical devices. Woodhead Publishing, pp 47–95
Sharma AK (2018) Semin Cancer Biol 52:iii–iv
Simon S, Ezan E (2017) Bioanalysis 9:753–764
Solanki S, Pandey CM, Gupta RK, Malhotra BD (2020) Biotechnol J 15:1900279
Soler M, Belushkin A, Cavallini A, Kebbi-Beghdadi C, Greub G, Altug H (2017) Biosens Bioelectron 94:560–567
Strimbu K, Tavel JA (2010) Curr Opin HIV AIDS 5:463–466
Sun X, Li B, Qi A, Tian C, Han J, Shi Y, Lin B, Chen L (2018) Talanta 178:426–431
Suntornsuk W, Suntornsuk L (2020) Electrophoresis 41:287–305
Tobiszewski M, Mechlińska A, Namieśnik J (2010) Chem Soc Rev 39:2869–2878
Torjesen I (2020) Br Med J 369:m1799
Vashist SK, Mudanyali O, Schneider EM, Zengerle R, Ozcan A (2014) Anal Bioanal Chem 406:3263–3277
Vashist SK, Luppa PB, Yeo LY, Ozcan A, Luong JHT (2015) Trends Biotechnol 33:692–705
Wang J, Li W, Ban L, Du W, Feng X, Liu B-F (2018) Sensors Actuators B Chem 254:855–862
World Health Organization (2020) WHO director-General's opening remarks at the media briefing on COVID-19—11 March 2020. World Health Organization, Geneva. https://www.who.int/dg/speeches/detail/who-director-general-s-opening-remarks-at-the-media-briefing-on-covid-2019%2D%2D-2011-march-2020. Accessed 15 July 2020
Xing Y, Chaudry Q, Shen C, Kong KY, Zhau HE, Chung LW, Petros JA, O'Regan RM, Yezhelyev MV, Simons JW, Wang MD, Nie S (2007) Nat Protoc 2:1152–1165
Yang G, Xiao Z, Tang C, Deng Y, Huang H, He Z (2019) Biosens Bioelectron 141:111416
Yu J, Wang J, Zhang Y, Chen G, Mao W, Ye Y, Kahkoska AR, Buse JB, Langer R, Gu Z (2020a) Nat Biomed Eng 4:499–506
Yu Y, Nyein HYY, Gao W, Javey A (2020b) Adv Mater 32:1902083
Zhang D, Liu Q (2016) Biosens Bioelectron 75:273–284
Zhang Y, Park S, Liu K, Tsuan J, Yang S, Wang T-H (2011) Lab Chip 11:398–406
Zhu Z, Park HS, McAlpine MC (2020) Sci Adv 6:eaba5575

Printed in the United States
by Baker & Taylor Publisher Services